Soft Matter

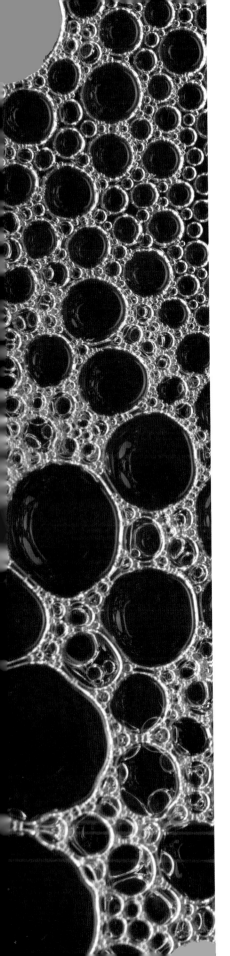

Soft Matter

Concepts, Phenomena, and Applications

Wim van Saarloos
Vincenzo Vitelli
Zorana Zeravcic

Princeton University Press
Princeton and Oxford

Published by Princeton University Press
41 William Street, Princeton, New Jersey 08540
99 Banbury Road, Oxford OX2 6JX

press.princeton.edu

All Rights Reserved

Names: van Saarloos, Wim, author. | Vitelli, Vincenzo, author. | Zeravcic, Zorana, author.
Title: Soft matter : concepts, phenomena, and applications / Wim van Saarloos, Vincenzo Vitelli, Zorana Zeravcic.
Description: Princeton : Princeton University Press, [2024] | Includes bibliographical references and index.
Identifiers: LCCN 2023030322 (print) | LCCN 2023030323 (ebook) | ISBN 9780691191300 (acid-free paper) | ISBN 9780691251691 (ebook)
Subjects: LCSH: Soft condensed matter—Textbooks. | BISAC: SCIENCE / Physics / General | SCIENCE / Life Sciences / Biophysics
Classification: LCC QC173.458.S62 .V36 2023 (print) | LCC QC173.458.S62 (ebook)
| DDC 530.4/1—dc23/eng/20231002
LC record available at https://lccn.loc.gov/2023030322
LC ebook record available at https://lccn.loc.gov/2023030323

British Library Cataloging-in-Publication Data is available

Editorial: Ingrid Gnerlich and Whitney Rauenhorst
Production Editorial: Natalie Baan
Text and Jacket Design: Wanda España
Production: Jacquie Poirier
Publicity: William Pagdatoon
Copyeditor: Bhisham Bherwani

Jacket image: Henrik Sorensen/Getty Images

This book has been composed in Palatino, Pazo Math, Latin Modern, and newpxmath

This book has been composed in LaTeX
The publisher would like to acknowledge the authors of this volume for acting as the compositor for this book.

Printed in China

10 9 8 7 6 5 4 3 2 1

Contents

II SOFT MATTER PHASES   161

Figures

Introduction

Chapter 1

Chapter 2

Chapter 3

Chapter 4

Chapter 5

Chapter 6

Chapter 7

Chapter 8

Chapter 9

Chapter 10

Preface

This book grew out of our experience teaching introductory courses on soft matter in Leiden and Chicago. The challenge in Leiden was to develop a course aimed at first-year master's students, students who just have completed a three-year bachelor degree in physics or a related field. They have a diverse background and will choose their specialization and decide whether to go into theoretical or experimental physics only some time after taking the course. The Chicago course targets beginning graduate students, but with similarly diverse backgrounds and interests.

Many colleagues we consulted about teaching soft matter from a physics perspective struggled with the same dilemma we faced: how to develop a course which introduces some of the basic concepts developed in the previous century, but which at the same time gives a feel for some of the exciting research questions these days, as well as for the revolutionizing new opportunities offered by modern visualization techniques and digital analysis. Moreover, for many of us the charm of soft matter is its diversity, the fact that it cannot simply be treated on the basis of a single overarching theoretical framework, and that it pays to have an intuitive understanding of many different approaches and materials. How can we bring across the necessity, power, and fun of being able to shift perspectives and to bring knowledge from various disciplines to bear on a problem? We found ourselves combining bits and pieces from several classical introductions to the field and from books focused on a particular phase of soft matter, excerpts from literature on applications and present-day research topics, and our own lecture notes.

This book reflects our teaching approach and philosophy: it is intended to be essentially the type of book we would have liked to have available as a basis for the courses we developed. In short, we have tried to write a somewhat different introductory textbook on the basic concepts of soft matter. Its aim is to give advanced undergraduate and beginning graduate students an introductory overview of the various soft matter phases and their rheology, and the conceptual framework to analyze them. We have attempted to choose our approach and topics in such a way that students who specialize in other sub-disciplines will acquire a good overview of the field, and get familiar with concepts and treatments that have broader application. Moreover, as students and researchers nowadays are motivated more than ever to pay attention to possible applications of their insights and methods, in both science and technology, we pay attention to the large range of applications. For students who continue in soft matter research, the book should be a stepping stone for further specialization, while for students whose main research focus is in biomatter or at the interface of physics with biology, this book should give them the necessary background to understand the application of soft matter physics concepts in biology. We have made an effort to include links between soft matter and biomatter throughout the book.

A distinctive feature of our treatment, especially when compared to most other introductory soft matter physics books, is its focus on the power of phenomenology and the hydrodynamic approach. The book reviews the main soft matter classes and their rheology with embedded explanations of key concepts and methods (scaling,

Landau approach, bifurcations, correlation functions, renormalization group, scattering approach, etc.) without assuming detailed previous knowledge of continuum mechanics. We do assume some background in statistical physics and some elementary knowledge of phase transitions, though. Quite a few concepts appear several times in different chapters and examples, as this deepens the students' understanding and stimulates them to explore how the various topics are interrelated. Through this, we hope to develop students' intuition and give them a kind of intellectual 'agility' in reasoning their way through complex soft matter phenomena.

Our approach is to develop many such embedded concepts 'on the fly,' rather than in separate appendices or boxes, mirroring how we ourselves often pick up new concepts while doing research, or from talks. The same holds for some of the modern topics we touch on only briefly with a short paragraph, a figure, or a note in the margin. We realize that, as a result, such topics are typically not developed in as much depth or as systematically as they would be were separate sections or appendixes devoted to them. But our own students appreciate this more informal style, which is closer to how science is actually often done in practice. Moreover, they find it stimulates them to realize and explore connections between topics that in the beginning of their studies were treated as separate subjects. We have also experienced that it helps to promote their agility and to overcome their hesitance to work with a concept they have not mastered completely. We routinely give pointers to literature where interested students can find more information.

We support this style and approach with our layout and use of references. We develop the main story line in the text as much as possible and without interruptions, and we reserve notes in the margin to point out connections or to draw the student's attention to important side issues. We view these margin notes, which often also contain references to relevant papers or to more detailed treatments in other textbooks, as an integral part of our approach. Numbered endnotes are used for backing up some of the assertions in the text, or for drawing attention to subtleties or connections to other works. These endnotes are intended for students who are eager to learn even more; sometimes they provide answers to subtle questions which might emerge from studying the main text. We imagine a reader skipping these endnotes when studying a topic for the first time.

The philosophy sketched above is also reflected in the organization of the chapters. They always start with a focus on introducing and explaining the basic concepts; we envision a lecturer wanting to treat these sections in detail if the book is used as the basis for a course. Toward the end, most chapters shift to descriptions of interesting examples and applications, which students should be able to study by themselves. There are, of course, ample opportunities for lecturers to highlight a few of these topics and expand on them, depending on the focus of the course and the interests of the students. But lecturers are advised to make a selection here and encourage the students to study the other material by themselves to enhance their understanding of the field and its breadth. We have attempted to provide sufficient references to the literature in all these later sections, which could also be used as a basis for student presentations.

Introducing well-established concepts which are part of a traditional field or of the soft matter canon, and connecting them with present-day developments, has forced us repeatedly to make tough choices about examples. We have tried to pick representative experiments or results from topics which are likely to continue to be actively explored in the coming years, and to include references to reviews that will give a suitable entry to the topic to students who would like to know more. Inevitably, the interests and knowledge of the authors introduce an element of bias in these choices.

We have split the problems which come with every chapter as much as possible into small, concrete steps. Here, too, we have been led by our experience with undergraduate students and the feedback we have received from them. As much as possible, the problems have been designed so that if a student has difficulty with one particular step, they should be able to move on to the next. The step-by-step format of the problems should also make them particularly suitable for active learning and reverse classroom settings. Similarly, instructors can easily transform these problems into advanced lectures by integrating mathematical details into the more qualitative introductions we provide in the main text. We have successfully adopted this approach ourselves when teaching the material in graduate classes. We hope that the step-by-step solutions available in the instructor manual will help other instructors achieve this goal. The more advanced problems are marked with an asterisk, the most challenging ones with two asterisks.

Students are also encouraged to deepen their understanding of the various topics by simulating simple processes on a computer. In order to facilitate updating and downloading of code, and to include links to relevant other material, we have made suggestions for coding problems available on the website www.softmatterbook.online complementing this book.

The topics to treat if this book is used for a course will naturally depend on the background and level of the students. The chapters in part I of the book have been included for students like most of our own, who have not yet had an introduction to fluid dynamics and elasticity theory, and who would like a short refresher on fluctuations. Even though the introductory parts of these chapters could be skipped by some students, the more advanced parts connect the classical fields with more modern developments that may be new even to some professors. So we recommend paying attention to these extensions. Part II contains the core material of the book; of this we suggest studying at least chapters 4–6, and time permitting also chapter 7. Whether or not any of the advanced topics of part III are included will depend very much on the background and interests of the students and the number of hours available. They can be left out of an introductory course without harm. The material in these chapters (possibly supplemented by selected readings from earlier chapters or from introductory textbooks on dynamical systems) could form the basis of an advanced graduate course emphasizing non-equilibrium aspects of soft matter physics. We had positive experiences teaching parts of this advanced material in summer schools also attended by postdocs and colleagues. We end the book in part IV with a brief perspective on new frontiers in soft matter research. Unlike the previous chapters, the one in this part is much less in the style of a textbook—it primarily gives a glimpse of emerging new directions, mostly by way of examples. These examples and corresponding pointers to the literature provide plenty of inspiration for students to pick end-of-course projects aimed at independently studying papers and presenting them in active learning sessions. The projects can complement our problems as a more dynamic way of getting students engaged and facilitating their transition to research.

This book of course reflects our own understanding of soft matter, as well as our own specific interests and style. Both have been shaped by our own teachers and by interactions with many colleagues worldwide who shared their knowledge and passion with us. WvS would like to take this opportunity to express his indebtedness to two former colleagues at Bell Labs, John Weeks and the late Pierre Hohenberg. VV would like to thank David Nelson for allowing him to see beauty in condensed matter physics through his eyes. ZZ would like to thank Sid Nagel, Martin van Hecke, and Michael Brenner for their long-lasting mentorship and collaboration. Hopefully this book reflects how each of them, in his own way, set an inspiring example for our careers, for how to approach physics, and for writing with passion and clarity.

Over the years, we have had the privilege of interacting and collaborating with many wonderful colleagues who have shared their insights with us. Our understanding of the topics treated in this book has benefited in particular from discussions and collaborations with Daniel Aalberts, Alexander Abanov, Andrea Alù, Ariel Amir, Denis Bartolo, Katia Bertoldi, José Bico, Daniel Bonn, Mark Bowick, Erez Braun, Michael Brenner, Carolina Brito, Jasna Brujic, Christiane Caroli, Mike Cates, Paul Chaikin, Hugues Chaté, Pat Cladis, Adam Cohen, Itai Cohen, Corentin Coulais, Chiara Daraio, Olivier Dauchot, Benny Davidovitch, Juan De Pablo, Martin Depken, Zvonimir Dogic, Marileen Dogterom, Ute Ebert, Wouter Ellenbroek, Nikta Fakhri, Alberto Fernandez-Nieves, Daan Frenkel, Joost Frenken, Michel Fruchart, Margaret Gardel, Luca Giomi, Paul Goldbart, Nigel Goldenfeld, Ray Goldstein, Ramin Golestanian, Ming Han, Silke Henkes, Martin Howard, David Huse, William Irvine, Heinrich Jaeger, Randy Kamien, Nathan Keim, Kinneret Keren, Daniela Kraft, Ludwik Leibler, Stan Leibler, Henk Lekkerkerker, Dov Levine, Peter Littlewood, Andrea Liu, Detlef Lohse, Teresa Lopez-Leon, Tom Lubensky, Andy Lucas, Tony Maggs, Lakshminarayanan Mahadevan, Vinny Manoharan, Cristina Marchetti, Alexander Morozov, Arvind Murugan, Sid Nagel, David Nelson, Peter Palffy-Muhoray, Deb Panja, Ji-woong Park, Jayson Paulose, Joey Paulsen, David Pine, Wilson Poon, Patrick Oakes, Sriram Ramaswamy, Pedro Reis, Olivier Rivoire, Ben Rogers, Benoit Romain, Chris Santangelo, Sri Sastri, Michael Schindler, Jim Sethna, Boris Shraiman, Jacco Snoeijer, Ellák Somfai, Anton Souslov, Francesco Stellacci, Kees Storm, Sebastian Streichan, Shashi Thutupalli, Brian Tighe, John Toner, Federico Toschi, Ari Turner, Suri Vaikuntanathan, Jan-Willem van de Meent, Willem van de Water, Martin van Hecke, Hans van Leeuwen, Brian Vansaders, Dave Weitz, Max Welling, Tom Witten, and Mathieu Wyart. We suspect virtually all of them will be able to identify particular choices, viewpoints, or wordings which they recognize as reflecting our interactions—we owe you a big thanks!

In addition, WvS would like to thank Luca Giomi for graciously sharing his notes from an earlier soft matter course when WvS started teaching the course which eventually stimulated his writing this book, and Zhihong You and Ludwig Hoffmann who as teaching assistants developed several problems for the course; some of these found their way to this book. Similarly, VV would like to thank Vinzenz Koning, Richard Green, Tali Khain, Noah Mitchell, Colin, Scheibner, Jonathan Colen and Luca Scharrer for serving as teaching assistants in the courses he taught at Leiden and Chicago and helping in preparing problem sets, solutions, and lecture notes. We thank Luca Scharrer and Ege Eren for preparing typeset solutions of the problems for the instructor manual. Finally, we would like to thank Yael Avni, Chase Broedersz, Sujit Datta, John Devany, Marjolein Dijkstra, Daan Frenkel, Michel Fruchart, Tali Khain,

Daniela Kraft, Henk Lekkerkerker, Detlef Lohse, David Martin, Alexandre Morin, Alexander Morozov, Michael Schindler, Daniel Seara, Kees Storm, Sebastian Streichan, and Martin van Hecke, who provided input or feedback during the writing process, for their help and their advice and Andrej Mesaros for his generous help, support, and advice throughout the whole process.

We would also like to express our gratitude to the great many colleagues who were kind enough to provide us with high-resolution images or plots from their earlier work. Their names are given in the credit list at the end of the book.

Finally, we would like to thank several staff members of Princeton University Press for their warm, dedicated, and eminent support: Ingrid Gnerlich for stimulating us to write this book, and for advising and guiding us through the application, writing, and review procedure; Whitney Rauenhorst for her help and advice on the figures; Natalie Baan for overseeing and coordinating the production; Dimitri Karetnikov for invaluable advice on finalizing the art; and copyeditors Bhisham Bherwani and Will DeRooy for meticulously going through the manuscript to preserve consistency of style and presentation, and ensure use of proper English.

You will be able to find supplementary material and coding problems for each chapter on our book's website www.softmatterbook.online. We will also keep a list of errata on this website and will be grateful to readers who send us any comments on the material and the way we present it, or suggestions for additional computer simulations. You can contact us via this website.

Leiden, Chicago, and Paris
Wim van Saarloos, Vincenzo Vitelli, Zorana Zeravcic
September 2023

Soft Matter

The Challenges, Relevance, and Fun of Soft Matter | Introduction

Welcome to our exploration of the challenges, relevance, and fun of soft matter. We hope that this introduction will entice you to join us on this fascinating journey.

The term "soft matter" refers to the subfield of condensed matter physics focused on matter which is soft in the sense that it is easily deformed under mechanical stresses. In the simplest cases the ensuing deformations of a large sample are elastic and reversible, but more often than not, structural changes in the material take place under the applied stress and the response is non-reversible. The material might even flow like a liquid—often it depends on the time scale of the perturbation whether the response is more solid-like or more liquid-like. Defined this way, "soft matter" is a very generic term. But a key element of what we typically gather under the name "soft matter" is that it consists of well-defined or recognizable structural units, e.g., long chains of molecules called polymers that themselves consist of very many atoms, but whose interactions are weak enough that their collective response is soft and often not captured by linearized or traditional theories.

Typically, thermal fluctuations also play an important role on the scale of these structural units. This means that some of the relevant interactions on the scale of the structural building blocks, either between them or within them, are on the order of a few k_BT, the thermal energy at room temperature. Many biomaterials are soft materials, and thermal fluctuations typically play an important role on the molecular and cellular scales of living matter.

Though many of the systems that we will encounter have long been around, interest in them in physics is more recent. Indeed, it was gradually realized that soft materials pose their own problems and challenges. Soft materials are often characterized by the confluence of three elements that make their physics interesting but also complicated. First, their response is often *nonlinear* due to the large deformations at play. Second, they are often *out of equilibrium* due to their propensity to flow easily. Third, their structure is often *amorphous* or disordered, unlike the crystals typically studied in solid state physics.

The confluence of these traits gives soft matter problems a particular kind of intellectual appeal, and the field lends itself to progress through playfulness, minimalist tabletop experiments, as well as virtuoso chemical synthesis and state-of-the-art experimentation, and intellectual agility (besides formal mathematical theories). It is

Figure I.1. Pierre-Gilles de Gennes, who in 1991 received the Nobel Prize in Physics "for discovering that methods developed for studying order phenomena in simple systems can be generalized to more complex forms of matter, in particular to liquid crystals and polymers." Picture courtesy of Marc Fermigier.

The French term for 'soft matter,' *matière molle,* reportedly was proposed in 1970 by de Gennes's collaborator Madeleine Vieyssié.

1

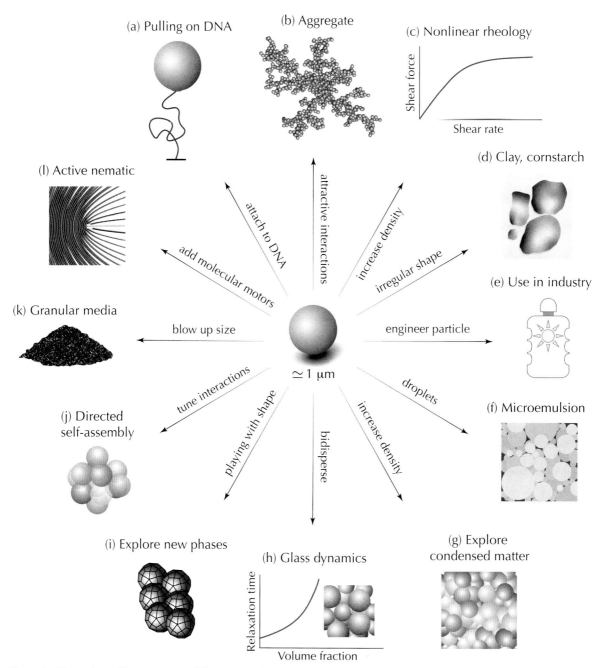

Figure I.2. Illustration of how we can modify a single colloidal particle of about a micrometer in size in different ways to create a plethora of soft matter systems. Some examples are relevant to applications, others play a role in bioscience or daily life, and yet others are nice model systems that allow us to pursue fundamental questions. All the examples are discussed in more detail in section I.1.

also a great playground for illustrating general concepts and connections between seemingly disconnected fields of science. Gradually, since the 1970s, accelerated by the awarding of the Nobel Prize in Physics to de Gennes in the 1990s (see figure I.1), soft matter gained recognition as a field in its own right, with its own richness and attraction. Ever since, the field has gained enormous traction, also due to its relevance to industry, its connections with bioscience and complex systems, new optical imaging techniques, and the rapidly increasing computer power for imaging, data analysis, and simulation.

I.1 Inspiration from an example

What makes soft matter special and inspiring to us is illustrated in figure I.2. Here, we start with a very simple building block at the center of the figure and let it explode into a diverse range of complex systems. We choose as our basic building block a so-called colloidal particle, a solid spherical particle with a diameter of about $1\,\mu\text{m}$ (one micrometer), say, dissolved in a fluid such as water. Such a particle is small enough to undergo Brownian motion visible through a microscope, due to the random kicks it experiences from the solvent molecules. This behavior was observed in experiments by Jean Perrin over 100 years ago when he tested Einstein's theory of Brownian motion; see figure I.3.

You probably won't consider a single Brownian particle in a fluid as soft matter per se. But it is one of its important building blocks. As figure I.2 illustrates, when we take this colloidal particle as a starting point and explore all kinds of extensions and modifications, we inevitably land in the realm of soft matter. We would like to start our journey with a kaleidoscope of examples that illustrate this point. Do not be disheartened if some of the systems or terms are new to you. Most of them will return in more detail later in the book.

a. Using a colloid to pull on DNA

A small micron-sized colloidal particle can nowadays easily be manipulated in so-called optical traps. If you attach the particle to DNA, you can use the trapped colloid to pull on the DNA and thus measure its elastic properties, by plotting curves of the force against the extension. Figure I.4 illustrates simple examples of pulling on DNA. The pulling force, sketched with red lines in the upper panel, is decreased in time, and the resulting change of the extension of the DNA with time is depicted with blue lines in the lower panel.

The figure illustrates experiments under two different conditions. The dashed lines indicate behavior in a solution with no salt added.

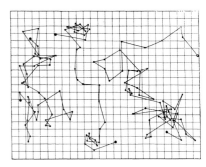

Figure I.3. Perrin's original drawing showing the displacement of three emulsion droplets. Perrin marked positions of each droplet at regular time intervals, and then joined the dots to illustrate the trajectories.[1] Brownian motion is discussed in chapter 3.

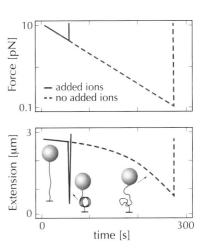

Figure I.4. Schematic rendering of experiments by Besteman et al., 2007 on pulling double-stranded DNA with a colloidal particle in an optical trap, starting from a completely stretched configuration. Depending on the solution in which the experiment is done, two types of behavior are observed as the pulling force is lowered. With ions added to the solution (solid line), the repulsive electrostatic forces between DNA segments are screened and the DNA coils up tightly when the force decreases. When the forces are not screened (dashed lines) the DNA does not coil up and one measures the force-extension curve resulting from the DNA behaving like a random coil. The width of the shaded area indicates the size of fluctuations.[2] Such types of experiments are discussed in section 3.6.1 and section 5.4.4.

As the lower panel shows, when the pulling force becomes low, the polymer behaves like a random coil (as sketched in the bottom panel): fluctuations in the extension (indicated by the shaded area) are large. The solid lines illustrate what happens when ions are added to the solution. Electric charges on the DNA that keep it straighter become less important as they are 'screened' by the ions, so the DNA reduces its extension by coiling up. We shall discuss in this book how thermal fluctuations make the DNA behave as an elastic spring, and how ions can screen electric forces, and thus affect biopolymers like DNA. Note that these forces are really tiny, on the order of piconewtons (pN). Other experiments of this type allow one to probe the effects of twisting and curling the DNA, or even to explore how the DNA sequence is reflected in the force variations needed to rip the two DNA strands apart.[3] Such studies are not limited to DNA; one can nowadays also attach particles to molecular motors and even observe fluctuations in their steps. This takes us into the realm of biomatter.

b. Formation of fractal aggregates due to attractive interactions

When we have a dilute solution of colloidal particles with attractive interactions, the particles aggregate slowly and form very open, floppy structures; see figure I.5. These structures are fractals, objects whose dimensionality is not an integer. This so-called fractal dimension depends on the processes that affect how fast the structures form. It could be either the diffusion rate of the particles or the reaction rate with which the particles stick to one another. The fractal dimension of such aggregates is typically determined experimentally by scattering light. We'll see that dilute polymers (i.e., long chains of molecules) in solutions also form very open structures with a fractal dimension. This dimension can be measured with neutrons rather than light (section 5.5.1.a).

c. Shear thinning and shear thickening of dense suspensions

Imagine you start with a dilute suspension, i.e., a heterogeneous mixture, of noninteracting colloidal particles in water, and you start increasing their concentration. For low concentrations, the suspension flows like water, albeit with a somewhat enhanced resistance to motion called viscosity. We say that the suspension behaves like a Newtonian fluid, a fluid whose flow is characterized by a viscosity. However, by the time the volume fraction occupied by the particles has become as large as about 30%, an interesting effect occurs, as illustrated in figure I.6. The suspension shows *shear thinning*, meaning that the effective viscosity decreases with increased shear rate. The shear rate denotes the rate of change of the velocity at which a fluid layer flows with respect to an adjacent layer. As depicted in the figure, the origin of shear thinning is the layering of particles. A fluid with a shear-dependent viscosity is denoted as 'non-Newtonian' or *complex*. When the concentration of

Figure I.5. Fractal aggregate grown in a petri dish. For so-called diffusion-limited aggregation in two dimensions, the fractals exhibit a fractal dimension $d_f \approx 1.71$.[4] We treat fractal structures in section 4.5.1, and we treat the light-scattering technique used to probe the fractal dimension in section 3.7.

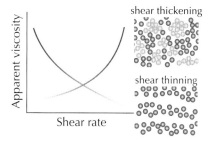

Figure I.6. Illustration of shear thinning behavior for small concentrations due to layering of particles and shear thickening at larger concentrations due to clustering of colloidal particles. The issue will be discussed in section 4.6.[5]

An everyday example of a shear-thinning fluid is blood, due to the presence of the red blood cells (which of course are not perfect spheres, and which deform under strong shear).

particles is increased even more, to around 60%, the opposite effect typically occurs: the effective viscosity now increases with shear rate, because obstructing structures form during flow. Almost all soft matter systems behave as *non-Newtonian* or *complex fluids* when flowing. Rheology is the time-honored field that studies how these materials deform and flow.

The main features and consequences of the complex fluid rheology of polymers are discussed in section 5.10.

d. Particles with irregular shapes: Even more complex rheology

When we change the particles' shape from a nice round sphere, we move quite close to what everyday materials like clay and cornstarch look like microscopically; see figure I.7. Indeed, these strongly exhibit complex fluid behavior (sometimes also due to charges on the particles)—just watching one of the many YouTube videos of people walking on cornstarch should be sufficient to convince you of its amazing and surprising rheology: on short time scales it resists deformations and holds your weight like an elastic solid, but on longer scales it flows like a viscous liquid.

e. Applications like sunscreen cream and paint

Let's move for a moment into a more application-oriented direction: sunscreen creams are essentially dense suspensions of particles engineered to block UV light, while when you zoom in on paint, as figure I.8 illustrates, you discover that it consists of many irregular particles. These particles spontaneously organize themselves into layered structures and are made to crosslink, i.e., stick to each other, when the solvent evaporates upon drying. A lot of science (chemistry!) and empirical know-how goes into optimizing the surface properties of paint particles.

f. Droplets rather than solid particles: Microemulsions

Let us imagine we change the constituent particles from solid to liquid: we obtain what is technically known as a microemulsion of droplets of one phase, say oil, in another, say water. Milk, mayonnaise, and yoghurt are typical everyday emulsions. In milk the suspension is not so dense, so milk behaves for most purposes as a regular Newtonian fluid like water. But mayonnaise is a dense suspension of oil droplets of various sizes; see figure I.8. As you know from daily life, mayonnaise is a *complex fluid* with abnormal flow properties. It is said to have a *yield stress*, which means in simple terms that it stays put on a spoon. The yoghurt you eat in the morning also has interesting rheological properties: it *ages*! If you leave it undisturbed it gradually shows solid-like behavior, but if you stir it for a moment it flows again quite easily. Microemulsions play a role in many industries, as the large surface-to-volume ratio speeds up many processes like, e.g., extraction of dissolved molecules from solution. The tragic Covid-19 crisis brought many soft matter phenomena to the fore, such as the breakup and evaporation of virus-bearing mucus droplets during coughing.

Figure I.7. Upper row: cornstarch (left) and clay (right) consist of very irregular particles. Below: two images illustrating that one can walk on cornstarch.[6] This is a vivid example of the remarkable properties that complex fluids can exhibit.

Figure I.8. On the scale of microns, paint (left) consists of particles which organize into layered structures that crosslink when the solvent evaporates. Mayonnaise (right) is a stable emulsion of oil droplets. Mayonnaise only starts to flow under a sufficiently large force, so it is an example of soft matter with a yield stress.[7] Droplets are discussed in section 1.15, yield stress and rheology of emulsions and colloidal suspensions in section 4.6. Emulsion image courtesy of Angus McMullen and Jasna Brujic.

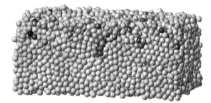

Figure I.9. With modern imaging techniques, one can follow the precise location of all the colloidal particles. In this case, local response to an applied strain of particles in a disordered dense packing is indicated with a color code, as detailed in figure 4.14.c. Image courtesy of Peter Schall.[8] Section 4.5.4 puts such experiments in context.

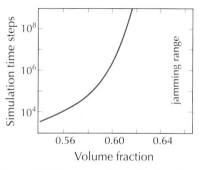

Figure I.10. Rendering of the relaxation time (measured in terms of the number of Monte Carlo steps) in simulations of a system of polydisperse spheres. Redrawn from "Equilibrium Sampling of Hard Spheres up to the Jamming Density and Beyond" by Berthier et al. (Phys. Rev. Lett. 116, 238002).[9] In section 4.5.4 we discuss why colloids are studied in the context of the glass transition.

g. A model system with which to probe the dynamics of disordered solids

Let us return to our spherical solid particles in suspension, in the absence of external forces or imposed shear deformations. The particles still explore phase space due to thermal fluctuations. So if we change their density we can explore how they form crystals or disordered solids, and how these respond to forces. We can in addition play with particle shapes and interactions. Thanks to recent advances in optical imaging and digital analysis, we have powerful experimental model systems with which to study basic questions of condensed matter, from crystalline to disordered phases and their formation. Indeed, experimentalists can nowadays follow all individual particles, which play the role of the constituent atoms of solid state systems. Figure I.9 shows an example from such a study.

h. A model system with which to study the glass transition

One of the salient features of glasses is that, as the temperature is lowered, the viscosity rises rapidly by many orders of magnitude: motion slows down to the point that it becomes essentially unnoticeable on human time scales. This reflects the fact that the constituent particles increasingly get arrested. Whether there is an underlying phase transition—and, if so, what its nature is—remains a matter of debate and ongoing research.

Soft matter models and ideas have recently thrown new light on these issues. Indeed, a collection of hard colloidal particles which is sufficiently polydisperse (i.e., the particles have different sizes) will not easily crystallize. As a result, when the volume fraction of the particles increases, one can follow how the system gets 'arrested,' i.e., how particles get stuck. Figure I.10 shows an example from a computer simulation of such a system. The relaxation time (i.e., the typical length of time the system takes to relax back into an unperturbed state after a disturbance) is plotted as a function of particle volume fraction: it rises by more than five orders of magnitude as the critical packing fraction at which the system jams is approached.

i. Playing with the shape of the hard core particles

If we think again about particles which experience Brownian fluctuations but have no interaction except for their hard core repulsion (meaning that the particles have infinite interaction when they overlap), we can consider changing their shapes, and explore what types of phases they form. It has become clear in recent years that a plethora of phases can be formed from hard polyhedra. A few simple phases found in simulations are shown in figure I.11.

In finite temperature (T) simulations with hard core interactions, the ordered phases minimize the free energy $F = U - TS$ by maximizing the entropy S. The thermodynamically stable phase is the

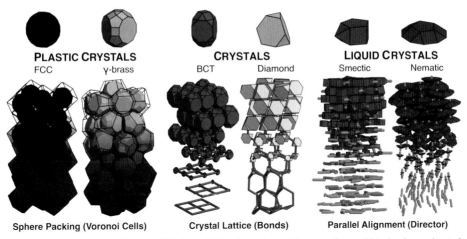

Figure I.11. If one experiments with their form, polyhedra with hard core interactions can form all kinds of complicated phases. Image from Damasceno et al., 2012.[10] The question of the formation of nontrivial phases of particles with hard core repulsion will return in section 4.2.1 when we discuss colloids and in section 6.1.1 in the context of our discussion of liquid crystals.

one where the particles have the most room to wiggle around, thus maximizing their entropy, while the internal energy U remains zero as nonoverlapping hard core particles don't interact.

j. Tuning the interactions: Toward programmable matter

Instead of changing particle shape, let us go back to micron-sized monodisperse spherical particles, but tune the interactions between them, i.e., make the particles distinguishable. In practice this can be done by coating the surface of the particles with (different) short single strands of DNA. For example, a particle of type A can strongly attract B, moderately attract C, not attract D, etc. Doing this for all the pairs allows us to define an *interaction matrix* that specifies how all the different particle types interact.

Figure I.12 shows an example of how one can tune the interactions of dozens of particles so as to make specific desired structures favored. As we will discuss in chapter 10, some basic questions concerning artificial life can be studied with such types of 'programmable matter.'

k. Blowing up the size: From colloids to granular matter

Imagine we increase the size of our micron-sized particle a thousand-fold and consider a large enough volume fraction so that many spheres inevitably touch—we have thus constructed a nice model system for granular media. In a more realistic granular medium, such as a heap of salt or the flowing grains shown in figure I.13, the particle shapes are irregular and frictional forces play an important role too. It has been realized in recent years that these athermal granular media—the particles are so heavy that

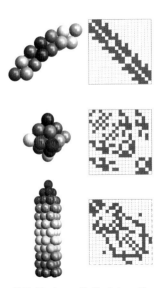

Figure I.12. Playing with the interactions between all pairs of particles allows one to explore how certain structures are favored and formed by thermal fluctuations. The top structure is a 19-particle chiral chain, the middle one a 19-particle square bipyramid, and the bottom one a 69-particle tower. Corresponding interaction matrices are shown to the right of each structure (the one for the tower structure is only partially shown). Each blue element in the symmetric matrix M_{ij} indicates the presence of an attractive interaction between particles i and j.[11] The topic is treated in chapter 10.

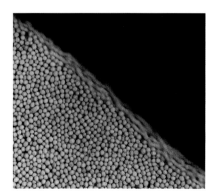

Figure I.13. A granular medium consisting of irregular millimeter size grains. The photo was taken with a relatively long exposure time, so that the grains flowing near the inclined surface are blurred. Clearly, the flow is localized to the top layer of a few grains. Image courtesy of Sid Nagel.[13] Granular media are discussed in section 2.10.

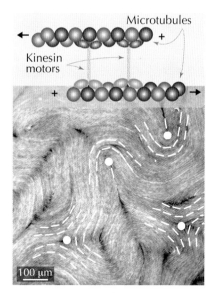

Figure I.14. A still from a continuously changing pattern obtained by microtubules and molecular motors sketched in the top. White lines are superimposed on the image to highlight the nematic order while defects in the nematic order are marked with white dots. Figure courtesy of Sattvic Ray and Zvonimir Dogic.[15] Active nematics are the subject of section 9.5.

thermal fluctuations don't matter—display striking similarities to thermal systems of Brownian spheres which are worth exploring (see section 2.10).

Interestingly, granular media research has yielded key insights for glass research, and vice versa. Theory and experiments on dense packing of shaken disks have convincingly demonstrated the importance of capturing density inhomogeneities in glasses using higher order correlation functions (e.g., the density is evaluated at three points in space), rather than the pair correlation functions traditionally employed in statistical physics.[12]

l. Self-propelled particles and active matter

We end this list of examples by mentioning one more research direction: the field of active matter, the subject of chapter 9, in which one studies collections of entities which can move by themselves. Swimming bacteria are a good example. It has long been known that sufficiently dense suspensions of bacteria naturally show collective swarming effects. This is essentially due to the tendency to align their swimming direction via hydrodynamic interactions.[14] More recently, particle-based systems composed of self-propelled colloids have been developed that mimic the swarming behavior of fishes and birds, which also tend to align the direction in which they swim or fly to that of their neighbors.

In microbiology, we encounter another interesting but less intelligent type of active matter, molecular motors, which play an important role for transport in cells. An interesting twist in active matter research is that of mixing microtubules (pretty stiff long biofilaments) with molecular motors and ATP (adenosine triphosphate) molecules, which the motors need as fuel. Figure I.14 shows a snapshot from the continuously swirling type of flow patterns one observes in this system. As the long microtubules tend to line up, they exhibit so-called nematic liquid crystalline order. However, the patterns are very much dominated by defects in the alignment of the filaments that are specific to nematics. In short, we have active matter with defect-driven liquid-crystal-like patterns!

I.2 Our view of soft matter and our approach

We could do an exercise very similar to what we did in figure I.2, starting from long molecular chains, the characteristic building blocks of polymers or liquid crystals, but we hope that this kaleidoscope of examples, reflected in the way we approach the field in this book, already gives you a glimpse of the appeal of soft matter.Some subdisciplines in physics have a clear and well-defined theoretical framework that forms the canon of the discipline. Soft matter, on the other hand, cannot be captured in terms of a single framework or a few basic concepts. As a result, it pays to approach problems from various perspectives and to take advantage of similarities that transcend traditional boundaries of (sub)disciplines, the more so since different concepts are relevant on different scales.

In short, we intend to portray soft matter as we perceive it: a diverse and somewhat iconoclastic field that attracts people with different backgrounds and training, and we invite you to join us in this tour.

Soft matter is a field which resonates at every level with the basic theme of Phil Anderson's famous "More Is Different" article (Anderson, 1972). It is a strong plea to acknowledge that each level of natural phenomena poses its own set of challenges that require new concepts to be introduced for understanding them.

I.2.1 Our approach in this book

The aim of this book is to serve as a textbook on the basic phenomena and concepts of soft matter for advanced undergraduate and beginning graduate students: we aim to present an introductory overview of the various soft matter phases and some of the key conceptual frameworks used to analyze them. The book should be a stepping stone for further specialization, while students who specialize in other subdisciplines will acquire a good overview of the field, and get familiar with concepts and a way of thinking that has broad applications. We also intend to give students whose main research focus is in biomatter or biophysics the necessary background to understand the application of soft matter concepts in life sciences. Therefore, throughout the book, examples of soft matter applications or concepts in biomatter are emphasized whenever possible. The book also pays attention to applications of the topics treated in science and technology.

The examples and discussion of the previous section illustrate the challenge of writing such an introductory book. How do we, without becoming encyclopedic, do justice to a field which branches out in many directions, and which thrives on cross-fertilization and interactions with other subdisciplines and applications? Our choice has been to focus on the main concepts and phenomena from a physics perspective (our own background), and to try to select applications which are interesting and instructive, in that they enhance one's understanding of important concepts.

In view of our conviction that it pays off not to draw sharp boundaries with other fields, we deliberately try to be open-minded in our choice of topics too: some examples and applications that you might justifiably not consider soft matter per se have been included because they are sufficiently illustrative and interesting for a soft matter scientist to become familiar with. Nonetheless, a number of important subjects have only been touched upon.

The syntheses of polymers, liquid crystals, and colloids are separate and well-developed subdisciplines, so there are many introductory books for them, each with its own focus and approach. For a starting point, we mention three popular books which have been revised and extended over the years: Carraher Jr., 2017 for polymers, Collings and Goodby, 2019 for liquid crystals, and Shaw, 2013 for colloid and interface chemistry.

For example, macromolecular chemistry is a subdiscipline of chemistry that lies at the basis of making the building blocks of soft matter; it is a large and important field in itself that we do not cover in detail by choice. Likewise, surface chemistry is important for manipulating the surface of colloids so as to tune them to the desired properties or to optimize industrial processes in using colloids or more generally dispersed media. As a result of our physics-oriented approach, we take the building blocks of soft matter and their properties mostly as given, while we refer you to other sources for detailed explanations on how to make them.

I.2.2 The hydrodynamic perspective

We started this chapter with a bottom-up approach aimed at illustrating how complex soft matter phases and phenomena arise starting from microscopic building blocks as familiar as colloidal particles or droplets; see figure I.2. We now provide a glimpse into a complementary, top-down approach that is often employed to systematically model soft matter systems. As before, we will start with a pictorial representation of the key ideas as illustrated in figure I.15.

The top row lists three examples of (soft) condensed matter *phases*: a crystal in which point particles are regularly arranged on a square lattice, a so-called nematic liquid crystal in which elongated molecules are aligned along a common axis, and finally a flock of birds all moving along the same direction. Despite their obvious differences, these three systems (and many more!) can be analyzed and modeled through a unified approach built around common concepts listed in the gray column of figure I.15. The second row shows that the three phases can be classified according to what symmetries they possess and, more crucially, what symmetries they break. The third row introduces a quantity called order parameter, which measures how ordered each of the three systems is. The last two rows reveal how each of the three systems can be perturbed by smooth periodic modulations of the ordered phase called hydrodynamic modes (fourth row), of which long-wavelength sound modes are a familiar example, and by even more drastic distortions called topological defects (fifth row).

As we said before, do not get discouraged if aspects of figure I.15 and the accompanying discussion seem abstract or if the associated terminology comes across as technical. Plenty of examples will follow naturally in the remainder of this book. For now, we refer to figure I.15 as the *hydrodynamic perspective* and sketch its main features. Here the word *hydrodynamics* is used to indicate a phenomenological approach that goes beyond the study of water, coffee, or any other fluid.

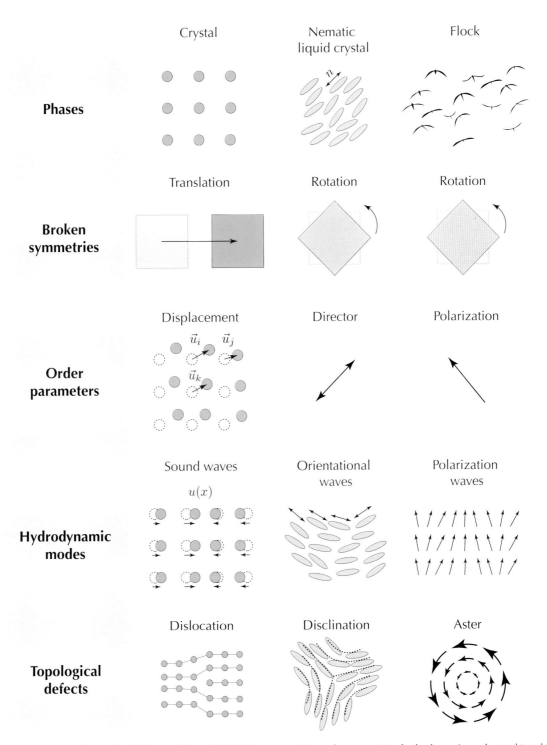

	Crystal	Nematic liquid crystal	Flock
Phases			
Broken symmetries	Translation	Rotation	Rotation
Order parameters	Displacement	Director	Polarization
Hydrodynamic modes	Sound waves	Orientational waves	Polarization waves
Topological defects	Dislocation	Disclination	Aster

Figure I.15. Soft matter phases classified according to their broken symmetries, order parameters, hydrodynamic modes, and topological defects represented as rows. The first column represents crystalline phases, the second column nematic liquid crystals (see chapter 6), and the third column flocking (chapter 9). The orientational order parameter of the nematic phase, the director $\hat{n}$, is drawn as a double-headed vector to express that the orientations $\hat{n}$ and $-\hat{n}$ are equivalent (see section 6.1.4 for further discussion).

A coarse-grained equation is a mathematical description of a many-body system in which some of its fine details, e.g., individual particle coordinates or velocities, have been smoothed over in favor of continuous fields.

The hydrodynamic approach is a prescription to construct 'coarse-grained' equations governing the macroscopic behavior of any physical system, starting from symmetries and conservation laws, without assuming or even needing detailed knowledge of its microscopic building blocks. It is pretty much the opposite approach to how we constructed figure I.2.

a. Spontaneously broken symmetries and associated order parameters

In the next chapter, we will show in detail how to derive the so-called Navier-Stokes equations that describe the evolution of the velocity fields of a fluid like water using only conservation of mass, momentum, and energy. When viewed from the perspectives of symmetries, a fluid is a very simple state: it is invariant with respect to arbitrary rotations and translations. However, this is not the case for other soft matter phases. For example, a crystal is invariant only under translations by a lattice spacing; see the first column of figure I.15. Hence, during a phase transition from a liquid to a crystal, the symmetry is lowered—the crystal is said to be a *spontaneously broken-symmetry phase*.

In the example considered here of a fluid freezing into a solid, translational symmetry is said to be *spontaneously* broken because the microscopic physical laws remain symmetrical throughout the process, even if the crystal phase manifestly breaks translational symmetry.

In order to quantify the amount of order in a spontaneously broken-symmetry phase, we need to define an additional field called an order parameter. For a crystal, the order parameter is the displacement that measures the deviation of the atoms from a perfect lattice; see the first column in figure I.15. Note that what really matters here are gradients in displacement because the crystal is invariant under constant particle translations. In order for the atoms to remain in their periodic arrangements and resist mechanical stresses or thermal fluctuations, the crystal must be endowed with a *rigidity* (i.e., the ability to resist deformations) whose strength is controlled by an elastic modulus. This is a generic feature of all phases with a spontaneously broken symmetry, not only the crystal, as shown in the last two columns of figure I.15, where a similar analysis is illustrated for the nematic and flocking phases.

Our discussion in this section of *broken symmetries* and the associated *order parameters* is very much inspired by that of Anderson, 1997.

b. Hydrodynamic modes and topological defects

Long-wavelength (also known as hydrodynamic) sound modes (or waves) are periodic modulations of the particle displacements (or, more generally, the order parameter) that nearly approach a uniform translation. Since their appearance restores the broken translational symmetry of the fluid, they cost only a nearly vanishing elastic energy. When the wavelength of the sound modes goes to infinity, the interparticle distance is left nearly constant.

More drastic distortions of the crystal, called dislocations, are created by removing a whole row of atoms. They have an energy cost that diverges as the size of the system is taken to infinity. These imperfections are examples of a broad class of objects called topological defects. The name *topological* stems from the fact that the presence of an isolated topological defect will not go unnoticed even far away from it, because any path (no matter its shape) encircling its center will reveal the existence of a missing row of atoms: there is a missing step equal to an integer multiple of the lattice spacing. Like hydrodynamic modes, topological defects are a ubiquitous feature of broken-symmetry phases, and they can be classified (using the mathematical language of group theory) according to the symmetry of the relevant order parameter; see the last two columns of figure I.15 for illustrations of the corresponding entries for nematic and flocking phases.

c. Classifying (soft) condensed matter phases by their broken symmetries

In the second column of figure I.15, we consider phases with broken rotational symmetry called nematic liquid crystals, familiar for their use in computer displays. As we will discuss in chapter 6, in a nematic phase, interactions between elongated molecules favor their alignment along a common orientation that singles out a specific direction in space, hence breaking rotational symmetry. But, unlike in a crystal, the molecules are not constrained to lie on a lattice: there is no translational order, only orientational order. Here, the order parameter, called a director, is a double-headed vector field (i.e., a line field) that measures the local deviations of the orientation of the molecules from their average direction. This director is the order parameter whose time evolution and orientational elasticity, i.e., the tendency of the molecules to resist gradients in their orientation of alignment, are crucial to understanding the physics of nematic liquid crystals. Also in this case, the broken orientational order can be restored by long-wavelength orientational waves. Similarly, topological defects called disclinations exist (last panel of the second column of figure I.15), and they can be classified according to the angle by which the director rotates around the defects' center; see chapter 6 for more details. This angle must be an integer multiple of π because the order parameter is a double-headed vector invariant under a rotation by π.

As a last example of this approach, consider the flocking phase that describes the alignment of self-propelled agents such as birds or fishes. In this case, in addition to a spontaneously broken rotational symmetry, Galilean invariance is also explicitly broken. What that means is that there is a preferred frame (set by the air or water) with respect to which the self-propelled agents like birds and fish move. In this case, the order parameter is a vector, called polarization, tracking the average velocity field of the self-propelled particles; see the third column of figure I.15. Hence both the hydrodynamic modes and the topological defects take new forms compared to nematics, but the strategy used to model them is the same as in the previous examples.

The flocking behavior of active media is discussed in section 9.2.

Rotational symmetry is spontaneously broken during the transition from a disordered phase (where the birds interact so weakly compared to noise that they fly in random directions) to the flocking phase (birds spontaneously pick a direction of alignment). Galilean invariance, on the other hand, is explicitly broken because each bird self-propels.

In the remainder of this book, we introduce the main classes of soft matter phases from this perspective, embedding explanations of key concepts and methods (scaling, correlation functions, Landau approach, renormalization group, scattering, etc.) without assuming previous knowledge of hydrodynamic theories or advanced statistical mechanics. We have made an effort to have concepts appear several times in different chapters and examples, as this will help deepen your understanding of soft matter and help you see how the various topics are interrelated.

I.2.3 A field relevant to society

Beyond its appeal from a basic science perspective, soft matter is a field extremely *relevant* to society with many real-world applications. Here, again, a range of scales plays a role. While in some cases (think of the process industry) its relevance stems from the ability to manipulate soft matter at the mesoscopic scales, in other cases its societal relevance stems directly from the macroscopic scale at which the properties of soft materials manifest themselves. Even if it can be traced down to molecular structure, the way soft materials deform or flow is often very tangible in daily life.

Think of accidentally spilling coffee, causing a stain. The tendency of particles to move to the rim of the spill, where most of the evaporation takes place, causing the familiar look of coffee stains, is explained in section 1.13.2. This phenomenon has direct relevance to the ink-jet printing industry.

For an intellectually stimulating and tasty introduction to soft matter science and cooking, we recommend the book and culinary creations of M. P. Brenner et al., 2020.

Soft matter research has helped launch major industries—think of liquid crystals, polymers and glasses—but you will also see that soft matter inventions play an increasingly important role in startups and consumer products, such as ones based on mechanical metamaterials.

As developments of new concepts and applications often go hand in hand, we have made the deliberate choice to integrate applications in the text where appropriate, rather than to list these at the end as an afterthought. In order to give some idea of the relevance and range of applications, let us list some examples. As most food is soft matter, the food industry relies on processing and controlling the different phases, stabilizing mixtures, minimizing degradation, and optimizing uptake.[16] Something similar holds for the process, cosmetics, painting and printing industries.[17] Polymer chemistry and physics are behind most consumer products made of plastic. The plastics industry is now in particular exploring new lightweight materials for insulation and the aerospace industry, biodegradable plastics, and soft electronics.[18] The pharmaceutical industry explores controlling amphiphile phases for optimizing drug delivery.[19] Covid-19 posed many soft matter questions, including virus spreading through aerosols.[20] Controlling droplet properties can help us minimize unwanted spreading of pesticides.[21] Glass with switchable properties and LCDs rely on the ability to switch liquid crystals with small electric fields.[22] Self-organized patterns create new material properties.[23] Soft matter robots, cloaking and auxetic metamaterials, origami, and programmable and designer matter provide new opportunities whose range of applications is only slowly becoming clear.[24]

I.3 Outline of the book and how to use it

Since coarse-graining and the hydrodynamic approaches are so much part of our description, and since flow, elasticity, and fluctuations are integral elements of the behavior of soft matter, we start our presentation by giving a brief introduction to the essentials from these topics in part I of this book. Even though we start each chapter with basic classical results, we deliberately end every chapter with some examples of modern directions. So even if you are familiar with the classical fields, or if these are not treated when this book is used for a course, we hope you will glance at those sections of the chapters in part I that convey present research directions, because they connect with the topics treated in later sections. Part II forms the heart of the book, and here we zoom in on the most salient examples of soft matter: colloids, polymers, surfaces, interfaces and membranes, and liquid crystals. In part III the focus is on two advanced topics: pattern formation and active matter. The latter chapter extends the material treated in part II to active matter. We end the book in part IV with a brief perspective on new frontiers in soft matter research. Unlike the previous chapters, the last one is much less in the style of a textbook—it merely gives a glimpse of emerging directions and opportunities.

A number of problems are included at the end of each chapter 1–9; we have split each problem into several small steps. This should make it possible to move on to the next step should you get stuck at a particular point. In many cases, the problems provide mathematical details that support the qualitative treatment of the material presented in the main text or offer real-world examples of soft matter research questions.

There are quite a few interesting videos on the web that illustrate or expound many of the topics discussed in this book. Because their URLs may change, we have not included these in the main text. Instead, we invite you to consult the website complementing this book: www.softmatterbook.online; it will be kept up-to-date. Computer simulations that are fun and instructive are also available via the website.

While our understanding of the formation of patterns in non-equilibrium systems was developed mostly in the last decades of the previous century, there are few textbook-style introductions to this topic. Since it plays a role in a large variety of soft matter, active matter and biomatter problems, we have included it in this book.

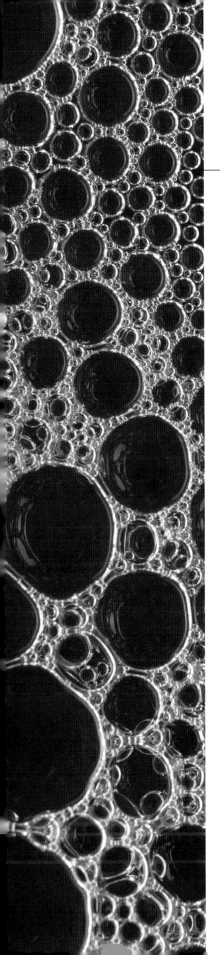

Part I

GROUNDWORK

FROM CLASSIC RESULTS TO SOFT MATTER TODAY

Fluid Dynamics | 1

As we sketched in the introductory chapter, soft materials often have unusual properties when they flow. They are indeed often examples of *complex fluids*. This term is used to set them apart from *ordinary* or *simple fluids*, like water, air, alcohol or glycerol. It is therefore useful to first discuss the continuum description of ordinary fluids, because it allows us to pinpoint in which sense the complex fluids behave differently from these in the later analysis of soft matter. A second reason why a good understanding of the behavior of classical fluids is important is that quite a few soft matter systems involve small particles or droplets immersed in and interacting through a simple fluid; also surface tension gradients at the interface of a droplet, resulting from temperature or concentration gradients, often cause interesting and surprising effects.

Fluid dynamics started as the study of the flow of water, hydrodynamics. It is an interesting twist of history that this word and the concepts underlying it have witnessed a revival in the last decades: the term 'hydrodynamic approach' is nowadays often used more generally for a coarse-grained description based on an appropriate set of conserved or slowly evolving variables.

1.1 The relevance and attractiveness of a continuum description of fluids

It is actually not a trivial task to define what we mean by simple fluids, as they are most easily defined once we can refer to the equations describing them, and the steps we took in their derivation. A normal liquid consists of a dense phase of molecules without any particular type of long-range order or broken symmetry. The simplest molecular-scale picture that comes to mind is the one in figure 1.1.a: it shows a snapshot from a computer simulation of 'atoms' interacting through the so-called Lennard-Jones potential (which we specify in section 4.3.1). Similarly, panel b shows a sketch of H_2O water molecules in the liquid phase. It is in general useful to have such pictures in mind when you think of simple fluids and the analysis below. However be aware of the scales here! You are zooming in on the atomic scale of nanometers (10^{-9} m) and the time scale of the vibrational motion of molecules which is typically less than a picosecond (10^{-12} s), while essentially all flows relevant to daily life, even those in micron-sized biological cells, vary

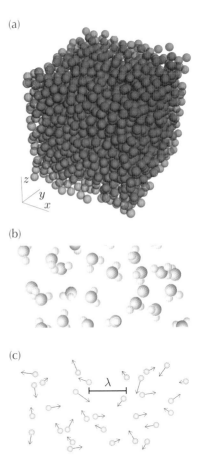

(a)

(b)

(c)

Figure 1.1. (a) Snapshot of a computer simulation of the liquid phase of atoms interacting through a Lennard-Jones potential, which is a good model for noble gases like argon. (b) Artist's impression of water in the liquid phase. (c) In a gas, what matters is the mean free path λ between collisions: hydrodynamics is an accurate description for variations on length scales much larger than λ and time scales much longer than the collision time. For air at room temperature, $\lambda \approx 70$ nanometers (nm).[1]

on much longer length and time scales. For these we use the fluid dynamics description.

It is also important to stress that fluid dynamics equally applies to the motion of gases like air, illustrated in panel c of figure 1.1, provided the variations in density and velocity are small on the spatial scale of the mean free path of the gas molecules, and slow on the time scale of their collisions. In air at room temperature the mean free path is about 70 nanometers and the mean collision time about 140 nanoseconds, small enough to leave a lot of room for fluid dynamics to apply to most practical situations.

We will pursue a continuum description of these fluids, i.e., we want to describe them with continuum fields which vary continuously as a function of space and time. As we just stressed, real matter is composed of particles, so at small enough spatial and time scales this description breaks down. We should thus think of these fields as *coarse-grained fields*—continuum fields obtained by averaging the behavior of the molecules over spatial distances and time scales that are (i) large enough that we obtain well defined and smooth continuum variables, but (ii) small compared to the overall length and time scales on which the flow varies. Figure 1.2 illustrates how we imagine the fluid dynamic equations to emerge from coarse-graining over molecular scales, but coarse-graining is actually much easier said than done. A practical approach to obtaining continuum fields is to use the Boltzmann equation for dilute gases.[2]

There are various reasons why an analysis of a problem with a continuum approach is very useful. First of all, analysis of the dynamics of all the particles is not only much too complicated, but it would in fact amount to an enormous overkill. We simply don't want to know all these details! Secondly, we often have to deal with large molecules (macromolecules) or colloidal particles embedded in a fluid consisting of much smaller molecules, and then it is often very useful—even if the behavior of the macromolecule or colloid is the object of study—to treat the surrounding fluid using continuum fluid dynamics. Thirdly, and most importantly, formulating a continuum approach will, as we shall see, in a very direct way illustrate that only very few effective parameters matter—e.g., a viscosity and a compressibility. So using a continuum theory allows us to zoom in very quickly on the essential ingredients that matter. Finally, using a continuum theory also brings the power of mathematics to bear on our problems, through the theory of (partial) differential equations, and the methods of solving or simulating them on a computer.

We will see this last point very clearly illustrated when we discuss the Reynolds number, which is an important dimensionless quantity distinguishing various flow regimes: at large Reynolds numbers, the flow is strongly nonlinear (and often turbulent at

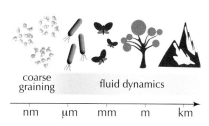

coarse graining   fluid dynamics

nm   μm   mm   m   km

Figure 1.2. Illustration of how coarse-graining on scales of a few tens of atomic diameters yields hydrodynamic equations which can be used for flow phenomena on all larger scales.

sufficiently large values), while at small Reynolds numbers, the flow is essentially laminar and governed by linear equations.[3] Especially for biomatter, this distinction is important—on the cellular scale we are in the limit of small Reynolds numbers, so this approach luckily suffices for understanding most behavior, as we will see.

Only in very simple cases (such as dilute gases composed of simple atoms) can parameters like the viscosity be calculated from the atomic interactions. For our approach, however, we consider them as *phenomenological parameters* that can be measured relatively simply in experiments or that can be looked up in published tables. It is important to keep in mind that *the equations do have a very wide range of validity*. The fact that a derivation from first principles is often lacking is not really a drawback—actually, it forces us to develop a good intuitive understanding of their range of validity.

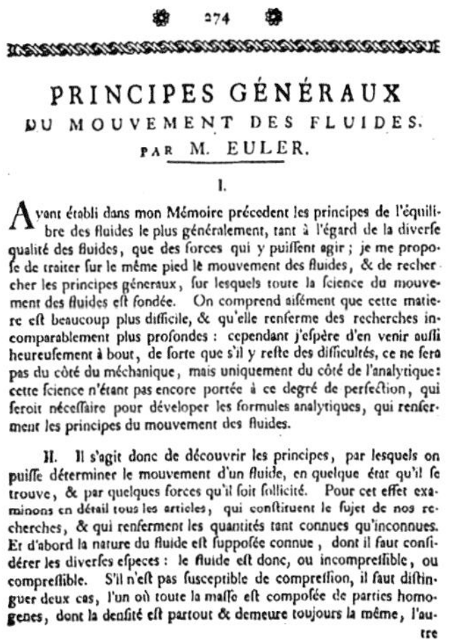

274

PRINCIPES GÉNÉRAUX
DU MOUVEMENT DES FLUIDES.
PAR M. EULER.

I.

Avant établi dans mon Mémoire précedent les principes de l'équilibre des fluides le plus généralement, tant à l'égard de la diverſe qualité des fluides, que des forces qui y puiſſent agir ; je me propoſe de traiter ſur le même pied le mouvement des fluides, & de rechercher les principes géneraux, ſur lesquels toute la ſcience du mouvement des fluides eſt fondée. On comprend aiſément que cette matiere eſt beaucoup plus difficile, & qu'elle renferme des recherches incomparablement plus profondes : cependant j'eſpère d'en venir auſſi heureuſement à bout, de ſorte que s'il y reſte des difficultés, ce ne ſera pas du côté du méchanique, mais uniquement du côté de l'analytique: cette ſcience n'étant pas encore portée à ce degré de perfection, qui ſeroit néceſſaire pour développer les formules analytiques, qui renferment les principes du mouvement des fluides.

II. Il s'agit donc de découvrir les principes, par lesquels on puiſſe déterminer le mouvement d'un fluide, en quelque état qu'il ſe trouve, & par quelques forces qu'il ſoit ſollicité. Pour cet effet examinons en détail tous les articles, qui conſtituent le ſujet de nos recherches, & qui renferment les quantités tant connues qu'inconnues. Et d'abord la nature du fluide eſt ſuppoſée connue, dont il faut conſidérer les diverſes eſpeces : le fluide eſt donc, ou incompreſſible, ou compreſſible. S'il n'eſt pas ſusceptible de compreſſion, il faut diſtinguer deux cas, l'un où toute la maſſe eſt compoſée de parties homogenes, dont la denſité eſt partout & demeure toujours la même, l'autre

Figure 1.3. The birth of hydrodynamics: the article by Leonhard Euler on ideal fluid mechanics published in 1757 in the *Mèmoires de l'acadèmie des sciences de Berlin*. The term *hydro* comes from the Greek word for water.[4]

In fact, the line of reasoning based on identifying the proper hydrodynamic or slow variables is often also very powerful for more complicated problems.

Hydrodynamics has a long history—see figure 1.3—focused mostly on water motion, but the term 'hydrodynamic description' is increasingly being used more generically for a description based on the slow modes of the system. Such modes often emerge due to a broken symmetry that occurs at a phase transition, and which is associated with a particular field, called an order parameter, in this context. We already illustrated this in figure I.15. A good example of the existence of a nontrivial order parameter associated with a symmetry breaking occurs in the nematic liquid crystal phase, where the so-called director field is the 'slow hydrodynamic variable' which we can treat with a continuum theory. (The phase field in a superconductor is another, admittedly much less familiar, example.) The form of the appropriate equations in such a phenomenological hydrodynamic approach can often be determined, guessed or postulated by understanding the symmetry of the relevant variables and by thinking in terms of a 'gradient expansion' of the equations. So even though our discussion will focus on the derivation of the fluid hydrodynamic equations, the line of approach has broader implications that we will try to bring across.

One sees this beautifully at work in the classification of dynamical critical phenomena (Hohenberg and Halperin, 1977).

We comment in passing on the term *emergent phenomenon* that researchers sometimes use in connection with hydrodynamics.[5] It refers to situations where on larger scales (or levels of description) qualitatively new and often surprising behaviors arise from the interactions among the entities on the small scale. In other words, the larger entities exhibit properties the building blocks themselves do not clearly exhibit. The term is most often used in connection to the collective phenomena and symmetry breaking such as in superconductivity, or the emergence of living matter from molecules. It is, however, sometimes also used in connection with the 'emergence' of the fluid dynamics equations out of the underlying molecular dynamics.

1.2 Hydrodynamics as a balance equation of fluid elements

You may also have come across the idea of coarse-graining in a discussion of how to arrive at a Landau approach or the field-theoretic analysis of critical phenomena. There, the focus is on equilibrium phenomena, the critical behavior that describes the asymptotic behavior upon approaching a critical point. In that type of coarse-graining approach a lot is thrown out (such as so-called irrelevant variables), but this is fine as one focuses on extracting the asymptotic critical exponent only. Here, however, we allow for non-equilibrium situations, and we extract the equations for the conserved quantities provided they vary on large length and time scales compared to the molecular ones.

Let us illustrate the idea of a small-gradient expansion as follows. If you consider a velocity change of $1\,\text{m/s}$ over $1\,\text{m}$, it corresponds to a shear gradient $\partial v_x / \partial y$ of $1\,\text{s}^{-1}$. Imposed shear rates are simulated on the molecular scale by periodic boundary conditions by replicating the boxes of the upper panel of figure 1.1, and sliding the boxes in the various layers sideways with the imposed shear rate (Frenkel and Smit, 2023). But for realistic parameters leading to collision times of picoseconds you would never notice a shear rate of order $1\,\text{s}^{-1}$ in the simulations. You can easily make the imposed shear rate many orders of magnitude larger and still be in the realm of local equilibrium with a small imposed shear gradient.

The original equations describing the molecular motion are Hamiltonian equations which are time-reversible. Nevertheless, the molecular motion leads to dissipative processes at the coarse-grained scale. When these effects are included, equations are time-irreversible. How irreversibility at the coarse-grained scale emerges from reversible behavior at the microscopic scale is a classic problem of statistical physics, but we focus here simply on how to implement dissipative effects in our equations. As we shall see, dissipative processes lead to entropy production. For an introduction to and overview of non-equilibrium thermodynamics, see the book by de Groot and Mazur, 1964.

As we already remarked above, rigorously deriving continuum equations from the underlying microscopic description is a daunting task, if not impossible. But luckily the underlying picture and the program for how to convincingly arrive at the equations by coarse-graining is actually quite clear in practice.

The intuitively most appealing way to arrive at the hydrodynamics is to imagine that the fluid consists of identifiable fluid elements, and that we can follow an element as it flows and deforms in time. This is called the Lagrangian description of fluid dynamics. In the spirit of the above discussion, we imagine that the element is large enough to contain a sufficient number of atoms or molecules (a thousand, say) that average quantities are well defined—an element that looks like the one in panel a of figure 1.1 will certainly do in practice.

Now, what are the slow variables? Since individual collisions conserve the number of atoms or molecules, momentum and energy, the total mass, momentum and energy of a fluid element can only change by what is flowing in and out of it at its boundary.[6] That makes the coarse-grained counterparts—*mass, momentum and energy densities*—indeed slow variables in space and time. Even in a non-equilibrium situation when there are variations in the coarse-grained density, velocity and temperature, the net in- and outflow of a fluid element will depend on the small gradient of these coarse-grained densities between neighboring fluid elements. Other quantities which you could consider, like a special type of correlation which is not conserved in individual collisions, will in contrast decay quickly on the collision time scale, on the order of picoseconds.

Because of the enormous separation of time scales of the slow variables and the molecular collisions, the molecules within the fluid element will be *locally* in equilibrium, which means that the usual thermodynamics relations hold between quantities like the internal energy, density and temperature, but then taken at their instantaneous local values. This idea of local equilibrium in the presence of small gradients is actually the basis of what is called the assumption of *non-equilibrium thermodynamics*. In this approach, so-called dissipative processes, like viscous dissipation or heat diffusion, are treated as small perturbations.

(a) Lagrangian description   (b) Eulerian description   (c) Material derivative   (d) Volume change of a fluid element

moving fluid element   fixed volume element

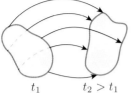
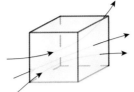
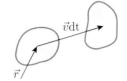
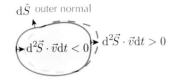

$$\frac{\mathrm{d}}{\mathrm{d}t} = \frac{\partial}{\partial t} + \vec{v} \cdot \vec{\nabla}$$

$$\int_S \mathrm{d}^2\vec{S} \cdot \vec{v}\, \mathrm{d}t = \int_V \mathrm{d}^3\vec{r}\, \vec{\nabla} \cdot \vec{v}\, \mathrm{d}t$$

$\mathrm{d}\hat{S}$ outer normal

$\mathrm{d}^2\vec{S} \cdot \vec{v}\mathrm{d}t < 0$ $\mathrm{d}^2\vec{S} \cdot \vec{v}\mathrm{d}t > 0$

$\vec{v}\mathrm{d}t$

$\vec{r}$

t_1 $t_2 > t_1$

Figure 1.4. (a) In the Lagrangian description, we follow a fluid element as it flows and distorts in time. (b) In the Eulerian description, we consider a fixed volume and analyze the balance of what flows in and out through the sides, as indicated by the dashed arrows. (c) The material derivative gives the time derivative of a physical quantity when following a fluid element. (d) The change in volume $\mathrm{d}V$ of a fluid element in a small time interval $\mathrm{d}t$ can be expressed as a volume integral of $\vec{\nabla}\cdot\vec{v}$.

1.3 Derivation of the equations

It is due time to derive the equations explicitly step-by-step.

1.3.1 The material or convective derivative

Conceptually, the most natural description of fluid dynamics is to follow fluid elements as they flow. This is illustrated in figure 1.4.a and is called the Lagrangian description.

Although we will use it below as a starting point for deriving the conservation equations, in practice this formulation is not convenient for doing calculations. In particular in numerical simulations we need to solve the equations for fixed boundaries, and following a moving coordinate system is very tricky. Therefore in practice, one specifies the fluid fields in the so-called Eulerian description relative to a fixed reference frame with position vectors $\vec{r}$ (see figure 1.4.b).

Suppose we want to follow some physical quantity a of a fluid element. How is $\mathrm{d}a/\mathrm{d}t$, the change of a with time while following the fluid element, written when we express $a(\vec{r}, t)$ in the fixed Eulerian frame with spatial coordinates $\vec{r}$? Well, if the fluid element is initially at position $\vec{r}$ and has a velocity $\vec{v}$ in the fixed frame, it moves to $\vec{r} + \vec{v}\mathrm{d}t$ in a time $\mathrm{d}t$, so we can write

$$\frac{\mathrm{d}a}{\mathrm{d}t} = \lim_{\mathrm{d}t\to 0} \frac{a(\vec{r}+\vec{v}\,\mathrm{d}t, t+\mathrm{d}t) - a(\vec{r}, t)}{\mathrm{d}t} = \frac{\partial a}{\partial t} + \sum_i v_i \frac{\partial a}{\partial x_i},$$

$$= \frac{\partial a}{\partial t} + \vec{v}\cdot\vec{\nabla} a. \tag{1.1}$$

This is summarized by defining the so-called material derivative, also called convective derivative,

Unless specified otherwise, we will in this book assume position vectors are defined in three dimensions, and we will indicate vectors like $\vec{r}$ with an arrow. The components are indicated as x_i, with $i = 1, 2, 3$, where unless specified otherwise we use the convention $x_1 = x$, $x_2 = y$, $x_3 = z$. Furthermore r denotes the length of the vector ($r^2 = |\vec{r}|^2$). Throughout we use a central dot $\cdot$ to denote a contraction of two vectors, so $\vec{a}\cdot\vec{b} = \sum_i a_i b_i$. Symbols and notation are summarized in appendix A.

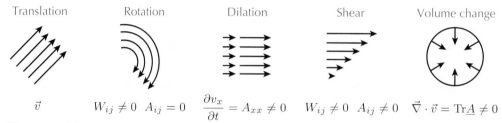

Translation   Rotation   Dilation   Shear   Volume change

$\vec{v}$ $W_{ij} \neq 0 \quad A_{ij} = 0$ $\dfrac{\partial v_x}{\partial t} = A_{xx} \neq 0$ $W_{ij} \neq 0 \quad A_{ij} \neq 0$ $\vec{\nabla} \cdot \vec{v} = \mathrm{Tr}\underline{A} \neq 0$

Figure 1.5. Illustration of the various types of flow components as defined in equation (1.3). A shear flow is a flow in which adjacent layers of a fluid move relative to each other. Note that what is often called a simple shear flow $v_x(y)$ in the fourth panel is actually a combination of shear and rotation.

$$\frac{\mathrm{d}}{\underbrace{\mathrm{d}t}_{Lagrangian}} = \underbrace{\frac{\partial}{\partial t} + \vec{v} \cdot \vec{\nabla}}_{Eulerian}, \tag{1.2}$$

which, as illustrated in figure 1.4.c, gives the change in time of a quantity when following a fluid element. As we shall see, the convective term $\vec{v} \cdot \vec{\nabla}$ is the most important source of nonlinearity of the fluid dynamics equations.

1.3.2 Separating out the various components of flow

For the forthcoming discussion it is illuminating to distinguish the various types of flow associated with flow gradients, with the aid of figure 1.5. When you have a general flow, nearby fluid elements will have different speeds, and this difference is generally a combination of a trivial local rotation and a shearing motion of fluid layers over each other. When you are rotating a bucket of water steadily, the water indeed executes a solid body–like rotation with the bucket, which clearly does not give rise to friction and dissipation. We can separate out effects like these locally by defining the symmetric and asymmetric part of the velocity gradients as

> With a trivial rotation we mean a solid body–like rotation as in the second panel of figure 1.5. In such a rotation, different layers do not shear relative to each other. The local solid body–like rotation of a velocity field is given by the vorticity $\vec{\nabla} \times \vec{v}$.

$$A_{ij} = A_{ji} = \frac{1}{2}\left(\frac{\partial v_i}{\partial x_j} + \frac{\partial v_j}{\partial x_i}\right), \quad W_{ij} = -W_{ji} = \frac{1}{2}\left(\frac{\partial v_i}{\partial x_j} - \frac{\partial v_j}{\partial x_i}\right). \tag{1.3}$$

As figure 1.5 illustrates, $\underline{W}$ selects components of the local solid body–like rotation $\vec{\nabla} \times \vec{v}$, while the matrix $\underline{A}$ filters out solid body–like rotation. Off-diagonal components of A describe shear of layers of fluid over each other, while diagonal components A_{ii} give the local stretching or shrinking, depending on their sign.

> Tensors of rank 2 can be thought of as the extension of vectors to objects with two components. Their components can be represented by a matrix.[7] Tensors like $\underline{W}$ are indicated in this book with an underscore. In particular, the stress tensor introduced below is $\underline{\sigma}$, with components σ_{ij}. Like we do for vectors, we indicate the contraction of two matrices with a dot $\cdot$, e.g., $(\underline{\sigma} \cdot \underline{\tau})_{ij} = \sum_k \sigma_{ik}\tau_{kj}$.

Finally, if we consider the trace $\mathrm{Tr}\underline{A} \equiv \sum_i A_{ii} = \vec{\nabla} \cdot \vec{v}$, this gives the relative rate of change of the volume of a fluid element. To see this, consider the volume element sketched in figure 1.4.d. If $\mathrm{d}\hat{S}$ denotes the outward pointing normal vector of the surface, then the contribution to the change in volume associated with the motion of the surface element $\mathrm{d}\vec{S}$ in a time $\mathrm{d}t$ equals $\mathrm{d}\vec{S} \cdot \vec{v}\,\mathrm{d}t$ (if $\vec{v}$ and $\mathrm{d}\vec{S}$

point in the same direction, the volume increases; if they point in opposite directions, the volume decreases). Hence for the total rate of change in volume dV of the element we have

$$\frac{dV}{dt} = \int_S d^2 \vec{S} \cdot \vec{v} = \int_V d^3 \vec{r} \, \vec{\nabla} \cdot \vec{v} = V \, \vec{\nabla} \cdot \vec{v}, \qquad (1.4)$$

Throughout the book, we will indicate intregrals over a two-dimensional surface by $\int d^2$ and integrals over a three-dimensional volume by $\int d^3$.

where we used the Gauss divergence theorem[8] to convert the surface integral to a volume integral, and the fact that $\vec{v}$ changes slowly on the scale of the small volume element. So $\vec{\nabla} \cdot \vec{v}$ simply gives the relative rate of volume change of a fluid element. If it is positive, the fluid element expands; if it is negative it contracts.

1.3.3 Conservation of mass

Let us now analyze the change in time of the mass density ρ of the fluid. For a small fluid element, ρ is simply the ratio of its mass M to its volume V,

$$\rho = \frac{M}{V}. \qquad (1.5)$$

By the definition of a fluid element its mass M does not change in time upon following the element, so ρ only changes as a result of the volume change of the element, which we just analyzed. Indeed

$$\frac{d\rho}{dt} = \frac{d}{dt}\frac{M}{V} = -\frac{M}{V^2}\frac{dV}{dt} = -\frac{M}{V}\left(\frac{1}{V}\frac{dV}{dt}\right) = -\rho\vec{\nabla} \cdot \vec{v}, \qquad (1.6)$$

where we used equation (1.4). To translate this back into the fixed Eulerian frame, we use equation (1.2) and the chain rule for differentiation to get

$$\frac{\partial \rho}{\partial t} = -\rho\vec{\nabla} \cdot \vec{v} - \vec{v} \cdot \vec{\nabla}\rho = -\vec{\nabla} \cdot \rho\vec{v}. \qquad (1.7)$$

We could also have arrived at this equation directly in the Eulerian description, in which we consider the mass balance in a small fixed volume V'. As figure 1.4.b illustrates, the mass density in the volume element is determined by the net in- and outflow of the mass; so we have via an argument similar to that for the volume change in equation (1.4):

$$V'\frac{\partial \rho}{\partial t} = -\int_{S'} d^2 \vec{S} \cdot \rho\vec{v} = -\int_{V'} d^3 \vec{r} \, \vec{\nabla} \cdot \rho\vec{v} = -V' \, \vec{\nabla} \cdot \rho\vec{v}, \qquad (1.8)$$

which reduces to equation (1.7),

$$\frac{\partial \rho}{\partial t} = -\vec{\nabla} \cdot \rho\vec{v}. \qquad (1.9)$$

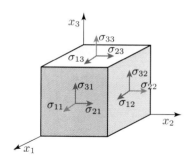

Figure 1.6. Illustration of the various components of the stress tensor. σ_{ij} is the force in the ith direction on a surface whose normal is pointing in the jth direction.

If $\underline{\sigma}$ is not symmetric, it models situations where the volume element has internal torques.[9] As we will discuss in section 9.6.2, such effects can occur in active matter. Note also that in the literature, e.g. in the book by de Groot and Mazur, 1964 on non-equilibrium thermodynamics, the opposite convention—that σ_{ij} is the force in the jth direction on a surface whose normal is in the ith direction—is used as well, but because of the symmetry of $\underline{\sigma}$ this does not matter.

In this equation, $\vec{v}\vec{v}$ is treated as a tensor with components $(\vec{v}\vec{v})_{ij} = v_i v_j$. Hence

$$\vec{\nabla} \cdot \rho\vec{v}\vec{v} = \sum_i \nabla_i (\rho v_i v_j).$$

A tensor $\vec{a}\vec{c}$ formed from two vectors $\vec{a}$ and $\vec{c}$ is called a dyadic.

1.3.4 Conservation of momentum

The derivation of the conservation of momentum equation proceeds along similar lines. According to Newton's acceleration law, the change of momentum equals the sum of forces. In the Lagrangian description the mass of a fluid element is fixed, so the acceleration of a small fluid element of volume V is simply given by $M\mathrm{d}\vec{v}/\mathrm{d}t = V\rho\,\mathrm{d}\vec{v}/\mathrm{d}t$. The total force on the fluid element is the sum of the forces exerted by the surrounding fluid on the element and the external body forces. In the presence of a gravity field $\vec{g}$, the external force per unit volume is simply $V\rho\vec{g}$.

Because both the force and the normal of the surface onto which the force acts are vectors, the forces exerted by the surrounding fluid are encoded in a stress *tensor* $\underline{\sigma}$, with two components. As illustrated in figure 1.6, σ_{ij} is the ith component of the force per unit area on a surface of the fluid element whose normal is pointing outward in the j direction. It can be shown that $\underline{\sigma}$ must be symmetric, i.e., $\sigma_{ij} = \sigma_{ji}$. The diagonal components of the tensor σ_{ii} are called normal stresses. An ideal (frictionless) fluid only has normal stresses, namely the pressure, but for a real viscous fluid there are also off-diagonal dissipative contributions to the stress tensor, as we will see below.

With this definition of the stress tensor, the force exerted by the outer fluid on the fluid element is

$$\int_S \underline{\sigma} \cdot \mathrm{d}^2\vec{S} = \int_V \mathrm{d}^3\vec{r}\,\vec{\nabla} \cdot \underline{\sigma}^T = V\,\vec{\nabla} \cdot \underline{\sigma}^T = V\,\vec{\nabla} \cdot \underline{\sigma}, \tag{1.10}$$

where we assumed, as before, that the volume element is small and where in the last step we used the fact that, since $\underline{\sigma}$ is symmetric, the transpose $\underline{\sigma}^T$ is equal to $\underline{\sigma}$.[10]

Taking all the terms together, the momentum equation that we obtain from Newton's acceleration law becomes

$$\rho\frac{\mathrm{d}\vec{v}}{\mathrm{d}t} = \vec{\nabla} \cdot \underline{\sigma} + \rho\vec{g}. \tag{1.11}$$

If we work out the material derivative equation (1.2) and use also the conservation of mass equation (1.9) we can write this in a nice form in Eulerian coordinates,

$$\frac{\partial\rho\vec{v}}{\partial t} = -\vec{\nabla} \cdot \left(\rho\vec{v}\vec{v} - \underline{\sigma}\right) + \rho\vec{g}. \tag{1.12}$$

By now you probably have a feel for the structure of the balance equations in the Eulerian formulation: this equation expresses that the change of momentum density in a fixed volume element is due to the external forces, plus the net balance of in- and outflow of momentum density across the boundaries—the local flow of momentum at the boundary is the velocity $\vec{v}$ times the momentum density $\rho\vec{v}$.

1.3.5 Conservation of energy

We now consider the balance equation for the energy density per unit mass, $\epsilon = v^2/2 + u$, with u the internal energy per unit mass. Its derivation follows the same lines of argument, and in the Eulerian reference frame the equation has a form which should have a familiar structure by now,

$$\frac{\partial \rho\epsilon}{\partial t} = -\vec{\nabla} \cdot \left(\rho\epsilon\vec{v} - \underline{\sigma} \cdot \vec{v} + \vec{J}^q \right) + \rho\vec{v} \cdot \vec{g}. \tag{1.13}$$

Indeed, the first term on the right is the balance of energy convected in and out of the element with the fluid flow. As energy generated per unit time has the form of force times velocity, we recognize in the second and last term the net energy flow due to the stress force and the change of energy due to the bulk force. The new term is the energy or heat current $\vec{J}^q$: in the absence of convection, there can be a flow of energy due to a temperature gradient—an example of a dissipative process that we will discuss in more detail below.

1.4 Once more: Reflections on the underlying picture

Equations (1.9), (1.12) and (1.13) are the five balance equations for a simple fluid—five, because the mass density ρ and the energy per unit mass ϵ are scalars, while $\vec{v}$ is a three-dimensional vector. They have the general form

$$\frac{\partial(\text{density})}{\partial t} = -\text{div (flux)} + \text{sources}, \tag{1.14}$$

because, as we discussed earlier, mass, momentum and energy are conserved on the microscopic scale in molecular collisions. As a consequence these quantities can only change within the fluid element if there is a net in- or outflow at the boundaries, or if there is a source or sink term, like an external force which changes the potential energy if the fluid element moves.

A few more remarks are in order. First of all, these are the *only* slow quantities for a simple fluid consisting of one type of molecules, as mass, momentum and energy are the only conserved quantities in molecular collisions at the microscopic scale. In contrast, if you would decide to analyze some unusual quantity, for example the correlation function of three constituent molecules forming a particular geometric configuration, this is *not* a conserved quantity. Therefore such a correlation decays on the time scale of the collisions, say in picoseconds. But if you have a mixture of various types of molecules, one has to extend the formalism to include appropriate concentration variables, as both concentrations are then slow variables since the molecules keep their identity during collisions.

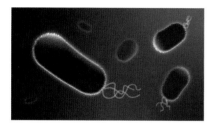

Figure 1.7. Artist's impression of a fluid with swimming bacteria as an example of an active fluid, for which the standard hydrodynamic equations don't hold. Active matter is the subject of chapter 9. As we will discuss in section 9.4, the swimming bacteria can actually help the flow so much that the fluid effectively flows like a frictionless fluid.

Secondly, it is clear that if you consider active matter, like bacteria—see figure 1.7—which consume ATP as fuel to swim, momentum is not conserved. Equations for active media indeed do not include momentum balance, while an equation of conservation of bacterial density is appropriate on time scales where division and death of bacteria can be ignored.

Thirdly, underlying the derivation is the idea that the fluid elements are in local equilibrium. We expressed this before by stating that we coarse-grain in time and space so that in each fluid element we have a sufficient number of molecules which reach equilibrium given the 'instantaneous' values of their local parameters, like density and temperature. This also means that thermodynamic quantities like the energy per unit mass ϵ, the pressure p or temperature T are related through their usual thermodynamic relations, expressed locally. At the same time, irreversible dissipative processes due to viscous friction or heat flow are treated as small perturbations, in view of the smallness of the gradients on the scale over which we coarse-grain. This is the essence of what is referred to as the non-equilibrium thermodynamics approach.[11]

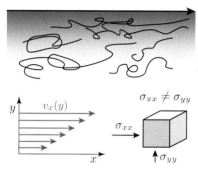

Figure 1.8. Illustration of long polymers (top) in simple shear flow (bottom left). The polymers are strongly stretched and oriented due to the flow—it is as if we have a fluid with many small stretched rubber bands in it. As a result the fluid is elastic and anisotropic, and as the polymers relax very slowly, the internal relaxation time is long. The anisotropy of the fluid gives rise to anisotropic normal stresses (bottom right). These normal stress differences actually give rise to many instabilities of polymer flows. See section 5.10.1 for further details.

To drive this last point home, imagine for a moment a dilute suspension of long polymers. In a simple shear flow as sketched in figure 1.8 these get stretched and oriented. The longer the polymers are, the stronger this effect is, and the more time it takes for the polymers to adapt to changes in the shear—the separation of time scales assumed above clearly breaks down, as the polymers don't 'instantaneously' follow their local environment: local equilibrium breaks down. As we will discuss in section 5.10, to model polymer rheology one therefore has to resort to more ad hoc phenomenological equations not based on non-equilibrium thermodynamics.

Interestingly, the small effect of thermal fluctuations on longer length scales can be analyzed theoretically with continuum versions of the Langevin-type equations (Landau and Lifshitz, 1987) discussed in chapter 3, and they can be probed with light-scattering measurements like those shown in figure 3.16 and discussed in section 3.7.2. Such measurements can be used to extract fluid parameters like the viscosity accurately.

Finally, the hydrodynamic equations are deterministic equations of first order in time: in principle, their evolution is completely determined once the initial conditions and boundary conditions are completely specified. At small enough scales the deterministic picture breaks down due to fluctuations inherent to the molecular nature of matter.

1.5 The dissipative terms: Onsager reciprocity relations

The balance equations for mass, momentum and energy are not a closed set of equations as long as the stress tensor $\underline{\sigma}$ and the heat current $\vec{J}^q$ are not yet specified. This is where the phenomenological analysis of dissipative effects comes in.

In the absence of dissipative effects, an ideal flow is reversible and the stress tensor is very simple: there is only the pressure field, which purely acts in the normal direction. According to our convention the stress tensor describes the forces exerted by the outer fluid onto the surface of a fluid element, counted positive if it is pointing in the same direction as the outer normal of the surface. As the pressure exerts an inward force, we clearly have for the stress tensor $\underline{\sigma}^{\mathrm{r}}$ in the *reversible* case of an ideal dissipationless fluid

$$\sigma_{ij}^{\mathrm{r}} = -p\,\delta_{ij}, \tag{1.15}$$

where p is the pressure and δ_{ij} the Kronecker delta, which is 1 if the indices are equal and zero otherwise.

In order to analyze the dissipative effects, it makes sense to introduce the dissipative contribution to the stress tensor $\underline{\sigma}$ as the difference,

$$\underline{\sigma}^{\mathrm{d}} = \underline{\sigma} - \underline{\sigma}^{\mathrm{r}}. \tag{1.16}$$

Now, according to the second law of thermodynamics, the entropy of an *isolated* system can only increase due to dissipative processes. Given the local equilibrium picture, this should be true for a fluid element in the Lagrangian description, which in turn implies that it should hold in the Eulerian description too provided we properly separate out what convects in and out through the boundaries of the volume element. If we introduce the entropy per unit mass s, and apply the usual thermodynamic relations, we can derive from the above equations the following equation for s:

$$\frac{\partial \rho s}{\partial t} + \vec{\nabla} \cdot \left(\rho \vec{v} s + \frac{\vec{J}^q}{T} \right) = \frac{\underline{\sigma}^{\mathrm{d}} : \vec{\nabla}\vec{v}}{T} + \vec{J}^q \cdot \vec{\nabla}\left(\frac{1}{T} \right). \tag{1.17}$$

See problem 1.2 for a step-by-step derivation of equation (1.17) from the energy conservation equation (1.13), using the so-called Gibbs-Duhem thermodynamic relation between u, s and ρ in addition to the conservation equations for mass and momentum.

Here T denotes the temperature and the colon : denotes the double contraction of $\underline{\sigma}$ and the tensor $\vec{\nabla}\vec{v}$, in other words, $\underline{\sigma} : \vec{\nabla}\vec{v} = \sum_{ij} \sigma_{ij} \partial v_i / \partial x_j$. The requirement that the terms on the right-hand side generally lead to an increase of the local entropy restricts the possible form of the dissipative stress tensor $\underline{\sigma}^{\mathrm{d}}$ and the heat current $\vec{J}^q$, as we now show.

This assumption is relaxed in our treatment of active fluids in section 9.7.2, where we explicitly consider viscosities that do not result in entropy production.

The general theory of dissipative processes is based on writing the entropy production term on the right-hand side of (1.17) as:

$$\frac{2R}{T} = \sum_{\alpha} \vec{F}^{\alpha} \cdot \vec{J}^{\alpha}, \qquad (1.18)$$

in terms of so-called thermodynamic forces $\vec{F}^{\alpha}$ that drive the dissipative processes (e.g., a temperature or velocity gradient), and the dissipative currents $\vec{J}^{\alpha}$ (a heat flow or momentum current) that these thermodynamic forces give rise to; R is called the dissipation function. The thermodynamic forces are thought of as vectors, as they are gradients of a physical field like the temperature. Then so are the currents, as is illustrated by the heat current $\vec{J}^{q}$ that a temperature gradient gives rise to. However, the thermodynamic force $\vec{\nabla}\vec{v}$ is actually a tensor, and so is the stress $\underline{\sigma}$. We will not go into the subtleties associated with this,[12] and we will also ignore the notational complications associated with anisotropic media—with the prime exception of liquid crystals, most soft matter phases we will discuss are indeed isotropic.

In the small gradient expansion it is assumed that the currents are linear functions of the forces, so generally

$$\vec{J}^{\alpha} = \sum_{\beta} L_{\alpha\beta} \vec{F}^{\beta}, \qquad (1.19)$$

where $L_{\alpha\beta}$ is a matrix of linear coefficients, the so-called Onsager coefficients. These relations between fluxes and forces are phenomenological relations, i.e., they are not derived but assumed, based on the premise that if no particular symmetry forbids a linear coupling, the dependence is expected to be present and linear. Note that when this expression is combined with the one for the entropy production, we get

$$\frac{2R}{T} = \sum_{\alpha\beta} L_{\alpha\beta} \vec{F}^{\alpha} \cdot \vec{F}^{\beta}. \qquad (1.20)$$

The requirement from non-equilibrium thermodynamics that the entropy production term R is indeed positive for general forces (as the entropy of a closed system can only increase) implies that $\underline{L}$ has to be a positive definite matrix. This implies that the common transport coefficients (like viscosities and diffusion coefficients) must be of a particular sign. In fact, already about a century ago Onsager derived an even stronger result, namely that $\underline{L}$ must be symmetric as a result of the reversibility of the underlying microscopic equations of motion:

$$L_{\alpha\beta} = L_{\beta\alpha}. \qquad (1.21)$$

This symmetry relation is particularly interesting when one considers so-called cross effects, like a temperature gradient giving rise to an electric field (thermo-electric effect or Seebeck effect) or to diffusion (thermo-diffusion or Soret effect)—the Onsager relations essentially express that an effect and its inverse are related.

So when the right-hand side of (1.17) is interpreted in the form of an entropy production term (1.18), $\vec{\nabla}\vec{v}$ plays the form of a thermodynamic driving force for the dissipative stress tensor $\underline{\sigma}^{d}$ (a 'momentum current'), while the temperature gradient $\vec{\nabla}T^{-1}$ is the thermodynamic force driving the heat current $\vec{J}^{q}$.

Lars Onsager was awarded the 1967 Nobel Prize in Chemistry for developing the reciprocity relations (Onsager, 1931). The form given here is valid in the absence of a magnetic field $\vec{B}$ or an overall rotation $\vec{\Omega}$ of a system. In their presence, the so-called Onsager reciprocity relations read $L_{\alpha\beta}(\vec{B}, \vec{\Omega}) = L_{\beta\alpha}(-\vec{B}, -\vec{\Omega})$. As discussed by de Groot and Mazur, 1964, the symmetric expression (1.21) is only valid for currents and forces which are both even or both odd expressions of the microscopic velocities. $L_{\alpha\beta}$ is antisymmetric for cross effects between an even and an odd quantity.

This is illustrated in figure 1.9 for the thermo-electric effect.

1.6 The stress tensor and heat current for a Newtonian fluid

1.6.1 Stress tensor and heat current

Let us return to our discussion of a simple fluid. For these, the phenomenological Onsager relations for the dissipative part of the stress tensor $\underline{\sigma}^{\mathrm{d}}$ and for the heat current $\vec{J}^q$ have the following form in terms of the local gradients,

$$\sigma_{ij}^{\mathrm{d}} = \eta \left(\frac{\partial v_j}{\partial x_i} + \frac{\partial v_i}{\partial x_j} - \frac{2}{3} \delta_{ij} \, \vec{\nabla} \cdot \vec{v} \right) + \zeta \, \delta_{ij} \, \vec{\nabla} \cdot \vec{v}, \qquad (1.22)$$

and

$$\vec{J}^q = -\kappa_T \vec{\nabla} T. \qquad (1.23)$$

For a fluid, the expression for $\underline{\sigma}^{\mathrm{d}}$ can be taken as the defining equation of a Newtonian fluid. This expression is conveniently written as the sum of a traceless tensor—the first term in (1.22)—and the identity tensor with components δ_{ij}. We already identified in equation (1.4) and figure 1.4(d) the term $\vec{\nabla} \cdot \vec{v}$ with the compression rate of volume elements, hence the coefficient ζ is usually called the bulk viscosity (sometimes it is also referred to as the compression viscosity). The first term in the expression of $\underline{\sigma}^{\mathrm{d}}$ is traceless and this removes the terms of the velocity gradients associated with the total compression $\vec{\nabla} \cdot \vec{v}$; this term therefore measures true shear in the fluid, and the coefficient η is called the dynamic viscosity (more loosely referred to as the shear viscosity). As will become clear in section 1.9, η is often the only viscosity that matters for the type of soft matter applications we consider, as the velocities are small enough that we can treat the flow as incompressible.

Finally, the relation (1.23) simply expresses the well-known fact that the heat flow is proportional to the gradient of the temperature and oriented from hot to cold regions, i.e., opposite to the gradient. The coefficient κ_T is called the thermal conductivity. It is formulated here for fluids, but the fact that heat current is proportional to the temperature gradient holds quite generally, also for solid matter.

1.6.2 The resulting hydrodynamic equations

Now that the stress and heat current are specified, we have a closed set of equations. Let us, for ease of reading, put them together here.

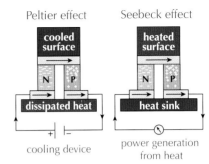

Figure 1.9. The Peltier effect refers to the phenomenon where a temperature difference is induced by a voltage difference. Its reciprocal is the Seebeck coefficient: here a temperature difference gives rise to a voltage. The Onsager relations imply that the two coefficients expressing these relations are the same. Both effects are used in applications, the Peltier effect for cooling with a setup as sketched in the left panel and the Seebeck effect to generate a voltage from a temperature difference with conceptually the same setup (the right panel). In these figures, N and P mark semiconductors with, respectively, negative mobile charges (so-called N-type) and positive mobile charges (so-called P-type). Peltier elements are used widely to control the temperature, especially in small-scale applications like chips.[13] The Seebeck effect is used to measure temperature differences at small scales, but it is even used to generate power from waste heat.[14]

The general expression for the dissipative stress tensor in equation (1.22) can be derived as follows. First, no dissipation occurs if the entire fluid flows with the same velocity v, which means that $\underline{\sigma}^{\mathrm{d}}$ depends only on space derivatives of the velocity. Second, under the assumption of small velocity gradients, the stress tensor can be written as a linear combination of first order derivatives of the velocity, $\partial v_i / \partial x_j$. Finally, noting that no dissipation should occur if the fluid rotates with the same angular velocity $\vec{v} = \vec{\Omega} \times \vec{r}$, further constraints $\underline{\sigma}^{\mathrm{d}}$ should depend only on combinations of $\partial v_i \partial x_j + \partial v_j \partial x_i$ (which vanish for global rotation). These arguments, along with the assumption of isotropic media, lead to equation (1.22).[15]

Having a closed set of equations means that we have exactly the right number of equations to determine, in principle, their solution. In this particular case, the phrase means that we have five equations that specify the time evolution of the five variables, the mass density, the three components of the velocity, and the energy density, in terms of the variables themselves. In other words, they are fully specified and can be implemented on a computer to study the temporal evolution, given appropriate initial and boundary conditions.

So $(\vec{\nabla}\vec{v})^{0s} = \frac{1}{2}(\partial_{x_j}v_i + \partial_{x_i}v_j - \frac{2}{3}\delta_{ij}\vec{\nabla}\cdot\vec{v})$.

Keep in mind that underlying the hydrodynamic equations is also the equation of state which relates variables like density, pressure and temperature.[17] As stressed in section 1.4, these thermodynamic relations are a consequence of the local-equilibrium assumption.

The heat diffusion equation is mathematically the same as the particle diffusion equation. See section 3.3.5 on fundamental solutions of this equation.

We have the mass conservation equation (1.9)

$$\frac{\partial \rho}{\partial t} = -\vec{\nabla}\cdot\rho\vec{v}, \tag{1.24}$$

and if we substitute the above expressions in the momentum balance equation (1.12) we get

$$\frac{\partial \rho\vec{v}}{\partial t} = -\vec{\nabla}\cdot(\rho\vec{v}\vec{v}) - \vec{\nabla}p + \rho\vec{g}$$
$$+ \eta\nabla^2\vec{v} + \left(\frac{1}{3}\eta + \zeta\right)\vec{\nabla}(\vec{\nabla}\cdot\vec{v}), \tag{1.25}$$

while the balance equation (1.13) written for the internal energy $u = \epsilon - v^2/2$ per unit mass of the fluid becomes

$$\frac{\partial \rho u}{\partial t} = -\vec{\nabla}\cdot(\rho u\vec{v}) - p\vec{\nabla}\cdot\vec{v}$$
$$+ \kappa_T\nabla^2 T + 2\eta(\vec{\nabla}\vec{v})^{0s} : \vec{\nabla}\vec{v} + \zeta(\vec{\nabla}\cdot\vec{v})^2, \tag{1.26}$$

where $(\vec{\nabla}\vec{v})^{0s}$ denotes the traceless symmetric part of the tensor $\vec{\nabla}\vec{v}$.[16] We have written the dissipative terms in these equations on the second line. In the last equation, the term $p\vec{\nabla}\cdot\vec{v}$ is the work done by the pressure in compressing a volume element.

1.6.3 Heat diffusion equation

For a fluid at rest, $\vec{v} = 0$, equation (1.26) simply reduces to

$$\rho c_v\frac{\partial T}{\partial t} = \kappa_T\nabla^2 T, \tag{1.27}$$

where $c_v = (\partial u/\partial T)_v$ is the specific heat per unit mass at constant volume. By introducing the heat diffusion coefficient $D_T = \kappa_T/\rho c_v$ we obtain the temperature diffusion equation,

$$\frac{\partial T}{\partial t} = D_T\nabla^2 T. \tag{1.28}$$

This equation is of the form of a diffusion-type equation for the temperature; this form arises quite ubiquitously, as heat transport is generally proportional to the gradient of the temperature.

If temperature effects due to heating and viscous dissipation play a role, one has to work with the complete set of equations for mass, momentum and energy (or temperature). Luckily, as we shall see, in most situations relevant to soft matter, the fluid flow can be taken to be *incompressible*, and then the entropy or internal energy equation is not needed explicitly. In order to be able to show this, we make a short digression to discuss sound, which is nothing but a compression wave.

1.7 Sound waves

It is instructive (and fun) to study sound waves in a fluid in the absence of gravity. This will prepare the ground for discussing under which conditions we can consider the flow of a fluid as incompressible. To keep the analysis relatively simple, we will assume that the thermal conductivity is negligible, so that there is no heat transport. This implies that we can treat the pressure variation with density as adiabatic (i.e., without heat exchange and so at constant entropy). For water, thermal diffusivity changes the damping of the waves only by $\sim 10\%$, so it is actually a pretty good approximation to ignore the thermal diffusion. We will focus on small-amplitude waves.

The full formula for the damping of the sound modes is given in equation (8.4.63) of Chaikin and Lubensky, 1995.

1.7.1 The equation for sound propagation

Sound waves in fluids are pressure waves. The pressure induces a density modulation, hence sound modes are alternating compression/expansion waves. In order to treat small-amplitude sound modes, we take the hydrodynamic equations and linearize them about the quiescent state $\rho = \rho_0 = constant$, $\vec{v} = 0$, by treating deviations $\Delta\rho \equiv \rho - \rho_0$ and the velocities $\vec{v}$ as small quantities. For the density equation we then get to linear order

$$\frac{\partial \Delta\rho}{\partial t} = -\rho_0 \, \vec{\nabla} \cdot \vec{v}. \qquad (1.29)$$

For the velocity equation, linearization means that in equation (1.25) the term proportional to v^2 can be neglected, so that we simply get

$$\rho_0 \frac{\partial \vec{v}}{\partial t} = -\vec{\nabla} p + \eta \nabla^2 \vec{v} + \left(\frac{1}{3}\eta + \zeta\right) \vec{\nabla}(\vec{\nabla} \cdot \vec{v}). \qquad (1.30)$$

In view of the fact that we ignore thermal heat exchange due to diffusion, a fluid element changes adiabatically under pressure, so we can write for the small changes in pressure

$$\Delta p \approx \left(\frac{\partial p}{\partial \rho}\right)_s \Delta\rho, \qquad \left(\frac{\partial p}{\partial \rho}\right)_s > 0, \qquad (1.31)$$

where $\Delta p = p - p_0$, with p_0 the pressure of the fluid in the quiescent state. The index s refers to the adiabatic condition of constant

Due to their finite stiffness, solids also support transverse waves in addition to the compression waves, which are called longitudinal waves in solids (see section 2.6). Transverse waves have the atoms move in a direction perpendicular to the direction of propagation. In seismology, such transverse or shear waves are called secondary waves or S-waves.

Note $\vec{\nabla} \cdot \rho\vec{v} = \vec{\nabla} \cdot (\rho_0 + \Delta\rho)\vec{v} \approx \rho_0 \vec{\nabla} \cdot \vec{v}$ as terms $\vec{v}\Delta\rho$ are of second order.

entropy s. Now if we take the time derivative of the density equation (1.29) and the divergence of the velocity equation (1.30), we obtain

$$\frac{\partial^2 \Delta\rho}{\partial t^2} = -\rho_0 \, \vec{\nabla} \cdot \frac{\partial \vec{v}}{\partial t}$$
$$= \left(\frac{\partial p}{\partial \rho}\right)_s \nabla^2 \Delta\rho - \left(\frac{4}{3}\eta + \zeta\right) \nabla^2 \vec{\nabla} \cdot \vec{v}, \tag{1.32}$$

which upon rearranging, and using the linear equation (1.29) for the density once more, becomes

$$\frac{\partial^2 \Delta\rho}{\partial t^2} - \left(\frac{\partial p}{\partial \rho}\right)_s \nabla^2 \Delta\rho = \rho_0^{-1}\left(\frac{4}{3}\eta + \zeta\right) \nabla^2 \frac{\partial \Delta\rho}{\partial t}. \tag{1.33}$$

To get a feel for this equation, let us consider simply a wave $\Delta\rho(x,t)$ propagating in the x direction so that we can replace ∇^2 by ∂_x^2. If we ignore for the moment the damping terms on the right, we obtain

$$\frac{\partial^2 \Delta\rho}{\partial t^2} - \left(\frac{\partial p}{\partial \rho}\right)_s \frac{\partial^2 \Delta\rho}{\partial x^2} = 0. \tag{1.34}$$

An equation of the form (1.34) is called a wave equation, as its solutions $\Delta\rho = f(x \pm c_s t)$ describe traveling waves of arbitrary waveform f and speed

Note that the subscript s of c_s refers to the speed of *sound*, while as a subscript of the density derivative of the pressure, it refers to the derivative taken at constant entropy s.

$$c_s = \sqrt{\left(\frac{\partial p}{\partial \rho}\right)_s}. \tag{1.35}$$

Clearly, $\Delta\rho = f(x \pm c_s t)$ describes sound waves propagating with speed c_s in the positive and negative x direction. In our case, c_s given by equation (1.35) is the *speed of sound*.

The term on the right-hand side of equation (1.33) is a damping term, as one might guess both from the fact that it involves a first order time derivative, and from the fact that it involves viscosities. We shall confirm this explicitly in the next subsection. Note also that the combination

$$\nu = \eta/\rho_0, \tag{1.36}$$

Table 1.1. Viscosities of water and air at $20°$ C. Many tables in the literature give the dynamic viscosities in centipoise $= 10^{-2}$ g/cm $\cdot$ s $= 10^{-3}$ kg/m $\cdot$ s $= 10^{-3}$ Pascal seconds (Pa $\cdot$ s).

which appears on the right in equation (1.33), is called the kinematic viscosity; we shall see below that it is the only viscosity that plays a role for incompressible flows.

Finally, for reference, in table 1.1 we give a few important values of the viscosities of and speed of sound in water and air at room temperature. The two cases shown illustrate that for a given fluid the dynamic viscosity and the bulk viscosity of fluids are typically comparable (although for water ζ is almost three times as great as η). However, when we compare different liquids, their dynamic viscosities can differ quite a lot. Compared to water, blood is about

$\eta \left[\dfrac{\text{kg}}{\text{m}\cdot\text{s}}\right]$	$\zeta \left[\dfrac{\text{kg}}{\text{m}\cdot\text{s}}\right]$	$\nu \left[\dfrac{\text{m}^2}{\text{s}}\right]$	$c_s \left[\dfrac{\text{m}^2}{\text{s}}\right]$	
$1.0 \cdot 10^{-3}$	$2.9 \cdot 10^{-3}$	$1.0 \cdot 10^{-6}$	1484	water
$1.8 \cdot 10^{-5}$	$1.6 \cdot 10^{-5}$	$1.5 \cdot 10^{-5}$	340	air

3 to 4 times, glycerol about 1400 times, and honey typically about 10^4 times as viscous.

1.7.2 Analysis of the equation with damping

Let us make a small digression to treat the sound waves with damping in some detail, as this will be a good example of how it can efficiently be done by 'complexifying' the linear equation. We will use a trick that is often implemented in analysis of linear partial differential equations, for example those that arise in linear response calculation or linear stability analysis.

The density modulation $\Delta\rho$ is a quantity that takes on real values only, as it is a physical field. In the presence of damping, depending on the problem at hand various cases could be considered. For example, one may have a case sketched in the upper panel of figure 1.10 where the sound waves are continuously excited at some point in space, and where the amplitude is decaying away from this point due to damping. Or a case like in the lower panel, in which a wave is initially excited with a spatially constant amplitude, which then because of the damping decays in time. Or even mixed cases. We could analyze equation (1.33) for these various cases separately—by taking combinations of real exponential and trigonometric functions as particular solutions—but it is usually much simpler to do it all in one fell swoop by complexifying the field through writing

$$\Delta\rho^{\mathrm{C}} = A\,e^{ikx-i\omega t}, \tag{1.37}$$

where the amplitude A, the 'wavenumber' k and the 'frequency' ω are all allowed to be complex. We use the names *wavenumber* and *frequency* here loosely as only their real parts play these roles, while the imaginary parts of k and ω play the role of damping in space and time, as figure 1.10 illustrates. For these reasons, at the end of the calculation, the physical fields (in this case the physical density modulation) are obtained by taking the real part of the complexified field,

$$\Delta\rho = \mathrm{Re}\left(\Delta\rho^{\mathrm{C}}\right). \tag{1.38}$$

In practice, though, you will hardly ever make the distinction between physical and complexified fields explicit.

Let us now introduce the ansatz equation (1.37) into the wave equation (1.34). You will notice that all the terms in the resulting equation are proportional to the same overall factor $A\,e^{ikx-i\omega t}$, due to the fact that the linear equation (1.34) is translation-invariant in space and time.[18] This factor can therefore be divided out, yielding the simple relation $\omega = \pm c_{\mathrm{s}}\,k$ between ω and k. This linear relation, a simple example of a so-called dispersion relation $\omega(k)$, describes

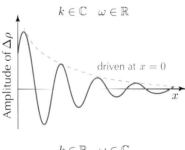

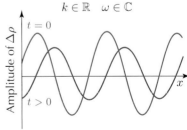

Figure 1.10. Illustration of various types of situations one may want to consider in analyzing the wave equation with damping. In the upper one, the wave is excited continuously at $x = 0$, and the amplitude decays in space. This is described by taking k complex and ω real in equation (1.37). The lower panel illustrates a case in which a wave of constant amplitude is initially created at $t = 0$, and whose amplitude then decays in time. This is modeled by taking k real and ω complex.

waves propagating to the right for k and ω real and of the same sign, and to the left when they are of opposite sign.

If we substitute the ansatz into the full equation (1.33) that includes damping, all terms will again have a common factor that can be divided out. This leaves us with a complex equation for the dispersion relation connecting ω and k,

$$-\omega^2 + c_s^2 k^2 = \left(\frac{4}{3}\nu + \zeta/\rho_0\right) i\omega k^2. \tag{1.39}$$

We will work out in more detail the case sketched in the upper panel of figure 1.10—a wave generated by a constant source with real frequency $\omega > 0$ at $x = 0$, focusing on the right-moving wave with the real part of k positive. By writing k as a function of ω we get

You may wish to play for yourself with the analysis of the temporal decay of a mode $\exp(ikx)$ with k real, by allowing ω to be complex.

$$\begin{aligned} k &= \omega/c_s \left(\frac{1}{1 - (4\nu/3 + \zeta/\rho_0)\,i\omega/c_s^2}\right)^{1/2}, \\ &\approx \omega/c_s \left(1 + \frac{i}{2}(4\nu/3 + \zeta/\rho_0)\,\omega/c_s^2 + \dots\right), \end{aligned} \tag{1.40}$$

where in the second step we have expanded the square root as the imaginary part is small. The real part of k gives the usual linear dispersion relation of sound modes from the wave equation, and the imaginary term corresponds to a small spatial damping $\sim e^{-\mathrm{Im}\,(k)x}$ of the modes, corresponding to a

$$\text{decay length} = \frac{1}{\mathrm{Im}\,(k)} = \frac{c_s}{\omega}\,\frac{2c_s^2}{(4\nu/3 + \zeta/\rho_0)\,\omega}. \tag{1.41}$$

For water and a frequency of 160 Hz (angular frequency of order 10^3 Hz), using the values of the viscosities given in table 1.1 we can estimate

For estimates like those in equation (1.42), we aim at order of magnitude only, so we deliberately put $4\pi \approx 10$, $4/3 \approx 1$, etc.

$$\text{decay length} \simeq \frac{1.4 \cdot 10^3\,\frac{\mathrm{m}}{\mathrm{s}}}{10^3\,\frac{1}{\mathrm{s}}}\,\frac{2\,(1.4 \cdot 10^3)^2\,\frac{\mathrm{m}^2}{\mathrm{s}^2}}{4 \cdot 10^{-6}\,\frac{\mathrm{m}^2}{\mathrm{s}}\,10^3\,\frac{1}{\mathrm{s}}}, \tag{1.42}$$

$$= \mathcal{O}(10^9\,\mathrm{m}).$$

This makes sonar possible and enables whales—some species use sounds in the range 31 Hz–31 kHz—to communicate over long distances. Since the decay length varies as ω^{-2}, at their highest frequencies the decay length reduces to something of order 25 km—still large and more than sufficient in practice, but not enough to cross the ocean.

Clearly, the decay length is so ridiculously large that damping of such waves does not play any role in practice. For sound emerging from a point source, attenuation of the amplitude from the spherical spreading and through scattering of waves off objects or for instance the sea bottom, is very much more important than damping.

1.8 When can we treat a flow as incompressible?

We now come to an important point relevant to most soft matter problems, namely the realization that in most cases the relevant velocities, length and time scales are such that we can treat the flow as incompressible, and that this allows us to reduce the equations to a conceptually much simpler form, namely that of the Navier-Stokes equations (1.49) below.[19]

To start the analysis for a steady time-independent flow, note that when we compare the first two terms on the right of equation (1.25), the nonlinear velocity term and the pressure gradient term, we see that pressure will vary with the flow as

$$\Delta p \simeq -\frac{1}{2}\rho_0 v^2. \tag{1.43}$$

The pressure difference leads to density changes $\Delta\rho$ of order

$$\Delta\rho \simeq (\partial p/\partial\rho)_{\mathrm{s}}^{-1}\frac{1}{2}\rho_0 v^2 \simeq \frac{1}{2}\rho_0 v^2/c_{\mathrm{s}}^2, \tag{1.44}$$

where we used equation (1.31) and the earlier result (1.35) for the speed of sound c_{s}. Since c_{s} is normally quite large, over 1,400 m/s in water, this means that in many situations of practical interest in which the flow speeds are much less than the speed of sound, the relative density variations $\Delta\rho/\rho_0$ due to variations in the flow speed are very small. They can therefore be ignored.

The above argument is only valid for steady flows. Let us now extend it to time-dependent flows, which vary on a time scale τ and length scale ℓ. Pressure changes Δp induced by changes in the flow speed due to the $\partial\vec{v}/\partial t$ term in the velocity equation (1.12) (or momentum conservation (1.25)) are estimated by comparing the order of magnitude of these two terms, according to

$$v/\tau \simeq \Delta p/(\ell\rho_0), \qquad \text{so } \Delta p \simeq \ell\rho_0 v/\tau, \tag{1.45}$$

which with (1.31) leads to density changes of order

$$\Delta\rho \simeq \ell\rho_0 v/\tau c_{\mathrm{s}}^2. \tag{1.46}$$

Now we go back to the density conservation equation and compare the time derivative of the density variations $\partial\rho/\partial t \simeq \Delta\rho/\tau$ to the compression term $-\rho_0\nabla\cdot\vec{v} \simeq \rho_0 v/\ell$. Upon combining this with (1.46), we find that the density time-derivative term is smaller than the compression term by a factor of

$$\frac{\mathcal{O}(\partial\rho/\partial t)}{\mathcal{O}(\rho_0\nabla\cdot\vec{v})} = \frac{(\ell/\tau)^2}{c_{\mathrm{s}}^2}. \tag{1.47}$$

The origin of the factor $1/2$ in equation (1.43) may not immediately be transparent from equation (1.25) for the momentum density. Indeed, it is obtained most clearly from the velocity equation (1.11) for steady flows. By writing the $\vec{v}\cdot\vec{\nabla}\vec{v}$ term from the convective derivative as $\vec{\nabla}v^2/2 - \vec{v}\times(\vec{\nabla}\times\vec{v})$ using a standard identity from vector calculus, one sees that the second term is pointing perpendicular to $\vec{v}$; hence it does not vary along streamlines which are the flow lines of fluid-elements and which hence are parallel to $\vec{v}$ (streamlines are defined in problem 1.3). From this, equation (1.43) arises for pressure and velocity variations along streamlines. Equation (1.43) is essentially the Bernoulli equation $p + \rho v^2/2 = const.$, which is considered in problem 1.3. For a fuller account see, e.g., section 5 of Landau and Lifshitz, 1987.

The Navier-Stokes equations are obtained from equation (1.48) and equation (1.11) for $\vec{v}$, together with the expressions (1.16), (1.15) and (1.22) for the stress tensor $\underline{\sigma}$. Note that the prefactor of the nonlinear term $\vec{v} \cdot \vec{\nabla} \vec{v}$ is precisely 1. This reflects Galilean invariance of the Navier-Stokes equations: if we make a transformation from the frame $\vec{r}$ to a frame $\vec{r}' = \vec{r} - \vec{U}t$ moving with velocity $\vec{U}$ with respect to the frame $\vec{r}$, then velocities transform as $\vec{v}' = \vec{v} - \vec{U}$. As $\partial/\partial t|_{\vec{r}} = \partial/\partial t|_{\vec{r}'} - \vec{U} \cdot \vec{\nabla}'$ we see that the terms on the left-hand side of the velocity equation transform as $\partial \vec{v}/\partial t + \vec{v} \cdot \vec{\nabla} \vec{v} \to \partial \vec{v}'/\partial t + \vec{v}' \cdot \vec{\nabla}' \vec{v}'$. Hence the Navier-Stokes equations are the same in a frame moving with a constant velocity $\vec{U}$, i.e., they are Galilean-invariant.

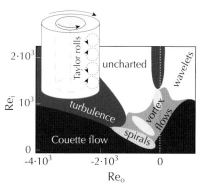

$2 \cdot 10^3$

$\mathrm{Re_i}$ 10^3

0

$-4 \cdot 10^3$ $-2 \cdot 10^3$ 0

$\mathrm{Re_o}$

Taylor rolls

uncharted

wavelets

turbulence

vortex flows

Couette flow

spirals

Figure 1.11. Schematic phase diagram of so-called Taylor-Couette flow, fluid between rotating cylinders. The experimental setup with two cylinders is shown on the upper left, and the phase diagram is drawn as a function of $\mathrm{Re_o}$ and $\mathrm{Re_i}$, the Reynolds numbers associated with the outer and inner cylinders.[20] For $\mathrm{Re_o} < 0$ the two cylinders are counterrotating. The basic instability from laminar flow, the dark region labeled 'Couette flow,' to Taylor vortex flow sketched in the inset, is driven by the Coriolis force. Regular Taylor vortices only occur in the bright blue region near the vertical axis where the outer cylinder is hardly rotating. Only major flow regions are indicated; the full phase diagram includes at least 18 different types of flow patterns. Figure adapted from Grossmann et al., 2016, which in turn was adapted from Andereck et al., 1986. [21]

In other words, if the velocity associated with the typical length and time scales is much less than the speed of sound, we can ignore the $\partial \rho/\partial t$ term and consider the fluid flow as *incompressible*. In other words, we can then take

$$\vec{\nabla} \cdot \vec{v} = 0. \qquad (1.48)$$

Since the speed of sound is so large, this condition is satisfied in many cases of practical interest, also in air.

1.9 The Navier-Stokes equations

Because of their importance, we highlight the so-called Navier-Stokes equations for incompressible flows without gravity,

$$\vec{\nabla} \cdot \vec{v} = 0,$$
$$\frac{\partial \vec{v}}{\partial t} + \vec{v} \cdot \vec{\nabla} \vec{v} = -\vec{\nabla} p/\rho_0 + \nu \nabla^2 \vec{v}. \qquad (1.49)$$

Here ρ_0 is the (constant) density of the fluid. Note that these equations only contain one phenomenological parameter, the kinematic viscosity ν defined in (1.36).

These equations are the basis of a vast part of fluid dynamics of practical interest. It is remarkable how rich in behavior the second equation is, with only one nonlinear term $\vec{v} \cdot \vec{\nabla} \vec{v}$, as a glance at figure 1.11 makes clear: the flow between two rotating cylinders exhibits many different regimes, some with regular 'coherent' patterns, others with turbulence. Even a basic problem like this one is still actively studied today, both experimentally and theoretically. The same holds for the transition to turbulence in pipe flow, first studied by Reynolds (see figure 1.12).

We are now finally able to explain the remark made at the end of section 1.6.3, that for incompressible flows the thermodynamic relations and the entropy equation are not needed anymore. When we deal with the full case of compressible flow, we have to specify how the pressure p depends on the thermodynamic variables (we saw an example of this when we studied sound modes). But once we study the Navier-Stokes equations for an incompressible fluid, we do not need to specify such a relation: instead p then is a quantity which is determined by the flow equations—thermodynamics does not explicitly play a role anymore.[22]

For analyzing flow in any practical situation, one has to specify proper boundary conditions. These depend a bit on the situation one considers. For flow past a static solid object that can withstand shear (i.e., does not easily deform or flow itself), one normally uses so-called no-slip boundary conditions, expressing that the fluid velocity vanishes at the surface of the fixed object. But for instance at the free surface of a fluid (e.g., water in contact with air), an appropriate boundary condition is that the free surface cannot withstand shear forces, so then the proper boundary condition is that the tangential stresses at the surface vanish. But as we will see in section 1.13, if there are surface tension gradients along the interface due to concentration gradients, the boundary condition is that the viscous forces parallel to the interface balance the forces due to the surface tension gradient.

1.10 The dimensions of physical quantities, dimensionless numbers, and similarity

In a phenomenological approach, it is good to check the *dimensions* of a quantity, as, for an equation to make sense physically, the various terms need to have the same dimension. For example, if the term on the left-hand side is an energy, then terms on the right-hand side have to have a dimension of energy as well.

Even though this is a rather obvious truth, paying attention to the dimensions is useful. First of all, if we introduce phenomenological coefficients, like the viscosities, their dimension is determined by the above requirement that the dimensions work out. Secondly, it is an important step toward identifying the proper dimensionless quantities or the proper dimensionless combinations, based on physical insight into what is relevant and what is not.

1.10.1 Dimensions of physical quantities

To make the first point concrete, we will in this book denote the dimensions of a quantity A with $[A]$. The dimensions of the mass density ρ and velocity v of a fluid are clearly

$$[\rho] = \text{kg}/\text{m}^3, \qquad [v] = \text{m}/\text{s}. \qquad (1.50)$$

For convenience, we have chosen the mksA units, but a dimensional analysis does not depend on what your favorite units are. Now, as the dimension of the time derivative is $[\partial/\partial t] = \text{s}^{-1}$ and $[\nabla^2] = \text{m}^{-2}$, the requirement that the viscous damping term in the Navier-Stokes equations (1.49) has the same dimension as the time derivative term on the left, implies that we need to have

$$[\nu] = \text{m}^2/\text{s}, \qquad [\eta] = [\nu\rho_0] = \text{kg}/\text{m s}, \qquad (1.51)$$

in agreement with the dimensions stated in table 1.1 for ν and η.

Figure 1.12. Osborne Reynolds at work experimenting on flow through a pipe (1883). By injecting ink Reynolds was able to visualize the flow, and he discovered that at large enough Reynolds numbers the flow is turbulent. The nature of this transition has only become essentially understood recently.[23]

Probably the most famous illustration of the power of dimensional analysis is how G. I. Taylor, one of the giants of fluid dynamics, upset the US military by determining the classified energy E_{blast} of the very first atomic bomb explosion simply from published photos. Taylor assumed that the only relevant parameters are the radius R of the fireball, the detonation energy E_{blast}, the density of air ρ_0, and the time t since the detonation. Their dimensions are $[E_{blast}] = \text{kg m}^2/\text{s}^2$, $[R] = \text{m}$, $[\rho_0] = \text{kg}/\text{m}^3$, $[t] = \text{s}$. The only law that can be formed from these and which is dimensionally correct is

$$R \simeq A \left(\frac{E_{\text{blast}}}{\rho_0}\right)^{1/5} t^{2/5},$$

with A a dimensionless number. By plotting the radius of the fireball as a function of $t^{2/5}$ and assuming A to be of order unity, Taylor estimated E_{blast} to be about 17 kilotons. The fact that he got this number simply from the published pictures at various times after the detonation, upset the military quite a bit—the real value which was published later was 20 kilotons, so Taylor's estimate was actually very accurate![24] For a modern application of Taylor's analysis to the 2020 Beirut explosion, see Díaz, 2021.

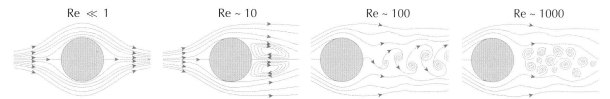

Re ≪ 1   Re ~ 10   Re ~ 100   Re ~ 1000

Figure 1.13. Schematic illustration of the flow past a cylinder as the Reynolds number $\mathrm{Re} = UD/\nu$ increases (U is the velocity far ahead and D the diameter of the cylinder). The flow is from left to right. At very small Reynolds number, the flow is laminar; nearby streamlines are parallel and smooth. For $\mathrm{Re} \gtrsim 5$, coherent vortices form behind the cylinder, but these stay behind the cylinder. At high enough Reynolds numbers ($\mathrm{Re} \gtrsim 47$), these vortices shed from the cylinder: they form a so-called Von Kármán vortex street. At very large Reynolds numbers, a turbulent wake forms behind the cylinder. Note that the flow is fully characterized by the Reynolds number only, we do not need to know the size of the cylinder or the flow speed.[25]

We now illustrate the importance of dimensionless numbers with the example of the Reynolds number.

1.10.2 The Reynolds number

The Reynolds number for a flow with far away velocity U around an object with linear size L is defined as

$$\text{Reynolds number} \qquad \mathrm{Re} = \frac{UL}{\nu}. \tag{1.52}$$

The important point is that for a flow around a single body of given shape, the Reynolds number is the *only* quantity specifying the flow, apart from the measures setting the length and time. This means that flow pattern past objects of the same shape but different size are the same if their Reynolds number is the same. In figure 1.13 we sketch the flow past a cylinder for increasing Reynolds numbers—the particular values of the size, flow speeds and viscosity are immaterial, only the Reynolds number matters.

1.10.3 Dimensionless numbers and similarity

We assume here that the proper time scale of variations in the flow is L/U. This is, e.g., true for a flapping flag, but it need not always be the case. For example, the flagella of some bacteria, as in figure 1.7, turn on a smaller time scale than this one. The analysis can easily be extended to such cases. The Reynolds number always measures the relative importance of the nonlinear term.

We can formalize this a bit by writing the Navier-Stokes equations (1.49) in dimensionless spatial and time variables defined as

$$\tilde{t} = tU/L, \quad \tilde{\vec{r}} = \vec{r}/L, \quad \tilde{\vec{v}} = \vec{v}/U. \tag{1.53}$$

In these scaled coordinates, the equations then become

$$\tilde{\vec{\nabla}} \cdot \tilde{\vec{v}} = 0,$$

$$\mathrm{Re} \left[\frac{\partial \tilde{\vec{v}}}{\partial \tilde{t}} + \tilde{\vec{v}} \cdot \tilde{\vec{\nabla}} \tilde{\vec{v}} \right] = -\tilde{\vec{\nabla}} \tilde{p} + \tilde{\vec{\nabla}}^2 \tilde{\vec{v}}, \tag{1.54}$$

where $\tilde{p} = pL/(\rho_0 U \nu)$ is a dimensionless pressure.

In essence, the Reynolds number measures the relative importance of the nonlinear $\vec{v} \cdot \vec{\nabla} \vec{v}$ term in the Navier-Stokes equations. For $\mathrm{Re} \ll 1$ the left-hand side in the velocity equation can be ignored, and the equations are linear (even if the time-derivative term is kept, if high-frequency motion is involved). This is the regime of low Reynolds number hydrodynamics. As the Reynolds number increases, the nonlinear term becomes increasingly important. We will come back to this presently.

The scale free equation (1.54) expresses that such problems obey a similarity law: flow past bodies of the same shape but different dimensions are the same apart from an overall rescaling,

$$\vec{v} = U f(\vec{r}/L, tU/L; \mathrm{Re}). \tag{1.55}$$

This similarity law underlies the use of scale models in engineering: we can, e.g., study how air will flow around a building by putting a scale model of the design in a wind tunnel, and by adjusting the speed in the tunnel so as to get the same Reynolds number as one expects in practice.

We will encounter other dimensionless numbers in different situations; typically a dimensionless number is the ratio of one physical effect over another, so that its magnitude is an indication of which effects are the dominant ones.[26]

We have just scratched the surface of dimensionless numbers and similarity analysis, but we will in passing touch on aspects of it later, for example in the context of the discussion of the diffusion equation and its so-called similarity variables.

A very nice introduction and overview can be found in the book *Scaling, Self-Similarity, and Intermediate Asymptotics* (Barenblatt, 1996) or its shorter and earlier version (Barenblatt, 1987).

1.11 From small to large Reynolds numbers

1.11.1 Low Reynolds number hydrodynamics

The Navier-Stokes equations (1.49) illustrate the difference between the two limits of small and large Reynolds numbers quite clearly. When $\mathrm{Re} \ll 1$, we can ignore the velocity terms on the left compared to the one on the right, and we arrive at the following equations for *low Reynolds number hydrodynamics* (in the original non-scaled coordinates):

$$\vec{\nabla} \cdot \vec{v} = 0, \quad \eta \nabla^2 \vec{v} = \vec{\nabla} p. \tag{1.56}$$

These are *linear equations*, and therefore they are *much* easier to analyze than the full nonlinear Navier-Stokes equations. Note that because of the incompressibility condition p satisfies the Laplace equation $\nabla^2 p = 0$ in the fluid. Almost anything can be calculated in the limit where this low Reynolds number approximation applies; the challenges are essentially only technical.[27]

Just take the divergence of the second equation of (1.56) to show that $\nabla^2 p = 0$.

The most well-known result[29] is that of the Stokes drag, the viscous drag force $\vec{F}_{\mathrm{drag}}$ on a sphere of radius R moving with velocity $\vec{U}$,

$$\text{Stokes' drag for } \mathrm{Re} \ll 1: \quad \vec{F}_{\mathrm{drag}} = -6\pi \eta R \vec{U}. \tag{1.57}$$

— time →

Figure 1.14. If one makes the Reynolds number small by using a fluid with a very large viscosity, one can illustrate the perfect flow reversal for Re ≪ 1 on a macroscopic scale with a fluid between two cylinders. In this demonstration, the letter 'C' is initially 'written' with ink in the whitish viscous fluid (left picture). The inner cylinder is then slowly turned a couple of times until the C has completely disappeared (second and third picture). Thereafter the inner cylinder is slowly rotated in the opposite direction. Note that in the fifth picture the letter C is already starting to appear faintly. After precisely the same number of turns, the fluid essentially returns to the original starting state, and the letter C reappears (sixth photo). Courtesy of Sid Nagel.[28]

The minus sign shows that the force is pointing in the direction opposite to the velocity.

Physically, the low Reynolds number regime is the regime in which viscous forces dominate over the inertial terms. The Reynolds number $Re = UL/\nu$ is small when the velocity U or the length scale L is small, or when both are. The discussion of colloids in chapter 4 will illustrate that this regime is relevant to most of soft matter on the scale of the building blocks. It is relevant as well in biology at the cellular level, where both the velocities and the scales are small.

Let us illustrate this with the *E. coli* bacterium: its size is of order 1μm, typical propulsion speeds are of order 30 μm/s, so with $\nu \approx 10^{-6}\,\mathrm{m^2/s}$ one gets

$$Re_{E.\ coli} \approx \frac{30.10^{-6}\ 10^{-6}}{10^{-6}} = 3.10^{-5}, \tag{1.58}$$

which is very small indeed.

An important point about low Reynolds number hydrodynamics is that if the velocities in the linear equations (1.56) are reversed, so are all the viscous forces. This implies that creatures living at low Reynolds numbers cannot propel themselves by making a simple back and forth motion in which they reverse their shape change, as then all forces just reverse sign, so that the net force of one cycle is zero. This result is often referred to as the *scallop theorem* as a scallop can't move by slowly opening and closing its shells. As figure 1.14 illustrates, the flow reversal can also easily be demonstrated at a macroscopic scale.

This question was first raised by Purcell in an article entitled "Life at Low Reynolds Numbers" (Purcell, 1977) which is still worth reading.[30]

So how do simple bacteria swim? Well, many use flagella, corkscrew-like tails which they rotate. By rotating it, they get around

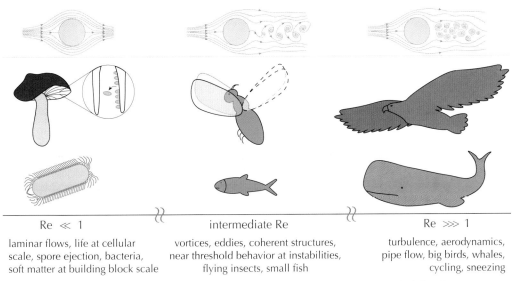

Figure 1.15. Illustration of the various flow regimes as the Reynolds number increases. Keep in mind that there are not necessarily sharp transitions between the regimes, and that the value of the Reynolds number where the crossover occurs depends strongly on the problem one considers. Note that the low Reynolds number hydrodynamics regime Re ≪ 1 described by the linearized Navier-Stokes equations applies to much of soft matter at the scale of the building blocks, such as colloids, membranes and polymers. It also applies to the flow in small channels used in microfluidics. To show the connection with the development of flows past a cylinder as a function of increasing Reynolds number, three representative sketches from figure 1.13 are included at the top.

the scallop theorem, which states that a back and forth motion does not work, but which does not apply to a rotating tail.[31]

When we go up in scale, the sizes and the flow speeds we encounter in biology increase, and hence so does the Reynolds number. Indeed, blood flow in the aorta is not at all in the small Reynolds number limit—rough estimates suggest that the corresponding Reynolds number is only moderately below the critical value for the transition to turbulence; it is reported that blood flow is normally laminar, but that especially in the downflowing aorta it can sometimes become turbulent.[32]

Figure 1.16. A cloud pattern resulting from the Kelvin-Helmholtz instability. Red Buffalo Studios/Shutterstock.[33]

1.11.2 Intermediate Reynolds numbers

As the Reynolds number increases, the inertial nonlinear term $\vec{v} \cdot \vec{\nabla} \vec{v}$ becomes more and more important, and one often gets a transition to more complicated flows. The typical scenario was already quite well illustrated by figure 1.13 for the case of flow past a cylinder: upon increasing the Reynolds number, we first observe vortices just behind the cylinder, while for even larger Re a regular 'vortex street' is shed from the cylinder. Clearly nonlinearities play an important role, but the structures or patterns one finds are still quite coherent. For very high Reynolds numbers, the wake behind the cylinder is turbulent.

For $47 \lesssim \mathrm{Re} \lesssim 80$ the vortex street does not interact noticeably with the vortices just behind the cylinder; for $\mathrm{Re} \gtrsim 80$ the vortex street forms right behind the cylinder. The formation of such vortices can actually make cables resonate, an issue of relevance for building suspension bridges.[34]

Poiseuille pipe flow (see figure 1.12) is different. In this case the laminar flow state is linearly stable for all Reynolds numbers but nonlinearly unstable. The transition is driven by a competition between laminar and turbulent domains. Inspired by this scenario, the question was brought up some years ago whether channel flow at small Re of viscoelastic polymer fluids, which is also linearly stable, can become nonlinearly unstable due to elastic effects. The issue was only settled convincingly by experiments in micro-channels after ten years. Indeed there is a nonlinear transition to elastic turbulence when elastic effects are large enough.[35] See section 5.10.4.

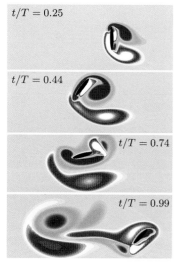

$t/T = 0.25$

$t/T = 0.44$

$t/T = 0.74$

$t/T = 0.99$

Figure 1.17. Vorticity field created by a two-dimensional idealized dragonfly wing motion at Re = 157. The four panels show the calculated vorticity generated by the wing (black) during the downstroke (upper two panels) and the upstroke (lower two panels). Blue represents clockwise vorticity; red represents counterclockwise vorticity. A pair of vortices is created in each wing beat. Image courtesy of Jane Wang[36] (Z. J. Wang, 2000).

Even though the details do depend very much on the problem at hand, as figure 1.15 illustrates, this sequence from smooth laminar flow via regular patterns dominated by clearly identifiable coherent structures to turbulence or chaos, as the driving is increased, is actually quite typical for many nonlinear problems. This is particularly the case if the patterns arise from a linear finite wavelength instability.

Instabilities are discussed from a more general perspective in chapter 8. Moreover, two common fluid dynamic instabilities, the Kelvin-Helmholtz instability—two fluids with different velocities parallel to the interface separating them—and the Rayleigh-Taylor instability—a light fluid pushing into a heavier fluid—are analyzed explicitly in problem 1.10. As that analysis shows, except in special circumstances like those shown in figure 1.16, these are not finite wavelength instabilities which typically lead to a well-defined scale; instead these instabilities usually extend over a large range of wavenumbers; as a result of this they typically lead quickly to strongly incoherent patterns.

It is interesting to note that vortices are an integral part of the aerodynamics of flying of insects: vortex formation is crucial for lift, as figure 1.17 illustrates. The smallest flying insects have a Reynolds number of order 10, while for the largest ones the Reynolds number can go up to about 10,000.

1.11.3 Very large Reynolds numbers

The term $\nu\nabla^2\vec{v}$ in the Navier-Stokes equations describes dissipation. As it is stabilizing and the highest order spatial derivative term in the equation, it tends to smoothen the field at the smaller scales. Roughly speaking, the intermediate range of Reynolds numbers we discussed above is the regime where the Reynolds number is large enough that the inertial nonlinearity $\vec{v}\cdot\vec{\nabla}\vec{v}$ plays a crucial role, while this smoothening effect is still strong enough that coherent structures like vortices are dominant: the interplay of the two terms is key. Figure 1.17 illustrates this with simulations of flow around a flapping wing of an insect.

At very large Reynolds numbers, however, there is a whole range of spatial scales ℓ for which the convective term is by far the dominant one, so for which this balance is lost. A large portion of studies of well-developed turbulence is focused on analyzing the statistical properties on this intermediate range of scales, between the large length scale ℓ_0 where energy is pumped in and the small scale, traditionally called η (don't confuse it with the dynamic viscosity!), where the energy dissipation kicks in. This intermediate range of scales $\eta \ll \ell \ll \ell_0$ is called the inertial subrange, as the statistical properties are dominated by the nonlinear interaction of modes due to the inertial term from the convective derivative. The Kolmogorov scale η is essentially the scale where the Reynolds number is of order unity.

You may still wonder to what extent and why one may ignore the viscous term on the right-hand side for many statistical aspects of well-developed turbulence. After all, energy is clearly dissipated. The answer is once more that it is a question of scales. For example, when we drive a car, the Reynolds numbers associated with the airflow around the car are extremely large. Then indeed, from the scale ℓ_0 of the car down to much smaller scales, the dynamics is essentially governed by the non-dissipative equation, obtained by ignoring the viscous term on the right. But the energy has to be transfered to smaller scales where it can finally get dissipated, so a crucial aspect of this picture is that there is an 'energy cascade' from large scales to small scales—the classic intuitive picture is that larger 'eddies' (circular currents) break up into smaller ones, and these again into even smaller ones until they reach a size where they are stable and where dissipation takes over. Since the importance of the viscous term grows when we go down in scale as it involves a second order partial derivative with respect to the spatial coordinates, it will eventually take over, and this sets the energy dissipation scale η. Clearly, in order that we can really distinguish a sufficiently large inertial subrange, the Reynolds number has to be extremely large.

1.12 Lubrication approximation for thin film flow

There are many situations in nature and in soft matter which involve the flow of a thin fluid film—the flow of a thin layer of paint immediately comes to mind, but the slime layer under crawling snails and adhesive films are in essence also thin film problems. But there are less obvious examples as well. Think for instance of situations in which droplets or (colloidal) particles in a fluid are moving toward each other, and where in the final stage the two get so close that the outflow of the thin layer of fluid in between them gives a significant counterpressure. Other examples are the thin layer of vapor that supports a droplet on a hot plate, and the final stages of the breakup of droplets. In fact, lubrication theory, which describes the motion of thin films, owes its name to the fact that humankind since antiquity has reduced wear and friction of moving parts with a thin layer of lubricating fluid. We now introduce the main ingredients for the case of a Newtonian fluid, leaving aside complications associated with the interaction with deformable media or with the non-Newtonian nature of the fluid, even though both are important for soft and living matter.

It is interesting to see that the basic scaling behavior of the Kolmogorov approach, which is still the reference frame of any statistical analysis of well-developed turbulence, can essentially be obtained from dimensional arguments. The important quantity is the energy per unit mass and time ε which is generated at the outer scale ℓ_0, $\varepsilon = u_0^2/\tau_0 = u_0^3/\ell_0$. With $[\nu] = \mathrm{m}^2/\mathrm{s}$ for the kinematic viscosity and $[\varepsilon] = \mathrm{m}^2/\mathrm{s}^3$ for the energy rate, the assumption that ε is the only quantity that matters in the inertial subrange suggest on dimensional grounds the Kolmogorov scales $\eta = (\nu^3/\varepsilon)^{1/4}$ for length, $u_\eta = (\varepsilon\nu)^{1/4}$ for the velocity at that scale, and $\tau_\eta = (\nu/\varepsilon)^{1/2}$. These scalings also imply $\eta/\ell_0 \sim \mathrm{Re}^{-3/4}$ and similarly for the other quantities. See, e.g. the book *Turbulent Flows* (Pope, 2000) for details and further discussion, and see problem 1.9 for a derivation of the logarithmic law of wall turbulence from dimensional arguments.

With a proper design, flow need not be turbulent immediately: aerodynamics (a field in itself) is based on controlling the flow over the wings of a plane so as to maximize lift. Incidentally, note that modern airplanes use winglets to reduce the generation of vortices at the edge of the wing, as vortices and turbulence reduce flight efficiency. The introduction of winglets is said to have been inpired by birds.

Lubrication theory goes back to Reynolds, 1886. A nice elementary treatment is given by Lautrup, 2005; his book treats many other interesting flow problems as well. For modern introductions to the lubrication approximation, see the overviews of Oron et al., 1997, Craster and Matar, 2009 and Langlois and Deville, 2013. A review of breakup of droplets and free-surface flows is given by Eggers, 1997, while applications in living matter can be found in the papers by Feng and Weinbaum, 2000 and Chan et al., 2005.

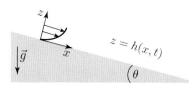

Figure 1.18. Schematic situation used to derive the lubrication approximation: a plane inclined with respect to gravity, with a thin layer of fluid of thickness h which varies on length scales much larger than the thickness. θ is the angle between the normal to the surface and the z direction.

The Laplace pressure is the pressure a curved interface exerts on the inner fluid. Think of a soap bubble.

As sketched in figure 1.18, we consider a situation of a thin layer of fluid of density ρ_0 (which we take to be constant) and height h above a surface at $z = 0$, inclined with respect to gravity. Variations of h in the x direction along the plane are on a much longer scale than the thickness h, and we assume that h does not vary in the y direction perpendicular to the x and z directions. Because of the separation of scales, the flow profile within the film will in lowest order of approximation depend on z and the local value of $h(x,t)$ only (we can make this more precise with a formal separation of scales expansion method[37]). In terms of h, the conservation of mass equation then becomes

$$\rho_0 \frac{\partial h}{\partial t} = -\frac{\partial J}{\partial x}. \qquad (1.59)$$

Here J is the total height-integrated mass flow in the x direction, per unit distance in the y direction, so dimensionally $[J] = \mathrm{kg/m\,s}$.

How does J depend on height? Well, as figure 1.18 illustrates, the flow down the plane is driven by three effects: first of all, when the plane is inclined with an angle θ with respect to gravity, there is a force $\rho_0 g \sin\theta$ along the plane in the downward direction. The second effect is that within the layer, because of the slow flow and the separation of scales, the hydrostatic pressure is proportional to the thickness of the layer. Finally, the pressure just below the free surface on the liquid side is affected by the curvature of the interface through the Laplace pressure $\gamma\kappa$, where κ is the curvature of the interface and γ the surface tension of the interface. For the nearly flat interface $\kappa \approx \partial^2 h / \partial x^2$ and the Laplace pressure is positive if the surface curvature is concave when viewed from the liquid side. We can put all these ingredients together by writing for the pressure in the fluid

$$p - p_0 \approx \rho_0 g \cos\theta \left(h(x,t) - z \right) - \gamma \partial^2 h / \partial x^2, \qquad (1.60)$$

where p_0 is the air pressure above the layer. Hence, a gradient of h leads to a pressure gradient $\partial_x p = \rho_0 g \cos\theta \, \partial_x h - \gamma \partial^3 h / \partial x^3$. Clearly, these two terms, the component of gravity parallel to the interface and the pressure gradient, are the driving forces.

Now how does the total current depend on these? Let us first do a dimensional analysis. If we realize that J must be proportional to the above three driving forces, and inversely proportional to the kinematic viscosity ν, it is easy to see on dimensional grounds that the total flow due to these forces must be proportional to h^3, in other words that

$$J = C \frac{\rho_0 g}{\nu} \left(\sin\theta - \cos\theta \frac{\partial h}{\partial x} + \frac{\gamma}{\rho_0 g} \frac{\partial^3 h}{\partial x^3} \right) h^3, \qquad (1.61)$$

where C is some dimensionless number. Physically, the fact that J is proportional to h^3 arises from the fact that for a parabolic profile (see the sketch in figure 1.18), the average flow speed is proportional to h^2. Hence, the total flow through a layer of thickness h is proportional to h^3. (That in Poiseuille pipe flow the average speed is proportional to the diameter squared is derived explicitly in problem 1.5.) The explicit calculation, done in problem 1.12, gives $C = \frac{1}{3}$, so by combining equations (1.59) and (1.61) we get for the dynamical equation in the lubrication approximation

$$\frac{\partial h}{\partial t} = -\frac{g}{3\nu}\frac{\partial}{\partial x}\left[\left(\sin\theta - \cos\theta\,\frac{\partial h}{\partial x} + \frac{\gamma}{\rho_0 g}\frac{\partial^3 h}{\partial x^3}\right)h^3\right]. \qquad (1.62)$$

As the above derivation shows, the nonlinear h dependence comes directly from the nonlinear thickness dependence of the average speed of the thin film flow.

It is clear that equation (1.62) can only be dimensionally correct provided

$$\ell_c = \sqrt{\gamma/\rho_0 g} \qquad (1.63)$$

has the dimension of a length. It is called the capillary length and it sets the length scale beyond which gravity plays a role. For the water-air interface, the capillary length is about 2.7 mm. Figure 1.19 illustrates a measurement of the thin air layer which gets temporarily formed under a droplet impacting on a surface, and whose dynamics can be analyzed with the lubrication approximation in the spirit of the above analysis.

As we discuss below and illustrate in figure 1.20, in the presence of a surface, a liquid can either wet the surface or form a droplet with a finite contact angle on the surface. Roughly speaking, in the latter case surface-droplets of size less than about ℓ_c form essentially a spherical cap, while larger droplets get deformed through gravity, as figure 1.21 illustrates.

1.13 Contact angle, coffee stains, and Marangoni flow

1.13.1 Contact angle and wetting

When a droplet is in contact with a surface, three interfaces with their respective interfacial tensions are involved, as sketched in figure 1.20.a: a liquid-vapor interface with surface tension γ_{lv}, a solid-liquid interface with tension γ_{sl} and a solid-vapor interface (γ_{sv}). The figure illustrates how for the three interfaces to be in

In short: from the above analysis the first term between brackets in (1.61) comes from the component of gravity parallel to the surface, the second one from the gradient of the hydrostatic pressure with a change in height, and the third one from the gradient of the Laplace pressure.

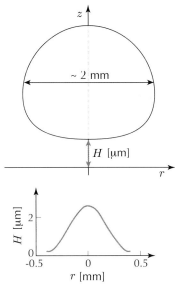

Figure 1.19. When a large enough droplet impacts a surface with sufficiently large velocity, a thin air layer can get trapped temporarily underneath. The situation is illustrated in the top panel, which is not to scale. The dynamics of this thin layer is described by the lubrication approximation. The lower panel shows an instantaneous thickness profile, as determined by light interference from the bottom. Note the micron scale.[38] Another application of the lubrication approximation is to the thin cushion of vapor on which little droplets can 'float' on a hot stove—the so-called Leidenfrost effect.[39]

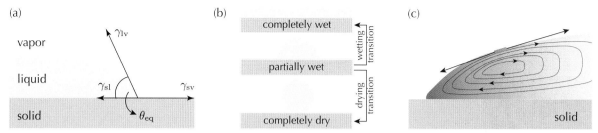

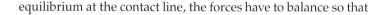

Figure 1.20. (a) A droplet on a surface forming a finite contact angle θ_{eq}. The arrows illustrate the forces resulting from the three surface tensions when the droplet is in equilibrium. (b) In the case of a droplet in contact with a solid surface and a vapor, there are three wetting regimes, depending on the relative surface tensions (see text). (c) Illustration of the fact that a concentration gradient in a droplet induces a surface tension gradient, which in turn induces flow. The arrows indicate the surface tension pulling from the left and right on a small element of the interface, resulting in a net force in the direction of the largest surface tension. For nice reviews of wetting and spreading, see Bonn et al., 2009a and de Gennes et al., 2004. As discussed there, the contact angle θ_{eq} has to be thought of as a macroscopic contact angle, beyond the scale of molecular forces, like the Van der Waals forces discussed in chapter 5.

Figure 1.21. Water droplets on the leaves of a plant. Michal Bednarek/Shutterstock. The leaves are obviously quite hydrophobic, as the contact angle is large. Note that the larger sizes get deformed, while the smaller ones are very spherical, in agreement with the fact that the capillary length ℓ_c defined in equation (1.63) is about 2.7 mm.

equilibrium at the contact line, the forces have to balance so that

$$\gamma_{sv} = \gamma_{sl} + \gamma_{lv} \cos \theta_{eq}, \tag{1.64}$$

where θ_{eq} is the equilibrium contact angle. This relation, which is nothing but a force balance equation, is referred to as Young's equation. Clearly, for a given choice of droplet fluid and substrate, as the cosine varies between -1 and 1, this equation only has finite contact angle θ_{eq} solutions provided γ_{sl} lies in the range

$$\text{finite } \theta_{eq}: \quad \gamma_{sl} - \gamma_{lv} < \gamma_{sv} < \gamma_{sl} + \gamma_{lv}. \tag{1.65}$$

At the upper end of this range, when $\gamma_{sv} \to \gamma_{sl} + \gamma_{lv}$, we have $\theta_{eq} \to 0$; when γ_{sv} is larger, $(\gamma_{sv} > \gamma_{sl} + \gamma_{lv})$ the liquid wets the solid surface, i.e., spreads over the surface. If γ_{sv} approaches the lower end $\gamma_{sl} - \gamma_{lv}$ of the finite contact angle range, $\theta_{eq} \to \pi$; when γ_{sv} is even smaller ($\gamma_{sv} < \gamma_{sl} - \gamma_{lv}$) a droplet on a surface avoids contact with the surface. In the wetting terminology, this is called the completely dry regime. These possibilities are illustrated in figure 1.20.b. Interestingly, with modern lithography techniques one can nowadays make surfaces with little micro-pillars, which have such a low surface tension that they favor the completely dry regime.

The capillary length introduced above is an important length scale for film spreading-, wetting-, and dewetting-type problems, whose dynamics can often be described by equation (1.62).[40]

1.13.2 Coffee stains resulting from enhanced evaporation at the rim of a droplet

Quite often, the meniscus of a droplet is pinned in place on a surface, due to dirt and inhomogeneities. An advancing or receding

Figure 1.22. When coffee is spilled on a table, upon drying most evaporation takes place at the rim of the spilled droplet and all dissolved particles are swept toward the rim by the resulting flow in the droplet. The figure illustrates the effect with a controlled experiment. Left panels: Quasi-two-dimensional colloidal gel, sedimented on a glass slide in a closed cell. Since there is no evaporation, the colloidal gel is homogeneous upon coarse-graining. Right panels: Quasi-two-dimensional layers of colloids, observed at the boundary of an evaporating droplet, illustrating how all colloids have moved to the rim during the drying process. The outermost images show enlargements as indicated, with a color coding for the height ranging from 0.5 particle diameter (dark blue) to five particle diameters (yellow). In both cases colloids are 1.5-micron-diameter polystyrene particles dissolved in water, and observations are made with a confocal microscope. Images courtesy of Olivier Dauchot. The coffee stain effect was highlighted in particular by Deegan et al., 1997, to whom we refer you for a detailed discussion of the effect.

contact line then typically moves by fits and starts; but the pinning can easily be strong enough to fix the position of the contact line in place, for example when a droplet is drying. This, in combination with the strong enhancement of evaporation near the rim, induces quite a few interesting effects that we also know from daily experience: coffee stains. Indeed, the evaporation of the droplet can be so dominated by the region near the rim that it induces a significant flow within the droplet toward the contact, which sweeps most of the dissolved particles to the outer rim. You may find it intriguing that for a finite contact angle the diffusion field of the evaporated vapor has a power law singularity and that this induces a power law growth of the mass accumulated at the ring. As the photos before and after drying in figure 1.22 illustrate, the effect is strong enough to explain the typical appearance of dried coffee stains, with their dark outer rim.

1.13.3 Marangoni convection

There are many non-equilibrium situations in which a third substance is dissolved in the droplet. If there is a concentration gradient, the surface tension will normally depend on this concentration and hence vary along the surface. As sketched in figure 1.20.c, the resulting surface tension gradient gives rise to a net force along the interface in the direction of the increasing surface tension. This induces a flow in the droplet from regions where the surface tension is small to regions where it is larger, so the flow is such that it lowers the extra energy associated with the surface. This surface tension gradient–induced flow is called Marangoni convection. We shall see in section 1.15 that it plays an important role in systems with three components or more. As any gradient in the surface tension will drive Marangoni convection, it can also be induced by temperature gradients, but in multicomponent systems Marangoni convection is usually driven by concentration gradients.

1.14 Bubble oscillations

The study of bubbles in fluids is important in fluid dynamics and engineering. It has a long history; indeed Lord Rayleigh already analyzed bubble dynamics over a century ago[41] because of the damage cavitation of bubbles did to ship propellers. Modern applications include the deliberate injection of bubbles under the hull of a ship to reduce ship drag and the use of bubbles for aeration or purification of water. Made possible by the advances in instrumentation and modeling of the last decade, micro-bubbles are nowadays also increasingly being used at small scales, not only as a diagnostic tool of the embedding material, but also for drug delivery and micro-surgery. As this is now even being used to probe soft matter response, we briefly introduce here the Rayleigh-Plesset equation for air bubbles, on which it is based.

A nice overview of modern research and applications of bubbles is given by Lohse, 2018 and by Lohse and X. Zhang, 2015, who focus on surface nanobubbles and droplets.

Bubbles much smaller than the capillary length are spherical, so the bubble and the flow pattern due to bubble oscillations can be taken to be spherical. Moreover, lengths and frequencies are such that we may treat the fluid flow as incompressible; in spherical coordinates the incompressibility condition $\vec{\nabla} \cdot \vec{v} = 0$ translates into $r^{-2} \partial (r^2 v_r) / \partial r = 0$, where v_r is the radial velocity. If we denote the bubble radius by R, the boundary condition at the surface of the bubble becomes $v_r(R) = \dot{R}(t)$ (the dot indicates a time derivative), so the solution for the flow field is $v_r = R^2 \dot{R}/r^2$.

For our discussion it is useful to keep initially the dissipative part of the stress tensor explicit in the velocity equation of the Navier-Stokes equations (1.49), so that its radial component reads

The dissipative part was defined in equation (1.16) and for a Newtonian fluid $\underline{\sigma}^{\mathrm{d}}$ is given in equation (1.22).

$$\rho_0 \left(\frac{\partial v_r}{\partial t} + v_r \frac{\partial v_r}{\partial r} \right) = -\frac{\partial p}{\partial r} + (\vec{\nabla} \cdot \underline{\sigma}^{\mathrm{d}})_r. \tag{1.66}$$

where $(\vec{\nabla} \cdot \underline{\sigma}^{\mathrm{d}})_r$ denotes the radial component of $\vec{\nabla} \cdot \underline{\sigma}^{\mathrm{d}}$. Since the radial dependence of the velocity field is known, we can integrate equation (1.66) to get

$$\rho_0 \left(R\ddot{R} + \frac{3}{2}\dot{R}^2 \right) = p(R) - p_\infty + \int_R^\infty \mathrm{d}r (\vec{\nabla} \cdot \underline{\sigma}^{\mathrm{d}})_r. \tag{1.67}$$

Here $p(R)$ is the pressure at the fluid side of the bubble interface. Now, the balance of forces at the interface of the bubble is expressed by

$$p_g = p(R) - \sigma_{rr}^{\mathrm{d}} + \frac{2\gamma}{R}, \qquad (1.68)$$

where p_g is the pressure in the gas bubble and where the last term on the right is again the Laplace pressure resulting from the total curvature of the interface, which equals $2/R$ for a sphere, and γ denotes the surface tension of the liquid-gas interface. If we use this in equation (1.66), we get

$$\rho \left(R\ddot{R} + \frac{3}{2}\dot{R}^2 \right) = p_g - p_\infty - \frac{2\gamma}{R} + \sigma_{rr}^{\mathrm{d}}(R) + \int_R^\infty \mathrm{d}r (\vec{\nabla} \cdot \underline{\sigma}^{\mathrm{d}})_r. \qquad (1.69)$$

In general, the last term on the right depends on the response behavior of the (complex) fluid outside the bubble; it therefore gives access to probing this behavior by studying the bubble response to oscillations.

To get a feel for this, let us first specify to the case of a bubble in a normal Newtonian fluid. In this case, $\sigma_{rr}^{\mathrm{d}} = 2\eta \partial v_r / \partial r |_R = -4\eta\dot{R}/R$, while $(\vec{\nabla} \cdot \underline{\sigma}^{\mathrm{d}})_r = 0$.[42] The equation for the bubble radius then becomes

$$\rho \left(R\ddot{R} + \frac{3}{2}\dot{R}^2 \right) = p_g - p_\infty - \frac{2\gamma}{R} - \frac{4\eta\dot{R}}{R}. \qquad (1.70)$$

This is the so-called Rayleigh-Plesset equation, which, in combination with an equation for the radial dependence of the gas pressure in the bubble, is an explicit nonlinear equation for the bubble radius. The equation of state for the pressure of the gas bubble is usually taken to be of the form $p = p_0 (R_0/R)^{3\kappa}$, where for adiabatic compression the exponent κ is equal to the ratio of the heat capacities at constant pressure and volume.

The relevance of this equation becomes clear when we consider the case of small-amplitude driving, the response of the bubble to a small driving of the pressure far away, $p_\infty = p_0 + \Delta p \sin \omega t$, with $\Delta p \ll p_0$. By linearizing the Rayleigh-Plesset equation about the static solution p_0, R_0 by assuming that the relative amplitude $x = \Delta R/R_0 \ll 1$, i.e., that the radial variations are small, one obtains for x the equation

$$\ddot{x} + 2\beta\dot{x} + \omega_0^2 x = -\Delta p \sin \omega t / \rho R_0^2, \qquad (1.71)$$

where

$$\omega_0^2 = \frac{1}{\rho R_0^2} \left(3\kappa p_0 + \frac{2(3\kappa - 1)\gamma}{R_0} \right), \qquad \beta = \frac{2\eta}{\rho R_0^2}. \qquad (1.72)$$

Equation (1.71) is simply the equation for a harmonic oscillator with natural frequency ω_0 and damping β. This shows that by acoustically exciting the pressure far away, one can make the bubble

It is important to stress that we consider an air bubble here. Because molecular diffusion is slow, the diffusion of air molecules into the liquid is ignored. For vapor bubbles, the situation is different: as vapor can condense, one then has to take into account the transport of the heat of condensation into the fluid. In a liquid the heat diffusion coefficient is normally much larger than the molecular diffusion coefficient. This is because the molecular diffusion involves actual displacement of molecules, while exchange of energy can happen without it. For a review of vapor bubbles, see Prosperetti, 2017.

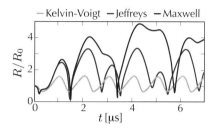

— Kelvin-Voigt — Jeffreys — Maxwell

Figure 1.23. Example of the behavior of bubble oscillations as determined from simulations of the equations given by Warnez and Johnsen, 2015 for the Rayleigh-Plesset equation (1.69) with various viscoelastic models for the embedding fluid (see section 2.5 for a brief introduction to such models). The response of a bubble embedded in a so-called Kelvin-Voigt fluid is indistinguishable from that of a bubble in a Newtonian fluid. The figure illustrates that the response depends sensitively on the viscoelastic behavior of the embedding material. An overview of probing soft matter and biomatter with bubbles, including the case of bubbles with a viscoelastically coated interface, is given by Dollet et al., 2019.

Our understanding of and ability to model the physicochemical hydrodynamics behavior of non-equilibrium droplets has in recent years increased dramatically, thanks to experimental and numerical advances similar to those which advanced soft matter research, as we sketched in the introduction. For a very nice perspective on these advances as well as the applications see Lohse and X. Zhang, 2020.

oscillate and study its resonance behavior. This acoustic resonance is for instance used in ultrasonic cleaning.

If the driving amplitude increases enough that the radius variations of the bubbles are so large that one has to use the full nonlinear equation (1.70), the temperature during the compression part of the oscillation cycle can get so large that the bubble gives off light. This is called sonoluminescence.[43] It is a phenomenon that allows one to uncover a lot of interesting physics.

An interesting new twist to this story is that small-amplitude bubble resonances are now used to probe the viscoelastic behavior of soft matter and biomatter in which they are embedded. This method employs the fact that the last integral term on the right in (1.69) depends on the non-Newtonian rheology of the material outside the bubble. As illustrated in figure 1.23, for a given viscoelastic model one can work out the response $(\vec{\nabla} \cdot \underline{\sigma}^{\mathrm{d}})_r$ of the soft matrix and then compare these with experimental measurements of the resonance curves as a function of frequency. It turns out to be a surprisingly sensitive local probe of soft matter and bio-tissues, which promises to become more important as bubbles are increasingly being used in biomedical applications, such as ultrasound imaging, drug delivery and microsurgery.

1.15 Droplets

Droplets in contact with a surface have already been touched upon in section 1.13. We now turn briefly to droplets in liquid phases, a topic of great importance for applications and which is so far-ranging that we can not properly do justice to it here. Indeed, as we already mentioned in the introductory chapter and illustrated in figure I.8, microemulsions of little droplets in suspension are common in the kitchen and in the food industry. But droplets play an equally important role in industrial bulk processing, chemical analysis, tracer extraction to detect pollutants, drug production and purification, inkjet printing etc. For instance, a common problem in industry is to extract a substance, say A, from the water in which it is dissolved. The so-called liquid-liquid extraction method is based on adding substance B, which forms droplets in water, and which has a higher solubility for A than water. So A dissolves from the water into the B-droplets; after dissolution one then subsequently extracts droplets B, taking out A along the way. Naturally, one wants to do this with small droplets as this increases the speed and efficiency of this process—both count in industry and this is therefore one of the drivers for the ubiquity of such processes in industry.

Clearly, we are here typically in the realm of non-equilibrium processes of multicomponent systems, with many competing effects: flow, composition gradients, demixing and nucleation of droplets, thermal effects, capillary forces, and gravity can all play a role, even though they may act on different time and length scales. This is best illustrated by the overview in box 1, in section 1.17, of the plethora of dimensionless numbers that help distinguish which physical effects and time scales are involved and competing.

Let us illustrate the richness with two simple examples in which the Marangoni effects which we discussed in section 1.13.3 play a dominant role. In the first one, sketched in figure 1.24.a, the Marangoni effect competes with gravity. We consider a droplet (black) of oil, for instance anise oil, in a concentration gradient of alcohol and water. The lighter alcohol-rich phase at the top is indicated by orange, the heavier water-rich phase below by blue. The oil is not miscible with alcohol or water, but alcohol and water are miscible so there is a smooth concentration gradient, as the figure indicates; the densities are chosen such that the oil is lighter than water but heavier than alcohol. Typically, oil and water don't mix well, and as a result the surface tension of the oil droplet with the water-alcohol mixture is lower the larger the alcohol concentration is. In the gradient setup of figure 1.24.a this means that a Marangoni flow is induced from the top of the droplet at A to the bottom at B.

If the droplet is released on the lighter alcohol side at the top, the first thing that happens is that the droplet sinks downward in the density gradient. Ethanol-rich liquid is entrained downward during this motion, and this effect is enhanced by the Marangoni flow downward along the drop's surface. At the same time, the Marangoni flow downward along the surface is counteracted by the flow around the droplet as it is sinking down, which is directed upward in the frame of the droplet. The buoyancy forces drive the droplet to the layer where the density of the alcohol-water mixture matches that of the anise oil, but as the droplet gets settled there, the counteracting flow from the sinking motion decreases and the entrained alcohol diffuses out, with the result that the Marangoni forces start to increase. The effect of this flow is to push the surrounding alcohol-water mixture down, and hence the droplet moves up again. In the case of anise it is reported that this Marangoni flow takes over after about a minute and that the droplet then shoots up again. The process then starts all over again—in experiments, such consecutive jumps have been observed to continue for more than six hours, as the process only stops once the stratified alcohol-water mixture has been sufficiently mixed by the droplet and by diffusion. This remarkable phenomenon has been modeled in detail; Marangoni convection, gravity and diffusion all play a role, so relevant dimensionless numbers from box 1, in section 1.17, are the Marangoni, the Rayleigh and the Péclet number.

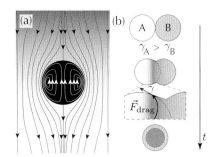

Figure 1.24. Two examples of Marangoni-induced effects in multicomponent droplet systems. (a) A droplet of oil in a vertically stratified alcohol-water mixture starts to oscillate up and down due to competition of Marangoni flow and buoyancy forces. See text. (b) Side view of two droplets of different immiscible fluids. The figure illustrates that the one with lower surface tension can engulf the one with larger surface tension in time. The condition under which this happens is reminiscent of the condition that a droplet wets a surface, illustrated in figure 1.20.b. The lower part illustrates the net capillary force and the resulting flow during the dynamical process.[44]

The droplet motion is reminiscent of a microswimmer in an active medium of the type discussed in chapter 9. Swimming droplets are reviewed by Maass et al., 2016.

The repetitive up and down motion of the droplet nicely illustrates experience from dynamical systems, namely that oscillatory limit-cycle like behavior is usually favored by competing phenomena with different time scales.

The second example, driven by surface tension differences, is depicted in figure 1.24.b: when two droplets of different immiscible fluids immiscible fluids come in contact, the relative surface tensions can be such that the droplet with the lowest surface tension with respect to the solvent engulfs the other one. As the lower panel illustrates, the process is driven by the capillary forces, which balance with viscous forces during the dynamical process. This balance determines the speed of the encapsulation front.

1.16 What have we learned

In this chapter we have introduced the classical equations of fluid dynamics which are based on the conservation of mass, momentum and energy at the molecular scale: the fact that these are the conserved quantities of simple fluids identifies them as the slow variables governed by hydrodynamic conservation laws. The derivation, based on coarse-graining together with a focus on the slow variables, also yields an important lesson for introducing hydrodynamic equations for more complicated problems: a generalized hydrodynamic formulation is likewise based on this physical picture of coarse-graining and the identification of the appropriate slow variables.

We discussed under what conditions the flow of a fluid can be treated as incompressible, and introduced the Reynolds number as a measure for the importance of the nonlinearities in the equations. In soft matter and the life sciences, relevant flow often occurs at small scales, so the Reynolds number is small and the flow equations are linear. The nonlinear behavior of the equations for flow of thin layers, obtained in the lubrication approximation, arises from the strong dependence of the flow on layer thickness.

The fluid dynamic equations also provide a good reference point for discussing flow of soft matter. Simple Newtonian fluids are characterized by a linear relation between the local stress and the instantaneous local shear rate. A complex fluid is a fluid for which such a simple proportionality does not hold. Most soft matter behaves like a complex fluid during flow—for instance, the stress may depend nonlinearly on the shear rate, or memory and relaxation terms may need to be included. We will encounter various examples in later chapters.

Soft matter and life at the scale of the cell often involve building blocks like colloidal particles or droplets dissolved in a fluid. In many non-equilibrium situations of practical interest, temperature or composition-induced gradients in the surface tension give rise to transport driven by Marangoni flow.

1.17 Box 1: Key dimensionless parameters

> **Box 1. Key dimensionless parameters of fluid dynamics relevant to soft matter**
>
> ▶ $\mathrm{Re} = UL/\nu$, the **Reynolds number**. Well-known dimensionless number defined in equation (1.52). Expresses the ratio of inertial forces to viscous forces, for flows characterized by a velocity U around a body of size L. ν is the kinematic viscosity defined in (1.36). See section 1.11 for discussion of Re dependence of flows.
>
> ▶ $\mathrm{Pe} = \dot{\gamma} R^2/D$, the **Péclet number**. Expresses the competition between advection and diffusion in a shear gradient $\dot{\gamma}$ for particles of radius R and with diffusion coefficient D. Introduced in equation (4.25) in our discussion of colloidal rheology.
>
> ▶ $\mathrm{Pr} = \nu/D_T$, the **Prandtl number**. As the ratio of the kinematic viscosity ν over the thermal conductivity D_T it expresses the competition between viscous damping and thermal diffusion. In typical simple fluids, Pr is just somewhat larger than 1, expressing that momentum diffusion is a bit faster than thermal diffusion. Plays a role in Rayleigh-Bénard convection; see section 8.3.1.
>
> ▶ $\mathrm{Ra} = \alpha g d^3 \Delta T/\nu D_T$, the **Rayleigh number**. Measures the importance of buoyancy effects induced by a thermal gradient, relative to stabilizing viscous damping and thermal diffusion. As discussed in section 8.3.1 on Rayleigh-Bénard convection, α measures the thermal expansion, g is the gravitational acceleration, d the thickness of the Rayleigh-Bénard cell, and ΔT the temperature across it. Defined in equation (8.11).
>
> ▶ $\mathrm{Sc} = \nu/D$, the **Schmidt number**. As molecular diffusion, characterized by the diffusion coefficient D, is typically slower than momentum or thermal diffusion, the Schmidt number is large, typically of order 10^3, for simple fluids. This disparity of the time scales of viscous damping and molecular diffusion gives rise to many physicochemical hydrodynamic effects, but it complicates numerical and experimental studies.[45]
>
> ▶ $\mathrm{Le} = D_T/D$, the **Lewis number**. Important when both molecular and thermal diffusion play a role. As $\mathrm{Le} = \mathrm{Sc}/\mathrm{Pr}$, the Lewis number is typically large.
>
> ▶ $\mathrm{Ca} = \eta U/\gamma$, the **capillary number**. With η the dynamic viscosity and γ the surface tension, the capillary number measures the ratio of viscous to capillary forces. See e.g. the wetting front analysis in problem 7.2.
>
> ▶ $\mathrm{Bo} = \rho g R^2/\gamma$, the **Bond number**. Measures the ratio of gravitational to capillary forces for a droplet of radius R and density ρ. Note that $\mathrm{Bo} = R^2/\ell_c^2$, with ℓ_c the capillary length defined in equation (1.63). The change of shape with radius R of droplets on a leaf in figure 1.21 is due to the varying Bond number.
>
> ▶ $\mathrm{We} = \rho U^2 R/\gamma$, the **Weber number**. Measures the ratio of inertial to capillary forces, e.g., during the deformation of a droplet of radius R hitting a surface with velocity U.
>
> ▶ $\mathrm{Ma} = R\Delta\gamma/\rho\nu D$, the **Marangoni number**. Compares the Marangoni forces discussed in section 1.13.3 with the stabilizing viscous forces and stabilizing mass diffusion. Note that the surface tension difference $\Delta\gamma$ can arise from concentration gradients and thermal gradients.
>
> ▶ $\mathrm{Oh} = \eta/(\rho\gamma R)^{1/2} = \mathrm{We}^{1/2}/\mathrm{Re}$, the **Ohnesorge number**. Gives the ratio of the time scale of viscous damping to the time scale of capillary oscillations.
>
> ▶ $\mathrm{Wi} = |\sigma_{xx} - \sigma_{yy}|/\sigma_{xy}$, the **Weissenberg number**. As discussed in section 5.10.2, for viscoelastic polymer fluids the Weissenberg number is a ratio of the (first) normal stress difference over the shear stress for a simple shear flow. In the UCM model (5.80) for polymer rheology, $\mathrm{Wi} = 2\dot{\gamma}\lambda$, where $\dot{\gamma}$ is the shear rate and λ is the relaxation time of the fluid; see equation (5.81). This ratio of the relaxation time to the time scale of the flow in polymer flows is also called the Deborah number De.
>
> ▶ The **Damköhler number** Da is in the presence of chemical reactions defined as the ratio of the chemical reaction rate and the diffusive or convective mass transport rate.

Relevant coding problems and solutions for this chapter can be found on the book's website www.softmatterbook.online under Chapter 1/Coding problems.

1.18 Problems

Problem 1.1 *Derivation of equation (1.12) in the Eulerian frame*

Derive the conservation of momentum equation directly in the Eulerian frame.

Problem 1.2 *Derivation of equation (1.18)*

In this exercise we derive the conservation equation (1.18) for the entropy from the energy conservation equation (1.13).

a. Equation (1.13) is the balance equation for the energy density per unit mass ϵ. The first step is to write the equation for the internal energy per unit mass $u = \epsilon - v^2/2$, already mentioned above equation (1.13). Use equation (1.11) for the material derivative of $\vec{v}$ together with the conservation equation for ϵ and the conservation of mass equation (1.7) to derive the following equation for the material derivative of the internal energy u:

$$\rho \frac{du}{dt} = \sum_{ij} \sigma_{ij} \frac{\partial v_i}{\partial x_j} - \vec{\nabla} \cdot \vec{J}^q. \tag{1.73}$$

b. Next, we use the basic assumption of non-equilibrium thermodynamics that a fluid element is in local equilibrium, so that the internal energy u, density ρ and entropy *per unit volume* s obey the usual thermodynamic relations. This means, for instance, that their differentials obey the so-called Gibbs-Duhem relation[46]

$$T\,ds = du + p\,d\rho^{-1}, \tag{1.74}$$

where p, as usual, is the pressure. Hence

$$T\frac{ds}{dt} = \frac{du}{dt} - \frac{p}{\rho^2}\frac{d\rho}{dt}. \tag{1.75}$$

c. Use (1.75) together with (1.73) and the mass conservation equation (1.7) to derive the following equation for ρs:

$$\frac{\partial \rho s}{\partial t} + \vec{\nabla} \cdot \left(\rho \vec{v} s + \frac{\vec{J}^q}{T} \right) = \sum_{ij} \frac{\sigma_{ij} + p\delta_{ij}}{T} \frac{\partial v_i}{\partial x_j} + \vec{J}^q \cdot \vec{\nabla}\left(\frac{1}{T}\right). \tag{1.76}$$

d. The above equation contains the total stress tensor $\underline{\sigma}$. Use (1.16) and (1.15) to show that (1.76) reduces to equation (1.18).

Problem 1.3 *Derivation of Bernoulli's law*

The common form of Bernoulli's law for a steady, inviscid, and incompressible flow, valid at any point along a streamline, is

$$\frac{\rho v^2}{2} + \rho g z + p = const, \tag{1.77}$$

where in line with the notation used earlier v is the fluid flow speed at a point on a streamline, g the gravitational acceleration, z the vertical distance, p the pressure at the chosen point, and ρ the density of the fluid. The constant on the right-hand side of the equation depends only on the streamline chosen. We derive the equation in this problem.

Streamlines are defined for steady (i.e., time-independent) flows as the line along which a small fluid element moves. In equations: for a velocity field $\vec{v}(\vec{r})$ the streamlines are defined as the lines along which dx, dy and dz obey

$$\frac{v_x}{dx} = \frac{v_y}{dy} = \frac{v_z}{dz}. \tag{1.78}$$

A set of streamlines can form a tubular surface called a streamtube; see figure 1.25. From equation (1.78), the fluid velocity is parallel to the streamline, so by construction there is no flow across the sidewalls of the streamtube. Assume an infinitesimal volume element ΔV moving within a streamtube with the geometry shown in figure 1.26. The light blue regions represent the volume element, and $\Delta V = \Delta s_1 A_1 = \Delta s_2 A_2$.

a. Show that the total work done by surrounding fluid of this volume element, ΔV, from point 1 to point 2, is $(p_1 - p_2)\Delta V$.

b. Derive Bernoulli's law (1.77) from energy conservation.

c. Also derive the Bernoulli equation for steady flows from the Navier-Stokes equation for incompressible ideal flow (so $\eta = \zeta = 0$), making use of the vector identity $\vec{v} \cdot \vec{\nabla}\vec{v} = \vec{\nabla}v^2/2 - \vec{v} \times (\vec{\nabla} \times \vec{v})$ by integrating the equation along a streamline.

d. Illustrate how Bernoulli's law can be used to argue for the lift of an airplane when the wing design causes streamtubes to narrow over the upper surface.

Problem 1.4 *Dimensional analysis and the drag on a car*

a. Estimate the order of magnitude of the Reynolds number for a car driving at 100 km/h.

b. Give the dimensions of the drag force F_{drag}, the speed and the air density.

c. Use dimensional analysis to argue that $F_{\text{drag}} = \rho U^2 A$, where $[A] = \text{m}^2$.

d. Argue that it makes sense to write $A = C_d A_{\text{cross}}$, where A_{cross} is the head-on cross-sectional area of the car, and where C_d is a dimensionless drag coefficient.

What we now call Bernoulli's law was first proposed by Daniel Bernoulli when he was investigating the relationship between the speed of blood flow and its pressure. As a side product, he invented the first sphygmomanometer, a blood pressure gauge. Although Bernoulli deduced that pressure decreases when the flow speed increases, it was Leonhard Euler who derived Bernoulli's equation in its usual form in 1752.

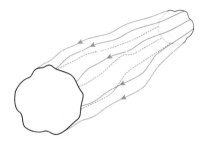

Figure 1.25. Illustration of streamlines and streamtube.

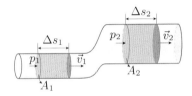

Figure 1.26. Illustration of volume element moving within a streamtube.

We here consider flow in a rigid pipe, with no-slip boundary conditions, i.e. $\vec{v} = 0$ at the walls. Interestingly, when the walls are made of deformable soft material (Louf et al., 2020), or when the walls are covered by a deformable hair bed (Alvarado et al., 2017), then the relation between flow and pressure is quite nontrivial and nonlinear.

e. Look for values of C_d of various cars on the internet, and convince yourself that air resistance a Porsche experiences at 200 km/h is comparable to that of a Hummer at 100 km/h, illustrating that cross-section and design matter.

Problem 1.5 *Poiseuille pipe flow*

In this exercise we analyze the laminar regime of pipe flow, called Poiseuille flow. We consider the steady laminar flow in a cylindrical pipe with length L and radius R ($R \ll L$) in the absence of gravity. A pressure difference $\Delta p = p_0 - p_L$ is applied across the pipe and no-slip boundary conditions apply at the wall. Because of the rotational symmetry of the system, the flow profile can be described simply by a radial function $v(r)$ for $0 \leq r \leq R$.

a. Write down the quantities governing the flow, and their dimensions.

b. We introduce Q, the amount of mass per unit time, flowing through the pipe. Write down its dimension.

c. Anticipate that for laminar flow Q is linear in Δp. Then use dimensional arguments to write down how Q will depend on the parameters of the pipe and the fluid. Pay particular attention to the subtlety associated with dealing with lengths.

Hint: Argue that Q should be linear in the combination $\Delta p / L$.

d. Take the Navier-Stokes equation for this case, write it in cylindrical coordinates, and show that it leads to a linear differential equation for the flow profile $v(r)$.

You should get a parabolic velocity profile.

e. Calculate $v(r)$ from the differential equation, using the proper boundary conditions.

f. Calculate the exact value of the rate of mass flow Q.

Hint: Integrate $v(r)$ over the cross section of the pipe.

g. Compare your initial guess from dimensional arguments with the exact calculation.

Problem 1.6 *Oscillatory motion of a plate*

This exercise is an example of a calculation which is most easily done using complexification of the equation, discussed in section 1.7.2. It also illustrates that the penetration depth of an oscillatory motion is determined by the frequency and the viscosity (can you guess the form of the result from dimensional considerations?).

We consider the oscillatory motion of a plate in the xy plane at $z = 0$, and we calculate the fluid motion in the incompressible and laminar fluid for $z > 0$. The plate oscillates with velocity $U_x = A \cos(\omega t)$, and we have no-slip boundary conditions at the plate.

a. Consider the symmetries, and show that the continuity equation implies that $v_z = 0$ in the fluid.

b. Show that the nonlinear convective term in the Navier-Stokes equations is zero. Hence the resulting equation for the fluid motion is linear.

c. Show that the pressure p is constant in the fluid.

d. Write down the linear equation governing v_x in the fluid.

e. Make the ansatz $v_x = Ae^{ikz-i\omega t}$ and calculate the exact functional form of v_x by using the proper boundary condition.

f. Show that the *viscous penetration depth* δ is given by $\delta = \sqrt{2\nu/\omega}$.

g. How large is the viscous penetration depth in water for 10 Hz and 10 kHz?

Values for the viscosities are given in table 1.1.

Problem 1.7 *Damping of a sound mode in time*

Suppose a standing sound mode of frequency ω has been excited in a resonator in air.

a. Calculate the decay time of the mode once the driving is turned off.

Hint: Use equation (1.39), keep k real but allow ω to be complex.

b. What is the decay time for a frequency of 100 Hz and for a frequency of 1 kHz?

Problem 1.8 *Estimation of Reynolds numbers*

To get a feel for Reynolds numbers, make estimates of the Reynolds number for biking and for swimming.

Values for the viscosities are given in table 1.1.

Problem 1.9 *The logarithmic law for the mean velocity near a wall in turbulence*

In this problem you will derive in a few simple steps the logarithmic law for the streamwise velocity near the wall in the case of high-Reynolds turbulent flow past a plate or in a pipe.

Very close to the wall, there is a small layer where the flow is laminar. For large enough Reynolds numbers, one will have well-developed turbulence on outer scales. We derive here the Von Kármán logarithmic wall law from dimensional arguments in the intermediate range between the viscous layer and the outer scale. We consider a flat wall at $y = 0$ with mean flow in the x direction, so y is measuring the distance from the wall.

a. We denote the wall stress by σ_w, and the fluid density by ρ_0 (be aware that in most of the literature the wall stress is indicated as τ).

As always, ν is the kinematic viscosity. Write down the dimensions $[\sigma_{\mathrm{w}}]$, $[\rho_0]$, $[y]$ and $[\nu]$.

b. Introduce the velocity scale $u_\sigma = \sigma_{\mathrm{w}}/\rho_0$. Show that this has indeed the dimension of velocity, and show that ν/u_σ has the dimension of length.

c. We postulate that the mean flow $U(y)$ can only depend on the quantities considered in step a in the intermediate range between the viscous layer and the outer scale. Write down the only possible law for $\mathrm{d}U(y)/\mathrm{d}y$ that is allowed on dimensional grounds.

d. Integrate the law derived in c. to derive the logarithmic law

$$U^+(y) = \frac{1}{\kappa}\ln(y^+) + A, \qquad (1.79)$$

where $U^+ = U/u_\sigma$ and $y/(\nu/u_\sigma)$ are the dimensionless mean velocity and distance from the wall, and where κ is a dimensionless constant, referred to as the Von Kármán constant.

This logarithmic law has been extensively confirmed in experiments and simulations. For smooth walls, it turns out that $A \approx 5.0$ and $\kappa \approx 0.4$—see Smits et al., 2011 for an overview.

Problem 1.10* *The Kelvin-Helmholtz and Rayleigh-Taylor instabilities*

This exercise illustrates how one does a stability calculation for the case of two ideal incompressible fluids flowing with different speeds relative to each other. The situation is sketched in figure 1.27. The analysis will demonstrate the essence of both the Kelvin-Helmholtz instability (two fluids flowing in opposite direction) and the Rayleigh-Taylor instability (a heavier fluid on top of a lighter fluid).

The calculation is broken up into many small steps, so as to make them doable. Some intermediate results are simply stated, but if you are ambitious you can derive these results explicitly with the information provided. If you do get stuck, we suggest that you go to the next step where you should be able to pick up.

a. We simplify the situation by considering the case of potential flow, for which

$$\begin{aligned}\text{upper half } z > 0: \quad &\vec{v}_1 = \vec{\nabla}\Phi_1; \\ \text{lower half } z < 0: \quad &\vec{v}_2 = \vec{\nabla}\Phi_2.\end{aligned} \qquad (1.80)$$

We take the fluid flow to be incompressible. Show that this implies that Φ_1 and Φ_2 obey the Laplace equation

$$\nabla^2\Phi_{1,2} = 0 \qquad (1.81)$$

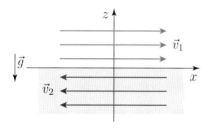

Figure 1.27. Illustration of the reference state for the stability calculation of the Rayleigh-Taylor and Kelvin-Helmholtz instability.

With this exercise, we intend to show you explicitly how a calculation of a linear instability is done in practice. Even though there are quite a few terms to keep track of and even though the dispersion relation cannot always be solved easily analytically, conceptually the steps are relatively straightforward: one linearizes the equations and boundary conditions and then substitutes a Fourier mode ansatz for all the fields.

in their respective domains.

b. The governing equation for an incompressible, inviscid flow is the Euler equation in both domains

$$\frac{\partial \vec{v}_i}{\partial t} + \vec{v}_i \cdot \vec{\nabla}\vec{v}_i = -\frac{1}{\rho_i}\vec{\nabla}p_i - gz\hat{e}_z, \quad i = 1, 2. \tag{1.82}$$

Here $\hat{e}_z$ is a unit vector in the z direction. Starting from the Euler equation, derive the Bernoulli equation for time-dependent potential flow

$$\frac{\partial \Phi_i}{\partial t} + \frac{1}{2}v_i^2 + gz + \frac{p_i}{\rho_i} = const., \quad i = 1, 2. \tag{1.83}$$

Note that the constant on the right-hand side is constant in each fluid, but could be different for the two fluids.

c. In the unperturbed case the flow is parallel to the x-axis and the surface separating the fluids is completely flat (see figure 1.27). Furthermore, the unperturbed pressures are simply given by the hydrostatic pressure with a linear variation in z (why?). In order to do the linear stability analysis, we assume small-amplitude perturbations of the surface and of the flow velocities, by writing

$$\Phi_1 = U_1 x + \phi_1, \quad \Phi_2 = U_2 x + \phi_2, \tag{1.84}$$

and considering ϕ_1 and ϕ_2 as small quantities. Clearly, these also satisfy the Laplace equation.

We write the small-amplitude variations of the interface as $z - \xi(x, t)$. By requiring that the total time derivative $d\xi/dt$ must equal the z-component of the velocity in both fluid 1 and fluid 2, one then gets, to linear order, the two boundary conditions

$$\frac{\partial \xi}{\partial t} + \frac{\partial \xi}{\partial x}U_1 = \frac{\partial \phi_1}{\partial z}, \quad \frac{\partial \xi}{\partial t} + \frac{\partial \xi}{\partial x}U_2 = \frac{\partial \phi_2}{\partial z}. \tag{1.85}$$

at the interface.

d. Now derive the boundary condition for the two pressure fields by writing in the perturbed case as $p_i = p_i^0 + \delta p_i$ where p_i^0 is the pressure field for the unperturbed straight interface and δp_i the change in pressure when the interface is perturbed. Show that p_i^0 vary linearly in z, and derive the equation which the δp_i obey to linear order in the perturbations from the fluid equation (1.83), linearized in the fields $\phi_{1,2}$. Show that this gives

$$\rho_1\left(\frac{\partial \phi_1}{\partial t} + \frac{1}{2}\left[U_1 + \frac{\partial \phi_1}{\partial x}\right]^2 + gz\right) + p_1 = \frac{1}{2}\rho_1 U_1^2 + p_1^0 \tag{1.86}$$

for phase 1, and similarly for phase 2.

The Rayleigh-Taylor instability arises for two fluids of different densities which occurs when the lighter fluid is pushing the heavier fluid. It is essentially buoyancy-driven and can occur in fluids in the lab, but it can also occur in a stratified atmosphere when the density is inverted. Other examples include the behavior of thin oil films suspended below a wafer (Limat et al., 1992), mushroom clouds like those from volcanic eruptions and atmospheric nuclear explosions, supernova explosions in which expanding core gas is accelerated into denser shell gas, instabilities in plasma fusion reactors and inertial confinement fusion.

An example of the Kelvin-Helmholtz instability is wind blowing over water: The instability manifests in waves on the water's surface. More generally, clouds, the ocean, Saturn's rings, Jupiter's Great Red Spot, and the sun's corona show this instability. The flapping of a flag is also due to it.

These equations are valid in both fluids, and we have to match the results at the interface; doing so gives the boundary conditions for the fields in fluid 1 and 2. One does so by analyzing the three equations at the perturbed interface to linear order, which involve the fields $\phi_{1,2}$ and the perturbed interface position.

Since mass cannot accumulate at the interface between the two fluids, the normal component of the velocity has to be continuous and equal to the normal velocity of the interface. Since here the unperturbed interface is flat, we can approximate the normal component by the z-component.

Note that as ξ, ϕ_1 and ϕ_2 are themselves small quantities, we can to lowest order write the boundary conditions at $z = 0$, as the shift to $z = \xi$ is a higher order correction.

The boundary condition at the interface is that the pressure on both sides of the interface is the same, hence that $p_1^0(z = \xi)$ $+ \delta p_1(z = \xi) = p_2^0(z = \xi) + \delta p_2(z = \xi)$, where again to lowest order we can evaluate the fields δp_i at $z = 0$ rather than at $z = \xi$.

e. Show that continuity of pressure leads, in linear order, to the boundary condition

$$\rho_1\left(\frac{\partial \phi_1}{\partial t} + U_1\frac{\partial \phi_1}{\partial x} + g\xi\right) = \rho_2\left(\frac{\partial \phi_2}{\partial t} + U_2\frac{\partial \phi_2}{\partial x} + g\xi\right). \quad (1.87)$$

at the interface.

Hint: Start from equation (1.86) and note that since ϕ_i is small, you can ignore terms quadratic in ϕ_i.

f. We are now going to solve these equations. For our small-amplitude pertubations we assume as in equation (1.37) Fourier modes of the form $\exp[i(kx - \omega t)]$ for the x and temporal behavior and use complexification as discussed in section 1.7.2. So we write

$$\phi_1 = \phi_1(z)\exp[i(kx - \omega t)], \quad \phi_2 = \phi_2(z)\exp[i(kx - \omega t)], \quad (1.88)$$

and solve for $\phi_1(z)$ and $\phi_2(z)$, imposing that these functions vanish for $z \to \infty$ and $z \to -\infty$, respectively. Verify that the ϕ_i obey the Laplace equation, and verify that we get

$$\frac{\partial^2 \phi_i}{\partial z^2} = k^2\phi_i, \quad (1.89)$$

with solutions $\phi_1(z) \propto \exp(-kz)$ and $\phi_2(z) \propto \exp(kz)$ for $k > 0$. We take $k > 0$ from here on.

g. Likewise, assume $\xi(x, t) = \hat{\xi}\exp[i(kx - \omega t)]$. Show that with this ansatz, boundary conditions (1.85) and (1.87) lead to

We now have all the linear equations and are going to solve them by substituting Fourier modes. This will lead in a few steps to equations (1.90).

$$-i\omega\hat{\xi} + U_1 ik\hat{\xi} = -k\hat{\phi}_1, \quad -i\omega\hat{\xi} + U_2 ik\hat{\xi} = k\hat{\phi}_2,$$
$$\rho_1\left((-i\omega + U_1 ik)\hat{\phi}_1 + g\hat{\xi}\right) = \rho_2\left((-i\omega + U_2 ik)\hat{\phi}_2 + g\hat{\xi}\right), \quad (1.90)$$

where $\hat{\phi}_{1,2} = \phi_{1,2}(0)$.

Hint: Simply substitute the ansatz into equations (1.85) and (1.87).

We now have the three linear equations for the three amplitudes, for given ω and k. The dispersion relation (1.91) is essentially the condition that this set of linear equations has a nontrivial solution.

h. Eliminate $\hat{\phi}_{1,2}$ from the first two equations and substitute this into the third one. Show that this leads to the dispersion relation.

$$\omega^2(\rho_1 + \rho_2) - 2\omega k(\rho_1 U_1 + \rho_2 U_2)$$
$$+ k^2(\rho_1 U_1^2 + \rho_2 U_2^2) + (\rho_1 - \rho_2)kg = 0. \quad (1.91)$$

We have the dispersion relation and are now ready to harvest!

i. With reference to the discussion in section 1.7.2, verify that with our mode ansatz $\exp[i(kx - \omega t)]$, a positive imaginary part of ω corresponds to an exponentially growing mode in time, i.e., an unstable mode, and verify that modes with a negative imaginary part of ω decay, i.e., correspond to stable modes.

j. We are now going to explore the implications of this dispersion relation for two different situations. Consider first the case of fluids at rest, $U_1 = U_2 = 0$. Show that the dispersion relation reduces to

$$\omega^2 = \frac{(\rho_2 - \rho_1)kg}{\rho_1 + \rho_2}. \tag{1.92}$$

k. Show that the dispersion relation implies that, when a heavier fluid is resting on top of a lighter fluid, this leads to instability for all modes $k > 0$. This instability is called the Rayleigh-Taylor instability.

l. Now, set gravity to zero, $g = 0$, so that we have a case of pure shear flow between different fluids. Show that in this case the solutions of the quadratic dispersion relation are

$$\omega_{\pm} = \frac{k(\rho_1 U_1 + \rho_2 U_2)}{\rho_1 + \rho_2} \pm ik\frac{\sqrt{\rho_1 \rho_2}}{\rho_1 + \rho_2}|U_1 - U_2|. \tag{1.93}$$

Discuss how this expression implies that one of these modes is unstable, and one is stable. Since there is always an unstable mode, the flat interface in this case is unstable.

This last case which you have analyzed is the *Kelvin-Helmholtz* instability. It too is a very common instability.

Note that according to the above analysis the instability occurs for all wavenumbers k in both cases. This is not realistic, of course, and it comes from the fact that we have studied ideal fluids without damping, and that we have ignored also any surface tension: surface tension will stabilize the instability at short enough wavelengths (We extend the analysis to include surface tension in problem 1.11 below). In natural systems like clouds like those in figure 1.16, the finite thickness of the layers or of the inhomogeneous region may stabilize sufficiently small wavenumber perturbations. Similarly, the finite width of a container in the lab or of the simulation domain in numerical calculations will cut off the instability for small wavenumbers. Nevertheless, both instabilities in practice usually exist over a large range of wavenumbers. As a result of this, both instabilities tend to result quickly in incoherent patterns and hence in strong mixing of the two layers, as figure 1.28 clearly demonstrates. The figure also illustrates how the Rayleigh-Taylor instability often leads to shear layers which subsequently exhibit a Kelvin-Helmholtz instability.[48]

Problem 1.11 *Capillary-gravity waves*

In the previous exercise, we did not include the stabilizing effects of the surface tension. We now return to the stable situation of a surface between two fluids at rest with the heavier fluid below ($\rho_2 > \rho_1$), like the fluid-air interface of water in a pond. Equation

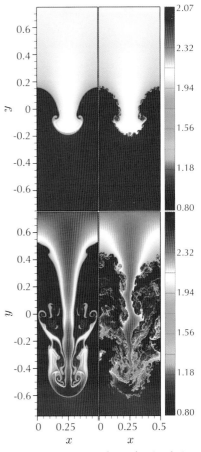

Figure 1.28. Four snapshots of a simulation of the Rayleigh-Taylor instability in two dimensions. The heavier fluid (yellow) is accelerated downward by gravity. Note that the instability quickly leads to interfacial shear layers which in turn exhibit a Kelvin-Helmholtz instability, and to structures at increasingly smaller length scales. Copyright 2019 by S. M. Rahman and O. San, CC BY 4.0.[47]

(1.92) already gave the dispersion relation of gravity waves, wavelengths long enough that the surface tension does not play a role.

a. Return to the result (1.86) in the previous derivation. We derived (1.87) by requiring continuity of the pressure. Argue that inclusion of the Laplace pressure implies that for small pressure changes we have $\delta p_2 = \delta p_1 - \gamma \partial^2 \xi / \partial x^2$ and that inclusion of the surface tension of the interface therefore amounts to the change $\rho_2 g \xi \to \rho_2 g \xi - \gamma \partial^2 \xi / \partial x^2$ in (1.87).

Note that in the previous exercise we took $k > 0$ for simplicity. We write the result here for arbitrary k.

b. Show from (1.92) that the above replacement leads to the following dispersion relation of capillary-gravity waves:

$$\omega^2 = |k| \left(\frac{(\rho_2 - \rho_1)g}{\rho_1 + \rho_2} + \frac{\gamma}{\rho_1 + \rho_2} k^2 \right). \qquad (1.94)$$

The dispersion relation $\omega \sim \sqrt{|k|}$ of gravity waves implies that the group velocity $d\omega/dk$ is half the phase velocity ω/k of waves. This result is responsible for the characteristic V-shape of the wake created by boats and swimming ducks.

c. For many cases, like a water-air interface, ρ_1 can be neglected in comparison with ρ_2. Show that in this case the above result can be written as

$$\omega^2 = |k| g \left(1 + \ell_c^2 k^2 \right), \qquad (1.95)$$

where ℓ_c is the capillary length introduced in equation (1.63). This expression confirms that the capillary length sets the scale separating short-wavelength surface tension dominated capillary waves from long-wavelength gravity waves.

d. The form (1.95) is unusual, since waves obeying a normal wave equation like (1.34) have $\omega^2 \sim k^2$. Argue that the special form of surface waves originates from the fact that the driving force is only at the surface, while the mass involved in the motion scales as the penetration depth of the wave, i.e., as k^{-1}.

e. Demonstrate that when the heavier fluid is above the lighter fluid, the Rayleigh-Taylor instability is cut off at a wavelength of order of the capillary length defined in equation (1.63) times a dimensionless factor determined by the density contrast.

Problem 1.12 *Derivation of the thin-flow equation (1.62)*

The thin film equations are used in problem 7.1 to study the instability of a cylindrical fluid layer and in problem 7.2 to study a wetting front.

In this equation, we derive equation (1.62) step-by-step. See figure 1.18 for the setup and coordinate system.

a. Argue that the gradient of quantities in x is much smaller than that in z, i.e. $\partial f / \partial x \ll \partial f / \partial z$, for any quantity f of interest. *Hints*: Rescale x as $\tilde{x} = L^{-1}x$ and z as $\tilde{z} = h_0^{-1}z$ where h_0 is a typical value of height h, and make use of the fact that $L \gg h_0$.

Argue next that for this particular situation, the dominant terms of the x-component of the hydrodynamic equations give

$$-\frac{1}{\rho_0} \frac{\partial p}{\partial x} + g \sin \theta + \nu \frac{\partial^2 v_x}{\partial z^2} = 0. \qquad (1.96)$$

b. Likewise, argue that the z-component reduces to

$$-\frac{1}{\rho_0}\frac{\partial p}{\partial z} - g\cos\theta = 0. \tag{1.97}$$

Note that this is essentially the hydrostatic equation for the pressure within the fluid layer.

c. The upper surface at $z = h(x,t)$ is a free surface. Argue that the appropriate boundary condition here is

$$\text{at } z = h(x,t): \quad \frac{\partial v_x}{\partial z} = 0, \quad p = p_0 - \gamma\frac{\partial^2 h}{\partial x^2}, \tag{1.98}$$

where p_0 is the constant atmospheric pressure, γ is the surface tension, and the last term expresses the Laplace pressure for a weakly curved interface. At the surface $v_x(z=0) = 0$.

d. Integrate the equation (1.97) in the z direction and obtain the expression for p with the boundary condition.

e. Substitute the obtained p into the equation (1.96), and integrate it in the z direction to obtain $v_x(z)$.
Hint: You should get terms linear and quadratic in z.

f. Calculate the total flow J in the x direction, $J = \rho_0 \int_0^h \mathrm{d}z\, v_x$.

g. Substitute your final result into (1.59) to get equation (1.62).

Problem 1.13* *Lubrication approximation for the hydrodynamic interaction between two spheres at close approach*

In this exercise we calculate the hydrodynamic interaction between two spheres which closely approach each other using the lubrication approximation. This is relevant to our discussion of colloids in chapter 4, as it determines the force between two colloidal particles upon close approach.

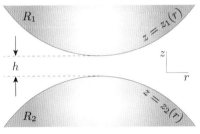

Figure 1.29. Two close spheres approaching each other. The gap is h.

We assume that $h \ll R_1$ and $h \ll R_2$, where h denotes the gap between the two spheres with radii R_1 and R_2—see figure 1.29. We use cylindrical coordinates. The lower sphere is fixed, while the upper one moves downward with a velocity V.

a. Show that to lowest order in r we can write $z_1 = (h + r^2/R_1)/2$ and $z_2 = -(h + r^2/R_2)/2$.

b. Show that this setup implies the following boundary conditions in the small gap limit

$$\begin{aligned}
v_r = 0, \quad v_z = -V \quad &\text{at} \quad z = z_1(r), \\
v_r = 0, \quad v_z = 0 \quad &\text{at} \quad z = z_1(r), \\
\partial p/\partial r = 0 \quad &\text{at} \quad r = 0, \\
p \to p_0 \quad &\text{for } r \to \infty.
\end{aligned} \tag{1.99}$$

c. Argue that the analysis of section 1.12 and problem 1.12 implies that in the lubrication approximation we can write the governing equations in cylindrical coordinates as

$$0 = -\frac{\partial p}{\partial r} + \eta \frac{\partial^2 v_r}{\partial z^2},$$

$$0 = -\frac{\partial p}{\partial z},$$

$$0 = \frac{1}{r}\frac{\partial(r v_r)}{\partial r} + \frac{\partial v_z}{\partial z}.$$

(1.100)

d. Integrate the first equation of (1.100) in the z direction and apply the boundary conditions to show that

$$v_r = \frac{1}{2\eta}\frac{\partial p}{\partial r}\left[z^2 - (z_2 + z_1)z + z_1 z_2\right].$$

(1.101)

e. Substitute expression (1.101) into the last equation of (1.100) which expresses incompressibility, integrate the resulting equation first with respect to z and use the boundary conditions, then integrate the resulting equation twice with respect to r to get

$$p = p_0 + \frac{3\eta V R_1 R_2}{h^2(R_1 + R_2)}\left(1 + \frac{(R_1 + R_2)}{2h R_1 R_2}r^2\right)^{-2}.$$

(1.102)

f. Integrate the pressure over the surface of sphere 1 to get for the total force on this sphere

$$F_z = -\frac{6\pi\eta V}{h}\frac{R_1^2 R_2^2}{(R_1 + R_2)^2}.$$

(1.103)

The matching to the hydrodynamic interaction if the spheres are far apart is discussed in section 1.8 of Russel et al., 1989, while Goddard et al., 2020 discuss the full solution of the two-sphere interaction for any separation in every detail. Such hydrodynamic interactions are incorporated in the Stokesian dynamic simulation of the flow of colloids discussed in section 4.6 and reviewed by Brady and Bossis, 1988. Be aware that hydrodynamic interactions also include terms involving sets of three, four, or even more particles, so that the hydrodynamic interactions cannot be written as a sum of two-particle terms.

This force diverges as $1/h$ as it becomes increasingly difficult to push the fluid out of the narrow gap. Note that, since we start from the linear hydrodynamic equations, the force is the same in magnitude if the spheres move away from each other: we need a large force to pull the spheres away if h is small.

Elasticity | 2

Fluids deform indefinitely—they continue to flow—when a stress is applied. A solid, on the other hand, can withstand a stress with a finite deformation. For small enough stresses a solid typically deforms elastically in response to the stress applied: no permanent changes take place within the medium, and when the stress is relieved, the material resumes its original shape. In the elastic regime the local stress in the material is a function of the local strain, which is a measure of the elastic deformation of the medium.

The main aim of this chapter is to introduce those elements of elasticity theory that provide the necessary background for our later discussion of soft materials and of ways to explore new phenomena and applications, like those illustrated in figure 2.1. We concentrate on the linear elastic response of isotropic soft matter. This is followed by a discussion of buckling and examples illustrating modern developments: we introduce metamaterials with unusual deformation properties and with memory, the concept of a marginal solid, a particularly ingenious and detailed soft matter experiment of the two-dimensional melting process, and topological mechanical materials.

2.1 Elasticity: A time-honored subject with a twist

We can well imagine that you wonder why we include in this book an elementary introduction to the most basic concepts of elasticity—after all, the birth of 'elastica' goes back to the eighteenth-century giants Euler and Bernoulli, and even more than fluid mechanics, elasticity theory is by outsiders considered the realm of engineers who design bridges and other structures. There are good reasons to treat elasticity. One trivial reason is that soft matter easily deforms, so we better learn how to describe elastic deformations. The most important reason, however, is illustrated already by figure 2.1: one can nowadays take advantage of the versatility of soft matter to design functional materials which exhibit surprising effects and allow new applications. Elementary concepts of elasticity, like the strain and stress fields in a bending sheet or rod, or the buckling transition, often underlie these developments.

We shall encounter several examples of these in our discussion of soft designer materials in chapter 10, but figure 2.2 and figure

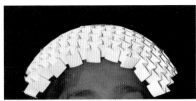

Figure 2.1. Illustration of how one can play with the structure to create materials with unusual properties: designer matter. As we discuss in section 2.9 below, upon stretching normal materials contract in the transverse directions, but so-called auxetic materials then expand in the transverse directions as well. The child illustrates how such a material naturally wraps around his head, while a conventional material does not. Elastic designer materials are discussed in more detail in chapter 10. Images courtesy of Joseph Grima.

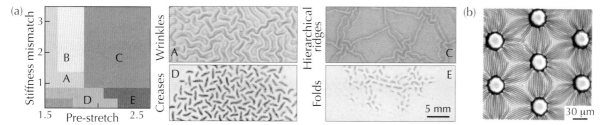

Figure 2.2. (a) Buckling patterns in experiments by Al-Rashed et al., 2017 on a bilayer system in which the substrate layer is initially prestressed. After releasing the stress of the substrate the other layer, consisting of rubbery-type polymers (elastomers), forms a variety of buckling patterns, depending on the amount of prestress of the substrate, and the mismatch in stiffness of the substrate and the polymer layer.[1] The typical length scale of the patterns is a few millimeters. Image courtesy of Pedro Reis. (b) Buckling pattern observed in an elastic PDMS layer which has been oxidized at the top. The oxidation leads the top layer to expand, and this leads to buckling. In this image the hexaogonal patterning of the starting layer leads to an ordered buckling pattern. From Bowden et al., 1999.[2]

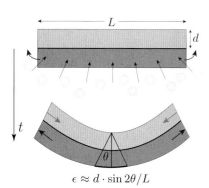

$$\epsilon \approx d \cdot \sin 2\theta / L$$

Figure 2.3. When one starts with the bilayer system sketched at the top, and induces one component to swell by a relative amount ε (the strain), for instance by having an appropriate solvent diffuse into it as illustrated, the bilayer spontaneously curves as sketched below. A simple geometric analysis shows that the radius of curvature R of the bent bilayer equals d/ε, with d the thickness of each layer. The arrows in the curved bilayer indicate the local elastic forces.[3]

Figure 2.4. Drying of the flesh of an apple results in a compressive stress on the skin, causing wrinkling. LanKS/Shutterstock.

2.3 already provide illustrative examples of how new structures and patterns can be created simply with a bilayer of two different types of soft materials, one of which is made to shrink or expand. Undoubtedly, you will actually have done an uncontrolled version of such an experiment unintentionally yourself when you left an apple too long uneaten: as figure 2.4 shows, when the flesh shrinks as it dries out, the skin starts to wrinkle as it is too stiff to conform to the contraction.

In addition, elasticity theory is a crucial ingredient of understanding the spontaneous deformations and fluctuations of small soft matter and biomatter structures. So an elementary discussion of the ease of bending thin objects will provide the necessary groundwork for later discussions of bending deformations of biopolymers in chapter 5, and for the analysis in chapter 7 of how viruses develop an icosahedral shape, and of the crumpling of sheets. Finally, our discussion will provide the proper reference frame for presenting the fact that granular materials have put our understanding of elasticity in a new light. Indeed, as we will discuss, close to the so-called jamming point, granular media are marginal solids; these are solids that are easily deformed and which exhibit anomalous behavior of the elastic constants. They also have many more low-energy vibrational modes than a classical elastic medium.

2.2 The strain tensor

So let's bite the bullet and introduce the concept of strain deformation. In chapter 1 we saw that the introduction of the fluid dynamics equations was conceptually clearest in the Lagrangian frame moving with the fluid elements, even though this frame is hardly ever used for concrete calculations—we are normally interested in flow around fixed or moving bodies, and not in the position of the individual fluid elements which are difficult to track anyhow.

Hence, in practice, the Eulerian description is always the favored and easiest one for analysis and numerical computations in fluid dynamics.

The situation is actually more mixed in the case of elasticity. Again, the Lagrangian picture is the most natural one for formulating the basics of elasticity theory. However, for calculating the small elastic response of stiff materials of fixed shape, especially in three dimensions, the Eulerian description is usually the most natural one again. But global shape deformations of lower-dimensional objects in three dimensions, like those of thin sheets, skin wrinkles, bacterial flagella, DNA or thin rods, can easily be large and complicated. For these it is usually more convenient and natural to use a Lagrangian description attached to the deformed shape, not a fixed Eulerian reference frame. We will not go into the details of such large deformations of lower-dimensional objects, but it is good to realize that the overall shape and dynamics of such elastic objects can be very complex and nonlinear. For instance, even the calculation of the shape of a simple elastic rod under force is a highly nontrivial problem: the mathematical description of such shapes is intricate, both because of the differential geometry needed to describe shapes, and because shape and stress are intimately coupled: the elastic stresses depend on the shapes, while the shapes depend on the stress balance.

To introduce the concept of the strain tensor, which is central to elasticity theory, let us consider a material element of the elastic solid with position vector $\vec{r}$ in the unstressed state. As before, we denote the components of $\vec{r}$ by x_i. When the solid deforms under stress two things happen: first these material elements in the Lagrangian picture move to different positions $\vec{r}'$: $\vec{r} \to \vec{r}'$. As a result of the deformation the *relative* distances between nearby material elements typically change. Elasticity theory is based on assuming that the elastic energy density is quadratic in the local change in relative distance between such nearby points.

To make this precise, let us write $\vec{u}$ as the shift in position of the material point which was originally at $\vec{r}$: $\vec{u} = \vec{r}' - \vec{r}$, and let us analyze how the relative distance between two nearby material elements initially at $\vec{r}$ and $\vec{r} + d\vec{r}$ changes. When the stress is applied, they are at $\vec{r}'$ and $\vec{r}' + d\vec{r}'$, with $d\vec{r}' = d\vec{r} + d\vec{u}$; see figure 2.5. So for the distance between the two points before and after the deformation we have with $d\ell^2 = (d\vec{r})^2$ and $d\ell'^2 = (d\vec{r}')^2$

$$(d\ell')^2 = (d\vec{r}')^2 = (d\vec{r} + d\vec{u}) \cdot (d\vec{r} + d\vec{u})$$

$$= \left(dx_i + \frac{\partial u_i}{\partial x_k} dx_k \right) \left(dx_i + \frac{\partial u_i}{\partial x_l} dx_l \right) \quad (2.1)$$

$$= (d\vec{r})^2 + 2\frac{\partial u_i}{\partial x_j} dx_i dx_j + \frac{\partial u_k}{\partial x_i} \frac{\partial u_k}{\partial x_j} dx_i dx_j,$$

For a more complete introduction, see *Theory of Elasticity* by Landau and Lifshitz, 1970. A modern account, with many applications and attention to numerical examples, is given by Sadd, 2014.

Quite often, we even pick such a local reference frame without thinking about it. For instance, when we will treat the bent rod in figure 2.13, we introduce an appropriate local coordinate system in the center of the rod.

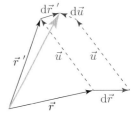

$$\vec{r}' + d\vec{r}' = \vec{r}' + d\vec{r} + d\vec{u}$$
$$= \vec{r} + d\vec{r} + \vec{u} + d\vec{u}$$

Figure 2.5. Illustration of the change in distances upon elastic deformation of a solid. Two nearby material points of the solid at $\vec{r}$ and $\vec{r} + d\vec{r}$ move to $\vec{r}' = \vec{r} + \vec{u}$ and $\vec{r}' + d\vec{r}' = \vec{r}' + d\vec{r} + d\vec{u}$.

Figure 2.6. A nice illustration of the importance of the nonlinear strain term for elastic sheets is provided by graphene. For a two-dimensional sheet oriented along the x, y directions and with vertical height $h(x, y)$, the nonlinear terms in (2.3) generate a coupling between the height and the in-plane strain fields, as $u_{ik} = 1/2(\partial u_i/\partial x_j + \partial u_j/\partial x_i + \partial h/\partial x_i \partial h/\partial x_j)$ for $i, j = x, y$. We shall discuss in section 7.4.1 that via this coupling the elastic fields effectively generate a long-range interaction between height variations, which strongly suppresses fluctuations of elastic sheets. As a result, graphene is much flatter than you might naively have thought.

The assumption that the elastic energy can be written as a quadratic form in the strain is less innocuous than it may appear. As we shall discuss in section 9.6.2, for active matter this assumption does not hold.

where we have used the Einstein summation convention for repeated indices. We can put this in the form

$$(\mathrm{d}\ell')^2 - (\mathrm{d}\ell)^2 = 2u_{ij}\mathrm{d}x_i\mathrm{d}x_j, \tag{2.2}$$

where u_{ij} is the symmetric *strain rate tensor*

$$u_{ij} = \frac{1}{2}\left(\frac{\partial u_i}{\partial x_j} + \frac{\partial u_j}{\partial x_i} + \frac{\partial u_k}{\partial x_i}\frac{\partial u_k}{\partial x_j}\right). \tag{2.3}$$

The fact that the strain rate tensor is itself already nonlinear at a basic level illustrates how nonlinearities can play a subtle role in elasticity, because distances are affected by strains or deformations in all directions. With figure 2.6 we illustrate how sheets of graphene are much flatter than one might naively expect due to the coupling of height fluctuations and elastic fields, via this nonlinear term.

General elasticity theory proceeds by assuming that the elastic energy density $\mathcal{E}_{\mathrm{elas}}$ is a quadratic form in terms of the strain rate tensor $\underline{u}$ in the Lagrangian frame $\vec{r}$,

$$E_{\mathrm{elas}} \equiv \int \mathrm{d}^3\vec{r}\, \mathcal{E}_{\mathrm{elas}}(\vec{r}) = \frac{1}{2}\int \mathrm{d}^3\vec{r}\, K_{ijkl}u_{ij}u_{kl}. \tag{2.4}$$

As $\underline{u}$ is a symmetric tensor ($u_{ij} = u_{ji}$), K_{ijkl} can be taken to be symmetric in its first and last pair of indices, but this in general still allows for many coefficients. For a crystalline material these can be reduced on the basis of the symmetry of the crystal.[4] But as we already stated, we will sidestep this issue below by concentrating right away on materials which are isotropic—for these there are only two elastic constants, the compression modulus and shear modulus discussed in the next section.

We will also focus the analysis below on cases in which the strains are relatively small, so that one may ignore the nonlinear terms in equation (2.3) to get[5]

$$u_{ij} \approx \frac{1}{2}\left(\frac{\partial u_i}{\partial x_j} + \frac{\partial u_j}{\partial x_i}\right). \tag{2.5}$$

Keep in mind that this approximation is not always justified, especially for two- and one-dimensional materials or the metamaterial with holes in them which we will discuss in section 2.9. These are indeed more susceptible to large strains because they are less constrained. Moreover, soft matter and biological matter are by their very nature much more prone to large strain deformations.

Two final remarks are in order. First of all, note that we *assume* in our discussion that the material responds elastically. We will come back to this issue in section 2.10, but it is good to be aware that this is an issue which is more subtle than it may sound for soft matter consisting of particles.

Secondly, while in most of the discussion in this chapter temperature will not explicitly play a role, quite often the models we will discuss do provide the basis for analyzing thermal fluctuations of soft matter at the (cellular) microscales, like the bending fluctuations of biopolymers, microtubules or membranes. Some of these examples will come back in later chapters. Temperature will even play a crucial role in the discussion of defect-mediated melting at the end of this chapter.

Defects interact through elastic fields. As we shall discuss in section 2.11, in two dimensions temperature fluctuations drive a melting transition associated with unbinding of defects pairs. A prime experimental realization will be seen to occur in colloidal systems.

2.3 The linear stress-strain relation

For small strain deformations the relation between the local stress tensor $\underline{\sigma}$ and strain tensor $\underline{u}$ is linear—this is essentially the extension of Hooke's law that you know from continuum mechanics stating a linear relation between the force and the extension or compression of a spring.

As in our discussion of fluids, σ_{ij} is the ith component of the outside surface force per unit area on a infinitesimal piece of surface whose outward normal points in the j direction, and $\underline{\sigma}$ is symmetric. Remember also that we denote tensors with an underscore.

As already stated, for simplicity we will confine the analysis to the elastic response of soft matter and biomatter phases which are isotropic on the coarse-grained scale, e.g. because they are disordered. For these, the linear stress-strain relation simplifies in three dimensions to

It is common, also for active matter, to define elasticity through the linear relation between stress and strain, but for systems considered here one can also introduce the stress components with (2.4) as the derivative of the elastic energy density, $\sigma_{ij} = \partial \mathcal{E}_{\text{elas}} / \partial u_{ij} = K_{ijkl} u_{kl}$.[6]

$$\text{Isotropic media:} \quad \sigma_{ij} = K \, u_{kk} \, \delta_{ij} + 2\mu \left(u_{ij} - \frac{1}{3} u_{kk} \, \delta_{ij} \right). \quad (2.6)$$

Here K is the *compression modulus* and μ the *shear modulus*. In the applied literature μ is sometimes also referred to as the rigidity modulus, but because in the physics community the word *rigidity* is used more generally for the stiffness modulus of an appropriate order parameter, we prefer to use the name *shear modulus*.

Note the similarity of equation (2.6) with the stress–shear rate relation (1.22) for a Newtonian fluid: the first term describes elastic compression, just as $\vec{\nabla} \cdot \vec{v}$ described the rate of compression of a fluid, while the second term is traceless, and hence it measures true shear-type strains without volume change. In view of these similarities it is important to stress that while under many circumstances of practical interest fluid flows can be considered incompressible, for elastic deformations compression often plays an important role—typically K and μ are of the same order.

For example, the bending moduli of the wormlike chain introduced in section 5.4.1, or of the membranes discussed in section 7.2, are commonly referred to as a bending rigidity. As mentioned in section I.2.2.b, in the framework of broken symmetries, the term *rigidity* is often used for the stiffness of the order parameter associated with the broken symmetry. For instance, in superconductivity or in the context of pattern formation the term *phase rigidity* is used for the stiffness of the phase (see section 8.5.2 for our use of *phase* in our discussion of pattern formation).

Just as for a harmonic spring with force $F = -kx$, the energy associated with it is $1/2 \, kx^2$, the elastic energy density $\mathcal{E}_{\text{elas}}$ associated

Note that $\sigma_{ij} = \partial \mathcal{E}/\partial u_{ij}$. The lack of the usual minus sign in the force comes from the definition of the stress as the force exerted from the outside on the material element, while $\mathcal{E}$ is the energy density of the material element itself.

with the strain and stress fields is

$$\mathcal{E}_{\text{elas}} = \tfrac{1}{2}\,\sigma_{ij}u_{ij}, \tag{2.7}$$

where we have used again the Einstein summation convention for the indices. Upon substitution of the stress-strain relation (2.6) in the expression for the energy density we get

$$\mathcal{E}_{\text{elas}} = \tfrac{1}{2}\lambda u_{ii}^2 + \mu u_{ij}^2, \qquad \lambda = K - 2\mu/3, \tag{2.8}$$

where written this way λ and μ are referred to as the Lamé coefficients. Whether the compression modulus K and shear modulus μ or the two Lamé coefficients λ and μ are the more natural parameters depends on the problem at hand.

As always, it is useful to consider the dimensions of the various quantities. We have

$$[u_{ij}] = 1, \quad [\sigma_{ij}] = \frac{\text{force}}{\text{area}} = \frac{\text{kg m/s}^2}{\text{m}^2} = \frac{\text{kg}}{\text{m s}^2}, \tag{2.9}$$

and μ and K simply have the dimension of σ

$$[\mu] = [K] = \frac{\text{force}}{\text{area}} = \frac{\text{kg}}{\text{m s}^2}. \tag{2.10}$$

Relations of the type (2.6) and (2.11) are the only linear isotropic forms one can write for two tensors and the Kronecker delta. The specific coefficients in (2.11) are most easily obtained from (2.6) by considering first the off-diagonal term, and then the diagonal term.

For future reference it is handy to also write the strain in terms of the stresses. It is easy to invert the relation (2.6) between σ_{ij} and u_{ij},

$$u_{ij} = \frac{1}{9K}\sigma_{kk}\,\delta_{ij} + \frac{1}{2\mu}\left(\sigma_{ij} - \frac{1}{3}\sigma_{kk}\,\delta_{ij}\right). \tag{2.11}$$

2.4 The Poisson ratio

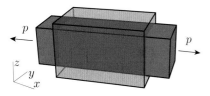

Figure 2.7. Illustration of the Poisson ratio. When a force per unit area p is applied to the long sides of an elastic bar the long sides extend. For common materials the transverse directions contract. The figure illustrates both the original shape of the bar (light magenta) and the deformed one (blue). The Poisson ratio ν is the ratio of the transverse contraction strain to the longitudinal extensional strain.

Now consider a bar of an isotropic elastic medium, on which we pull on both sides in the x direction with a force per unit area (hence with dimension of pressure) p—see figure 2.7. With our convention for the stress tensor, this means that $\sigma_{xx} = p$. As there are no other forces, this means that

$$\sigma_{xx} = p, \quad \sigma_{ij} = 0 \quad \text{for all other components } ij \neq xx. \tag{2.12}$$

From equation (2.11) we get immediately for the strains

$$u_{xx} = p\left(\frac{1}{9K} + \frac{1}{3\mu}\right), \tag{2.13}$$

$$u_{yy} = u_{zz} = p\left(\frac{1}{9K} - \frac{1}{6\mu}\right), \tag{2.14}$$

while all off-diagonal elements of the strain tensor are zero. The coefficient that relates the extension u_{xx} to the pressure is the so-called Young's modulus E_Y,

$$u_{xx} = \frac{p}{E_Y}. \tag{2.15}$$

The expression for E_Y in terms of the elastic moduli simply follows from equation (2.13) as the reciprocal of the term between brackets,

$$E_Y = \frac{9\mu K}{\mu + 3K}. \tag{2.16}$$

The larger the Young's modulus, the stiffer the material.

Now consider the contraction in the transverse direction given by (2.14) as a result of the extensional force in the x direction

$$u_{yy} = u_{zz} = -p \frac{3K - 2\mu}{18\mu K} = -\nu \, u_{xx}, \tag{2.17}$$

where the *Poisson ratio* ν is given by[7]

$$\text{Poisson ratio} \quad \nu \equiv -\frac{u_{yy}}{u_{xx}} = \frac{3K - 2\mu}{2(3K + \mu)}. \tag{2.18}$$

We see that the Poisson ratio is a measure of the change in the transverse direction relative to the change in the main stretching or compression direction. Both elastic constants are positive, therefore as the ratio K/μ is varied, the Poisson ratio can vary between -1 and 0.5, with $\nu \approx 0.5$ in the limit $K \gg \mu$ and $\nu \approx -1$ for $K \ll \mu$.

For future reference, we also quote here the expressions for K and μ in terms of the Poisson ratio and Young's modulus, which can be obtained straightforwardly from the above equations (2.16) and (2.18),

$$K = \frac{E_Y}{3(1 - 2\nu)}, \qquad \mu = \frac{E_Y}{2(1 + \nu)}. \tag{2.19}$$

Materials with $\nu > 0$ contract (stretch) in the transverse directions if stretched (squeezed) in one direction. This is the common behavior of materials—figure 2.8 illustrates why cork, with a Poisson ratio of about zero, is so exceptional. In fact, typical materials, including many metals and ice, have $\nu \approx 0.33$, so these contract in the transverse directions by about a third of the amount with which they are stretched. What does this mean for the volume change? Well, for the relative volume change $\Delta V/V$ of the bar in figure 2.7 we have

$$\frac{\Delta V}{V} = u_{xx} + u_{yy} + u_{zz} = u_{xx}(1 - 2\nu). \tag{2.20}$$

Note that in this section we consider an elastic bar in three dimensions. The analysis in two dimensions proceeds along the same lines, and it leads to similar relations between the elastic coefficients. Only the numerical factors are slightly different.

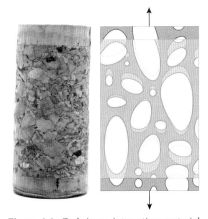

Figure 2.8. Cork is an interesting material with many cells of size of about a tenth of a millimeter filled with air, which give rise to a Poisson ratio $\nu \approx 0$. The illustration at right shows the deformation of cork under stretching. This property makes cork such a nice material to use to stopper wine bottles: when one pushes the cork into the wine bottle, the cork hardly expands in the transverse direction. If it were to expand, it would be very difficult to insert into the bottle. This is the disadvantage of using a plastic 'cork': it is difficult to insert, because when you push on it, it expands in the transverse direction.[8] Left: Yarbeer/Shutterstock.

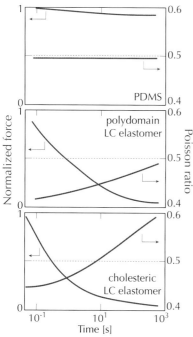

Figure 2.9. Trends of the time-dependent Poisson ratios for various so-called elastomers, rubber-like polymers. The Poisson ratio is shown as a function of time. Red lines indicate the force as a function of time normalized to the force for small times (left axis), and the blue lines indicate the corresponding Poisson ratio on the right axis. Top: rendering of the experimental data for the elastomer PDMS.[9] The Poisson ratio is close to 0.5 and is independent of time. Middle: the Poisson ratio of a polydomain liquid crystal elastomer, which is time-dependent and which approaches 0.5 for long times. Bottom: curves for a cholesteric liquid crystal elastomer—see chapter 6 for more on cholesteric liquid crystals—with a Poisson ratio that approaches about 0.6 for long times, due to anisotropy. Figure adapted from measurements of Pritchard et al., 2013.

We will encounter examples of these effects in colloids (section 4.6), polymers (section 5.10) and liquid crystals (section 6.7).

This equation shows that when the bar is stretched in one direction, its change in volume for materials with positive ν is reduced significantly by the contraction in the transverse directions. Materials with $K \gg \mu$ leading to $\nu \approx 0.5$ in fact hardly change volume at all. Rubber and rubbery synthetic polymers, so-called elastomers, are examples of soft matter with Poisson ratio close to 0.5. The upper panel of figure 2.9 illustrates this for PDMS, an elastomer which is used in the food and cosmetics industries, but also as a soft material in which to grow cells in chips. We will come back to the time dependence of the measurement in the next section.

It is important to keep in mind that the Poisson ratio is defined here for isotropic materials. Many materials are anisotropic, and these can actually have Poisson ratios larger than 0.5 in some directions.[10] The lower panel of figure 2.9 illustrates this for a liquid-crystal-type elastomer at long times.

What about materials with a negative Poisson ratio, which have the curious property that they expand in the transverse direction when stretched in the primary direction? For a long time it was thought that such materials did not exist or were limited to artificial engineering structures with rods and hinges, until soft matter opened up the field of so-called metamaterials. We will come back to this in section 2.9, after we have discussed buckling instabilities, as these play an important role for metamaterials.

2.5 Frequency-dependent generalization of the shear modulus

In our discussion of fluids, we stressed that Newtonian fluids are characterized by a stress which is proportional to the local and *instantaneous* values of the velocity gradients, while many complex fluids exhibit strong memory effects, shear rate-dependent effective viscosities, etc. The situation is analogous for elasticity: the classic approach sketched above expresses the elastic stress in terms of the instantaneous local value of the strain. Elastic energy which is stored in an elastic medium by loading it with a force is completely recovered upon unloading the force: the behavior is reversible. In soft matter, relaxation effects are often important too; in such cases the response is characterized by a response time and by dissipation. In addition, in solid materials large strains can induce plastic deformations on a long time scale; these are due to structural rearrangements.

You actually know this from daily experience: if you stretch a rubber band, the elastic force is initially stronger than it is after some time—a rubber band which has been stretched for a long time can become quite weakened. Likewise, when you release the force, the band typically only snaps back to part of its original length and then slowly relaxes back completely. This behavior is illustrated in the middle and lower panels of figure 2.9, which show significant changes of the Poisson ratio of liquid crystal elastomers over a range of time scales, from a fraction of a second to about 15 minutes.

For materials with a time-dependent response, it is experimentally often convenient to probe the linear response of the material as a function of frequency. Typically, one then applies a small oscillatory shear strain $u_{xy}e^{-i\omega t}$, with a given angular frequency ω, and measures the resulting shear stress $\sigma_{xy}(\omega)$ to determine the response $G(\omega)$,

$$\sigma_{xy}(\omega) = G(\omega)u_{xy}(\omega), \qquad \left(\text{fields} \sim e^{-i\omega t}\right). \qquad (2.21)$$

These Fourier amplitudes of the shear strain and stress are allowed to be complex (we effectively complexify the fields, as discussed in section 1.7.2), and in fact the measured dynamic modulus $G(\omega)$ generally has a real part G' and an imaginary part G''. The first, G', is called the storage modulus, and the imaginary part G'' the loss modulus. This is because for a normal elastic material with a time-independent response, G is real and equal to the shear modulus for all frequencies, so 'storage' refers to the fact that the strain deformation stores elastic energy which can be recovered upon releasing the strain. The imaginary part is due to energy dissipation, hence the name *loss modulus*; it results from stress and strain being out of phase, very much like in the case of the driven harmonic oscillator with damping—the damping term leads to driving and response being out of phase.

As an example of this type of data, we show in figure 2.10 the measured storage and loss modulus of an ultrasoft elastic polymer network, developed especially for mechanobiological applications— the low frequency value of the storage modulus G' matches that of brain tissue, so that the material can for instance be used to simulate a brain in an automotive crash test. At low frequencies, this polymer gel has a predominantly elastic response with little loss, but at frequencies ω above about $1\,\mathrm{s}^{-1}$, loss and storage moduli are comparable. In this regime the gel network responds viscoelastically.

Two very simple models which are often useful as a reference in an analysis of response are illustrated in figure 2.11. The Maxwell model consists of a damper (sometimes referred to as a dashpot) and a spring, arranged in series. If we denote the total strain by ε and the total stress by σ, in the Maxwell model the strain of the spring and damper add up, while the stresses on the two are the same. This gives a relation of the type:

The real part and the imaginary part of $G(\omega)$ are actually not independent, as they satisfy the so-called Kramers-Kronig relations, which express causality in the time domain: a response at time t can only depend on the cause in the past, i.e. on times up to t, and not on what might happen in the future.[11]

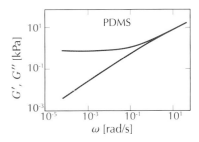

Figure 2.10. The storage and loss modulus of an ultrasoft polymer gel, made of PDMS, the material whose Poisson ratio is shown in figure 2.9. At low frequencies, the network responds elastically, with very little loss, but at high frequencies the loss and the storage modulus are comparable. Adapted from Heinrichs et al., 2018.

We discuss viscoelastic behavior of polymers in section 5.10.

Maxwell model

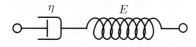

Kelvin-Voigt model

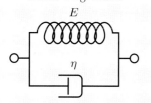

Figure 2.11. Illustration of two simple models for time-dependent response, the Maxwell model with a damper and spring in series, and the Kelvin-Voigt with a damper and spring in parallel. The elastic spring characterizes an elastic solid with linear stress-strain relation $\sigma_e = E\varepsilon$, and the damper a viscous material with $\sigma_v = \eta\dot{\varepsilon}$. When the two elements are put in parallel, the stresses add up, $\sigma_{tot} = \sigma_e + \sigma_v$ while $\varepsilon_{tot} = \varepsilon_e = \varepsilon_v$, while when put in series, the stresses we have $\sigma_{tot} = \sigma_e = \sigma_v$ and $\varepsilon_{tot} = \varepsilon_e + \varepsilon_v$. Equations (2.22) and (2.23) can easily be derived from this; see problem 2.1.

$$\text{Maxwell:} \quad \sigma + \frac{\eta}{E}\dot{\sigma} = \eta\dot{\varepsilon} \quad \Leftrightarrow \quad G(\omega) = \frac{-i\eta\omega}{1 - i\eta\omega/E}. \qquad (2.22)$$

In the Maxwell model, when a system is put rapidly under strain and then held constant, the stress initially follows the increase and then relaxes exponentially to zero. The Maxwell model is the basis of the most common rheological model for viscoelastic polymer flow, which we will discuss in section 5.10.

The Kelvin-Voigt model can be thought of as a spring and damper in parallel. In this case the forces from the two add up, which translates to the relation

$$\text{Kelvin-Voigt:} \quad \sigma = E\varepsilon + \eta\dot{\varepsilon} \quad \Leftrightarrow \quad G(\omega) = E - i\eta\omega. \qquad (2.23)$$

This model describes how when a material is put under constant stress, it deforms at a decreasing rate; likewise, when the stress is released, the material relaxes back to its undeformed state. It is therefore most often used as a starting point for describing creep in materials. As discussed in problem 2.1, a combination of the two models exhibits both creep and viscoelastic behavior.

We stress that these are only very idealized models, and that there is almost always more structure in the frequency dependence of the dynamics modulus than these models can capture. The structure in $G(\omega)$ typically provides a clue for understanding the physics at the scale of the building blocks of soft matter.

2.6 A brief foray into elastodynamics

Let us now discuss briefly the effect of elasticity on dynamics. As we shall see, the restoring elastic forces can lead to sound waves in solids.

2.6.1 Sound waves in solids: Continuum approach

We will proceed very much along the reasoning we adopted to study sound waves in fluids in section 1.7. Sound modes in solids are a familiar example of the hydrodynamic modes we have invoked in the general framework of figure I.15 in our introduction to the hydrodynamic approach. The dynamics of the displacement field u_i can be described by

$$\rho\frac{\partial^2 u_i}{\partial t^2} + \gamma\frac{\partial u_i}{\partial t} = f_i = \partial_j \sigma_{ij}, \qquad (2.24)$$

with γ a friction coefficient and ρ the mass density of the medium.

Using the stress-strain relations expressed in terms of the compression modulus K and shear modulus μ, we find that the elastodynamics equation explicitly reads

$$\rho \frac{\partial^2 \vec{u}}{\partial t^2} + \gamma \frac{\partial \vec{u}}{\partial t} = \left(K + \frac{\mu}{3}\right) \vec{\nabla}(\vec{\nabla} \cdot \vec{u}) + \mu \Delta \vec{u}. \qquad (2.25)$$

By performing a Fourier transform of equation (2.25), or simply by considering modes $e^{ikx - i\omega t}$ as in (1.37), we find

$$(\omega^2 \rho + i\gamma\omega) \begin{pmatrix} u_{\parallel} \\ \vec{u}_{\perp} \end{pmatrix} = k^2 \begin{pmatrix} K + \frac{4\mu}{3} & 0 \\ 0 & \mu \end{pmatrix} \begin{pmatrix} u_{\parallel} \\ \vec{u}_{\perp} \end{pmatrix}, \qquad (2.26)$$

in which we have defined the longitudinal component $u_{\parallel} = \hat{k} \cdot \vec{u}$ and transverse component $\vec{u}_{\perp}$ (so $\vec{u}_{\perp} \cdot \hat{k} = 0$) of the displacement. In an overdamped system, where the (first) inertial term in (2.25) can be neglected, the (passive) elastodynamic equations are $\omega = -ik^2 \frac{K + \frac{4\mu}{3}}{\gamma}$ and $\omega = -ik^2 \frac{\mu}{\gamma}$. Vice versa, when the inertial term dominates over friction, we recover the wave equation just as in our discussion of sound waves in fluids—compare equation (1.39).

2.6.2 Dynamical matrices: Microscopic description of elasticity

We now consider an elastic solid from a discrete perspective rather than in terms of displacement fields and their dynamics. We describe it as a collection of particles labeled by α that interact through two-body interactions. The displacement of particle α from its equilibrium position is denoted by $\vec{u}^\alpha$. The force $\vec{f}^\alpha$ on a particle from all the other particles is given by the dynamical matrix D as

$$f_i^\alpha = -D_{ij}^{\alpha\beta} u_j^\beta, \qquad (2.27)$$

where i, j label the Cartesian components of the displacements and forces. In a regular crystal, the dynamical matrix can be Fourier transformed to obtain a momentum-space version of the matrix $D_{ij}^{mn}(q)$, where q is the momentum and m, n label particles in a unit cell of the crystal.

Newton's law of motion is then

$$m\partial_t^2 \vec{u}^\alpha = \vec{f}^\alpha, \qquad (2.28)$$

where m is the mass of the particles. To connect this microscopic description with elasticity, recall that the continuum equivalent of the equation above is

$$f_i = \partial_j \sigma_{ij} \quad \text{with} \quad \sigma_{ij} = K_{ijk\ell}\partial_\ell u_k, \qquad (2.29)$$

Problem 2.6 will guide you step-by-step through this derivation. Note also that in the chapter on active matter the generalization of this approach to active solids is presented in section 9.6.2 on odd elastodynamics, where the presence of additional active elastic moduli allows waves to propagate even in overdamped solids.

With our conventions, a negative imaginary frequency implies that a wave is attenuated—compare the discussion in section 1.7.2. The behavior $\omega \sim -ik^2$ is often called diffusive, as it is the same as that obtained from a diffusion-type equation, such as equation (1.28) for the temperature.

We refer the reader unfamiliar with vibrational modes in crystals to a solid state physics textbook, such as Simon (2013).

A derivation of equation (2.30) can be found in the classic book by Born and K. Huang, 1954. The formula has reemerged in the context of computer simulations (Lutsko, 1989) and in recent work on odd elasticity (Fruchart and Vitelli, 2020; Scheibner et al., 2020); discussions of the relation can also be found in these papers. The formula hides some subtleties that are explored in detail in problem 2.7.

We shall call a three-dimensional object which has *one* small dimension a *sheet* and an object that has *two* small dimensions a *rod*. In daily life we often rather use the words *plates* and *beams* for thicker objects with similar aspect ratio. The analysis below equally applies to these when they are weakly bent with a radius of curvature much less than their thickness. The word *shell* is typically used for a sheet or plate which is pre-curved.

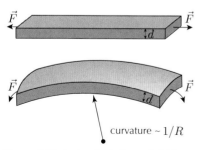

Figure 2.12. A stretched sheet of thickness d and a sheet bent with curvature $\kappa = 1/R$. The sheet is taken to be homogeneous across the thickness.

As the straight rod or sheet is supposed to be a lowest energy configuration, there cannot be a term linear in the stretching.

where f is a force acting on the solid continuum. In terms of the dynamical matrix, the stiffness tensor can be expressed as

$$K_{ijk\ell} = \left[\frac{1}{2} \frac{\partial^2 D}{\partial q_i \partial q_k} - \frac{\partial D}{\partial q_i} D^{-1} \frac{\partial D}{\partial q_k} \right]_{j\ell}. \qquad (2.30)$$

This formula is a bridge from microscopic models of crystals represented by the dynamical matrix to the macroscopic stiffness tensor which contains all the allowed elastic moduli. In problem 2.7 you will learn how to coarse-grain triangular and honeycomb lattices of masses and springs to obtain their elastic moduli. In addition, by looking at the spectrum of the dynamical matrix, one can determine the density of vibrational modes, $D(\omega)$, at a given frequency both for regular structures and for amorphous ones, as we shall see when we discuss jamming later in this chapter.

2.7 Bending is the low-energy deformation of sheets and rods

We know from daily experience that thin objects like sheets, rods and rulers are much easier to bend than to stretch. And if we look through a microscope at biomatter, we encounter many bending deformations, from those of membranes to flagella and microtubules. In fact, bending is the low-energy deformation of thin objects, and it is important to understand in some more detail from elasticity theory why this is so.

2.7.1 Scaling with thickness: Dimensional analysis

Let us first approach this with a dimensional argument, by comparing the stretching and bending energy of a sheet of thickness d, as sketched in figure 2.12. How will the stretching energy per unit area scale with d?

First note that when we deform a straight sheet or rod by stretching it, the energy will increase quadratically in the strain. Likewise, if we bend it the energy must increase quadratically in the bending curvature. Indeed, if there were a linear term, the original object would have been able to lower its energy by spontaneously stretching, contracting or bending. So the deformation energy must be quadratic in the strain and in the curvature.

Since strain is dimensionless, and since (as discussed in section 2.3) the dimension of Young's modulus (2.16) is force per unit area or energy per unit volume, dimensionally the stretching energy must be

$$\frac{E_{\text{stretch}}}{area} \simeq E_Y \, (strain)^2 \, d, \qquad (2.31)$$

that is, it must scale linearly with the thickness d. This is what we know from daily experience—two sheets are twice as strong as one. But intuitively we also know that the force needed to bend a sheet or a rod depends much more on the thickness. Indeed, as the bending energy will be proportional to the square of the curvature (with dimension 1/length) we must have on dimensional grounds

$$\frac{E_{\text{bend}}}{area} \simeq E_Y \, (curvature)^2 \, d^3 \simeq E_Y \, \frac{d^3}{R^2}, \qquad (2.32)$$

with R the radius of curvature. You are undoubtedly roughly familiar with this effect from daily experience: if you screw or glue two wooden beams together, they are not twice as resistant to bending deformations, but eight times as resistant.

The important lesson from this comparison is that sufficiently thick sheets and rods for that matter are very rigid and they resist bending, while bending is the low-energy deformation of thin ones.

2.7.2 Analysis of the strain and energy of a bent sheet

Before we explore this in more detail, it is good to derive this result more systematically by zooming in on what the bending implies for the elastic distortions within a sheet or rod. The assumption that these objects are thin means that we assume that the local radius of curvature R is much larger than their thickness, i.e., $d/R \ll 1$. Moreover, as long as the curvature varies on length scales much larger than d, we can simply do the analysis locally at a point where the local radius of curvature is R, and then allow R to vary along the bent object later. Note that for simplicity, we assume that there is just one radius of curvature, in other words that the object is bent in one direction only.

Figure 2.13 sketches what happens in a cross section of such a bent object; bending implies that the outer half of a sheet or rod is stretched, while the inner half is compressed. For an isotropic and homogeneous medium, there is clearly a 'neutral surface' halfway through the sheet (rod) which is neither stretched nor compressed. Let us take for simplicity a local coordinate system with x parallel to the neutral surface in the direction of bending, and y in the direction perpendicular to it, with y increasing outward as in the figure (remember that we assumed there is no curvature in the z direction). Then the relative stretching of a little material element in the x direction along the sheet or rod is clearly proportional to

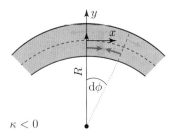

Figure 2.13. Illustration of a bent sheet or rod. The dashed line in the center denotes the neutral line or surface, which is neither bent nor stretched; its radius of curvature is R.[12] On the outer side of the neutral surface, the sheet or rod is stretched; on the inner side it is compressed. The amount of compression or stretching is proportional to the distance from the neutral line, and it is inversely proportional to the local radius of curvature; see equation (2.33). The purple arrows indicate the forces on a surface element of the cross section. Note that these elastic forces will also give rise to a torque on the cross section. The curvature κ for this situation is taken as negative, in accord with the fact that the curve $y(x)$ parameterizing the neutral line has negative curvature.

the increase in length of each circular element,

$$\frac{\partial u_x}{\partial x} \approx \frac{(R+y)\,\mathrm{d}\phi - R\,\mathrm{d}\phi}{R\,\mathrm{d}\phi} = \frac{y}{R}, \tag{2.33}$$

which confirms that we have stretching on the outer side $y > 0$ and compression on the inner side $y < 0$.

As we saw above, when a medium is stretched or compressed in one direction, so does (do), depending on the Poisson ratio, the transverse direction(s). Since the surfaces of the sheet are free (no forces), we must also have $\sigma_{yy} \approx 0$ in the sheet or rod. By imposing this on expression (2.6), we then get

The prefactor in equation (2.34) is different from ν in equation (2.18) as there both transverse directions contracted while here we consider a case where the z direction out of the plane does not strain. See problem 2.2.

$$u_{yy} \approx -\frac{\nu}{1-\nu}\frac{y}{R}, \tag{2.34}$$

which shows that indeed for a normal material with positive Poisson ratio the sheet or rod is compressed toward the neutral zone on the upper side and expanded away from the zone on the lower side.

You may already like to convince yourself that a material with negative Poisson ratio tends to bend in the other direction as well. We will discuss this in section 2.9.

Let us now analyze the elastic bending energy E_{bend} associated with the bending. We saw already in equation (2.7) that the elastic energy density equals $1/2\,\sigma_{ij}u_{ij}$. With this expression we get for a little element of the sheet or rod for the energy per unit area parallel to the neutral surface

$$\frac{E_{bend}}{area} \simeq \tfrac{1}{2}\int_{-d/2}^{d/2}\mathrm{d}y\,\sigma_{ij}u_{ij},$$

$$\simeq \tfrac{1}{2}\,E_{\mathrm{Y}}\,f_b(\nu)\int_{-d/2}^{d/2}\mathrm{d}y\left(\frac{y}{R}\right)^2,$$

$$= \frac{E_{\mathrm{Y}}\,d^3}{24(1-\nu^2)}\kappa^2. \tag{2.35}$$

Here E_{Y} is the Young's modulus introduced before, and $\kappa = 1/R$ denotes the local curvature of the sheet or rod. Moreover, in the second line $f_b(\nu)$ is a simple function of the Poisson ratio, which is calculated to be $1/(1-\nu^2)$ in problem 2.2 where the analysis is done in more detail; this result has been used in the third line. The variation of the integrand as y^2 with the height y from the neutral surface in the second line comes from the fact that all nonzero elements of the elastic deformations u and stresses σ vary linearly in y. This variation is clearly very general, as it follows directly from the geometry of a weakly bent object.

Above, we analyzed the contribution to the elastic energy with local curvature $\kappa = 1/R$ (varying in one direction only). For a deformed sheet or rod, κ will vary along the sheet. So based on the above analysis, one concludes that the total elastic bending energy of a

deformed sheet or bar is

$$E_{bend} \simeq \frac{E_Y d^3}{24(1-\nu^2)} \int d^2S \, \kappa^2, \qquad (2.36)$$

where the integral is taken along the neutral surface. If one considers a rod of rectangular cross section bent only in the long direction, the integral reduces to the width W times an integral along the neutral line,

$$E_{bend}^{rod} = \tfrac{1}{2}B \int ds \, \kappa^2, \qquad (2.37)$$

where

$$B = \frac{E_Y \, W \, d^3}{12(1-\nu^2)}. \qquad (2.38)$$

Here s is the coordinate along the neutral line, so essentially along the rod: in practice we can identify very thin objects with the geometric neutral surface or line that characterizes them.

We stress that we have in this analysis assumed that the bending is in one direction only, and that variations in the other direction along the sheet can be ignored. The expression for the more general case of a distorted sheet with bending in both directions, the so-called Föppl-Von Kármán equations,[13] is more complicated, as a result of the fact that forces and strains in the directions tangential to the surface are coupled via the stress-strain relation (2.6). Nevertheless, the structure and order of magnitude of the terms is very much the same, so for the purpose of our discussion the expression (2.36) for E_{bend} suffices.

2.7.3 Implications

The important lesson for us is that the energy of bending-type deformations of sheets go as d^3 and for rods with width $W \approx d$ as d^4. For large d, this means that it is extremely difficult to bend a thick beam, due to the d^3 dependence—an important rule in engineering! The flipside of the coin is that thin objects easily bend: when we stretch a sheet or plate, the elastic stretching energy $E_{stretch}$ will vary *linearly* with the thickness d. As E_{bend} varies as d^3, bending energies of thin rods and sheets are much smaller than stretching energies. Stated differently, *bending deformations are the low-energy deformations at long length scales* for sheets and rods, as playing with a sheet of paper immediately confirms—see also figure 2.14. These observations apply in particular to biomatter, as membranes abound at the cellular scale; as we shall discuss in chapter 7, shape changes of such membranes are indeed typically analyzed in terms of an effective model based on bending deformations. A common

We will see later that expression (2.37) is the energy used in the wormlike chain model for biopolymers whose statistical properties are discussed in section 5.4.

Figure 2.14. Since a thin sheet easily bends, it spontaneously bends around a droplet if this lowers the surface energies (see section 1.13.1). The photos from the experiments show a square (left) and a triangular (right) polymer sheet wrapped around a water droplet. This effect causes soft contact lenses to attach to your eye. Left: Reproduced with permission from "Capillary Origami: Spontaneous Wrapping of a Droplet with an Elastic Sheet" by Charlotte Py et al. (*Phys. Rev. Lett.* 98, 156103). © 2007 The American Physical Society. Right: Image courtesy of José Bico.

Keep in mind, also, that the statement that bending is the low-energy deformation is only true for bending on scales much longer than d, so for $\kappa d \ll 1$. Crumpling a sheet of paper typically gives rise to folds, localized regions with $\kappa d = \mathcal{O}(1)$ where stretching and bending balance (Witten, 2007).[14]

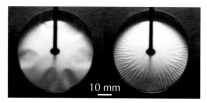

Figure 2.15. Upper images: sandwiching elastic plates into a spherical mold results in radial wrinkles at the outer edge, with wavelength decreasing with the separation of the mold parts (Hure et al., 2012).[15] Image courtesy of José Bico.[16] Lower image: example of a simple elastic sheet, like the plastic wrap used in a kitchen, which is stretched and as a result develops wrinkles in the transverse direction. The wavelength is determined by a balance of stretching and bending terms; see problem 7.10. Image courtesy of Enrique Cerda.[17]

Figure 2.16. Illustration of the ability to support a weight with a paper sheet by bending it. If you have ever tried to eat a large pizza slice with one hand, you probably did so, without thinking about it, by bending it slightly as on the right. Photo by Nynke Smits.

problem in biomatter is also that of a thin membrane or sheet attached to a thick elastic medium—examples from daily life are provided by your own skin, which gets wrinkles when you grow old, or the skin of the apple in figure 2.4, which starts to wrinkle as the flesh of the apple dries out. Such problems have only become addressed rather recently. Because soft matter is easy to deform, it provides an ideal playground for studying the interplay between wrinkling and geometry, as figure 2.15 illustrates.

Be aware that the equations governing general bending deformations of thin sheets—even those for the static shapes only—are much less benign than the above discussion might suggest. Not only are deformations in the two directions along the sheet coupled and are the equations of high order in the derivatives, but also the nonlinear coupling of the deflection with the stresses and strains bites back and leads to intricate nonlinearly coupled partial differential equations.[18]

Note, finally, that there is also an interesting interplay between bending and structural stability, which is illustrated by figure 2.16: a paper sheet easily bends, but is strong enough to support a weight when bent. Understanding this requires considering bending in two directions.

2.8 Static shapes and buckling of rods

In this section we discuss the full equations and buckling of rods with length L and thickness d. We will sometimes use the terms *long rods* and *thin rods* interchangeably to stress that $L \gg d$. Moreover the fact that they are weakly bent means that the radius of curvature R of the bending is also much longer than d, $R \gg d$. After summarizing the expressions of the most important geometrical quantities for small deflections, we discuss buckling, and then we derive the general force and torque balance equations and their simplifications in the limit of small deflections.

2.8.1 Geometrical quantities for small deflections

As we remarked above, we identify the neutral line of a thin rod with the rod itself; we assume that the rod lies in the xy plane and use s as a coordinate along the rod. It is easy to see from the simple geometric construction illustrated in figure 2.17, that we then have for the unit vector $\hat{t}$ tangent to the curve

$$\frac{\mathrm{d}\hat{t}}{\mathrm{d}s} = \kappa\,\hat{n}, \tag{2.39}$$

where $\hat{n}$ is the normal to the curve, pointing in the positive y direction. Furthermore for a vector $\vec{r}$ in the xy plane on the curve parameterized by $y(x)$ we have

$$\vec{r} = x\,\hat{x} + y\,\hat{y}, \quad d\vec{r} = dx(\hat{x} + y'\,\hat{y}), \quad ds^2 = \left(1 + (y')^2\right)\,dx^2. \tag{2.40}$$

Now consider small deflections of the rod (curve) about the x axis. To lowest order we then have

$$\hat{t} = \frac{d\vec{r}}{ds} = \frac{d}{ds}(x\,\hat{x} + y\,\hat{y}) = \frac{\hat{x} + y'\,\hat{y}}{\sqrt{1 + (y')^2}}, \tag{2.41}$$

$$\frac{d\hat{t}}{ds} = \kappa\,\hat{n} \approx \frac{d}{dx}(\hat{x} + y'\,\hat{y}) = y''\,\hat{y}, \quad \Leftrightarrow \quad \kappa \approx y''. \tag{2.42}$$

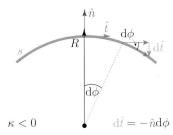

Figure 2.17. Since $\hat{t}$ is a unit vector along the curve, $d\hat{t}$ is parallel to the normal $\hat{n}$ of the curve and of length $|\hat{t}|d\phi = d\phi$. The length ds of the line element is $ds = R\,d\phi = d\phi/\kappa$. Upon eliminating $d\phi$ we get the result (2.42). For a concave curve, with negative curvature as shown here and in figure 2.13, $d\hat{t}$ is pointing down.

2.8.2 Buckling of a long rod

When a long straight rod is compressed by exerting an external force, it starts to bend when the force exceeds a threshold value; this phenomenon is called buckling. We will analyze the full general equations for the shape of bent rods under the influence of external forces. But it is illuminating to first understand the buckling instability on the basis of the following energy argument for the case sketched in figure 2.18: a rod of length L which is free to pivot at its ends, and compressed by an external force F^{ext}. When the rod bends, the total energy is increased by the bending energy (2.37) and lowered by the displacement of the force over a length Δx,

$$E^{\text{total}} = E^{\text{rod}}_{\text{bend}} - F^{\text{ext}}\Delta x = \tfrac{1}{2}B\int ds\,\kappa^2 - F^{\text{ext}}\Delta x. \tag{2.43}$$

Close to threshold, the bending deflections will be small, so we can use the approximations derived above to write $\kappa \approx y''$. Moreover, for Δx we have from (2.41) $\Delta x = L - \int_0^L dx\,\hat{t}\cdot\hat{x} \approx L - \int_0^L dx\,[1 - (y')^2/2] = \int_0^L dx\,(y')^2/2$, so to lowest order in the buckling amplitude y

$$E^{\text{total}} = \tfrac{1}{2}\int_0^L dx\,[B(y'')^2 - F^{\text{ext}}(y')^2]. \tag{2.44}$$

This expression nicely shows that the buckling transition results from the competition of two terms: the energy increase associated with bending the rod and the energy decrease associated with the inward displacement of the endpoint. Indeed, an analysis of the full equations derived in the next subsection shows that near threshold the buckling deformation mode which becomes unstable first is[19]

$$y(x) = A\sin\left(\frac{\pi x}{L}\right), \tag{2.45}$$

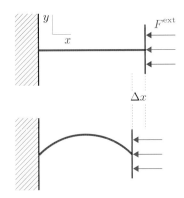

Figure 2.18. A rod of length L compressed by an external force F^{ext} starts to buckle when the applied force is large enough. The upper panel shows the original rod, the lower one the buckled state. The buckling results when the energy decrease $F^{\text{ext}}\Delta x$ is larger than the energy increase due to the bending.

which upon substitution gives

$$E^{\text{total}} = \frac{A^2 L}{4} \left(\frac{\pi}{L}\right)^2 \left[B \left(\frac{\pi}{L}\right)^2 - F^{\text{ext}} \right]. \qquad (2.46)$$

When the external force F^{ext} exceeds the critical threshold value

$$F_{\text{c}}^{\text{ext}} = B\pi^2 / L^2, \qquad (2.47)$$

the buckled state with $A \neq 0$ is the lowest energy state of the rod. For forces below this value the straight rod with $A = 0$ is stable as this state has the lowest energy—see the upper row in figure 2.19. The precise values of the buckling amplitude above threshold, associated with the minima of the total energy, cannot be determined by this analysis to lowest order in the deflections. However, we can have a pretty good guess for the behavior near threshold. Indeed, for a perfectly symmetric rod and symmetric boundary conditions one expects that for small amplitudes close to threshold the total energy can be written as an expansion in even powers of A with a stabilizing fourth order term,

$$E^{\text{total}} = \frac{A^2 L}{4} \left(\frac{\pi}{L}\right)^2 \left[F_{\text{c}}^{\text{ext}} - F^{\text{ext}} \right] + g_4 A^4 + \mathcal{O}(A^6), \qquad (2.48)$$

with g_4 positive. The coefficient g_4 and the fact that it is positive can be derived from an explicit calculation (as in problem 6.7 in the context of liquid crystals). But irrespective of the precise value of g_4, one concludes from this expression that above the threshold the buckling amplitude grows as

$$A \sim \sqrt{\frac{F^{\text{ext}} - F_{\text{c}}^{\text{ext}}}{F_{\text{c}}^{\text{ext}}}}, \qquad (2.49)$$

as sketched in the lower row of figure 2.19.

In the context of systems which are described by deterministic equations without fluctuations, the buckling transition is an example of a *bifurcation*. Bifurcations are sudden changes in the qualitative behavior of a system as a parameter is varied. There are several types of bifurcations. The one we encounter here, the so-called supercritical or forward *pitchfork bifurcation*, is one of the most common ones. At a pitchfork bifurcation, a new nontrivial state bifurcates continuously: the amplitude grows continuously from zero as a control parameter is increased above the threshold value where a stable state becomes unstable.[21] In systems where the base state is associated with the minimum of an energy-type functional the scenario is usually as sketched in the left column of figure 2.19. But you should be aware that such supercritical bifurcations are also found in many pattern forming instabilities

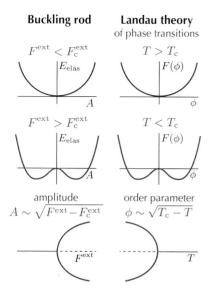

Figure 2.19. Left: the elastic energy of the buckled rod as a function of the buckling amplitude A for compression forces below the critical value (upper figure) and above the critical value (middle figure). As the figures on the right illustrate, this behavior is very reminiscent of the behavior of the free energy density $F(\phi)$ as a function of the order parameter ϕ in the Landau theory of phase transitions (a short summary of Landau theory is given in box 2, in section 2.14).[20] The lower figures illustrate the resulting behavior of the buckling amplitude A and the order parameter ϕ as a function of the external force and temperature. This behavior is an example of a supercritical pitchfork bifurcation (see e.g. Strogatz, 2015 and V. I. Arnold, 1973). This name refers to the form of the graph resembling a farmer's pitchfork—we discuss the bifurcation also in section 8.2.1.

not governed by a (free) energy functional. We will discuss this in detail in chapter 8.

Furthermore, if you are familiar with the Landau mean-field theory of phase transitions, you will surely also have recognized the similarity of the above expression and line of argument with that of the Landau coarse-grained free energy $F(\phi)$ for a simple order parameter ϕ with up-down symmetry, as one has for instance for an Ising-type spin system. This similarity is also highlighted in figure 2.19.

Avoiding buckling is an important design principle of engineering, as buckling of a beam can lead to a snowball effect resulting eventually in the structural collapse of a building. But as we shall see, buckling can also be used to one's advantage in designing soft metamaterials with unusual nonlinear response.

In passing, it is fun to point out that the simple buckling analysis above is already rich enough to give some insight into the scaling of tree heights. In order for a tree of height H and diameter d not to buckle, its gravitational force $Mg \simeq \rho_{\text{wood}} d^2 H g$ should not exceed the critical buckling force $F_c^{\text{ext}} \simeq E_Y d^4 / H^2$ (we take $W \approx d$). This shows that the maximum tree height $H_{\max}$ should scale with the tree diameter as $d^{2/3}$. Figure 2.20 shows that trees conform to this scaling law quite well. Figure 2.21 illustrates that da Vinci was already aware of some of the scaling behavior of thin bent rods some 500 years ago.

2.8.3 The general force and torque balance equations for static rods

In order to illuminate the underlying effects of the torques involved with a bent rod, we now derive the general equations for force and torque balance governing the static shape of a thin rod of rectangular cross section in the absence of torsion. These equations, which allow for external forces to be applied along the rod, will also enable us to analyze more general situations, such as the bending of the tip of an atomic force microscope (see problems 2.3 and 2.4). In line with our discussion above, we identify this thin rod with the neutral line that characterizes its shape when we zoom out. We will use a coordinate system in which the undistorted rod is along the x axis, and the bent rod lies in the xy plane as in figure 2.13; as before, we count the curvature negative for a concave shape with $\mathrm{d}^2 y / \mathrm{d} x^2 < 0$.

In the static case, the total forces and torques have to add up to zero everywhere along the rod. We will allow for an external force $\vec{f}^{\text{ext}}$ per unit length being applied to the rod; when this force is not parallel to the local rod direction, this force gives rise to a torque

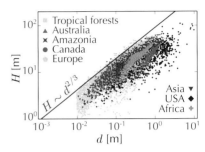

Figure 2.20. Scatter plot of tree height H versus trunk diameter d for a large number of species from various continents. The maximum tree height conforms well to the scaling $H_{\max} \sim d^{2/3}$, with prefactors consistent with the density and Young's modulus of wood. Image courtesy of Fabian Brau.[22]

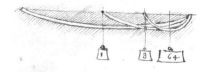

Figure 2.21. As this sketch from one his notebooks[23] shows, around 1500 da Vinci already found that the weight at the center needed to make a rod of length L deflect a particular vertical distance, goes as $1/L^3$: when the rod's length is halved, an eight times heavier weight is needed. The result follows directly from the equations for small deflections of section 2.8.4, as shown in problem 2.3.

as well. Clearly, when we consider a rod element of infinitesimal length $\mathrm{d}s$, force balance implies that the external force on this element is balanced by the net result of the elastic stresses at both ends of the rod element in figure 2.13; similarly, the external torque should balance with the net torque from the elastic stresses on both sides of the rod element.

Let us denote the total elastic stress integrated over the cross section of the rod by $\vec{F}_{\mathrm{el}}$, and, as in figure 2.17, the unit vector tangent to the (neutral line of the) rod by $\hat{t}$. With our convention for the stress tensor that $\underline{\sigma} \cdot \hat{t}$ is the force density on the cross section normal to $\hat{t}$, we thus have

$$\vec{F}_{\mathrm{el}} = \int_{\Sigma} \mathrm{d}^2 S \, \underline{\sigma} \cdot \hat{t}, \tag{2.50}$$

where the integral is taken over the cross section Σ of the rod normal to $\hat{t}$.

In figure 2.22 we reproduce the same bent rod as in figure 2.13, but now with a focus on the torques in the two cross sections as a result of the elastic stresses σ_{xx}. If we consider for simplicity the situation as sketched in the figure in which the normal of the cross section is locally pointing in the x direction and the curvature is counted negative, we get for the elastic torque

$$\begin{aligned} \vec{M}_{\mathrm{el}} &\approx - \int_{\Sigma} \mathrm{d}^2 S \, \sigma_{xx} \, y \, \hat{z}, \\ &\approx \frac{E_{\mathrm{Y}}}{1-\nu^2} \, W \int_{-d/2}^{d/2} \mathrm{d}y \, y^2 \, \kappa \, \hat{z}, \\ &= \frac{E_{\mathrm{Y}} \, W \, d^3}{12(1-\nu^2)} \, \kappa \, \hat{z} = B \, \kappa \, \hat{z}. \end{aligned} \tag{2.51}$$

Here W is the width of the rod in the direction normal to the bending plane, B was introduced before in equation (2.38), and the factor $E_{\mathrm{Y}}/(1-\nu^2)$ arises for the same reasons as in equation (2.35). So we see that the internal stresses lead to a torque which is proportional to the curvature of the rod and which has a sign depending on the way the rod is curved; very much to what we saw before, the size of the torque stiffness B is proportional to the width W and to the cube of the thickness, and of course the Young's modulus.

After these preliminaries on external and elastic forces, it is straightforward to write the force and torque balance equations. Force balance on a little element of length $\mathrm{d}s$ along the rod amounts to

$$(\vec{F}_{\mathrm{el}} + \mathrm{d}\vec{F}_{\mathrm{el}}) - \vec{F}_{\mathrm{el}} + \vec{f}^{\,\mathrm{ext}} \, \mathrm{d}s = 0,$$

so

$$\frac{\mathrm{d}\vec{F}_{\mathrm{el}}}{\mathrm{d}s} + \vec{f}^{\,\mathrm{ext}} = 0. \tag{2.52}$$

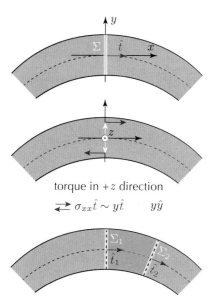

torque in $+z$ direction

$\sigma_{xx}\hat{t} \sim y\hat{t}$ $y\hat{y}$

Figure 2.22. Focus on the behavior of torques of the bent rod of figure 2.13. Top: a cross section Σ on which the elastic torque is integrated. Middle: the torque is in the $+z$ direction on both the stretched and the compressed part of the rod, since both the force and the position change sign between these two halves. Bottom: the total elastic torque on the shaded volume element is $\vec{M}_2 - \vec{M}_1$ because the outward normal on its left surface is $-\hat{t}_1$.

Turning to the torque balance, the elastic force gives rise to a torque $\mathrm{d}\vec{s} \times \vec{F}_{\mathrm{el}} = \mathrm{d}s\,\hat{t} \times \vec{F}_{\mathrm{el}}$ on the rod element, while the external force would give a contribution of order $\mathrm{d}s^2$, so the total torque balance equation becomes

$$\frac{\mathrm{d}\vec{M}_{\mathrm{el}}}{\mathrm{d}s} + \hat{t} \times \vec{F}_{\mathrm{el}} = 0. \tag{2.53}$$

These three equations, (2.52) and (2.53) together with $\vec{M}_{\mathrm{el}} = B\,\kappa\,\hat{z}$ according to (2.51) are the equations describing the static shape of a rod under the application of external and boundary forces.

We can combine these equations by differentiating the torque balance equation (2.53) and then combining it with the force balance equation (2.52) and the torque-curvature relation (2.51) to get

$$B\,\kappa''\,\hat{z} + \hat{t}' \times \vec{F}_{\mathrm{el}} - \hat{t} \times \vec{f}^{\mathrm{ext}} = 0. \tag{2.54}$$

Here a prime indicates differentiation with respect to s along the curve, and $B = E_{\mathrm{Y}}W d^3 / [12(1 - \nu^2)]$ as defined in (2.38).

This equation illustrates for the case of a thin rod the remarks made earlier, that the shape equation is much less benign than it might first appear. The shape and the solution are intimately coupled, as the derivatives with respect to s are not with respect to a fixed coordinate but with respect to a coordinate which itself is determined by the shape one seeks to determine. The interest in nonlinear solutions has been revived by the connection with DNA coiling, and some solutions are indeed known.[24] The corresponding equations for a two-dimensional sheet are, however, known to be prohibitively difficult to solve.

2.8.4 Equations in the small deflection approximation

The static equilibrium equations for a rod in the small deflection limit can be obtained by combining the above result (2.54) with the small deflection approximations (2.41) and (2.42) to get

$$B\,y''''\,\hat{z} + y''\,\hat{y} \times \vec{F}_{\mathrm{el}} - (\hat{x} + y'\,\hat{y}) \times \vec{f}^{\mathrm{ext}} = 0, \tag{2.55}$$

which finally leads to

$$B\,y'''' - y''\,F_{\mathrm{el},x} + y'\,f_x^{\mathrm{ext}} - f_y^{\mathrm{ext}} = 0. \tag{2.56}$$

As a simple application, let us return to the case of a long rod compressed by a force F^{ext}, which we considered in the discussion of the buckling instability. In this case the external force $\vec{f}^{\mathrm{ext}}$ applied along the rod is zero, while equation (2.52) shows that $\vec{F}_{\mathrm{el}}$ is constant along the rod and hence, because of the boundary condition,

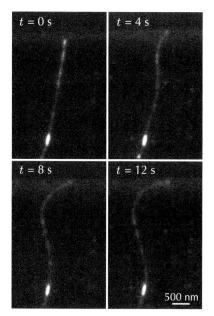

Figure 2.23. Four successive images of a fluctuating microtubule. Microtubules, which act as the rails along which micromotors walk in cells, can be modeled as thin rods with a bending stiffness B. This bending stiffness can be measured via analysis of the thermal fluctuations, along the lines discussed in section 3.6.3, and the force they exert can be determined via analysis of their buckling when they hit a wall during growth in a confined cell. The wall is visible in the upper part of each image, while the bright spot marks the place where the microtubule is attached.[25] Image courtesy of Maurits Kok and Amol Aher.[26] A simple dynamical equation for microtubules is discussed in problem 2.5.

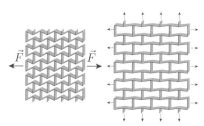

Figure 2.24. Simple illustration of auxetic response of a two-dimensional regular structure. When the material on the left is stretched in the horizontal direction, the structural elements also expand in the transverse direction, as the right panel shows. Note that this structure is anisotropic, while our discussion of elasticity in this chapter is limited to isotropic materials.

equal to $-F^{\text{ext}}\,\hat{x}$. So equation (2.56) then reduces to

$$B\,y'''' + y''\,F^{\text{ext}} = 0. \tag{2.57}$$

It is easy to check that the expression (2.45) for $y(x)$ does indeed solve this equation at the critical force; since the torque is proportional to the curvature, the fact that $y''(0) = y''(L) = 0$ expresses that the rod experiences no torques at its ends, in other words, it is free to pivot.

We will use this equation to analyze two applications in problems 2.2–2.4. Figure 2.23 shows two pictures of microtubules, which can be modeled as a rod with a bending stiffness B.

2.9 Auxetics: Metamaterials with negative Poisson ratio

As we discussed in section 2.4, under stretching normal materials contract in the directions transverse to the stretching. This means their Poisson ratio is positive. We have seen that the Poisson ratio of isotropic materials can vary between -1 and $1/2$—see equation (2.18)—and no principle excludes materials with negative Poisson ratio. In other words no principle excludes the existence of materials whose transverse directions expand or contract in unison with the expansion or contraction of the longitudinal direction. Although crystals can have negative Poisson ratios in some directions,[27] for a long time common wisdom had it that negative Poisson ratio materials were an anomaly. But as was already shown back in 1987[28] for a foam, mesoscopic structures can easily be cleverly designed to build materials whose macroscopic elastic response is an example of an *auxetic*, the name used for negative Poisson ratio materials. Figure 2.24 illustrates this for a regular structure in two dimensions. We are entering the realm of designer matter here, a topic which is put in a more general perspective in chapter 10.

Auxetic materials have an interesting property, namely, that when bent in one direction, they have a natural tendency to bend in the other direction as well. This can most easily be seen from figure 2.13: the bending shown there in one direction leads to a compression of the material on the inner side and an extension on the outer side. But the negative Poisson ratio of an auxetic implies that the material or structure will show a concomitant expansion and contraction in the orthogonal direction out of the page as well, and this automatically induces bending in this direction. As figure 2.1 illustrates, auxetic materials therefore naturally wrap around curved surfaces, without the tendency to form folds as normal

Figure 2.25. Metameterials built with rubbery elastic matter (elastomers). Panel (a) shows how when such a material with a two-dimensional pattern of holes is compressed, buckling-type instabilities lead to a pattern of deformed holes and to auxetic behavior: the material also contracts in the transverse (horizontal) direction. Image courtesy of Katia Bertoldi.[30] (b): When a material is made with building blocks as sketched in (c), which can buckle when compressed in one particular direction, neighboring elements influence each other. This makes it possible to design structures which respond in surprising but prescribed ways under compression (Coulais et al., 2016), as the images illustrate. This structure consists of 10^3 of these building blocks. Image courtesy of Corentin Coulais and Martin van Hecke.

materials do. Clearly, such behavior is attractive for use in clothes and footwear, and some running shoes indeed already have auxetic sole structures.

Such structures are examples of mechanical metamaterials. The word *metamaterial* was introduced in optics for materials—often designed by cleverly exploiting patterned structures of building blocks like resonators smaller than the wavelength of light—with exceptional properties, such as a negative index of refraction for light. This allows for spectacular Harry Potter–like applications such as an invisibility cloaking, the fact that an object is made invisible because electromagnetic waves bend around it.[29] We will come back to cloaking and designer matter in chapter 10. In a similar spirit the term *mechanical metamaterials* is nowadays used for materials with remarkable mechanical properties. Like their optics counterpart, mechanical metamaterials are often based on clever design and use of elementary building blocks. Soft matter plays a central role in this field.

Figure 2.25 illustrates this nicely. Panel a shows a material made of rubber-like elastic polymers (elastomers), which has been made with a two-dimensional square pattern of square-shaped holes. When this structure is compressed, the initial response is linear and elastic, but as the right-hand image shows, as the strain is increased at some point, depending on their thickness, the bridges between the holes start to buckle and the holes start to deform. Because a single buckling event affects the two neighboring holes, the resulting pattern of the hole deformations has a staggered structure. The image clearly shows that the structure contracts in the horizontal direction when compressed in the vertical direction: the structure is auxetic.

If one goes beyond regular arrays of building blocks of holes, one enters the realm of programmable materials. Figure 2.25.c shows an interesting building block made of elastic polymers: under vertical compression of this structure along the longitudinal axis, the four vertical sides can buckle in or out, as illustrated. As

For an introduction to and overview of the field of flexible mechanical metamaterials, see Kadic et al., 2013 (with both optical and mechanical examples) and Bertoldi et al., 2017. Notable also are origami-inspired metamaterials, made of stiff materials with flexible joints, such as solar panels built to unfold gently in space from the space load of the launch vehicle.

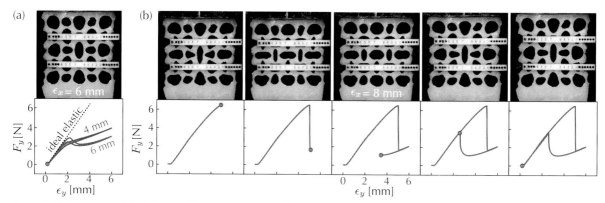

Figure 2.26. A soft material with a two-dimensional pattern of holes with two different sizes behaves as a metamaterial. The images show the behavior for various prestrains ϵ_x imposed with two horizontal constraints (Florijn et al., 2016). Panel (a) shows the force-strain curves, the measured compression force F_y as a function of the vertical strain ϵ_y, for the two cases of not too large prestrain, $\epsilon_x = 4$ mm and 6 mm.[32] For increasing ϵ_y the material softens, while for a prestrain $\epsilon_x = 6$ mm the force-strain curve has a maximum and exhibits some hysteresis. (b) For $\epsilon_x = 8$ mm the deformed holes in the material flip orientation at large enough ϵ_y (compare the second image with the first one, just before the reorientation of the holes), resulting in a sudden drop in the vertical force. Upon subsequent reduction of the strain, the response follows a lower branch (third image) until the two branches merge (fourth image). The images show that the sudden changes in the effective stiffness result from transitions between differently oriented patterns of the hole-deformations, while the hysteresis illustrates that such compression patterns can be bistable. Image courtesy of Martin van Hecke.

the building block itself has overall the shape of a cube, one can build large three-dimensional structures out of them, with at each point the freedom to choose the orientation of the longitudinal axis of the building block. Since buckling affects neighboring elements—for example when one buckles out the neighboring element has to buckle in—each stacking leads to a different response under compression. As figure 2.25.b shows, designing for a particular type of response behavior is indeed possible. Since there are so many ways to stack the building blocks, the design of such structures poses an interesting combinatorial problem of reverse engineering.[31]

One need not be surprised that once the bridges between the holes buckle, the response of the material rapidly becomes quite nonlinear. Moreover, nonlinear behavior can be enhanced and manipulated by using holes of different sizes which result in switches between different pattern orientations. To make this concrete, consider the material with a regular pattern of holes of different sizes, shown in figure 2.26. When this material is compressed in the vertical y direction, the large holes tend to deform into an oval-like shape, with the short axis in the y direction. Panel a shows that the force-strain response of the system is relatively linear as the holes continuously deform, although some weakening is noticeable (note that the curve flattens) at strains larger than about 5%. The case shown in a is for when the material is prestrained by 12% in the horizontal direction by applying the two horizontal clamps visible in the photo. The stress-strain response in this case is very nonlinear: there is a regime where the vertical force even decreases by increasing the compression—it is as if the material has a negative compression modulus. This behavior results from the fact that the prestrain brought the system into a state in which the large deformed holes have their short axis in the x direction: the nonlinear behavior amounts to switching from this state to the one where their short axis is rotated 90 degrees. This transition is also weakly hysteretic: the force upon releasing the strain is slightly different from upon compression. Panel b shows that this switching between states becomes very abrupt when the prestrain is increased to about 16%. At the same time, the hysteresis becomes very large: the system stays for a very long time in the 'wrong' state until it switches.

These examples illustrate the enormous range of possibilities of varying the mechanical response of soft metamaterials. It is also interesting to see that while buckling is typically avoided in hard materials and engineering, buckling instabilities underlie many of the design principles of soft metamaterials and similarly of soft robots.[33]

We finally note that while most research on mechanical metamaterials has focused on materials which are patterned on the mesoscopic scale, a recent advance is the discovery of liquid crystal elastomers and protein crystals which show auxetic response at the molecular level.[34]

2.10 Packings of particles *jammed* together: Beyond standard elastic behavior

We often take it for granted that solid materials respond elastically when a force is applied to them, and elasticity theory is indeed developed on the assumption of a stress-strain relation which is linear for small strains—see equation (2.6). But soft matter forces us to rethink under what conditions such elastic response might cease to apply.

As illustrated in figure 2.27, interatomic forces are typically quite smooth, and they have a strong attractive tail. These attractive tails are responsible for the formation of molecules, and for the attractive Van der Waals interaction that we will discuss in chapter 4 in the context of colloids. Moreover, the smoothness of the potential ensures the linear relation between stresses and strains, when a solid is deformed slightly.

As we already saw in the introductory chapter and as is illustrated again in figure 2.28, many soft matter phases, on the other hand, consist of small particles with a well-defined radius. These particles, be they grains, colloids, droplets or bubbles, hardly

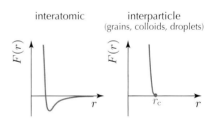

Figure 2.27. While interatomic forces typically exhibit an attractive tail, forces between small particles are typically dominated by a strong repulsive force up to some radius r_c; beyond r_c the force is essentially negligible.

Figure 2.28. Many soft matter systems consist of grains, particles, droplets or bubbles which hardly interact until they touch, and which strongly repel when we push them into each other.[35]

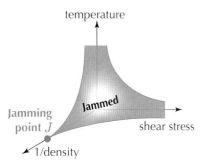

Figure 2.29. The jamming phase diagram for particles with a finite radius and strong repulsive forces. J marks the jamming point. A jammed phase can be unjammed by reducing the density, applying a shear force (as you know from playing with sand) or raising the temperature. The idea to think of heaps of particles jammed randomly together as the phase of a system with a critical point, and to explore the analogy with the phase diagram of thermal systems, is attributable in particular to A. J. Liu and Nagel, 1998. Introductory reviews of the field of jamming are given by van Hecke, 2010 and A. J. Liu et al., 2011. Figure adapted from A. J. Liu et al., 2011.

interact until they touch, while as soon as they touch, they repel strongly—very much like marbles or pebbles. This virtually nonanalytic behavior of the interparticle force as a function of their separation has stimulated rethinking the response of packings formed from them.

In the absence of an attractive tail, dense phases of such particles only form under the action of an external force or by controlling the density of the packing. Moreover, when the particles are sufficiently large, as they are in granular media, interaction energies are much larger than thermal energies $k_B T$, so collections of such particles do not explore phase space like thermodynamics systems do: these are disordered *athermal* systems in which temperature does not play a role. Nevertheless, the question naturally emerges whether we can unify their most prominent behavior in a generic phase diagram, very much like an equilibrium phase diagram of molecular systems.

2.10.1 The jamming phase diagram

These considerations have stimulated the introduction of the so-called jamming phase diagram of figure 2.29 for particles exhibiting harsh repulsive potentials with a finite range. For athermal systems in the absence of an external shear stress, the important variable is the density of the particles. Note that in the jamming phase diagram, the *inverse density* is plotted along one of the axes, so that the high density phase where particles are 'jammed' into a packing, very much like cars stuck in a traffic jam, is on the inside near the origin.

Figure 2.30 illustrates the idea of jamming of weakly compressible particles with a finite range as the volume fraction ϕ (the fraction of volume taken up by the particles) increases, and at zero temperature. At some value ϕ_J it becomes unavoidable that a finite fraction of the particles touches; beyond this value most particles exert repulsive forces.

The interesting finding of the jamming studies is that the jamming point J marked in figure 2.29 does indeed play the role of a critical point for the model system of frictionless particles with repulsive forces of finite range. We will demonstrate this in sections to follow, and we will discuss experimental obervations of this critical behavior in microemulsions in section 4.6.3. For the discussion below it is good to stress that since the system is athermal, the jamming point is not a thermodynamic critical point—packing arguments instead of the theory of critical phenomena are needed to understand it, and the exponents we will encounter are classical.

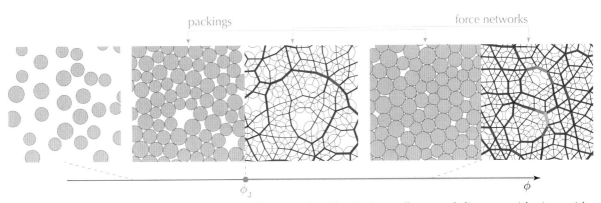

Figure 2.30. Illustration of the basic idea of jamming. The first, second and fourth pictures illustrate polydisperse particles, i.e. particles of different radii, for various packing fractions ϕ. For ϕ less than the jamming volume fraction ϕ_J, it is possible to fill space without having the particles touch. When packed randomly, they start to touch at ϕ_J—this marks the jamming point. Above ϕ_J the particles are compressed and the number of contacts increases with increasing volume fraction. We note that when the particles all have the same radius, ϕ_J coincides with the so-called random close packing density of disks/spheres, $\phi_{RCP} \approx 0.88$ (Zaccone, 2022) in two dimensions and $\phi_{RCP} \approx 0.64$ in three dimensions. We stress that we consider an athermal system (the temperature is zero), so the pressure of the packing vanishes for $\phi \leq \phi_J$. The force networks of the two jammed packings are shown in the third and fifth panels; here the thickness of each line is proportional to the force between two particles.

2.10.2 Counting argument for frictionless spheres

The simplest setting for jamming is to place frictionless compressible spheres in a box, as illustrated in figure 2.30. Let us explore how this packing behaves as we increase the volume fraction (or density) of the particles.[36] For low volume fraction all particles can be placed without having them touch, but at some point it becomes unavoidable to have almost all of them touch—this marks the volume fraction ϕ_J where they get jammed. As we increase the volume fraction beyond ϕ_J, the particles start to get compressed and the number of contacts steadily increases. How does the number of contacts evolve in this process?[37]

Let us start with asking ourselves when z, the average number of contacts per particle, is high enough that the system can form a rigid structure. There is a simple argument attributed to Maxwell which shows us that there is an intriguing simple formula for the critical average coordination number z_c,

$$z_c = 2d, \tag{2.58}$$

which plays the role of the minimum value needed for a structure to be stiff. Here d is the dimension of the system. The argument is as follows.

Just at the jamming threshold, there is a precise balance between the number of degrees of freedom and the number of constraints. Let us start by considering the N particles which experience contacts.[39] In d dimensions, there are d degrees of freedom associated with the coordinates of the center of each of these particles.

In this section we focus on how the response properties of packings evolve if we approach the jamming point from the large density side. This line of research is different from the so-called rigidity percolation problem, which has a long history. Rigidity percolation concerns the question of how structures of flexible units joined together acquire rigidity as connected structures grow sufficiently large and with a sufficient number of crosslinks to form infinite structures which can resist shear stresses. Rigidity percolation is of importance for the formation of colloidal gels and has even been suggested to play a role during embryonic development.[38]

Thus the total number of degrees of freedom of all these particles amounts to dN. Now, there must be a sufficient number of repulsive contact forces to constrain all particles. Because each contact is shared between two particles, for a given coordination number z, there are $Nz/2$ contacts. So in order that there are indeed enough contact forces to fix the dN translation degrees of freedom, we need to have

$$\text{enough contact forces to fix coordinates:} \quad z \geq 2d. \quad (2.59)$$

Now change perspective, and focus on these contact forces. Infinitesimally close to threshold, all particles just touch—a finite amount of compression would lead to large internal forces, and hence to rearrangements. This means that for all pairs of spheres of radius R_i and R_j which just touch, their centers need to be *exactly* a distance $R_i + R_j$ apart at jamming. So associated with each contact there is a constraint on the distances of the centers, and hence in total there are $Nz/2$ such constraints. In order that the dN coordinates of the centers of the spheres provide enough degrees of freedom to satisfy these constraints, we need to have

$$\text{enough coordinates to fix contact distances:} \quad 2d \geq z. \quad (2.60)$$

The above two constraints imply that the critical coordination number z_c at jamming equals

$$z_c = 2d. \quad (2.61)$$

Note that the first condition (2.59), $z \geq 2d$, always needs to hold, while the second condition (2.60) only holds *at* jamming: above the jamming density particles are pressed into each other, so they are less constrained.

This Maxwell criterion (2.61) therefore leads to the picture that for $z > z_c$ the material is rigid as it has more contacts ('bonds') than the minimum needed. The critical point z_c is often referred to as the *isostatic* point. In computer studies of random packings of compressible frictionless spheres it is indeed found that z increases continuously from z_c if the packings are compressed and the volume fraction increases beyond ϕ_J.

A packing with $z < z_c$ cannot be stable, as it has too few contacts to constrain all the particle coordinates, and so the packing at jamming with z_c contacts per particle is at the edge of instability: it is a *marginal solid*. Indeed, if we remove a few contacts from it, the packing does not have enough contacts to constrain all the particles in place, so it can then be deformed without any energy cost. These zero-energy deformation modes are disordered but extended: as we shall see below they typically involve many particles. On the other hand, when we compress the packing beyond the jamming density, additional new contacts form because particles are pressed

It is important to realize that there are subtleties associated with the fact that the details of the scenario are somewhat protocol-dependent. This is best understood for computer generated packings as these are made according to well-defined protocols, but also the detailed response of experimental packings tends to depend on how they were made.

into each other, and the packing becomes less fragile. We illustrate this picture of packings as marginal solids that are close to the jamming point with the data below.

2.10.3 Scaling of the ratio of elastic constants

A strong signal that the properties of packings change upon approaching point J from the jammed side comes from the behavior of the elastic moduli. The compression modulus K and shear modulus μ typically both vanish in this limit, but the way they do depends on the particular force law. However, their ratio is independent of the force law but linearly dependent on the distance from the jamming point: as figure 2.31 illustrates, μ/K scales linearly with the excess number of contacts relative to the isostatic one,

$$\frac{\mu}{K} \sim \Delta z, \qquad \text{with } \Delta z = z - z_{\mathrm{c}}. \qquad (2.62)$$

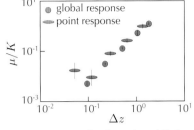

Figure 2.31. Data for the ratio μ/K from three-dimensional computer generated packings, showing that it varies linearly in $\Delta z = z - z_{\mathrm{c}}$. Adapted from Ellenbroek et al., 2009.[40]

2.10.4 Excess of low-frequency modes

A good way to probe the emergence of an excess number of low-energy modes as the jamming point J is approached from the jammed side is by gradually reducing the pressure in a simulation of a computer model of frictionless compressible spheres. Once a packing has equilibrated, one then linearizes the dynamical equations of the spheres given the configuration and studies the spectrum of eigenmodes.

Long-wavelength modes of an elastic medium, discussed in section 2.6.1, have a linear dispersion $\omega \approx ck$ for small k (long wavelength), very much like the sound modes in a fluid we studied in section 1.7. Standard results for the phase space density of k modes going as k^{d-1} in d dimensions[41] then imply that the density of states $D(\omega)$ ($D(\omega)\mathrm{d}\omega$ gives number of eigenmodes with frequency between ω and $\omega + \mathrm{d}\omega$) goes as ω^{d-1} for an elastic medium, hence as ω^2 in three dimensions.

The data presented in figure 2.32 for $D(\omega)$ demonstrate that the particle packings do indeed exhibit more and more low-frequency modes as $\phi \downarrow \phi_J$. Especially the lower panel, where data for a large number of two-dimensional packings are plotted on a log-log scale, shows that $D(\omega)$ exhibits the ω^{d-1} scaling of elastic modes for low frequencies, with dimension $d = 2$; however as the packing density is reduced toward ϕ_J the crossover to elastic behavior moves to lower and lower frequencies, and $D(\omega)$ becomes constant: gradually more and more excess low-frequency modes emerge. This is a salient signature of the fact that one approaches a marginal solid, which at the jamming point has no elastic modes anymore.

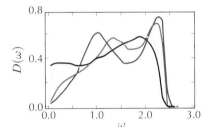

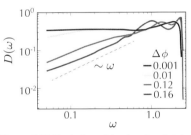

Figure 2.32. Evolution of the density of states $D(\omega)$ of two-dimensional packings of frictionless polydisperse compressed particles upon approaching the jamming point. Upper panel: rendering of data on a linear scale for four different volume fractions. As $\Delta\phi = \phi - \phi_J$ decreases, more and more low-frequency modes emerge. Lower panel: on a log-log scale, one clearly sees the development of a large frequency-independent plateau in $D(\omega)$ upon approaching the jamming point. Away from the jamming point, the lowest-frequency data are consistent with the $D(\omega) \sim \omega$ scaling characteristic of elastic modes in two dimensions.[42]

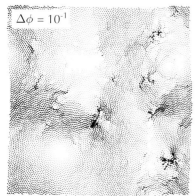

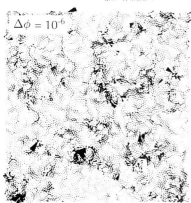

Figure 2.33. The lowest-frequency eigenmodes for a packing of 10,000 bidisperse disks in two dimensions far above the jamming point (top) and extremely close to the jamming point (bottom). The arrows indicate the direction and amount of displacement of each particle in the mode. Although there is some disorder, the lowest-frequency mode above the jamming point is clearly reminiscent of a smooth elastic mode. The eigenmode at the jamming point spans the whole packing (which has periodic boundary conditions) but is very disordered. Some particles make very large moves.[43]

The emergence of a flat density of states upon approaching point J is understood along the lines of the argument below for the crossover length scale beyond which elastic behavior arises.

2.10.5 The crossover length scale

To give an idea of the type of arguments used to understand marginal solids, we determine the length scale ℓ^* which characterizes the crossover from elastic behavior at long scales to floppy mode behavior at shorter scales.

Imagine we start with a very large computer generated compressed packing in three dimensions with average contact number per particle $z = z_c + \Delta z$, where $\Delta z \ll z_c$. We can imagine it was created by starting with an isostatic packing of frictionless spheres at the jamming density and then subsequently compressing the packing slightly. This compression creates the extra contacts needed to bring the average contact number z above z_c and ensures elastic response at sufficiently long length scales. The total number of contacts between spheres is $Nz/2 = N(z_c + \Delta z)/2$.

We now do the following thought experiment. Imagine we focus on all the particles in a large cubical portion of the overall system. The cube has linear size L (measured in units of the average sphere radius). So the volume is L^3, and there are $N_L \simeq L^3$ particles in the cube. The center of each spherical particle has three spatial degrees of freedom, so the total number of degrees of freedom of the spheres in the imaginary cube grows as $3N_L \simeq 3L^3$. Furthermore, the number of excess constraints due to the additional contacts of the particles within the cube is of order $N_L \Delta z/2 \simeq 3L^3 \Delta z/2$. Furthermore, the number of contacts crossing the boundary grows with their surface, so as L^2.

Now imagine that we release these contacts across the surface of the cube, turning them into degrees of freedom we can play with to push or pull on the spheres in the cube near the surface, and that we monitor whether the packing within the cube resists these perturbations or distorts as a marginal solid. Clearly, the crossover between these two regimes happens at the length scale ℓ^* which is such that the number of artificial degrees of freedom created this way at the surface, which is of order L^2, becomes comparable to the excess number of constraints which is of order $3L^3 \Delta z/2$. In other words, at $L \simeq \ell^*$ we must have $(\ell^*)^2 \simeq (\ell^*)^3 \Delta z$, so that

$$\ell^* \sim \frac{1}{\Delta z}. \tag{2.63}$$

In summary, ℓ^* is the crossover scale from elastic response of the packing at longer length scales to marginal solid type of behavior at smaller length scales. As figure 2.33 illustrates, the spatial structure of the eigenmodes in this latter regime is very disordered and typically involves many particles.

The above argument can be extended to show that the crossover frequency ω^* separating the two regimes in the density of states scales linearly as $\omega^* \sim \Delta z$.[44] Detailed analysis of the density of states $D(\omega)$ confirms this scaling.

We finally note that the jamming point also strongly affects the flow rheology of sheared frictionless spheres. Microemulsions are a good experimental realization of this system; in section 4.6.3 we'll see that data for sheared emulsions are indeed nicely consistent with the scaling behavior found in computer simulations of compressible spheres near the jamming point.

2.10.6 Jammed packings versus disordered crystals

The discussion in this section has been based on disordered packings of spheres as an idealized model for jammed systems. This amounts to taking random close packings of spheres as the reference model. A natural question then is how relevant this limit is for partially ordered packings.

To phrase it differently, suppose we start at the other extreme with a computer model of ideal crystalline packing of compressible spheres, introduce point defects such as vacancies or interstitials, and let the system relax to a nearby minimum energy state. We then ask the question: how long does the response of these defected crystalline packings remain close to that of the ideal crystalline system upon our increasing the density of defects? Likewise, when does the response start to exhibit significant features of random packings? The surprising answer from numerical studies of this type[46] is that indeed already for quite a small density of point defects the response of weakly disordered crystal packings resembles that of jammed random packings much more than that of an ideal crystalline packing. Hence the jamming behavior discussed above appears more robust than one might have expected at first glance.

2.10.7 Toward designer granular matter

We will discuss designer matter from a broader perspective and in more detail in chapter 10, but it is illuminating to illustrate already here how, armed with these insights, it is a small step to start exploring designer granular matter, granular matter which is tuned so as to exhibit specific desirable properties. We briefly mention two examples.

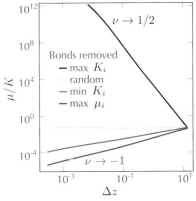

Figure 2.34. Evolution of the ratio μ/K of a three-dimensional network as various types of bonds are pruned. The thickness of the lines is proportional to the density of data points in the numerical studies of Goodrich et al., 2015. The network is obtained by starting from a compressed packing with $\Delta z = 0.13$. This packing is then translated into a network by the replacement of all the particle centers with nodes and the replacement of the contacts between the spheres with springs of unit stiffness. Subsequently, the network is pruned via removal of particular types of bonds. This reduces Δz and the elastic constants of the network. When bonds which contribute most to the energy under compression are removed (indicated as 'max K_i removed'[45]) K rapidly decreases while the shear modulus μ remains largely unchanged, so that μ/K rapidly increases. Likewise 'max μ_i removed' indicates data for networks obtained by removing bonds which contribute most to the energy under shear. Clearly, the Poisson ratio ν can be tuned via selective pruning of bonds, while random removal of bonds leaves the ratio of elastic constants unchanged. Reproduced and adapted with permission from "The Principle of Independent Bond-Level Response: Tuning by Pruning to Exploit Disorder for Global Behavior" by Carl P. Goodrich, Andrea J. Liu, and Sidney R. Nagel (*Phys. Rev. Lett.* 114, 225501). © 2015 The American Physical Society.

Particle shape optimization

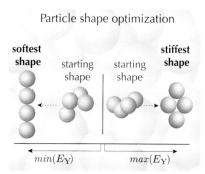

Figure 2.35. Summary of the results from simulations by Miskin and Jaeger, 2013 of the shape of granular motifs consisting of four spherical particles fused together, using an evolutionary algorithm starting from a random shape of the motif. Shapes that maximize (right) or minimize (left) the Young's modulus E_Y starting from random packings made of initial four-particle motifs as sketched are found iteratively. Miskin and Jaeger, 2013 also show experimental data of packings made from such simple motifs.

A nice introductory overview of two-dimensional melting is the lecture notes of Nelson, 1980, delivered soon after the birth of the field, while overviews of the subsequent developments are included in the later reviews by Strandburg, 1988 and Ryzhov et al., 2017. The renormalization group analysis of the Kosterlitz-Thouless transition is treated in section 6.10.

a. Tunable Poisson ratio networks

In numerical studies of the response of packings, it is straightforward to analyze the contribution each bond makes to the shear and compression response of a packing. As it turns out, these are largely independent: there are bonds which contribute mostly to the compression response but not much to the shear response, and the other way around. As illustrated in figure 2.34, by removing one of these types of bonds, one can create computer-generated networks of nodes connected by bonds that have either $\mu/K \to \infty$ and hence $\nu \to 1/2$ for the Poisson ratio, or have $\mu/K \to 0$ and hence $\nu \to -1$. To our knowledge, this approach has not yet been exploited experimentally.

b. Playing with granular motifs of various shapes

Another interesting research direction that has emerged from jamming studies is to play with the shapes of the constituent particles so as to optimize a particular property of the packing. This is illustrated in figure 2.35 for 'motifs' consisting of four spheres which are fused together in particular shapes. An evolutionary algorithm is used to optimize the shape of the motifs so as to give the highest Young's modulus E_Y of a packing of the motifs, or the smallest value of E_Y. In this particular case, the fact that chain-like configurations give the smallest Young's modulus may be not so surprising (as this promotes liquid-crystal-like ordering where sliding of strings is easy), but the fact that a flat configuration of four spheres gives the largest elastic modulus is already less obvious. Optimal shapes of motifs of five or more particles are even more nontrivial. Experimental studies of packings consisting of such motifs are in good agreement with the predicted behavior, and at the same time they illustrate the power of using evolutionary algorithms for optimizing disordered designer matter.

2.11 Dislocations and defect-mediated melting

So far, in this chapter we have not explicitly considered the effects of thermal fluctuations. In this section, however, we do want to discuss an important example where the interplay of the elastic interactions and statistical properties of defects leads to a basic thermal phase transition in two dimensions. The transition emerges in many corners of condensed matter, and soft matter systems provide an attractive platform for detailed experimental studies.

We aim to describe a two-dimensional melting phase transition which is mediated by lattice defects called dislocations. The peculiar feature of a dislocation is that it introduces an extra row of atoms or particles which originates at the dislocation center, as illustrated in figure 2.36. This feature is topological, meaning that it cannot be removed by smooth local deformations, and this topological signature is captured by the Burgers vector, $\vec{b}$. This vector quantity is analogous to an electric charge and it can be sensed far from the dislocation site. Any loop like the one in the figure with a dashed line, drawn around a dislocation with an equal number of steps back and forth in each lattice direction (as measured by lattice sites traversed), will not close, and the difference between the ending site and starting location is the Burgers vector $\vec{b}$. At the core of the defect are a pair of fivefold and sevenfold coordinated lattice points.

To move a step toward the theory of two-dimensional melting, we need to assess the energetic cost of a dislocation. In section 2.3 we showed that the energy can be written according to equations (2.7)–(2.8) as

$$E_{\text{elas}} = \int \mathrm{d}^2\vec{r}\left(\frac{1}{2}\lambda u_{ii}^2 + \mu(u_{ij})^2\right),\tag{2.64}$$

where λ and μ are the two-dimensional variants of the Lamé coefficients already introduced in (2.8) in the context of the discussion of the elastic energy.

Note that this expression is subject to the same assumptions that applied when we wrote down the constitutive relation, including that the material is isotropic. Luckily, in two dimensions most particles form a triangular lattice in the solid phase, and the (elastic) properties of two-dimensional triangular lattices are in fact isotropic.

So what is the strain field associated with such a defect? For a full calculation, we have to take this energy functional (2.64) and solve the two-dimensional partial differential equation for the strain field obtained by minimizing the energy, and then substitute that strain field into the elastic energy expression to determine the energy of the defect. Luckily, for the argument we present below we only need the strain behavior far away from the defect. This allows us to bypass solving the full problem by using essentially dimensional analysis plus the very definition of the Burgers vector of the dislocation.

If we integrate the displacement around a loop encircling the defect, we have as illustrated in the top panel of figure 2.36

$$\oint \mathrm{d}u_i = \vec{b} = \oint \frac{\partial u_i}{\partial x_j}\mathrm{d}x_j.\tag{2.65}$$

What does this equation tell us? Note that this line integral is the perimeter of the loop, which goes as $\sim r$, times the strain. This

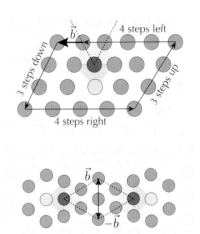

Figure 2.36. Upper panel: construction of the Burgers vector (black arrow) for a dislocation in a triangular lattice. The two dashed lines indicate two possible ways of including an extra row of atoms to create this type of dislocation. Note at the core of the dislocation there is a pair of a fivefold and a sevenfold coordinated lattice point. Lower panel: a pair of dislocations with opposite Burgers vectors indicated by the black arrows. Note the extra rows of atoms extend only in the space between the two dislocations indicated by dashed lines. Bringing the dislocations closer limits the resulting elastic distortion. Hence dislocations with opposite Burgers vectors experience an attractive force.

immediately implies that the strain $\partial u_i / \partial x_j$ must fall off as $\sim 1/r$. In particular,

$$|u_{ij}| \propto \frac{b}{r}. \tag{2.66}$$

To go from this scaling to an energy, we use this in the expression (2.64) for the elastic energy to obtain schematically

$$E_{\text{elas}} \simeq c\, E_Y \int_a^R \frac{b^2}{r^2} r\, \mathrm{d}r = c\, E_Y\, b^2 \ln\left(R/a\right), \tag{2.67}$$

where as before E_Y is the Young's modulus and c is a constant of order unity that depends on the Poisson ratio ν (and thus on the ratio of the Lamé coefficients) and the integrated angular dependences of u_{ij}. Note that we have picked integration limits to ensure that the energy does not diverge at short and long distances. The natural choices here are readily given by the system: at short scales, our continuum elastic description is cut off at the length scale of the lattice spacing, A, while at long scales, we use our system size, R.

Even with these cutoffs, notice that the elastic energy does indeed diverge in the large system limit $R/a \to \infty$, making the presence of dislocations unfavorable unless one applies mechanical strains that force them in. If we do this calculation more carefully, we find that the constant of order unity is $c = 1/8\pi$, so we have for the leading $\ln R/a$

$$E_{\text{elas}} = \frac{E_Y}{8\pi} b^2 \ln\left(R/a\right). \tag{2.68}$$

Let us now turn temperature on, which means that we need to consider the *free energy*

$$F = U - TS, \qquad U = E_{\text{elas}}, \tag{2.69}$$

which accounts for elastic as well as entropic contributions. To estimate the entropy, S, we need the number of microstates $N_{\text{microstates}}$ available for the defect. If we have a dislocation in a lattice with n^2 sites, lattice spacing A, and size R, we have

$$N_{\text{microstates}} \simeq \left(\frac{R}{a}\right)^2, \tag{2.70}$$

up to factors of order unity. Then

$$S \simeq k_B \ln N_{\text{microstates}} \longrightarrow TS \approx 2k_B T \ln(R/a). \tag{2.71}$$

Note from comparing (2.68) and (2.71) that the same divergent factor appears in both U and TS so that the free energy becomes

$$F = \left(\frac{E_Y b^2}{8\pi} - 2k_B T\right) \ln\left(\frac{R}{a}\right). \tag{2.72}$$

Of course, there are modifications to the $1/r$ dependence (2.66) close to the defect; these give corrections to (2.67), but these do not alter the leading $\ln R/a$ divergence with R, and therefore they do not alter the argument below.

In problem 2.8 different types of defects in two-dimensional triangular crystal lattices are analyzed in detail.

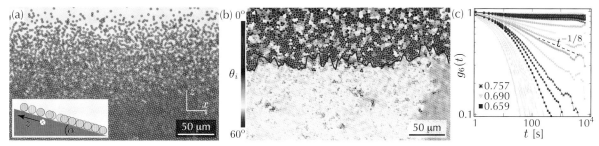

Figure 2.37. Experiment of Thorneywork et al., 2017 probing the details of the transitions between the solid, hexatic and liquid phases in a two-dimensional layer of hard core colloidal particles of diameter 2.79 μm. Do note the extremely precise data and the fact that the correlation function in panel c is followed over four orders of magnitude in time! (a) Image and illustration of the sample. As sketched in the inset, the sample is slightly slanted under a variable angle α ($\alpha = 0.56°$ in the photo), resulting in a small density gradient due to gravity. (b) Color plot of the orientational order parameter of each particle for a situation in which the lower part of the sample is hexatic, as illustrated by the almost uniformly yellow color, and the upper liquid part where the orientational order is gone. The dark line indicates the sharp interface resulting from the liquid-hexatic interface. The fluctuations of this interface are so-called capillary waves, and from detailed fits of these waves, the capillary length ℓ_c defined in (1.63) is inferred; this provides an additional check on the density difference between the coexisting hexatic and liquid phase. (c) Plot of the sixfold bond orientational order parameter g_6 as a function of time for volume fractions ranging from 0.659 (blue, liquid phase) to 0.757 (red, solid phase). While most earlier work on two-dimensional layers is focused on the spatial correlations of the bond angles in the various phases, in this experiment the temporal correlations are analyzed: $g_6(t)$ probes for each particle the time dependence of the bond angle correlation with neighboring particles.[47] In the solid, when the packing stays almost perfectly triangular, each particle has six neighbors and bonds make close to a 60-degree angle relative to one another, so g_6 is long ranged and close to 1. In the liquid phase (gray curves), g_6 decays exponentially to zero, while the hexatic phase is characterized by a power law decay $g_6(t) \sim t^{-\eta_6/2}$, with the exponent $0 < \eta_6 < 1/4$. At intermediate values of the packing fraction (orange to light green curves) this power law decay is indeed observed, showing that the phase is hexatic. In the experiment, such data are combined with various other detailed measurements to show that the transition from solid to hexatic is continuous, while the transition from hexatic to liquid is first order, with a hexatic-liquid coexistence region in ϕ ranging from 0.68 to 0.70. Images courtesy of Thorneywork and Dullen.[48]

This means that F changes sign at the critical temperature

$$T_c = \frac{E_Y b^2}{16\pi k_B}. \tag{2.73}$$

Below T_c, the free energy is enhanced by the presence of the defect. Above T_c, however, the entropic contribution to the free energy overcomes the strain energy, giving rise to a plasma of dislocations that are responsible for the loss of translational order—recall that dislocations disrupt perfect lattices by introducing extra rows of atoms. This mechanism is usually referred to as the Kosterlitz-Thouless (or topological) phase transition.

While our discussion was done in the context of a single dislocation, a full theory includes interactions of defects and their nucleation. In fact, defects are usually created by unbinding pairs of opposite Burgers vectors which experience an attractive force like oppositely charged particles, as illustrated in figure 2.36. Interestingly, the expression for T_c in the full theory is still of the simple form (2.73), albeit with renormalized elastic constants.[49] Moreover, an intermediate so-called hexatic phase intervenes between the solid and the liquid, in which the translation order of the solid is already lost, but bond orientational order still exists. The transition from the hexatic to the liquid phase is associated with the unbinding of

The discoverers of the Kosterlitz-Thouless transition received the Nobel Prize (together with Haldane) in 2016 "for theoretical discoveries of topological phase transitions and topological phases of matter." We follow the main steps of their renormalization group analysis in section 6.10.1. The application of the theory to two-dimensional melting of crystals was in particular developed further by Halperin and Nelson, and by Young, who showed that melting in two dimensions is a two-step process via an intermediate hexatic phase. In the broader context the theory is therefore often recognized as the KTHNY-theory, or even, in recognition of Berezinskii's early contributions, the BKTHNY-theory (Ryzhov et al., 2017).

the sevenfold and fivefold coordinated lattice points that together form the core of the defects, as shown in figure 2.36.

The two-dimensional melting transitions are subtle, and it took some years before unequivocal evidence for the intermediate hexatic phase was obtained in computer simulations or experiments. From the start, soft matter has played an important role in probing the predicted transitions experimentally. Experiments on smectic liquid crystals and colloidal layers already strongly supported the scenarios soon after the theory was developed,[50] while modern observation and preparation techniques nowadays allow us to prepare large colloidal systems and to probe correlation functions—even dynamical ones—that were heretofore not accessible. Figure 2.37 illustrates this with some of the detailed data from one of the most sophisticated colloidal experiments to date. Such data illustrate the power of contemporary soft matter to probe fundamental physics using modern visualization techniques and data processing. We will illustrate in figure 4.14 how colloids can be used to study the glassy or plastic behavior of solids.

2.12 Topological mechanics

Topology is a branch of mathematics that describes the properties of objects that are preserved under continuous deformations of their shapes. As an example, consider the smooth deformation of a doughnut into a mug (with a handle). While the shape of the object changes during this transformation, the number of handles is preserved provided no cuts are introduced. This integer number, the so-called genus illustrated in figure 7.4, is an example of a topological invariant. An important application of the theory of topological invariants is the characterization of topological defects like the dislocations studied in the last section. The presence of such defects is said to be topologically protected because an isolated dislocation cannot be removed by continuous deformations of the underlying lattice structure.

A similar mathematical mechanism ensures the robustness of so-called edge modes in topological insulators, a class of materials first explored in the context of electron physics. In recent years, these edge modes have been identified in suitably designed mechanical structures. Like their optical and electronic counterparts, topological mechanical systems can exhibit classical states such as vibrations, free motions or load-bearing states, that owe their existence to topological invariants.[51] These invariants are integers that are preserved under continuous changes in the physical parameters of the material (rather than under continuous deformations of the underlying structure).

Actually, the analysis of capillary waves in figure 2.37.c is a rare but nice example of the clever use of the analysis of interfacial properties to extract information about the bulk. Another interesting soft matter example of using interface properties to probe bulk behavior is the analysis of the dynamical properties of the interface between the nematic and smectic-A phases of liquid crystals (Anisimov et al., 1990) to show that the transition is weakly first order, as was predicted theoretically.

For an introduction to and overview of topological insulators, see Fruchart and Carpentier, 2013; Hasan and Kane, 2010; and Shankar et al., 2022.

2.12.1 Topological waves

We start with mechanical media that display 'topologically protected' waves like the lattice of coupled gyroscopes in figure 2.38.

The robustness of these waves can be traced to features of their band diagram. In general, a band diagram is a plot of frequency versus wavevector that describes the frequency range at which waves (e.g., sound modes or equivalently phonons) are allowed to propagate, along with the way the system vibrates at a given frequency. The phononic band diagram of the gyroscope lattice exhibits band gaps, frequency intervals in which the mechanical waves cannot propagate in the bulk of the material. But this obstruction does not necessarily exist in a finite sample. Right in the band gaps, we find a special class of waves, called chiral edge modes, that can propagate unidirectionally only at the edge of the sample. These phononic edge states are topologically protected: they will not be scattered even if the shape of the boundary is modified, e.g. by introducing sharp corners. Nonetheless, these robust systems do retain an element of tunability: upon changing geometric parameters of the lattice one can control the penetration depth of such phononic edge states. This paradigmatic example raises an important question: where does the topological robustness come from?

The answer lies in a principle known as the bulk-edge correspondence. The phononic eigenvectors and eigenvalues of the gyroscope lattice calculated with periodic boundary conditions (i.e., in the bulk) can be used to define an integer valued topological invariant called a Chern number, as long as the system remains gapped. Since the invariant is an integer, it cannot smoothly change across space and in particular not at the sample's edge that separates a topological material from the outside. Instead what happens is that the gap "locally closes" at the edge (so that the Chern numbers are no longer defined). As a result of the topological nature of the bulk-band structure, gapless edge modes appear in a finite system no matter what the shape of the boundary is. In the gyroscopic lattice, the gap arises from a combination of the geometry of the underlying structure and broken time-reversal symmetry: the gyroscopes all spin with one chirality that is only inverted by flipping the arrow of time. As a result, the wave propagation at the edge will be chiral (or unidirectional) and immune to backscattering because the edge modes can only move forward.

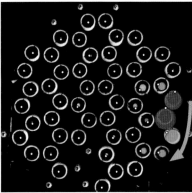

Figure 2.38. Upper panel: honeycomb lattice of coupled suspended gyroscopes, photographed from below, in the experiments by Nash et al., 2015. The bottom image shows the gyroscopes with their phase angle indicated with colors. Note that upon touching a gyroscope at the edge a vibrational mode starts propagating clockwise along the boundary of the sample. Image courtesy of Lisa Nash, Dustin Kleckner, and William Irvine.

For an example of a topological number see our discussion of topological defects in liquid crystals in section 6.6.1, where we discuss the topological invariant characterizing those defects.

2.12.2 Topological zero-energy modes

Topologically protected mechanical states need not be waves that exist at finite frequency. They can also be zero-energy (or equivalently zero-frequency) modes, i.e. deformations of a mechanical

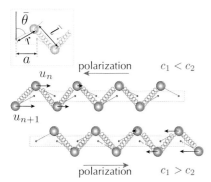

Figure 2.39. Mechanical structure of masses connected with springs that harbors a topological zero-energy mode localized either on the left (top panel) or on the right (bottom panel) (B. G. Chen et al., 2014; Kane and Lubensky, 2014). The projection of the displacement on the x-axis is labeled by u_n where n labels the nth mass. See problem 2.9 for a more detailed analysis of this structure and its edge modes.

Redundant constraints are tensions or compressions applied to the bonds that do not result in net forces on the nodes. Note that we already introduced the Maxwell counting argument in section 2.10.2 in the context of jamming.

structure that do not entail an increase in its elastic potential energy. One of the simplest examples is illustrated schematically in figure 2.39, which shows a structure composed of a chain of rigid rotors with masses attached to their end and connected by springs. Note that this structure displays a peculiar breaking of inversion symmetry: it is not invariant under reflection. Like many mechanical structures, this chain can be viewed as a network composed of N_S sites (the point masses) connected by N_b central force bonds (the springs and rigid rotors).

To count the number of zero modes we use again the simple criterion invented by Maxwell.[52] For a generic structure in d dimensions, the number of degrees of freedom associated with the coordinates is given by dN_S, and the total number of constraints is simply the number of bonds N_b. The number of zero-energy modes n_m is then given by the relation

$$dN_S - N_b = n_m - n_{ss}, \qquad (2.74)$$

where n_{ss} denotes the number of redundant constraints. The redundant constraints should not have any effect in reducing the total number of degrees of freedom dN_S. In other words, the effective number of bonds is $N_b - n_{ss}$. The redundant constraints are technically known as states of self-stress. If we apply the Maxwell criterion to the structure in figure 2.39 we find that there is one zero mode.[53]

Where is this zero-energy mode located in the chain? The answer depends on the chain configuration. The condition for a zero-energy mode is that the springs are neither stretched nor compressed. One can prove that the zero-energy deformation satisfies the linearized constraint equation

$$c_2 u_{n+1} - c_1 u_n = 0. \qquad (2.75)$$

where u_n is the horizontal displacement of the nth node connecting adjacent bars and springs, while the structural parameters c_1 and c_2 are geometrical projection factors which depend on the geometry of the unit cell.[54] Upon solving this equation we obtain

$$u_{n+1} = \frac{c_1}{c_2} u_n. \qquad (2.76)$$

If c_1/c_2 is greater (less) than 1, the displacements are amplified (reduced), respectively. This means that the zero-energy mode is localized at the right edge of the chain for $c_1 > c_2$ or at the left edge for $c_1 < c_2$, depending on which of the two directions the rotors lean. The breaking of left-right symmetry in the unit cell is captured by a polarization P whose orientation depends on whether the rotors point to the right or the left. The two cases are illustrated in figure 2.39.

The same principle has been used to (i) create topological origami that fold more easily from one side; (ii) elucidate how topological defects such as dislocations can localize topological zero-energy modes and states of stress; or (iii) demonstrate topological control of material failure in three-dimensional cellular metamaterials.

See the review by Bertoldi et al., 2017 for further discussion of these issues.

2.13 What have we learned

Elasticity theory is based on the linear relation between stress and strain in an elastic medium. For most soft matter which is isotropic on the coarse-grained scale, this linear relation is characterized by two elastic constants, a compression modulus and a shear modulus; their ratio determines the Poisson ratio. Most materials have a positive Poisson ratio, which expresses that when stretched, they contract in the transverse directions.

Bending is the low-energy deformation mode of thin rods and sheets. When thin rods are compressed, they buckle at some well-defined threshold value of the compression force. The elastic description of thin rods also provides the basis for the continuum formulation of stiff polymers.

The response of collections of almost incompressible particles pressed together only becomes elastic on relatively long length scales. The jamming approach shows how up to some crossover length scale the static response is strongly disordered and characterized by an excess of low-energy modes. This is reflected in an excess of low-frequency modes in the dynamical response of packings of particles.

Modern (3D printing) techniques nowadays allow us to build structures of soft matter which behave as elastic metamaterials or programmable matter. Examples are auxetics, materials with a negative Poisson ratio, or materials which deform strongly nonlinearly and hysteretically, or whose mesoscale response to a global strain deformation is programmable. At the same time soft matter throws new light on long-standing fundamental questions concerning the elastic behavior of solids, thanks to the fact that the positions of all individual particles in a colloid can be tracked with incredible precision.

Topological concepts also throw new light on the mechanical behavior of materials.

2.14 Box 2: Summary of Landau theory

Box 2. Summary of Landau theory of phase transitions

▶ The Landau theory of phase transitions[55] is formulated in terms of a free energy for a coarse-grained order parameter. For the simplest case where the order parameter ϕ is a scalar and where the system has $\pm\phi$ symmetry, as when ϕ describes the magnetization of Ising spins, the coarse-grained free energy F reads

$$F = \int d^d\vec{r}\,\mathcal{F} = \int d^d\vec{r}\,\left[\tfrac{1}{2}\xi_0^2\,(\vec{\nabla}\phi)^2 + \tfrac{1}{2}r(T)\phi^2 + \tfrac{1}{4}u\phi^4\right]. \tag{2.77}$$

▶ Landau theory assumes $u > 0$ and a smooth temperature dependence of u and r. In particular, it is assumed that r changes sign smoothly at some temperature T_c, so that $r(T) = r'(T - T_c)$. It is customary to assume $r' > 0$ so that the high-temperature phase is the disordered phase. The temperature dependence of u is not important and is therefore ignored.

▶ Consider first homogeneous phases where ϕ is space-independent. The free energy density then becomes $\mathcal{F} = \tfrac{1}{2}r'(T - T_c)\phi^2 + \tfrac{1}{4}u\phi^4$. Its behavior is sketched in figure 2.19: it has a single minimum at $\phi_{\min} = 0$ for $T > T_c$ and double well structure for $T < T_c$. T_c thus marks the transition temperature from a disordered phase to an ordered phase with nonzero order parameter $\phi_{\min}$ corresponding to the minimum.

▶ Below the critical temperature T_c the order parameter corresponding to the minimum of the free energy is

$$\phi_{\min} = \left[r'(T_c - T)/u\right]^{1/2} \sim (T_c - T)^{1/2} \qquad (T < T_c). \tag{2.78}$$

▶ In critical phenomena the relevant quantities typically have a power law scaling with distance from T_c, for instance the order parameter goes as $|T_c - T|^\beta$ where β is called a critical exponent. The above result (2.78) shows that $\beta = 1/2$ in Landau theory.

▶ If the order parameter is not a conserved quantity, the associated time-dependent equation is[56]

$$\tau_0 \frac{\partial \phi}{\partial t} = -\frac{\delta F}{\delta \phi} = \xi_0^2 \nabla^2 \phi - r'(T - T_c)\phi - u\phi^3. \tag{2.79}$$

▶ Small fluctuations of the order parameter about $\phi = 0$ in the disordered state for $T > T_c$ can be studied by linearizing the above equation about $\phi = 0$ to get $\tau_0 \partial\phi/\partial t = \xi_0^2 \nabla^2 \phi - r'(T - T_c)\phi$. According to this equation, fluctuations decay over a length scale ξ (the correlation length) and time scale τ given by

$$\xi = \frac{\xi_0}{[r'(T - T_c)]^{1/2}} \sim \frac{1}{(T - T_c)^{1/2}}, \qquad \tau = \frac{\tau_0}{r'(T - T_c)} \sim \frac{1}{T - T_c}, \qquad (T > T_c). \tag{2.80}$$

So the length and time scale diverge with characteristic power law behavior when approaching T_c.[57] By linearizing about $\phi_{\min}$ one gets the same expressions for $T < T_c$, with r' replaced by $2r'$ (cf. problem 6.3).

▶ Landau theory is essentially a mean-field theory of phase transitions, and the power law behaviors (2.78) and (2.80) define the mean-field exponents. Field-theoretic Renormalization Group Theory is based on starting from the coarse-grained F and integrating out more and more short-length scale fluctuations[58] (see problems 4.11 and 4.12).

▶ A cubic term ϕ^3 in $\mathcal{F}$ breaks the $\pm\phi$ symmetry and leads to a (weakly) first order transition. An example where this happens is the isotropic-to-nematic transition in liquid crystals, as discussed in section 6.2.

▶ A complex version of (2.79) emerges as the Ginzburg-Landau equation of superconductivity and as the amplitude equation for stationary patterns in non-equilibrium systems—see section 8.5.2.

2.15 Problems

Relevant coding problems and solutions for this chapter can be found on the book's website www.softmatterbook.online under Chapter 2/Coding problems.

Problem 2.1 *Derivation of the equations for the Maxwell model and the Kelvin-Voigt model, and extension to the Standard Linear Solid Model.*

In this exercise we derive equations (2.22) and (2.23) for the Maxwell model and the Kelvin-Voigt model.

    a. In the caption of figure 2.11, it is explained what the schematic diagrams of the two models mean. Implement these equations and show that (2.22) and (2.23) directly follow from them.

    b. Evaluate the storage modulus $G'(\omega)$ and loss modulus $G''(\omega)$ and plot these on a log-log scale.

Problem 2.2 *Derivation of equation (2.34) and equation (2.35) for a sheet bent in one direction*

In this problem we back up some of the steps of section 2.7.

    a. Argue for the situation sketched in figure 2.13 that for a sheet bent in the x direction only, the fact that there is no external pressure applied to the surface of the sheet implies that at the upper and lower side $\sigma_{yi} = 0$ for $i = x, y, z$.

    b. Argue that since in particular $\sigma_{yy} = 0$ at the upper and lower sides, to lowest order in the curvature we must have $\sigma_{yy} = 0$ everywhere within the sheet.

    c. Use the linear stress-strain equation (2.6) for $\sigma_{yy} = 0$ to derive equation (2.34).

Hint: It is easiest to express K and μ immediately in terms of E_Y and ν using equation (2.19).

    d. Show with these results that $\sigma_{xx} = E_Y y/(1-\nu^2)R$ and that $\sigma_{ij} u_{ij} = E_Y y^2/(1-\nu^2)R^2$, as is used in equation (2.35).

Problem 2.3 *Cantilever rod with transverse tip load in the small deflection approximation*

The following problem is relevant to the tip of an *atomic force microscope*: a thin rod clamped at $x = 0$ and parallel to the x-axis, bent due to a perpendicular force $F^{\text{ext}} \, \hat{y}$ direction at $x = L$. We leave out the gravitational force (you can practice by including this yourself).

Horizontal clamping at $x = 0$ implies the boundary conditions $y(0) = 0, \quad y'(0) = 0$.

    a. Argue that for the cantilever tip, i.e., at $x = L$, we have L we have the boundary conditions $\vec{F}_{\text{el}}(L) = -F^{\text{ext}} \, \hat{y}$, and $M_{\text{el}}(L) = 0$, so that $y''(L) = 0$.

b. Show that $\vec{F}_{el}$ is constant along the cantilever and hence that $\vec{F}_{el}(x) = -F^{ext}\,\hat{y}$.

c. Show that our basic equation (2.56) now reduces to $y'''' = 0$.

d. Seek a general solution in the form of a third order polynomial $y(x) = a_0 + a_1\,x + a_2\,x^2 + a_3\,x^3$. Show that the boundary conditions imply $a_0 = a_1 = 0$.

e. Apply the boundary condition for the vanishing torque at the end to express a_2 in terms of a_3.

f. Show that the torque balance equation (2.53) applied to the rod element at the very tip implies in the weak deflection approximation $B\,y'''(L) - F^{ext} = 0$.

g. Apply this result to show that $a_3 = F^{ext}/(6B)$, with the final result

$$y(x) = \frac{F^{ext}}{6B}\,x^2(x - 3L). \tag{2.81}$$

h. To analyze the case drawn by da Vinci shown in figure 2.21, we take the center of the rod at $x = 0$; the total rod length is L. Argue that the starting equation (2.56) for this case reduces to $By'''' + W\delta(x) = 0$, where W is the weight attached at $x = 0$. Show that this gives the condition $By'''(0^+) - By'''(0^+) = -W$, while lower order derivatives should be continuous at $x = 0$.

i. Show that you can solve the equation with the polynomial ansatz introduced in step d, with a_3^+ for $x > 0$ and $a_3^- = -a_3^+$ for $x < 0$, and with $a_1 = 0$.

j. Solve the equation with the boundary condition $y''(\pm L/2) = 0$ derived in step a, and show that the total downward deflection of the center of the rod equals

$$y(0) - y(\pm L/2) = -\frac{WL^3}{48B}. \tag{2.82}$$

Problem 2.4 *Onset of buckling of a rod clamped on one side*

In this problem we analyze the buckling threshold for a rod clamped on one side. We will consider the particular case where the rod is clamped at $x = 0$, meaning that its direction is fixed there, while it is free to pivot at L (so there is not torque). Hence the boundary conditions are $y(0) = 0, y'(0) = 0, y''(L) = 0$. As in the example discussed in section 2.8.4 there is an external compression force F^{ext} in the x direction.

a. Show that in the small deflection approximation the shape $y(x)$ obeys the differential equation (2.57).

b. Show that in order for $y(x)$ to satisfy, in the small deflection limit, the torque balance (2.53) (from which our basic equation

(2.56) was obtained by differentiation) we need to have

$$B\,y''' + F^{\text{ext}}\,y' = 0. \tag{2.83}$$

c. Integrate the above equation once, and show that the resulting equation is of the form of the harmonic oscillator equation and that the integration constant obtained in the last step can be thought of as a shift of the equilibrium position of the harmonic oscillator. What is the role of x in the harmonic oscillator analogy?

d. Formulate the present buckling problem in analogy with the harmonic oscillator where the original boundary conditions play the role of two initial conditions and a condition for the behavior at 'time' $x = L$.

e. Write the solution $y(x)$ as a sum of a $\sin(kx)$ and $\cos(kx)$ term. Give the relation between k and the parameters in the equation.

f. Apply the boundary conditions at $x = 0$ and $x = L$ to determine the constants in the solution obtained in the previous step. Show that nontrivial (buckled) solutions are possible when

$$kL = (1 + 2n)\,\pi/2 \quad (n = 0, 1, 2, \ldots), \tag{2.84}$$

and that this translates into critical force values

$$F_{\text{c},n}^{\text{ext}} = \frac{(1 + 2n)^2\,\pi^2 B}{4L^2}. \tag{2.85}$$

g. Argue that $F_{\text{c},0}^{\text{ext}}$ for $n = 0$ will in practice be the true critical value for buckling. Why is this critical value different from the one given in equation (2.47)? What do the values $F_{\text{c},n}^{\text{ext}}$ with $n \neq 0$ correspond to?

Problem 2.5 *A dynamical equation for a microtubule or flagellum*
Microtubules and flagella can be modeled as thin rods with a bending stiffness B. We use the result (2.56) for the y deflection of a flagellum oriented along the x-axis to get a simple approximation for its oscillations.

a. Argue that you can approximate the hydrodynamic friction experienced by the flagellum by a term $f_y^{\text{ext}} = -\zeta \partial y / \partial t$ in equation (2.57) and that $F_{\text{el},x} = 0$ in this equation.[59]

b. Consider an oscillating flagellum in a fluid, such that $y(x = 0) = A \cos(\omega t)$. Analyze the decay of the oscillations along the flagellum in the linear approximation from the equation derived in step a.

Hint: Assume $y(x, t) = A e^{ikx - i\omega t}$; see the discussion on complexification of the equation in section 1.7.2.

c. For a given frequency there are four different solutions for the complex wavenumber k. Which two are relevant in the case of a long flagellum?

d. Show that the results imply a characteristic relation between the wavelength of the oscillations along the flagellum and the decay length of the amplitude of the oscillations.

Problem 2.6 *Derivation of the equation for sound waves in solids*
Substitute the stress-strain relations (2.6) for an isotropic lattice into the dynamical equation (2.24), and fill in the steps to derive the wave (2.25) for sound propagation in solids.

Problem 2.7 *Elastic moduli from coarse-graining*
In this problem, we ask this question: how do you determine the macroscopic elastic properties of a material, such as its Lamé coefficients, if you are given a microscopic model of the material? For simple models, such as a network of masses connected by springs, the coarse-graining procedure can be carried out analytically.

We consider what is potentially the simplest model solid: a two-dimensional network of masses connected by springs. We will study both the *triangular lattice* shown at the top of figure 2.40, for which the unit cell has only one type of particle, as well as the *honeycomb lattice* shown at the bottom. In both cases, only nearest neighbor sites are connected by springs. Since the triangular case is the simplest, we will first analyze this case. After step i, we turn to the honeycomb case, which is more complicated as it has two sublattices (red and blue in the figure) and, correspondingly, a unit cell with two atoms.

Before deformation, each particle n sits at an equilibrium position $\vec{x}_n$, which we think of as labeling the particle. After deformation, each particle has moved to a new position $\vec{u}(\vec{x}_n) + \vec{x}_n$, where $\vec{u}(\vec{x}_n)$ is the displacement field. Each of the nearest neighbors are connected by Hookean springs with spring stiffness k which have a rest length A. To help navigate the lattice, we will introduce the three nearest neighbor bond vectors

$$\vec{a}^{(1)} = a \begin{pmatrix} \sqrt{3}/2 \\ 1/2 \end{pmatrix}, \quad \vec{a}^{(2)} = a \begin{pmatrix} -\sqrt{3}/2 \\ 1/2 \end{pmatrix}, \quad \vec{a}^{(3)} = a \begin{pmatrix} 0 \\ -1 \end{pmatrix},$$
(2.86)

which point between particles when the lattice is undeformed. See figure 2.40. For ease of notation, we will from now on suppress the index n on $\vec{x}_n$.

Triangular lattice
a. First, convince yourself that the spring force on the particle at $\vec{x}$ by its neighbor at $\vec{x} + \vec{a}^{(\alpha)}$ is given by

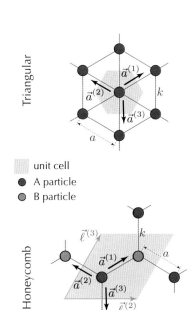

Triangular

unit cell
A particle
B particle

Honeycomb

Figure 2.40. A triangular lattice and honeycomb lattice of Hookean springs that are coarse-grained in problem 2.7 to derive the elastic constants of continuum theory.

$$\vec{F}(\vec{x}, \vec{x} + \vec{a}^{(\alpha)}) = k\left(\left|\delta\vec{u} + a^{(\alpha)}\right| - a\right)\frac{\delta\vec{u} + \vec{a}^{(\alpha)}}{\left|\delta\vec{u} + \vec{a}^{(\alpha)}\right|}, \qquad (2.87)$$

where $\delta\vec{u} = \vec{u}(\vec{x} + \vec{a}^{(\alpha)}) - \vec{u}(\vec{x})$.

Hint: The first factor sets the magnitude and the second factor sets the direction of the force.

b. Linear elasticity theory describes solids under mild deformation. Mathematically, mild deformation means that relative displacements between neighbors are small compared to their equilibrium spacings, i.e., $|\delta\vec{u}| \ll a$. To this end, show that $\vec{F}(\vec{x}, \vec{x} + \vec{a}^{(\alpha)})$ to linear order in $\delta\vec{u}$ is given by

$$\vec{F}(\vec{x}, \vec{x} + \vec{a}^{(\alpha)}) \approx k\left(\hat{a}^{(\alpha)} \cdot \delta\vec{u}\right)\hat{a}^{(\alpha)}, \qquad (2.88)$$

where $\hat{a}^{(\alpha)} = \vec{a}^{(\alpha)}/a$.

c. Next, convince yourself that the total force on the particle at $\vec{x}$ is given by

$$\vec{F}(\vec{x}) = \sum_{\alpha=1}^{3}\left[\vec{F}(\vec{x}, \vec{x} + \vec{a}^{(\alpha)}) + \vec{F}(\vec{x}, \vec{x} - \vec{a}^{(\alpha)})\right], \qquad (2.89)$$

and substitute the linearized force (2.88) into equation (2.89) to obtain

$$\vec{F}(\vec{x}) = k\sum_{\alpha=1}^{3}\hat{a}^{(\alpha)}\hat{a}^{(\alpha)} \cdot \left[\vec{u}(\vec{x} + \vec{a}^{(\alpha)}) + \vec{u}(\vec{x} - \vec{a}^{(\alpha)}) - 2\vec{u}(\vec{x})\right].$$
$$(2.90)$$

d. In the previous step, we have made only a small displacement approximation. To derive macroscopic properties, we will also need to make a small gradient approximation, i.e., that $\vec{u}(\vec{x})$ changes slowly in space. To do this, we will write $\vec{u}(\vec{x})$ as a Taylor series,

$$u_m(\vec{x} + \vec{a}^{(\alpha)}) = u_m(\vec{x}) + a_j^{(\alpha)}\partial_j u_m\bigg|_{\vec{x}} + \frac{1}{2}a_j^{(\alpha)}a_n^{(\alpha)}\partial_j\partial_n u_m\bigg|_{\vec{x}} + \dots$$
$$(2.91)$$

and we will keep only the lowest gradients possible. Substitute equation (2.91) into equation (2.90), and show that the total force is approximately

$$F_i(\vec{x}) \approx ka^2\sum_{\alpha=1}^{3}\hat{a}_i^{(\alpha)}\hat{a}_j^{(\alpha)}\hat{a}_m^{(\alpha)}\hat{a}_n^{(\alpha)}\partial_j\partial_n u_m\bigg|_{\vec{x}}. \qquad (2.92)$$

e. Next, we convert this force-displacement relationship into a stress-strain relationship. Recall that $\partial_j\sigma_{ij} = f_i$, where $\vec{f} = \vec{F}/A$

is the force per unit area and $A = \sqrt{3}a^2/2$ is the area of a single unit cell in the triangular lattice. Show that equation (2.92) can be rewritten as

$$\partial_j \sigma_{ij} = \partial_j \left[K_{ijmn} \partial_m u_n \right],\tag{2.93}$$

where

$$K_{ijmn} = \frac{2k}{\sqrt{3}} \sum_{\alpha=1}^{3} \hat{a}_i^{(\alpha)} \hat{a}_j^{(\alpha)} \hat{a}_m^{(\alpha)} \hat{a}_n^{(\alpha)}\tag{2.94}$$

is an object known as the elastic modulus tensor. It encodes macroscopic properties, such as the Lamé coefficients, that relate stress to strain.

f. Verify explicitly the components of the elastic modulus tensor:

$$
\begin{aligned}
K_{xxxx} &= K_{yyyy} = \frac{3\sqrt{3}k}{4}, \\[4pt]
K_{xxyy} &= K_{yyxx} = \frac{\sqrt{3}k}{4}, \\[4pt]
K_{xyxy} &= K_{yxyx} = K_{xyyx} = K_{yxxy} = \frac{\sqrt{3}k}{4}, \\[4pt]
K_{xxxy} &= K_{xxyx} = K_{xyxx} = K_{yxxx} = 0, \\[4pt]
K_{yyyx} &= K_{yyxy} = K_{yxyy} = K_{xyyy} = 0.
\end{aligned}
\tag{2.95}
$$

Using this, check that K_{ijmn} can be written in the form

$$K_{ijmn} = \lambda \delta_{ij}\delta_{mn} + \mu(\delta_{im}\delta_{jn} + \delta_{in}\delta_{jm}),\tag{2.96}$$

where λ and μ are the Lamé coefficients. Verify that $\lambda = \mu = \sqrt{3}k/4$ for the triangular lattice.

g. There is another way of deriving the elastic modulus tensor from microscopics, which will be generalized when we analyze the honeycomb lattice. Show that equation (2.90) can be written as

$$F_i(\vec{x}) = -\sum_{\vec{x}'} D_{ij}(\vec{x} - \vec{x}')u_j(\vec{x}'),\tag{2.97}$$

where

$$D_{ij}(\vec{x}) = -k \sum_{\alpha=1}^{3} \hat{a}_i^{(\alpha)} \hat{a}_j^{(\alpha)} \left(\delta(\vec{x} + \vec{a}^{(\alpha)}) + \delta(\vec{x} - \vec{a}^{(\alpha)}) - 2\delta(\vec{x}) \right)\tag{2.98}$$

is the dynamical matrix.

h. The Fourier transform of the dynamical matrix is defined as

$$D_{ij}(\vec{q}) = \sum_{\vec{x}} e^{-i\vec{q}\cdot\vec{x}} D_{ij}(\vec{x}).\tag{2.99}$$

Note that the obtained elastic tensor is fully isotropic: the lattice axes did not introduce preferred directions for elasticity. This is not an artifact of the spring model, but holds in general as a consequence of the threefold rotational symmetry of the lattice in two dimensions. Hence you will find the same feature in the honeycomb case too.

Also note that the expression for Lamé coefficients in terms of K and μ given in (2.8) is valid only in three dimensions. Check for yourself that in two dimensions $\lambda = K - \mu$.

Show that

$$D_{ij}(\vec{q}) = 2k \sum_{\alpha=1}^{3} \hat{a}_i^{(\alpha)} \hat{a}_j^{(\alpha)} \left[1 - \cos(\vec{a}^{(\alpha)} \cdot \vec{q}) \right], \qquad (2.100)$$

and verify that

$$K_{ijmn} = \frac{1}{2A} \frac{\partial D_{im}}{\partial q_j \partial q_n} \bigg|_{\vec{q}=\vec{0}}. \qquad (2.101)$$

i. Commonly, when in two dimensions, one recasts the stiffness tensor and equation (2.93) in the basis defined with the matrices

$$\tau^0 = \begin{pmatrix} 1 & 0 \\ 0 & 1 \end{pmatrix}, \qquad \tau^1 = \begin{pmatrix} 0 & -1 \\ 1 & 0 \end{pmatrix},$$

$$\tau^2 = \begin{pmatrix} 1 & 0 \\ 0 & -1 \end{pmatrix}, \qquad \tau^3 = \begin{pmatrix} 0 & 1 \\ 1 & 0 \end{pmatrix}, \qquad (2.102)$$

by transforming the stress tensor and the strain according to

$$
\begin{aligned}
\sigma^0 &= \tau_{ij}^0 \sigma_{ij} & &\text{pressure,} \\
\sigma^1 &= \tau_{ij}^1 \sigma_{ij} & &\text{torque density,} \\
\sigma^2 &= \tau_{ij}^2 \sigma_{ij} & &\text{shear stress 1,} \\
\sigma^3 &= \tau_{ij}^3 \sigma_{ij} & &\text{shear stress 2,} \\
u^0(\vec{x}) &= \tau_{ij}^0 \partial_i u_j(\vec{x}) & &\text{dilation,} \\
u^1(\vec{x}) &= \tau_{ij}^1 \partial_i u_j(\vec{x}) & &\text{rotation,} \\
u^2(\vec{x}) &= \tau_{ij}^2 \partial_i u_j(\vec{x}) & &\text{shear strain 1,} \\
u^3(\vec{x}) &= \tau_{ij}^3 \partial_i u_j(\vec{x}) & &\text{shear strain 2,}
\end{aligned}
\qquad (2.103)
$$

and by transforming the four-component stiffness tensor into the two-component stiffness tensor,

$$K^{\alpha\beta} = \frac{1}{2} K_{ijmn} \tau_{ij}^{\alpha} \tau_{mn}^{\beta}. \qquad (2.104)$$

Demonstrate that for the two-dimensional triangular lattice, the two-component stiffness tensor is given by

$$K^{\alpha\beta} = 2 \begin{pmatrix} K & 0 & 0 & 0 \\ 0 & 0 & 0 & 0 \\ 0 & 0 & \mu & 0 \\ 0 & 0 & 0 & \mu \end{pmatrix}, \qquad (2.105)$$

with $K = 2\mu = \sqrt{3}k/2$. K is the *compression* or *bulk modulus* and μ the *shear modulus* introduced in section 2.3.

Honeycomb lattice
We now turn to the generalization of (2.101), which becomes especially relevant when we have more than one particle per *unit cell*.

The lattice geometry for which we will treat this case is the honeycomb lattice, shown at the bottom in figure 2.40, together with the relevant unit cell.

To proceed, we will make use again of the lattice vectors introduced in equation (2.86). The unit cell is defined by the unit cell vectors $\vec{\ell}^{(\alpha)}$, which tell you which unit cell you enter by traversing a bond in the $\vec{a}^{(\alpha)}$ direction from the A particle, colored red in figure 2.40. In this problem, we use $\vec{x}$ to define the location of a unit cell. Therefore, $\vec{u}^\mu(\vec{x})$ denotes the displacement of the type μ particle in the unit cell given by the position $\vec{x}$.

j. Show that the unit cell vectors are given as

$$\vec{\ell}^{(\alpha)} = \vec{a}^{(1)} - \vec{a}^{(\alpha)}. \tag{2.106}$$

Note that $\vec{\ell}^{(1)} = \vec{0}$, which is not shown in figure 2.40.

k. Argue why the linearized force on particle type μ due to the displacement of particle ν given by

$$\vec{F}^\mu(\vec{x}, \vec{x} + \vec{a}^{(\alpha)}) \approx k(\hat{a}^{(\alpha)} \cdot \delta\vec{u}^{\mu\nu})\hat{a}^{(\alpha)}, \tag{2.107}$$

with $\delta\vec{u}^{\mu\nu} = \vec{u}^\nu(\vec{x} + \vec{\ell}^{(\alpha)}) - \vec{u}^\mu(\vec{x})$, must have $\mu \neq \nu$.

l. Convince yourself that the force on particle A (red) at a position $\vec{x}$, $\vec{F}^A(\vec{x})$, and the force on particle B (blue) at a position $\vec{x}$, $\vec{F}^B(\vec{x})$, are given by

$$\vec{F}^A = \sum_{\alpha=1}^{3} \vec{F}^A(\vec{x}, \vec{x} - \vec{a}^{(\alpha)}), \qquad \vec{F}^B = \sum_{\alpha=1}^{3} \vec{F}^B(\vec{x}, \vec{x} + \vec{a}^{(\alpha)}). \tag{2.108}$$

Insert equation (2.107) to find

$$\vec{F}^A = k\sum_{\alpha=1}^{3} \hat{a}^{(\alpha)} \left[\hat{a}^{(\alpha)} \cdot \left(\vec{u}^B(\vec{x} - \vec{\ell}^{(\alpha)}) - \vec{u}^A(\vec{x}) \right) \right],$$

$$\vec{F}^B = k\sum_{\alpha=1}^{3} \hat{a}^{(\alpha)} \left[\hat{a}^{(\alpha)} \cdot \left(\vec{u}^A(\vec{x} + \vec{\ell}^{(\alpha)}) - \vec{u}^B(\vec{x}) \right) \right]. \tag{2.109}$$

m. Show that equation (2.109) can be written as

$$F_i^\mu(\vec{x}) = -\sum_{\vec{x}'} D_{ij}^{\mu\nu}(\vec{x} - \vec{x}')u_j^\nu(\vec{x}'), \tag{2.110}$$

where

Here, $\varepsilon^{AB} = 1 = -\varepsilon^{BA}$ and $\varepsilon^{AA} = \varepsilon^{BB} = 0$ are components of the two-dimensional Levi-Civita tensor and $\delta^{\mu\nu}$ is the Kronecker delta.

$$D_{ij}^{\mu\nu}(\vec{x}) = k(2\delta^{\mu\nu} - 1)\sum_{\alpha=1}^{3} \hat{a}_i^{(\alpha)}\hat{a}_j^{(\alpha)}\delta(\vec{x} - \varepsilon^{\mu\nu}\vec{\ell}^{(\alpha)}). \tag{2.111}$$

n. Show that the Fourier transform of (2.111) is

$$D_{ij}^{\mu\nu}(\vec{q}) = k(2\delta^{\mu\nu} - 1) \sum_{\alpha=1}^{3} \hat{a}_i^{(\alpha)} \hat{a}_j^{(\alpha)} e^{-i\varepsilon^{\mu\nu}\vec{q}\cdot\vec{\ell}^{(\alpha)}}. \qquad (2.112)$$

Perform the sum over α to show

$$\underline{D}(\vec{q}) = \begin{pmatrix} \underline{D}^{AA}(\vec{q}) & \underline{D}^{AB}(\vec{q}) \\ \underline{D}^{BA}(\vec{q}) & \underline{D}^{BB}(\vec{q}) \end{pmatrix}, \qquad (2.113)$$

where

$$D_{ij}^{AA} = D_{ij}^{BB} = k\frac{3}{2}\delta_{ij},$$

$$D_{11}^{AB} = -\frac{3k}{4}\left(1 + e^{-i\sqrt{3}aq_x}\right),$$

$$D_{12}^{AB} = D_{21}^{AB} = -\frac{\sqrt{3}k}{4}\left(1 - e^{-i\sqrt{3}aq_x}\right), \qquad (2.114)$$

$$D_{22}^{AB} = -\frac{k}{4}\left(1 + e^{-i\sqrt{3}aq_x} + 4e^{-i\frac{\sqrt{3}a}{2}\left(\sqrt{3}q_y + q_x\right)}\right),$$

$$\underline{D}^{BA} = \left[\underline{D}^{AB}\right]^{\dagger},$$

where $\underline{A}^{\dagger}$ is the conjugate transpose of a square matrix $\underline{A}$.

o. Equation (2.113) is in the so-called particle-type basis. Its upper indices index over the two particles in the unit cell. In order to make progress, we move to a *center-of-mass* basis. Transform our force equation in momentum space to

$$\widetilde{F}_i^{\mu}(\vec{q}) = \widetilde{D}_{ij}^{\mu\nu}\widetilde{u}_j^{\nu}, \qquad (2.115)$$

with $\widetilde{F}_i^{\mu} = U^{\mu\sigma}F_i^{\sigma}$, $\widetilde{u}_j^{\nu} = U^{\nu\rho}u_i^{\rho}$ and $\underline{\widetilde{D}}^{\mu\nu} = U^{\mu\sigma}\underline{D}^{\sigma\rho}(U^{-1})^{\rho\nu}$, with the matrix

$$\underline{U} = \frac{1}{2}\begin{pmatrix} 1 & 1 \\ 1 & -1 \end{pmatrix}. \qquad (2.116)$$

The result will be the tensor

$$\underline{\widetilde{D}}(\vec{q}) = \begin{pmatrix} \underline{\widetilde{D}}^{SS}(\vec{q}) & \underline{\widetilde{D}}^{SF}(\vec{q}) \\ \underline{\widetilde{D}}^{FS}(\vec{q}) & \underline{\widetilde{D}}^{FF}(\vec{q}) \end{pmatrix}. \qquad (2.117)$$

The superscripts S and F denote 'slow' and 'fast' degrees of freedom. The slow degree of freedom is the center of mass of the two particles in the unit cell, $\widetilde{u}_i^S = (u_i^A + u_i^B)/2$. The fast degree of freedom is the orthogonal vector, $\widetilde{u}_i^F = (u_i^A - u_i^B)/2$. These are also called normal modes.

While $\underline{\widetilde{D}}$ is a complicated matrix, you should find

$$\underline{\widetilde{D}}(\vec{q}=\vec{0}) = \begin{pmatrix} \underline{0} & \underline{0} \\ \underline{0} & 3k\mathbb{1} \end{pmatrix}, \qquad (2.118)$$

where $\underline{0}$ is a 2×2 matrix with all zeros, and $\mathbb{1}$ is the 2×2 identity matrix.

The subtleties alluded to in the margin next to (2.30) refer to the following facts: we must consider the dynamical matrix in the center-of-mass basis; the derivatives are evaluated at $\vec{q} = 0$; and we need the factor of n/V in front.

p. We are finally ready to revisit equation (2.30), written properly as

$$
K_{ijk\ell} = \frac{n}{V}\left[\frac{1}{2}\frac{\partial^2 \widetilde{\underline{D}}^{\mathrm{SS}}}{\partial q_j \partial q_\ell} - \frac{\partial \widetilde{\underline{D}}^{\mathrm{SF}}}{\partial q_j}\left(\widetilde{\underline{D}}^{\mathrm{FF}}\right)^{-1}\frac{\partial \widetilde{\underline{D}}^{\mathrm{FS}}}{\partial q_\ell}\right]_{ik}\Bigg|_{\vec{q}=0}, \quad (2.119)
$$

where n is the number of particles in the unit cell of volume V. Here, $n = 2$ and $V = 3\sqrt{3}a^2/2$. Take the derivatives to find

$$
K_{ijkl} = \begin{cases} \dfrac{k}{2} & ijkl = \{xxxx, xxyy, yyxx, yyyy\}, \\[2mm] 0 & \text{otherwise.} \end{cases} \quad (2.120)
$$

q. Transform the four-component stiffness tensor $K_{ijk\ell}$ to the two-component tensor $K^{\alpha\beta}$ as in step i to find

$$
K^{\alpha\beta} = 2\begin{pmatrix} K & 0 & 0 & 0 \\ 0 & 0 & 0 & 0 \\ 0 & 0 & 0 & 0 \\ 0 & 0 & 0 & 0 \end{pmatrix}, \quad (2.121)
$$

where $K = k/2\sqrt{3}$.

Problem 2.8* *Defect energetics in two-dimensional crystals*

In this exercise, we will determine the energy associated with different types of defects in a two-dimensional triangular crystal lattice. The analysis below builds on the idea that defects act as sources of large strain, as illustrated in figure 2.41. To perform the analysis we first take a few preliminary steps.

a. The energy in a crystal due to some strain field $\vec{u}(\vec{r})$ is given by equation (2.64), reprinted here for convenience as

$$
E_{\mathrm{elas}} = \frac{1}{2}\int \mathrm{d}^2\vec{r}\,\sigma_{ij}u_{ij} = \frac{1}{2}\int \mathrm{d}^2\vec{r}\left(2\mu u_{ij}^2 + \lambda u_{kk}^2\right), \quad (2.122)
$$

where $u_{ij} = \frac{1}{2}\left(\partial_j u_i + \partial_i u_j\right)$ is the strain rate tensor, which is purely linear when confined to two dimensions. Take the variation of the energy functional to show that the lowest energy state satisfies the force balance equation:

$$
\partial_i \sigma_{ij} = 0. \quad (2.123)
$$

b. We can rewrite the stress tensor in terms of the so-called Airy function $\chi(\vec{r})$, defined by

Here ϵ_{ij} is the two-dimensional antisymmetric Levi-Civita tensor. $\epsilon_{xy} = -\epsilon_{yx} = 1$, $\epsilon_{xx} = \epsilon_{yy} = 0$.

$$
\sigma_{ij} = \epsilon_{ik}\epsilon_{jl}\partial_k\partial_l\chi, \quad (2.124)
$$

Prove that this definition of the stress tensor automatically satisfies the force balance equation.

c. Hooke's law, written as

$$\sigma_{ij} = 2\mu u_{ij} + \lambda \delta_{ij} u_{kk}, \tag{2.125}$$

gives us a linear relation between the stress and strain rate tensors. Invert this equation to find u_{ij} in terms of σ_{ij}, and then use the definition of the Airy function to find the strain rate in terms of χ. From this, show that χ satisfies the biharmonic equation

$$\frac{1}{Y}\Delta^2 \chi = \epsilon_{ik}\epsilon_{jl}\partial_k\partial_l u_{ij}, \tag{2.126}$$

where $Y = \frac{4\mu(\mu+\lambda)}{2\mu+\lambda}$ is the two-dimensional Young's modulus, and $\Delta^2 = (\nabla^2)(\nabla^2)$ is the biharmonic operator.

d. Show that, up to boundary terms, the energy functional can be expressed in terms of the Airy function as

$$E_{\text{elas}} = \frac{1}{2Y}\int d^2\vec{r}\,[\nabla^2\chi(\vec{r})]^2. \tag{2.127}$$

e. The right-hand side of equation (2.126) can be thought of as a source term for the biharmonic, as it is generally zero everywhere except in the immediate vicinity of defects. For the rest of this problem, we will write

$$\rho(\vec{r}) = \epsilon_{ik}\epsilon_{jl}\partial_k\partial_l u_{ij}(\vec{r}). \tag{2.128}$$

Different types of defects contribute different terms to the source function $\rho(\vec{r})$. Dislocations, as discussed in section 2.11, are characterized by the Burgers vector $\vec{b}$, defined by

$$\oint du_i = b_i, \tag{2.129}$$

where the integral is performed along a closed loop encircling the dislocation. Show that the differential form of this equation for an isolated dislocation at position $\vec{r}_\alpha$ can be written as

$$\epsilon_{ij}\partial_i\partial_j u_k = b_k\delta(\vec{r}-\vec{r}_\alpha). \tag{2.130}$$

Then show that this results in the following source term on the right-hand side of (2.126) for a collection of dislocations at positions $\{\vec{r}_\alpha\}$ with Burgers vectors $\{\vec{b}_\alpha\}$:

$$\rho_{\text{disloc}}(\vec{r}) = \sum_\alpha \epsilon_{ij}b_{\alpha i}\partial_j\delta(\vec{r}-\vec{r}_\alpha). \tag{2.131}$$

f. Show that, for a single dislocation at the origin of a circular patch of crystal of radius R, the resulting Airy stress function can be written as

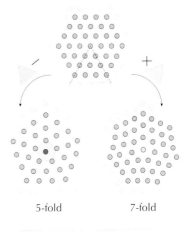

5-fold   7-fold

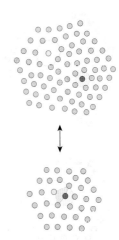

Figure 2.41. A sevenfolded disclination is obtained by adding a $\pi/6$ areal sector to a triangular lattice, while for a fivefolded disclination the same sector is removed. This is known as the Volterra construction. A dislocation (bottom panel) is a tight pair of five and seven folded disclinations. Dislocations disrupt translational order, because they are constructed by inserting additional lattice lines.

$$\chi(\vec{r}) = \frac{Y}{4\pi} b_i \epsilon_{ij} r_j \ln\left(\frac{r}{a}\right), \tag{2.132}$$

where a denotes the lattice spacing. Substitute this expression into equation (2.127) and show that the total elastic energy in the crystal can be written as

$$E_{\text{elas}}^{\text{disloc}} = \frac{Y b^2}{8\pi} \ln\left(\frac{R}{a}\right) + \text{const.} \tag{2.133}$$

g. A second type of topological defect that can occur in crystals is called *disclination*. Disclinations will be covered in more detail in the context of liquid crystals in section 6.6. In a two-dimensional crystal, they correspond to points with a different coordination number than the standard six for a triangular lattice, usually either five or seven. Disclinations are characterized by the following equation

$$\oint d\theta(\vec{r}) = s \frac{2\pi}{6}, \tag{2.134}$$

where $\theta(\vec{r}) = \frac{1}{2}\epsilon_{kl}\partial_k u_l$ is the local bond angle on the lattice. s is the topological strength of the defect, denoting how many extra or fewer nearest neighbors the disclination has (i.e., $s = 1$ corresponds to seven nearest neighbors, $s = -1$ corresponds to five). Using a similar technique as in step e, show that disclinations introduce a source term of the form

$$\rho_{disc} = \frac{\pi}{3} \sum_{\beta} s_\beta \delta(\vec{r} - \vec{r}_\beta) \tag{2.135}$$

for a collection of dislocations at positions $\{\vec{r}_\beta\}$ with topological strengths $\{s_\beta\}$.

h. Repeat your calculation from step f, but now for a disclination of strength s placed at the origin. You should find the Airy function to be

$$\chi(\vec{r}) = \frac{Ys}{8\pi} r^2 \left[\ln\left(\frac{r}{a}\right) + \text{const.}\right], \tag{2.136}$$

and for the total energy

$$E_{\text{elas}}^{\text{disc}} = \frac{Ys^2}{32\pi} R^2. \tag{2.137}$$

Compare this result with the energy associated with a single dislocation. Which is less energetically favorable in the thermodynamic limit?

i. Finally, let $\Gamma_B(\vec{r}, \vec{r'})$ be the Green's function of the biharmonic operator. Show that the energy functional can be rewritten as an interaction energy between defects:

As figure 2.41 illustrates, a single dislocation can be thought of as a tight dipole of two disclinations of opposite strength; this should be intuitive by comparing the different types of source terms induced by dislocations and disclinations. We use the result (2.137) in our discussion of virus shapes in section 7.3.

$$E_{\text{elas}} = \frac{Y}{2} \int d^2\vec{r} \int d^2\vec{r}' \rho(\vec{r}) \Gamma_B(\vec{r}, \vec{r}') \rho(\vec{r}'). \qquad (2.138)$$

Hint: Try integrating by parts.

Problem 2.9 *A mechanical topological insulator*

In this problem, we quantitatively study the edge modes for the one-dimensional topological chain illustrated in figure 2.39. Let $l_{n,n+1}$ be the length of the bond connecting the masses n and $n + 1$. Let θ_i being the angle of the ith rotor. Note that θ_i is measured with respect to the vertical and is taken to be positive when the rotor leans to the right. Let r be the length of each rotor, and let A be the spacing between the base of the rotors.

a. Derive the relationship:

$$\begin{aligned} l_{n,n+1}^2 = a^2 &+ 2ar(\sin\theta_{n+1} - \sin\theta_n) \\ &+ 2r^2(1 + \cos(\theta_n + \theta_{n+1})). \end{aligned} \qquad (2.139)$$

b. Write $\theta_n = \bar{\theta} + \delta\theta_n$, where $\bar{\theta}$ is the initial angle of the rotors when the all the bonds are unstrained, and $\delta\theta_n$ is a small perturbation about the initial angle. Expand equation (2.139) to linear order in $\delta\theta_n$. You should obtain

$$\begin{aligned} l_{n,n+1}^2 = a^2 &+ 2ar(\delta\theta_{n+1} - \delta\theta_n)\cos\bar{\theta} \\ &+ 2r^2[1 + \cos 2\bar{\theta} - (\delta\theta_n + \delta\theta_{n+1})\sin 2\bar{\theta}]. \end{aligned} \qquad (2.140)$$

c. Let $\bar{l}$ be the unstrained length of the bonds and let $\delta l_{n,n+1} = l_{n,n+1} - \bar{l}$. Notice that $l_{n,n+1}^2 = (\bar{l} + \delta l_{n,n+1})^2 \approx \bar{l}^2 + 2\bar{l}\delta l_{n,n+1} + \mathcal{O}(\delta l^2)$. Using this fact, show that

$$\delta l_{n,n+1} = q_+ \delta\theta_n + q_- \delta\theta_{n+1}, \qquad (2.141)$$

where

$$q_\pm = -\frac{r\cos\bar{\theta}}{\bar{l}}(2r\sin\bar{\theta} \pm a) \qquad (2.142)$$

d. You can now compute the penetration depth of a zero mode localized to one edge of the topological chain. Start from the condition $\delta l_{i,i+1} = 0$ for all i and show that $\delta\theta_n = e^{-\frac{an}{l}}\delta\theta_0$, for some l. Write l in terms of r, A, and $\bar{\theta}$, and physically interpret how the sign of l depends on the sign of $\bar{\theta}$.

e. Build the mechanical topological insulator following the instructions on the website of the book, www.softmatterbook.online. Then try to check whether the topological edge mode is actually localized only on one end of the chain. What happens to the topological mode (initially localized at the edge) when you tilt the chain?

f. Now let's make the connection to topology a bit more explicit. To get started, we will express displacements and elongations of the chains in terms of Fourier modes:

$$\delta\ell(k) = \sum_n e^{-ikn}\delta\ell_{n,n+1}$$

$$\delta\theta(k) = \sum_n e^{-ikn}\delta\theta_n. \tag{2.143}$$

In terms of Fourier modes, show that equation (2.141) reads

$$\delta\ell(k) = C(k)\delta\theta(k), \tag{2.144}$$

where $C(k) = q_+ + q_-e^{ik}$.

Hint: Use the relationships

$$\delta\theta_n = \int_0^{2\pi} \frac{dk}{2\pi} e^{ikn}\delta\theta(k) \tag{2.145}$$

and

$$\delta_{mn} = \int_0^{2\pi} \frac{dk}{2\pi} e^{ik(m-n)}, \tag{2.146}$$

where δ_{mn} is the Kronecker delta.

g. Here, $C(k)$ is an example of an object known as a compatibility matrix. (In this case, a 1-by-1 matrix). Given this matrix in Fourier space, we can define the winding number

$$\nu = \oint_0^{2\pi} \frac{dk}{2\pi i}\frac{d}{dk}\log\det C(k). \tag{2.147}$$

The winding number ν is an example of a topological index. How does ν depend on q_+ and q_-? How does its value relate to the presence or absence of a localized mode?

For a more comprehensive introduction, we refer you to Mao and Lubensky, 2018.

Brownian Motion, Thermal Fluctuations, and Diffusion | 3

In the introductory chapter we took a small micron-sized particle in a fluid as the starting point for illustrating the breadth of soft matter physics. Such a particle executes Brownian motion: as illustrated in figure 3.1, the small kicks from the surrounding fluid molecules give rise to small, rapidly varying, virtually random forces on the particle, and this results in the particle displacement being a succession of uncorrelated tiny movements. The reproduction of Perrin's observations in figure I.3 illustrates the resulting Brownian motion nicely.

Since so much of soft matter and biomatter consists of intermediate-scale building blocks that exhibit thermal fluctuations, we introduce in this chapter the concepts most frequently used to account for thermal fluctuations on the coarse-grained scale. We discuss both the description in terms of a stochastic differential equation and the equivalent formulation in terms of a probability distribution.

Thermal fluctuations are crucial for the building blocks of soft matter to form equilibrium phases, but the rate at which equilibration happens depends on their scale. In addition, thermal fluctuations can also be used as a probe: we discuss several examples of how by observing the magnitude of fluctuations one can infer effective force or stiffness constants of a single biomolecule or a membrane.

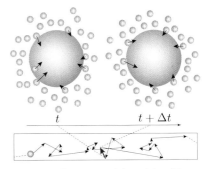

Figure 3.1. Illustration of the origin of Brownian motion of a large particle in a bath of small ones. The left panel shows a snapshot of the particle at a time t; the velocities of all particles colliding with it within a time Δt are indicated by arrows. The Brownian particle is seen to receive predominantly kicks from the left. The right panel illustrates how a moment later, it receives mostly kicks from particles on its right. Below is a sketch of the random trajectory of the particle at successive times.[1]

For instance, colloids can form colloidal crystals, but the time scale on which this happens is large, and flow drives them quickly far out of equilibrium. Common liquid crystal molecules, on the other hand, are small enough that they equilibrate rapidly and that local equilibrium is maintained during flow.

3.1 A matter of scales and description

The extent to which thermal fluctuations play an important role in soft matter depends on the scale of the phenomena we are interested in. Indeed, as we already mentioned in the introduction and section 2.10, granular media research is focused on the analysis of heaps of grains which are so large that thermal fluctuations do not play a role: granular media are essentially a-thermal systems. But when the grains or particles are small enough—as we shall see, typically about a micron in diameter or less—thermal fluctuations make them disperse (distribute) throughout the fluid; then we are in the realm of colloids. Likewise, the statistical properties of polymers and the force-extension curve of biopolymers like DNA that we will derive in chapter 5 result from thermal fluctuations. On the other hand, most of the peculiarities of viscoelastic polymer

time goes to zero by writing

$$\overline{L(t)L(t')} = 2\gamma\,\delta(t-t'). \tag{3.8}$$

Here $\delta(t)$ is the delta function and γ is a parameter for the force strength which will be determined below. Taking the noise to be 'delta-correlated' in time expresses that the force at different times is uncorrelated, while it still results in nonzero mean square fluctuations of the stochastic quantities that we will study below.[4]

As usual, it pays to consider the dimensions. Since L has the dimension of force divided by mass, $[L(t)] = \mathrm{m/s^2}$, and since the dimension of $\delta(t)$ is s^{-1}, we have

> An easy way to remember that the dimension of a delta function is the inverse of the dimension of its argument is to note that $\int \mathrm{d}x\,\delta(x) = 1$.

$$[\gamma] = \frac{\mathrm{m}^2}{\mathrm{s}^3}. \tag{3.9}$$

3.2.3 Mean square variations of velocity and position: Diffusion

We now analyze equation (3.6). By taking the stochastic average, we simply get for the stochastically averaged velocity

$$\frac{\mathrm{d}\overline{V}}{\mathrm{d}t} = -\Gamma\overline{V}, \quad \rightarrow \quad \overline{V} = V_0\,e^{-\Gamma t}, \tag{3.10}$$

where V_0 is the velocity at time $t = 0$. For the fluctuations around the average, we get

> It is easy to check the solution (3.11) by differentiating it, or to derive it from the Langevin equation (3.6) by the variation of constants method.

$$\Delta V(t) = V(t) - \overline{V}(t) = \int_0^t \mathrm{d}t_1\, e^{-\Gamma(t-t_1)}L(t_1). \tag{3.11}$$

Due to the relaxation of the velocity on a time scale $\tau = \Gamma^{-1}$, the memory of the initial velocity essentially fades away in this time τ. From then on the instantaneous velocity is determined by the (weighted) sum of the random forces over the last time interval of order τ.

From the previous result and (3.8), we can analyze the mean square fluctuations of the velocity

> This equation illustrates that velocity correlations decay during a time scale $\tau = \Gamma^{-1}$, as anticipated in the discussion leading to equation (3.5).

$$\overline{(\Delta V(t))^2} = \int_0^t \mathrm{d}t_1 \int_0^t \mathrm{d}t_2\, e^{-\Gamma(t-t_1)-\Gamma(t-t_2)}\,\overline{L(t_1)L(t_2)}$$

$$= \int_0^t \mathrm{d}t_1\, e^{-2\Gamma(t-t_1)}\,2\gamma = \frac{\gamma}{\Gamma}\left(1 - e^{-2\Gamma t}\right). \tag{3.12}$$

The analysis of the mean square displacement $\overline{(\Delta X(t)^2)}$ for the Langevin equation proceeds along the same lines; we relegate it to problem 3.2, and we simply quote the large-time result

$$\sigma_X^2(t) = \overline{(\Delta X(t))^2} \approx \frac{2\gamma}{\Gamma^2}t, \qquad t \gg \Gamma^{-1}. \tag{3.13}$$

This equation expresses the *essence of diffusion: the mean square displacement increases linearly in time*. In general, the strength of the diffusion is measured by the *diffusion coefficient D*, defined as the proportionality constant of the linear growth of of the mean square displacement,

$$\sigma_X^2(t) = \overline{(\Delta X(t))^2} = 2Dt \quad \text{for large enough } t. \tag{3.14}$$

For our Langevin equation we have from (3.13)

$$D = \frac{\gamma}{\Gamma^2}. \tag{3.15}$$

Figure 3.3 illustrates the diffusion behavior of micron-sized particles of various diameters. We will discuss the diffusion in more detail after taking the final step by including the effect of temperature.

3.2.4 The Stokes-Einstein equation for the diffusion coefficient

So far, the results are quite general, and they also apply to non-thermodynamic systems. Intuitively, it is clear that in thermal systems, the noise strength should increase with increasing temperature, as the strength of the kicks which the particle experiences will be greater the higher the temperature is.

To make this more concrete, note that since the surrounding molecules act as a heat bath for the Brownian particle, the average thermal kinetic energy associated with each velocity component of the Brownian particle is simply $1/2k_\mathrm{B}T$,

$$\frac{1}{2}M\langle V^2 \rangle = \frac{1}{2}k_\mathrm{B}T. \tag{3.16}$$

Here $\langle \cdot \rangle$ denotes a thermal average.

So far, we did not take into account that the fluctuating forces in our Langevin equation are coming from fluid molecules in thermal equilibrium. Let's do that now. In order that our Langevin equation result (3.12) for the mean square velocity fluctuations approaches the proper thermal average value (3.16), we should require

$$\frac{1}{2}M\,\overline{V^2(t)} \quad \overset{t\to\infty}{\Longrightarrow} \quad \frac{1}{2}k_\mathrm{B}T. \tag{3.17}$$

Since for long times $\overline{V^2} = \overline{(\Delta V)^2} = \gamma/\Gamma$ according to equation (3.12), combining these two results gives

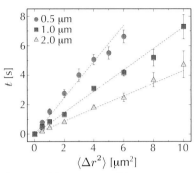

Figure 3.3. Measurement of the mean square displacement of colloidal particles of various diameters $2R$ in a modern version of Perrin's experiment. The slope of the line gives the diffusion coefficient D. Note that the smaller the particles, the faster the diffusion, in accord with the Stokes-Einstein expression (3.19) for D. Data courtesy of Joseph Peidle.[5]

3.3 The Fokker-Planck equation for the probability distribution

The Langevin stochastic differential equation description is especially useful if the deterministic equations of a system in the absence of noise are linear; adding the noise terms and requiring that the mean square of all linear modes of the system be consistent with equipartition of energy is then often the most efficient way to analyze the fluctuations.

This Langevin program also works for spatially extended systems. For instance, Landau and Lifshitz, 1980 discuss fluctuating hydrodynamics along these lines.

In this section we describe an equivalent description in terms of a Fokker-Planck equation for a probability distribution. Analyzing the effect of fluctuations in terms of an appropriate probability distribution often offers more physical insight and facilitates analytical calculations and approximations. Moreover, it provides a more natural language for interpreting experimental data: one can measure a probability distribution, but not a Langevin equation.

A very accessible introduction to and overview of the Fokker-Planck equation and its relation to the Langevin equation (and the master equation for Markov processes discussed in section 3.4) is given by van Kampen, 2009 and Gardiner, 2004. The equivalence stated here for one variable extends to many variables and to situations where the coefficient b in (3.26) is a function $b(X)$.[6]

3.3.1 The Fokker-Planck equation: Equivalence to a Langevin equation

When we use a probability distribution $\mathcal{P}(X, t)$ to analyze the evolution of a stochastic quantity X, we can think of X as forming the phase space and $\mathcal{P}$ as being the probability density of an ensemble of systems: $\mathcal{P}(X, t)\mathrm{d}X$ is the probability of finding the system with an interval $\mathrm{d}X$ around the value X at time t. Each point in phase space represents a particular realization of the system, and we can think of $\mathcal{P}$ as the density of the points representing the ensemble. This is illustrated in figure 3.4. The notation $\mathcal{P}(X, t | X_0, t_0)$ is often used to denote the conditional probability of finding the system at X at time t, given that it was at X_0 at time t_0.

We simply state here without proof that each Langevin equation with delta-correlated noise of strength 2γ, $\overline{L(t)L(t')} = 2\gamma\,\delta(t - t')$ as in (3.8), is *completely equivalent* with a so-called Fokker-Planck equation for the dynamical evolution of $\mathcal{P}$. To be concrete, the equivalence is expressed by

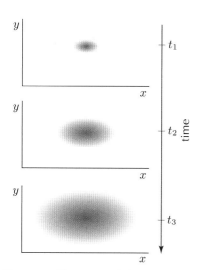

Figure 3.4. Illustration of the evolution of the probability distribution $\mathcal{P}(X, Y, t)$ in a two-dimensional phase space. Each point in phase space corresponds to a realization of the system, and the density of points corresponds to the probability density $\mathcal{P}$. The three snapshots illustrate how as time progresses downward, the distributions spread, while the density of points diminishes.

$$\text{Langevin eq.:} \quad \frac{\mathrm{d}X}{\mathrm{d}t} = a(X) + bL(t),$$

$$\Updownarrow \qquad\qquad (3.26)$$

$$\text{Fokker-Planck eq.:} \quad \frac{\partial \mathcal{P}}{\partial t} = -\frac{\partial\,[a(X)\mathcal{P}]}{\partial X} + \gamma\frac{\partial^2\,[b^2\mathcal{P}]}{\partial X^2}.$$

The second derivative term in the Fokker-Planck equation gives rise to the spreading of the distribution due to the fluctuations, illustrated in figure 3.4.

3.3.2 The Fokker-Planck equation for the velocity of the Brownian particle

The above equivalence immediately applies to the Langevin equation (3.6) for the velocity V. In this case V plays the role of the variable X in (3.26), and $a(X)$ becomes the term $-\Gamma V$, while $b = 1$. Thus the Fokker-Planck equation becomes

$$\frac{\partial \mathcal{P}(V,t)}{\partial t} = \frac{\partial}{\partial V}\left(\Gamma V + \gamma \frac{\partial}{\partial V}\right)\mathcal{P}(V,t). \qquad (3.27)$$

We can check for consistency with the earlier result by imposing that the Maxwell-Boltzmann equilibrium distribution

$$\mathcal{P}_{eq}(V) \sim \exp(-\tfrac{1}{2}MV^2/k_BT) \qquad (3.28)$$

is indeed a stationary solution of this equation. Requiring that $\partial_t \mathcal{P}_{eq} = 0$ in (3.27) gives

$$\left(\Gamma V + \gamma \frac{\partial}{\partial V}\right)\mathcal{P}_{eq} = 0 \quad \Longleftrightarrow \quad \gamma = \frac{\Gamma k_B T}{M} = \Gamma^2 D, \qquad (3.29)$$

in full accord (3.18) for γ and with the expression (3.19) of γ in terms of the diffusion coefficient D.

3.3.3 The Fokker-Planck equation for the position of a Brownian particle in an external potential

For studying diffusion, the probability density $\mathcal{P}(V,t)$ is hardly relevant: the relaxation time $\tau = \Gamma^{-1}$ is typically so short that the velocity relaxation is simply invisible; moreover one is normally interested in the effects of diffusion in position space. We can in principle determine the spatial diffusion by augmenting the phase space with the positional coordinate X and extracting the diffusion from the evolution of the full distribution $\mathcal{P}(X,V,t)$.[7] But it is simpler and more insightful to go straight to the distribution $\mathcal{P}(X,t)$ by using the earlier shortcut which showed that if we coarse-grain over times much longer than $\tau = \Gamma^{-1}$, we can effectively describe the diffusion by a Langevin equation of the form (3.22) for the spatial coordinate X. The presence of the potential U translates in this case into the function $a(X) = -U'/(\Gamma M)$ in (3.26), while $b = \Gamma^{-1}$. The appropriate Fokker-Planck equation thus becomes

$$\frac{\partial \mathcal{P}(X,t)}{\partial t} = \frac{\partial}{\partial X}\left(\frac{U'\mathcal{P}(X,t)}{\Gamma M}\right) + D\frac{\partial^2 \mathcal{P}(X,t)}{\partial X^2}, \qquad (3.30)$$

where we have made use again of the fact that $\gamma/\Gamma^2 = D$, according to (3.15).

Again, it is simple to check the self-consistency by noting that the thermal equilibrium distribution

$$\mathcal{P}_{\text{eq}}(X) \sim \exp(-U(X)/k_{\text{B}}T) \tag{3.31}$$

should be a stationary solution of this Fokker-Planck equation for $\mathcal{P}(X,t)$. Indeed

$$\left(\frac{U'}{\Gamma M} + D \frac{\partial}{\partial X} \right) \exp(-U/k_{\text{B}}T) = 0$$

$$\iff D = \frac{k_{\text{B}}T}{\Gamma M} = \frac{k_{\text{B}}T}{6\pi\eta R}. \tag{3.32}$$

The equation on the right is simply the Stokes-Einstein relation (3.19) for D, which demonstrates the consistency.

3.3.4 The diffusion equation and its Gaussian solution

This equation is written here for a probability density of a single particle, but since the equation is linear we can for the case of a finite density of noninteracting particles simply multiply the equation by the total number of particles to obtain the same equation for the particle density n.[8]

In the absence of a potential the Fokker-Planck equation (3.30) reduces to the simple *diffusion equation*

$$\frac{\partial \mathcal{P}(X,t)}{\partial t} = D \frac{\partial^2 \mathcal{P}(X,t)}{\partial X^2}. \tag{3.33}$$

Suppose at time $t = 0$ the particle is at the origin, so that we need to solve this equation with the initial condition $\mathcal{P}(X, t=0) = \delta(X)$. The solution is the well-known Gaussian

One can for instance derive this equation by taking a spatial Fourier transform of (3.33). Each mode e^{ikx} is then found to decay as $e^{-Dk^2 t}$. Converting back to real space then gives a Gaussian k-integral that yields (3.34).

$$\mathcal{P}(X,t) = \frac{1}{\sqrt{4\pi Dt}} e^{-X^2/4Dt}. \tag{3.34}$$

The exponent illustrates the familiar spreading of the diffusion with a width growing as $\sqrt{t}$. Since this solution reduces to the delta-function initial condition as $t \downarrow 0$, we can write the solution to any general initial distribution $\mathcal{P}(X, t=0) = \mathcal{P}_0(X)$ as

$$\mathcal{P}(X,t) = \int_{-\infty}^{\infty} \mathrm{d}X_1 \frac{1}{\sqrt{4\pi Dt}} e^{-(X-X_1)^2/4Dt} \mathcal{P}_0(X_1). \tag{3.35}$$

This expression nicely illustrates that if the initial distribution is nonzero in some localized region centered around X_0, then at long times the probability distribution approaches a Gaussian centered around X_0.[9]

The above expression means that the solution (3.34) is actually the Green's function of the equation (3.33), and it illustrates that the long time Gaussian solution (3.34) emerges very robustly in diffusion, irrespective of the details of the initial condition.

3.3.5 Self-similarity and self-similar solutions

The Gaussian solution (3.34) of the diffusion equation (3.33) is an example of a so-called self-similar solution, as it depends on the spatial coordinate X only through the combination $X/t^{1/2}$. This means that when viewed at different times these solutions are the same, provided that the spatial coordinate is rescaled.

More generally, self-similar solutions of a partial differential equation in one spatial dimension X are solutions of the form

$$t^{-\alpha} G(\xi), \quad \text{with } similarity \ variable \ \xi = X/t^{\beta}. \tag{3.36}$$

In our particular case of the Gaussian solution, we have $\alpha = \beta = 1/2$, but to illustrate how these exponents can already be obtained from the form of the equation, let us pretend for the moment that we do not know this solution yet and follow the typical line of analysis.

In the spirit of a similarity analysis, we assume a solution of the form (3.36) by writing $\mathcal{P}(X, t) = t^{-\alpha} G(\xi)$. We first note that the original diffusion equation (3.33) conserves the integrated probability, as the particle has to be somewhere with probability 1: indeed for solutions which vanish at $\pm\infty$, integration of the equation shows that

$$\int_{-\infty}^{\infty} \mathrm{d}X \, \mathcal{P}(X, t) = const, \quad \text{so} \quad t^{-\alpha + \beta} \int_{-\infty}^{\infty} \mathrm{d}\xi \, G(\xi) = const., \tag{3.37}$$

which implies $\alpha = \beta$. Secondly, upon substituting the ansatz (3.36) into equation (3.33), we get after dividing out a common factor $t^{-(\alpha+1)}$

$$-\beta G' \xi - \alpha G = t^{1-2\beta} G''. \tag{3.38}$$

Here a prime denotes differentiation with respect to ξ. The term on the right-hand side shows that as $G(\xi)$ is time-independent, we need to have $\beta = 1/2$, and hence as $\alpha = \beta$ also $\alpha = 1/2$. We are now left with an ordinary differential equation for $G(\xi)$, whose solutions can be obtained using standard methods for ordinary differential equations.

Solutions of the type (3.36) are called similarity solutions or self-similar solutions: the solutions at different times are mathematically the same, upon rescaling the spatial coordinate X properly with a time-dependent factor, so that the similarity variable ξ is unchanged—in a way, this is a generalization of the concept of *similarity laws* which we discussed in section 1.10.3.

Self-similar solutions emerge in many problems as 'attractors' of the dynamics, the solutions to which the dynamics converges irrespective of the details of the initial conditions or of the equation. Crucial for the understanding of these similarity solutions is the

For a very nice introduction to and overview of self-similarity, see Barenblatt, 1996. The example we discuss here, in which the exponents are determined by conservation laws and dimensional analysis, are often referred to as self-similar solutions of the first kind. Such considerations don't suffice for self-similar solutions of the second kind: for these the equations have to be solved explicitly to determine the exponents.

In section 7.4 we briefly discuss how similarity solutions play a role in analyzing crumpling as a fragmentation process. The analysis of the integro-differential equation is given in problem 7.4.

Quite interestingly, the similarity solution (3.39) plays a crucial role in the intermediate asymptotics of the problem of front propagation into unstable states, as it happens to be the dominant term in the 'leading edge' of the front (Ebert and van Saarloos, 2000; van Saarloos, 2003).[10]

fact that they typically determine the 'intermediate asymptotics' of a problem—the 'intermediate' is used to stress that we are focusing on the intermediate time scale, large enough for the similarity behavior to be dominant, but before the solution dies away completely as $t \to \infty$. And as the example illustrated, the similarity solutions are determined by solving an ordinary differential equation, which is easier than solving a full partial differential equation.

Finally it is worth noting that when one has a solution $\mathcal{P}(X, t)$ of the diffusion equation (3.33), also $\partial \mathcal{P}(X, t) / \partial X$ is a solution. This extends to the similarity solutions: it implies that even though the Gaussian form (3.34) is the most commonly encountered solution, its derivative

$$\frac{X}{(4\pi D t)^{3/2}} e^{-X^2/4Dt} \tag{3.39}$$

is an exact dipole-type similarity solution. One can generate a series of such solutions by taking more and more derivatives.

3.3.6 The Kramers problem: Fluctuation-driven escape over a barrier

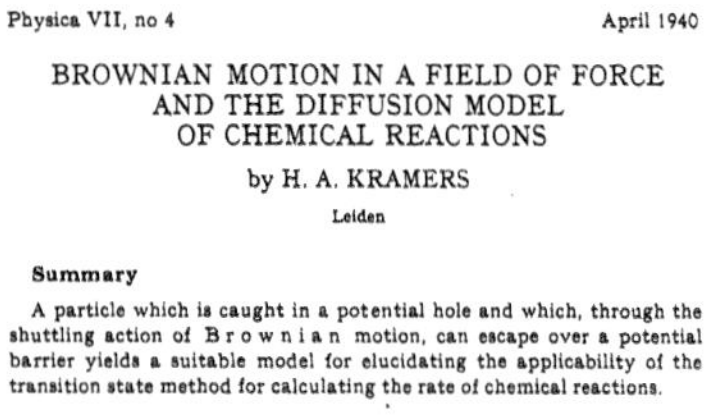

Let us return to the Fokker-Planck equation for a Brownian particle in a potential U, and analyze what is often referred to as the Kramers problem, following the 1940 paper by Kramers, illustrated in figure 3.5. We consider a potential $U(X)$ with a deep potential well of the type sketched in figure 3.6; a deep well means that the barrier height $E_b \gg k_B T$.

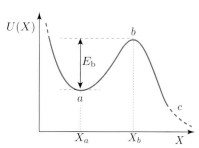

Physica VII, no 4   April 1940

BROWNIAN MOTION IN A FIELD OF FORCE
AND THE DIFFUSION MODEL
OF CHEMICAL REACTIONS

by H. A. KRAMERS

Leiden

Summary

A particle which is caught in a potential hole and which, through the shuttling action of B r o w n i a n motion, can escape over a potential barrier yields a suitable model for elucidating the applicability of the transition state method for calculating the rate of chemical reactions.

Figure 3.5. Title and abstract of Kramers's original paper from 1940.[11]

Imagine that our Brownian particle is placed near X_a in this well. The thermal fluctuations make the particle explore mostly the configurations in the well, but there is a small probability that they drive the particle uphill toward the peak at b, and have it cross the barrier at X_b. Once it passes the barrier, it is driven down the potential toward large X. We are interested in calculating the small rate at which the particle escapes from the well and over the barrier, due to thermal fluctuations. This is the Kramers escape over a barrier problem.

The Fokker-Planck equation (3.30) provides the basis for an explicit calculation, which we include as problem 3.3. As the result is central to many problems, we quote it here: the escape rate r over the barrier, due to thermal fluctuations, is

$$r = \frac{D}{2\pi k_B T} [|U''(X_a)U''(X_b)|]^{1/2} e^{-E_b/k_B T}. \tag{3.40}$$

Figure 3.6. The potential $U(X)$ considered in the analysis of Kramers of the rate of escape of a particle from a potential well, due to thermal fluctuations.

The result that the transition rate goes as $e^{-E_b/k_B T}$ is essentially the so-called Arrhenius exponential behavior that ubiquitously arises as the dominant temperature and energy dependence of chemical reactions and of localized rearrangements.[12] An important part of the question of many applications concerns the most favorable reaction pathway of such a local event in a high-dimensional phase space.

In addition, the calculation shows the prefactor of this dominant exponential to depend on the curvature of the potential at the well and the barrier, $U''(X_a)$ and $U''(X_b)$: the narrower the well or the steeper the barrier, the larger the transition rate. Quite interestingly, the prefactor decreases with increasing temperature, but of course the overall temperature dependence of the transition rate is dominated by the exponential term.

Examples of this Arrhenius behavior include the conformational change of a biomolecule, such as diffusion controlled interchain reactions in polymers, the binding of ligands to receptor molecules or protein folding, chemical reactions, or a rearrangement of atoms or particles in a solid or colloidal glass.

In a multi-dimensional phase space, the barrier looks like a saddle or mountain pass; the prefactor of the transition rate also depends on the curvature of the potential in these other directions.[13]

3.4 The master equation

We have so far focused on the Brownian particle with a continuum position coordinate. When a stochastic process concerns a discrete set of states with label n, a common description is in terms of the discrete equation

$$\frac{dp_n(t)}{dt} = \sum_{n'} \left[W_{nn'} p_{n'}(t) - W_{n'n} p_n(t) \right]. \tag{3.41}$$

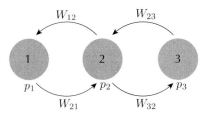

Figure 3.7. Illustration of the master equation (3.41). $p_1, \ldots, p_n$ denote the probabilities of the states $1, \ldots, 5$ indicated with circles and $W_{nn'}$ the transition probabilities between these states.

The equation is illustrated in figure 3.7; p_n is the probability of the process being in state n, and the $W_{nn'} \geq 0$ are the transition probabilities for the probability per unit time for the system to make a change from state n' to n. Thus, the first term on the right-hand side describes the gain in probability p_n due to transitions from other states n' to n; the second term describes the loss in probability due to transitions to other states n'. This equation, which is usually called the master equation, describes a so-called Markov process, a stochastic process without memory: the change in the p_n at time t only depends on the probabilities at that same time. In some cases, it is convenient to make the time a discrete variable too.

Note that the Fokker-Planck equation also describes a Markov process, a stochastic process without memory.

There is in principle a lot of freedom to choose the transition probabilities $W_{nn'}$; when n labels physical configuration states a natural choice is to take the $W_{nn'}$ such that they are compatible with the so-called detailed balance condition for the equilibrium probabilities p_n^{eq},

$$\text{Detailed balance:} \qquad W_{nn'} \, p_{n'}^{\mathrm{eq}} = W_{n'n} \, p_n^{\mathrm{eq}}. \tag{3.42}$$

This equation expresses that in equilibrium the transitions also balance locally. If an energy E_n is associated with state n, then a natural choice is to take

$$W_{nn'} = e^{-(E_n - E_{n'})/k_{\mathrm B}T} \qquad \Longleftrightarrow \qquad p_n^{\mathrm{eq}} \sim e^{-E_n/k_{\mathrm B}T}. \qquad (3.43)$$

so that in the equilibrium the states are weighed with their appropriate Boltzmann factors.

When the n label sites and the $W_{nn'}$ are the probabilities of making hops to neighboring sites, as suggested by figure 3.7, the equation describes a discrete diffusion process. At long times the discrete character is washed out because the diffusion length $\sqrt{Dt}$ becomes large, and as a result the probabilities p_n approach a Gaussian distribution of the form (3.34). Problems 3.4 and 3.5 give examples of the calculation of the diffusion coefficient in such a case of a discrete hopping process. The versatility to switch between a discrete and continuum description, because most details are irrelevant on the long time and large length scales relevant to diffusion, can often be used to one's advantage to pick the most convenient model. In a similar spirit, we took advantage of coarse-graining over the relaxation time τ of a Brownian particle to write the Fokker-Planck equation (3.30) as a Markov process in the space variable. Additional examples of how one derives a continuum description starting from a master equation description can be found in problems 3.7–3.9, while problems 3.10 and 3.11 introduce a general method of deriving hydrodynamic equations from a system of Langevin equations.

3.5 Size matters for diffusion and dispersion of Brownian particles

Let us return to our Brownian particle of radius R in a fluid and explore the implications of their size dependence. We will see that a micron is an important crossover scale both for diffusion and for the dispersion of the particles through a container.

3.5.1 Diffusion

Brownian motion varies more strongly with radius than it may appear at first sight from the $1/R$ dependence. To see this, let us estimate the time t_{diff} it takes for a particle in water at room temperature to diffuse by a distance on the order of $2R$. We get

$$(2R)^2 \approx 2D\, t_{\mathrm{diff}} = 2\frac{k_{\mathrm B}T\, t_{\mathrm{diff}}}{6\pi\eta R} = 2\frac{4 \cdot 10^{-21}\mathrm{m}^3/\mathrm{s}}{2 \cdot 10^{-2}\, R}\, t_{\mathrm{diff},}$$
$$\Longrightarrow \qquad t_{\mathrm{diff}} \simeq 10\, R^3 (\mathrm{s}/\mu\mathrm{m}^3). \qquad (3.44)$$

For a particle of radius 1 μm, this gives a diffusion time of about 10 seconds, which is just about the time scale where you can still reasonably see a particle move by eye through a microscope. But for particles a few times larger, the time already becomes uncomfortably large because of the increase as R^3 with radius R. At the same time, particles ten times smaller diffuse their own size on the scale of 10 milliseconds. In other words, because of this rapid crossover, 1 μm is a convenient scale where on the one hand particles are easily detected optically with great precision, while on the other hand diffusion and equilibration is not prohibitively slow.

3.5.2 Dispersions versus granular media

For a Brownian particle in the presence of gravity, the gravitational potential is $U = M_{\text{eff}}gh$ where h is the height, and where M_{eff} is the effective mass due to buoyancy (Archimedes's law). The equilibrium distribution then gives the usual exponential law,

$$\mathcal{P}_{\text{eq}} \sim \exp(-M_{\text{eff}}gh/k_{\text{B}}T). \tag{3.45}$$

Let us determine the particle radius at which thermal fluctuations lift the particle by a distance of order its radius. In other words, what is the radius R for which the exponent is about unity for $h = R$? This is the case when

$$\tfrac{4}{3}\pi R^3 \Delta\rho\, gR \approx k_{\text{B}}T \iff R \approx \left(\frac{3k_{\text{B}}T}{4\pi\Delta\rho\, g}\right)^{1/4}, \tag{3.46}$$

where $\Delta\rho$ is the density difference between the particle and the fluid. It is easy to check that for any reasonable density difference this also gives a crossover radius of order 1 μm. In other words, because of the rapid crossover due to the R^4 term on the left-hand side of (3.46), for particles somewhat larger than about a micron the thermal fluctuations are unable to lift the particles appreciably: they settle at the bottom. On the other hand, smaller particles are strongly affected by thermal fluctuations and little affected by gravity: they disperse throughout the container.

Interestingly, both for diffusion and for dispersion, the crossover length is about a micron. In other words, a micron is about the crossover length distinguishing granular media—particles large enough that thermal fluctuations and Brownian motion are not important—from dispersions like colloidal systems consisting of particles small enough that thermal fluctuations and Brownian motion make them disperse throughout the container and allow them to be viewed as a statistical-physical system. This latter regime is clearly important in biological matter at the cellular level.

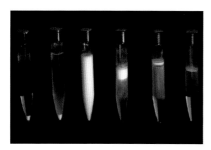

Figure 3.8. Photo of six vials with colloidal suspensions with fluid density increasing to the right, obtained after centrifuging the samples for one hour at 1000 g to speed up the settling process.[14] When the density of the colloid and fluid match best, as in the third flask from the left, the colloids nicely disperse throughout the cell. When the fluid is lighter, as in the two leftmost flasks, the colloids settle on the bottom, while the colloids float to the top when the fluid is heavier (the three flasks on the right).[15] Experiment and image courtesy of Ali Azadbakht.

The invention of *optical tweezers*, for which Arthur Ashkin was awarded the Nobel Prize in 2018, was a breakthrough for research on soft matter and biomatter at the scale of the building blocks and cells. Optical tweezers make use of the fact that when one or more laser beams are focused very tightly, the radiation pressure that tends to repel objects from the light is overcome by the intense radiation field that produces a gradient force to the center of the beam. Via movement of the beams, small objects like micron-sized beads, viruses and cells can be moved around and hence manipulated. Interestingly, for a trapped bead one can probe the curvature of the radiation-induced potential well from fluctuation measurements, as we discuss in section 3.6.1. From these data one can then infer the force with which a bead pulls on a molecule attached to it. See Grier, 2003 and Moffitt et al., 2008 for introductions to the power of optical tweezers to manipulate and explore soft matter and biomatter. Overviews are given in many of the books dedicated to the technique and its applications.

In biophysical systems, density differences $\Delta\rho$ are often small enough that particles about a micron in size still disperse appreciably. In the lab, one often uses polystyrene beads for the colloid. Polystyrene is only 5% denser than water, so for these particles $\Delta\rho$ is already small. Moreover, by adding appropriate salts to the water, the densities can be made to be precisely the same: gravity effectively is switched off in the experiment by density matching, as figure 3.8 illustrates. We will discuss colloids in detail in chapter 4.

3.6 Probing fluctuations and taking advantage of them as a probe

Now that we are armed with the classic description of fluctuations and Brownian motion, it is time to look at some applications. Modern experimental techniques, in particular optical traps and optical tweezers, have made it possible to probe and manipulate matter at the small scales where fluctuations are prominent. Moreover, in combination with the massive data acquisition and processing possibilities offered by digital cameras and computers, these optical techniques nowadays allow researchers to follow (many) fluctuating building blocks over a long enough time and in enough detail to probe the individual or collective behavior of soft (bio)matter by tracking the fluctuations.[16] The colloidal experiments on two-dimensional melting shown in figure 2.37 actually already illustrated this.

3.6.1 Measuring force constants of biomatter experimentally

It is instructive to first illustrate with concrete data how one puts Brownian motion to work to infer the shape of the potential well associated with an optical trap of a bead: one just observes the thermal position fluctuations. Indeed, the equilibrium fluctuations of a bead around the minimum U_0 of a two-dimensional well are obtained from the equilibrium distribution (3.31) by expanding the potential around the minimum, to get

$$\mathcal{P}_{\mathrm{eq}} \sim \exp(-U''(X^2+Y^2)/2k_{\mathrm{B}}T), \qquad (3.47)$$

where X and Y measure the distances from the minimum of the well. Since the bead is in a thermal bath, it performs a Brownian motion in the potential well, so one can just infer the curvature of the potential by tracing the bead's positions and fitting the Gaussian distribution (3.47) to it. Panels (a) and (b) of figure 3.9 illustrate

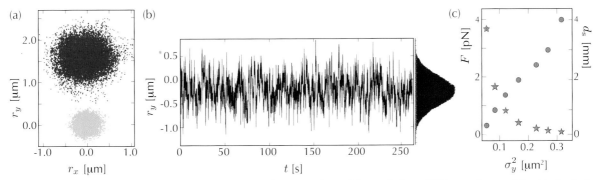

Figure 3.9. Data from experiments by Vilfan et al., 2009 with a magnetic particle attached to a DNA strand, as explained in the text. Panel (a) shows the cloud of measured positions for two different cases of the force exerted by the DNA (the dark green cloud of observations is shifted upward for visibility). Panel (b) shows a sample measurement, illustrating the Brownian motion of the bead in the potential well, with on the right the fit to the Gaussian (3.48). Panel (c) shows the relation between the mean square variation of the particle position, the force F_{DNA} (stars) and with blue circles the distance d_s between the magnet and the sample (the farther away the magnet, the less force it exerts on the particle). Adapted from Vilfan et al., 2009.[17]

this with real experimental data from the experiment sketched in figure 3.10.

In these experiments, a small paramagnetic bead is attached to a DNA strand. The magnetic field is arranged such that there is a magnetic gradient in the vertical direction only: the horizontal dashed lines in figure 3.10 are lines of constant magnetic potential. So with the magnetic force, one can pull the bead vertically up and down, while the bead remains free to move in the horizontal plane. In addition to the magnetic force, the bead experiences a force from the DNA, which is attached to the bead: the DNA acts like an elastic spring on the particle which pulls mostly downward but which also has a small component pointing inward to the center at $X = Y = 0$ if X and Y measure the horizontal coordinates from the minimum-energy position. Now, if we denote the length of the DNA at $X = 0, Y = 0$ by ℓ_0, then this particular configuration implies for the change in length $\Delta\ell$ of the DNA strand $\Delta\ell \approx (X^2 + Y^2)/(2\ell_0)$ for small excursions within the horizontal plane. As a result, the equilibrium probability distribution becomes

$$\mathcal{P}_{\mathrm{eq}} \sim \exp\left(-\left[U_{\mathrm{DNA}}(X, Y) - U_{\mathrm{DNA}}(0,0)\right]/k_{\mathrm{B}}T\right),$$

$$\sim \exp\left(-\frac{F_{\mathrm{DNA}}(X^2 + Y^2)}{2\ell_0\,k_{\mathrm{B}}T}\right), \qquad F_{\mathrm{DNA}} = \left|\frac{\mathrm{d}U_{\mathrm{DNA}}}{\mathrm{d}\ell}\right|. \tag{3.48}$$

So indeed we find a Gaussian distribution for the fluctuations of the bead around the minimum-energy position, while in this particular case the width of the distribution gives immediately F_{DNA}/ℓ_0, where $F_{\mathrm{DNA}} = |\mathrm{d}U_{\mathrm{DNA}}/\mathrm{d}\ell|$ is the force with which the DNA pulls on the bead. Figure 3.9.c shows the relation between the force and the width of the probability distribution of the bead fluctuations measured in these experiments.

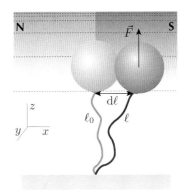

Figure 3.10. Sketch of experimental setup as used in experiments of Vilfan et al., 2009 of which data are shown in figure 3.9. A nucleic acid like DNA is attached to the orange bead which is paramagnetic. The magnetic field is such that there is only a gradient in the vertical direction: dashed lines indicate lines of constant magnetic field. N and S indicate the north and south poles of an additional magnet that allows rotation of the magnetic bead (which is never quite perfect and hence has a preferred magnetic moment due to anisotropy).

The interesting twist in these particular experiments is that by turning the magnetic particle by putting a torque on the magnetic moment with the help of magnets, we can wind or unwind the DNA strand. This can lead to twists or loops, termed supercoils. These are observed through changes in the effective force F_{DNA}; an example of such data is shown in figure I.4. Such fascinating types of experiments allow researchers to unravel the properties and response of DNA in great detail.

3.6.2 Directed Brownian motion of molecular motors

Molecular motors play an important role in transport in biological cells: these are proteins which move along actin or microtubules in cells, contracting a muscle or pulling their cargo, e.g. a vesicle, behind them. Molecular motors come in two classes: so-called processive and non-processive motors. Processive motors have the ability to make many steps along their track before detaching from it, so a single motor molecule can transport a cargo over a significant distance. In most cases, processive motors have two identical heads and alternately move these in a 'hand-over-hand' fashion illustrated in figure 3.11. Non-processive motors, on the other hand, detach from the track after each step; only by working cooperatively together can such motors transport loads over long distances. Figure 3.12.a illustrates a kinesin motor moving along a microtubule.

The motion of molecular motors had long been modeled as a directed Brownian walk, but it has become possible only rather recently to reveal the stochastic behavior explicitly experimentally, and to probe it in sufficient detail that the data allow us to understand the sequence of events underlying a step at the molecular scale. We illustrate this here with experimental observations on two processive motors, as conceptually these are closest to directed Brownian motion of single walker.

As molecular motors are used for transport or muscle contraction, they perform directed walks. The top panel in figure 3.12 illustrates the steps made by a particular myosin motor which makes large steps of about 40 nm, while pulling a bead along. The data confirm that there is a significant variation in the times between successive steps. Detailed analysis of the data gives evidence that since the steps of the myosin are relatively large, in this case the distribution of waiting times between steps is actually affected by the need for the bead to catch up through Brownian motion, after each step. In other words, the pulling along of the cargo significantly feeds back on the stochastic nature of the motor.

The bottom panel of figure 3.12 shows a trace of another single-headed kinesin motor which makes smaller steps and which is

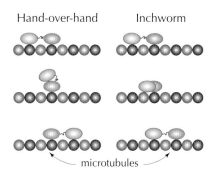

Hand-over-hand   Inchworm

microtubules

Figure 3.11. Illustration of the hand-over-hand and inchworm movement of a Brownian motor. In the hand-over-hand step, one of the heads overtakes the other in each step, while in the inchworm movement one of them is always in front.

attached to a bead in an optical trap. As the force with which the bead pulls the motor back increases, the trace shows that backsteps become more frequent. From the flattening of the curve for large forces, we can read off that the maximum force with which this motor can pull, its so-called stall force, is about 7 pN. Two detachments are also visible. These data contain enough information to estimate the energy barriers for forward and backward steps of a kinesin molecular motor.

3.6.3 Bending modulus or surface tension from shape fluctuation measurements

As we have discussed in section 2.7, bending is the small-energy deformation of thin rods and sheets. There are many examples where this is relevant in biomatter, because of the ubiquity of membranes at the cellular level.

The bending modulus of such small objects (we discuss the bending modulus in more detail in section 7.2) can nowadays be measured quite accurately by studying the thermal fluctuations. Figure 3.13.b shows an example of such a measurement of a giant vesicle with a diameter of about 15 microns. In this case, the shape fluctuations are made visible in a two-dimensional cross section with a thin sheet of laser light, which makes fluorescent probe molecules embedded in the membrane light up.

The analysis to extract the bending modulus from such fluctuation measurements essentially always proceeds along the following lines, as illustrated in figure 3.13.a and detailed in box 3, in section 3.9. One writes the energy of small-amplitude bending modulations in terms of normal modes and then uses the fact that in a thermal bath the average energy associated with each independent bending mode equals $\frac{1}{2}k_\mathrm{B}T$. Since mean square amplitudes of the normal modes can be inferred from the experimental pictures using modern interface tracking algorithms, one can immediately infer the bending modulus from such data.

A particular advantage of measuring properties of matter at a small scale by simply watching them fluctuate in their natural environment, is that it is a non-intrusive experiment. A nice illustration of this can be seen in the experiment of figure 3.13.c on nucleoli, membraneless organellae embedded in chromatin solution inside the cell nucleus (chromatin is the material of which the chromosomes are composed). Since there is no membrane the inner and outer part are separated by what is essentially an interface between two fluids governed by surface tension. The figure illustrates the experiments in vivo on the shape fluctuations and fusion kinetics of such human nucleoli. Panel c1 (top) shows two isolated nucleoli, while panel c2 (top) illustrates how their thermal shape

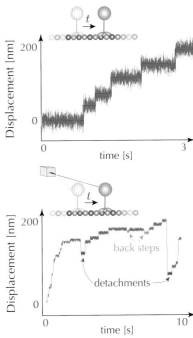

Figure 3.12. Impression of experimental observations of fluctuations in the position of a molecular motor with length and time scales typical for experiments on myosin like those of Yanagida et al., 2008. Top: typical observations for a molecular motor which pulls a bead along, as sketched at the upper left.[18] Note the variation in the time between successive steps. Bottom: when the bead which the motor pulls along is attached to a needle or stylus (or kept in an optical trap) which pulls back on the motor, one can extract the maximum force with which the molecular motor can pull, the so-called stall force. Typical stall forces are a few piconewtons. Also backsteps and detachment events are visible in the data.[19]

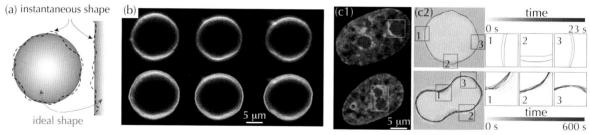

Figure 3.13. (a) Illustration of how one calculates thermal shape fluctuations of rods, flat surfaces or spherical vesicles or drops using the 'four commandments' detailed in section 3.9. For straight objects like a stretched polymer or the fluctuating microtubule of figure 2.23 (for which the bending modulus has also been extracted from fluctuations following these steps), the normal modes are simple Fourier modes. For circular objects like the vesicles (panel b) the normal modes are associated Legendre functions, while if one has access to the full three-dimensional traces, one uses spherical harmonics. The program of calculating mean square shape fluctuations from those of the modes is followed explicitly in section 5.4.4 in the calculation of the force-extension curve of the wormlike chain model. (b) Determination of the bending modulus of a giant vesicle from the experimental observations of the fluctuations by Loftus et al., 2013. Upper row: three images of a giant single lamellar vesicle. Fluorescent probe molecules are included in the vesicles so when these are illuminated by a thin sheet of laser light, the fluctuations are visible in a two-dimensional cross section of the vesicle. Lower row: the same photos with the traces of the edge in light blue, obtained by tracking the circumference digitally. Image courtesy of Raghu Parthasarathy.[20] (c) Determination of the surface tension and viscosities of human nucleoli from their fluctuations and fusion, as observed and analyzed by Caragine et al., 2018. As nucleoli have no membrane, the fluctuations and dynamics are analyzed in terms of two different fluids with a surface tension governing their interface. Panel (c1 top) shows a photo of two isolated nucleoli, while panel (c2 top) shows an example of the minute shape fluctuations which are used to determine the effective surface tension of the interface. The surface tension is found to be more than four orders of magnitude smaller than that of a water-air interface. The column on the right of (c2) shows enlargements of the areas labeled with numbers in the left part. Panel (c1 bottom) shows snapshots from a fusion event, while panel (c2 bottom) shows interface traces just after the fusion. Image courtesy of Alexandra Zidovska.[21]

fluctuations can be tracked with digital techniques to extract the very small surface tension of the interface along the lines sketched above. Interestingly, as the lower image and trace of panels c1 and c2 show, also the fusion behavior of two nucleoli can be tracked in detail, which allows comparison with the highly nonlinear dynamics of necking of fluid droplets. From these measurements also the relative viscosity of the fluids can be determined. The viscosity of the outer fluid is found to be very high, even much larger than that of honey!

3.6.4 Thermal fluctuations in a buckling colloidal chain

Traditionally, the onset of buckling is discussed in the context of macroscopic objects, objects large enough that thermal fluctuations can be neglected and that the appropriate equations are (partial) differential equations. We also followed this route in section 2.8.2. In such an approach, the transition from the unbuckled to the buckled state is sharp. However, buckling effects occur at the smaller scale of the building blocks of soft matter as well: for instance under deformation in polymers, filaments and actin networks.[22] For structures and building blocks of such small size, thermal fluctuations do play a role. At the same time, they are large enough that modern experimental techniques allow us to probe them.

Figure 3.14 shows an illustration of the buckling of a one-dimensional chain of colloidal particles. The top and the bottom colloids are held in place, and as the upper colloid is pushed downward the chain is compressed. At some point, buckling is observed. The system is small enough that thermal fluctuations are significant, as the overlays in the top panel illustrate.

As we illustrated in figure 2.19, there is a strong similarity between transitions like buckling and the Landau theory of phase transitions (see box 2, in section 2.14, for a summary of Landau theory). Near phase transitions, the strength of the fluctuation and their correlation time grows, as they diverge at the phase transition. Based on the analogy with the Landau theory, one expects the same to happen near the buckling transition. The lower panel of figure 3.14 shows that this is indeed borne out by the observations on the buckling chain. A more detailed analysis of the growth of the fluctuation strength is in reasonable agreement with the power law divergence of the Landau theory. It is quite remarkable that the data are precise enough to make such a comparison, given the small system size and the experimental error at such a scale. Numerical simulations of a model for chains of beads with angular-dependent forces and noise allow one to probe the fluctuation behavior in more detail, and they support the expected behavior and experimental observations.[24]

3.7 Probing soft matter with scattering techniques

Scattering techniques play an important role in probing both the structure and the dynamics of soft condensed matter phases. While the principles of X-ray scattering for determining the crystal structure is normally introduced in courses on solid state physics, little attention is usually paid to the use of scattering techniques for probing disordered soft matter, even though the underlying principles are quite the same. Because we will encounter several examples of this throughout the book, we sketch the essentials here.

3.7.1 Essentials of scattering experiments

When one wants to probe atomic structure or arrangements, for instance to determine a crystalline structure, the dependence of the scattering intensity of the X-rays (which mostly scatter off the electron clouds) or neutrons (which scatter primarily off the atomic nuclei, or from magnetic moments via their spin) on the different atoms may be important. This scattering strength translates into

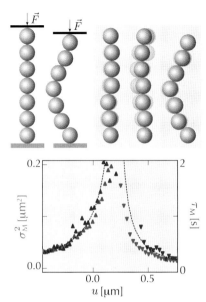

Figure 3.14. Illustration of the buckling of a colloidal chain under compression as observed in the experiments by Stuij et al., 2019. Upper panel: A colloidal chain under compression exhibits buckling as sketched on the left. In the experiment the upper and lower colloid are kept in place with an optical trap, as indicated with the horizontal lines. The position of the upper one is lowered in successive frames, compressing the chain. At sufficient compression the chain is found to buckle as illustrated. Upper right: sketches of three observed chain configurations in the experiments; the fluctuations are indicated by plotting also the chain configuration about a second earlier in light gray. The later ones are shown in dark gray. Some of the beads clearly show significant fluctuations. Lower panel: the mean square fluctuations σ_M^2 and correlation time τ_M of the buckling mode as a function of strain u of the chain, as observed in the experiments. The data are nicely consistent with the behavior of the continuum Landau model behavior summarized in box 2 of section 2.14: as indicated by the black dashed lines, Landau theory predicts a divergence of the fluctuations and the relaxation time at the transition. It is remarkable that this continuum behavior is already observed for such a short colloidal chain. Data credits Simon Stuij and Peter Schall.[23]

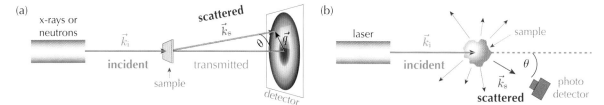

Figure 3.15. Illustration of the fact that an X-ray- or neutron-scattering experiment and a light-scattering experiment are conceptually the same. The scattering wavevector $\vec{q} = \vec{k}_s - \vec{k}_i$ can be changed by changing the scattering angle θ. Panel (a) shows the setup of an X-ray or neutron-scattering experiment, panel (b) the one for a light-scattering experiment.

different so-called form factors of different atoms. Even though one can sometimes take advantage of the different scattering strength of different atoms to single out a particular effect,[25] we will be interested mostly in coarse-grained quantities, like the variation of the density on length scales longer than those of the building blocks. For simplicity, we ignore these form factors, which enter mostly as overall prefactors.

The essence of the treatment of static X-ray scattering that you are probably familiar with, is that waves scatter off the atoms, and that the *scattering intensity* $S(\vec{q})$ is of the form

$$S(\vec{q}) \propto \frac{1}{N} \Sigma_{j,k} \langle \exp(i\vec{q} \cdot \vec{r}_j) \exp(-i\vec{q} \cdot \vec{r}_k) \rangle, \qquad (3.49)$$

where N is the total number of particles. Here $\vec{q}$ is the scattering wavevector, which is related to the wavenumbers $\vec{k}_i$ and $\vec{k}_s$ of the incoming and scattered wave by

$$\vec{q} = \vec{k}_s - \vec{k}_i. \qquad (3.50)$$

> One can understand the origin of this equation as follows. The incoming field at an atom or particle at position $\vec{r}_i$ varies as $e^{i\vec{k}_i \cdot \vec{r}_i}$. Waves scattered from it with wavevector $\vec{k}_s$ expand spherically; at large distances $\vec{r}$, this behavior can be approximated as a planar wave, so the scattered wave varies as $e^{i\vec{k}_i \cdot \vec{r}_i} e^{i\vec{k}_s \cdot (\vec{r} - \vec{r}_i)} = e^{-i\vec{q} \cdot \vec{r}_i} e^{i\vec{k}_s \cdot \vec{r}}$. Formula (3.49) essentially determines the total scattering intensity as the constructive interference of the sum of all the waves scattered off the particles in the sample.

This is represented in figure 3.15, which also illustrates that conceptually X-ray, neutron and laser light scattering are similar, although the apparatus needed and the length scales which are probed are in each case different.

When we consider a crystal with the atoms forming a regular lattice, expression (3.49) implies that the scattering intensity is only nonzero on Bragg spots, where the scattered waves interfere constructively. For disordered matter there are no Bragg spots, but the scattering intensity $S(\vec{q})$ then still gives interesting information on longer scales about the density correlation function. To see this, let us write the microscopic particle density as $\rho(\vec{r}) = \Sigma_j \delta(\vec{r} - \vec{r}_j)$ so that we can write

$$S(\vec{q}) \propto \frac{1}{N} \int d^3\vec{r}_1 \int d^3\vec{r}_2 \, \langle \rho(\vec{r}_1)\rho(\vec{r}_2) \rangle \, \exp[i\vec{q} \cdot (\vec{r}_1 - \vec{r}_2)]. \quad (3.51)$$

This is the general expression. When we probe a homogenous system or an aggregate or polymer on length scales where correlations only depend on the relative distance $\vec{r}_1 - \vec{r}_2$, we may write

$$S(\vec{q}) \propto \frac{\int d^3\vec{r}_1}{N} \int d^3\vec{r} \, \langle \rho(\vec{r}_1)\rho(\vec{r}_1 + \vec{r}) \rangle \exp(i\vec{q} \cdot \vec{r}). \qquad (3.52)$$

In words: the structure factor which one can measure in a scattering experiment is simply the Fourier transform of the density-density correlation function $\langle \rho(0)\rho(\vec{r}) \rangle$. This key result underlies essentially all scattering experiments. We will encounter applications of this formula when we come across the use of static light scattering to probe the fractal dimension of colloidal aggregates in section 4.5.1, the use of neutron scattering to determine the critical exponent of polymers in section 5.5.3, and measurements of the correlation length near the isotropic-nematic transition in liquid crystal in section 6.2.

Dynamic light scattering is conceptually very similar: instead of measuring the total scattered light,[26] one measures the spectrum of the scattered light as a function of frequency ω at each selected scattering vector $\vec{q}$. In direct analogy with (3.52), the scattering intensity $S(\vec{q}, \omega)$ is then the spatial and temporal Fourier transform of the density correlation,

$$S(\vec{q}, \omega) \propto \int d^3\vec{r} \int dt' \, \langle \rho(\vec{r}_1, t)\rho(\vec{r}_1 + \vec{r}, t + t') \rangle \exp(i\vec{q} \cdot \vec{r} - i\omega t'). \qquad (3.53)$$

In dynamical light-scattering experiments the frequency ω with which one probes the physical system according to this formula is the *frequency shift* relative to the frequency of the incoming light,

$$\omega = \omega_s - \omega_i, \qquad (3.54)$$

just as according to (3.50) the scattering wavevector $\vec{q}$ is the change in wavevector. Dynamic light scattering is also called inelastic light scattering, because of this change in the frequency (or energy) of the light. Figure 3.16 illustrates such a measured spectrum for the case of light scattering off the thermal density fluctuations in water. Nowadays, also inelastic or dynamic X-ray scattering is possible, but we won't encounter examples of this in this book, because we will be interested mostly in behavior on length scales much larger than the interatomic distances, which are probed by X-rays.

Indeed, in a scattering experiment the scattering wavenumber can be made small by making the scattering angle θ small, while the maximum scattering wavenumber q is about $2k_i$. Hence light scattering is very well suited for probing matter on length scales relevant to soft matter, on the order of the wavelength of light and up. But we will come across neutron scattering as well: with modern neutron sources which can produce sufficiently mono-energetic beams, one can use neutrons to probe length scales on the order of

Dynamic light scattering has a long history and an enormous range of applications in soft matter research, so there are many books introducing the field. The 1976 classic by Berne and Pecora, 2000 still provides a nice entry and overview. A very accessible later account is the book by Schmitz, 1990. For an introductory overview of the various probes of spatial structure we refer you to section 2.3.2 of Witten and Pincus, 2004, while section 2.2. of Chaikin and Lubensky, 1995 gives more background on scattering.

Note that the frequency shifts of the light are typically quite small: for the sound waves probed in figure 3.16 one has $\omega = c_s q$ with c_s the speed of sound given in equation (1.35) while for the laser light $\omega_i = \mathcal{O}(ck_i)$ with c the speed of light. In water $c_s / c \approx 4 \cdot 10^{-6}$ so the frequency shifts are only on the order of parts per million.

a few tenths of a nanometer. So neutrons are often used to probe structure at this length scale; this is relevant to probing the conformational scaling of polymers in solution in section 5.5.3.

3.7.2 Probing small fluctuations in continuum systems with laser light scattering

Laser light scattering is a powerful and versatile tool with which to explore the propagation and decay of small fluctuations at the length scales of light. Hence, such experiments yield information about coarse-grained behavior governed typically by continuum equations. It is a technique which is quite a bit older than that of optical traps: it came to fruition relatively soon after the invention of the laser.

With dynamical scattering techniques one essentially probes according to (3.53) the space-time Fourier transform of the density-density correlation function, and we saw that by picking a scattering angle one selects the scattering wavevector $\vec{q}$ with which one probes the sample. The temporal behavior of the density correlation function determines the frequency spectrum of the scattered light: exponential damping of a mode gives essentially a Lorentzian line in the spectrum, with width proportional to the damping rate. Moreover, a finite frequency shift $\Delta\omega$ in the spectrum relative to the carrier frequency indicates a propagating density mode of the type $\cos(qx - \Delta\omega t)$: it is as if the mode is Doppler shifted by this propagating mode.

Even though fluctuations are typically small in a fluid, at least away from the critical point, minute density fluctuations are still measurable with light scattering. The theoretical calculation of small thermal fluctuations in such cases essentially proceeds along lines that can be thought of as the extension to continuum systems of the Langevin-equation analysis for a Brownian particle in section 3.2.

In figure 3.16 we show the spectrum obtained with laser light scattering from a simple fluid. The two peaks with a finite freqency shift indeed correspond to scattering off thermally excited sound waves. For a given wavenumber q this shift is simply $\Delta\omega = c_s q$, with c_s the speed of sound discussed in section 1.7; the width of these peaks gives the damping of such waves.[28] Such scattering techniques are nowadays part of the standard toolbox of soft matter researchers, to explore soft matter phases, for instance the effect of adding surfactants to a surface or of additives to a fluid.[29]

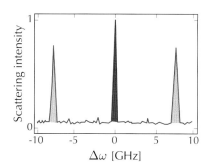

Figure 3.16. Example of a light-scattering spectrum of a fluid like water. The two peaks at finite frequency correspond to scattering of light from thermally excited sound waves which propagate with the speed of sound c_s; the central peak corresponds to scattering off thermal fluctuations which diffuse out as described by equation (1.28).[27]

As fluctuations are small, the coarse-grained dynamics of fluctuations is governed by the linear equations obtained by linearizing the continuum equations about the stationary state. We can view these (partial differential) equations as the generalization of the friction law (3.10), $\mathrm{d}\overline{V}/\mathrm{d}t = -\Gamma\overline{V}$, for the average velocity of a Brownian particle. For a translation-invariant system, the different spatial Fourier modes are the proper normal modes, so the mean square amplitudes of the modes are obtained in accord with the 'four commandments' of section 3.9 by writing the energy in these modes and then equating the thermal average energy of each mode with $\frac{1}{2}k_B T$ per degree of freedom. The temporal behavior of each mode is governed by the behavior of the linearized equations—their structure determines whether the mode is propagating or not, and their damping.

3.8 What have we learned

On the mesoscopic scale the effect of thermal fluctuations on a system can conveniently be modeled with Langevin equations. The strength of these fluctuating forces is determined by the fluctuation-dissipation theorem, which relates fluctuation phenomena and the dissipative terms. Specialized to a Brownian particle in a fluid, this expression amounts to the Stokes-Einstein relation which gives its diffusion coefficient in terms of the temperature, the fluid viscosity and the particle radius.

For particles in water, the crossover scale that distinguishes small particles which disperse throughout the container from large ones that settle to the bottom is about one micron.

The Langevin formulation is equivalent to a Fokker-Planck equation for the probability distribution. The Langevin approach is often most convenient for cases in which the equations governing the dynamics of the fluctuations can be taken to be linear; the Fokker-Planck approach compares more directly to experimental measurements of probability distributions and is more convenient for analyzing nonlinear situations, like the Kramers problem of the rate of escape over a high barrier.

The master equation is a suitable description for the evolution of the probability distribution from the transition rates between the states of the underlying process. It models these changes as a Markov process, in which transitions only depend on the present state, not on its history. The Langevin and Fokker-Planck description effectively also describe Markov processes.

The possibility of observing and manipulating microscopic beads with optical tweezers, and of measuring effective forces from the thermal fluctuations of these beads, has provided new avenues to observe the elastic response of soft matter and biomatter at the small scale, even of single molecule properties of DNA and other biomolecules. Likewise the bending moduli or surface tensions of biomembranes and surfaces are accessible through simple observation of their fluctuations.

Scattering techniques, like laser light scattering, are powerful methods to probe soft matter at suitable length scales. Such techniques essentially allow us to measure the Fourier transform of the density-density correlation function.

3.9 Box 3: Calculating thermal averages

Box 3. The four commandments for calculating averages of thermal fluctuations

Whenever you need to calculate the thermal average of small-amplitude fluctuations of an inherently translation- or rotation-invariant extended system, you should essentially follow step-by-step the following 'four commandments':

1. Write the free energy to quadratic order in the amplitude of the fluctuations. In the base state the energy is minimal, so there are no linear terms.

2. Write the fluctuation amplitudes in terms of Fourier modes[30] (or spherical modes for spherically symmetric problems), to find that these diagonalize the free energy—in other words, different modes are treated as independent, i.e. as uncoupled Gaussian variables.

3. Use equipartition of energy to calculate the thermal average of each mode: the average energy of each mode has energy $1/2 \, k_\mathrm{B} T$.

4. Revert to real space to calculate the thermal average of any quantity you want.

It is good to be aware that there is actually a subtlety underlying the analysis. The Fourier components $A(q)$ and $A(-q)$ of a real physical quantity $A(x)$ are actually *not* independent, as $A(-q) = A^*(q)$, where the star denotes complex conjugation. In other words, the modes q and $-q$ are actually not independently fluctuating quantities even though at first it looks as if they are treated as independent. A more careful analysis is based on treating the real and imaginary terms separately; these turn out to be proper, independent Gaussian-distributed variables. The analysis shows that in the end the handwaving argument obtained by assigning each mode q an energy $1/2 k_\mathrm{B} T$ actually does give the right answer.[31]

The general approach is discussed implicitly or explicitly in many books, among them Landau and Lifshitz, 1980. Hydrodynamic fluctuations are treated in detail in chapter 9 of Landau and Lifshitz, 1981b. Note in addition that the linear equations governing the dynamics of fluctuations in spatially extended systems can be turned into Langevin-type equations by the addition of fluctuating force terms which are delta-correlated in space and time. Conceptually, this is a straightforward generalization of the Langevin equation for a Brownian particle in this chapter.

We encounter this program in a large variety of problems:

▶ the fluctuations of tubules shown in figure 2.23.[32]

▶ the shape fluctuations of a droplet or cell as in figure 3.13.b; in this case, the relevant modes are not Fourier waves, but spherical harmonics, on account of the spherical symmetry.[33]

▶ the calculation of the density fluctuations of fluids or of thermally excited capillary waves at a fluid-air interface discussed in the context of light scattering in section 3.7.2.[34]

▶ the calculation of the force-extension curve of the wormlike chain model for biopolymers in section 5.4.4.

▶ the orientational fluctuations in nematic liquid crystals treated in problem 6.12, and of those in smectic and columnar phases in problem 6.13.

▶ the analysis of the effect of fluctuations on the nematic-to-smectic-A transition treated in problem 6.4.

3.10 Problems

Problem 3.1 *Review of probability concepts*

The probability distribution for a Gaussian variable X with mean μ and variance σ^2 is given by

$$p(x) = \frac{1}{\sqrt{2\pi}\sigma} \exp\left[\frac{(x-\mu)^2}{2\sigma^2}\right]. \tag{3.55}$$

For this exercise, you may use without proof the evaluation of the Gaussian integral:

$$\int_{-\infty}^{+\infty} dx \exp(-ax^2) = \sqrt{\frac{\pi}{a}}. \tag{3.56}$$

a. If x_1 and x_2 are independent Gaussian variables with means μ_1, μ_2 and variances σ_1^2, σ_2^2, respectively, show that the variable $y = x_1 + x_2$ is also Gaussian with mean $\mu_1 + \mu_2$ and variance $\sigma_1^2 + \sigma_2^2$, as follows. The probability density for y is given by

$$
\begin{aligned}
p(y) &= \int dx_1 dx_2 \, p(x_1, x_2) \, \delta(y - x_1 - x_2) \\
&= \int dx_1 dx_2 \, p_1(x_1) p_2(x_2) \, \delta(y - x_1 - x_2) \\
&= \frac{1}{2\pi\sigma_1\sigma_2} \int dx_1 \exp\left[-\frac{1}{2\sigma_1^2}(x_1 - \mu_1)^2 \right. \\
&\quad \left. -\frac{1}{2\sigma_2^2}(y - (\mu_1 + \mu_2) - (x_1 - \mu_1))^2 \right].
\end{aligned}
\tag{3.57}
$$

b. Complete the square to show that

$$p(y) = \frac{1}{\sqrt{2\pi}\sqrt{\sigma_1^2 + \sigma_2^2}} \exp\left[-\frac{(y - \mu_1 - \mu_2)^2}{2(\sigma_1^2 + \sigma_2^2)} \right]. \tag{3.58}$$

c. Calculate the second and fourth moments of the Gaussian distribution relative to the mean, μ_2, and μ_4. Verify that the so-called kurtosis $\mu_4 / \mu_2^2 = 3$.

d. The generating function $G(z)$ of a discrete distribution is defined by $G(z) = \sum_0^\infty \langle z^n \rangle$. What is the generating function of the binomial distribution $p(n) = \binom{N}{n} p^n q^{N-n}$, where $q = 1 - p$?

e. Using this generating function, calculate the mean, variance, and skewness, $\gamma_1 \equiv \mu_3 / \mu_2^{3/2}$, of the binomial distribution.

Relevant coding problems and solutions for this chapter can be found on the book's website www.softmatterbook.online under Chapter 3/Coding problems.

The generating function $G(z)$ of some probability distribution over a random variable X can be constructed as $G(z) = \langle e^{zX} \rangle$. The nth moment of the distribution can then be found by evaluating $\frac{\partial^n G(z)}{\partial z^n}\Big|_{z=0}$. If you are not familiar with the binomial distribution, consult any introductory book on statistics or statistical physics, like Kestin and Dorfman, 1972 or Pécseli, 2000.

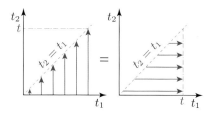

Figure 3.17. Illustration of the fact that we can rewrite the double integrals for $\Delta X(t)$ to get (3.59).

Problem 3.2 *Derivation of diffusion from the Langevin equation (3.6)*

a. Integrate equation (3.11) once to obtain the expression for the deviation $\Delta X(t) = X(t) - \overline{X(t)}$ from the mean position.

b. Rewrite the double integral that you got in the previous step with the help of figure 3.17 in terms of different integration variables, such that one of the integrations can be trivially performed, to obtain

$$\Delta X(t) = \frac{1}{\Gamma} \int_0^t dt_2 \left(1 - e^{-\Gamma(t-t_2)}\right) L(t_2). \tag{3.59}$$

c. Use the above expression to write $\sigma_X^2(t) = \overline{(\Delta X(t))^2}$ as a double integral and show that after two integrations this gives

$$\sigma_X^2(t) = \overline{(\Delta X(t)^2)} = \frac{\gamma}{\Gamma^3} \left(2\Gamma t - 3 + 4e^{-\Gamma t} - e^{-2\Gamma t}\right). \tag{3.60}$$

d. Expand the above result for small times, to show that $\overline{(\Delta X(t))^2}$ increases only as t^3. Can you explain the t^3 behavior intuitively, based on the earlier result derived for the mean square velocity fluctuations?

e. Analyze the above expression for large times to show that it reduces to (3.13).

Problem 3.3* *The Kramers problem: Derivation of the escape rate over a high barrier*

In this exercise, following Kramers, you will solve the Fokker-Planck equation for the case mentioned in section 3.3.6, the case of a Brownian particle in a one-dimensional potential well $U(x)$.

We assume that the barrier energy E_b is large and that the particle is initially within the well. For a high barrier, the particle has enough time to equilibrate within the well, in other words, we expect the particle to reach a close-to-equilibrium distribution within the potential well before crossing the barrier at b to a state c. The situation is sketched in figure 3.6. We start the analysis from the Fokker-Planck equation (3.30).

a. Demonstrate that (3.30) can be rewritten in the form

$$\frac{\partial \mathcal{P}(X,t)}{\partial t} = -\frac{\partial J}{\partial X}, \tag{3.61}$$

with the flux

$$J(X,t) = -De^{-U(X)/k_B T} \frac{\partial}{\partial X} \left[e^{U(X)/k_B T} \mathcal{P}(X,t)\right]. \tag{3.62}$$

b. After a very short initial relaxation, the particle at $X < X_b$ will reach a close-to-equilibrium state, and there is only a very small constant probability current J_0 across the barrier, i.e., $\partial \mathcal{P}/\partial t \approx 0$ for $X < X_b$, so that $J(x, t) \approx J_0$. Demonstrate that

$$J_0 = \frac{D e^{U(X_a)/k_B T} \mathcal{P}(X_a)}{\int_a^c e^{U(X')/k_B T} dX'},\tag{3.63}$$

where c is some point far beyond the barrier (on the dashed part of the curve in figure 3.6).

Hint: Use (3.62) with $J = J_0$ and integrate the equation from X_a to X_c, and the fact that $\mathcal{P}(X_c)$ is small (think about why this is so).

c. Note that, in the well, the distribution will essentially be the equilibrium distribution

$$\mathcal{P}(X) = \mathcal{P}(X_a) e^{-[U(X) - U(X_a)]/k_B T}.\tag{3.64}$$

Starting from this result, show that the probability of finding a particle within the well is

$$p = \mathcal{P}(X_a) \left[\frac{2\pi k_B T}{|U''(X_a)|} \right]^{1/2}.\tag{3.65}$$

Hint: Use the fact that the potential well is very deep, Taylor expand $\mathcal{P}(X)$ around a, and then integrate the resulting Gaussian integral. Note that the first derivative of $U(X)$ vanishes at a.

d. Show that the escape rate over the barrier, $r \equiv J/p$ can be expressed as

$$r = \frac{D}{2\pi k_B T} [|U''(X_a) U''(X_b)|]^{1/2} e^{-[U(X_b) - U(X_a)]/k_B T}.\tag{3.66}$$

Hint: Taylor expand J from (3.63) around b.

Problem 3.4 *The master equation for a particle on a one-dimensional lattice*

In this exercise we use the master equation (3.41) to analyze the stochastic hopping of a particle on a one-dimensional lattice, as sketched in figure 3.7. Lattice points are labeled by the integer numbers n, and the lattice spacing of the points is a. We consider the situation where the particle makes hops one step to the right (left) with probability r_1 (l_1) or hops two steps to the right (left) with probability r_2 (l_2).

a. Give all the nonzero transition probabilities $W_{nn'}$ for the master equation (3.41).

b. Write out the master equation (3.41) for the probability p_n.

c. Based on the fact that for long times the probabilities will approach values given by a smoothly varying probability distribution $P(X,t)$, associate X with point n and write $p_{n-1} = P(X-a,t)$, and similarly for other nearby points. Use Taylor expansion to show that the continuum equation for P can be written as

$$\frac{\partial P(X,t)}{\partial t} = -c\frac{\partial P}{\partial X} + D\frac{\partial^2 P(X,t)}{\partial X^2}. \tag{3.67}$$

d. Why is it allowed to cut the derivatives at second order?

e. Give the expression for c and D in terms of r_1, r_2, l_1 and l_2.

f. Show that c has the interpretation of the average speed with which the hopping particle is moving to the right or left. What determines whether it moves right or left?

Hint: See section 3.3.5.

Problem 3.5 *Random walk in a quadratic potential*

Note that in problem 3.4 we treated a particle on a homogeneous lattice, while in the present problem the hopping probability depends on the lattice position.

A particle, confined to a one-dimensional lattice along X, is subjected to random steps on a lattice $\{0, \pm 1, \pm 2, \ldots, \pm N\}$ in a potential $U = \frac{1}{2}kX^2$. In a time step $t \to t+1$, it takes *at most* one step to the right or left. At the point $X = n$, the probability of a move to the right, $n \to n+1$, is $(1 - \frac{1}{2}kn)D$ and the probability of a move left, $n \to n-1$, is $(1 + \frac{1}{2}kn)D$, where $kN \ll D$ and $N \gg 1$. At $t = 0$ the particle is placed at position $X = X_0$ and released.

a. Derive the master equation for the probability that the particle is at X_n at time $t+1$, $P(n, t+1)$, in terms of $P(n+1, t)$, $P(n,t)$, and $P(n-1,t)$.

b. Approximate the discrete master equation with continuous variables X and t and derive the Fokker-Planck equation:

$$\frac{\partial P(X,t)}{\partial t} = D\left[\frac{\partial^2 P(X,t)}{\partial X^2} + k\frac{\partial}{\partial X}(XP(X,t))\right]. \tag{3.68}$$

c. Determine the probability distribution at equilibrium, $P_{eq}(X)$, explaining your reasoning.

d. Before equilibrium is attained, $P(X,t)$ takes the form

$$P(X,t) = \frac{1}{\sqrt{2\pi\sigma^2(t)}}\exp\left(-\frac{(X-\mu(t))^2}{2\sigma^2(t)}\right). \tag{3.69}$$

Using equation (3.68) and integrating by parts, show that:

$$\frac{d\mu}{dt} + Dk\mu = 0, \tag{3.70}$$

$$\sigma \frac{d\sigma}{dt} + Dk\sigma^2 = D. \tag{3.71}$$

e. Using these results, find μ and σ^2 as functions of t.

Problem 3.6 *Master and Fokker-Planck equations: Chemical reactions*

Consider the following chemical reaction system:

$$A \rightarrow B, \tag{3.72}$$
$$B \rightarrow A. \tag{3.73}$$

Reactions (3.72) and (3.73) occur at rates k_1 and k_2, respectively. Let $P(x, t)$ be the probability of the concentration of A, i.e., the number of molecules of species A in the reactor, equal to x at time t. The concentration of B follows from conservation of the total concentration of A and B, equal to M.

a. Derive the following difference equation for the temporal evolution of the probability distribution:

$$
\begin{aligned}
P(x, t + \Delta t) &= P(x + 1, t)\, k_1\,(x + 1)\,\Delta t \\
&\quad + P(x - 1, t)\, k_2\,(M - x + 1)\,\Delta t \\
&\quad + P(x, t)\,(1 - k_1 x\, \Delta t)\,(1 - k_2\,(M - x)\,\Delta t) \\
&\quad + P(x, t)\, k_1 x\, \Delta t\, k_2\,(M - x)\,\Delta t,
\end{aligned}
\tag{3.74}
$$

where you may assume that two or more reactions of the same type in the time interval Δt are negligible.

b. Derive the following master equation by taking the limit in which Δt becomes infinitesimally small, i.e., $\Delta t \rightarrow dt$,

$$
\begin{aligned}
\frac{\partial P(x, t)}{\partial t} &= P(x + 1, t)\, k_1\,(x + 1) - P(x, t)\, k_1 x + P(x - 1, t) \\
&\quad \times k_2\,(M - x + 1) - P(x, t)\, k_2\,(M - x). \tag{3.75}
\end{aligned}
$$

c. By applying Taylor's approximation to $f(x + 1) \equiv P(x + 1, t)\, k_1\,(x + 1)$ and $g(x - 1) \equiv P(x - 1, t)\, k_2\,(M - x + 1)$ to the second order, derive the Fokker-Planck equation from the chemical master equation.

d. In equilibrium, assuming that M is sufficiently large that both $xP(x)$ and $(M - x)P(x) \rightarrow 0$ as $x \rightarrow M$ and $x \rightarrow 0$, show (explaining your reasoning) that

$$
\begin{aligned}
k_1 x P(x) &+ \frac{k_1}{2} \frac{d}{dx}\,[x P(x)] \\
&= k_2(M - x)P(x) - \frac{k_2}{2} \frac{d}{dx}\,[(M - x)P(x)]. \tag{3.76}
\end{aligned}
$$

e. Now taking $k_1 = k_2 = k$, solve equation (3.76) to obtain the equilibrium probability distribution. What are its mean and variance?

Problem 3.7 *Population dynamics: The Lotka-Volterra model*

Consider a simple model of the populations of foxes and rabbits on a shared terrain. Rabbits feed on grass, of which there is unlimited supply, and breed at a constant rate μq_2 (where q_2 is the number of rabbits), but unfortunately all eventually die by being eaten by foxes, none dying of natural causes. Foxes feed only on rabbits; when a fox eats a rabbit, another fox is immediately born and this process occurs at a rate $\lambda q_1 q_2$, where q_1 is the number of foxes. Foxes die of natural causes at a rate σq_1.

a. Consider a time interval $(t, t + \delta t)$. Assuming that no more than one of the above processes may occur in this time interval, derive the master equation for the probability $P(q_1, q_2, t + \delta t)$ that there are q_1 foxes and q_2 rabbits at time $t + \delta t$:

$$
\begin{aligned}
P(q_1, q_2, t + \delta t) = {} & P(q_1, q_2 - 1, t)\mu(q_2 - 1)\delta t \\
& + P(q_1 + 1, q_2, t)\sigma(q_1 + 1)\delta t \\
& + P(q_1 - 1, q_2 + 1, t)\lambda(q_1 - 1)(q_2 + 1)\delta t \\
& + P(q_1, q_2, t)(1 - \mu q_2 \delta t - \sigma q_1 \delta t - \lambda q_1 q_2 \delta t).
\end{aligned}
\tag{3.77}
$$

b. Derive the Fokker-Planck equation:

$$
\begin{aligned}
\partial_t P = {} & \sigma \left[\partial_1 + \frac{1}{2}\partial_1^2 \right] q_1 P + \mu \left[-\partial_2 + \frac{1}{2}\partial_2^2 \right] q_2 P \\
& + \lambda \left[\partial_2 + \frac{1}{2}\partial_2^2 - \partial_1 + \frac{1}{2}\partial_1^2 + \partial_1 \partial_2 \right] q_1 q_2 P.
\end{aligned}
\tag{3.78}
$$

See section 3.3.5 and problem 3.4 for a discussion on why approximating the finite differences by derivatives is justified.

You may use the following continuous derivative replacements, valid for any function $A(q_1, q_2)$:

$$
\begin{aligned}
\partial_1 A(q_1, q_2) &= \frac{1}{2}\left[A(q_1 + 1, q_2) - A(q_1 - 1, q_2) \right], \\
\partial_1^2 A(q_1, q_2) &= A(q_1 + 1, q_2) - 2A(q_1, q_2) + A(q_1 - 1, q_2), \\
& \left[(\partial_2 - \partial_1) + \frac{1}{2}(\partial_2 - \partial_1)^2 \right] A(q_1, q_2) \\
&= A(q_1 - 1, q_2 + 1) - A(q_1, q_2).
\end{aligned}
\tag{3.79}
$$

c. An alternative formulation of this problem is as a pair of coupled first order differential equations:

$$
\frac{\partial q_1}{\partial t} = -\sigma q_1 + \lambda q_1 q_2,
\tag{3.80}
$$

$$\frac{\partial q_2}{\partial t} = \mu q_2 - \lambda q_1 q_2. \tag{3.81}$$

Defining dimensionless variables (Q_1, Q_2, τ) through

$$q_1 = \frac{\mu}{\lambda} e^{\epsilon Q_1}, \qquad q_2 = \frac{\sigma}{\lambda} e^{Q_2/\epsilon}, \qquad t = \frac{\tau}{\sqrt{\mu\sigma}}, \tag{3.82}$$

with $\epsilon \equiv \sqrt{\frac{\sigma}{\mu}}$, show that

$$\frac{\partial Q_1}{\partial \tau} = -1 + e^{Q_2/\epsilon} \tag{3.83}$$

$$\frac{\partial Q_2}{\partial \tau} = 1 - e^{\epsilon Q_1}. \tag{3.84}$$

d. By looking at equations (3.80) and (3.81), determine what are the two trivial and one nontrivial fixed points in (Q_1, Q_2)-space.

e. We define the Hamiltonian

$$H = \frac{1}{\epsilon}(e^{\epsilon Q_1} - 1) - Q_1 + \epsilon(e^{Q_2/\epsilon} - 1) - Q_2. \tag{3.85}$$

Show that Q_1 and Q_2 form a canonical pair, i.e.,

$$\partial_\tau Q_1 = \partial_{Q_2} H, \tag{3.86}$$

$$\partial_\tau Q_2 = -\partial_{Q_1} H. \tag{3.87}$$

Calculate $\partial_\tau H$. Interpret this result in terms of the dynamics of the populations of foxes and rabbits.

f. For small values of (Q_1, Q_2), expand H to second order to determine the shape of the contours of constant H and plot these contours in (Q_1, Q_2)-space.

Problem 3.8 *Thermal fluctuations in a diode: Master equation*

This problem deals with Alkemade's diode, which consists of two electrodes of different materials with work functions W_1 and W_2, connected to a capacitor with capacitance C. The probability that an electron leaves electrode 1 in a time interval dt is given by

This problem is inspired by section 12.1 of the book by Pécseli, 2000.

$$P_1 dt = \frac{4\pi m}{h^3} (kT)^2 A e^{-\frac{W_1}{kT}} dt \equiv \xi dt, \tag{3.88}$$

where A is the area of the electrode. When there is an imbalance of N electrons between the two halves of the circuit, the probability $P_2 dt$ that an electron leaves electrode 2 toward electrode 1 (overcoming the potential difference) is

$$P_2 dt = \xi e^{-\frac{W_2 - W_1}{kT}} e^{-\frac{e^2}{kTC}\left(N + \frac{1}{2}\right)} dt. \tag{3.89}$$

a. Let $P(N,t)$ denote the probability of N excess electrons on electrode 1 at time t. Show that, at time $t + \Delta t$, $P(N, t + \Delta t)$ is given by

$$
\begin{aligned}
P(N, t + \Delta t) = {} & P(N,t)\left[1 - P_1 \Delta t\right]\left[1 - P_2(N)\Delta t\right] \\
& + P(N-1, t)\, P_2(N-1)\, \Delta t\left[1 - P_1 \Delta t\right] \\
& + P(N+1, t)\, P_1 \Delta t\left[1 - P_2(N+1)\Delta t\right] \\
& + P(N,t)\, P_1 \Delta t P_2(N)\, \Delta t,
\end{aligned}
\tag{3.90}
$$

where the probability of two or more electrons being emitted in Δt is negligible. Give an interpretation of each term.

b. Ignoring terms containing $(\Delta t)^2$ and using the explicit expressions for P_1 and P_2, obtain the following master equation:

$$
\begin{aligned}
\frac{1}{\xi}\frac{\partial P(N,t)}{\partial t} = {} & P(N+1,t) - P(N,t) \\
& + \alpha\left[e^{-(N-1)\epsilon}P(N-1,t) - e^{-N\epsilon}P(N,t)\right].
\end{aligned}
\tag{3.91}
$$

What are ϵ and α?

c. By rearranging terms in (3.91), show that the equilibrium solution $P_{eq}(N)$ satisfies:

$$
P_{eq}(N+1) - \alpha e^{-N\epsilon}P_{eq}(N) = \text{const., independent of N.} \tag{3.92}
$$

What is that constant? (Note that (3.92) holds also for $N \to \infty$ when $P_{eq}(N) \to 0$.)

d. Show by iteration that

$$
P_{eq}(N) = e^{-\epsilon\frac{N^2}{2} + \eta N}P_{eq}(0), \tag{3.93}
$$

and state the value of η. Treating $P_{eq}(0)$ as normalization, show that

$$
P_{eq}^{-1}(0) = \sum_{-\infty}^{+\infty} e^{-\epsilon\frac{N^2}{2} + \eta N} \approx \sqrt{\frac{2\pi}{\epsilon}}\, e^{\frac{\eta^2}{2\epsilon}}, \tag{3.94}
$$

approximating the sum with an integral.

e. Show that

$$
\langle N \rangle_{eq} = \frac{\eta}{\epsilon} = \frac{(W_1 - W_2)C}{e^2}, \tag{3.95}
$$

and that the mean square fluctuations of the voltage V are

$$
\left\langle (V - \langle V \rangle_{eq})^2 \right\rangle_{eq} = \frac{k_\mathrm{B}T}{C}. \tag{3.96}
$$

Note that the nonlinearity in the diode builds up a net voltage $-\frac{e}{C}\langle N\rangle_{eq} = -\frac{(W_1 - W_2)}{e}$.

f. Considering that $\frac{d\langle N\rangle}{dt} \equiv \frac{d\sum N P(N)}{dt}$, derive

$$\frac{1}{\xi}\frac{d\langle N\rangle}{dt} = -1 + e^{\eta - \frac{\epsilon}{2}}\left\langle e^{-\epsilon N}\right\rangle. \tag{3.97}$$

Equation (3.97) can be solved approximately in powers of ϵ by Taylor expanding $\left\langle e^{-\epsilon N}\right\rangle$ and doing a moment expansion, $\left\langle N^2\right\rangle$, $\left\langle N^3\right\rangle, \ldots$, but it's too difficult! Instead, consider the macroscopic features of the diode in the limit of vanishing fluctuations c.

g. Take the limit of (3.97) when $C \to \infty$, $N \to \infty$, and $\epsilon \to 0$ in such a way that the voltage $-\frac{eN}{C}$ is fixed. Upon introducing the current $I = e\frac{d\langle N\rangle}{dt}$, show that the standard diode characteristic is obtained as

$$I = e\xi\left(e^{\frac{eV}{k_B T}} - 1\right), \tag{3.98}$$

where

$$V = \frac{W_1 - W_2}{e} - \frac{eN}{C}. \tag{3.99}$$

Problem 3.9** *Fermi acceleration: Generating function*

Figure 3.18 shows the trajectory of a *massive* particle (or piston) moving with a sawtooth-like oscillatory motion (blue) with a constant velocity of magnitude either $+1/2$ or $-1/2$. The period of oscillation is taken to be $2t_0$ and the peak-to-peak amplitude is then $1/2t_0$. A *light* particle (red) is colliding with the massive one. Let the velocity of the light particle be u before impact. After the collision, its velocity will be $u + 1$ if it happens to be reflected head-on, or $u - 1$ in the case of an overtaking collision. Assume that the light particle arrives at the position of the heavy one with equal probability within any time interval $(t, t + \delta t)$.

This problem is based on chapter 13 of the book by Pécseli, 2000.

a. Show that the transition probabilities are

$$P(u \to u + 1) = \frac{1}{2} + \frac{1}{4u}\ , \quad P(u \to u - 1) = \frac{1}{2} - \frac{1}{4u}. \tag{3.100}$$

Hint: From figure 3.18, note that the transition probabilities $u \to u + 1$ and $u \to u - 1$ are in the ratio DC to AD. Note that the probabilities above are given that a reflection occurs.

b. Assume that the probability of a collision in a time interval dt is given by $2u\,dt$. Show that the probabilities relating a velocity u at time t to the velocity at $t + dt$ are

$$P(u + 1|u) = \frac{2u + 1}{2}dt,$$

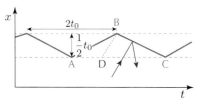

Figure 3.18. A space-time diagram illustrating Fermi acceleration.

$$P(u|u) = 1 - 2u\mathrm{d}t,$$

$$P(u-1|u) = \frac{2u-1}{2}\mathrm{d}t. \tag{3.101}$$

These equations describe a Markov process.

c. To simplify the problem, assume that the light particle starts out with velocity $\frac{1}{2}$. All later velocities are $\frac{1}{2} + k$, where k is an integer. Show that the probability of being in state k is

$$P_k(t + dt) = [1 - (2k+1)dt]P_k(t) + [k\,dt]P_{k-1}(t)$$
$$+ [(k+1)dt]P_{k+1}(t). \tag{3.102}$$

d. Derive the master equation

$$\frac{dP_k(t)}{dt} = -(2k+1)P_k(t) + kP_{k-1}(t)$$
$$+ (k+1)P_{k+1}(t), \tag{3.103}$$

with initial conditions $P_0(0) = 1$ and $P_{k>0}(0) = 0$.

e. To determine the evolution of $P_k(t)$, we have to determine the temporal evolution of $P_{k-1}(t)$, $P_{k-2}(t)$, etc. The complications associated with this hierarchy of equations can be circumvented by introducing the *moment generating function*

$$G(x,t) \equiv \left\langle x^k \right\rangle = \sum_{k=0}^{\infty} P_k(t)x^k. \tag{3.104}$$

Show that the master equation can be reformulated in terms of $G(x,t)$ as

$$\frac{\partial G(x,t)}{\partial t} = (1-x)^2 \frac{\partial G(x,t)}{\partial x} - (1-x)G(x,t), \tag{3.105}$$

for $0 \le x \le \infty$.

f. Show by substitution that the solution to equation (3.105) is

$$G(x,t) = \frac{1}{1 + (1-x)t}, \tag{3.106}$$

with $G(x,0) = 1$, as required by the normalization of P.

g. Show that

$$\left\langle \frac{k!}{(k-n)!} \right\rangle = n!\, t^n. \tag{3.107}$$

Hint: Apply $\frac{\partial^n G(x,t)}{\partial x^n}\big|_{x=1}$ to equations (3.104)–(3.106).

h. Use (3.107) to calculate $\langle k \rangle$ and $\langle k^2 \rangle - \langle k \rangle^2$. Will the light particle(s) attain an equilibrium velocity with a well-defined (finite) scatter?

i. Consider k as a continuous variable and derive the following equation:

$$\frac{\partial P(k,t)}{\partial t} = k \frac{\partial^2 P(k,t)}{\partial k^2} + \frac{\partial P(k,t)}{\partial k}. \tag{3.108}$$

j. What are the steady-state solutions of equation (3.108)? Can they be normalized? Compare your answers to your conclusions in step h for the discrete model. Are they consistent?

Problem 3.10** *The Dean equation*

This problem shows how to derive approximate hydrodynamic equations from a system of Langevin equations, following a method which is widely employed to derive continuum theories of active matter from microscopic models.

We follow the method introduced by Dean, 1996.

We assume our system has N particles each with n degrees of freedom $\vec{R}^\alpha(t) \in \mathbb{R}^d$, where $\alpha \in \{1, \ldots, N\}$ indexes the particles. The particles obey the following dynamics:

$$\frac{d\vec{R}^\alpha}{dt} = \sum_{\beta=1}^{N} \vec{F}\left(\vec{R}^\alpha(t) - \vec{R}^\beta(t)\right) + \vec{G}(\vec{R}^\alpha(t)) + \vec{\xi}^\alpha(t). \tag{3.109}$$

Here, $\vec{F}$ is a pairwise force with $\vec{F}(0) = 0$, and $\vec{\xi}^\alpha$ is Gaussian white noise with correlations $\langle \xi_i^\alpha(t) \xi_j^\beta(\tau) \rangle = 2T \delta_{ij} \delta^{\alpha\beta} \delta(t - \tau)$ and where G is an arbitrary position-dependent function (e.g., coming from an external potential). We define the one-particle density function as

$$\rho^\alpha(\vec{R}, t) = \delta(\vec{R}^\alpha(t) - \vec{R}), \tag{3.110}$$

and our goal is to get a closed equation for the total density

T is a temperature-like variable which sets the noise strength in this general formulation, very much as the temperature sets the noise strength in the case of a Brownian particle obeying the fluctuation-dissipation theorem.

$$\rho(\vec{R}, t) = \sum_{\alpha=1}^{N} \rho^\alpha(\vec{R}, t). \tag{3.111}$$

a. Consider an arbitrary function, $f(\vec{R})$. Show that

$$f(\vec{R}^\alpha(t)) = \int d^d\vec{R} \, \rho^\alpha(\vec{R}, t) f(\vec{R}). \tag{3.112}$$

b. Similarly, show that

$$\frac{df(\vec{R}^\alpha(t))}{dt} = \int d^d\vec{R} \, \rho^\alpha(\vec{R}, t) \frac{df(\vec{R})}{dt}. \tag{3.113}$$

Start with the noisy differential equation

$$\frac{\mathrm{d}x}{\mathrm{d}t} = -kx + \sqrt{2D}\xi(t) \qquad (3.126)$$

for the position x of a particle in one dimension. Here, ξ is Gaussian white noise with correlations $\langle \xi(t)\xi(t')\rangle = \delta(t-t')$. The quantities k and D are parameters. This stochastic process is known as an Ornstein-Uhlenbeck process. It represents an overdamped Brownian particle in a harmonic potential with stiffness k, an example of which is a colloid trapped by optical tweezers.

See section 3.7.2 for a discussion of the utility of optical tweezers in soft matter experiments.

a. Using the Dean equation but ignoring the noise term, show that

$$\partial_t \rho(x,t) \simeq D\partial_x^2 \rho(x,t) + \partial_x \left(kx\rho(x,t)\right). \qquad (3.127)$$

b. Show that you recover the standard diffusion equation when $k = 0$.

c. We now consider the case where $k \neq 0$. Verify that

$$\rho^{\mathrm{ss}}(x) = \frac{1}{\sqrt{2\pi\sigma^2}} \exp\left(-\frac{x^2}{2\sigma^2}\right), \qquad (3.128)$$

where $\sigma^2 = D/k$ is a stationary solution of (3.127).

d. What would happen to a Gaussian with a different standard deviation than the one in (3.128)? What if it were not centered at zero?

e. Verify that your answer matches the one derived using the Fokker-Planck equation in section 3.3.3. You will encounter nontrivial applications of the Dean equation in the active matter chapter and problems.

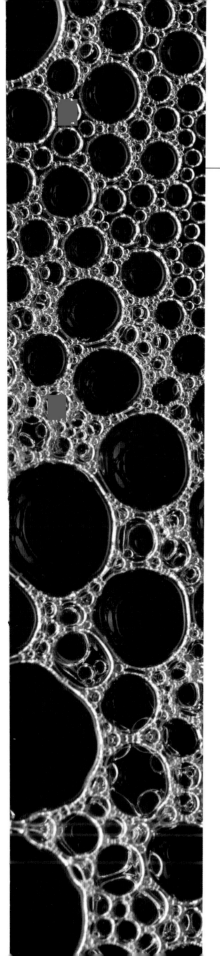

Part II

SOFT MATTER PHASES

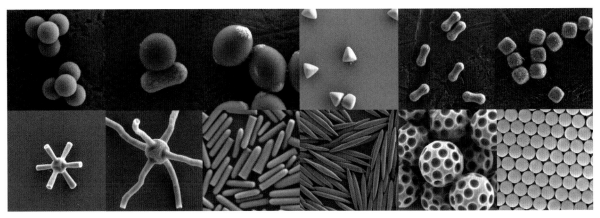

Figure 4.2. Collage of particles of various shapes and sizes which can nowadays be synthesized, illustrating the enormous control experimentalists have gained over the growth of such particles. Images courtesy of Stefano Sacanna.

4.1.1 Colloids: Fundamental studies

Note that we can even consider wet foams as a form of a heterogeneous phase of gas bubbles. While such wet foams indeed share some properties with emulsions, the behavior of dry foams is dominated by the surface tension trying to minimize the area of the surface separating them. See, e.g., Weaire and Hutzler, 2001.

Check out the website softmatterdemos.org for interactive demonstrations of self-assembly of a number of colloidal phases.

An old brute-force method of making colloids is to grind up macroscopic particles. It should come as no surprise that this leads to a large variety in shapes and sizes. Over time, researchers have been able to control the bottom-up synthesis of increasingly monodisperse colloids with specific shapes: initially spheres (Overbeek, 1982), then rods, then plates and nowadays even cubes as figure 4.2 shows. Polymers like polystyrene and PMMA (polymethylmethacrylate) are popular for making spheres, silica can be used for making rods with desired aspect ratio (Kuijk et al., 2011). The silica reaction is controlled by having the spheres grow within droplets in an emulsion!

Colloids and emulsions are of a great interest, both from a scientific point of view and from an applied point of view. Indeed, colloidal systems are very nice model systems with which to study a great variety of fundamental physical phenomena: properties like their size and shape, as well as their composition and interactions, can nowadays be engineered almost at will, as figure 4.2 beautifully illustrates. Moreover, colloidal particles are large enough to be optically detectable and small enough to execute Brownian motion and hence to explore phase space, and they are also easy to simulate on a computer. In combination with modern data processing techniques, which allow large sets of particles to be traced automatically, colloids are nowadays used to study defects in crystals, the formation of crystal phases, the properties of glasses, packing effects, defects, grain boundaries, phase transitions and active media. Colloidal systems are even being explored to study fundamental questions of biology, like replication, a topic touched on in chapter 10.

4.1.2 Colloids: Application perspective

Colloids and emulsions are ubiquitous in the natural world around us, and in applications. Just think of foods like mayonnaise, milk, or cornstarch, or of paint, clay, etc.—do have another look at figure I.7 and figure I.8 of the introductory chapter. Figure 4.1 shows a rod-shaped virus particle occurring in nature; there are in fact quite a few virus colloids with all kinds of curious shapes. But you can nowadays also put colloids to use in many ways—e.g., by irradiating plasmonic nanoparticles with a laser you can locally heat them, and use this for catalytic conversion, solar energy harvesting, biomedical imaging technique or cancer therapy. Moreover, as we mentioned already in section 1.15, emulsions are often used in industry for liquid-liquid extraction, which is one of the core processes in chemical industry to transfer the solute from one solvent to another. Of course, practically relevant systems are typically not as 'clean' as the model systems studied in the science labs, but in practice basic research and practical applications of colloids are closely intertwined. Nowadays there is a whole toolbox that can be applied to control their shape, size and types of interactions, and researchers in the field are very clever in engineering the desired properties.

One reason that colloids and emulsions are so important technologically is that they have a lot of surface: this speeds up processes like reactions. For instance, 5 liters of emulsion paint consists typically of about a kilogram of polymers (plus some pigment and other ingredients) dispersed in water in little spheres of about 200 nm radius. As a result, the total area of interface between water and polymer is about 15,000 m^2: colloid science is intimately connected with surface science.

4.2 Colloids as a thermodynamic system with effective interactions

It gradually became clear in the previous century that it pays to view colloids in a dispersion as a thermodynamic system of particles with effective interactions in its own right. In other words, we essentially view the surrounding fluid to simply play the role of a thermal bath which gives rise to the Brownian motion that makes the colloidal particles explore configuration space. This picture implies that once we understand the effective interactions between the colloids we can focus on understanding their thermodynamic phases by applying the tools and methods of equilibrium statistical physics to particles endowed with these effective interactions.

4.2.1 Hard core particles: Model system with entropic interactions

Historically, the interest in this approach was boosted by the fact that computer simulations of molecular systems or of particles with well-defined interactions became readily available in parallel with the experimental developments. A particular attractive model system for computer studies are hard core particles, such as spheres: particles with a well-defined shape (like spheres or rods) which do not interact until they touch, and which are impenetrable. So their interaction is infinite when they touch and zero otherwise. This means that hard core systems are governed by entropy alone— there is no energy scale, and their phase behavior can be studied with Monte Carlo simulations that sample the configuration space. Such simulations of hard core particles played an important role in showing that (liquid) crystal ordered phases can emerge from entropy effects alone: we can have 'order from disorder' if in the ordered phase particles have more configuration space to explore—

Hard core spheres also play a role in the field of jamming discussed in section 2.10. The jamming point at a volume fraction $\phi_J \approx 0.64$ is indeed the random close packing density of hard core spheres; the spheres then have no room to move anymore. In comparison: the packing density of a face-centered cubic (fcc) lattice of spheres is $\pi/3\sqrt{2} \approx 0.74$. So clearly at densities below this value, hard core spheres have a finite entropy when ordered fcc-like, while a random packing cannot exist.

the entropy is larger in the ordered than in a disordered phase. When particles are anisotropic, these effects can be very strong. In fact, simulations of hard core particles played an important role in the realization that the elongated shape of particles is sufficient to give many liquid crystal phases discussed in chapter 6: most if not all of these phases can be reproduced with properly shaped hard core particles without any attractive interactions. Indeed, figure I.11 in the introductory chapter already illustrated that almost any crystalline order can be obtained as the equilibrium phase of some properly designed hard core particles.

In colloid science, these developments stimulated a drive to control the interactions between the particles in experiments, in particular to make them as much as possible behave like hard core particles, and to reduce their polydispersity, that is the variation in their dimensions. The extent to which this can nowadays be achieved, not only for spherical particles but also rods, plates, cubes, and other shapes, is remarkable and one of the reasons why colloids are such a favorite model system and stepping stone for applications in new directions.

4.2.2 Colloids tend to attract

The present experimental control of the interactions is quite an achievement—and a bit of an art—because there are three general mechanisms which make colloidal particles naturally *attract* each other: the Van der Waals interaction, the depletion interaction, and the interaction resulting from the effect of their perturbation on their vicinity. Consequently, unless one takes special precautions, colloids tend not to form a stable dispersion: the attractive forces make them aggregate and stick together, and once the resulting clusters are sufficiently large, they settle to the bottom or flocculate (form flakes). Because of this natural tendency to cluster, in order to stabilize colloidal dispersions, one has to prevent the particles from sticking. We will first consider the origin of the attractive forces in the next section, and then discuss how one can use charges or little polymer bumpers to prevent aggregation. After we have laid the groundwork, we will discuss a number of applications.

We estimate the barrier needed to prevent coagulation in problem 4.1.

4.3 Naturally occurring attractive forces between colloids

In this section we review the three types of attractive forces that naturally occur between colloidal particles in a fluid.

4.3.1 The Van der Waals attraction

a. Van der Waals' attraction between two atoms

We will first show how if we have two simple atoms, the fluctuating dipole moments lead to an attractive force which falls off as $1/r_{12}^6$ as a function of their separation r_{12}. After discussing the origin of this *Van der Waals interaction*, we will proceed to analyze the resulting effective interaction between two bodies, like two spheres or two planar interfaces.

The origin of this general attractive potential lies in a fluctuation-induced dipole interaction. For a permanent dipole $\vec{\mu}$ in an electric field $\vec{E}$, the potential energy is

$$U = -\vec{\mu} \cdot \vec{E}. \tag{4.1}$$

The minus sign here expresses that the energy is lowest when the dipole is pointing in the same direction as the electric field. Now, if an atom does not have a permanent dipole, a field will generally induce a dipole moment

$$\vec{\mu}_{\text{ind}} = \alpha \vec{E}, \tag{4.2}$$

with α the *polarizability* of the atom. As a result, the energy associated with the induced dipole of the atom is

$$U_{\text{ind}} = -\int_0^E dE' \, \mu_{\text{ind}}(E') = -\tfrac{1}{2}\alpha E^2 = -\tfrac{1}{2}\mu_{\text{ind}}(E)\,E. \tag{4.3}$$

Let us now consider two identical atoms, a distance r_{12} apart, in the absence of an external field. Even if the atoms do not have a permanent dipole moment, due to quantum mechanical fluctuations they have a fluctuating dipole moment; the Van der Waals interaction originates from the interaction of each dipole with the field mediated by the other atom. For simplicity we will assume atom 1 to have at some instant a fluctuating dipole moment μ_1 pointing toward atom 2 as in the top panel of figure 4.3, $\vec{\mu}_1 \parallel \vec{r}_{12}$. At atom 2 the field E_2^{ind} generated by this fluctuating dipole moment is given by the usual dipole formula from electrostatics

$$E_2^{\text{ind}} = \frac{2\mu_1}{4\pi\epsilon_0 \, r_{12}^3}. \tag{4.4}$$

This induces a dipole $\mu_2 = \alpha E_2^{\text{ind}}$, and this again causes a small field

$$E_1^{\text{ind}} = \frac{2\mu_2}{4\pi\epsilon_0 \, r_{12}^3} = \frac{2\alpha}{4\pi\epsilon_0 \, r_{12}^3}\frac{2\mu_1}{4\pi\epsilon_0 \, r_{12}^3} \tag{4.5}$$

at atom 1 which points in the same direction as $\vec{\mu}_1$ and hence which lowers the energy according to (4.3). If we put all this together, we

For an introduction with focus on applications, see Cosgrove, 2010. This book also treats emulsions. Our discussion of the forces between colloids is motivated by the discussion of these by Witten and Pincus, 2004.[3] The book by Russel et al., 1989 and overview by Pusey, 1991 still provide a very accessible introduction to and overview of colloidal dispersions up to 1989. A modern overview of the various forces and the implications for numerical simulations of colloids is given by Dijkstra and Luijten, 2021. The book edited by N. J. Wagner and Mewis, 2021 provides a modern perspective on colloid rheology.

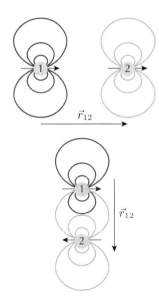

Figure 4.3. Illustration of how an instantaneous dipole moment of atom 1 generates a dipole field (blue) which induces a dipole at 2. The induced dipole at 2 creates in turn a dipole field (gray) which acts back on atom 1 in such a way that the energy is lowered. Top panel: when $\vec{\mu}_1 \parallel \vec{r}_{12}$ the dipole $\vec{\mu}_2$ points in the same direction as $\vec{\mu}_1$. Right panel: when $\vec{\mu}_1 \perp \vec{r}_{12}$ the induced dipole $\vec{\mu}_2$ points in the opposite direction from $\vec{\mu}_1$ but again the field generated at atom 1 by this dipole (gray) points in the same direction as $\vec{\mu}_1$: there is a small angular dependence due to the strength of the dipole field depending on the angle between $\vec{\mu}$ and $\vec{r}_{12}$, but the effect is always of the same order and such that it lowers the energy.

see this gives for the interaction energy $U_{1,2} = -\frac{1}{2}\mu_1 E_1^{\text{ind}}$ of the pair, due to these dipole fluctuations of atom 1

$$U_{1,2} = -2\alpha \frac{\mu_1^2}{(4\pi\epsilon_0)^2\, r_{12}^6}. \tag{4.6}$$

Of course, the dipole moment is fluctuating in arbitrary directions, so we still need to average over both the size of μ_1 and the direction in which it is pointing, and add $U_{2,1}$ from fluctuations of atoms 2 as well. But as the bottom panel of figure 4.3 illustrates for the case that $\vec{\mu}_1 \perp \vec{r}_{12}$, the energy is always lowered, and in order of magnitude we get an expression of the form (4.6) with μ_1^2 replaced by $\langle \mu_1^2 \rangle$ and with a slight correction from the angular average.

In order of magnitude the dipole fluctuations will be $\langle \mu_1^2 \rangle = \mathcal{O}(ea)^2$, where a is the size of the atom, and e as usual its charge, and the polarizability α is typically of $\mathcal{O}(4\pi\epsilon_0\, a^3)$. In any case, the upshot is that due to these dipole fluctuations, two atoms generally attract each other with a power law interaction

$$\text{Van der Waals' attraction:} \quad U_{12} = -C\left(\frac{a}{r_{12}}\right)^6, \tag{4.7}$$

where C is a positive constant with dimension of energy and order of magnitude comparable to the binding energy of an atom, several electron volts: $C = \mathcal{O}(\text{several eV})$.

Repulsive forces between two atoms typically become dominant at distances less than 2 to $3a$, so there will be a minimum in the total effective potential of order $U_{12}(2-3a) = \mathcal{O}(0.01-0.1\text{eV})$. This is in the range of $k_B T \approx 0.025$ eV, so thermal fluctuations will typically be able to drive the two particles out of this potential well.

For molecular dynamics simulations, in which the classical equations of motion are solved for a collection of atoms or molecules to study, e.g., the equilibrium phase diagram, one needs a good model-potential for the atomic interactions. A very popular model, which exhibits the Van der Waals r_{12}^{-6} power law behavior at long distances and a steep repulsion at small distances, is the so-called Lennard-Jones pair potential

$$U_{\text{LJ}}(r_{ij}) = \varepsilon \left[\left(\frac{d}{r_{ij}}\right)^{12} - \left(\frac{d}{r_{ij}}\right)^6 \right],$$

for every pair of atoms ij with distance r_{ij}. This potential, which is quite accurate for noble gases, just has two constants, ε which sets the energy scale, and d which sets the distance scale—it is the distance at which $U_{\text{LJ}} = 0$. The minimum of the potential is at $r/d = 2^{1/6}$.

So far, we considered two atoms in free space. In a solvent, C depends on the difference in polarizabilities and the dielectric constant, which is frequency-dependent. So a full analysis takes into account the frequency dependence of these quantities and involves an integral over all fluctuating frequencies. Even when all the details are accounted for, the effective potential is still of the form (4.7),[4] although the value of C between molecules in a solvent is different (typically smaller) from the one for molecules in free space.

b. Effective Van der Waals interaction between colloids

So far we considered the Van der Waals interaction between two molecules in free space or a solvent. If we consider colloids, we have objects consisting of lots of molecules that are different from the solvent molecules. We are then interested in the effective interaction $U(h)$ between two colloidal particles 1 and 2 a distance h apart. This effective interaction results simply from the sum total of the Van der Waals interaction

$$U(h) = -\Sigma_{\vec{r}_1} \Sigma_{\vec{r}_2} C \left(\frac{a}{r_{12}} \right)^6 \qquad (4.8)$$

of the constituent atoms. Here $\vec{r}_1$ and $\vec{r}_2$ denote the positions of the atoms or molecules in colloid 1 and colloid 2, respectively.

This effective Van der Waals interaction between objects can be worked out for various cases, by converting the sums to integrals on account of the fact that the Van der Waals interaction itself is slowly varying as r_{12}^{-6}. The case of two semi-infinite slabs separated by a distance h and with lateral area A is simplest. As we derive in problem 4.1, the effective interaction falls off as h^{-2},

$$\text{slabs:} \quad U_{\text{VdW}}(h) = -\frac{H\,A}{12\pi\,h^2}, \qquad (4.9)$$

where H is the so-called Hamaker constant. H has the dimension of energy; its expression in terms of the parameters of the Van der Waals interaction is derived in problem 4.2.

While for slabs the variation is simply h^{-2}, the variation for other objects depends of course on their shape and, for nonsymmetrical objects, even on their relative orientation. It can be worked out explicitly along the lines of the calculation of problem 4.1 for slabs, the integrals only become more intricate the more complex the shape of the objects is. For two spheres of radius R, the effective result for the interaction as a function of h, the minimum distance between their two surfaces, is

$$\text{spheres:} \quad U_{\text{VdW}}(h) = -\frac{H\,R}{12\,h}, \qquad (h \ll R) \qquad (4.10)$$

in the limit when the two spheres are very close, $h \ll R$—see problem 4.3. In the intermediate range $h \approx R$ the complete formula has to be used, which for very large distances $h \gg R$ crosses over to a h^{-6} behavior.

In conclusion, even though the precise functional form depends on the shapes and relative distance, the important result of this discussion is that mesoscopic objects generally attract each other with a slowly varying power law potential.

4.3.2 Depletion interaction

The depletion interaction[5] has a very simple origin. As figure 4.4 illustrates, when two large particles in a thermal bath of smaller ones approach one another a distance comparable to the size of the smaller particles, these smaller ones start to be excluded from a small zone in between the large particles. The effect is stronger

Table 1.5 of Kleman and O. D. Lavrentovich, 2003 lists the effective Van der Waals interaction for a number of relevant cases like spheres, rods, sphere and slab, and cylinders.

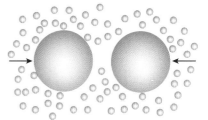

overlap volume

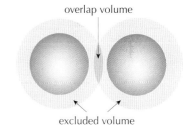

excluded volume

Figure 4.4. Illustration of the origin of the depletion interaction. Top panel: when the two large particles approach each other by a distance of order the size of the smaller particles, the small particles cannot enter the region in between them—this region is 'depleted.' As a result, the particles feel an effective attractive force. The effect is very much like an osmotic pressure. The bottom panel illustrates another way to think about this: around each particle, there is an excluded volume, which the small particles cannot enter due to their finite radius. When the two large particles come close together, the total excluded volume is decreased, as the overlap zone of the excluded volumes increases. A decrease of the excluded volume enhances the entropy of the small particles, and hence it induces an effective attraction.

the closer the big particles approach each other. The result of this is that when this happens, the large particles feel a net pressure inward from the small particles, as there are more particles that exert pressure on the outer sides. In other words, there is an effective attractive force with a range of order the size of the small particles. Note by the way that the smaller particles need not be colloidal particles—they can also be polymer blobs.[6]

Another way to view this is illustrated in the bottom panel in figure 4.4: around each large particle, there is an 'excluded volume' from which the small particles, due to their finite size, are excluded. When the larger particles near each other, the excluded volume zones overlap, and the total excluded volume decreases. Hence it is favorable for particles to be close to each other. For this reason, the depletion interaction is sometimes called an excluded volume interaction. It is an entropic interaction, since when the excluded volume decreases, the small particles have more room to explore entropically. For spherical colloids, the overlap volume is simple to calculate geometrically and one can then estimate the excluded volume force as the product of the osmotic pressure of the small particles, times the excluded volume—see problems 4.4 and 4.5. Excluded volume interactions are typically not very strong; typically the effective potential is of order $k_B T$.

4.3.3 Induced attractive interaction due to perturbations of the surrounding medium

Colloidal particles normally disturb their environment, in that in their neighborhood they change the surrounding packing, density, composition, the orientational correlations, etc. When these perturbations do fall off smoothly away from the colloidal particle instead of in an oscillatory fashion, this usually leads to an effective attractive interaction. Figure 4.5 illustrates how for colloidal particles at a fluid interface this is mediated by a change in shape of the meniscus. The origin of the effective attraction is that if a particle affects property W in its neighborhood, then clearly another particle of the same type will feel happier in the neighborhood of the original one, where property W is already affected in a way it likes. It thus leads to a perturbation-induced attraction.

Clearly, the precise nature of the attraction depends very much on the problem under consideration, so a general discussion of these attractive forces is hard to give. Generally, the functional form of the distance dependence of the attraction is set by the way in which small perturbations of the field mediating the interaction die out—in other words, it reflects the nature of the correlations and the response. For instance, if changes in the order parameter are well described by a Landau-type theory, in which perturbations of the

Figure 4.5. Colloids tend to be attracted to a fluid interface—the so-called Pickering effect—simply because they reduce the area of the fluid-air interface and hence the total interfacial area. The figure illustrates how two colloids attract each other through the deformation of the interface: one way of understanding this is as a lowering of the gravitational energy, as together they sink somewhat deeper. An alternative way to understand the attraction is based on the realization that the contact angles on both sides of the particle (red arrows; see section 1.13.1) make a slightly different angle with the horizontal, so that the resultant force drives the particles toward each other. The deflections of the fluid surface typically extend over a range of order the capillary length discussed in section 1.12, hence this sets the range over which the attractive force extends. For a general discussion of how perturbations of the surrounding medium give rise to attractions, see in particular Witten and Pincus, 2004; they refer to the effect as the *Perturbation-Attraction Theorem* to stress its general validity.

order parameter die off exponentially, so does the effective inter-action.[7] While if the interaction is mediated by elastic fields, it can be power-law like. Note that the interaction of defects discussed in section 2.11 is an example of a power law interaction mediated by the elastic field, and that in such cases defects with the same topo-logical charge actually repel each other, while those with opposite charge attract. This may serve as a warning that one cannot blindly apply the principle that perturbations of the local environment lead to attractive interactions. Returning to colloidal particles in a fluid, it is worth mentioning that interesting effects also occur close to a critical point, where the correlation length grows: the effective in-teraction mediated by the fluid perturbations then actually become very long ranged.

The conclusion from this discussion is that we should be aware that colloidal particles generally also experience attractive interactions induced by the way they perturb their local environment, but that the strength and range of the interaction depends very much on the situation at hand.

This effect, often called the critical Casimir effect in view of its similarity to Casimir forces resulting from quantum mechanical modes, was first predicted by M. E. Fisher and de Gennes, 1978. For an observation of colloidal aggregation induced by the attrac-tive force from critical Casimir fluctuations in a solvent close to its critical point, see Bonn et al., 2009b. For a general review see Krech, 1994 and Gambassi, 2009.

4.4 Repulsive forces

We have seen in the previous section that colloidal particles natu-rally tend to attract each other, and hence form aggregates or floc-culates instead of staying dispersed. In order to prevent this from happening, experimentalists have developed clever ways to engi-neer the particles with repulsive forces that keep them apart. We now discuss two important tricks with which to stabilize the disper-sions by endowing colloids with repulsive forces, either by charges or by coating the particles with a polymer bumper or brush.

4.4.1 Electrostatic stabilization

An important way to stabilize a colloidal dispersion is by effective charges on the particles; such surface charges often naturally occur because surface molecules ionize in solution. The effective inter-action between particles with charges of the same sign is however not a simple Coulomb $1/r$ repulsion, because their charges get *screened* by ions of opposite charge which are attracted to them. This effect, which is illustrated in figure 4.6, is especially important in polar solvents like water with a large dielectric constant. This for instance causes dissolved salt molecules like NaCl to split into ions which only weakly interact. Because of the strong Coulomb forces, the system should be essentially charge neutral in total, but this still allows for formation of a charge layer near the colloid which screens its effective charge over a distance which depends on the

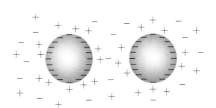

Figure 4.6. Illustration of charge screening in a fluid. In this case the two colloidal particles are negatively charged. As a result, positive ions are attracted to the colloids and *screen* the charge. This is often referred to as the formation of a *double layer*, as the negatively charged colloid is surrounded by a positively charged cloud. This double layer formation leads to a screening of the Coulomb poten-tial over a distance κ^{-1} given by equation (4.15). Surface charges often naturally oc-cur on colloids because in solution surface molecules become ionized, especially in po-lar solvents.[8]

concentration of ions. We treat the basic description of the effect here, the so-called Debye-Hückel screening.

Consider a solution of a monovalent salt like NaCl in a polar solvent in which the molecules are split completely into Na^+ and Cl^- ions. We denote the total NaCl concentration by c, the local charge density by ρ, the total anion concentration by c_+ and the cation charge concentration by c_-. From electrostatics we then have the following relation between the electric potential Φ and the charge density:

$$\epsilon_0 \epsilon_r \nabla^2 \Phi(\vec{r}) = -\rho(\vec{r}), \tag{4.11}$$

where ϵ_0 is the dielectric permittivity of vacuum and ϵ_r the relative dielectric permittivity. In the solution, the ions are mobile, and are able to rapidly respond to the forces they experience. It is therefore natural to assume that the average ion density will be the one in which the ions are in local equilibrium with the average potential that they feel, i.e., the one given by the Boltzmann distribution,

$$c_\pm(\vec{r}) = c_\pm^0 \exp(\mp e\,\Phi(\vec{r})/k_B T). \tag{4.12}$$

The overall constraint $2c = V^{-1} \int d^3\vec{r}\,(c_+(\vec{r}) + c_-(\vec{r}))$, with V the volume, fixes the constants $c_\pm^0$. If there is overall charge neutrality, as we will assume, and if the density of charged colloids is low, then the total number of ions in the screening layers is small leading to

$$c_\pm^0 = c. \tag{4.13}$$

Combination of (4.11) and (4.12) together with the fact that the total charge density $\rho = ec_+ - ec_-$ yields a self-consistent partial differential equation for the potential $\Phi(\vec{r})$,

$$\epsilon_0 \epsilon_r \nabla^2 \Phi(\vec{r}) = -ec\left[\exp(-e\,\Phi(\vec{r})/k_B T) - \exp(+e\,\Phi(\vec{r})/k_B T)\right]. \tag{4.14}$$

This is called the Poisson-Boltzmann equation. It is a nonlinear equation, as the potential appears in the exponent. However, for not too large potentials, we can expand the exponentials up to first order in the arguments, to get the so-called Debye-Hückel equation

$$\nabla^2 \Phi(\vec{r}) = \kappa^2 \Phi(\vec{r}) \quad \text{with} \quad \kappa^2 = \frac{2e^2 c}{\epsilon_0 \epsilon_r\, k_B T} = 8\pi \lambda_B c. \tag{4.15}$$

Here λ_B is the so-called Bjerrum length,

$$\lambda_B = \frac{e^2}{4\pi \epsilon_0 \epsilon_r\, k_B T}, \tag{4.16}$$

which is an intrinsic quantity of the solvent. For water at room temperature, this Bjerrum length is about 0.7 nm.

The term κ in (4.15) plays the role of an *inverse screening length*. Indeed, if you solve this equation in one dimension, the decaying

Note that this formulation in terms of the average quantities is a type of mean-field approximation in which we ignore all correlations, and which is valid if there are many ions within a screening length. In terms of the quantities defined below this is the case if $\kappa \lambda_B \ll 1$.

solutions are exponentials $\exp(-\kappa|x|)$, while the solution for a point charge in three dimensions is the so-called Yukawa potential of nuclear physics (see problem 4.3)

$$\Phi_{\text{DH}}(r) = \frac{e\,e^{-\kappa r}}{4\pi\epsilon_0\epsilon_r\,r}. \tag{4.17}$$

For a small salt concentration of about 1 millimolar in water, the charge screening length $\kappa^{-1} \approx 10$ nm at room temperature. Hence moderate salt concentrations are sufficient to screen out charge differences over a reasonably short length scale, about a hundredth of a typical colloid diameter.

The above expression is the expression for the potential of a point charge. It is easy to derive this expression and the Debye-Hückel potential around a spherical colloid with charge Ze:

$$\text{Sphere:} \quad \Phi_{\text{DH}}(r) = \frac{Ze}{4\pi\epsilon_0\epsilon_r(\kappa R + 1)}\frac{e^{-\kappa(r-R)}}{r}. \tag{4.18}$$

See problem 4.6. In practice, one may also have to take into account the fact that mobile counterions from the solvent tend to form a little layer around the colloid, making their effective charge smaller than their bare charge.

The essence of this discussion shows that by charging colloids in a polar solvent, one gets repulsive electrostatic forces that are screened out over distances of order κ^{-1}. Screening of electric charges also plays an important role for many biopolymers in solution; see section 5.4.3.

In figure 4.7 we show the effective interaction between colloids according to the so-called DLVO theory, in which the interactions are modeled by combining the attractive Van der Waals potential with the repulsive screened Debye-Hückel potential. This DLVO potential is often the starting point for the analysis of colloidal interactions.

4.4.2 Steric stabilization by grafting polymers onto the surface

One way to stabilize the colloids at short range against the attractive forces, is by grafting polymers onto the surface that act as a 'bumper'. This is illustrated in figure 4.8: the colloids effectively have a polymer bumper which keeps them from getting too close. For this to work, the polymers in the brush have to 'like' the solvent, which as we will discuss in chapter 5 means that the solvent is a 'good solvent' for the polymer. Typically uncharged hydrocarbon-based polymers do not like polar solvents such as water, so the

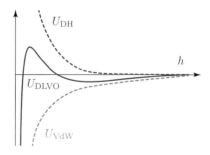

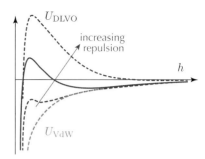

Figure 4.7. The starting point of most treatments of colloidal interactions is the DLVO theory, named after Derjaguin, Landau, Verwey and Overbeek. The DLVO interaction is the sum of the Van der Waals attraction of section 4.3.1 and the electrostatic repulsion between colloids. Upper panel: sketch of the DLVO potential (solid line) as a function of the distance h between the colloids. Lower panel: development of the DLVO potential as the repulsive charge interaction increases from curve 1 to 3. For further discussion see problem 4.7 and Lekkerkerker and Tuinier, 2011 and Israelachvili, 2011. Modern experiments have shown that deviations from DLVO theory typically only manifest themselves at distances below a few nanometers (Smith et al., 2020).

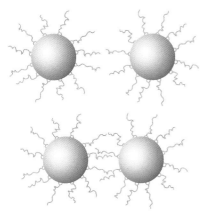

Figure 4.8. Colloids can also be endowed with a short-range steric repulsion, by grafting them with polymers. The coating of the particles by a polymer brush acts like a bumper which prevents the colloids from coming too close to one another. When the colloids are sufficiently separated as on the top, the polymer brushes don't interact, while if they get close the polymers interact, resulting in a steric repulsion. Beware that such cartoons are never drawn to scale: the polymer brushes are often about 10–20 nm, so they are relatively thin compared to the typical micron scale of the colloids.

An early example of the unexpected use of colloids which is fun to mention for historical reasons is the case of the so-called long time tails. Back in 1967 some of Alder and Wainwright's first molecular dynamics simulations of equilibrium fluids had shown that the velocity correlation function of particles decayed in d dimensions as a power law $t^{-d/2}$ with time, rather than as an exponential (Alder and Wainwright, 1967). This came as a big surprise at the time, and it gave rise to a flurry of theoretical work.[9] The phenomenon eluded experimental observation for quite a while, until it was demonstrated by precise light-scattering experiments of Paul and Pusey, 1981 and Ohbayashi et al., 1983 that the Brownian diffusion of colloidal particles has indeed the expected subdominant correction as a result of this long time tail.

interaction has to be tuned carefully by the choice of the polymers and by optimizing their length and the density of what is called a polymer brush. We will not go into these details here but will treat some of the ingredients of the analysis of polymer brushes in section 5.6.

Of course, by grafting the colloids with polymers can never completely undo the long-range attractive tail of the Van der Waals attraction, but as long as it prevents the colloids from getting so close that the attraction grows to the order of $k_B T$ or larger, the colloids can be made to behave to a good approximation as hard core particles.

4.5 Playing with colloids as model systems

We already encountered in section 2.11 an interesting example of how a well-controlled colloidal dispersion provides a great experimental model system with which to probe the two-stage melting process in two dimensions. By a slight tilting of the sample a small density gradient was created, which allowed probing of the various phases in different parts of the sample, to extract the equation of state of the various phases and even to create an interface between the various phases. In this section we further illustrate the versatility of colloids as interesting model systems with which to explore diverse basic physical phenomena.

4.5.1 Colloidal aggregates

We have seen that if one does not take special precautions, colloidal particles attract each other, in particular due to the relatively long-range Van der Waals interaction. As a result, they then tend to form aggregates: particles diffuse around until they hit one or more others, and then stick to these; so one gets ever-growing clusters of particles. Since in this case the diffusion of particles is the rate-limiting step, this process is referred to as diffusion-limited aggregation or DLA. An example of the type of tenuous structures one gets in this regime is shown in figure 4.9; it is generated from a computer simulation of 10^7 particles.

The DLA growth process is easily simulated on a computer[10] and this allows tracking of the growth history: the color darkness in the computer-grown cluster in figure 4.9 indicates when the particles joined the cluster: the darker ones are released first, the lightest ones the last. If you look carefully, you'll see that the light green ones have mostly stuck on the tips of the outer branches.

How to characterize these structures, which clearly do not fill space, but instead are open and very fluffy, with much open space? Indeed, they are nice examples of so-called fractal structures, which have an effective scaling dimension lower than the bulk dimension. To explain this concept, consider a particle at the center of the cluster, and denote by $\rho(r)$ the density of particles at a distance r from it. For an ordinary homogeneous bulk material, the density $\rho(r)$ is clearly constant as long as we are in the cluster, and hence the total mass $M(r)$ of the particles with a radius r goes as

$$\text{ordinary matter:} \quad M(r) = \int_{r' < r} \mathrm{d}^d r' \rho(r') \sim r^d. \quad (4.19)$$

Fractal aggregates are objects which are scale-invariant, and which have a particle density which falls off with distance r as a power law,

$$\text{fractals:} \quad \rho(r) \sim r^{d_\mathrm{f} - d} \quad \Longleftrightarrow \quad M(r) \sim r^{d_\mathrm{f}}, \quad (4.20)$$

which clearly generalizes (4.19) and justifies the name *fractal dimension* for d_f. This result confirms that fractals with $d_\mathrm{f} < d$ are very fluffy, as $\rho(r)$ and the total density of particles within a ball of radius r falls off as $M(r)/r^d \sim r^{d_\mathrm{f} - d}$, which becomes arbitrarily small as r grows. The concept of fractals, objects with an anomalous scaling dimension, is more general than this: these aggregates are random objects, but there are also regular fractals—see problem 4.8 for two examples.

For our purposes, it suffices to think of the mass or density scaling (4.20) as the key property of random fractals; the scaling $\rho(r) \sim r^{d_\mathrm{f} - d}$ will come back when we treat the configuration of polymers in section 5.5.3. Nevertheless, be aware that such aggregates have additional, more hidden characteristic properties, such as the 'hull' or outer boundary, each with their own fractal-like scaling: such clusters are actually *multifractals*[12] with a continuous spectrum of exponents. From computer simulations with many millions of particles, the fractal dimension of DLA clusters has been found to be ≈ 1.71 in two, and ≈ 2.2 in three dimensions.

What about measurements of the fractal dimension of real colloidal aggregates? As we discussed in section 3.7, the scattering intensity $S(q)$ measured in static scattering experiments is essentially the Fourier transform of the density-density correlation function. So such experiments are very well suited for determining d_f. Indeed, for a fractal structure with scaling (4.20) of the particle density, one finds that the structure factor $S(q)$ scales as[14]

$$S(q) \sim q^{-d_\mathrm{f}}. \quad (4.21)$$

Over which range of wavenumbers can one expect to see this scaling? If we denote the size of cluster by R_cl then for $qR_\mathrm{cl} \lesssim 1$ one can ignore the factor $i\vec{q} \cdot \vec{r}$ in the exponent in the expression (3.52)

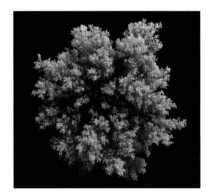

Figure 4.9. When no special precautions are taken, colloidal particles naturally attract each other and form clusters. In the DLA regime, particles stick to the cluster as soon as they touch it, so the diffusion of particles is the rate-limiting effect. The image shows an aggregate grown on the computer by letting successive particles diffuse around until they stick to the cluster. The darkness of the color indicates when the particles stuck to the cluster, with darker particles indicating earlier times. Simulation and rendered image by Michael Fogleman: https://www.michaelfogleman.com/. In simulations of two-dimensional DLA clusters, the fractal dimension d_f is about 1.71, whereas simulations of three-dimensional clusters find d_f of about 2.2.

The concept of a fractal was introduced and popularized by B. Mandelbrot. An extensive overview with a whole host of examples is given in his book *The Fractal Geometry of Nature* (Mandelbrot, 1982).[11]

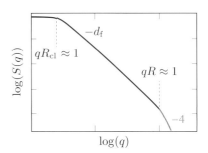

Figure 4.10. Schematic behavior of the scattering intensity $S(q)$ from colloidal aggregates as a function of the scattering wavenumber q on a log-log plot. For $qR_\mathrm{cl} \lesssim 1$, with R_cl the radius of the cluster, the scattering flattens out, while for $qR \gtrsim 1$ with R the radius of the colloids, the scattering intensity is determined by the surface of the particles[13] and falls off as q^{-4}. In between these regimes $S(q) \sim q^{-d_\mathrm{f}}$.

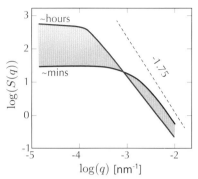

Figure 4.11. Schematic representation of scattering-intensity data from fractal clusters grown from aluminum oxide particles of diameter of about 0.4 μm as a function of time (Glover et al., 2000). Note that as time increases, the range over which the scaling $S(q) \sim q^{-d_f}$ extends becomes larger since the crossover to the flat behavior moves to smaller q. This shift of the crossover shows that the cluster size grows progressively larger. The measured fractal dimension is about 1.75 which is the value typically found for aggregating particles in solutions with low salt concentration, so that repulsive electrostatic interactions are important.[15]

for $S(q)$, so $S(q)$ is roughly q-independent. Hence the scattering flattens out for small q: at long wavelengths the scattering just reflects the presence of a small object but does not probe its structure. Likewise, for particles of size R, the scattering for $qR \gtrsim 1$ is determined by (the surfaces of) the particles. Therefore in the intermediate range $R_{cl}^{-1} \lesssim q \lesssim R^{-1}$ one expects to observe the fractal scaling (4.21), as is illustrated in figure 4.10 with d_f being the negative slope on a log-log plot.

Figure 4.11 summarizes experimental data from static light scattering of aggregating aluminum colloidal particles for various times. As time passes, the clusters grow in size, and this is visible in the data: the flattening of the curves happens at smaller and smaller q as the clusters grow, and so the range over which the fractal scaling is observed gradually increases—for the largest clusters in these experiments this range extends over about two decades, and the slope gives a fractal exponent d_f of about 1.75. This is significantly below the value $d_f \approx 2.2$ found in computer studies of DLA in three dimensions—as explained in the caption, this apparent discrepancy is due to the fact that in actual experiments the colloids repel each other through electrostatic forces, while the DLA computer simulations are based on particles that do not interact until they touch.

4.5.2 From spheres, rods, and plates to cubes and beyond

Figure I.11 in the introductory chapter illustrates that with computer simulations one can play endlessly with the shapes of hard core particles so as to produce almost any imaginable crystal structure. Likewise figure 4.2 gives an impression of the range of shapes and structures that can be synthesized in the lab.

The study of the equilibrium phases of well-defined colloidal particles started with studying spheres and evolved to more complicated shapes via rods and then plates. As an illustration of this line of research today, we show in figure 4.12 some of the cube-like shapes which can nowadays be made experimentally, and which are well described by the 'superball' shape formula

$$\left|\frac{x}{a}\right|^m + \left|\frac{y}{a}\right|^m + \left|\frac{z}{a}\right|^m \leqslant 1. \tag{4.22}$$

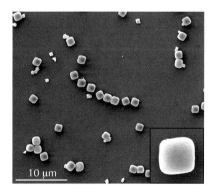

Figure 4.12. Illustration of colloids with the shape of a rounded cube. Image courtesy of Daniela Kraft. J.-M. Meijer et al., 2019 show examples of rounded cubes whose shape is well fitted by equation (4.22) with $m = 2.9$, 3.5, and 3.6.

which interpolates via rounded cubes from spheres for $m = 2$ to cubes as $m \to \infty$. With such well-defined shapes, the equilibrium phases can be explored in detail, and be compared to computer simulations of hard core rounded cubes. Some of the phases have remarkable properties that one does not often encounter in atomic crystals, like an extremely high vacancy density of 6%.[16]

4.5.3 The use of colloidal crystals to make optical bandgap materials

When spherical colloidal particles are sufficiently monodisperse, they easily form crystals with few defects. This property has found an interesting application, as a viable route to make optical bandgap materials.

As you might know from solid state physics, in crystals the periodicity of the atomic lattice leads to the formation of electronic bands. Gaps in such bands, ranges of energy where no band states exist, underlie the behavior of insulators and semiconductors. It has long been realized that if one were to build periodic structures with lattice spacing on the order of the wavelength of light, the propagation of light through such a crystal would be also characterized by bands, and this would create the possibility of creating band gaps. The existence of such a photonic bandgap implies that in a particular frequency range, no light can propagate in the lattice. This in turn provides a way to guide or confine the light.

Fabricating three-dimensional materials with photonic bands presented a challenge before the invention of three-dimensional printing, until it was realized in 1998 that self-assembled colloidal crystals could be used as a nice mold for fabricating them.[18] Since light does not propagate in most colloidal materials, the trick is to first grow a colloidal crystal, then have the spheres sinter (grow together) to provide some mechanical stability, subsequently fill the remaining space with a proper material and finally etch away the original colloids. Figure 4.13 illustrates an example of the very nice regular structures of 'air spheres' that can be fabricated this way: the spherical empty spaces were originally filled by the colloids which had self-assembled into a crystal structure.

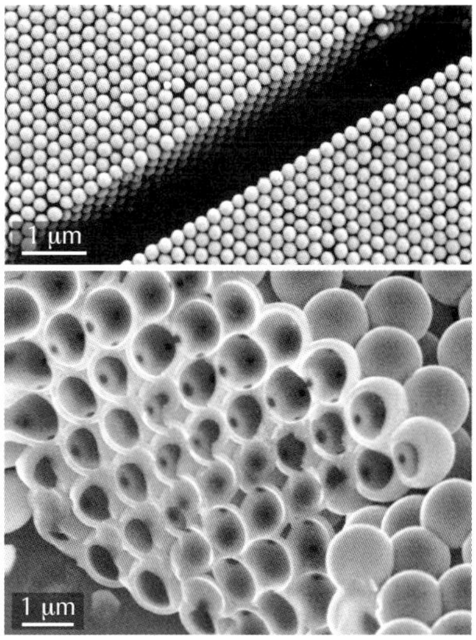

Figure 4.13. Upper panel: example of a colloidal crystal. Courtesy of Daniela Kraft. Lower panel: Example of a photonic bandgap material with 'air spheres' made from a self-assembled colloidal crystal. See text for an explanation of how such 'air crystals' are synthesized. Note the visibility of small 'necks' where the original colloids were fused together. Image courtesy of Dannis 't Hart and Job Thijssen.[17]

4.5.4 Colloidal glasses

In many experiments, especially those focused on the formation of crystal phases, it is important to reduce the polydispersity of the colloids as much as possible—in other words to reduce the variations in their size. The opposite is actually true if one wants to use colloids as an example of a glass-forming system. Indeed, when the polydispersity is significant, the formation of ordered phases is suppressed and the colloids tend to get stuck in some random configuration as their volume fraction is increased: we have an example of a colloidal glass. Figure 4.14.a shows a snapshot of a dense disordered phase in a binary system of colloidal particles.

What can colloids add to the study of glasses, which is a rich field in itself with a long history and a lot of active research? The answer to

The literature on glasses is enormous. A brief introduction and overview of the various approaches is given by Tarjus, 2011, while Lubchenko, 2015 gives a comprehensive pedagogical introduction to the structural glass transition. A nice short introduction to colloidal glasses is given by E. R. Weeks, 2016, while Gokhale et al., 2016; Hunter and E. R. Weeks, 2012; and Y. M. Joshi, 2014 give more complete overviews. The various chapters in the collection of Berthier et al., 2011 discuss dynamical heterogeneities in a variety of glassy systems.

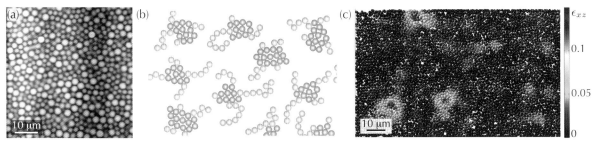

Figure 4.14. (a) Two-dimensional image of a dense binary system of colloidal particles used by Narumi et al., 2011 to study heterogeneous behavior of colloidal glasses. The ratio of the diameters of the particles is 1.3. Image courtesy of Eric Weeks. (b) Visualization of typical rearrangement events in the two-dimensional colloidal glass of Nagamanasa et al., 2015, made from a binary mixture of colloids with diameter ratio 1.3 at $\phi = 0.76$.[19] Only the most mobile 10% of particles during some time interval are shown, with light red indicating the core particles and blue those with a string-like rearrangement. (c) Identification of local strain changes in a slice of a three-dimensional colloidal glass after applying a shear strain during times on the order of tens of seconds in experiments of Chikkadi et al., 2011. Color coding gives the shear component of the strain, as indicated by the color bar. Note that rearrangements occur in strongly localized red regions, while in large fractions of the sample changes in the local strain are small (the spheres colored yellow, green, and light blue). Detailed analysis of the data as a function of time shows that a rearrangement favors successive ones in its vicinity. Another example from a similar experiment is shown in figure I.9 in the introductory chapter. Image courtesy of Peter Schall.

this question is again related to what we expressed in the introduction: large colloidal systems can nowadays easily be simulated and with present techniques experimentalists can track the behavior of individual particles in a sample. So unlike in molecular glasses, in colloidal systems one can follow what type of rearrangements dominate the dynamics and compare with simulation results.

To put this in perspective, we note that the quintessential behavior of glass-forming molecular materials is that when approached from the high-temperature (liquid) side, the viscosity is found to increase exponentially fast as the temperature is decreased. For instance, a typical law with which the viscosity is fitted is the empirical Vogel-Fulcher law

$$\frac{\eta}{\eta_0} = \exp[C\, T_0 / (T - T_0)], \qquad (4.23)$$

with C some constant. This expression suggests a divergence at a temperature T_0. But one is never able to get very close to the temperature T_0 where the viscosity would diverge, because once the viscosity reaches the order of 10^{12} Pa·s, or fifteen(!) orders of magnitude greater than that of water, the liquid does not flow anymore on a human time scale—from such a perspective a glass is essentially an arrested liquid, because the relaxation has just become prohibitively slow. The question of whether a true phase transition with divergent length scale at some finite temperature like T_0 underlies this behavior actually concerns much of the glass literature.

For colloids which behave to a good approximation like hard core spheres, temperature is irrelevant (see section 4.2). But for these, one observes similar behavior as a function of volume fraction ϕ:

as figure 4.15 illustrates, the viscosity rapidly rises as ϕ increases. In fact, as the solid line shows, the underlying data are well fitted by a law reminiscent of (4.23),

$$\frac{\eta}{\eta_0} = \exp[C\,\phi/(\phi_0 - \phi)]. \tag{4.24}$$

Such a strong viscosity increase with the volume fraction ϕ is also found in simulations of polydisperse hard core spheres; see figure I.10 in the introductory chapter. Also those simulation data can be fitted well with a law of the form (4.24).

The strong increase of the viscosity of a colloidal dispersion with volume fraction reflects the fact that as the density goes up, particles get more and more in each other's way. In other words, in order for a particle to move, several others have to move out of the way; this becomes increasingly difficult when the volume fraction approaches the one where hard core particles get stuck or, in the language of section 2.10, get jammed as the density approaches the jamming density ϕ_J.

Clearly, in this picture the dynamics in colloids becomes increasingly cooperative and heterogeneous as the volume fraction increases. Even though the details of this mechanism may differ for molecular glasses or spin glasses, also for these most theories are based on the idea that as the glass temperature is approached the dynamic rearrangements involve increasingly more cooperative 'moves' of particles, bond, spins, etc.

If we assume that there are indeed a number of general principles underlying the formation of dynamical heterogeneities, then colloidal glasses do present a special experimental opportunity to probe these, as one can nowadays trace the displacements of many colloids accurately over a long time. Panels b and c of figure 4.14 illustrate this from two different experiments. Panel b illustrates how the 10% most mobile particles during some time interval in a two-dimensional bidisperse colloidal glass tend to form well-defined clusters: rearrangements at this particular value of $\phi = 0.76$ are not single-particle events but a combination of string-like moves (blue) and blob-like clusters. From measurements at other volume fractions it is found that string-like rearrangements dominate at smaller ϕ, and blob-like clusters at larger ϕ. Panel c also shows how in a different experiment on a colloidal glass under strain the most important strain rearrangements occur in localized 'shear transformation zones', while rearrangements in large fractions of the sample are small. Moreover, rearrangements tend to be successive events that are highly correlated in space and time.

These data present of course only a small sample of properties which one can measure explicitly in colloidal glasses and compare to predictions from theories of glasses.

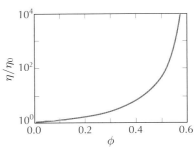

Figure 4.15. General behavior of the viscosity of a colloidal dispersion as a function of the volume fraction. Note the strong enhancement by over three orders of magnitude as the density of the dispersion increases, and the similarity with the data in figure I.10 for the relaxation time in a simulation of polydisperse hard spheres. Adapted from Hunter and E. R. Weeks, 2012, who combine experimental data from various groups.[20]

It is worth pointing out that colloidal dispersions have also been used as a model system for studying the transition to irreversible behavior as a function of strain, as well as memory formation effects related to this. See section 10.2 and in particular figure 10.18, which illustrates these types of experiments.

Colloidal motifs

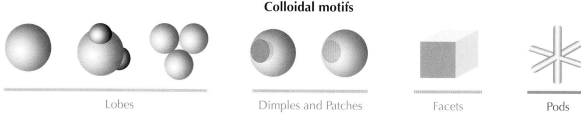

Lobes   Dimples and Patches   Facets   Pods

Figure 4.16. Sketch of the main classes of functional colloidal building blocks which can be engineered to aggregate into desired colloidal structures. The figure illustrates how by playing with the shape or by creating patches with specific surface chemical properties one can engineer colloidal motifs with preferred directional bonding (very much like the directional molecular bonds), as illustrated in figure 4.17. Colloids of this type can self-organize into new forms of colloidal matter: one turns colloids into a form of designer matter. For details and extensions, see Hueckel et al., 2021.

4.5.5 Colloidal motifs as the building blocks of designer matter

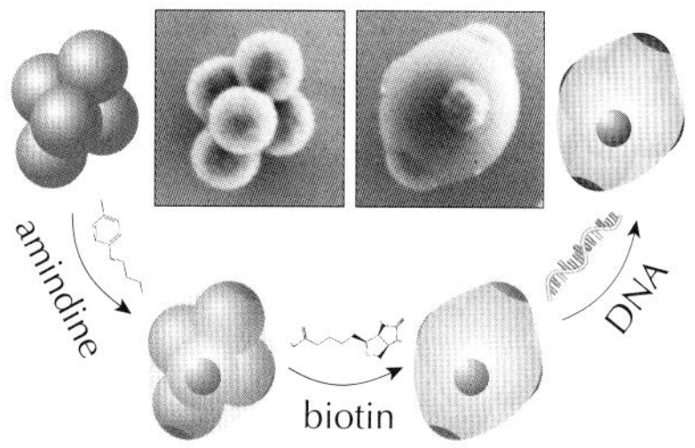

We have already encountered various illustrations of the fact that the interactions and shapes of colloids can be controlled to a remarkable degree. The sophistication with which this can be done nowadays has reached a level such that one can control both the shape of the colloids and the local surface chemistry. As figure 4.16 illustrates, this allows the creation of colloidal motifs which prefer to bond in specific directions, and hence the turning of colloids into self-assembling designer matter.

Designer matter is a modern topic that we will put in perspective in chapter 10; nevertheless, we like to illustrate the above remarks here with a simple example of colloids with patches of DNA which promote directional bonding. Figure 4.17 gives an impression of how this is done in practice, and of the resulting clusters that these colloidal particles self-organize into. This emerging field is expected to play an increasingly important role in the field of 'artificial life', as it provides versatile model systems for testing ideas.

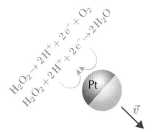

Figure 4.17. As shown by Y. Wang et al., 2012, colloids can nowadays be synthesized with patches of DNA, which endows them with very specific directional bonding. The figure illustrates the main steps in the fabrication of colloidal particles with patches of DNA. Electron micrographs courtesy of Yufeng Wang, Yu Wang and Dave Pine.

4.5.6 Colloids as active matter

We will discuss active matter in chapter 9. This field was motivated by examples from the life sciences, like the flocking of birds, the swarming of bacteria, and the interactions induced by molecular motors. But since it was discovered how colloidal particles can be made into microswimmers, it has become clear that active colloids provide a versatile model system for active matter. This field is rapidly developing, and we only sketch here how playing with colloids allows probing diverse aspects of active matter.

The simplest example of an active colloid, which has become a work-horse for many experiments, is sketched in figures 4.18 and 4.19 a—a colloidal particle coated on one side with platinum. Such

Figure 4.18. The workhorse of active colloids is a so-called Janus particle, a colloidal particle coated on one side with platinum. Platinum is a good catalyst, so in a solution with hydrogen peroxide (H_2O_2), the hydrogen peroxide reacts on the platinum side. The reaction pushes the particle forward.[21]

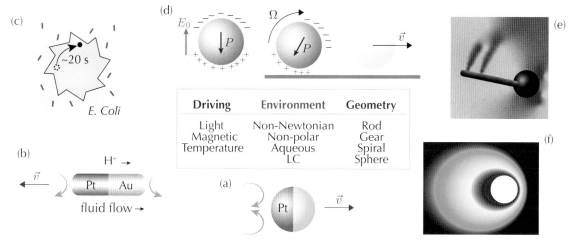

Figure 4.19. Active colloids can nowadays be made and manipulated in an enormous variety of ways. In addition, their interaction depends on the suspending medium. The central box of the figure[22] indicates how one can play with the shape of the colloids, how one can manipulate them with magnetic fields, electric fields or light, and how one can take advantage of specific effects of the suspending fluid. (a) The colloidal Janus particle of figure 4.18, which is used in many experiments on active matter. (b) Bimetallic nanorods favor one of the two reaction steps of hydrogen peroxide to water at each of their ends. Electrons flow through the metals, while H^+ flows through the fluid from the platinum to the gold side. This induced flow makes the rod self-propelling.[23] (c) A micromotor driven by bacteria.[24] In an equilibrium system such a symmetry breaking is impossible. (d) Schematic illustration of the so-called Quincke rotation mechanism of active colloids.[25] When an electric field E_0 is applied to an insulating sphere immersed in a conducting fluid, positive and negative charges accumulate at the surface in such a way that the sphere gets a dipole moment pointing opposite to the field. Above a well-defined critical field amplitude the situation is unstable and the sphere starts to rotate spontaneously (see problem 4.9). Due to the induced rotation, settled particles move along the bottom. We shall encounter various examples of active matter based on the Quincke rotation mechanism in chapter 9. (e) As we discussed in section 1.11.1, in a Newtonian fluid the scallop theorem expresses that creatures cannot propel themselves by a back and forth motion at small Reynolds number. When particles like the one shown in black are immersed in a viscoelastic fluid they can self-propel by a back and forth wiggling motion induced by an external field: the viscoelastic behavior of the immersion fluid breaks time-reversal symmetry and so allows circumvention of the scallop theorem.[26] (f) In section 1.13.3 we discussed the Marangoni effect, the motion of a droplet due to a surface-tension gradient induced by a concentration gradient. The picture illustrates the fact that also droplets can be self-propelled swimmers, due to a combination of the Marangoni effect and dissolution. The color coding shows the density concentration profiles associated with a droplet moving to the right. The self-sustained effect results from the coupling: the Marangoni effect gives flow, while the flow breaks the symmetry between front and back side. The transition to self-propelled states happens at sufficiently large Péclet number.[27]

particles are often called Janus particles, after the Roman god Janus who was often portrayed with two faces. Platinum is a well-known catalyst for many reactions. So when such a Janus particle is put in a solution of hydrogen peroxide (H_2O_2), the hydrogen peroxide reacts on the platinum coated side in two steps to oxygen and water, and this reaction drives the particles forward in the direction opposite the platinum side.

It should be clear that active colloids will have many interactions. In addition to the static and entropic interactions already discussed in the earlier in this chapter, active colloids interact through the hydrodynamic fields—a moving object disturbs the velocity field around it, which is felt by other particles—and through the concentration fields associated with the chemical reactions that propel them. Not only can the concentration gradients be manipulated with light, one can also play with the shapes of the particles or manipulate them with electric or magnetic fields. In addition one can

Recent reviews of active colloids approach the topic from various angles. J. Zhang et al., 2017 focus on new applications and research opportunities, including the interaction with non-Newtonian soft matter liquids. Figure 4.19 is based on this paper. Marchetti et al., 2016 focus on the role of mechanical interactions, Mallory et al., 2018 give particular attention to self-assembly of active colloids, and L. Wang and Simmchen, 2019 give an overview of the interaction with passive tracers. Bacterial hydrodynamics is reviewed by Lauga, 2016.

This section is inspired by N. J. Wagner and Brady, 2009, who give an enjoyable introduction to non-Newtonian effects and who highlight some of the applications of shear-thickening in sportswear, the automotive industry and body armor. Russel et al., 1989 still provide a very accessible entry to colloidal hydrodynamics in the regime where the dispersion behaves essentially as a Newtonian fluid. For modern developments, see N. J. Wagner and Mewis, 2021.

change the fluid in which they are dissolved into a complex fluid, to take advantage of its special properties. Clearly, the parameter space which one can explore is large, and we content ourselves here by presenting the kaleidoscope of possibilities in figure 4.19.

4.6 Non-Newtonian rheology of colloidal dispersions

How do colloidal dispersions flow? When a dispersion is dilute, its behavior is essentially that of a Newtonian fluid with an effective viscosity which is enhanced by the colloids. In fact we already showed that the viscosity of the dispersion rapidly increases with the volume fraction of colloidal particles—see figure 4.15. At the same time, since we discussed these data in the context of colloidal glasses, which are characterized by very long relaxation times due to particles getting stuck, it will probably come as no surprise that when the fluid dispersion becomes denser, non-Newtonian effects become important: the dispersion increasingly behaves like a complex fluid.

In this section we first discuss various aspects of the rheology of colloidal dispersions, in particular shear thinning and thickening, and the temporal transition in the viscosity of natural materials like clay and mud. We close this section with a brief comparison with related phenomena in emulsions and granular media, and their relation to the earlier discussion of jamming.

4.6.1 Shear thinning and shear thickening

One of the first ways in which non-Newtonian behavior often manifests itself is in terms of a shear-rate dependence of the effective viscosity. In such a discussion, one considers a simple shear flow[28] $\vec{v}(\vec{r}) = \dot{\gamma} y \hat{x}$ which has only one nonzero shear component $v_x = \dot{\gamma} y$ (we follow common practice to denote the shear rate by $\dot{\gamma}$). For a spherical colloid of radius R in such a flow, an important dimensionless parameter is the Péclet number

$$\mathrm{Pe} = \frac{\dot{\gamma} R^2}{D}, \tag{4.25}$$

where D is the diffusion coefficient of the particle, given by the Stokes-Einstein relation (3.19). In terms of the diffusion time t_{diff} introduced in (3.44), the time it takes the colloid to diffuse a distance on the order of its diameter, we can write the Péclet number as

$$\mathrm{Pe} = \dot{\gamma} \, t_{\mathrm{diff}} / 2. \tag{4.26}$$

Since $\dot{\gamma}$ is the local rotation rate of the fluid,[29] we see that the Péclet is essentially the ratio of the translation diffusion time over the rotation time of the particle: for small enough shear rate $\dot{\gamma}$ a particle diffuses around fast enough to restore a near-equilibrium configuration. But for shear rates such that the Péclet number is larger than 1, the rearrangements due to the flow are too fast to be undone by the diffusion. Indeed, the reorientation of an anisotropic colloidal particle or a cluster of particles, due to the flow-induced rotation, is significant for large Péclet numbers. It should be intuitively clear that this is likely to affect the rheology of a dispersion of strongly anisotropic particles, like the clay and cornstarch shown in figure I.7, if you picture the fluid microscopically as having many counter-rotating gear wheels. Indeed, already for moderate densities, changes in the rheology typically happen when the Péclet number becomes somewhat larger than 1.

The top panel of figure 4.20 shows the behavior of the viscosity (obtained as the ratio of the measured shear stress to shear rate) of latex colloids at various densities. For the lowest volume fractions the viscosity is essentially independent of the applied shear stress, hence the dispersion behaves very much like a Newtonian fluid. But as the volume fraction ϕ increases (middle curve), the viscosity initially decreases at sufficiently high shear stresses. The decrease of viscosity with increasing shear rate is termed shear thinning.

The uppermost curve drawn for a dense suspension shows another interesting effect: at high densities, the dispersion actually exhibits a *yield stress* at small shear stress: it behaves like a solid which can resist a finite amount of stress before it 'yields'. As the sample does not flow, the viscosity appears infinite for small stresses, while the viscosity rapidly falls once the sample yields and starts flowing.

A second striking effect we see in the data for the two larger volume fractions is an increase of the viscosity for the largest shear stresses: this is termed shear thickening.

The origin of these intriguing effects have been uncovered with the help of numerical simulations. As illustrated in figure 4.20 and figure 4.21, shear thinning is due to the fact that the colloids tend to organize (roughly) into layers which more or less slide over each other. The shear thickening, on the other hand, is due to the particles forming clusters large enough to obstruct each other in their motion and rotation. These effects were actually already alluded to in figure I.6 in the introductory chapter.

From such numerical simulations it was found that the attractive interactions play an important role in the formation of the clusters which give rise to shear thickening. So one expects that one can suppress the shear thickening by making the colloids behave more hard-sphere like, by grafting them with a polymer brush.

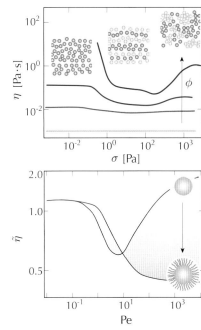

Figure 4.20. Illustration of the typical non-Newtonian rheology of colloidal dispersions, as it appears in the effective viscosity.[30] Top: schematic representation of the viscosity of dispersions of latex particles in water as a function of increasing shear stress σ, for four representative volume fractions covering the range from dilute to dense suspensions. Initially shear thinning is observed for increasing shear stress; for higher densities, eventually shear thickening is typically observed. Dense samples also show the existence of a yield stress through a divergence of η for small σ. The sketches at the top illustrate how shear thinning is due to layering of the particles, while shear thickening is due to particles forming clusters that obstruct each other—see figure 4.21 for more details. Bottom: By gradually reducing the attractive interactions in simulations of colloidal suspensions or in experiments by grafting a polymer brush on the particle (as discussed in section 4.4.2), the Péclet number dependence of the viscosity crosses over from that of the top curve (attractive interactions, strong shear thickening) to the bottom one (no attractive interactions, no shear thickening).[31] The fact that hard core particles without attractive interactions do not shear thicken confirms that attractions are crucial for shear thickening.[32] This is also consistent with the fact that emulsions, whose droplets behave as compressible spheres, only show shear thinning—see section 4.6.3.[33] Adapted from N. J. Wagner and Brady, 2009.

Pe ~ 100   Pe ~ 1000

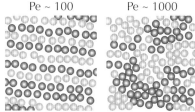

Figure 4.21. Illustration of the layering and clustering observed in rheological simulations by Melrose and R. C. Ball, 2004 of 2,000 spherical colloids with an interaction that models coating with a polymer brush (see section 4.4.2). Strong layer formation occurs for Pe = 100, causing shear thinning. Simulations for Pe = 300 show a combination of layering and cluster formation; the fact that clusters form causes shear thickening. Colors are introduced to guide the eye; all particles in the simulations are identical. Colloidal flow simulations—sometimes referred to as Stokesian dynamics (Brady and Bossis, 1988) and used in the simulations in this figure—are based on the coupled linear equations for the motion of the colloidal particles where the forces include: the external forces, hard core and other colloid-colloid interactions, the hydrodynamic interactions of the colloids, derived for general configurations in the low Reynolds number hydrodynamic regime[34] and matched to the lubrication forces when two particles get very close (see problem 1.13), and the Langevin forces modeling the Brownian dynamics.

This effect is illustrated in the bottom panel of figure 4.20: the more the attractive forces are reduced, the more the shear thickening is suppressed. The rheology of colloidal suspensions without attractive interactions becomes close to that of emulsions: as we will see below, these only exhibit shear thinning.

4.6.2 A temporal transition due to competition between aging and rejuvenation

So far, we have focused on the effective shear rate or shear stress dependence of dense colloidal dispersions. But as we have seen before, complex fluids typically show also a clear time or frequency dependence in their response. Indeed, as the colloidal density increases, the effective viscosity starts to develop a frequency dependence in the spirit of the discussion in section 2.5. Moreover, once the glass transition of section 4.5.4 is approached one starts to see a rapid broadening of the spectrum of relaxation times. A second illustration of a strong time-scale-dependent response is provided by cornstarch dispersions, whose response on short time scales is elastic enough that you can run on it, as figure I.7 illustrates, while you sink into it if you walk too slowly.

These examples are sufficient to illustrate that once we move away from ideal model systems, the temporal or frequency response of dispersions is often far from trivial. We cannot really do justice to the topic here[35], and so we content ourselves with drawing attention to an effect which is especially prominent in natural colloidal systems like clay, a temporal bifurcation in the viscosity.

You probably know from experience that Greek yoghurt can become quite stiff if left untouched for some time, while it flows reasonably well as a fluid (albeit with a small yield stress) once you stir it sufficiently: upon aging it becomes stiff, while sufficient stirring promotes flowing as a fluid. The competition between these two effects can actually play up in real time in natural colloids. Natural colloids like clay, whose micrograph is shown in figure I.7, typically consist of very irregular particles, some of them very plate-like. You can imagine that such particles can easily form clusters or gel-like networks that resist flowing, especially if additional forces due to charges or the Van der Waals interaction are important. This effect can actually be so strong that the competition between two effects—the breaking of clusters due to the shear forces or the mixing by the flow, and the 'aging' due to the preference to form spanning or percolating clusters or networks—leads to a sharp transition or bifurcation.

This scenario, which is sometimes referred to as a competition between 'aging' and 'shear rejuvenation', is illustrated by the data shown in figure 4.22 for the rheology of drilling mud as a function of time, for a given level of the shear stress. As the data show, in all cases, the effective viscosity starts out relatively small, although still a hundred times greater than that of water. For large levels of the shear stress, the viscosity increases over time, but after about a thousand seconds or so, it levels off to a finite value which depends somewhat on the stress. However, below some well-defined stress level (defined to within a few percent), the viscosity suddenly rises rapidly by a few orders of magnitude after a few hundred seconds, as if the sample suddenly gels. Eventually the system enters a regime in which the viscosity slowly increases with time, more or less independent of the shear stress applied. To give you a sense of it: during this phase the viscosity becomes over 100 times that of typical peanut butter.

4.6.3 Comparison with emulsions

When they are not driven (sheared), colloidal dispersions are thermal systems, due to the Brownian motion. But as we discussed above, once the shear rate is large enough that the Péclet number is much larger than 1, the thermal motion of the colloids is too small to have the colloids equilibrate—under these conditions sheared colloids are a far-from-equilibrium system. One would intuitively expect that under such conditions the difference between dense colloidal dispersions and sheared granular media becomes less important. Exploring this idea has indeed turned out to be fruitful.

Indeed, as we discussed briefly in section 2.10, even though granular media are inherently zero-temperature systems, the jamming approach to granular media has stimulated thinking of packings of compressible particles with a finite radius as a system with a critical point. This point is identified as the random close packing density at which the particles just touch. We identified this in section 2.10 as the jamming density or volume fraction ϕ_J. Many of the developments in the field of jamming have been stimulated by investigation of the properties of frictionless compressible spheres or disks which can slide past each other without friction force. Admittedly this may be quite an idealized model for most granular media consisting of solid state particles, especially if they have irregular shapes, which exert also tangential friction forces at their contacts. However this may be, from dynamical simulations of such models, it has emerged that in fact close to the jamming point, for $\Delta\phi = \phi - \phi_J$ small, the flows of frictionless spheres exhibits nice scaling behavior. In this sense, the jamming point acts truly like a critical point: all the data collapse well onto one curve if we plot the rescaled shear stress $\tilde{\sigma} = \sigma / |\Delta\phi|^{\Delta}$ versus the rescaled shear rate $\tilde{\dot{\gamma}} = \dot{\gamma} / |\Delta\phi|^{\Gamma}$. Note that if we are above the jamming point ($\Delta\phi > 0$), we expect the packings to have a finite yield

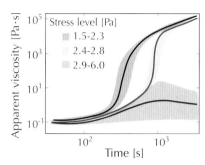

Figure 4.22. The rheology of drilling mud is an example of the scenario that depending on the stress level, the fluid either fluidizes or stiffens with time. The figure summarizes the experimental observations of Bonn et al., 2017 by clustering traces of the temporal evolution of the effective viscosity for three different ranges of stress levels. For small stress levels the effective viscosity rapidly increases with time over four orders of magnitude (gray band), while for large stress the mud 'rejuvenates' and the viscosity decreases in time (red band). The actual data show a very rapid crossover between these two regimes. Similar rheological behavior is reported for bentonite, a natural clay.[36] Adapted from Bonn et al., 2017.

For a more extensive discussion ranging between jamming, yield stress and rheology and review of the literature, see section II.E.3c of Bonn et al., 2017. Figure 4.23 is also put in a broader context in this review.

Such scaling collapse generally occurs near equilibrium phase transitions. Problems 4.10–4.12 introduce some essential features of critical points and the renormalization group.

See Forterre and Pouliquen, 2008 for a review of continuum-type granular flow stressing the importance of the dimensionless number In, which they refer to as the inertial number. The review by Kamrin, 2019 focuses on modeling thin shear zones like in figure I.13 with a continuum formulation.

You can understand this time scale as follows. The force F on a grain of diameter d is $\mathcal{O}(pd^2)$, its mass M is $\mathcal{O}(\rho d^3)$, so the force gives an acceleration $F/M = \mathcal{O}(p/\rho d)$. According to Newton's second law, the time it takes a grain to move a distance d as a result of this force is of order $\sqrt{d/(F/M)} = d\sqrt{\rho/p}$.

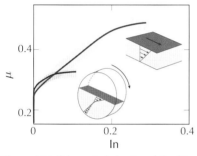

Figure 4.24. Schematic behavior of the friction coefficient μ on the dimensionless number In, as extracted from two-dimensional granular plane shear simulations at constant pressure (blue line) and rotating drum simulations (red line).[43] Note that the friction coefficient μ approaches a finite value as In $\rightarrow$ 0. Adapted from Forterre and Pouliquen, 2008.

For flow configurations with side shear zones, an important role is played by the dimensionless inertial number

$$\mathrm{In} = \frac{\dot{\gamma}d}{\sqrt{p/\rho}}, \tag{4.30}$$

where d is the diameter of the grains, p the pressure and ρ the density of the material the grains are made of. This dimensionless number can be thought of as the analogue of the Péclet number (4.25) for colloids and microemulsions: as $d/\sqrt{p/\rho}$ is the time scale it takes for a grain to fall into a hole of size d, In is the ratio of this rearrangement time and the time scale associated with the shear. Continuum theories are typically based on assuming that the so-called friction coefficient μ, defined as the ratio of the shear stress σ and the pressure p,

$$\sigma = \mu(\mathrm{In})p \tag{4.31}$$

is a function of In, $\mu = \mu(\mathrm{In})$. Figure 4.24 shows an example of the $\mu(\mathrm{In})$ behavior extracted from granular simulations of plane shear and a rotating drum. Note that the branch corresponding to nonzero flow rate approaches a finite value of μ as In $\rightarrow 0$. This reflects the fact that for granular media the tangential frictional forces between grains have to overcome a finite value in order for the contact to slide.

4.7 What have we learned

Colloidal particles naturally attract each other, in particular because of the Van der Waals interaction and the depletion interaction. In order to prevent colloidal particles from aggregating into clusters, colloidal dispersions are stabilized electrostatically or sterically. This is done by putting charges on the particles or by grafting polymers to their surface to serve as a bumper.

For increasing densities, colloidal suspensions behave increasingly as complex fluids with non-Newtonian rheology. Typically, above a shear rate determined by the Péclet number, their flow exhibits shear thinning due to layering of particles. But particles with appreciable attractive forces eventually exibit shear thickening for large shear, due to cluster formation. Experiments on microemulsions, whose behavior should be close to that of compressible spheres without friction forces, exhibit the nontrivial scaling behavior predicted by jamming approaches.

Colloids increasingly serve as model systems with which to explore the interface with other subdisciplines, such as the formation of unusual crystal phases, glass formation, artificial life and active matter.

4.8 Problems

Problem 4.1 *Estimating the energy barrier for coagulation*

Suppose we have a dispersion of particles in water at room temperature. If they interact with the potential sketched in figure 4.25, the energy barrier may prevent particles from crossing into the attraction regime and consequently aggregating. In this problem, we are going to estimate the height of the barrier needed in order to keep the particles dispersed for the duration of a day. Let us consider a 10% dispersion by weight of glass particles with radius of $R = 100$ nm.

a. The relative density of glass is 2.5. Calculate the volume fraction of this suspension and show that the typical distance between colloids is of order 500 nm.

b. Calculate the Stokes-Einstein diffusion coefficient given in equation (3.19), section 3.2.4 (see also the discussion in section 3.5.1).

c. Estimate the mean distance $\ell_1 \simeq \left(\langle V^2 \tau^2 \rangle\right)^{1/2}$ a bead moves during the time the velocity is correlated (as discussed above equation (3.21)) and show that it is much smaller than the radius R of the beads. What do you conclude from this?

d. Show that the mean time t_{coll} it takes two colloids to 'collide' is of order 0.12 s.

Hint: The conclusion you drew in c should convince you that the behavior of the colloids is diffusive also at the scale of microns, and hence that you can estimate the 'collision time' with the help of the results of b and a.

e. Requiring the probability of two particles overcoming the energy barrier upon collision to be less than 1/typical number of collisions, show that the height of the energy barrier should exceed about $13 k_{\mathrm{B}} T$.

Problem 4.2 *Derivation of Van der Waals interaction between two slabs*

We calculate here the effective Van der Waals interaction (4.9) between two semi-infinite slabs which extend over a large area A in the lateral directions x, y and which are separated by a distance h, as drawn in figure 4.26. The upper slab extends infinitely in the $+z$ direction and the lower one in the $-z$ direction. The positions of atoms in the lower slab are labeled by x_1, y_1, z_1 and those of atoms in the upper plate by x, y, z.

a. As the Van der Waals interaction is varying slowly, we can replace the sums in equation (4.8) by integrals. Show that in this approximation the expression (4.8) for the effective potential can

Relevant coding problems and solutions for this chapter can be found on the book's website www.softmatterbook.online under Chapter 4/Coding problems.

This problem is adapted from the textbook by Israelachvili, 2011, which provides a systematic introduction to intermolecular and surface forces.

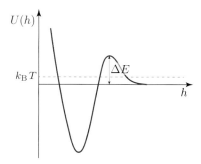

Figure 4.25. Typical interaction potential between colloids.

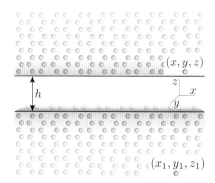

Figure 4.26. Two semi-infinite slabs a distance h apart in the calculation of the Van der Waals interaction. The atoms in the upper slab are labeled by x, y, z, those in the lower slab by x_1, y_1, z_1. For the calculation of the effective interaction per unit area, we have to perform the integrals over the coordinates x, y, z and z_1.

$$F(h) \approx 2\pi \left(\frac{R_1 R_2}{R_1 + R_2} \right) \int_h^\infty dw\, f(w) = 2\pi \left(\frac{R_1 R_2}{R_1 + R_2} \right) u(h),$$
(4.40)

where $u(h)$ is the potential per unit area. Show that for the case of the Van der Waals interaction between two spheres this gives

$$\text{different spheres:} \quad U_{\text{VdW}}(h) = -\frac{H}{6\,h} \left(\frac{R_1 R_2}{R_1 + R_2} \right).$$
(4.41)

This is the generalization of (4.10) to spheres of different sizes.

The full expression for arbitrary h is given in table 1.5 of Kleman and O. D. Lavrentovich, 2003.

f. The Derjaguin approximation holds for $h \ll R$ and $z \ll R$. Interpret the physical meaning of these assumptions.

g. The above calculation focuses on Derjaguin's argument in terms of effective area. We can actually derive this result also systematically by retracing the last few steps from problem 4.2, by changing the slabs to two spheres with radius R_1 and R_2 for $h \ll R$, where, as above, h is the minimum separation between the two surfaces. In this limit, we can view the spheres as weakly curved slabs, with total vertical separation $h + \frac{1}{2}r^2(1/R_1 + 1/R_2)$, where $r^2 = x_1^2 + y_1^2$ in terms of the coordinates of figure 4.26. Show that in this limit we can use the result (4.34) with A replaced by the two-dimensional integral over x_1 and y_1, and h replaced by $h + \frac{1}{2}r^2(1/R_1 + 1/R_2)$. Perform the two-dimensional integration in terms of the variable r and show that this yields the result (4.41).

Problem 4.4 *Calculation of the excluded volume and the resulting depletion interaction*

In this exercise, we back up the discussion of section 4.3.2 with an explicit calculation of the excluded volume and the resulting depletion interaction, using the two illustrations in figure 4.4.

A somewhat similar calculation can be found in section 4.3.5 of Jones, 2002.

We consider two large spheres with diameter D, surrounded by N small spherical particles with diameter d in a box with volume V. The magnitude of the attractive depletion force F experienced by the large spheres is equal to the derivative of the Helmholtz free energy with respect to the distance r between their centers $(r > D)$, $F(r) = -dA/dr$, where A is the Helmholtz free energy of the whole system. As all particles are taken to have hard cores, so that they cannot overlap with each other, the Helmholtz free energy of the small surrounding particles is then $A \equiv -TS = -k_B T \ln Z$, where

$$Z = \frac{V_a^N}{N! \lambda^{3N}}$$
(4.42)

is the canonical partition function. Here, V_a is the total volume available for small particles and λ is the de Broglie wavelength.

a. Calculate the available volume $V_a(r)$ for small spheres for $D < r < (D + d)$.

b. Derive the force $F(r)$ and sketch it as a function of r.

c. In the previous steps we have calculated the attractive force $F(r)$ from the entropic point of view. Now let us do it from the point of view of the pressure. The random kicks from small particles which exhibit Brownian motion can be characterized by the osmotic pressure: $p_o = N k_B T / V$. Calculate the resultant force due to the osmotic pressure on the big spheres, $F_o(r)$, as a function of their distance r, and compare it with $F(r)$, calculated from the Helmholtz free energy in step b.

Note that when the density of the small particles is high, the depletion interaction is never large, but of order $k_B T$.

Problem 4.5 *Lock and key colloids*

In this problem we determine the depletion interaction, discussed in section 4.3.2 and the previous problem, for the colloids illustrated in figure 4.28.

a. Consider the so-called 'lock and key' colloids. The 'key' is a spherical particle (red) that fits precisely into the spherical cavity of a 'lock' particle (gray). In the presence of polymer molecules in solution, these colloids experience an attractive interaction that favors binding of the key in the lock site. Argue that the depletion interaction is at play here.

b. Argue that the interaction potential between the lock and key colloids is

$$U = -n_p k_B T \Delta V, \qquad (4.43)$$

where n_p is the number density of polymers in solution, and ΔV is the change in volume from which polymers are excluded.

c. Consider a key particle with radius R_1 and a lock particle with radius R_2. In the preferred lock-key configuration, the two fit perfectly together, and sit with a distance d between their centers. Given that the radius of the polymer molecules is R_p, show that the maximum excluded volume is approximately

$$\Delta V_{\max} = 4\pi R_1^2 \left(1 + \frac{R_2^2 - R_1^2 - d^2}{2 R_1 d} \right) R_p. \qquad (4.44)$$

Problem 4.6 *Debye-Hückel potential of a point charge and a charged sphere*

In this exercise we derive the potential for a point charge and a sphere with radius R and charge Ze from the Debye-Hückel equation (4.15).

a. Write the Debye-Hückel equation (4.15) in spherical coordinates.

For further reading, see Sacanna et al., 2010 and section 1.4.4 in Lekkerkerker and Tuinier, 2011.

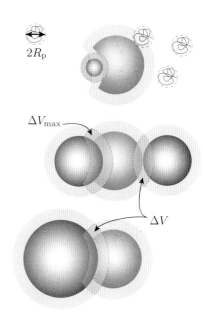

Figure 4.28. Interaction between lock and key colloids. See Sacanna et al., 2010 for further details and images from these experiments.

b. Make the following substitution $\Phi(r) = W(r)/r$, and solve the differential equation for $W(r)$, given that $W(r)$ should vanish as $r \to \infty$.

c. Show that the solution for a single point charge is given by (4.17).

d. We now turn to the sphere. The surface charge density σ on the sphere is $\sigma = Ze/(4\pi R^2)$. Apply the boundary condition from electrostatics,

$$\sigma = -\epsilon_0 \epsilon_{\mathrm{r}} \left. \frac{\partial \Phi}{\partial r} \right|_{r=R}, \tag{4.45}$$

to derive the Debye-Hückel solution (4.18).

Problem 4.7 *The DLVO theory and the Schultz-Hardy rule*

The interaction between colloidal particles depends on the solution of salt in which they are dispersed. In this exercise, we derive the DLVO potential between spheres, illustrated in figure 4.7, and use it to predict the critical salt concentration at which colloids begin to aggregate.

For further reading, see section 12.18 of Is-raelachvili, 2011 and chapter 7C of J. C. Berg, 2010.

a. We start with the interaction potential per unit area between two parallel charged surfaces a distance h apart,

$$U(h) = \frac{64 k_{\mathrm{B}} T c \gamma^2}{\kappa} e^{-\kappa h}. \tag{4.46}$$

Here, c represents the total electrolyte concentration in the solution, $\gamma = \tanh\left(Ze\psi_0/4k_{\mathrm{B}}T\right)$, and ψ_0 is the potential difference between the two surfaces. Applying the Derjaguin approximation (see problem 4.3), show that the potential between two charged spheres of radius R is

$$U(h) = \frac{64\pi k_{\mathrm{B}} T R c \gamma^2}{\kappa^2} e^{-\kappa h}. \tag{4.47}$$

b. As discussed in the caption of figure 4.7, the DLVO potential between two spherical particles consists of the above repulsive term and an attractive Van der Waals term (4.10):

$$U(h) = \frac{64\pi k_{\mathrm{B}} T R c \gamma^2}{\kappa^2} e^{-\kappa h} - \frac{HR}{12h}. \tag{4.48}$$

Sketch the different regimes that can occur as the strength of the two terms above is varied, and verify that you reproduce figure 4.7 qualitatively. How many zeros can $U(h)$ have? In which regime will colloid coagulation (aggregation) occur?

c. At the critical aggregation concentration c_*, the function $U(h)$ has exactly one zero. By definition, this zero occurs when $U = 0$

and $dU/dh = 0$. Using these conditions, show that

$$\frac{\kappa^6}{c_*^2} \propto \left(\frac{k_\mathrm{B} T \gamma^2}{H} \right)^2 . \tag{4.49}$$

d. As $\kappa^2 \propto Z^2 e^2 c / k_\mathrm{B} T$ according to (4.15), the above implies

$$(Ze)^6 c_* \propto \frac{(k_\mathrm{B} T)^5 \gamma^4}{H^2} . \tag{4.50}$$

Using the definition of γ from a, show that the critical aggregation concentration scales as $c_* \propto 1/(Ze)^6$ for large potential difference ψ_0 and as $c_* \propto 1/(Ze)^2$ for small ψ_0. These scalings, which are successfully described by DLVO theory, were first discovered empirically and are known as the Schultz-Hardy rule.

Problem 4.8 *Scaling of regular fractals*
In this exercise we analyze the scaling of two regular fractals, the so-called Koch curve and the Sierpinski gasket illustrated in figure 4.29. We first consider the Koch curve. The starting stage is shown at the top at level $k = 1$. We have four line elements, each of a length $L_0/3$. In each successive stage, the line is divided into four pieces, each with a length a third of that of the previous stage. Clearly the total line gets longer in each stage of the process.

   a. Give the expression for the length r_k of a line segment in stage k of the process.

   b. Write from this expression k as a function of L_0 and r_k.

   c. Write the total length L_k of the curve in stage k.

   d. Use the previous two results to write L_k as a function of r_k and L_0. Show that $L(r) = L_0^{d_\mathrm{f}} r_k^{1-d_\mathrm{f}}$ and use it to calculate d_f. Note that this is similar to equation (4.20), as $\rho(r) \sim 1/L(r)$.

   e. How does the number of line segments $N(r)$ scale with r_k?

   f. Now repeat the steps for the scaling of the total 'mass' $M(r_k)$ (the shaded portion of each triangle) of Sierpinski gasket.

Problem 4.9 *Motorized colloids: Quincke rollers*
Quincke rotation, illustrated in figure 4.19 d, is an electro-hydrodynamic phenomenon in which an insulating sphere immersed in a conducting fluid begins to rotate under an applied electric field, $\vec{E}_0$. In this exercise, we will derive this angular velocity $\vec{\Omega}$.

   a. Let us begin with a simpler situation: a conducting fluid with charge density ρ in the absence of a sphere. The electric current $\vec{j}$ in the fluid is related to the electric field $\vec{E}$ through Ohm's law,

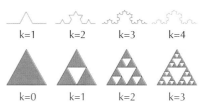

Figure 4.29. Illustration of four stages of regular fractals, the Koch curve (top), and the Sierpinski gasket (bottom).

Quincke rotation provides a popular experimental route to studying active media. You will encounter examples of this in figure 9.4 and figure 9.7.

$\vec{j} = \sigma\vec{E}$. Using Gauss' law, $\nabla \cdot \vec{E} = \rho/\epsilon$, and charge conservation, show that ρ relaxes to zero in a typical time $\tau = \epsilon/\sigma$.

b. If the fluid flows with nonzero velocity $\vec{v}$, then the electric current is $\vec{j} = \sigma\vec{E} + \rho\vec{v}$ due to advection of the charge. In this case, show that the modified charge conservation equation is

$$\partial_t \rho + \frac{1}{\tau}\rho + \nabla \cdot (\rho\vec{v}) = 0. \tag{4.51}$$

c. Now, if an insulating sphere is immersed in such a conducting fluid, electric charges build up on the surface of the colloidal particle to generate a dipole moment, $\vec{P}$. The dynamical equation for the dipole moment $\vec{P}$ is given by

$$\frac{d\vec{P}}{dt} + \frac{1}{\tau}\vec{P} = -\frac{1}{\tau}2\pi\epsilon_0 R^3 \vec{E}_0 + \vec{\Omega} \times (\vec{P} - 4\pi\epsilon_0 R^3 \chi^\infty \vec{E}_0), \tag{4.52}$$

where R is the radius of the sphere and $\vec{E}_0$ is taken to be uniform in space and constant in time. The remaining constants are defined as $\chi^\infty = \frac{\epsilon_p - \epsilon_l}{\epsilon_p + 2\epsilon_l}$, with ϵ_p the dielectric permittivity of the sphere and ϵ_l the permittivity of the fluid, and $\tau = \frac{\epsilon_p + 2\epsilon_l}{2\sigma_l}$ is the Maxwell-Wagner time, with σ_l the conductivity of the fluid. By comparing this equation with that for the charge density ρ in step b), provide an interpretation for the different terms in this equation.

d. It is convenient to split the dipole moment into two terms, $\vec{P} = \vec{P}^\epsilon + \vec{P}^\sigma$, with $\vec{P}^\epsilon = 4\pi\epsilon_0 R^3 \chi^\infty \vec{E}_0$. In terms of these variables, show that the dipole moment equation reduces to

$$\frac{d\vec{P}^\sigma}{dt} + \frac{1}{\tau}\vec{P}^\sigma = -\frac{1}{\tau}4\pi\epsilon_0 R^3 \left(\chi^\infty + \frac{1}{2}\right)\vec{E}_0 + \vec{\Omega} \times \vec{P}^\sigma. \tag{4.53}$$

e. Show that if $\vec{\Omega} = 0$, there exists a steady solution for the dipole moment (i.e., $d\vec{P}/dt = 0$) in which the dipole moment anti-aligns with the applied electric field, given by $\vec{P} = -2\pi\epsilon_0 R^3 \vec{E}_0$.

f. Due to its surface charge distribution, the colloid can experience an electric torque that may cause it to rotate. When this torque is balanced with viscous drag in the fluid, the colloid rotates with angular velocity $\vec{\Omega} = \mu_r \frac{\epsilon_l}{\epsilon_0}\vec{P} \times \vec{E}_0$, where $\mu_r = (8\pi\eta R^3)^{-1}$ and η is the viscosity of the fluid. In this case, show that the equation of motion takes the form

$$\begin{aligned}\frac{d\vec{P}^\sigma}{dt} + \frac{1}{\tau}\vec{P}^\sigma = &-\frac{1}{\tau}4\pi\epsilon_0 R^3 \left(\chi^\infty + \frac{1}{2}\right)\vec{E}_0, \\ &- \mu_r\frac{\epsilon_l}{\epsilon_0}(\vec{P}^\sigma \cdot \vec{E}_0)\vec{P}^\sigma + \mu_r\frac{\epsilon_l}{\epsilon_0}|\vec{P}^\sigma|^2\vec{E}_0.\end{aligned} \tag{4.54}$$

g. Notice that the sphere rotates only if the dipole moment and the applied electric field are misaligned. Let us further split $\vec{P}^\sigma$ into a component along the direction of $\vec{E}_0$, which we will call $P_\parallel^\sigma$, and a component perpendicular to it, which we will call $P_\perp^\sigma$. Show that the equation above reduces to the following set of equations:

$$\frac{dP_\parallel^\sigma}{dt} + \frac{1}{\tau}P_\parallel^\sigma = -\frac{1}{\tau}4\pi\epsilon_0 R^3\left(\chi^\infty + \frac{1}{2}\right)E_0 + \mu_r\frac{\epsilon_1}{\epsilon_0}(P_\perp^\sigma)^2 E_0,$$

$$\frac{dP_\perp^\sigma}{dt} + \frac{1}{\tau}P_\perp^\sigma = -\mu_r\frac{\epsilon_1}{\epsilon_0}E_0 P_\parallel^\sigma P_\perp^\sigma. \qquad (4.55)$$

h. We now look for a steady state solution for which $P_\perp^\sigma \neq 0$. Show that if $\chi^\infty + \frac{1}{2} > 0$ and $E_0 > E_Q$, where $E_Q = (4\pi R^3\epsilon_1\tau\mu_r(\chi^\infty + \frac{1}{2}))^{-1/2}$, there exists a steady state given by

$$P_\parallel^\sigma = -\frac{\epsilon_0}{\epsilon_1\tau\mu_r}\frac{1}{E_0}, \qquad P_\perp^\sigma = \frac{\epsilon_0}{\epsilon_1\tau\mu_r}\frac{1}{E_0}\sqrt{\frac{E_0^2}{E_Q^2} - 1}. \qquad (4.56)$$

i. Show that in this steady state, the rotation speed is

$$\Omega = \frac{1}{\tau}\sqrt{\frac{E_0^2}{E_Q^2} - 1}. \qquad (4.57)$$

Note that the rotation axis can be any direction perpendicular to $\vec{E}_0$.

j. Based on the terms in the equations in g, argue that the nonrotating state is unstable. That is, if we start near the nonrotating steady state, with $P_\parallel^\sigma \approx -4\pi\epsilon_0 R^3\left(\chi^\infty + \frac{1}{2}\right)E_0$ and with a small misaligned $P_\perp^\sigma$, the dipole moment evolves to misalign further.

If rotating colloids are sedimented onto a substrate, they begin to roll. Aligning interactions between the colloids cause them to flock, or move unidirectionally, forming an ordered phase (see Bricard et al., 2013). To learn more on this go to the active matter chapter! See in particular figure 9.4 and figure 9.7.

Problem 4.10 *Gelation of milk and percolation*

Milk is an example of a colloidal suspension, specifically water interspersed with small protein particles called casein micelles (micelles are illustrated in figure 6.24). Most of the time, these micelles do not stick together because their surfaces are hydrophilic, i.e., they want to be surrounded by water (see section 6.9.4). However, when an enzyme such as chymosin is added to milk, the suspension begins to thicken into a gel: certain parts of the surface of the micelles are hydrolyzed (i.e., broken down by water) and want to form bonds with other micelles. For this reason, groups of micelles start clumping together. Once enough bonds are formed, the once-fluid milk becomes a gelatinous network of bonds that can resist deformations.

Here, we view this process through the lens of the percolation transition. For simplicity, imagine that the casein micelles are not

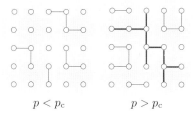

$p < p_c$ $p > p_c$

Figure 4.30. The percolation transition. The micelles floating in milk are represented as white circles. The lines represent bonds that form with probability p because of the enzyme called chymosin. If $p < p_c$ the bonds form disconnected clumps. If $p > p_c$, there is on average at least one clump (red) that spans the entire system.

floating around in water, but rather are arranged on a lattice as shown in figure 4.30. We assume that each micelle forms a bond with its neighbor with probability p, which depends on the amount of chymosin added to the solution. As you can imagine, if p is very small, then the lattice will consist mostly of small disconnected clumps. However, if p is sufficiently large, then there will on average be at least one clump (red) that spans the entire size of the system. The value $p = p_c$ at which the first system-spanning clumps form is known as the critical value, and the behavior of the system near this point is known as the percolation transition. Imagine trying to 'stretch' milk in a glass: if $p < p_c$, then the system will flow since there is nothing holding the small clumps together. If $p > p_c$, then the system will resist deformation because certain connected clumps span the volume of the glass.

To warm up to quantitative calculations, we will begin with a one-dimensional glass of milk as shown in the top row of figure 4.31. Each micelle is separated from its neighbor by a lattice spacing a. As you can see, in one dimension, the critical value is $p_c = 1$, since all the bonds must be occupied in order to form a clump that spans the entire system. To get a feel for the physics near this critical value, try the following two exercises.

For more details on percolation we recommend the book by Stauffer and Aharony, 2003, whose material we have greatly condensed to construct this problem and the next two.

We recommend the paper by Tokita, 1989 for an introduction to gelling as a percolation problem.

a. Let's define an object known as the *correlation function* $g(r)$ as the probability that two micelles a distance r apart are connected by an unbroken chain of bonds. Consider a lattice model of bonds; bonds are treated as independent, and each bond is present with probability p. Show that $g(r)$ can be written as

$$g(r) \sim \exp(-r/\xi), \tag{4.58}$$

with $\xi = -a/\log p$. The length scale ξ is called the *correlation length*.

b. Show that near the percolation transition $p_c = 1$, the correlation length ξ diverges in the form

$$\xi \propto \frac{a}{|p_c - p|^\nu}, \tag{4.59}$$

with $\nu = 1$. This exponent ν is an example of a *critical exponent* (compare the margin note opposite equation (5.4)).

Hint: Note that in this case $p_c = 1$ and expand the logarithm.

Percolation is a very general concept that goes well beyond milk. It has found applications in the study of turbulence, insulator-conductor transitions, and fluid flow through porous media.

Figure 4.31. Coarse-graining a one-dimensional model. A one-dimensional lattice is coarse-grained by retaining every third lattice site. Points are connected in the coarse-grained lattice if the points were connected in the original lattice.

Problem 4.11 *Real space renormalization group analysis for percolation in one dimension*

The percolation transition from the previous problem has features that are actually ubiquitous throughout physics: near a phase transition, there exists a correlation length ξ that diverges as a power law in the parameter $|p - p_c|$ that measures the distance from the critical point. These are very general features of a broader class of transitions known as second order, or continuous, phase transitions. Other examples of second order phase transitions include boiling of water, melting of a two-dimensional crystal, and the onset of ferromagnetism. The exponent ν is often called universal, since many systems with seemingly dissimilar microscopic components have the same critical exponent. Systems that share the same exponents are said to be in the same *univerality class*.

It was discovered in the mid-twentieth century that the key ingredient behind these second order phase transitions is the notion of *self-similarity*. The percolation transition is a perfect example. Suppose $p < p_c$; then, microscopically, you may see small chunks of micelles sticking together, but when you zoom out you see that the system is disconnected. If $p > p_c$, then microscopically, you might see pockets of water, but when you zoom out you will see that chains of micelles span the entire system. If $p = p_c$, no matter how far you zoom out, you won't be able to tell which phase you are in. In other words, no matter which length scale you look at, the system should be statistically self-similar. As an illustration, figure 4.32 shows a simulation of two-dimensional percolation with $p = p_c$. The three successive zoom-ins display the self-similarity: on average, they each seem to have similar statistical properties, like the relative amounts of black and white.

A framework for understanding these transitions is the *renormalization group* (RG). The basic idea is that you start with a model of your system, then you *coarse-grain* it, which means that you smear out some microscopic information, in order to obtain a new model that describes physics on a larger scale. To see what this means in practice, and to see why it is useful, we are going to look at one-dimensional percolation again. We'll pretend that we don't already know p_c and ν, and we will compute them in a completely different way. As shown in figure 4.31, the main idea is to start with a microscopic lattice with a given bond probability p, then build a new lattice by keeping every bth lattice site from the original. In the illustration, $b = 3$, a bond is drawn in the new lattice if the two points were connected in the original lattice. Our goal now is to understand the statistical properties of the new lattice in terms of the properties of the old one.

    a. Let p' be the probability of a bond existing between two neighbors in the new lattice. Argue that $p' = p^b$.

    b. At the critical point, the old lattice and the new lattice are statistically identical. Using this fact, argue that $p_c = 1$.

    c. The correlations lengths for the new and the old lattices are $\xi' = |p' - p_c|^{-\nu}$ and $\xi = |p - p_c|^{-\nu}$. Since the two lattices are

For an introduction to the theory of critical phenomena, see, e.g., Goldenfeld, 1992, Herbut, 2007, Nishimori and Ortiz, 2010, or Honig and Spalek, 2018.

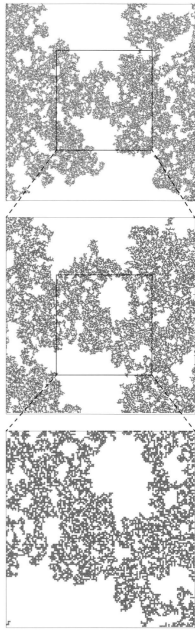

Figure 4.32. Self-similarity in the percolation transition. A simulation of two-dimensional percolation with $p = p_c$ is shown. Each successive zoom-in of the simulation looks statistically similar to the previous one.

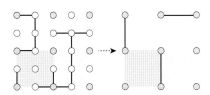

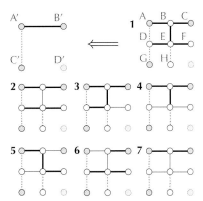

Figure 4.33. Renormalization group for 2D percolation. The top row illustrates a coarse-graining procedure in which every other row and column is eliminated. At bottom we illustrate seven different microscopic bond configurations (labeled 1 to 7) that would result in a horizontal connection between sites A′ and B′ in the coarse-grained lattice. The sites A′ and B′ are considered connected if in the original lattice A and/or D are connected (by thick solid lines) to C and/or F. We omit configurations obtained by vertical mirroring or by exchange of upper two rows. The dashed bonds do not affect the A′B′ horizontal bond so are not considered.

statistically identical at the critical point, the correlation length in the new lattice ξ' is simply a rescaling of the old correlation length: $\xi' = \xi/b$. Use this fact to show that

$$\frac{1}{\nu} = \frac{\log \lambda}{\log b}, \tag{4.60}$$

where $\lambda = (p' - p_c)/(p - p_c) = \frac{dp'}{dp}$ as $p \to p_c$.

d. Using the fact that $p' = p^b$ and $p_c = 1$, evaluate (4.60) in the limit $p \to p_c$, and show that $\nu = 1$.

That is it! We did an RG calculation: we started with a model, smeared out the microscopic data, related the new model and the old model, then used the property of self-similarity to obtain useful, measurable quantities like ν and p_c. In one dimension, the problem was simple enough that we didn't really need RG. In the next problem, we will do percolation in two dimensions, for which the situation is more complex.

Problem 4.12* *Renormalization group analysis for bond percolation in two dimensions*

We will in this problem repeat the calculation from the previous problem but now in two dimensions. Note that, unlike for the one-dimensional problem, which could be treated exactly, we have to resort to an approximate procedure that will yield a value for the critical exponent which is close to, but not exactly the same as, the value obtained by exact calculations.

Consider the square lattice sketched in the top of figure 4.33. We coarse-grain it by drawing a new square lattice that retains only every other site in each direction. The rule for drawing bonds in the new lattice (i.e., for considering neighboring sites in the new lattice to be connected) is illustrated in the lower part of the figure. Let's focus on the process for the horizontal bonds only, as the vertical ones will work analogously. For example, the yellow square labeled '1' shows the area of the microscopic lattice that will be coarse-grained into the square A′B′C′D′, with all bonds belonging to it marked by solid or dashed lines. Two horizontally adjacent sites in the new lattice A′ and B′ are taken as connected if there is a percolating path (thick solid lines) on the original lattice connecting A and/or D with C and/or F. Bonds AD, DG, and EH are marked with dashed lines as their presence or absence is immaterial for determining whether there is a connection between sites A′ and B′ in the new lattice.

a. The lower part of figure 4.33 shows seven fundamentally different possibilities of how a horizontal bond between A′ and B′ can occur in the coarse-grained lattice. Show that the probability for each of the seven cases is given by

$$p_1 = p^5, \qquad\qquad p_4 = p_5 = p_6 = p^3(1-p)^2,$$
$$p_2 = p_3 = p^4(1-p), \qquad p_7 = p^2(1-p)^3. \tag{4.61}$$

b. In addition, we have to take into account the multiplicity M_i of each of the graphs that is formed by the bonds indicated by solid lines. The multiplicity M_i is the number of ways in which each graph i can be drawn on the sublattice formed by sites A–F. By enumerating all the possible ways (draw them, finding the ones we omitted in the figure!) of getting a bond in the new lattice, show that the multiplicities are

$$M_1 = M_2 = 1, \quad M_3 = M_6 = 4, \quad M_4 = M_5 = M_7 = 2. \tag{4.62}$$

c. Combine the results of the previous two steps to show that probability p' for the bond between A′ and B′ occuring in the new coarse-grained lattice is

$$p' = p_1 + p_2 + 4p_3 + 2p_4 + 2p_5 + 4p_6 + 2p_7,$$
$$= 2p^5 - 5p^4 + 2p^3 + 2p^2. \tag{4.63}$$

d. Apply self-similarity: set $p' = p = p_c$ in (4.63) to show that $p_c = \frac{1}{2}$.

e. As in the previous problem, use the relationships $\lambda = \frac{dp'}{dp}|_{p=p_c}$ and

$$\frac{1}{\nu} = \frac{\log \lambda}{\log b} \tag{4.64}$$

to show that $\lambda = 13/8$, and that this implies $\nu \approx 1.428$.

Here, $b = 2$ since the side length of the new lattice is twice that of the old. It is worth mentioning that this coarse-graining procedure is actually too simple in the sense that it throws out too much information about the underlying lattice. As a consequence the value of ν differs from the result of exact calculations, which yield $\nu = 4/3$.

See section 4.2 of Stauffer and Aharony, 2003 for further examples and details and for discussion of the renormalization group flow implied by equation (4.63). This type of RG treatment in real space is an example of the block spin transformation originally developed by Leo Kadanoff for spin systems (Kadanoff, 1966). For an introduction to this method, which was a forerunner of RG theory, see section 9.1 of Goldenfeld, 1992.

5 | Polymers

Polymers, long chain molecules, play an important role in the natural and life sciences and in technology. As a result, polymer science is a rich and vast field, ranging from very fundamental to applied and from soft matter to biomatter. We discuss some of the classical results concerning the statistical properties of polymers as a function of chain length, and the effects of self-avoidance interactions to the scaling behavior. Particular attention is paid to the worm-like chain formulation of the ideal chain model and the resulting force-extension curve, as this has become a reference model for stretching experiments on DNA and other biopolymers.

Our overview includes the scaling behavior of polymers in solutions; the mean-field theory for polymer mixtures and the effects of fluctuations on the phase diagram; polymer networks; polymer rheology; and the intriguing effects and instabilities associated with polymer solutions and melts being complex fluids whose rheology exhibits strong viscoelastic effects.

5.1 The ever-broadening field of polymer science

Sir Sam Edwards addressed the scaling of a polymer coil in three dimensions back in 1965 (Edwards, 1965). His formulation (Doi and Edwards, 1986) has become the basis for field-theoretic calculations of polymer statistics and dynamics. See Goldenfeld, 2015 for a warm account of Edwards's foundational contributions to soft matter, including granular matter and spin glasses.

The books *Scaling Concepts in Polymer Physics* by de Gennes, 1979, *Statistical Mechanics of Chain Molecules* by Nobel laureate Flory, 1969, and *Theory of Polymer Dynamics* by Doi and Edwards, 1986 are still inspiring to read. Note also figure I.1 for de Gennes's Nobel citation. For a pedagogical modern introduction, we recommend the textbook by Rubinstein and Colby, 2003.

A brief history of the field is in order, given its scope. The study of properties of polymers received an enormous boost during the first half of the previous century, when it became possible to synthesize many new polymers at industrial scales (see figure 5.1). Gradually it became clear that when polymers are very long, i.e., when their degree of polymerization N is large, a number of their important statistical properties can be captured in simple statistical-physical models. This realization was exploited by Edwards, de Gennes, and their collaborators, who enriched our understanding by bringing insights from critical phenomena, in particular, those of scaling ideas, to bear on both the static and dynamical properties of polymers. Their way of thinking has become so much part of our present understanding that it is hardly ever stated anymore. We include in this chapter some examples where we see the power of their approach at work. Edwards's and de Gennes's ideas played an important role in stimulating the interest of the physics community in polymers, and in our accepting polymers as the cornerstone of the newly emerging field of soft matter.

In the last few decades, new enrichments have come first and foremost from the possibility of studying the properties of single

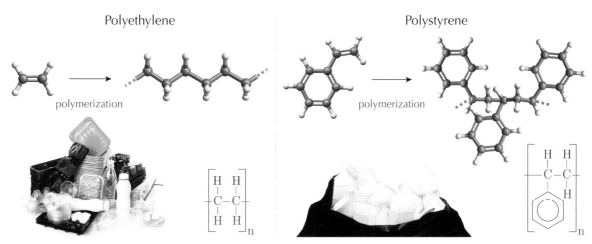

Figure 5.1. Polyethylene and polystyrene. Polyethylene is made by polymerization of the ethylene molecule 2H-C=C-2H. It is one of the simplest polymers and because of the single C-C bond it is very flexible (its Kuhn length, defined in equation (5.2), is $\ell_K = 1.4$ nm. Typical degrees of polymerization are $n = 10^3 - 10^5$. Polyethylene is the most widely used plastic in the world; you'll often encounter it in food wraps, shopping bags, detergent bottles, and automobile fuel tanks, for instance. Polystyrene is a hard plastic which is well known from food packaging and coffee cups; when combined with various colorants, additives, or other plastics, polystyrene is also used to make appliances, electronics, automobile parts, toys, etc. The benzene rings affect the possible local orientations somewhat, but polystyrene is still quite flexible ($\ell_K = 1.8$ nm); they also inhibit crystallization, so solid polystyrene is glassy. Left photo JasminkaM/Shutterstock; right photo Kabardins/Shutterstock.

biopolymers and branching out from there into biopolymer networks and beyond. This has broadened the field and has inspired many life scientists to embrace polymer science. It has also become increasingly clear that polymer rheology is more than an engineering topic: it poses new and interesting fundamental challenges. This fact, together with the modern experimental opportunities offered by microfluidics, flow visualization, and tracking techniques, has brought together communities at the interface of polymer science, non-equilibrium physics, and fluid dynamics.

Nowadays, the importance of the strong link between polymer science and the life sciences hardly needs explanation, since biopolymers such as RNA, DNA, microtubules, and proteins are primary building blocks of life. We will in this book not cover biopolymers like proteins or enzymes, whose detailed structures and ways of folding determine their functions. But we do treat the main statistical properties of semiflexible polymers like DNA, whose properties are widely probed experimentally, as we saw in section 3.6.1.

The ubiquity of plastics in modern society illustrates the importance of polymers in applications, and the impact of polymer research in industrial labs. As we saw for colloids, researchers today have many tricks at their disposal, which they can apply to polymers, and which feed back into fundamental science. The colloids coated with patches of DNA shown in figure 4.17 are just one example of the new opportunities created by DNA nanotechnology (we shall see more in section 5.4 and chapter 10), while the

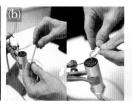

Figure 5.2. Within chemistry, an important recent development is the possibility of synthesizing supramolecular polymers. Instead of the traditional chain polymers which we will study and which are based on strong covalent bonds, supramolecular polymers are formed from molecules which bond together by secondary interactions, like hydrogen bonds, or that can rearrange their topology by so-called exchange reactions without depolymerization. These interactions are much more dynamic than covalent bonds. This creates new possibilities, such as for self-healing rubbers and biodegradable plastics. (a) The images illustrate the process of self-healing of a piece of supramolecular rubber which initially is cut into two.[1] (b) Permanently crosslinked polymers whose networks can change topology by thermally activated bond exchange are called vitrimers. Like silica, by being heated locally and without the use of molds, vitrimers can be worked and welded to make complex objects, as illustrated in the sequence of images.[2] (c) Self-healing materials with good mechanical properties like fast elastic return even in the cold and resistance to fracture are finding their way to the field of soft robotics. Image (a) © François Tournillac / Ludwik Leibler / CNRS Photothèque; (b) © Cyril Fresillon / ESPCI / CNRS Photothèque; (c) courtesy of Jakob Langenbach.[3] A nice account of the supramolecular polymer field is given by Aida and E. W. Meijer, 2020.

supramolecular polymers illustrated in figure 5.2 hold promises for biodegradable and self-healing materials.

In this chapter we will first focus on the basic statistical properties of polymers as long chain molecules, including the differences and similarities between synthetic polymers and biopolymers. After discussing polymers in solution and polymer brushes, we briefly treat the Flory-Huggins mean-field theory, biopolymer networks, and the rheology of dense polymer solutions.

Many topics treated in the first sections of this chapter are treated in more detail in the textbook on polymer physics by Rubinstein and Colby, 2003, and by Strobl, 1997, who puts more emphasis on rubber response and fracture. Witten and Pincus, 2004 give a brief introduction to the variety of polymers and the polymerization process, and they throw additional light on several of the topics we cover.

5.2 Polymers: Long chain molecules with many accessible conformations

For an introduction to polymer chemistry, see, e.g., Carraher Jr., 2017 and Billmeyer, 1984.

Since we will focus on statistical-physical properties which are characterized by only a few effective parameters, we will not go into the chemistry and microscopic properties of polymers here. For us, the essence of polymers is that they are long chain molecules, made by repetition of the basic monomeric unit. The most common and simplest ones are hydrocarbons, made of carbon and hydrogen. Polyethylene, illustrated in figure 5.1, is prime example of this. Many polymers with different properties, like the polystyrene shown too, are obtained by replacing one or more of the hydrogens with a different molecule or group. The varieties, uses, and properties of so-called synthetic polymers—polymers made in the lab or on an industrial scale—are overwhelming.

We will in this chapter focus on the behavior of long linear polymers. But you should be aware that polymer science is not limited to chain molecules. For instance, diblock copolymers consist of two different linear chains attached together at their ends; they can form all kinds of ordered phases if the two ends don't like each other, and so have a preference for demixing—just peek ahead at figure 5.24! Star polymers consist of linear chains attached all to one central knot; branched polymers have, as the name suggests, many branches; dendrimers are like star polymers that continue to branch out like trees; etc.

Polymerization is the process by which such chain molecules are grown. The typical description of polymerization reactions is in terms of rate equations for the reactants. We will not go into details, but even though the polymerization reactions can nowadays be controlled extremely well so as to produce polymers with the desired properties and lengths, it should be clear that the growth process is inherently a stochastic process. Nevertheless, the relative fluctuations Δn in the degree of polymerization in polymers of large degree of polymerization n are small: because of the law of large numbers, Δn grows as $\sqrt{n}$, so the relative fluctuations vanish as $\Delta n / n \sim 1/\sqrt{n}$ for large n. We will therefore in our treatment ignore effects associated with polydispersity, the fact that, in reality, one will have polymers with a distribution of the degree of polymerization.

The important property of chain polymers which underlies the statistical description is their great conformational freedom. As figure 5.3 illustrates, when you consider the plane spanned by two neighboring carbon bonds, the next C-C bond has the lowest energy in the so-called trans configuration, which lies in the same plane. But alternative low-energy configurations are the gauche conformations when the bond makes a 120 degree angle with the plane. The energy difference ΔE between gauche and trans conformations is low, only about $0.8 k_B T$ for polyethylene, so the thermal population of gauche configurations is quite significant. Of course, the values of these energies differ for each polymer, as these are affected, among other things, by the single/double nature of the carbon bonds and the steric hindrance effects from sidegroups. Nevertheless, the conclusion is that long polymers can statistically explore many different conformations. Figure 5.4, where polymers are projected to a plane, illustrates how a few local transformations change the overall conformation of a relatively short chain quite dramatically. For studying the statistical properties, we will simply view polymers as long flexible chains.

The above discussion is focused on synthetic polymers, whose flexibility becomes noticeable on the molecular scale. Many biopolymers, on the other hand, are stiff enough that it is better to think about them as thin bendable rods. As figure 5.5 illustrates for DNA, their flexibility only becomes noticeable on a sufficiently long length scale, the 'persistence length', which, as we shall see, is set by the ratio of the bending stiffness to the thermal energy. This is the basis of the so-called wormlike chain model, which is

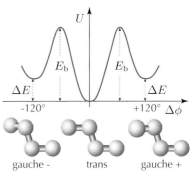

Figure 5.3. Illustration of the rotational freedom of neighboring bonds between carbon molecules of a hydrocarbon polymer such as polyethylene. Bottom: four carbon atoms of the backbone of the polymer. Relative to the plane spanned by the bonds of the right three atoms, the left atom has three possible orientations. The zigzag-type conformation with the leftmost bond lying in the plane is called the trans conformation, while the two out-of-plane orientations with $\Delta\phi = \pm 120°$ are called gauche orientations. In polyethylene the bond angle θ is about $112°$. The graph shows the energy U of the bond as a function of the rotation angle $\Delta\phi$. Trans is the lowest energy configuration. The energy difference ΔE with the gauche conformations is small for polyethylene, about $0.8\ k_B T$, while the energy barrier E_b is about $6\ k_B T$.

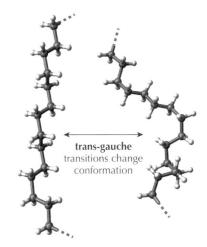

Figure 5.4. Illustration of different conformations of a short piece of a polymer, in this case polyethylene. The figure illustrates how a simple trans-to-gauche transformation changes the conformation of the polymer considerably.

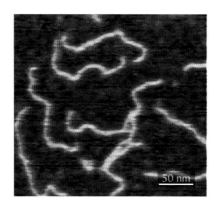

Figure 5.5. Atomic force microscope image from Lysetska et al., 2002 of 538 base pair DNA fragments deposited on mica. The scale on which the DNA is bent, the so-called persistence length of DNA, is clearly comparable to the scale bar of 50 nm.[4] Persistence lengths of various biopolymers are given in table 5.2 below.

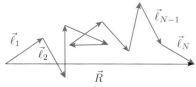

Figure 5.6. Illustration of the ideal chain model. The chain is the sum of vectors $\vec{\ell}_i$ of the same length whose orientations are uncorrelated.

We shall see in (5.31) that the Kuhn length is twice the persistence length.

so important for interpreting data on biopolymers that we devote a separate discussion to it in section 5.4. On length scales beyond the persistence length, its statistical scaling properties are the same as those of the ideal chain model, which we now discuss.

5.3 Ideal chains, excluded volume effects, and the Flory argument

The fact that most synthetic polymers are very floppy implies that their low-energy behavior is governed by their large *conformational entropy*. In this section we first introduce the simplest model for this, the noninteracting ideal chain model. We then discuss the effects of excluded volume interactions, first in terms of self-avoiding walks and then in the Flory approximation.

5.3.1 The ideal chain model

a. The ideal chain model

The ideal chain model is the fundamental noninteracting model describing the statistics of a floppy chain. An ideal chain can be thought of to consist of the vector sum of N vectors $\vec{\ell}_i$ of equal length, with no correlation in the directions of the different vectors—see figure 5.6. We consider the model in three dimensions.

The total end-to-end vector $\vec{R}$ after the N steps is

$$\vec{R} = \sum_{i=1}^{N} \vec{\ell}_i, \tag{5.1}$$

and the assumption that the vectors or steps are completely uncorrelated is expressed by the averages

$$\langle \vec{\ell}_i \rangle = 0,$$
$$\langle \vec{\ell}_i \cdot \vec{\ell}_j \rangle = \langle |\vec{\ell}|^2 \rangle \, \delta_{ij} = \ell_K^2 \, \delta_{ij}. \tag{5.2}$$

Here, ℓ_K is an effective length, the so-called Kuhn length. It is important to stress that Kuhn length is not the monomer size, but an effective length obtained from comparison of the results below with experiments: roughly speaking, it is the effective length after which successive sections can be thought to be directionally uncorrelated. For instance, as indicated in figure 5.1, the Kuhn length of polyethylene is about 1.4 nm, which is about nine times the C-C bond length. For most common simple synthetic polymers, the Kuhn length is between 1 and 2 nanometers, but, as we will see for biopolymers, the Kuhn length is comparable to the persistence length, which figure 5.5 already showed is of order 50 nm for DNA, but which is much bigger for many other biopolymers.

A similar remark concerns N, which is proportional to but not equal to n, the actual number of monomers. Indeed, N is the number of statistically independent units of length ℓ_K in this ideal chain representation. Even though the two differ typically by a factor of order 10 for synthetic polymers, as the Kuhn length is a factor of this size larger than the bond length of the C-C backbone,[5] we will follow standard practice of referring loosely to N as the degree of polymerization.

For the expectation value of $\vec{R}$, we straightforwardly get, from (5.2),

$$\langle \vec{R} \rangle = 0,$$

$$\langle |\vec{R}|^2 \rangle = \sum_{i,j} \langle \vec{\ell}_i \cdot \vec{\ell}_j \rangle = \sum_{i=1}^{N} \ell_K^2 = \ell_K^2 \, N. \tag{5.3}$$

In polymer physics, one commonly focuses on the root mean square end-to-end distance $R_0 = \langle |\vec{R}|^2 \rangle^{1/2}$. Since we will often focus on how properties scale with N without caring about the precise prefactor, we will often intuitively associate R_0 with the typical radius of the polymer coil. More explicitly, we write for the ideal chain

$$R_0 = \ell_K \, N^\nu \quad \text{for ideal chains: } \nu = 1/2. \tag{5.4}$$

> In problem 5.2 we derive that the radius of gyration R_g, the root mean square radius of the ideal chain, equals $R_0/\sqrt{6}$.

Note that introducing the exponent ν may look like overkill at this point, but it will pay off to think of it as the critical exponent of the noninteracting ideal chain model.

Since $\vec{R}$ is the sum of many uncorrelated random vectors, the probability distribution $\mathcal{P}(N, \vec{R})$ is a Gaussian, on account of the central limit theorem; in d dimensions, we have

$$\mathcal{P}(N, \vec{R}) \overset{N \gg 1}{\approx} \left(\frac{1}{2\pi R_0^2/d} \right)^{d/2} e^{-dR^2/2R_0^2}, \tag{5.5}$$

> Near critical points of a thermodynamic system, properties such as the correlation length ξ typically diverge as a power of the distance from the critical point, as $\xi \sim |T - T_c|^{-\nu}$ with T_c the critical temperature. The exponent ν is referred to in the theory of critical phenomena[6] as a *critical exponent*. Note the similarity of this expression with (5.4) for the mean end-to-end distance. This is no coincidence: in both expressions a length scale is related to the distance from criticality, where for the polymers the limit $N \to \infty$ is like that of the approach to the critical point, $T \to T_c$.[7]

with R_0 given by (5.4). The consistency of this formula is checked in problem 5.1.

b. Relation between the ideal chain model and random walks

By now you will probably have realized that the problem we are considering here is a random walk in a slightly different language. We can think of a configuration of an ideal chain as the path of a random walker which makes a step of fixed length $\vec{\ell}$ every time unit τ. In this analogy N is the number of steps, i.e., the time measured in units of τ: $N = t/\tau$. Indeed, the connection of the behavior of the ideal chain model and diffusion is quite precise, as the end-to-end distance of the ideal chain is just the distance a walker has moved from its starting point. As the Brownian motion diffusion coefficient D gave, according to (3.14), the mean square displacement in one dimension, for three dimensions we have the mapping

$$\text{ideal chain: } R_0^2 \simeq \ell_K^2 N = \frac{\ell_K^2}{\tau} t \iff \text{diffusion: } \langle \vec{R}^2 \rangle \simeq 3 \cdot 2Dt$$

(5.6)

for large N and large times. The ideal chain expression on the left illustrates again nicely that, in general, we can estimate a diffusion coefficient D_{diff} of a process which makes steps ℓ in time τ as

$$D_{\text{diff}} \simeq \frac{\ell^2}{\tau}.$$

(5.7)

Keep in mind that such an estimate is most accurate if the successive steps are essentially in random directions. In cases in which successive steps tend to be roughly in the same direction, the diffusion coefficient is enhanced, while if there is a tendency to backtrack upon successive steps, the diffusion coefficient is smaller than this estimate suggests.[8] A distribution of step sizes will also affect the estimate.

We used this insight to obtain the Stokes-Einstein relation for the diffusion coefficient of a particle in equation (3.21).[9]

c. The conformational entropy of the ideal chain model

The shape of a polymer is usually termed its conformation. If we denote by $\Omega(N, \vec{R})$ the number of conformations (the number of shape configuations) associated with a given end-to-end distance of a polymer of size N, the probability distribution $\mathcal{P}(N, \vec{R})$ introduced above is nothing but the fraction of all conformations which have an end-to-end vector between $\vec{R}$ and $\vec{R} + d\vec{R}$,

$$\mathcal{P}(N, \vec{R}) = \frac{\Omega(N, \vec{R})}{\int d^d \vec{R} \, \Omega(N, \vec{R})}.$$

(5.8)

If we use the Gaussian expression (5.5) for $\mathcal{P}$ in the Boltzmann formula of statistical physics for the entropy of this state,

$$S(N, \vec{R}) = k_B \ln \mathcal{P}(N, \vec{R}),$$

(5.9)

we see that the only $\vec{R}$ dependence comes from the term in the exponent, so that

$$\Delta S(N, \vec{R}) \equiv S(N, \vec{R}) - S(N, 0) = -\frac{dk_B}{2} \left(\frac{R}{R_0} \right)^2.$$

(5.10)

ΔS is the change in the conformational entropy with increase in the end-to-end vector $\vec{R}$. It decreases with increasing R, as the chain has fewer states available the more it is stretched. Note also that this expression is valid only for small stretching—as the chain is stretched to its maximum length, there is only one available state. We will come back to the stretching force for finite polymer chains in the discussion of the wormlike chain model.

The free energy is generally defined as $F = U - TS$, and so we get, from (5.10), for the conformational contribution to F

$$\Delta F(N, \vec{R}) = \frac{dk_B T}{2} \left(\frac{R}{R_0} \right)^2 = \frac{dk_B T R^2}{2\ell_K^2 N}. \qquad (5.11)$$

The quadratic increase of the free energy with R shows that at equilibrium, the ideal chain responds like an elastic spring whose energy grows quadratically with the extension, and hence whose force is linear in the extension. This is relevant to the types of experiments discussed in section 3.6.1, in which one pulls on a long coiled DNA molecule: the linear behavior for small extensions is visible in the small extension regime in the inset of figure 5.15, below. We will recover this ideal chain behavior for small extensions in our discussion of the wormlike chain model.

d. Polymer scaling and fractal dimension

We have seen that ideal chains are not compact objects, but have the conformation of random walks. This means that, just as we found the colloidal aggregates to be fractal objects, so it is with polymer configurations. Indeed, if we write (5.4) as $N \sim R_0^{1/\nu}$ and compare with (4.20), this suggests the identification of the fractal dimension d_f with the exponent $1/\nu$:

> polymers are fractal with fractal dimension $d_f = 1/\nu$. $\qquad$ (5.12)

As the chain is like a random walk on all scales beyond the Kuhn length, the comparison of the scaling of R_0 with N can immediately be extended to show that for the ideal chain the number of units within a ball of radius r scales as $N(r) \sim r^{1/\nu}$, in complete analogy with (4.20).

We will come back to measurement of the exponent ν from neutron-scattering experiments in section 5.5 below.

5.3.2 Excluded volume interaction and self-avoiding walks

The ideal chain model is clearly a very idealized model, as it leads to many configurations with different parts of the chain very near each other, or even on top of each other when the polymer configuration is considered as a chain or as a walk on a lattice. Stated differently, the model does not take into account the fact that around each monomer there is a region from which other parts of the chain are excluded on account of the short-range repulsive interactions.

How should we account for this excluded volume effect? Let us take what may appear to be a bold step. We still think of chain configurations as random walks, as we did before, but now put the random walk on a lattice. Then a natural model to take into account the excluded volume interaction is one that requires that the walk never visit the same site twice, or, in other words, that we have a *self-avoiding walk*. This is illustrated in figure 5.7.b for a square lattice in two dimensions.

Clearly, in one dimension, a self-avoiding walk is trivial: the walk cannot turn back, so the distance between the starting point and

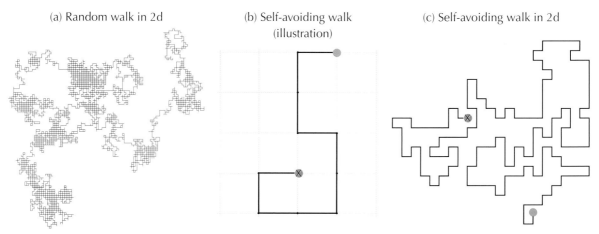

(a) Random walk in 2d   (b) Self-avoiding walk   (c) Self-avoiding walk in 2d
                                    (illustration)

Figure 5.7. Illustration of a random walk and self-avoiding walks on the square lattice in two dimensions. (a) Example of a random walk. Note that many points are visited more than once. (b) Illustration of the short piece of a self-avoiding walk on the square lattice. Note that from the endpoint, indicated by a cross, one cannot step farther, illustrating that this configuration cannot be part of a longer walk simulation. (c) Example of a self-avoiding walk simulation. Note that the structure is much more open than that of the random walk in (a). This illustrates the size scaling as $R_0 \sim N^{\nu_{\text{SAW}}}$ with $\nu_{\text{SAW}} = 3/4$ in two dimensions, while the random walk of panel (a) has $\nu = 1/2$.

the end point (the end-to-end distance, in polymer language) grows linearly in N: $\nu_{\text{SAW}} = 1$ in $d = 1$. In two dimensions, $d = 2$, the self-avoidance constraint is also quite severe: most regular random walk configurations, like the one in figure 5.7.a, are then not allowed either because the walk visits the same site more than once, or because it gets trapped in pockets like the one indicated by a cross in figure 5.7.b. So, self-avoiding walks swell much faster with N than ordinary random walks, as the example in figure 5.7.c illustrates: from exact results we know that their end-to-end distance grows as $R_0 \sim N^{\nu_{\text{SAW}}}$ with $\nu_{\text{SAW}} = 3/4$.[10]

In three dimensions, $d = 3$, the self-avoidance intuitively appears to be already less constraining than in $d = 2$, since the walk can often step up or down to avoid other parts of the chain. The self-avoidance constraint is indeed less important the larger the dimension is. From simulations, the $d = 3$ exponent is found to be $\nu_{\text{SAW}} \approx 0.588$.

It is important to keep in mind that for the chain or walk, the self-avoidance constraint is a nonlocal constraint that affects monomers which are very far apart along the chain just as much as it affects monomers which are close together on the chain. In the language of the walk: the self-avoiding walker never forgets where it has been before; every site which it has ever visited before is excluded. It never loses its memory.

Of course, real polymers don't live on a lattice, and their interactions are much more complex than those represented by the simple self-avoidance constraint. Why is this discussion nevertheless relevant? The answer is that we should think of the $N \to \infty$ limit as similar to approaching a critical point. Just as the behavior near a critical point is characterized by scaling exponents and scaling functions, so is the scaling behavior of polymers for $N \gg 1$.

This insight also implies that we should think of universality classes for the critical behavior. In ordinary equilibrium critical phenomena, most details of the interactions don't matter and the exponents are essentially determined by the dimension and the symmetry of the order parameter. In much the same way, we should think of the exponent $\nu = 1/2$ as the critical exponent of the noninteracting ideal chain model, and ν_{SAW} as the proper exact critical exponent of real polymers with short-range strongly repulsive interactions, no matter their precise form. This understanding will also play a role in our discussion of the scaling of polymers in solution in section 5.5, and of the osmotic pressure as a function of concentration in section 5.5.2. This way of thinking has indeed brought all the tricks and power of the theory of critical phenomena to bear on the polymer problem.

For an introduction to the theory of critical phenomena, see, e.g., Goldenfeld, 1992, Herbut, 2007, Nishimori and Ortiz, 2010, or Honig and Spalek, 2018. See also problems 4.11 and 4.12 on percolation as an example of a critical phenomenon.

5.3.3 The Flory argument for the excluded volume interaction

Already long before the modern theory of critical phenomena and the renormalization group were developed, Flory developed a simple but insightful argument for taking the repulsive interactions of polymers into account.

Paul Flory received the Nobel Prize in Chemistry in 1974 "for his fundamental achievements, both theoretical and experimental, in the physical chemistry of the macromolecules." His book *Statistical Mechanics of Chain Molecules* (Flory, 1969) is still worth consulting.

We already know that the ideal chain behaves like an elastic spring whose free energy (5.11) increases quadratically with end-to-end distance R: the entropic elastic force favors the end-to-end distance R being minimal, as the polymer has fewer configurations available the more it is stretched. Since any reasonable measure of the radius of the polymer blob scales the same as R (see problem 5.2 for an analysis of the radius of gyration), let us somewhat loosely associate R also with the radius of the polymer, and (5.11) with the elastic free energy that minimizes the polymer radius.

We phrase Flory's approach as follows. While the entropic forces prevent the polymer volume from becoming too large, the repulsive forces prevent the polymer from contracting into too small a volume. To estimate this effect, we imagine that the repulsive interaction of each monomer is proportional to the average number of monomers within some ball around it, as sketched in figure 5.8. As the monomer density in d dimensions scales as $n_m = N/R^d$, the average number of monomers each monomer interacts with is n_m times the volume of the ball. And as the number of monomers also scales as N, we estimate the total repulsive interaction as

$$U_{\text{int}}/k_{\text{B}}T = v\,N\,n_m = v\,N^2/R^d. \tag{5.13}$$

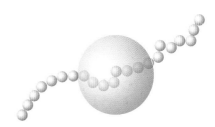

Figure 5.8. To estimate the repulsive energy in the approximation by Flory, we imagine each monomer to have a repulsive interaction with the monomers in the gray ball around it. The total average number of such monomers is the volume of the ball times the monomer density $n_m = N/R^d$. Note that this approximation is a mean-field-type argument which ignores all correlations, and which works best if each monomer has a repulsive interaction with a large number of other monomers. We will come back to the volume of the ball in section 5.4.3.

Here v measures the strength of this effective interaction. Notice that all correlations are ignored here and that the mean-field-type argument works best if the size of the ball which marks the radius of the interaction is rather large, so that there are on average quite a few monomers within it. In practice this is not very realistic—the repulsive forces are rather short-ranged.

The total free energy in Flory's approximation is taken as the sum of the interaction term (5.13) and the entropic term (5.11) derived for the ideal chain,

$$\frac{F}{k_\mathrm{B} T} = v\frac{N^2}{R^d} + \frac{d R^2}{2N\,\ell_\mathrm{K}^2}. \tag{5.14}$$

This behavior is sketched in figure 5.9. The free energy is large for small R due to the fact that then many monomers get close, while it also gets large with R large, as then we have to perform work against the configurational entropy. Minimizing with respect to R yields

$$-\frac{d\,v N^2}{R^{d+1}} + \frac{dR}{N\,\ell_\mathrm{K}^2} = 0, \quad \Rightarrow R = \left(v\ell_\mathrm{K}^2\,N^3\right)^{1/d+2} \sim N^{3/(d+2)}. \tag{5.15}$$

In summary, the Flory expression for the excluded volume or self-avoiding walk exponent ν_SAW is

$$\nu_\mathrm{Flory} = \frac{3}{d+2}. \tag{5.16}$$

Excluded volume effects are especially important for synthetic polymers: their radius scales with the exponent ν_SAW. The statistical behavior of biopolymers, on the other hand, is much less affected by excluded volume effects: these tend to behave much more like ideal chains. We will discuss why this is so in section 5.4.3.

5.3.4 Taking stock

The Flory expression (5.16) for the exponent ν turns out to be remarkably accurate, more than one has a priori reason to believe given the simplicity of the argument. We compare the values in table 5.1 for various dimensions. It is correct in $d=1$, where it predicts $\nu=1$, in accord with what we already concluded above. Secondly, it turns out to be exactly right in $d=2$, where $\nu=3/4$ is known from exact results for lattice models (note how our understanding of universality plays a role here).[11] Moreover, in $d=3$, the actual value $\nu_\mathrm{SAW}=0.588$ is remarkably close to the Flory value of $3/5$—the difference is too small to be relevant experimentally, so the Flory value $3/5$ is in practice used for most theoretical arguments and for interpreting experimental data. Finally the Flory

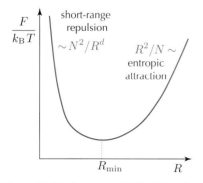

Figure 5.9. The free energy (5.15) of the polymer as a function of the radius R as determined by Flory.

Table 5.1. Comparison of the exponents ν_Flory and ν_SAW in various dimensions.

	ν_Flory	ν_SAW
$d=1$	1	1 (exact)
$d=2$	3/4	3/4 (exact)
$d=3$	3/5	0.588 (sim., ε-exp.)
$d=4$	1/2	1/2 (upper crit. dim.)

In renormalization group theory of critical phenomena, it is found that when the dimension d is taken as a continuous variable, the critical behavior, as determined by the critical exponents, is that of a free noninteracting model above a so-called upper critical dimension. For ordinary critical phenomena the upper critical dimension equals 4. This implies that above four dimensions, the large N scaling behavior of self-avoiding walks and 'polymers' is given by the ideal chain exponent $\nu = 1/2$.

expression gives the ideal chain result $\nu = 1/2$ in $d = 4$. This is actually also right, as in the language of critical phenomena, $d = 4$ is the upper critical dimension above which the ideal chain scaling with noninteracting exponent $\nu = 1/2$ is correct.

Flory's theory does better than one could have reasonably expected, as two effects more or less compensate for each other.[12] On the one hand, the conformational entropy used in the estimate is that of an ideal chain; this largely overestimates the entropy of a self-avoiding walk, which has many fewer configurations available to it than an ideal chain. On the other hand, the mean-field-type argument for the repulsive interaction neglects correlations; hence the argument also overestimates the repulsive potential for many local configurations in which chain segments get near each other without interacting significantly.

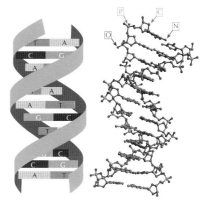

Figure 5.10. Illustration of the DNA molecule.

5.4 The wormlike chain model for biopolymers

The ideal chain model and the self-avoiding walk models which we discussed above are good models for most synthetic polymers, which are rather floppy, as they have a Kuhn length a few times the monomer size. Most biopolymers, on the other hand, are not nearly as flexible on the molecular scale. For instance, the two connected strands of DNA illustrated in figure 5.10 give the molecule quite a bit of rigidity. Indeed, figure 5.5 illustrated already that the typical scale on which DNA is bent is of order 50 nm. The scale on which the microtubules of figure 2.23 are curved is even much larger (see table 5.2 below). Such biopolymers are often referred to as semi-flexible, and it is better to think of them as thin bendable rods.

This motivates describing biopolymers with the so-called wormlike chain model, often abbreviated as WLC. This simple noninteracting continuum model is motivated by our discussion in chapter 2, which showed that the lowest energy deformations of thin rods are bending deformations.

As we shall see, the wormlike chain model is characterized by one important length scale, the persistence length. This is the length scale on which the energy of bending modes is of order $k_B T$. For long biopolymers, the persistence length plays the role of the Kuhn length (we'll see they differ by a factor of 2), in that it is the length scale over which orientational correlations get lost. Indeed, the statistical scaling properties of very long wormlike chains in equilibrium are essentially those of the ideal chain model.

There are several reasons why it is worth treating the wormlike chain model in quite a bit of detail. First of all, we have already

As mentioned in the caption of figure 5.1, the Kuhn length of polyethylene is $\ell_K \approx 1.4$ nm; for polystyrene $\ell_K \approx 1.8$ nm.

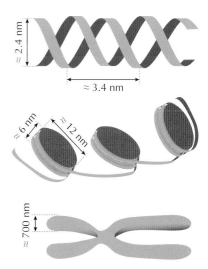

Figure 5.11. Illustration of the organization of DNA in the cell at various scales via the winding on histones (the disk-like structures in the middle panel) to the formation of chromatin (the 'beads of histones on a string'), and via the organization of chromatin structures to the chromosomes on the largest scale. See, e.g., Schiessel, 2013 for an introduction to and Parmar et al., 2019 for a recent review of the folding of DNA.

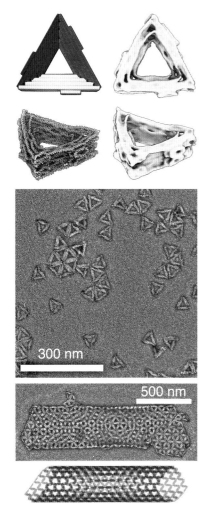

Figure 5.12. Top panel: design and realizations of a DNA origami building block for tubule assembly. The top left image shows the design in which each colored side has a complementary interaction. Going from here in the clockwise direction, we first show the single-particle reconstruction of the resulting triangular building blocks as observed in cryo-TEM from two different angles.[14] Finally, the image in the lower left corner shows a pseudo-atomic model reconstruction of the resulting structure.[15] Middle panel: image of such DNA origami building blocks. The bottom panel shows an image of a tubule assembled from these DNA origami triangles together with an illustration of the corresponding tubule.[16] Images courtesy of Daichi Hayakawa, Thomas Videbaek, and Ben Rogers. See Hayakawa et al., 2022 for more details.

seen in section 3.6.1 that it is nowadays possible to probe a variety of single-molecule properties experimentally by pulling on the molecule with beads held in an optical trap. When one pulls hard enough on biomolecules like DNA, the force-extension relation crosses over from the small entropic elastic force regime of ideal chains to the regime where the molecule is almost completely stretched. The wormlike chain model[13] is able to describe both of these regimes and their crossover, and it has therefore become the standard reference frame for interpreting these types of experiments. As we shall see (figure 5.16), the deviations from the wormlike chain behavior can, for instance, give signs of the underlying molecular structure of DNA. Secondly, in view of our earlier discussion of the importance of self-avoidance effects for synthetic polymers, we need to discuss why such effects play much less of a role in the statistical properties of biomolecules, or, in other words, why they behave like ideal chains to a good approximation.

The wormlike chain model is also a convenient starting point for modeling the elastic properties of biomolecules on the scale below the persistence length, for instance, to analyze how DNA winds around the spools, the so-called histones, to form chromatin—see figure 5.11. We do not have room in this book to discuss such applications or the vast opportunities offered by what is referred to as DNA technology, illustrated in figure 5.12.

5.4.1 The wormlike chain model and its persistence length

In section 2.7 we saw that the low-energy deformations of thin sheets and rods are bending deformations. The wormlike chain model for a free chain of length L amounts to using the bending energy expression (2.37) for a thin rod as the defining energy

$$E_{\text{WLC}} = \int_0^L \mathrm{d}s\, \tfrac{1}{2}\,\alpha\,\kappa^2. \tag{5.17}$$

Here s is a coordinate along the chain, modeled as a smooth curve in three dimensions with local curvature κ. Note furthermore that the expression for E_{WLC} does not include an energy penalty for self-intersections of the curve. So the wormlike chain model is essentially a continuum version of an ideal noninteracting chain. The parameter α, finally, clearly plays the role of a bending stiffness and has dimensions of energy times length. Hence the quantity

$$\text{persistence length WLC:} \quad \ell_{\text{p}} = \frac{\alpha}{k_{\text{B}}T} \tag{5.18}$$

has dimensions of length. It is called the persistence length: as we will show below, ℓ_{p} is the length over which the orientations of a smooth free curve decorrelate. It is clearly related to the Kuhn length ℓ_K; indeed, we shall find that $\ell_K = 2\ell_p$.

To analyze the orientational correlations, we will for simplicity consider two dimensions, so that the wormlike chain is a curve in the plane. Denoting as before in section 2.8.1 the unit vector tangent to the curve by $\hat{t}$ and the angle $\hat{t}$ makes with the x-axis by θ, we have[17]

$$\hat{t}_x(s) = \cos\theta, \quad \hat{t}_y(s) = \sin\theta, \quad \frac{\mathrm{d}\theta}{\mathrm{d}s} = \kappa. \tag{5.19}$$

To analyze the fluctuations, we will follow a route which makes contact with our earlier discussion of diffusion and with similar types of analyses for freely fluctuating interfaces. The analysis is inspired by the observation that if we think of the coordinate s along the curve as time and realize that $\kappa = \mathrm{d}\theta/\mathrm{d}s$, the problem is that of a random walker making steps in θ-space.

According to the energy expression (5.17), the probability distribution of a whole chain configuration is simply given by

$$\mathcal{P}[\theta] \sim \exp\left[-\frac{1}{k_B T}\int_0^L \mathrm{d}s\, \tfrac{1}{2}\alpha\left(\frac{\mathrm{d}\theta}{\mathrm{d}s}\right)^2\right]. \tag{5.20}$$

We have written $\mathcal{P}[\theta]$ to stress that $\mathcal{P}$ depends in this formulation on the whole shape function $\theta(s)$. The connection with a random walker suggests that we should concentrate rather on the probability $\mathcal{P}(\theta, s)$ that the chain has angle θ at position s, for which we have

$$\mathcal{P}(\theta, s + \mathrm{d}s) = \int \mathrm{d}\theta'\, W(\theta, \theta')\mathcal{P}(\theta', s), \tag{5.21}$$

where $W(\theta, \theta')$ is the transition probability for the angle of changing from θ' to θ within $\mathrm{d}s$. From (5.17) we have explicitly

$$W(\theta, \theta') = C\exp\left[-\frac{(\Delta\theta)^2}{2\ell_p^{-1}\mathrm{d}s}\right], \quad \Delta\theta = \theta - \theta'. \tag{5.22}$$

Here $C = 1/\sqrt{2\pi\,\ell_p^{-1}\,\mathrm{d}s}$ is the constant needed to ensure that $\int \mathrm{d}\theta\, W(\theta, \theta') = 1$, expressing that the curve definitely makes a transition to some new angle. Now, since for small $\mathrm{d}s$, $W(\theta, \theta')$ is strongly peaked around $\Delta\theta = 0$, we can expand the integrand on the right-hand side of (5.21) about θ,

$$\mathcal{P}(\theta, s + \mathrm{d}s) = \int \mathrm{d}\Delta\theta\, W(\Delta\theta)$$

$$\times \left(1 + (\Delta\theta)\frac{\partial}{\partial\theta} + \tfrac{1}{2}(\Delta\theta)^2\frac{\partial^2}{\partial\theta^2} + \cdots\right)\mathcal{P}(\theta, s). \tag{5.23}$$

Compare with our discussion of Markov processes in section 3.4 to see why we think of $W(\theta, \theta')$ as a transition probability of going from angle θ' at s to angle θ at $s + \mathrm{d}s$. By the way: our analysis is very much in the spirit of transfer matrix formulations of the partition function of statistical physics problems. In this formulation,[18] the partition function is written as the product of matrices, the so-called transfer matrices; for large system sizes the partition function is then simply determined by the largest eigenvalue of the transfer matrix.

Note that higher order terms in the expansion in $\Delta\theta$ in (5.23) give higher order terms in ds, which vanish in the limit d$s \to 0$.

The integrals over $\Delta\theta$ are, with the expression (5.22) for W, just standard Gaussian integrals which are straightforward to perform. The term linear in $\Delta\theta$ integrates to zero, and upon taking the limit d$s \to 0$ we arrive at the good old Fokker-Planck equation

$$\frac{\partial \mathcal{P}(\theta, s)}{\partial s} = D_\theta \frac{\partial^2 \mathcal{P}(\theta, s)}{\partial \theta^2}, \qquad (5.24)$$

with θ-diffusion coefficient

$$D_\theta = \frac{1}{2\ell_{\mathrm{p}}}. \qquad (5.25)$$

The above derivation makes the connection with a random walk quite explicit: the wormlike chain model in two dimensions implies a diffusive motion in the θ-space. Indeed, if we take $\theta(0) = 0$ without loss of generality, the solution, analogous to the Gaussian solution (3.34) of the diffusion equation (3.33), is

$$\mathcal{P}(\theta, s) = \frac{1}{\sqrt{4\pi D_\theta s}} e^{-\theta^2/(4 D_\theta s)}. \qquad (5.26)$$

This expression implies a mean square displacement $\langle \theta^2(s) \rangle = 2D_\theta s$.

We now have all the ingredients to calculate the correlation function of the tangent vectors. Taking for convenience again $\theta(s = 0) = 0$, we have

$$\langle \hat{t}(0) \cdot \hat{t}(s) \rangle = \langle \cos\theta(s) \rangle = \langle e^{i\theta(s)} + e^{-i\theta(s)} \rangle / 2. \qquad (5.27)$$

The Gaussian integrals resulting from (5.26) are easy to perform (see problem 5.3), and we find that this correlation function falls off as

in 2 dimensions: $\quad \langle \hat{t}(0) \cdot \hat{t}(s) \rangle = e^{-s/2\ell_{\mathrm{p}}}. \qquad (5.28)$

This confirms that in two dimensions, $2\ell_{\mathrm{p}}$ is the length over which angular correlations decay, in other words, the persistence length equals $2\ell_{\mathrm{p}}$. In three dimensions, the bending in the other direction gives a similar contribution to the decay of angular correlations, so the persistence length equals

in 3 dimensions: $\quad \langle \hat{t}(0) \cdot \hat{t}(s) \rangle = e^{-s/\ell_{\mathrm{p}}}, \qquad (5.29)$

Table 5.2. Typical values for the persistence length of various biopolymers. Keep in mind that as discussed in section 5.4.2, the persistence length of molecules like DNA is affected by charges, and hence it is dependent on the ionic concentration when in solution; see equation (5.32). The value quoted is a typical value for ionic concentrations of about 10^{-2} mol/L.

biopolymer	persistence length
DNA	$\ell_{\mathrm{p}} \approx 50$ nm
dsRNA	$\ell_{\mathrm{p}} \approx 64$ nm
collagen	$\ell_{\mathrm{p}} \approx 100$ nm
actin	$\ell_{\mathrm{p}} \approx 1$ μm
microtubules	$\ell_{\mathrm{p}} \approx 1$ mm

with $\ell_{\mathrm{p}} = \alpha/k_{\mathrm{B}}T$ according to (5.18).

In table 5.2 we list typical values for the persistence length extracted from various experiments. The broad array of values illustrates that the word *semiflexible* is used for quite a wide range of flexibilities!

Now that we have the correlation function (5.29) in three dimensions, it is straightforward to calculate the end-to-end correlation function as follows:

$$\langle R^2 \rangle = \int_0^L \mathrm{d}s_1 \int_0^L \mathrm{d}s_2 \, \langle \hat{t}(s_1) \cdot \hat{t}(s_2) \rangle,$$

$$= \int_0^L \mathrm{d}s_1 \left[\int_0^{s_1} \mathrm{d}s_2 \, e^{-(s_1-s_2)/\ell_\mathrm{p}} + \int_{s_1}^L \mathrm{d}s_2 \, e^{(s_1-s_2)/\ell_\mathrm{p}} \right],$$

$$= \ell_\mathrm{p} \int_0^L \mathrm{d}s_1 \left[(1 - e^{-s_1/\ell_\mathrm{p}}) + (1 - e^{(s_1-L)/\ell_\mathrm{p}}) \right],$$

$$= 2\ell_\mathrm{p} L \left[1 - \frac{\ell_\mathrm{p}}{L} \left(1 - e^{-L/\ell_\mathrm{p}} \right) \right]. \tag{5.30}$$

For $L \ll \ell_\mathrm{p}$ this reduces to $\langle R^2 \rangle = L^2$: for lengths much smaller than the persistence length, the polymer essentially behaves like a rigid rod. For polymers much longer than ℓ_p, we have $\langle R^2 \rangle = 2\ell_\mathrm{p} L = (2\,\ell_\mathrm{p})^2 (L/2\ell_\mathrm{p}) = \ell_\mathrm{K}^2 (L/\ell_\mathrm{K})$: they behave like ideal chains with Kuhn length $\ell_\mathrm{K} = 2\ell_\mathrm{p}$ and $N = L/\ell_\mathrm{K}$ Kuhn segments,

$$\text{WLC: } L, \ell_\mathrm{p} \iff \text{ideal chain: } \ell_\mathrm{K} = 2\ell_\mathrm{p}, N = L/\ell_\mathrm{K}. \tag{5.31}$$

The persistence length and the Kuhn length do not need to be precisely the same. In the continuum wormlike chain model, angles decorrelate smoothly over a distance of order ℓ_p, while the ideal chain model is a discrete model with completely uncorrelated successive links.

5.4.2 Charge effects on the persistence length

In our discussion of colloids, we already noted that colloids in a polar solvent can become charged when surface molecules split into two ions, one of which stays charged at the surface while the other one goes in solution. A similar effect happens for so-called polyelectrolytes, polymers which become charged in solution due to ion formation. As for colloids, such charges can have strong effects: the charges typically lead not only to a repulsive interaction between the polymers but also to an enhancement of the stiffness of the polymer itself. This is because bending of the polymer tends to bring like charges closer together. Both effects are affected by the Debye-Hückel screening of these charges by ions in solution, discussed in section 4.4.1.

The charge-induced stiffening of a polymer means that the persistence length in solution depends on the ion concentration in the solvent, through the concentration dependence of the screening length. If the polyelectrolyte is modeled as a thin rod with elementary charges e at every distance A along the rod, then the effective persistence length is shown in problem 4.4 to be

Equation (5.32) for the effective persistence length in the presence of charges was derived by Odijk, 1977 and Skolnick and Fixman, 1977. This length is sometimes referred to as the Odijk length.[10]

$$\ell_\mathrm{p} = \ell_\mathrm{p}^0 + \frac{\lambda_\mathrm{B}}{4A^2\kappa^2}, \tag{5.32}$$

where λ_B is the Bjerrum length introduced in (4.16) and κ the inverse screening length given in (4.15). In this expression, the second term is via κ inversely proportional to the ion concentration in

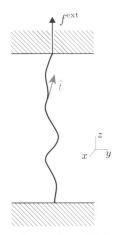

Figure 5.14. Sketch of the DNA at large pulling force in the calculation of the force-extension curve at large forces.

There is actually a subtlety underlying treating different q-modes as independent—see the discussion below box 3, in section 3.9.

the perpendicular component $\vec{t}_\perp = t_x\hat{x} + t_y\hat{y}$, we can write

$$t_z = \sqrt{1 - \vec{t}_\perp^2} \approx 1 - \tfrac{1}{2}\vec{t}_\perp^2. \tag{5.36}$$

Following *commandment 1*, we need to expand the free energy to quadratic order in the fluctuation amplitudes. To lowest order, we indeed obtain from (5.34) the following expression for the energy, which is quadratic in the transverse vector $\vec{t}_\perp$,

$$E = \int_0^L ds\, \tfrac{1}{2}\left(\alpha \left|\frac{d\vec{t}_\perp}{ds}\right|^2 + f\,\vec{t}_\perp^2 \right) - fL. \tag{5.37}$$

According to *commandment 2*, in translation-invariant situations like this, the energy is diagonalized by decomposing the fields in Fourier modes. In this case, as we have a finite s-interval of length L, we write (using periodic boundary conditions)

$$\vec{t}_\perp = \sum_q \vec{t}_\perp(q)e^{iqs}; \quad q = \pm\frac{2\pi}{L}, \pm\frac{4\pi}{L}, \cdots. \tag{5.38}$$

By virtue of Parseval's theorem, which simply states that[22]

$$\frac{1}{L}\int_0^L ds\, g^2(s) = \sum_q |g(q)|^2, \tag{5.39}$$

we can write the energy as

$$E = \tfrac{1}{2}L \sum_q \left(\alpha q^2 + f \right) \left|\vec{t}_\perp(q)\right|^2. \tag{5.40}$$

Indeed, in accord with our commandments, we see that modes are now decoupled; in other words, each mode with a given q is independent of modes with a different wavenumber. As a result, $\exp(-E/k_BT)$ becomes a product of independent Gaussians.

Following *commandment 3*, we now use equipartition for the average value of each mode,

$$\langle |t_x(q)|^2 \rangle = \langle |t_y(q)|^2 \rangle = \frac{k_BT}{L(\alpha q^2 + f)}. \tag{5.41}$$

We now proceed to apply *commandment 4*, which states we are ready to calculate any desired quantity. For the extension, we need to calculate, according to (5.36),[23]

$$\tfrac{1}{2}\int_0^L ds\, \langle |\vec{t}_\perp|^2 \rangle = \sum_q \frac{k_BT}{(\alpha q^2 + f)} = \frac{L}{2\pi}\int dq\, \frac{k_BT}{\alpha q^2 + f}$$

$$= \frac{L}{2} \frac{k_B T}{\sqrt{f\alpha}} = \frac{L}{2} \sqrt{\frac{k_B T}{f\ell_p}}. \tag{5.42}$$

In the last step we used $\alpha = k_B T / \ell_p$ from (5.18). From this result and (5.36) we finally get for the force-extension curve at large forces

$$\langle R_z \rangle = L \langle t_z \rangle = L \left(1 - \frac{1}{2} \sqrt{\frac{k_B T}{f\ell_p}} \right). \tag{5.43}$$

If we invert the relation, we can express f as a function of the extension as

$$\frac{f\ell_p}{k_B T} = \frac{1}{4} \left(1 - \frac{\langle R_z \rangle}{L} \right)^{-2}. \tag{5.44}$$

This expression is valid only in the large force regime. However, both in this expression and in the weak force expression (5.35) the combination $f\ell_p / k_B T$ is given in terms of the relative stretch $\langle R_z \rangle / L$, so it is relatively simple to interpolate between the two limits. A popular empirical interpolation formula which gives the right asymptotic small and large force limits, and which interpolates nicely in the crossover regime, is[24]

$$\frac{f\ell_p}{k_B T} = \frac{\langle R_z \rangle}{L} + \frac{1}{4(1 - \langle R_z \rangle / L)^2} - \frac{1}{4}. \tag{5.45}$$

Figure 5.15 shows the data from an illustrative experiment on the force-extension curve of DNA, together with the fit to the interpolation formula (5.45). The formula clearly fits the data quite well. The persistence length and the length of the DNA are immediately obtained from such a fit.

The versatility of such experiments is illustrated by the experiments of figure 5.16, in which DNA is suspended between two beads in optical traps. This configuration is used to study the response of (over)stretched DNA. For not too large forces, the force-extension curve of the DNA follows quite closely the line labeled WLC corresponding to that of the wormlike chain model, but strong deviations occur for very large forces. The data show that the DNA strands start to zip apart in the manner illustrated in panel a; the images in the bottom of this panel show the actual experiment with the single stranded DNA marked with the aid of fluorescence. Interestingly, the data of panel c show that this unzipping of the two DNA strands is not happening smoothly but happening in a sawtooth-type manner. This reflects the ACGT sequence of the DNA: there are parts which are more difficult and parts which are less difficult to unzip. Indeed, different double stranded molecules with the same ACGT sequence reproducibly exhibit the same sawtooth-type unzipping curve.[26]

This result illustrates the power of simply following the four commandments to calculate fluctuating quantities in equilibrium.

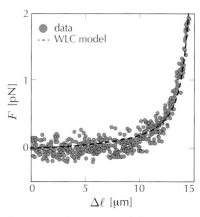

Figure 5.15. Comparison of the wormlike chain force-extension curve with DNA data, obtained by pulling on the DNA with optical tweezers in experiments by Rocha, 2015, along the lines discussed in section 3.6.1. The solid curve corresponds to (5.45) with persistence length $\ell_p = 50$ nm and length $L = 15.6$ μm. Image courtesy of Márcio Rocha.[25]

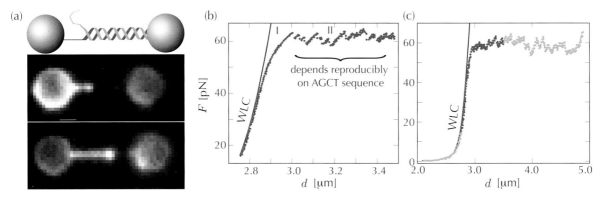

Figure 5.16. Experiments by Gross et al., 2011 on overstretching of DNA. Panel (a) illustrates how the DNA is attached to two beads in an optical trap, which allows us to stretch the DNA by pulling the beads apart. The sketch at the top illustrates how overstretching the DNA results in the DNA being peeled open. The two fluorescent images below it are from the actual experiments: a small amount of fluorescently labeled single-stranded binding protein is added to the solution, which only binds to those parts of the DNA strands which have been peeled open. So in these images only the left part of the DNA is peeled apart. Panel (b) exhibits the measured force-extension curve, the force as a function of extension of the DNA. In region I, strong deviations from the wormlike chain behavior start to occur. The sawtooth behavior observed in region II results from the stepwise increase in length as the two strands of the DNA are increasingly being torn apart; the sawtooth-like nature of the data shows that this tearing in the overstretching regime happens in bursts. Additional measurements show that these features of the curve are reproducible, and that they are correlated with the microscopic ACGT sequence of the DNA binding some parts of the strands more strongly than others. (c) The same data as in (b) with the WLC fit shown over a larger range; purple data points correspond to those shown in (b). Images and plots courtesy of Gijs Wuite.

5.5 Polymers in solution

Let us return to synthetic polymers and discuss their scaling behavior in solution in the three different regimes, the dilute, the semi-dilute, and the concentrated.

5.5.1 The dilute regime

So far, we have discussed the statistical mechanics of synthetic polymers in isolation, without paying attention to the interactions with a solvent. What effects do the polymer-solvent interactions and the interactions between polymers have? We first consider the situation in which the polymer concentration is so *dilute* that interactions between different polymers can be neglected. So the picture in this regime is that of isolated polymers interacting only with the solvent.

When the solvent is a so-called good solvent, a solvent into which the polymer easily dissolves, and when the Kuhn length is short, then effectively the self-avoidance effects are important and the radius grows with the Flory exponent $\nu_{\mathrm{Flory}} = 3/5$: $R \sim N^{\nu_{\mathrm{Flory}}}$. This implies (see section 5.3.1) that the statistical properties of the polymer are like those of a fractal with fractal dimension $d_{\mathrm{f}} = 1/\nu_{\mathrm{Flory}} = 5/3$.

Real data are never precise enough to distinguish between the self-avoiding walk exponents ν_{SAW} and ν_{Flory}, so in comparing with experimental data we will always use the Flory value in three dimensions.

a. Probing the scaling with neutron-scattering experiments

We saw in section 4.5.1 that scattering experiments can be used to study such behavior. Since the Kuhn length of typical synthetic polymers is only about 1–2 nm, the typical radius of gyration for degrees of polymerization of order 10^4–10^5 is only about 10 nm. Light has too large a wavelength to probe this length scale, but small-angle neutron scattering is very appropriate for scattering experiments at these scales.

In neutron-scattering experiments, one can actually extract how a few polymers behave in the dense environment provided by all the other ones. This is done by deuterating them. Deuteration means that the hydrogens of a polymer are replaced with deuterium. Because deuterium scatters neutrons much more strongly than hydrogen, if one uses a small concentration of deuterated polymers in a dense solution of the same polymers without deuteration, one can effectively filter out the single chain statistical properties of these 'stained' polymers.

b. Neutron-scattering data

The overall form of the scattering intensity $S(q)$ is found to be consistent with the schematic form sketched in figure 4.10, and from detailed fits to the crossover at small scattering wavenumbers, one obtains the radius of gyration R_g of the polymer. Curve A in figure 5.17 shows that the scaling of $S(q)$ for a dilute solution of polystyrene in a good solvent is indeed consistent with the expected scaling $S(q) \sim q^{-5/3}$ associated with self-avoidance effects. Modern data which probe $S(q)$ over about two and a half orders of magnitude of q, and include the flattening at small q, display the self-avoiding walk scaling $S(q) \sim q^{-d_f} \sim q^{-5/3}$ for a variety of polymers and solutions.[27]

On the other extreme, when the solvent is a bad solvent into which the monomers do not like to dissolve, then the polymer tends to form a compact coil (imagine a ball of wool) whose dimension scales as that of a compact three-dimensional object, $N \sim R^3$.

In between these two regimes something interesting happens: if you imagine being able to continuously tweak a solvent from a good solvent to a bad solvent, there is a point where the polymer solvent interactions, which favor the polymer's being compact, just about offset the polymer self-avoidance effects for the scaling properties. Such a solvent is called a theta-solvent. In such a theta-solvent the asymptotic polymer swelling is that of an ideal chain, $R \sim N^{1/2}$ with $d_f = 1/\nu = 2$! The 'theta-point' at which this happens thus also marks the transition from a coil (in good solvents) to a globule (in bad solvents). Curve B in figure 5.17 is taken at the theta-point of polystyrene in cyclohexane, and the scaling is indeed consistent with the expected behavior $1/S(q) \sim q^{d_f} = q^2$.

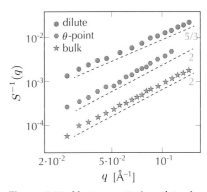

Figure 5.17. Neutron-scattering data for polystryrene and deuterated polystyrene of molecular weight of about 10^6 (degree of polymerization of about 10^4). Note that the inverse of $S(q)$ is plotted on a log-log scale. Circles: Dilute solution of deuterated polystyrene in carbon disulfide, a good solvent. The slope is consistent with a $q^{5/3}$ scaling. Hexagons: Nondeuterated polystyrene in deuterated cyclohexane is at the Θ point. The data are consistent with slope 2, hence with ideal chain scaling. Stars: Scaling for dense polymer solutions ('bulk') measured from deuterated polystyrene in a dense solution of undeuterated polystyrene. The scaling is also consistent with ideal chain behavior; why this is so is explained in section 5.5.3. Data from Cotton et al., 1974.

5.5.2 From semi-dilute to concentrated solutions

a. The crossover to the semi-dilute regime

We have already seen in our discussion of aggregates that objects with fractal dimension d_f less than the bulk dimension are very fluffy or relatively empty objects. This also holds for swollen polymers. As a result, when long polymers are dissolved in a good solvent, they quickly start to overlap and interact, even though we have a dilute solution as the concentration of polymers is still low. As illustrated in figure 5.18, the crossover concentration c^* where one enters this regime can be thought of as the monomer concentration at which polymer blobs start to touch. In other words, we can think of c^* as the concentration at which polymers of length N and volume about R^3 just about fill space, so that

$$c^* \simeq \frac{N}{R^3} \simeq \frac{N}{\ell_K^3 N^{3\nu_{Flory}}} \simeq \ell_K^{-3} N^{-4/5}. \qquad (5.46)$$

This confirms that for large degrees of polymerization N, the concentration c^* above which polymer interactions are important is quite small—in other words, the dilute regime covers a range of only very dilute concentrations. It is important to stress also that c^* is a crossover concentration rather than a concentration at which a transition occurs, and that it is its scaling with N which matters.

Polymer solvents with concentrations $c^* \lesssim c \ll 1$ are termed semi-dilute solutions: as sketched in figure 5.18, in this regime interactions between different polymers are important, but we are still far below the dense regime where it is essentially one polymer soup without much solvent.

b. The crossover length scale

So far, we have used the monomer concentration c, the number of monomers per unit volume. This depends on the units, and it is sometimes somewhat ambiguous. For many of the scaling arguments it is more convenient to use instead the dimensionless monomer volume fraction ϕ. For typical hydrocarbon polymers, ϕ^* is of order 1%.

In semi-dilute solutions, for $\phi \gtrsim \phi^*$, polymers interpenetrate, so another length scale comes in: the length scale on which the effect of the interaction with the other polymers becomes important. To make this more precise, let the average monomer volume fraction in the solvent be ϕ_0, and let's analyze the local density away from a monomer on a given polymer. The *local* volume fraction $\phi(r)$ of monomers is high near the starting monomer but will fall off as a function of the distance r away from it as $\phi(r) \simeq (A/r)^{d-d_f} = (A/r)^{4/3}$ in three dimensions. Here we take the Flory value for the exponent, and A is a length of order of the Kuhn length. As long as this value is larger than ϕ_0, the density is dominated, on

Here and in (5.48) below we for simplicity don't distinguish between the degree of polymerization n and the number of Kuhn segments N. For the scaling arguments here this is fine, but if one wants to compare, for instance, with expansion results for small concentrations, one of course has to be more careful.

For a more in-depth discussion of the semi-dilute regime we refer you to de Gennes, 1979 and Witten and Pincus, 2004.

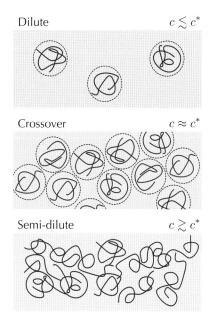

Dilute $\qquad c \lesssim c^*$

Crossover $\qquad c \approx c^*$

Semi-dilute $\qquad c \gtrsim c^*$

Figure 5.18. Illustration of the dilute and semi-dilute regime. At the crossover concentration c^*, the polymer blobs begin to touch.

average, by the enhanced monomer density resulting from the starting polymer, but at distances of order ξ_ϕ, defined by $\phi(\xi_\phi) \approx \phi_0$, there is a crossover to behavior influenced by average polymer density, hence by the other polymers. From this we get a crossover scale

$$\xi_\phi \simeq A\phi^{-3/4}. \tag{5.47}$$

Clearly, as the volume fraction ϕ increases, the crossover length decreases, and polymers overlap more and more. The segment with scale ξ_ϕ of a polymer is often called a blob. On the scale of each blob, the polymer properties are predominantly those of a single self-avoiding chain, but on the scale beyond ξ_ϕ, interactions with other chains do play as much of a role.

The interesting point is that for the statistical behavior of the chain beyond the scale ξ_ϕ, the interaction effects with other polymers are just as important as those of the polymer with itself. In other words, blobs experience no real self-avoidance effects—intuitively, you might say that beyond the scale ξ_ϕ, it feels for each polymer as if the other polymers interact just as much as the polymer does with itself, so that self-avoidance effects are canceled out. Hence, on the large scales, beyond ξ_ϕ, the statistical properties of a polymer are like those of a random chain of blobs, i.e., like an ideal chain without self-avoidance effects! This scenario is illustrated in figure 5.19.[29] Figure 5.20 shows neutron-scattering data for increasing values of the concentration above c^*, in which the crossover, obtained from detailed fits, is marked with a black arrow.

The picture of blobs is a very fruitful notion for understanding a number of scaling properties of polymer solvents in the semi-dilute regime. We will see an example of such a type of argument in our discussion of polymer brushes below.

c. Scaling analysis for the osmotic pressure

A quantity which is easy to measure for polymer solutions is the osmotic pressure Π: one measures the pressure difference across a membrane which is impermeable to the polymers. In the dilute regime $c \lesssim c^*$ the osmotic pressure from the polymers is in lowest order in the monomer concentration c given by the ideal gas expression

$$\Pi \approx \frac{c}{N}\, k_B T + \mathcal{O}(c^2), \tag{5.48}$$

as the concentration of polymers is c/N. We treat an example of this (so-called virial) expansion in powers of c in problem 5.8, but the expansion is immaterial here as we will focus below on a scaling argument which illustrates the power of thinking in terms of scaling concepts.

The insight that[30] the behavior of polymers for $N \gg 1$ is similar to the behavior near a critical point where universality is expressed by

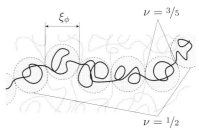

Figure 5.19. Sketch of the blobs and the crossover length ξ_ϕ in the semi-dilute regime.

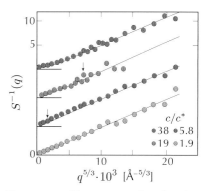

Figure 5.20. Neutron-scattering function from a very small concentration of deuterated polystyrene in undeuterated polystyrene in a solution of carbon disulfide (a good solvent) for increasing concentrations above the crossover value $c^* = 0.013\,\mathrm{g/cm^3}$. Data from different concentrations are displaced vertically for clarity. The inverse of $S(q)$ from the deuterated polymers is plotted versus $q^{5/3}$, so that data falling on a straight line are consistent with the self-avoiding walk exponent $\nu_{\mathrm{Flory}} = 3/5$. For concentrations just above c^* (lower curve), this behavior is observed for all values of q; for the largest concentrations the curves cross over at the value marked with a black arrow to ideal chain behavior with $S^{-1}(q) \sim q^2$ at the smallest q, i.e., at larger scales. The solid lines are fits to a function describing the full crossover (at the largest c/c^* it is outside the range of q values which are probed). The fact that gradual development of some curvature of the fits is hardly visible for the range of polymer length one typically has available illustrates the difficulty of distinguishing between the two regimes as the crossover from $q^{5/3}$ to q^2 behavior is smooth. Thus, the crossover wavenumber should be considered as an appropriate fit parameter, and not as a sharply defined value.[28]

scaling functions leads us to surmise the following scaling behavior for Π in good solvents:

$$\Pi = k_B T \frac{c}{N} f\left(\frac{c}{c^*}\right), \qquad (5.49)$$

with f some unknown scaling function. The expression (5.48) for the lowest order term of Π in powers of c then means that we need to have $f(x) \approx 1 + const.\, x$ for small arguments x.

Now, we saw that in the large N limit beyond a range ξ_ϕ, the single-chain properties do not matter anymore. This naturally leads to the idea that the osmotic pressure in the semi-dilute regimes must approach a behavior which depends on c, but which becomes independent of the degree of polymerization N as $N \to \infty$. In view of the prefactor N^{-1} of the scaling function f and the scaling of c/c^* as $cN^{4/5}$ according to (5.46), this implies that $f(c/c^*)$ *must* behave as a simple power for $c \gg c^*$,

$$f\left(\frac{c}{c^*}\right) \sim \left(\frac{c}{c^*}\right)^{5/4} \sim c^{5/4} N, \qquad (5.50)$$

as only then is the factor $1/N$ in the expression (5.49) for Π canceled. But this also predicts that the osmotic pressure Π should scale as $c\, c^{5/4} = c^{9/4} \sim \phi^{9/4}$. Experimental data are nicely consistent with this prediction; see figure 5.21. Moreover, the experiments confirm another important result of this analysis: upon properly rescaling the volume fraction and osmotic pressure, data in figure 5.21 coming from different polymers of various lengths fall on one curve. In the language of critical phenomena, we have scaling collapse.

5.5.3 Concentrated solutions

We saw above (equation (5.47)) that the crossover length ξ_ϕ decreases as the volume fraction increases. Concentrated solutions are those solutions for which ϕ is large enough that there is no proper separation of scales anymore between the monomer size and ξ_ϕ. In other words, there is no length scale anymore for which self-avoidance plays a role. Indeed, in view of our observation above that the polymer behaves like an ideal chain of blobs on the scale beyond ξ_ϕ, we conclude that in concentrated solutions the statistical behavior of a polymer is that of an ideal chain, with end-to-end length scaling as $R \sim N^{1/2}$.

Lower curve (stars) in figure 5.17 shows that this ideal chain behavior is indeed observed in neutron-scattering experiments with deuterated polymers.

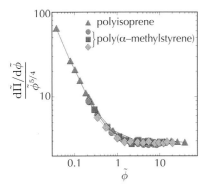

Figure 5.21. Scaling plot of the osmotic pressure as a function of the scaled volume fraction $\tilde{\phi} = \phi/\phi^*$ for various polymers and solvents, as determined by Witten and Pincus, 2004.[31] The plot contains data for one sample of polyisoprene in cyclohexane and three different lengths of a polystyrene-like polymer. The collapse of the various data onto a single curve is strong support for the scaling theory. The fact that the ratio $(\mathrm{d}\tilde{\Pi}/\mathrm{d}\tilde{\phi})/\tilde{\phi}^{5/4}$ becomes constant at large volume fractions confirms the scaling approach prediction that Π should scale as $\phi^{9/4}$ at large concentrations, and that the osmotic pressure is independent of the chain length N. Data courtesy of Tom Witten.[32]

5.6 Polymer brushes

In our discussion of colloids, we mentioned that polymer brushes are often used as bumpers on the colloids, which prevent the colloidal particles from aggregating. As an illustration of a scaling-type argument based on the existence of blobs, we now briefly discuss the description of a polymer brush in the high density regime. This means that we consider a case in which a surface is grafted with polymers with a surface density σ which is high enough that the polymers are in the semi-dilute regime. For a given grafting density σ, the typical distance ξ_0 between the chains on the surface is

$$\xi_0 = 1/\sqrt{\sigma}. \tag{5.51}$$

As illustrated in figure 5.22, we imagine that the total thickness or height of the polymer brush is h, and that the polymer density is rather uniform with height and then drops sharply to zero, i.e., polymers end abruptly.

The thickness of the brush is determined as the balance of two effects: the osmotic pressure which tends to push the brush out and an elastic term which resists the stretching of the polymers. For the osmotic repulsion, we can use the result derived above that the osmotic pressure Π scales as the volume fraction $\phi^{9/4}$ in the dilute regime. Since the volume occupied by the brush is proportional to the height, we have $\phi \sim h^{-1}$, and so

$$\Pi \sim h^{-9/4}. \tag{5.52}$$

To estimate the elastic stretching force, we note that according to the argument given above, we can view each polymer as an ideal chain of blobs of size ξ_ϕ (figure 5.19). Within a blob, polymer self-avoidance effects are dominant, so the number of monomers g in a blob goes as

$$\xi_\phi \simeq \ell_K g^{3/5} \implies g \simeq \left(\frac{\xi_\phi}{\ell_K}\right)^{5/3}. \tag{5.53}$$

As there are N/g blobs, and since the statistics of the blobs are like those of an ideal chain, the mean square displacement of a chain of blobs goes as

$$\langle R_{\text{blobs}}^2 \rangle \simeq \frac{N}{g}\xi_\phi^2 \simeq N\ell_K^{5/3}\xi_\phi^{1/3} \simeq N\,\ell_K^{5/3}A^{1/3}\,\phi^{-1/4}, \tag{5.54}$$

where we used (5.53) and (5.47). Upon comparing with the free energy expression (5.11), we see that the elastic energy term of a chain of blobs is like that of an h^2 harmonic potential with spring constant proportional to $1/\langle R_{\text{blobs}}^2 \rangle \sim \phi^{1/4} \sim h^{-1/4}$. Hence, the elastic stretching force goes as $h^{3/4}$. When we combine this result with

The theory of which only some elements are sketched here was developed by S. Alexander, 1977 and de Gennes, 1976 and de Gennes, 1981, and is therefore usually referred to as the Alexander–de Gennes theory. An elementary introduction is given by de Gennes, 1987, while Witten and Pincus, 2004 give a more extensive account and comparison with experimental results.

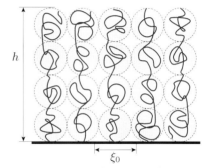

Figure 5.22. Sketch of a polymer of height h. The fact that the polymer blobs are more or less stacked on top of each other, as in the figure, is an outcome of the calculation. The abrupt ending at height h above the surface is not realistic. More advanced approaches account for a smoother density profile.

The first term between square brackets is a repulsive term which comes from the osmotic pressure; the second one is the attractive elastic stretching energy.

the repulsive osmotic pressure, we get for the brush the following expression for the force per unit area:

$$f \simeq \frac{k_\mathrm{B} T}{\sigma^{3/2}} \left[\left(\frac{h_e}{h} \right)^{9/4} - \left(\frac{h}{h_e} \right)^{3/4} \right], \tag{5.55}$$

where h_e is the equilibrium thickness of the brush. This formula shows that a brush is typically difficult to compress beyond its equilibrium thickness. In fact, if one follows the argument a bit more carefully, so as to keep track of the right orders of magnitude, one finds that the equilibrium thickness is such that the chain of blobs is essentially fully stretched, and that $\xi_\phi = \xi_0 = 1/\sqrt{\sigma}$. This means that each chain is essentially confined to a cylinder of cross section ξ_0, as sketched in figure 5.22, and that the thickness of the brush is simply

$$h_\mathrm{e} \simeq \xi_0 \frac{N}{g} = N \, \sigma^{1/3} \, \ell_\mathrm{K}^{5/3}. \tag{5.56}$$

See Milner et al., 1988 for a treatment with a smooth parabolic density profile and Milner, 1988 for comparison of this approach with experimental data.

Although the abrupt ending of the brush assumed in the model and as sketched in figure 5.22 is not physically realistic, more advanced approaches based on a smooth density profile are still consistent with this scaling behavior.

5.7 Flory-Huggins mean-field theory

5.7.1 Flory-Huggins approach

A nice introduction to the Flory-Huggins theory is given in chapter 4 of Rubinstein and Colby, 2003 and chapter 3 of Strobl, 1997.

The Flory-Huggins theory for polymer mixtures is in essence a mean-field-type approach which can be used to understand the main effects and trends of a number of relevant cases. Among these are the properties of polymers in solution or of polymer mixtures, the calculation of phase diagrams as a function of interaction parameter and degree of polymerization, the analysis of phase separation, the interpretation of small-angle neutron-scattering data, etc. Since you are therefore likely to encounter it, we briefly sketch its main ingredients here.

The starting point for the Flory-Huggins approach is the expression for the entropy of mixing of noninteracting molecules or monomers of type A with volume fraction ϕ_A and monomers of type B with volume fraction ϕ_B. The mixing entropy of ideal mixtures is nothing but the gain in entropy if we bring a volume V_A with species A and a volume V_B with species B together and allow the molecules to mix without taking interactions into account. Both types of monomers then have more position states available which they can occupy, so mixing increases their entropy. The expression for the mixing entropy, derived in problem 5.6, is then:

$$S_{\text{mix}} = -k_{\text{B}} \left[\frac{\phi_{\text{A}}}{N_{\text{A}}} \ln \phi_{\text{A}} + \frac{\phi_{\text{B}}}{N_{\text{B}}} \ln \phi_{\text{B}} \right]. \qquad (5.57)$$

Here each molecule of type A is taken to be of size N_{A}, and similarly for B. So the case with $N_{\text{A}} = N_{\text{B}} = 1$ describes the ideal mixing of a regular molecular solution, while $N_{\text{A}} \gg 1$, $N_{\text{B}} = 1$ describes a polymer in a simple molecular solution. Likewise, the case $N_{\text{A}} \gg 1$, $N_{\text{B}} \gg 1$ can be used for modeling mixtures of polymers of different length. Through these parameters, the degree of polymerization is thus built into the theory.

In the derivation of the above expression for the mixing entropy, interactions are ignored. In the Flory-Huggins theory, the interactions between A and B are accounted for in the spirit of a mean-field-type approach by a term $\phi_{\text{A}}\phi_{\text{B}}$ proportional to the volume fractions of both species $\phi = \phi_{\text{A}}$ and $\phi_{\text{B}} = 1 - \phi$ (note the similarity to the Flory treatment of excluded volume interactions in section 5.3.3). This then gives for the free energy of mixing

$$F_{\text{mix}} = k_{\text{B}}T \left[\frac{\phi}{N_{\text{A}}} \ln \phi + \frac{1-\phi}{N_{\text{B}}} \ln(1-\phi) + \chi\phi(1-\phi) \right], \qquad (5.58)$$

where χ is the effective interaction parameter. Written this way, you can think of the case $\chi = 0$ as the one in which the interaction between A and B is the same as that between like molecules. The larger χ is, the more unlike molecules repel each other.

This simple expression forms the basis of many discussions of phase diagrams of polymers in solutions (with $N_{\text{B}} = 1$) and of polymer blends (mixtures). Note that this free energy expression essentially contains only two parameters: if we pull out the length $N = N_{\text{A}}$ from the term between square brackets, the combination $k_{\text{B}}T / N$ sets the energy scale while the term between square brackets depends only on the length ratio $N_{\text{A}}/N_{\text{B}}$ and the effective interaction parameter χN.

As the parameter χN is a measure of the repulsion of two species, for large enough values of this parameter the theory predicts a demixing of the two phases. Figure 5.23 illustrates this for the case of a mixture of two polymers of equal length. For $\chi N > 2$, the free energy has a two-well structure, so that a mixture with composition ϕ between the values of the well is predicted to separate into an A-rich phase and a B-rich phase. For details of this calculation we refer you to problem 5.7.

5.7.2 Flory-Huggins as a mean-field theory

As was already mentioned, the Flory-Huggins theory is often the starting point for discussions of polymer properties like the phase

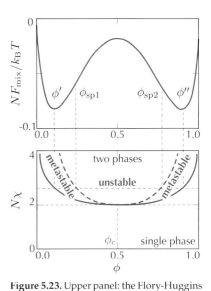

Figure 5.23. Upper panel: the Flory-Huggins free energy of a symmetric blend for $\chi N = 2.7$. Lower panel: the phase diagram for a symmetric blend with $N_{\text{A}} = N_{\text{B}} = N$. The solid line marks the line above which the mixture phase separates, while the dashed line marks the so-called spinodal line above which the homogeneous state is linearly unstable. In between these two lines the homogeneous mixture is metastable. Formation of the new phase in this region occurs through nucleation and growth. See also problem 5.7 and Rubinstein and Colby, 2003 for further discussion of this plot and its implications.

diagrams of mixtures—we will illustrate this with a nontrivial example below. Its success is related to the fact that it is a mean-field theory which is relatively easy to evaluate and which, in spite of its shortcomings, quickly gives overall trends, especially for understanding the various scaling properties with polymer length.

It is useful to throw light on the validity of a mean-field approach for polymers from another angle as well. Mean-field theories generally work well in condensed matter when the bare correlation length, the correlation length away from a critical point, is large. The most well-known example of this is the BCS theory of classical superconductors.

The BCS framework works well because the so-called coherence length of classical superconductors like lead and tin is so large, 100 nanometers or more, that the number of quasiparticles (electrons, Cooper pairs) within one coherence length is extremely large. Deviations from mean-field BCS behavior therefore only become apparent extraordinarily close to the transition temperature, so close that they are not of practical value.

As we have stressed in the introduction to this chapter, the large-N limit of polymers is analogous to approaching a critical point. This motivates the question of how important deviations from mean-field behavior are for large N. In other words, could mean-field theory possibly work so well because for $N \gg 1$ the large polymer size is the large length scale in the problem which makes mean-field theory so effective? We throw light on this issue from two different perspectives.

a. The Ginzburg criterion applied to polymers

First of all, as the case of superconductivity which we mentioned illustrates, an important clue is to look for when deviations from mean-field theory become dominant near a phase transition. When one approaches an equilibrium phase transition of a system which is reasonably well described by a mean-field approach, at some point near the transition fluctuation effects become important. The general criterion identifying when this happens is called the Ginzburg criterion. This criterion involves several quantities, most notably the 'bare' correlation length away from the critical point, but also a susceptibility type term and a parameter associated with the strength of the order parameter in the ordered phase.

See Als-Nielsen and Birgeneau, 1977 or Landau and Lifshitz, 1980 for an introduction to the Ginzburg criterion. The critical range in polymer systems is reviewed by Mortensen, 2001, who concludes, "the statement that polymer blends in principle are meanfield systems is therefore not correct."

The crossover to the fluctuation-dominated non-mean-field regime has been investigated in detail experimentally for polymer blends. The conclusion is that even though the region with deviations from mean-field behavior shrinks as $N \to \infty$, for realistic values of N of order 10^4 non-mean-field fluctuation effects are significant over quite a range. Polymer blends with N of order 100 can have an even bigger critical range than small-molar-mass systems!

b. Effect of fluctuations on the phase diagram of block copolymers

Theory and experiments on the phase diagram of diblock copolymers are reviewed by Bates and Fredrickson, 1990. Interestingly, the same fluctuation-driven first order transition occurs in other systems with rotational

A second interesting perspective is provided by a set of results on diblock copolymers. These are polymers which consist of two different polymer strands A and B attached together. Now, if these two chain ends don't mix well, and if the strands are roughly

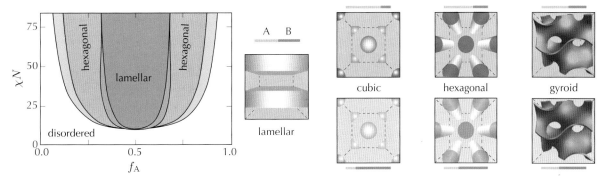

Figure 5.24. The phase diagram of diblock copolymers as a function of the fraction f_A of the total chain length composed by the A-part, as obtained from a Flory-Huggins mean-field approach based on an extension of (5.58) to account for inhomogeneous phases. Along the vertical axis the Flory-Huggins parameter χ times the chain length N is plotted. The cubic phase forms a body-centered cubic (bcc) lattice and exists in the thin green zones of the phase diagram between the disordered and hexagonal phases. The hexagonal phase consists of a hexagonal lattice of cylinders, while for polymer blocks of almost equal lengths a layered lamellar phase is formed. The gyroid phase is an ordered bicontinuous double diamond structure, and it exists only in the thin light red-brown slivers of the phase diagram between the hexagonal and lamellar regions. The disordered phase denotes the homogeneous mixture. The phase diagram is reflection-symmetric, with A and B interchanging roles as $f_A \to 1 - f_A$. All these phases have also been observed experimentally. Adapted from Matsen and Bates, 1996.

of comparable length, then such block copolymers form ordered periodic structures with A-rich domains and B-rich domains. This is illustrated in figure 5.24. Plotted along the horizontal axis is the fraction f_A of the chain which is of type A (shown in light ochre; the B-rich regions are shown in blue). The effective Flory-Huggins interaction parameter χN modeling repulsion between the two ends is plotted along the vertical axis. The figure is based on an extension of the Flory-Huggins theory to include the spatial composition modulations associated with the two different strands.

The figure illustrates the power of the mean-field Flory-Huggins approach to predict the formation of various types of nontrivial phases which have subsequently all been observed experimentally. Quite remarkably, the gyroid phase is a complex bicontinuous phase in which both the A and the B domains are continuous and span the whole sample.

For these diblock copolymer systems, the fluctuation effects have been investigated in detail. It has been found that as $N \to \infty$, the phase diagram approaches that of the Flory-Huggins mean-field approach, but that the convergence is rather slow, namely as $N^{-1/3}$. Indeed, for realistic values of N, fluctuation effects alter the phase diagram significantly. Figure 5.25 illustrates this for $N = 10^4$: there is a finite range of values of f_A where there is a direct transition from the homogenous phase to the lamellar phase. Moreover, while the mean-field transition is continuous (second order), the actual transition was predicted to be a fluctuation-induced first order transition. It took a combination of various types of experiments (including small-angle neutron-scattering and rheological

symmetry of the order parameter, e.g., weakly anisotropic antiferromagnets, liquid crystals with nematic-to-smectic-C transition, and convection patterns near the Rayleigh-Bénard transition. But the copolymer case is the only example where the fluctuation strength can be tuned so nicely by changing N.[33]

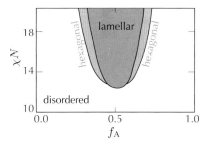

Figure 5.25. The phase diagram of diblock copolymers for $N = 10^4$ with fluctuation effects included. In the central region of the phase diagram the transition from the disordered phase to the lamellar phase does not go via an intervening hexagonal phase, as the mean-field theory displayed in figure 5.24 predicts. Moreover, the transition is (weakly) first order, instead of second order as in mean-field theory. Experiments reviewed by Bates et al., 1990 have indeed established the first order nature of the transition. Adapted from Bates et al., 1990, as based on the results of Fredrickson and Helfand, 1997.

measurements) to settle that the transition is weakly first order indeed.

This short overview serves to illustrate that while the Flory-Huggins theory is a powerful tool with which to calculate various polymer properties, including quite complicated phase diagrams like those of figure 5.24, one should at the same time be aware that deviations from mean-field behavior can be significant, even at relatively large degrees of polymerization.

5.8 Response of biopolymer networks

5.8.1 Biopolymer networks

Networks of biopolymers play an important role in living systems: the cytoskeleton serves as an intracellular scaffold, while fibrous matrices consisting of the protein collagen are principal structural components at the extracellular scale in your body. For instance, bones are strong lightweight materials made of a collagen matrix. But also many softer and deformable organs rely on collagen for their structure. Figure 5.26.a shows an example of a collagen network, while 5.26.c illustrates the very nonlinear stress-strain response of several of such biopolymer networks.

We have already seen in section 5.4 that most biopolymers are semiflexible polymers with a persistence length ℓ_p which ranges from about 50 nm for DNA or 100 nm for collagen to about 1 mm for microtubules. A glance at the collagen network of figure 5.26.a immediately shows that the collagen strands are pretty straight or only smoothly bent. This indicates that typical length ℓ_{crlk} between crosslinks is smaller than or at most of order ℓ_p: $\ell_{crlk} \lesssim \ell_p$. This in particular sets them apart from the behavior of gels, networks of synthetic polymers with typically $\ell_{crlk} \gg \ell_p$.

For an introduction to the physics of gels, see, for instance, part III of Rubinstein and Colby, 2003.

We will build on the earlier discussion of the wormlike chain model to treat here the concept of a thermal fluctuation-driven force-extension curve of biopolymer strands. We then show how this implies a strain stiffening like that seen in many networks beyond some typical crossover value of the strain. You should keep in mind that we cannot do justice to this broad and rich topic here. Apart from the fact that athermal models of crosslinked fibers which can both bend and stretch show similar trends,[34] what model is most appropriate for a particular experimental realization will depend on the scale and properties of the fibers and the network.

This section is stimulated by the review of the physics of biopolymer networks by Broedersz and MacKintosh, 2014. You will find many concepts in that paper which resonate with topics you encounter in this book.

An important advance that has helped unravel the various effects underlying the response properties of biopolymer networks has been the possibility of purifying biomolecules like actin and

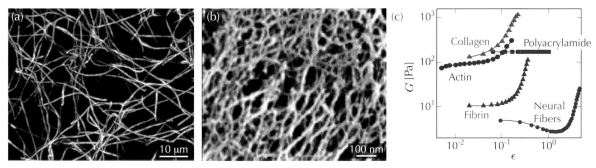

Figure 5.26. (a) Confocal microscopy image of a fluorescently labeled collagen network. The persistence length of collagen is of order 100 nm. Image courtesy of Stefan Münster and Chase Broedersz. (b) Electron micrograph of a fixed and rotary shadowed filamin-F-actin network with average filament length 15 μm. From Broedersz and MacKintosh, 2014. (c) The differential elastic shear modulus G as a function of strain[35] of various biopolymer networks as reported by Storm et al., 2005. Note that a normal elastic medium has constant shear modulus, so most networks show a very strong shear stiffening for strain values beyond about 0.1. For comparison, also shown is the shear modulus of the synthetic polyacrylamide gel. Graph courtesy of Chase Broedersz and Fred MacKintosh.[36]

microtubules and reconstituting them into bundles whose properties can be tuned and then used as building blocks of networks. Figure 5.26.b shows an example of such a reconstituted actin network. These have served as model systems with which to both test and stimulate recent approaches.

As figure 5.26.c shows, the elastic response is reported in terms of the differential shear modulus $G = d\sigma/d\varepsilon$, the derivative of the shear stress σ with respect to the strain ε. In the linear elastic regime this differential modulus is constant, so a nonlinear elastic response is visible as a deviation from this constant value. The crucial ingredient for understanding the elastic properties of such biomolecular networks is understanding the elastic response of a biomolecule stretched between the two nodes where it is crosslinked. We show here how we can derive this response by building on the earlier results for the wormlike chain model.

In chapter 2, we denoted the shear modulus by μ, but we follow the literature on biopolymer networks here by denoting the differential shear modulus by G. Some experimental results are obtained at a small frequency, so that what is actually reported is the real part $G'(\omega)$ of the response for small ω—see the discussion in section 2.5.

5.8.2 The slack or thermal-fluctuation-induced contraction

In section 5.4 we derived the force-extension curve for long polymers for large external pulling force f, and we calculated the relation between the contraction, relative to the maximal extension, due to the thermal fluctuations. We now consider the opposite limit $\ell_{\text{crlk}} \ll \ell_{\text{p}}$, where we think of a biopolymer in a network as a wormlike chain or thin rod being 'clamped' between two nodes which pull on it with a force f. We then have, from (5.42), for the thermal-fluctuation-induced contraction,

$$\langle \Delta\ell \rangle_f = k_{\text{B}}T \sum_q \frac{1}{\alpha q^2 + f}. \tag{5.59}$$

Here $\langle \Delta \ell \rangle_f$ is the contraction of the polymer due to its thermal fluctuations in the presence of the force f. As in the earlier discussion, this is the contraction measured relative to the total length of the polymer between the two clamped ends.

In the earlier analysis we converted the sum over modes q to an integral; this was appropriate for large forces, but for the general case of a fluctuating chain clamped on two sides the sum runs over wavenumbers[37] $q_n = n\pi/\ell_{\text{crlk}}$, with $n = 1, 2, \ldots$. Note that for small forces, this sum is dominated by its lowest order $n = 1$ term, in other words, by the fluctuations of the lowest order bending mode. This motivates defining a reference zero force contraction due to fluctuations:

We here use the mathematical result

$$\sum_{n=1,2,\cdots} \frac{1}{n^2} = \frac{\pi^2}{6}.$$

$$\langle \Delta \ell \rangle_0 = k_{\text{B}} T \sum_{q_n} \frac{1}{\alpha q_n^2} = \frac{k_{\text{B}} T \, \ell_{\text{crlk}}^2}{\alpha \, \pi^2} \sum_{n=1}^{\infty} \frac{1}{n^2} = \frac{\ell_{\text{crlk}}^2}{6\ell_{\text{p}}}. \tag{5.60}$$

This quantity is sometimes referred to as the *slack* of the filament. Note that the combination

$$f_{\text{c}} = \frac{\alpha \pi^2}{\ell_{\text{crlk}}^2} \tag{5.61}$$

is precisely the critical buckling force of a rod with bending modulus α—compare this result to equation (2.47). For forces roughly below this value, the sum in (5.59) is dominated by the lowest order term $q_1 = \pi/\ell_{\text{crlk}}$. In other words, for small forces the lowest order bending mode dominates the contraction, while as the force increases beyond the buckling value, more and more modes contribute, and nonlinearity in the contraction versus force response sets in.

To bring this to the fore, it is convenient to analyze the deviation in the contraction $\langle \Delta \ell_f \rangle$ from the zero-force reference value,

$$\delta\ell(f) = \langle \Delta \ell \rangle_0 - \langle \Delta \ell \rangle_f = \frac{\ell_{\text{crlk}}^2}{\pi^2 \ell_{\text{p}}} \sum_{n=1}^{\infty} \frac{\tilde{f}}{n^2 (n^2 + \tilde{f})}, \tag{5.62}$$

with

$$\tilde{f} = \frac{f \ell_{\text{crlk}}^2}{\alpha \pi^2} = \frac{f}{f_{\text{c}}}. \tag{5.63}$$

above summation is known,[38] and this gives for the relative extension

$$\epsilon = \frac{\delta\ell}{\langle \Delta \ell \rangle_0} = 1 - 3 \frac{\pi \sqrt{\tilde{f}} \coth\left(\pi \sqrt{\tilde{f}}\right) - 1}{\pi^2 \tilde{f}}. \tag{5.64}$$

Note that ϵ measures the extension relative to the zero-force value, so it varies from zero for small force to 1 in the limit of large stretching forces. In the linear regime, for small forces, one finds, from the expansion of the hyperbolic cotangent in (5.64),

$$\delta\ell = \langle\Delta\ell\rangle_0 \frac{\pi^2 \tilde{f}}{15} = \frac{\ell_{\text{crlk}}^2}{90\ell_{\text{p}}} \frac{\ell_{\text{crlk}}^2 f}{\alpha} = \frac{\ell_{\text{crlk}}^4}{90\ell_{\text{p}}\alpha} f. \qquad (5.65)$$

If we think of this as the linear expression of a spring, the effective spring constant is $90\ell_{\text{p}}\alpha/\ell_{\text{crlk}}^4$. As the persistence length *decreases* with increasing temperature, this spring weakens as the temperature increases. This is in contrast to what we showed before with equation (5.35), namely that a wormlike chain behaves for small forces as an entropic spring which *stiffens* with increasing temperature. The resolution of this apparent discrepancy is that we consider two different cases. The entropic elastic behavior we discussed before is that of a long wormlike chain with $L \gg \ell_{\text{p}}$, which coils up like an ideal chain. Here, however, we analyze the temperature-induced bending mode fluctuations of a polymer with $\ell_{\text{crlk}} \lesssim \ell_{\text{p}}$ between the two crosslinks—a bit like a clamped microrod at a scale where temperature fluctuations are important. In this case, with increasing temperature the mode fluctuations increase the zero-force contraction $\langle\Delta\ell\rangle_0$; as the extension is measured relative to this point, it is as if the spring has weakened.

Figure 5.27 shows the force-extension curve behavior as given by the full expression (5.64), both for compression and extension. Note the large asymmetry between compression and extension: under compression the polymer increasingly buckles but the force behaves smoothly,[39] while upon extension the force is found from (5.64) to diverge as

$$\frac{f\ell_{\text{p}}}{k_{\text{B}}T} = \frac{\ell_{\text{crlk}}^2}{4\left(\langle\Delta\ell\rangle_f\right)^2} \qquad (5.66)$$

as ϵ approaches 1. This is in agreement with the expression (5.44) derived before for large forces.

5.8.3 The stress-strain response of a network

It is easy to see that these simple results immediately imply the type of crossover of the differential shear modulus G shown in figure 5.26.c. To make this concrete, it is easiest to imagine the network to have a mesh size ℓ_{crlk} and to assume that under a strain ε, the network deforms affinely. This means that the deformation is the same, and hence equal to the applied global strain ε, throughout the system.

As sketched in figure 5.28, under an affine strain $\varepsilon = \partial u_x / \partial y$ the polymers in the network change length by an amount $\varepsilon\ell_{\text{crlk}}$ times a geometric factor. For strands lying in the xy plane, as in the figure, this factor is $\sin\theta\cos\theta\sin(\pi/2)$; this factor is maximal when the strands are oriented along the diagonal. Now, even if a polymer

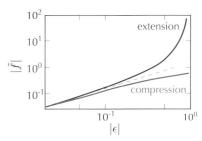

Figure 5.27. The dimensionless force $\tilde{f}$ as a function of the relative extension ϵ according to equation (5.64), for both compression ($\tilde{f} < 0$) and extension, corresponding to $\tilde{f} > 0$. The nonlinearities in the force behavior become important for $|\tilde{f}|$ at about 1, when the strength of the force is comparable to the buckling value f_c. Adapted from Broedersz and MacKintosh, 2014.

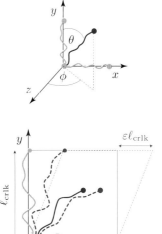

Figure 5.28. Sketch of polymers in a network under an affine strain $\varepsilon = \partial u_x / \partial y$ in the x direction. Only a two-dimensional cross section in the xy plane is shown. The dots mark the points where the polymers of length ℓ_{crlk} are crosslinked. As indicated, a polymer whose orientation makes an angle θ relative to the vertical direction is stretched by $\varepsilon\ell_{\text{crlk}}\sin\theta\cos\theta\sin\phi$ with θ and ϕ the spherical angles shown on the right. The full analysis requires an average over the orientation of the strands, by averaging over ϕ and θ.

was originally weakly stretched, $\epsilon \ll 1$, it is clear that polymers oriented roughly along the diagonal direction get strained into the nonlinear stress-strain regime $\epsilon \gtrsim 0.5$ if

$$\varepsilon \ell_{crlk} \gtrsim \langle \Delta \ell \rangle_0 \quad \Rightarrow \quad \varepsilon \gtrsim \varepsilon_c \equiv \frac{\langle \Delta \ell \rangle_0}{\ell_{crlk}} = \frac{\ell_{crlk}}{6\ell_p}. \tag{5.67}$$

For strains below this value, most polymers will respond to the imposed strain according to the linear force-strain result (5.65). If we denote the length of polymer strands per unit volume (or equivalently the number of strands per unit area crossing a plane) by ρ, then we expect, for the shear stress σ_{xy}, the linear stress-strain relation

$$\sigma = \rho \frac{90\ell_p \alpha}{\ell_{crlk}^3} C_{av} \, \varepsilon. \tag{5.68}$$

Here C_{av} denotes the average over all polymer orientations of the contribution each stretched polymer makes to the shear stress σ_{xy}. Just as each polymer stretching involves an angular term, as sketched in figure 5.28, the contribution of each polymer to the shear force σ involves the same angular factor. It is then easy to show[40] that $C_{av} = 1/15$. Hence, in the linear regime

$$\sigma = \rho \frac{6\ell_p \alpha}{\ell_{crlk}^3} \varepsilon \qquad (\varepsilon \lesssim \varepsilon_c). \tag{5.69}$$

This corresponds to an elastic shear modulus $G_0 = 6\rho \ell_p \alpha / \ell_{crlk}^3$. Moreover, for the shear stress at the strain ε_c where there is a crossover to the nonlinear regime, we get with (5.67) the surprisingly simple result

$$\sigma_c = \frac{\rho \alpha}{\ell_{crlk}^2}. \tag{5.70}$$

As the shear strain ε becomes larger than the crossover value ε_c, the stress-strain response will be dominated by the polymers which get strongly stretched, and for which f diverges as $f \sim 1/(\Delta \ell)_f)^2$, as derived in (5.66) and shown in figure 5.27. Experimental measurements are typically reported in terms of the differential shear modulus $G = d\sigma / d\varepsilon$, so for these the force law for large stresses implies

$$G = \frac{d\sigma}{d\varepsilon} \sim \frac{df}{d\ell} \sim \left(\frac{1}{\Delta \ell} \right)^3 \sim f^{3/2} \sim \sigma^{3/2}. \tag{5.71}$$

As figure 5.29 illustrates, this behavior is nicely observed at large stresses in actin filament networks. Further analysis confirms that upon rescaling the shear modulus by the small-strain plateau value G_0, and the stress by the crossover value σ_c, all data nicely collapse onto a single curve.[41] This is good support for the credence of the above analysis.

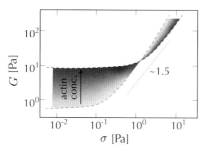

Figure 5.29. Rendering of the differential elastic modulus G as a function of applied stress in experiments by Gardel et al., 2004 on crosslinked and bundled actin filament networks at various concentrations. The dashed lines indicate theoretical predictions for each concentration determined from the single-filament response based on the wormlike chain model; actual data follows this behavior quite accurately. The dotted line indicates the $G \sim \sigma^{3/2}$ scaling predicted by expression (5.71). Upon rescaling the data by dividing G by its small-stress value and σ by the crossover value, the experimental values (not shown) all follow essentially the same universal behavior.[42]

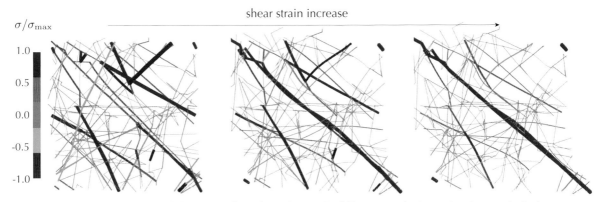

Figure 5.30. Three snapshots of the deformation of an athermal network of filaments under increasing shear strain, in the computer simulation of Žagar et al., 2011 of filaments which can stretch and bend. From left to right, $\epsilon = 0.05, 0.15$, and 0.20. The left picture corresponds to a deformation at the end of the linear regime, the middle one to that in the crossover regime, and the one on the right to that in the strain stiffening regime where $G \sim \sigma^{3/2}$. As the color bar indicates, filaments under tension are shown in blue, stretched filaments in red, with the thickness of the line proportional to the axial force, normalized to the maximum one in each simulation. Increasingly the filaments orient and the stress is concentrated in fewer filaments as the strain increases. Images courtesy of Patrick Onck and Erik van der Giessen.[43]

5.8.4 Beyond the simple approximation

In practice, the situation is much more complex—and interesting—than the above discussion might suggest. The assumption that the network deforms more or less affinely is not always justified. Indeed, in simulations of athermal networks one observes a smooth crossover from a predominantly affine deformation regime to a nonaffine deformation regime as the ratio of the bending rigidity to the stretching rigidity increases.[44] The bending-dominated regime which we analyzed above appears to extend over a smaller range than one might at first have expected.

In other regimes, different effects have to be included. Figure 5.30 illustrates this for an athermal model which includes stretching of the fibers: for increasing strain, the formation and subsequently reorientation of strong 'stress path' regions is found. Nevertheless, the overall stress-strain response of such networks is for some parameters very similar to that in figure 5.29, while for others it gives rise to a smooth crossover to $G \sim \sigma^{1/2}$ behavior.

Such non-affine deformations are very difficult to capture in analytical approaches. Moreover, the parameter space needed to describe biopolymer networks is large; it also includes the crosslinking density and the connectivity of the network. Even ideas from percolation and marginal networks (related to those of marginal solids, discussed in section 2.10) have been brought to bear on our understanding of their effect of connectivity and the crossovers between the various types of regimes. As a result, the phase diagram covering all the various possibilities and competing effects is quite complex.[45]

5.9 Reptation and the viscosity of polymer melts

A very accessible more in-depth discussion of reptation is given by Rubinstein and Colby, 2003. The treatments by Larson, 1999 and Doi, 2013 and the classic one of de Gennes, 1976 are also worth consulting.

We now return to synthetic polymers. We start with a brief discussion of the reptation model and the scaling of the viscosity with degree of polymerization, before we highlight some salient features of polymer rheology in the next section.

5.9.1 The polymer viscosity plays only a limited role in several relevant flow effects

The effect of polymers on the dynamical response and rheological properties of fluids is a very diverse topic, with only a limited role for viscosity. For instance, if one has a dilute solution of polymers, at flow rates small enough that the polymer conformation is not distorted significantly, the effective viscosity is increased in proportion to the polymer concentration—just as the viscosity of a dilute colloidal dispersion increases linearly with the volume fraction. However, we have already seen that the longer a polymer is, the easier it is to stretch it, as the entropic stiffness is inversely proportional to N. As a result, as we illustrated already in figure 1.8, long polymers get stretched and oriented at relatively small shear rates. As we will discuss in the next section, when this happens the solution behaves as a complex viscoelastic fluid: the stretching of the polymers gives rise to elastic, anisotropic, and relaxational effects, which in turn drive new types of rheological instabilities. From this perspective, the most remarkable features of polymeric fluids are not so much due to their enhanced viscosity in the zero shear limit as they are due to these non-Newtonian effects which happen at small shear rates for N.

But there are other surprises as well. A noteworthy counterintuitive effect is that adding dilute amounts of polymers to a turbulent flow can reduce the turbulent drag in pipes[46]—this is taken advantage of to reduce the drag in oil pipelines and fire hoses. Thinking of this in terms of a reduced viscosity is the wrong picture: these are high Reynolds number turbulent flows. The actual effect of the polymers is to give rise to the existence of long-lived recurrent transients, associated with the existence of edge states, that reduce the overall turbulent drag. Here too, the ability of polymers to store energy somewhere and release it elsewhere plays an important role: in regions of large shear and large drag they get stretched and in doing so reduce the turbulence, while they relax back and release their elastic energies in regions of low drag.

5.9.2 Reptation

We will content ourselves here with discussing the reptation process which determines the longest relaxation time, and so the viscosity. We already sketched in figure 5.19 how at sufficiently high concentrations a polymer can be thought of as a chain of blobs. For simplicity, we will immediately focus on a dense polymer melt, so that excluded volume effects are unimportant, as we argued in section 5.5.2. In this regime it is appropriate to think of the polymer as being constrained by all the other polymers to a 'tube.' Within the tube the polymer can move relatively unhindered, but in the direction perpendicular to the tube the polymer is constrained to a thickness of the order of the tube diameter d_{tube}.

Describing the dynamics and relaxation properties of a melt of long polymers is a baffling challenge. However, as sketched in figure 5.31, the easiest way in which the overall polymer configuration in a melt can relax is through a sliding diffusive motion through the tube: one side, the 'tail,' retracts, then the accumulated length diffuses through the tube, and finally this allows the 'head' on the other side to advance. Because of the similarity with how a reptile like a snake advances, the process is called reptation. The experimental observations of figure 5.32 confirm that this is actually a realistic picture of the dynamics: a long DNA molecule in a solution is seen to retract through the original tube, after being stretched with a bead pulled with optical tweezers.

In a dense melt, the excluded volume interactions are screened, as already mentioned, so also within a tube the polymer conforms to the ideal chain statistics discussed in section 5.3.1. The number N_{s} of Kuhn segments within a volume of size d_{tube}^3 is then

$$d_{\text{tube}} \simeq \ell_{\text{K}} \sqrt{N_{\text{s}}}. \tag{5.72}$$

If we think of the tube as consisting of N / N_{s} sections of size d_{tube} each with about N_{s} Kuhn segments, then the end-to-end distance R of the tube can be viewed either as a random walk of these sections or a random walk of the chain as a whole. With the above relation, we have

$$R \simeq d_{\text{tube}} \sqrt{\frac{N}{N_{\text{s}}}} \simeq \ell_{\text{K}} \sqrt{N}. \tag{5.73}$$

The average contour length $\langle L \rangle$ of the tube is the number of segments N / N_{s} times the size d_{tube}, so

$$\langle L \rangle \simeq d_{\text{tube}} \frac{N}{N_{\text{s}}} \simeq \frac{\ell_{\text{K}} N}{\sqrt{N_{\text{s}}}}, \tag{5.74}$$

i.e., $\langle L \rangle$ is a factor $1 / \sqrt{N_{\text{s}}}$ less than the total length of the polymer as each segment consists of a random walk of N_{s} Kuhn segments.

The arguments below can be extended to semi-dilute solutions by properly taking into account importance of excluded volume effects within blobs (so that the polymer statistics are given by the Flory exponent).

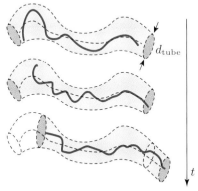

Figure 5.31. Stages of polymer reptation through which a polymer tube is renewed. From top to bottom, the 'tail' of the polymer retracts, propagates along the tube, and creates a new segment at the other side. This process is often referred to as tube renewal.

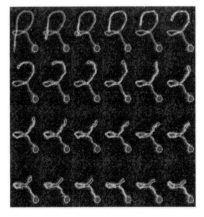

Figure 5.32. Illustration of tube-like reptation by Chu, 1998, along the lines of the experiments by Perkins et al., 1994a; Perkins et al., 1994b. The fluorescent DNA is originally pulled through a polymer solution with optical tweezers, creating a stretched polymer in the shape of an R at the upper left. The 24 images, from upper left to lower right—captured just seconds apart[47]—show the DNA retracting through the original tube while one side is held fixed with the bead.[48]

Of the many relaxation times associated with a polymer, the longest one is the time it takes the polymer to diffusively 'reptate' out the tube on one of the sides, as illustrated in figure 5.31. Let us estimate this 'reptation time' τ_{rep} for this tube renewal process. We note that the effective diffusion coefficient D_{c} for this diffusive motion of the constrained polymer along the curved but one-dimensional tube will scale with N as

$$D_{\mathrm{c}} = \frac{k_{\mathrm{B}}T}{N\zeta}. \tag{5.75}$$

Here ζ is a friction constant of a monomer. This expression is quite natural and similar to the Stokes-Einstein relation (3.19) for a single bead, if one thinks of the polymer as a string of beads connected by elastic springs as in the so-called Rouse model.[49] The reptation time τ_{rep} is then clearly the time it takes the polymer to diffuse a distance of order $\langle L \rangle$ in the tube, hence

$$\tau_{\mathrm{rep}} \simeq \frac{\langle L \rangle^2}{D_{\mathrm{c}}} \simeq \frac{\ell_{\mathrm{K}}^2 N^2}{N_{\mathrm{s}}} \frac{N\zeta}{k_{\mathrm{B}}T} = \frac{\zeta \ell_{\mathrm{K}}^2}{k_{\mathrm{B}}T} N_{\mathrm{s}}^2 \left(\frac{N}{N_{\mathrm{s}}} \right)^3. \tag{5.76}$$

In a similar way the reptation argument predicts the viscosity to grow as the cube of the degree of polymerization,

$$\eta \sim \tau_{\mathrm{rep}} \sim N^3. \tag{5.77}$$

As figure 5.33 shows, experimental observations are typically consistent with a larger exponent of about 3.4, rather than the reptation value 3. Detailed investigations have shown that this discrepancy is due to tube length fluctuations which are not accounted for in the above argument. For instance, simulations of a chain between fixed objects are consistent with a scaling of the relaxation time τ_{rep} with an exponent of around 3.4. For more details, see discussions in the recent books mentioned at the beginning of this section.

In any case, the rise of the viscosity with such a large power implies a steep rise of the viscosity with the degree of polymerization—indeed, the viscosity data in figure 5.33 span more than ten orders of magnitude—solutions and melts of very long polymers are very viscous!

5.10 Non-Newtonian rheology of polymer solutions and melts

So far we have mainly focused on properties of (bio)polymers and polymer solutions on the molecular scale. What if we look on much longer hydrodynamic scales and consider polymer rheology? The analysis in the previous section showed that solutions and melts of

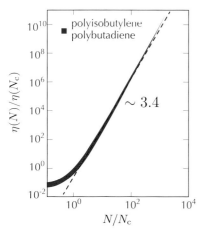

Figure 5.33. The viscosity of various polymer melts as a function of degree of polymerization N, combining data for three different polymers (polyisobutylene, polybutadiene, and hydrogenated polybutadiene), and with the thickness of the curve roughly comparable to the density of data points. After the rescaling of N by a crossover value N_{c} and $\eta(N)$ by $\eta(N_{\mathrm{c}})$ all data fall on the red curve, which is consistent with a scaling $\eta(N)/\eta(N_{\mathrm{c}}) \sim (N/N_{\mathrm{c}})^{3.4}$ for large N. A similar exponent is typically found for other polymers as well. Adapted from Rubinstein and Colby, 2003.[50]

For an introduction to polymer rheology, see the book by Larson, 1999, the introductory chapter of Morozov and Spagnolie, 2015, or the classic treatise by Bird et al., 1987a. The book by Joseph, 1990 focuses on the fluid dynamics aspects and treats several different flow geometries in detail. A historical overview is given by Tanner and Walters, 1998.

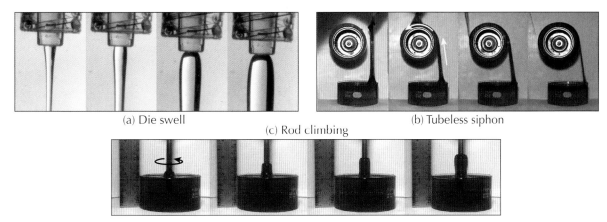

(a) Die swell

(c) Rod climbing

(b) Tubeless siphon

Figure 5.34. Stills from various videos of the MIT Non-Newtonian Fluid Dynamics Group[51] demonstrating surprising effects of the non-Newtonian rheology of polymer solutions. (a) *Die swell* refers to the remarkable swelling of a polymer fluid flowing out of a tube (in industrial processing, the device to press a metal or plastic into a mold is called a die (Rauwendaal, 2001)). The swelling is due to the fact that the polymers are stretched along the flow direction in the tube because of the shear, and they relax back to a spherical coil outside the die. As a result the fluid expands sideways. (b) Illustration of the fact that a rotating wheel can continue to draw a lint of polymer out of the container, without a tube. In a similar way, polymers can continue to flow over the rim of a siphon: a tubeless siphon. (c) Rod climbing refers to the fact that when a rod is rotated within a polymer solution, the polymers 'climb' up the rod. This results from the fact that the polymers are stretched and curved because of the circular streamlines. This gives rise to elastic forces, often called hoop stresses, pointing inward toward the tube. The effect is illustrated in figure 5.35.

long polymers are characterized by long relaxation times and very large viscosities. It should therefore come as no surprise that their rheology is very non-Newtonian. Figure 5.34 illustrates some of the remarkable counterintuitive features of polymer flows visible on the scale of daily life.

5.10.1 Importance of polymer stretching effects

What makes polymer rheology special among complex fluids is our finding that the longer polymers are, the easier they can be stretched. Remember that the free energy expression (5.11) showed that the effective spring constant of polymers, resulting from entropic effects, is inversely proportional to N. As we have already illustrated in figure 1.8, this implies that shear gradients readily cause long polymers in solution to stretch and orient along the streamlines. This is the root cause of most salient features of the non-Newtonian rheology of polymer solutions and melts.

For instance, the so-called die swell of figure 5.34.a of a polymer solution flowing out of a tube is due to the fact that, inside the tube, the polymers are stretched and oriented along the flow direction, because of the shear gradient. When they exit the tube, the shear is released and the polymers relax back to a spherical coil, pushing the fluid sideways. The drawing of a polymer lint out of a container by a rotating wheel shown in panel b is due to the strong entanglement of polymers at sufficiently high densities. As a result, polymers pull

The stretching of polymers in flow gradients is nicely demonstrated with fluorescent DNA solutions in cross-channel (also called cross-slot) flow. This is a setup in the form of a cross with inward flow through two channels on opposite sides and outward flow through the two channels oriented perpendicular to the inward flow channels. One can follow fluorescent DNA molecules as they flow through the cross and see them contract upon approaching the stagnation point where gradients are small and then see them stretch again when they flow away from it. There are many videos of this demonstration on the internet.

each other along and up when the flow rate is high enough that they do not have the time to disentangle. Finally, the rod climbing effect of panel c is due to the fact that when the rod in the center is rotated, polymers in the fluid get stretched and bent because of the shear gradients and the circular streamlines. As illustrated in figure 5.35, because of this bending there is an effective elastic force pointing inwards—it is as if in the fluid there are circular rubber bands pulling inward. These stresses, which are often referred to as hoop stresses, push the fluid toward and hence up the rod.

Based on these simple observations, it should be clear why polymer solutions are called viscoelastic. They are characterized by a large viscosity and, for sufficiently strong shear rates, by strong effects due to elasticity, anisotropy, and the slow response of the polymer conformation to changes in shear. We can think of a sufficiently sheared polymer solution as a fluid full of stretched and oriented little rubber bands, whose stretching adjusts slowly when the shear rate changes.

After we introduce the dimensionless Weissenberg number and discuss a common rheological model to capture these effects, we will briefly describe the type of instabilities one observes in many cases due to the interplay of elasticity and curved streamlines.

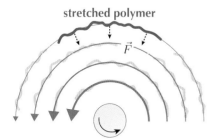

stretched polymer

$\vec{F}$

Figure 5.35. Illustration of the hoop stresses due to curved streamlines, sketched here as a top view of flow induced by a rotating rod. Due to the shear in the flow, the polymers are stretched and largely lined up with the circular streamlines. The polymer in red indicates how this gives an effective force directed inward. It is as if there are stretched rubber bands along the streamlines.

5.10.2 The dimensionless Weissenberg number

In a simple linear shear flow like the one sketched in figure 1.8, the dissipative stress (1.22) of a Newtonian fluid has only nonzero off-diagonal terms. However, the above picture of sheared polymers as oriented and stretched rubber bands implies that the normal stress components in a shear flow are nonzero and different from each other—in the case of a simple shear $v_x(y)$ as in figure 1.8, $\sigma_{xx} \gg \sigma_{yy}$. The dimensionless Weissenberg Wi number characterizes the importance of the effects as the ratio of the stress difference[52] over the shear stress

$$\text{Wi} = \frac{\text{normal stress difference}}{\text{shear stress}} = \frac{|\sigma_{xx} - \sigma_{yy}|}{\sigma_{xy}}, \qquad (5.78)$$

for simple shear $v_x(y)$. Since the normal stress difference increases quadratically with the shear rate, the Weissenberg number grows linearly in the shear rate for not too large shear.[53]

5.10.3 The Oldroyd-B and upper convected Maxwell model for polymer rheology

Most polymer flows are normally at low enough speeds that the flow can be considered incompressible. Therefore the description of the rheology of polymer solutions is based on the incompressible Navier-Stokes equations, with the important extension that the dissipative stress term $\underline{\sigma}^{\mathrm{d}}$ is written as the sum of a solvent and a polymer term,

$$\underline{\sigma}^{\mathrm{d}} = \underline{\sigma}_{\mathrm{s}}^{\mathrm{d}} + \underline{\sigma}_{\mathrm{p}}^{\mathrm{d}}. \tag{5.79}$$

Beware that in the polymer rheology literature the stress tensor is often written as $\underline{\tau}$ rather than $\underline{\sigma}$.

The solvent is simply treated as an incompressible Newtonian fluid modeled by equation (1.22), so the crux of the problem is to determine a proper *constitutive equation* for the polymer stress tensor $\underline{\sigma}_{\mathrm{p}}^{\mathrm{d}}$, a relation that expresses the polymer stress tensor in terms of the other relevant variable, like the shear gradient. Given the complexity of the problem, it should come as no surprise that one has to resort to effective phenomenological models which only capture the essentials. The starting point of most approaches in the field is models according to which the polymer stress tensor obeys the following constitutive equation:

$$\underline{\sigma}_{\mathrm{p}}^{\mathrm{d}} + \lambda \left[\frac{\partial \underline{\sigma}_{\mathrm{p}}^{\mathrm{d}}}{\partial t} + \vec{v} \cdot \vec{\nabla} \underline{\sigma}_{\mathrm{p}}^{\mathrm{d}} - (\vec{\nabla}\vec{v})^{\dagger} \cdot \underline{\sigma}_{\mathrm{p}}^{\mathrm{d}} - \underline{\sigma}_{\mathrm{p}}^{\mathrm{d}} \cdot (\vec{\nabla}\vec{v}) \right]$$
$$= \eta_{\mathrm{p}} \left[(\vec{\nabla}\vec{v}) + (\vec{\nabla}\vec{v})^{\dagger} \right]. \tag{5.80}$$

Here we follow common practice in the field of using λ for the parameter with the unit of time, η_{p} denotes a polymer viscosity, and the dagger indicates the transpose of a matrix.

Let us discuss the various terms in this equation. The first two terms on the left-hand side, $\sigma_{\mathrm{p}}^{\mathrm{d}}$ and $\lambda \dot{\underline{\sigma}}_{\mathrm{p}}^{\mathrm{d}}$, together with those on the right, are precisely the terms of the simple Maxwell model (2.22) for response discussed in section 2.5. It describes how the stress responds to shear with a time constant λ—this is because the polymers take some time to stretch or to relax in response to changes in shear. We can therefore think of λ as the longest relaxation time of the polymers which we discussed in the previous section. Note that λ is the *only* nontrivial parameter of the model! The term $\vec{v} \cdot \vec{\nabla} \underline{\sigma}_{\mathrm{p}}^{\mathrm{d}}$ in (5.80) we recognize as coming from the convective derivative (1.2) that emerges when we follow a fluid element.

Long polymers in a fluid have long relaxation times (even up to the order of seconds) because of the viscous friction with the solvent and the fact that the elastic constant of the polymer decreases with increasing N. As a result, local equilibrium is not established, so we cannot use the formalism of nonequilibrium thermodynamics to describe the dissipative processes. Instead we are forced to resort to effective models which capture only the essence of a polymer solution or melt, namely that it exhibits anisotropy, elasticity, and slow relaxation, while ignoring all the internal relaxation modes of the polymers. Such an effective approach is called phenomenological.

The other two terms between square brackets on the left-hand side of (5.80), which are bilinear in the velocity gradient and stress tensor, are new. They essentially take into account how the orientation and extension of a stretchable object like a polymer changes with time in the presence of velocity gradients—very much as, as we illustrated in figure 5.28, how affine strain changes the orientation and extension of biopolymers in a network. In fact, the terms between square brackets on the left of (5.80) form the so-called upper convected time derivative. This, together with the fact that the relaxation properties of the model are based on the simple Maxwell model, explains the name *upper convected Maxwell model*. This is the name used when the polymer stresses are dominant so that

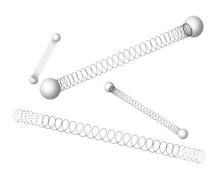

Figure 5.36. Illustration of the fact that the constitutive equation (5.80) can be derived exactly for a model of noninteracting dumbbells connected with infinitely extensible linear springs. The two ends are convected with the flow.

Luckily, so-called Boger fluids (James, 2009) have a rheology which conforms well to the Oldroyd-B model. For this reason these polymer solutions are favorite experimental model fluids. Boger fluids are smartly chosen combinations of solvents and polymers, whose shear stress is typically dominated by the solvent, while the normal stresses are due to the polymers only.

For an introduction to viscoelastic flow instabilities, see Larson, 1992. A complete review of the modern perspective is given by Datta et al., 2022.

the solvent viscosity can be ignored; the general case in which the solvent flow is retained is called the Oldroyd-B model.

An intuitive understanding of these models comes from the fact that, as figure 5.36 illustrates, the constitutive equation can be derived exactly for non-interacting dumbbells connected by infinitely extensible linear springs. Clearly, in reality, polymers have a finite extensibility[54] and do interact. Nevertheless, even though a particular extension may do better in particular circumstances, the common wisdom in the field is that the constitutive equation (5.80) captures the most essential gross features of polymer rheology. Moreover, in spite of its conceptual simplicity, the technical challenges associated with its tensor character and its numerical sensitivity[55] are often formidable enough to discourage us from going much further.

As derived in problem 5.9, for a uniform shear flow with shear rate $\dot{\gamma}$, the above constitutive equation gives a nonlinear stress difference which grows as $\dot{\gamma}^2$. As the shear stress remains linear in $\dot{\gamma}$, the Weissenberg number grows linearly in $\dot{\gamma}$. In fact,[56]

$$\text{UCM model:} \qquad \text{Wi} = 2\lambda\dot{\gamma}. \qquad (5.81)$$

As λ is the only parameter in the model and has dimension of time, the appearance of the combination $\lambda\dot{\gamma}$ need not come as a surprise.

5.10.4 Polymer flow instabilities due to hoop stresses

The fact that viscoelastic polymer flows often exhibit instabilities, or even turbulence, is due to the same hoop stresses that give rise to the rod climbing effect of figure 5.34.c. As figure 5.35 illustrates, the combination of shear gradients which stretch polymers and curve streamlines results in a force which pulls the fluid to regions of larger curvature and gradients where the effect is even greater. In other words, if this effect is strong enough to overcome stabilizing viscous forces, a small perturbation of the flow will tend to amplify more and more, and tend to lead to an instability of the base flow.

Whether this leads to an instability, however, also depends on the competition between the growth of perturbations and the response time of the polymers. If a polymer can follow changes in the flow almost instantaneously, not much will happen. In other words, for the elastic effects to drive the instability, the elastic energy which is 'charged' by the strong gradients at some point needs to be 'discharged' elsewhere. This requires that the polymer relaxation be sufficiently slow.

We have already seen in figure 5.33 that the viscosity of polymer melts scales as $N^{3.4}$ with the degree of polymerization, so that

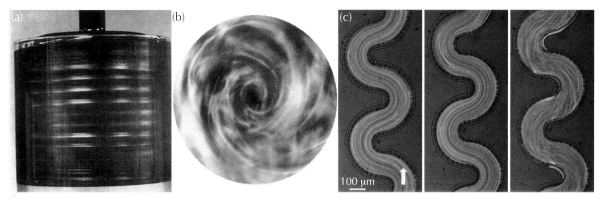

Figure 5.37. Various examples of viscoelastic flow. (a) Banded pattern observed by Larson et al., 1990 in viscoelastic flow between two rotating cylinders (a Taylor-Couette cell, as sketched in figure 1.11) for small Reynolds numbers. (b) Top view of turbulent viscoelastic flow between two rotating plates, as observed by Groisman and Steinberg, 2004. Note that there is still a spiral-like structure visible. The transition to such turbulent flow occurs at about Wi = 3.5, while the typical Reynolds number at that point is about 0.3. The Weissenberg number is defined in terms of the shear rate near the outer edge of the plate.[57] The measured shear stress as a function of shear rate is shown in figure 5.38. To make flow structures visible in such setups the fluids are typically fluorescently dyed (Zilz et al., 2012) or seeded with light-reflecting flakes (Groisman and Steinberg, 2004). (c) Snapshots of viscolastic flow in experiments by Zilz et al., 2012 with a so-called serpentine channel of width about 100 μm, designed for the study of viscoelastic flow in curved channels. The arrow indicates the flow direction. The panels show stable flow at Wi = 0.22 (left), slightly unstable flow close to onset of viscoelastic instabilities at Wi = 0.24 (middle), and unstable flow at Wi = 0.25 (right). Note that the transition is quite sharp. The Reynolds number in these experiments varies between 0.2 and 5. Image courtesy of Anke Lindner.

it becomes extremely large for long polymers. Likewise, polymer solutions can easily have a large viscosity in the semi-dilute regime when the polymers are long enough. The larger the viscosity, the smaller the Reynolds number (1.52). Since the Reynolds number can easily be made small, experimentalists can make sure that any instability they see is due to elastic effects and not due to the convective nonlinear terms in the Navier-Stokes equations that drive most instabilities of Newtonian fluids.

Since it was predicted and experimentally verified in 1990 that viscoelastic flow between rotating cylinders (the Taylor-Couette setup of figure 1.11) does indeed become unstable at small Reynolds numbers once the hoop stresses are large enough, it has gradually become clear that viscoelastic flow instabilities qualify as an interesting field in themselves, with their own particular set of challenges.

Figure 5.37 illustrates that when viscoelastic flow at small Reynolds numbers in curved geometries becomes unstable, one can observe coherent patterned flow as well as time-dependent or even turbulent flow. Indeed panel a shows a banded flow observed between two rotating cylinders. Panel c shows three snapshots of a curly serpentine channel in which one observes a sharp transition from laminar to time-dependent flow. Moreover, in the experiment of panel b, with a circular cell whose flow is driven by a rotating lid, one observes turbulent behavior at moderate Weissenberg numbers. And all that even though the Reynolds number is small!

While it was long realized that instabilities and the formation of vortices near nozzles and constrictions can be the rate-limiting effect in the process industry,[58] the systematic study of purely elastic instabilities was instigated by Larson et al., 1990. They predicted and observed such an instability for viscoelastic Taylor-Couette flow—see figure 5.37.a. A review of these effects and early follow-up work is given by Shaqfeh, 1996. This overview does not discuss the instability criterion (5.82), as this was published in the same year (Pakdel and McKinley, 1996). The News and Views article by Larson, 2000 has been instrumental in identifying viscoelastic turbulence as an important topic and challenge for research. The essay by Morozov and van Saarloos, 2007 may provide a simple starting point for you to learn more about this criterion and its implications. A more research-oriented review of the field is given by Datta et al., 2022.

The criterion is usually referred to as the Pakdel-McKinley criterion (Pakdel and McKinley, 1996), but sometimes also as the M-criterion.

From the investigation of a large variety of such setups, it has become clear that the threshold criterion for viscoelastic flow in curved geometries to become unstable can invariably be captured in the form[59]

$$\text{curved viscoelastic flow unstable if} \quad \text{Wi}\,\frac{\lambda\,U}{\mathcal{R}} > M^2. \quad (5.82)$$

Here λ is, as above, the typical relaxation time of the fluid, U is a typical flow velocity, $\mathcal{R}$ is the radius of curvature of the setup, and M is some number, typically larger than 1, whose precise value depends on the flow geometry.

This inequality (5.82) quantifies how the three effects combined—anisotropic elastic normal forces, curved streamlines, and sufficiently slow relaxation time—have to conspire to make viscoelastic flow unstable. Indeed, according to this empirical criterion, instability is favored by a large anisotropic and elastic force as measured by Wi, a large relaxation time λ and flow velocity U, and a large curvature of the streamlines, hence a small radius of curvature $\mathcal{R}$. In problem 5.10 we use this criterion to investigate the instability threshold for the experiments on the serpentine channel flow shown in figure 5.37.c.

A salient feature of many of these hoop stress–driven viscoelastic instabilities is that they tend to be highly nonlinear: once the instability occurs, nonlinearities tend to amplify the instability. As a result, once one has entered the unstable flow regime and then reduces the driving rate (such as the rate of rotation of a circular lid), the flow tends to remain in the patterned or turbulent state at values of the driving below the one where the instability originally sets in. In other words: one observes strong hysteresis, very much as near a first order phase transition or, in the language of bifurcations (see section 8.2.3), near a subcritical bifurcation.

Let us illustrate this for the setup shown in figure 5.37.b, which exhibits a rapid transition to turbulent flow. In figure 5.38 the shear stress, normalized to that in the laminar phase, is plotted on a logarithmic scale as a function of the shear rate. We see that quite soon after deviations from the laminar flow behavior become noticeable, there is a rapid transition to a branch where the shear stress is enhanced by a factor of 5 to 8. This is the regime of the self-sustained turbulent flow illustrated in figure 5.37.b. Moreover, as the arrows indicate, when the driving shear is slowly reduced the flow stays on this turbulent branch at shear rates much below where the instability originally set in. Clearly, once we are in the turbulent state, with polymers constantly being stretched in regions of swirls and large shear rates, with dumping of their elastic energy in more quiescent regions, the flow tends to get stuck in this self-sustained nonlinear turbulent regime. In other words, the flow does not

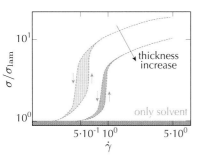

Figure 5.38. Measured stress, relative to that in the laminar state, as a function of shear rate in the experiments of Groisman and Steinberg, 2004 corresponding to figure 5.37.b for two runs at different cell thickness. The rapid increase upon increasing the shear rate marks the transition to the turbulent state. Note that the stresses rapidly increase by a large amount. The curves for increasing shear rate are different from those for decreasing shear rate, as indicated by the arrows, so there is a significant amount of hysteresis. The transition occurs for Wi of about 3.5, where Wi is defined in terms of the shear rate near the outer edge of the plate.

easily find its way back to the quiescent laminar state. As mentioned in a margin note in section 1.11.2, as a result, also a viscoelastic channel flow exhibits a nonlinear instability to a weakly turbulent state.

5.11 What have we learned

In the absence of interactions, the statistics of polymers for large degrees of polymerization N are given by the ideal chain model, whose scaling of the mean end-to-end distance R is similar to that of a random walk with time, $R \sim N^{1/2}$. Excluded volume interactions change this behavior to $R \sim N^\nu$, with $\nu > 1/2$ for physical dimensions d below 4. The Flory argument gives an excellent approximation, $\nu_{\text{Flory}} = 3/(d+2)$. This is especially relevant to synthetic polymers with a short Kuhn length.

Many biopolymers have a large Kuhn/persistence length. For these, excluded volume effects are small and their statistical behavior is well described by the wormlike chain model, a continuum variant of the ideal chain model. The wormlike chain model has become the reference point for interpreting experiments involving the force-extension curve of biopolymers.

In passing, our analysis of the force-extension curve of the wormlike chain model illustrated how to calculate thermal averages of fluctuating quantities by simply following the four commandments of box 3 in section 3.9.

The scaling behavior of synthetic polymers in a good solvent is governed by a concentration-dependent crossover length. For small length scales, single chain excluded volume effects dominate; however, for large scales, the difference between avoiding the same chain and others blurs, resulting in a crossover to free chain-type scaling.

Reptation theory gives a scaling $\eta \sim N^3$ for the viscosity of dense polymer solutions and melts, but experiments give an even faster growth of η with N.

The rheology of polymer solutions and melts is strongly non-Newtonian: the shear-induced stretching and orientation, together with the slow relaxation of long polymers, makes polymer flows viscoelastic and anisotropic. This results in various characteristic flow instabilities and the occurrence of elastic turbulence, even at small Reynolds numbers.

5.12 Problems

Relevant coding problems and solutions for this chapter can be found on the book's website www.softmatterbook.online under Chapter 5/Coding problems.

Problem 5.1 *The Gaussian expression for the ideal chain model*

In this problem we'll check the Gaussian expression (5.5) of $\mathcal{P}(\vec{R})$.

a. Write $\vec{R}$ in d dimensions as $\vec{R} = \sum_{i=1}^{d} \vec{x}_i$ and show that

$$\int d^d \vec{R}\, \mathcal{P}(\vec{R}) = \prod_i \left[\left(\frac{1}{2\pi R_0^2/d} \right)^{1/2} \int dx_i\, e^{-x_i^2/(2R_0^2/d)} \right]^d = 1.$$
(5.83)

b. Show along the same lines that

$$\int d^d \vec{R}\, \vec{R}^2 \mathcal{P}(\vec{R}) = R_0^2,$$
(5.84)

and convince yourself that this is consistent with the definition of R_0 just below equation (5.3).

Problem 5.2 *The radius of gyration of the ideal chain model*

In this problem we calculate the radius of gyration R_g of the ideal chain model for a case in which all the monomers have the same mass.

a. Let $\vec{R}_i$ denote the positions of the points joining the vectors $\vec{\ell}_i$ of the ideal chain model. In terms of the mean position $\vec{R}_m$ is

$$\vec{R}_m = \frac{1}{N} \sum_{i=1}^{N} \vec{R}_i,$$
(5.85)

and the radius of gyration is defined as the mean square distance from the center of mass,

$$R_g^2 = \frac{1}{N} \sum_{i=1}^{N} \langle (\vec{R}_i - \vec{R}_m)^2 \rangle.$$
(5.86)

Show that you can write the expression for R_g as

$$R_g^2 = \frac{1}{N^2} \sum_{i=1}^{N} \sum_{j=1}^{N} \langle (\vec{R}_i^2 - \vec{R}_i \cdot \vec{R}_j) \rangle.$$
(5.87)

b. Rewrite the above expression for the mean radius of gyration in the form

$$R_g^2 = \frac{1}{N^2} \sum_{i=1}^{N} \sum_{j=i}^{N} \langle (\vec{R}_i - \vec{R}_j)^2 \rangle.$$
(5.88)

c. Since we are interested in the large-N limit, we can switch to continuum coordinates s_1 and s_2 along the chain, which vary between zero and $N\ell_K$. Show that this allows us to write (5.88) as

$$R_g^2 = \frac{1}{N^2\,\ell_K^2} \int_0^{N\ell_K} \mathrm{d}s_1 \int_{s_1}^{N\ell_K} \mathrm{d}s_2 \, \langle (\vec{R}(s_1) - \vec{R}(s_2))^2 \rangle. \qquad (5.89)$$

In step a we have started for clarity from the discrete formulation, but if you already have some experience with such problems, you might have realized that you could immediately have written down expression (5.89).

d. Argue that for the ideal chain model, one can write

$$\langle (\vec{R}(s_1) - \vec{R}(s_2))^2 \rangle = \frac{s_1 - s_2}{\ell_K} \, \ell_K^2. \qquad (5.90)$$

e. Using (5.90) in (5.89), perform the s-integrals by going to the difference variable $s_1 - s_2$ to show that

$$R_g^2 = \frac{N\ell_K^2}{6} = \frac{R_0^2}{6}. \qquad (5.91)$$

Problem 5.3 *Evaluation of $\langle \hat{t}(0) \cdot \hat{t}(s) \rangle$ in equation (5.27).*

a. Show, by completing the square in the exponential, that one can write (5.27) with (5.26) for $P(\theta, s)$ as

$$\begin{aligned}
\langle e^{\pm i\theta} \rangle &= \frac{1}{\sqrt{4\pi D_\theta s}} \int_{-\infty}^{\infty} \mathrm{d}\theta \, \exp\left(-\frac{(\theta \mp 2iD_\theta s)^2}{(4 D_\theta s)} \right) \\
&\quad \times \exp\left(\frac{i^2(2D_\theta s)^2}{4D_\theta s} \right).
\end{aligned} \qquad (5.92)$$

b. Perform the Gaussian integral by shifting the integration contour up or down in the θ plane and show that this leads to equation (5.28).

Problem 5.4* *Derivation of the Odijk length for a polyelectrolyte*

In section 5.4.2, we discussed how for a polyelectrolyte (i.e., a charged polymer chain), the persistence length ℓ_p is also influenced by the electrostatic interactions between charges distributed along the chain. We derive here the expression (5.32) for the persistence length in the presence of charges, the so-called Odijk length. We do so by calculating the increase in energy due to the charges if a charged rod is bent. By adding this contribution to the elastic bending energy, we can then obtain the total persistence length.

The length was introduced first by Odijk, 1977.

For simplicity, we consider a semiflexible polymer (which we can think of as a thin rod) of length L which is bent in a circular fashion so that its radius of curvature R is constant and large; the radius of curvature can be taken to be so large that $L \ll \ell_p, R$. Here ℓ_p is the persistence length in the presence of charges. We take the charges to be uniformly distributed along the chain, with a line density $\rho = e / A$. Between two line elements $\mathrm{d}s_1$ and $\mathrm{d}s_2$ on the chain, located at s_1 and s_2, the charge interaction is taken to be the Debye-Hückel potential,

$$U_{DH}(r)\mathrm{d}s_1\mathrm{d}s_2 = \frac{\rho^2}{4\pi\epsilon_0\epsilon_{\mathrm{r}}r}e^{-\kappa r}\mathrm{d}s_1\mathrm{d}s_2, \qquad (5.93)$$

where $r = r(s_2 - s_1, R)$ is the distance between the line elements. s_1 and s_2 are measured along the contour from one end of the polymer. Furthermore, ϵ_{r} is the relative dielectric permittivity and κ^{-1} the Debye-Hückel screening length—see the discussion in section 4.4.1.

a. Write the total elastic energy E_{WLC} from (5.17) for this case with the polymer bent with constant radius of curvature R: the curvature brings distant pieces of the polymer closer together, hence increasing the repulsive energy.

In principle we would want to derive the total electrostatic energy

$$U_{DH}(R) = \int_0^L \mathrm{d}s_1 \int_{s_1}^L \mathrm{d}s_2\, U_{DH}(r) \qquad (5.94)$$

as a function of R. However, this expression is divergent for any R due to the $1/r$ term in expression (5.93). To circumvent these divergencies, we calculate $\Delta U_{DH}(R) = U_{DH}(R) - U_{DH}(R=\infty)$, the change in electrostatic potential due to the curvature.

b. As a first step toward implementing this, we calculate Δr $(R, s_2 - s_1) = r(R, s_2 - s_1) - r(\infty, s_2 - s_1)$. Show, with the help of figure 5.39, that

$$\Delta r(R, s_2 - s_1) = 2R\sin\theta/2 - (s_2 - s_1). \qquad (5.95)$$

c. Since $L \ll R$, the angle θ is very small. Use this to expand $\sin\theta/2$ in (5.95) to show that

$$\Delta r(R, s_2 - s_1) = -\frac{(s_2 - s_1)^3}{24R^2} = -\frac{r^3(\infty, s_2 - s_1)}{24R^2}. \qquad (5.96)$$

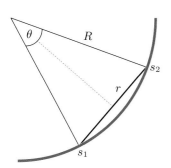

Figure 5.39. Sketch of the various quantities used to calculate the charge contribution to the persistence length of a polyelectrolyte. The distance between two points at positions s_2 and s_1 along the curved polymer is indicated by r, $\theta = (s_2 - s_1)/R$. Note that in the calculation $R \gg L$ is assumed so that θ is a small angle.

d. Show that for ΔU_{DH} we can write

$$\Delta U_{DH}(R) = \int_0^L \mathrm{d}s_1 \int_{s_1}^L \mathrm{d}s_2\, \frac{\rho^2}{4\pi\epsilon_0\epsilon_{\mathrm{r}}}\left[\frac{e^{-\kappa r(R)}}{r(R)} - \frac{e^{-\kappa r(\infty)}}{r(\infty)}\right],$$
$$(5.97)$$

where for simplicity we write $r(R)$ for $r(R, s_2 - s_1)$.

e. Now expand $r(R)$ about $r(\infty)$ to show that

$$\frac{e^{-\kappa r(R)}}{r(R)} - \frac{e^{-\kappa r(\infty)}}{r(\infty)} = -\left[\frac{1}{r^2(\infty)} + \frac{\kappa}{r(\infty)}\right]e^{-\kappa r(\infty)}\Delta r(R, s_2 - s_1).$$
$$(5.98)$$

f. Use (5.96) in expression (5.98), and perform the integrals to get

$$\Delta U_{DH}(R) = \frac{\rho^2}{24(4\pi\epsilon_0\epsilon_{\rm r})R^2}\frac{1}{\kappa^3} \tag{5.99}$$
$$\times\left[e^{-\kappa L}(8 + 5\kappa L + \kappa^2 L^2) + (-8 + 3\kappa L)\right].$$

g. Show that in the limit $\kappa L \gg 1$ this gives

$$\Delta U_{DH}(R) = \frac{\rho^2 L}{32\pi\epsilon_0\epsilon_{\rm r}\kappa^2 R^2}. \tag{5.100}$$

h. Combine this with the expression obtained in step a. to show that the total increase in energy for bending with radius of curvature R is

$$E_{\rm tot} = E_{\rm WLC} + \Delta U_{DH}(R) = \frac{L}{2R^2}\left[\alpha + \frac{\rho^2}{16\pi\epsilon_0\epsilon_{\rm r}\kappa^2}\right]. \tag{5.101}$$

i. Use $\rho = e/A$ and the expression (4.16) for the Bjerrum length $\lambda_{\rm B}$ to show that this finally yields expression (5.32) for the persistence length.

Problem 5.5 *Evaluation of the linear response of a wormlike chain*

In this exercise we calculate the response of the wormlike chain to a small force f pulling on its end. To do so, we start from the partition function Z for the energy (5.34)

$$Z = \sum_{\rm dof} e^{-E/k_{\rm B}T}, \tag{5.102}$$

where the sum is over all degrees of freedom.

a. Show from the energy expression (5.34) that

$$\langle R_z \rangle = k_{\rm B}T\,\frac{1}{Z}\frac{\partial Z}{\partial f}. \tag{5.103}$$

b. Show from the above expression that

$$\frac{{\rm d}\langle R_z \rangle}{{\rm d}f} = \frac{1}{k_{\rm B}T}\left(\langle R_z^2 \rangle - \langle R_z \rangle^2\right). \tag{5.104}$$

c. To get the linear response, you can evaluate the above result in the absence of a force, for $f = 0$. Argue that then $\langle R_z \rangle = 0$, and $\langle R_z^2 \rangle = \langle R^2 \rangle/3$.

d. Use (5.104) and (5.30) to derive the wormlike chain linear response result for $\langle R_z^2 \rangle$ and show that your result reduces to (5.35) in the limit $L \gg \ell_{\mathrm{p}}$.

Problem 5.6 *Derivation of expression (5.57) for the entropy of mixing*

As explained in the text, we consider two species of molecules A and B, consisting of chains of N_A and N_B monomers, respectively (if $N_B = 1$, the units of B can be thought of as the solvent molecules). Each monomer takes an amount of space v_0 independent of the type of species, and it is therefore easy to imagine monomers occupying the sites of a lattice. Molecules A originally occupy volume V_A, and molecules B volume V_B, so the numbers of monomers are $M_A = V_A / v_0$ and $M_B = V_B / v_0$. We aim to determine the change in entropy when the volumes are put together so that both species now occupy the total volume $V_A + V_B$. The two volume fractions of monomers after mixing are then $\phi_A = V_A / (V_A + V_B)$ and $\phi_B = V_B / (V_A + V_B) = 1 - \phi_A$.

a. Use the Boltzmann formula $S = k_B \ln \Omega$ to determine the entropy of a *single monomer* A before and after the mixing, and show that the difference is

$$\text{single monomer:} \quad \Delta S_A = -k_B \ln \phi_A. \tag{5.105}$$

Note that this expression is independent of the elementary volume v_0 used to count the number of states.

b. Assume the monomers are noninteracting and derive, starting from the result for ΔS_A and ΔS_B, the *total* entropy of mixing:

$$\text{all monomers:} \quad \Delta S_{\mathrm{mix}} = -k_B M_A \ln \phi_A - k_B M_B \ln \phi_B. \tag{5.106}$$

c. We now take into account that the molecules of species A are polymers consisting of N_A monomers and B polymers of N_B monomers. Argue, in analogy with the previous result, that the mixing entropy is

$$\text{polymers:} \quad \Delta S_{\mathrm{mix}} = -k_B \frac{M_A}{N_A} \ln \phi_A - k_B \frac{M_B}{N_B} \ln \phi_B. \tag{5.107}$$

d. Calculate from the above expression the total entropy of mixing per monomer and show that this gives expression (5.57).

Note that various approximations were used in the derivation you just did. The elementary volume is taken to be the same for both monomers, interactions are ignored, and polymers are taken as swarms of monomers moving together. In practice there may also be volume changes upon mixing. In spite of these approximations, the expression 5.107 captures the essence of mixing.

Problem 5.7 *Evaluation of the Flory-Huggins free energy*

In this exercise, we evaluate the Flory-Huggins free energy (5.58) for a symmetric polymer blend with $N_A = N_B = N$.

a. Show that the free energy is symmetric with respect to $\phi = 1/2$.

b. Evaluate $dF_{mix}/d\phi$ for the symmetric blend.

c. Show that for $\chi N < 2$ the free energy has one minimum $\phi = 1/2$ while for $\chi N > 2$ it has two minima symmetrically positioned with respect to $\phi = 1/2$.

d. Argue that the common tangent construction describing phase mixing is the horizontal line through the minima of F_{mix} for $\chi N > 2$.

e. Use the previous result to derive the equation for the phase boundary for demixing for $\chi N > 2$. This line is called the binodal line. It is indicated by a solid line in figure 5.23.

f. Derive the equation for the so-called spinodal line where $d^2 F_{mix}/d\phi^2 = 0$. It is indicated by a dashed line in figure 5.23.

g. Convince yourself that in between the binodal line and the spinodal line the second derivative is positive, indicating linear stability of the homogeneous state in this part of the phase diagram, while above the dashed line the homogenous mixture is unstable.

> Because of the linear stability of the homogeneous state between the binodal and spinodal lines, the system is metastable here. Phase change occurs through nucleation of the minority phase. Above the spinodal line, the instability immediately results in growth of inhomogeneities in the composition.

Problem 5.8* *Application of the Flory-Huggins theory to dilute suspensions of polymers*

In this exercise we apply the Flory-Huggins theory to a dilute suspension of polymers, so to the case $\phi \ll 1$, $N_B = 1$. The osmotic pressure Π is defined as the derivative of the total mixing free energy with respect to volume, $\Pi = -\partial F_{mix}^{tot}/\partial V\big|_{N_A}$, where $F_{mix}^{tot} = n F_{mix}$, determined in problem 5.6, with $n = M_A + M_B$ the total number of lattice sites, and where n_A is the number of polymers of size N_A. The first steps in the exercise below are necessary to transform the derivative with respect to V at constant n_A in terms of a derivative with respect to the volume fraction ϕ.

a. Show that, since $M_A = n_A N_A$, we can write the volume fraction ϕ of polymers as

$$\phi = \frac{v_0 n_A N_A}{V}, \tag{5.108}$$

with v_0 the elementary volume used in problem 5.6.

b. Show that since the derivative of F_{mix}^{tot} with respect to V is taken at constant n_A, we can write

$$\frac{\partial}{\partial V}\bigg|_{n_A} = \frac{1}{v_0 n_A N_A} \frac{\partial}{\partial(1/\phi)} = -\frac{\phi^2}{v_0 n_A N_A} \frac{\partial}{\partial \phi}. \tag{5.109}$$

c. Show that

$$\Pi = -\frac{\partial n F_{\text{mix}}}{\partial V}\bigg|_{n_A} = \frac{\phi^2}{v_0 n_A N_A} \frac{\partial (n_A N_A F_{\text{mix}}/\phi)}{\partial \phi}\bigg|_{n_A}$$

$$= \frac{\phi^2}{v_0} \frac{\partial (F_{\text{mix}}/\phi)}{\partial \phi}\bigg|_{n_A}.$$

(5.110)

d. Evaluate the derivative in (5.110) to obtain the following expansion of Π for small volume fractions ϕ (with $N_B = 1$):

$$\Pi = \frac{k_B T}{v_0}\left[\frac{\phi}{N_A} + \frac{\phi^2}{2}(1 - 2\chi) + \frac{\phi^3}{3} + \cdots\right].$$

(5.111)

This expression has the form of a so-called virial expansion—the name used for a small density expansion—for the osmotic pressure. Note that the quadratic term results from both the Flory repulsion term for low volume fractions and the entropic term. The prefactor of the quadratic term changes sign at $\chi = 1/2$. Below this value, the interactions increase the osmotic pressure, but for $\chi = 1/2$ the quadratic term vanishes. At that point, entropic terms just cancel the repulsive terms; this point therefore identifies, within the Flory-Huggins theory, the theta-point. Measuring the second derivative of the osmotic pressure moreover gives a direct handle for determining χ.

For more details and other applications, see chapter 4 of Rubinstein and Colby, 2003, where it is also discussed how the interaction parameter χ can be determined from small-angle neutron-scattering data.

Problem 5.9 *Determination of the Weissenberg number for the Oldroyd-B and upper convected Maxwell model*

In this exercise we determine the Weissenberg number (5.78) for the polymers from the constitutive equation (5.80) of the Oldroyd-B and upper convected Maxwell models, for the case of a simple shear flow $\vec{v} = \dot{\gamma} y \, \hat{x}$. We follow the notation common in the literature to denote the shear rate by $\dot{\gamma}$.

a. Check that $\dot{\gamma}$ has the dimension of a rate, or inverse time.

b. Write down the equation for $\underline{\sigma}_p$ when the velocity suddenly stops, and show that the relaxation time with which components of $\underline{\sigma}_p$ relax to zero is simply λ.

c. As a preparation for the next step, write out the tensors $(\vec{\nabla}\vec{v})$ and $(\vec{\nabla}\vec{v})^\dagger$ explicitly for our case of simple shear rate.

d. Now write out the constitutive equation (5.80) for the case of steady shear rate and solve explicitly for the elements of $\underline{\sigma}_p^d$. (NB: You should find just three nonzero elements.)

e. For Newtonian fluids the normal stresses are all the same. But here you find that there is a difference between the normal stresses. Determine the Weissenberg number (5.78) for this case. How does

Wi depend on $\dot{\gamma}$ and how does λ come in? Show that the result reduces to expression (5.81).

As discussed in section 5.10, instabilities typically occur when the Weissenberg number is somewhat larger than 1, so that the hoop stresses are larger than the shear stress.

Problem 5.10* *Application of the instability criterion (5.82) to serpentine flow*

In this exercise we apply the instability criterion (5.82) to the serpentine channel of figure 5.37.c, which consists of a sequence of circularly bent channel sections as sketched in figure 5.40.

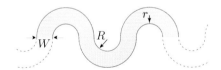

Figure 5.40. Illustration of a serpentile channel and the coordinates used by Zilz et al., 2012.

We follow the conventions illustrated in figure 5.40 to denote the radius of curvature of the inner sidewall of the channel by R, and to use r, in cylindrical coordinates, for the radial distance measured from the center line and counted positively in the direction pointing inward to the center. The width of the channel is W. We assume that the channel is deep enough that the flow profile can be taken as two-dimensional, i.e., that variations in depth can be ignored.

a. Assume that the radius of curvature R is very large ($R \gg W$), and ignore as a lowest order approximation the curvature of the channel in the flow profile (so essentially you analyze planar Poiseuille flow, similarly to the pipe Poiseuille flow problem treated in problem 1.5). Show that you can write the flow velocity in this approximation as

$$u_{\mathrm{s}} = (3/2)U \left[1 - \left(\frac{r}{W/2} \right)^2 \right], \qquad (5.112)$$

where u_{s} denotes the velocity along the circular streamline a distance r from the central line, and U is the average velocity in the channel.

b. Determine the shear rate $\dot{\gamma}$ as a function of r.

c. Use the previous result and the fact that for the upper convected Maxwell model, $\mathrm{Wi} = 2\lambda\dot{\gamma}$ according to (5.81), to show that the instability criterion (5.82) can be written in this case as

$$36 \left(\frac{\lambda U}{W} \right)^2 F(r) > M^2 \ \text{ with } \ F(r) = \frac{\frac{|r|}{W} \left[1 - 4 \left(\frac{r}{W} \right)^2 \right]}{\frac{R}{W} + \frac{1}{2} - \frac{r}{W}}. \quad (5.113)$$

d. Even though we do not know M (which follows from a full stability analysis of the equations), we can for given R/W determine the location r_c in the channel where the instability sets in first, by determining the position r where $F(r)$ is maximal. This can easily be done numerically. Here you are asked to simply derive analytically the following two results in the limits $R/W \to \infty$ (in which case the approximation in a is justified) and $R/W \to 0$ (where it is not):

$$R/W \to \infty : \mathrm{Wi}_c \to 0.38M\sqrt{\frac{R}{W}}, \quad \frac{r_c}{W} \to 0.29, \qquad (5.114)$$

$$R/W \to 0 : \mathrm{Wi}_c \to \frac{M}{\sqrt{72}}, \quad \frac{r_c}{W} \to \frac{1}{2}. \qquad (5.115)$$

The analysis given here follows the one by Zilz et al., 2012. The authors summarize the two asymptotic results in the following approximate interpolation formula:

$$\mathrm{Wi}_c = 0.38M\sqrt{0.1 + \frac{R}{W}}. \qquad (5.116)$$

Zilz et al., 2012 then argue that analysis of flow in a rectangular channel of depth corresponding to the one in their experiments leads to a similar result but with the prefactor 0.38 in this formula replaced by 0.27. Their comparison between theory simulations and experiments is shown in figure 5.41. The critical Weissenberg number is found to be of order unity and the above simple analysis is found to reproduce the overall behavior quite well with $M = 1.75$. This illustrates the validity of the instability criterion (5.82) and its usefulness for a back-of-the-envelope calculation.

Problem 5.11** *Rubber and the replica trick*

In this exercise, you will learn about the *replica trick*, which is a powerful mathematical tool to use when dealing with the *quenched disorder*, i.e., inhomogeneities frozen in the structure of a material. While the replica trick has found applications in the study of various disordered media, from spin glasses to jammed packings, here we will apply it to vulcanized rubber, following the seminal work of Sam Edwards, who invented the technique. Vulcanization is the process whereby crosslinks are introduced between the polymer chains that compose rubber so that it becomes more rigid and stable. An idealized picture of vulcanized rubber is reproduced in figure 5.42. In this system, the fixed random configuration of crosslinks is the quenched disorder.

The vulcanized rubber problem is complicated, and we will build up the analysis in steps. First, we describe a single polymer chain as a smooth path in space, a more detailed statistical treatment than addressing its end-to-end vector (see section 5.3.1). Second, we consider a number of C chains with steric repulsion. Third, we complete our system by adding the quenched disorder: a number of M fixed random crosslinks between the chains. We then arrive at the replica trick, which allows us to perform a physically meaningful quenched-disorder average, at the expense of introducing a

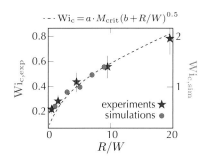

Figure 5.41. Comparison of the critical Weissenberg number Wi_c (left axis, data points marked by stars) as measured by Zilz et al., 2012 in their experiments on serpentine channels shown in figure 5.37.c with numerical simulations (right axis, circular data points) and the interpolation formula based on the instability criterion (5.116) (dashed line).

This problem is inspired by the treatment in the book *Stealing the Gold*, where the original article by Sam Edwards (Deam and Edwards, 1976) is accompanied by an insightful commentary by Paul Goldbart and Nigel Goldenfeld (Goldbart and Goldenfeld, 2004), who write: Edwards found the rubber problem significantly more complicated than his earlier forays into polymer science, and with characteristic ingenuity devised a variety of novel techniques and physical pictures to meet the challenge, all of which have survived as essential components of our current understanding, as well as being of general importance in statistical mechanics.

number of n copies of our entire system. The result is the *replicated partition function*, which represents an interacting system of chains without any randomness; yet it can provide the thermodynamics of our chain system with quenched disorder.

a. Let us start with a single ideal chain of length L having segments of Kuhn length ℓ_K. Coarse-grain this chain into path-segments of a certain length ℓ', assumed to be large enough, $\ell' \gg \ell_K$, so that each path-segment obeys the probability distribution of a long chain (see equation (5.4)), yet also assumed to be small enough, $\ell' \ll L$, so that the segments add up to a continuous position vector $\vec{R}(s)$ as a function of the path arclength s. Derive the partition function for this chain and check that it has the form

$$Z \sim \int \mathcal{D}\vec{R} \exp\left(-\frac{d}{2\ell_K L} \int_0^1 \left| \frac{\mathrm{d}\vec{R}(s)}{\mathrm{d}s} \right|^2 \mathrm{d}s \right), \qquad (5.117)$$

where the symbol $\mathcal{D}\vec{R}$ represents an integral over all possible realizations of the chain path $\vec{R}(s)$.

b. Generalize the result you obtained to a collection of C non-interacting polymer chains, denoting $\vec{R}_a(s)$ as the path of the ath chain for $a = 1, \ldots, C$.

c. In reality, different chain segments cannot overlap. This can be accounted for via the interaction potential

$$V = \frac{vL^2}{2\ell_K^2} \sum_{a,a'=1}^C \int_0^1 \mathrm{d}s\mathrm{d}s' \delta\left(\vec{R}_a(s) - \vec{R}_{a'}(s') \right), \qquad (5.118)$$

where v is an excluded volume parameter related to the steric repulsion between two chain segments, and δ is the Dirac delta function we encountered before. Show that the new partition function is

$$Z \sim \int \Pi_{a=1}^C \mathcal{D}\vec{R}_a \exp\left(-H_0(\mathcal{R}, \dot{\mathcal{R}}) \right), \quad \text{with}$$

$$H_0(\mathcal{R}, \dot{\mathcal{R}}) = \sum_{a=1}^C \frac{d}{2\ell_K L} \int_0^1 \mathrm{d}s \left| \frac{\mathrm{d}\vec{R}_a(s)}{\mathrm{d}s} \right|^2$$

$$+ \frac{vL^2}{2k_B T \ell_K^2} \sum_{a,a'=1}^C \int_0^1 \mathrm{d}s\mathrm{d}s' \delta\left(\vec{R}_a(s) - \vec{R}_{a'}(s') \right), \qquad (5.119)$$

where we introduce the shorthand notation $\mathcal{R}$ for the entire list of variables $\vec{R}_a$, and $\dot{\mathcal{R}}$ for the list $\dot{\vec{R}}_a$, as $a = 1 \ldots C$.

d. Next, we introduce M crosslinks which permanently connect path segments positioned at s_j and s_j' in the paths of chains a_j and

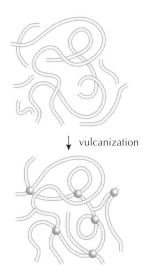

↓ vulcanization

Figure 5.42. Sketch of a polymer melt (top) that transforms into a crosslinked melt (bottom) in the process of vulcanization.

a'_j, respectively, for $j = 1, \ldots, M$. This introduces a fixed random set of M constraints of the form

$$\vec{R}_{a_j}\left(s_j\right) = \vec{R}_{a'_j}\left(s'_j\right), \text{ for } j = 1 \ldots M, \qquad (5.120)$$

defining the quenched disorder in the problem. Labeling this set by $\mathcal{M} = \{a_j, a'_j, s_j, s'_j\}_{j=1}^{M}$, show that the new Z is

$$Z\left(\mathcal{M}\right) = \int \Pi_{a=1}^{C} \mathcal{D}\vec{R}_a \ \Pi_{j=1}^{M} \delta\left(\vec{R}_{a_j}\left(s_j\right) - \vec{R}_{a'_j}\left(s'_j\right)\right)$$
$$\times \exp\left(-H_0(\mathcal{R}, \dot{\mathcal{R}})\right). \qquad (5.121)$$

e. In its present form, the partition function you obtained cannot be calculated, as it depends on quenched random variables (the crosslinks), which are not averaged over in the partition function as an annealed random variable would have been. However, we can average over the quenched random variables in the free energy: this amounts to dividing the material into large spatial regions and considering the regions as an ensemble of independent quenched-disorder realizations, then summing over the regions to recover the whole material using the fact that the free energy F is extensive (unlike the partition function!). We define the average of any variable $\mathcal{O}$ over a quenched-disorder ensemble as

$$\langle O \rangle_{\text{qd}} = \sum_{\mathcal{M}} P\left(\mathcal{M}\right) \mathcal{O}\left(\mathcal{M}\right)$$
$$= \sum_{M=0}^{\infty} \Pi_{j=1}^{M} \left[\sum_{a_j, a'_j = 1}^{C} \int_0^1 \mathrm{d}s_j \mathrm{d}s'_j\right] P\left(\mathcal{M}\right) \mathcal{O}\left(\mathcal{M}\right). \qquad (5.122)$$

Using the additive character of F, convince yourself that $Z \neq \langle Z \rangle_{\text{qd}}$, but $F = \langle F \rangle_{\text{qd}}$. Then use the identity

$$\lim_{n \to 0} \frac{x^n - 1}{n} = \ln x \qquad (5.123)$$

to obtain

$$\langle F \rangle_{\text{qd}} = -k_{\text{B}} T \lim_{n \to 0} \frac{\langle Z^n \rangle_{\text{qd}} - 1}{n}. \qquad (5.124)$$

f. We have now transformed the problem from calculating $\ln Z$ to calculating the *replicated partition function*, $\langle Z^n \rangle_{\text{qd}}$. Show that

$$\langle Z^n \rangle_{\text{qd}} = \int \Pi_{\alpha=1}^{n} \left[\Pi_{a=1}^{C} \mathcal{D}\vec{R}_a^{\alpha}\right] \exp\left(-\sum_{\alpha=1}^{n} H_0(\mathcal{R}^{\alpha}, \dot{\mathcal{R}}^{\alpha})\right)$$
$$\times \langle \Pi_{\alpha=1}^{n} \left[\Pi_{j=1}^{M} \delta\left(\vec{R}_{a_j}^{\alpha}\left(s_j\right) - \vec{R}_{a'_j}^{\alpha}\left(s'_j\right)\right)\right]\rangle_{\text{qd}}, \qquad (5.125)$$

where $\vec{R}_a^\alpha(s)$ denotes the path of the ath chain in the αth replica of the system, while $\mathcal{R}^\alpha$ is short for the entire list of variables in the αth replica, i.e., $\vec{R}_a^\alpha$, for $a = 1 \ldots C$, with α fixed (analogously for $\dot{\mathcal{R}}^\alpha$ and $\dot{\vec{R}}_a^\alpha$).

g. To proceed, we must assume something about the distribution of the crosslinks. Imagine the following procedure: the polymers are prepared at equilibrium without the crosslinks. Then, crosslinks are introduced in an instantaneous process, in which all pair segments which happen to be close to each other in that instant connect with some fixed probability μ. Show that the probability distribution for a specific crosslinks configuration is then proportional to

$$P(\mathcal{M}) \propto \frac{1}{M!} \mu^M Z(\mathcal{M}), \qquad (5.126)$$

where $Z(\mathcal{M})$ is the partition function in 5.121.

h. Using 5.122 and 5.126, show that

$$\langle \Pi_{\alpha=1}^n \left[\Pi_{j=1}^M \delta \left(\vec{R}_{a_j}^\alpha(s_j) - \vec{R}_{a_j'}^\alpha(s_j') \right) \right] \rangle_{\text{qd}}$$

$$\sim \int \Pi_{a=1}^C \mathcal{D}\vec{R}_a^0 \exp\left(-H_0(\mathcal{R}^0, \dot{\mathcal{R}}^0)\right) \qquad (5.127)$$

$$\times \exp\left(-H_I(\mathcal{R}^0, \ldots, \mathcal{R}^n, \dot{\mathcal{R}}^0, \ldots, \dot{\mathcal{R}}^n)\right),$$

where

$$H_I(\mathcal{R}^0, \ldots, \mathcal{R}^n, \dot{\mathcal{R}}^0, \ldots, \dot{\mathcal{R}}^n) =$$

$$- \mu \sum_{a,a'=1}^C \int_0^1 \mathrm{d}s \mathrm{d}s \, \Pi_{\alpha=0}^n \delta \left(\vec{R}_a^\alpha(s) - \vec{R}_{a'}^\alpha(s) \right). \qquad (5.128)$$

i. Show that the replicated partition function $\langle Z^n \rangle_{\text{qd}}$ is

$$\langle Z^n \rangle_{\text{qd}} \sim \int \Pi_{\alpha=0}^n \Pi_{a=1}^C \mathcal{D}\vec{R}_a^\alpha$$

$$\exp\left(-\sum_{\alpha=0}^n H_0(\mathcal{R}^\alpha, \dot{\mathcal{R}}^\alpha) - H_I(\mathcal{R}^0, \ldots, \mathcal{R}^n, \dot{\mathcal{R}}^0, \ldots, \dot{\mathcal{R}}^n) \right). \qquad (5.129)$$

We reached the end and managed to completely get rid of the quenched disorder—instead we obtained an effective Hamiltonian for the replicated system, where H_I contains interactions between the replicas. Particularly, our effective theory has an interplay between excluded volume that opposes overlapping chain segments and the interaction between replicas that favors overlap.

The replicated partition function can be approximated using, for example, the variational method. It can also be used to study the phase transition from a liquid to an amorphous solid state that occurs above a critical number of crosslinks. The phase transition is intimately related to the symmetry of the effective Hamiltonian under permutations of the replicas. This symmetry breaks at the liquid-solid transition (Goldbart and Goldenfeld, 1989; Goldenfeld, 1992). To learn more about the replica method and its broad applications in glassy systems, see Mézard et al., 1987 and Parisi et al., 2020.

6 | Liquid Crystals

Figure I.15 puts the nematic phase in the context of broken symmetries, while figure 6.1 illustrates the orientational order in the nematic, smectic, and cholesteric phases. Note that the smectic phases have partial positional order, due to the formation of layers.

Liquid crystals are intermediate between ordered crystals and disordered liquids, as they combine positional disorder with orientational order. They are nontrivial dense thermodynamic phases that have a broken rotational symmetry: they are characterized by an orientational order parameter and anisotropic properties.

We will focus our discussion on the nature and description of the most basic of all liquid crystal phases, the nematic phase. But the common feature of all nontrivial liquid crystal phases is the existence of an order parameter associated with the fact that locally the molecules point on average in a particular direction. Most characteristics and applications of liquid crystals are intimately connected with the slow spatial variation of this order parameter. Examples of modern developments which we explore are especially centered around the new phenomena and opportunities which arise at the interfaces with other fields.

6.1 Liquid crystals as mesophases

The books by de Gennes and Prost, 1993 and Chandrasekhar, 1992 give nice introductions to and overviews of the physics of liquid crystals. Most basic topics discussed in this chapter are discussed there in more detail. A different perspective is provided by Oswald and Pieranski, 2006a and Oswald and Pieranski, 2006b, who pay particular attention to experimental illustrations.

Since liquid crystals are neither fully positionally ordered crystals nor isotropic disordered liquids, the name *liquid crystal* is a bit of a contradiction in terms. To emphasize their intermediate nature between crystals and liquids, the more appropriate term *mesophase* has been introduced, but we will follow standard practice to use the generally accepted name *liquid crystal*.

6.1.1 A bewildering variety of liquid crystal phases

In figure 6.1 we sketch three of the most common mesophases: first of all the nematic phase, in which the molecules have an average orientation but no positional order. Secondly, the smectic one, in which the molecules form layers with the average orientation of the molecules parallel to the normal to the layers (smectic-A) or tilted with respect to it (smectic-C). And finally the cholesteric phase, in which the orientation rotates in a plane along the axis perpendicular to the planes. The artist's impressions of the phases in figure 6.1 illustrate how one commonly draws the molecules as smooth impenetrable objects, like hard ellipsoids. This is often very helpful, but it is important to keep in mind that it is a gross oversimplification, and that the actual molecular structure does matter for properties like the electric or magnetic polarizability[1]

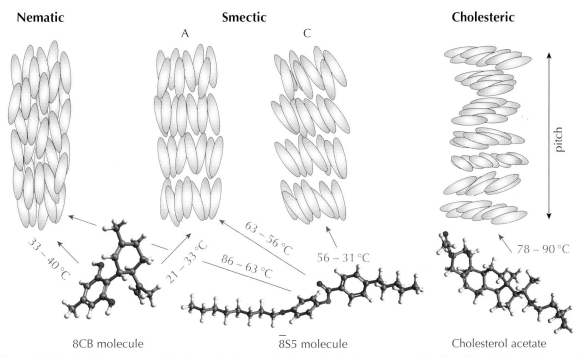

Figure 6.1. Sketch of three common liquid crystal phases, with the molecules drawn as smooth ellipsoidal objects. In the nematic there is average orientational order but no positional order. A smectic-A consists of layers; there is no order of the molecules within the layers but on average they point in the direction perpendicular to the layers (smectic-A) or are tilted with respect to the normal (smectic-C). In the cholesteric phase the average orientation rotates in the plane when we move in the direction perpendicular to the planes in which the molecules are oriented. The pitch is the distance after which the direction has rotated by an angle 2π. Also shown are two examples of molecules which exhibit one or more of the liquid crystal phases, together with the temperature range in which each phase is formed.[2]

and the preference to form particular phases. Likewise, the fact that the cholesterol acetate molecule shown in figure 6.1 exhibits a cholesteric phase with successive layers slightly rotated with respect to one another is a result of the slightly bent shape of the molecule. The relation between molecular architecture and the tendency to form particular phases is subtle!

We have already shown before examples of the fact that hard core particles can, depending on their shape, form various nontrivial thermodynamic phases—see figure I.11 and figure 4.12. So an aside about the relevance of studying the phase diagram of hard core particles may seem superfluous now. Nevertheless, for a long time it was not clear to what extent attractive molecular interactions were important or even necessary for the formation of liquid crystal mesophases. As far back as 1949, Onsager[4] predicted that long, thin needles exhibit a phase transition: at low densities they form a disordered liquid phase without orientational order (which in the liquid crystal context is called the isotropic phase), and at higher densities a nematic phase with orientational order. Because of the hard core interactions, there is no intrinsic energy scale, so this

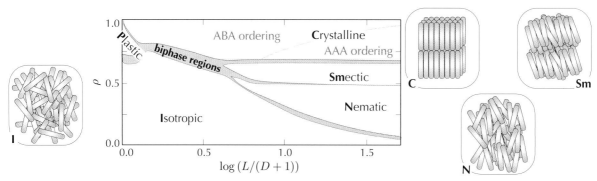

Figure 6.2. Phase diagram of hard core spherocylinders, cylinders of length L and diameter D capped on both sides with half a sphere, as a function of density and aspect ratio. For small densities and aspect ratios an isotropic disordered phase is found, while at sufficiently large aspect ratios the phase diagram displays large nematic, smectic-A, and crystalline phases. In the plastic phase at the upper left the slightly anisotropic particles form a crystal but remain orientationally disordered. The thin line through the crystalline region distinguishes between regions with ABA stacking and AAA stacking of successive planes. Adapted from Lekkerkerker and Vroege, 2013, based on computer simulations of Bolhuis and Frenkel, 1997.[3] Remember that for hard core interactions, temperature is an irrelevant parameter: all transitions are driven purely by entropy (see section 4.2).

In problem 6.1 you retrace Onsager's calculation.

transition is entropy-driven: at sufficiently high density, the needles have more room to explore (or more states available) when they are roughly oriented in the same direction. Since the 1980s, computer simulations of elongated hard-core particles like ellipsoids and disks have shown that hard-core interactions are sufficient to give many of the liquid crystal phases. For instance, as figure 6.2 shows, spherocylinders exhibit, besides the low-density disordered and nematic phases predicted by Onsager, a smectic phase and several crystalline phases at higher densities.

There are many more liquid crystal phases besides the three mentioned so far. Flat disk-shaped molecules like to form for columnar phases, as illustrated in figure 6.3; the tilt angle of a smectic-C phase can rotate around the axis normal to the layers to form a chiral phase denoted by smectic-C*; etc. Remarkably, the molecules can even form superstructures: in the so-called blue phases, an example of which is shown in figure 6.4, the molecules order chirally in columns, which themselves stack as ordered structures. When the length scale of these structures is on the order of the wavelength of light, the blue phases show Bragg spots in the visible.

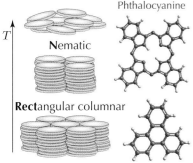

Figure 6.3. Flat disk-shaped molecules like the ones shown, examples of discotics, form nematic as well as hexagonal and rectangular columnar phases. Within the columns there can be additional order: the molecules can, for instance, stack in an ordered or a helical fashion, or behave plastically, meaning that they rotate about the disk axis. Adapted from Bushby and Lozman, 2002.

We will in this chapter focus on the nematic phase, since the discussion of this most common liquid crystal phase is a natural route to introducing the director field $\hat{n}$ which characterizes the orientational order. This field is an essential ingredient not only of the description of any liquid crystal phase but also of the new phenomena that emerge when liquid crystals are combined with colloids or active media. But it is important to be aware of the plethora of phases that liquid crystals can form.[5] Our discussion actually provides a good groundwork for exploring their enormous richness, should you be sufficiently intrigued by the example to do so.

6.1.2 Molecular liquid crystals versus colloidal liquid crystal phases

Historically, the tobacco mosaic virus which we showed in figure 4.1 was a convenient natural experimental model system for probing the thermodynamic phases of hard, long rods. But with its length at about 300 nm, it immediately brings up the question of what the distinction, if any, is between liquid crystal phases composed of rod-shaped molecules and those consisting of rod-shaped colloidal particles.[6]

There is actually no sharp distinction between the two as far as the formation of equilibrium phases is concerned. For, as we saw in chapter 4, we can think of colloids as particles endowed with effective interactions that, due to the Brownian motion, explore configuration space. They form equilibrium thermodynamic phases governed by these interactions, provided they are given enough time to equilibrate and explore phase space.

The main difference between the molecular and colloidal liquid crystal phases is associated with the time scales, and therefore also with their flow behavior. Rod-shaped colloids of roughly a micron in size diffuse and equilibrate on time scales of seconds, while true molecular liquid crystals, like the ones displayed in figure 6.1, form local equilibrium on the rapid time scale of molecular collisions. For this reason, molecular liquid crystals also remain essentially always close to local equilibrium, even when they flow. On the other hand, almost any flow drives a dense suspension of rod-shaped colloidal particles strongly out of equilibrium; as a result, the rheology of dense suspensions is almost immediately strongly non-Newtonian and governed by shear-induced jamming of anisotropic particles.

To make this more concrete, remember that we discussed in section 4.6.1 that the Péclet number

$$\mathrm{Pe} = \frac{\dot{\gamma}\,R^2}{D} = \frac{\dot{\gamma}\,6\pi\eta R^3}{k_\mathrm{B}T} \tag{6.1}$$

is the proper dimensionless number characterizing a colloidal particle of radius R with diffusion coefficient D in a flow with shear rate $\dot{\gamma}$. For flows with Péclet number larger than about 1, particles do not have the time to equilibrate, and the formation of shear-induced structures dominates. At small Péclet numbers, the shear is a small perturbation.

The factor R^3 in the above expression is crucial: for typical microns-sized colloidal rods, one is almost immediately in the large Péclet number regime for any reasonable shear rate; such flows are strongly out of equilibrium, so we cannot apply the concepts of non-equilibrium thermodynamics to capture this regime. On the other

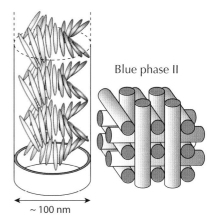

Blue phase II

~ 100 nm

Figure 6.4. The so-called blue phases consist of ordered stacks of columns (right) within which the molecules have chiral order (left). Illustrated here is the blue phase II, where the columns form a cubic superstructure. Because of the order and the large scale of the lattice constant, the blue phases can show Bragg reflections in the visible. Interestingly, blue phases are even explored for LCD technology and as photonic bandgap materials.[7]

The molecular examples we show in figure 6.1 are examples of so-called thermotropic liquid crystals, where phase changes occur as a function of temperature. In lyotropic liquid crystals, phase changes occur upon a change in composition. Colloidal liquid crystals, for which temperature does not play much of a role, are therefore more like lyotropic liquid crystals.

hand, for molecular liquid crystals with a scale of a few nanometers, virtually any flow is in the very small Péclet number regime. This means that the molecules have sufficient time to equilibrate, and that we can apply the framework of non-equilibrium thermodynamics of section 1.5 to formulate the dissipative hydrodynamic equations. We will follow this route in section 6.7.

6.1.3 The power of coarse-graining in the spirit of Landau

The above considerations concerning molecular liquid crystals are also the underlying reason why the standard approach to describing the nematic phase in terms of a coarse-grained free energy, expanded in gradients, works so well. In fact, the discussion of the Frank free energy below has wider pedagogical value as it is a beautiful illustration of how to approach a problem in terms of the appropriate order parameter and its associated coarse-grained effective free energy, in the spirit of the Landau theory of phase transitions and the concepts of hydrodynamics and broken symmetries introduced in section I.2.2: the director field is the order parameter associated with the breaking of rotational symmetry. Learning how to think this way is an investment worth the effort, as it is often a powerful way to write an effective phenomenological approach for a new problem you encounter. We'll see this illustrated again in chapter 9 on active matter, and even in chapter 8 on pattern formation.

6.1.4 The director field $\hat{n}$

As stated, we will focus our analysis on the nematic phase, in which there is only orientational order of the molecules. This means that *on average* the molecules point in a particular direction, which commonly is denoted by $\hat{n}$, a unit vector called the director. In the spirit of coarse-graining discussed in the context of hydrodynamics in section 1.2 and as illustrated in figure 6.5, we think of the director as a continuum field, obtained by coarse-graining over elements which are large enough that the average is well defined, but still small compared to the outer scale on which the director field varies.

For virtually all liquid crystals, the directions $\hat{n}$ and $-\hat{n}$ are equivalent, even though the molecules themselves are not up-down symmetric. The reason for this is that both packing considerations and electric dipole forces favor the fact that there are *on average* as many 'up' as 'down' molecules, as illustrated in figure 6.6: a configuration in which asymmetric molecules predominantly orient in one direction would result in a large strain and/or a significant dipole

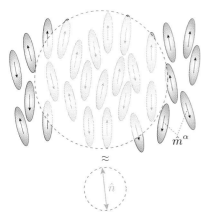

Figure 6.5. Illustration of the director $\hat{n}$ as an average orientation field of a number of molecules within the yellow region. Very much as we discussed for the derivation of the hydrodynamics equations in section 1.3, the director $\hat{n}$ is slowly varying in space and time; the average orientation is obtained by coarse-graining the molecular orientations in space over the spatial region indicated in yellow, and over a number of molecular collisions in time. The orientations of the molecules are indicated by the normal vectors $\hat{m}^\alpha$, which are used in expression (6.2) for the microscopic order parameter in Q_{ij}^m.

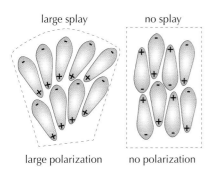

Figure 6.6. Illustration of why both packing considerations and electric dipole forces favor states with on average as many molecules 'up' as 'down', hence a $\hat{n} \leftrightarrow -\hat{n}$ symmetry. The left case is highly strained and has neighboring dipoles pointing in the same direction, while the configuration sketched on the right, with on average as many up as down molecules, has lower strain and electric energy. Nevertheless, molecules of the type caricatured here typically exhibit a large coupling between director deformations and electric fields—see figure 6.20.

interaction energy. Of course, there is no rigorous proof that this has to be so, and as usual nature provides an exception that proves the rule: asymmetric dipolar molecules exist which order in alternating up and down domains—if you are curious, just peek ahead to figure 6.19. If we coarse-grain over a length scale much larger than the width of these domains, we recover up-down symmetry, but then we mask some of the interesting physics of this material.

Almost all nematics are so-called uniaxial nematics, which means that only the direction of $\hat{n}$ is special, and that the two directions perpendicular to it are equivalent. In other words, on average there is symmetry under rotations around the direction $\hat{n}$.

To stress the $\hat{n} \leftrightarrow -\hat{n}$ symmetry, $\hat{n}$ is sometimes written as a two-headed vector $\overleftrightarrow{n}$.

6.2 Landau–de Gennes approach to the isotropic-nematic transition

Before we introduce the coarse-grained Frank elastic energy expression, we briefly discuss the Landau–de Gennes approach of the isotropic-nematic transition, as this is a nice pedagogical illustration of symmetry analysis and of a Landau-type argument.

a. Landau–de Gennes argument

The director $\hat{n}$ gives the preferred direction into which the molecules point, but it does not tell us how strong the order is—are molecules almost perfectly aligned with $\hat{n}$ or only weakly so? To construct a proper microscopic expression for the order parameter which upon averaging over many molecules yields the coarse-grained order parameter, we assign unit vectors $\hat{m}^\alpha$ to the molecules labeled by α, which point along their long axis—see figure 6.5. Because of the fact that on average there are as many molecules pointing up as down (the $\hat{n} \leftrightarrow -\hat{n}$ symmetry), terms linear in $\hat{m}^\alpha$ are not suitable candidates for the microscopic expression we look for: only expressions even in $\hat{m}^\alpha$ will do. Clearly a quadratic expression is the simplest nontrivial option, and this means we have to look at second order tensors built from $\hat{m}^\alpha$. Furthermore, the expression we look for should measure nematic order, so it should be zero in the isotropic phase when there is no preferred direction. Taking the average over the three directions amounts to taking the trace of a tensor whose components label the spatial directions. This then suggests constructing the microscopic expression for the nematic order parameter as the traceless tensor

$$Q_{ij}^{\mathrm{m}} = \frac{V}{N} \sum_\alpha \left(m_i^\alpha m_j^\alpha - \frac{1}{3}\delta_{ij} \right) \delta(\vec{r} - \vec{r}^\alpha), \qquad (6.2)$$

for N particles in a volume V.[8]

The argument sketched in this section was developed by Stinson and Litster, 1970 based on de Gennes's extension of a Landau argument to nematics. The topic is also treated briefly by de Gennes and Prost, 1993 and Chandrasekhar, 1992. Our presentation is inspired by Chaikin and Lubensky, 1995, and for ease of comparison, we follow their choice and notation in (6.5) below.

The fact that the proper order parameter is a tensor is also interesting from a broader perspective. The order parameters of the superfluid phases of helium-3 are in fact also tensors. But they are complex tensors, since the superfluid amplitude has an amplitude and a phase. The similarities between liquid crystals and ^{3}He stimulated various developments in condensed matter physics, such as the development of a topological classification of defects.[9]

A Landau argument is based on coarse-graining this microscopic expression to get the coarse-grained order parameter

$$Q_{ij} = \langle Q_{ij}^{\mathrm{m}} \rangle, \qquad (6.3)$$

where $\langle \cdot \rangle$ denotes the coarse-grained average. From the microscopic expression, Q_{ij} is seen to be traceless, and since $\hat{n}$ is the only vector quantity which characterizes the average orientation, it is clear that Q_{ij} can be expressed as

$$Q_{ij} = S\left(n_i n_j - \frac{1}{3}\delta_{ij}\right). \qquad (6.4)$$

Here S is precisely the parameter which measures the strength of the nematic order. The disordered isotropic phase corresponds to $S = 0$, whereas the perfectly ordered nematic phase with all molecules pointing in the same direction $\hat{n}$ corresponds to $S = 1$.

In a Landau approach, the free energy density $\mathcal{F}$ is expanded in powers of the coarse-grained order parameter field. Every term which is not forbidden by symmetry is assumed to be present. In this case Q_{ij} is a tensor, and since the free energy is a scalar, only invariant powers of Q_{ij} are allowed. This gives

Check the second line yourself by substituting (6.4).

$$\begin{aligned}\mathcal{F} &= \tfrac{1}{2}r\left(\tfrac{3}{2}\mathrm{Tr}\,\underline{Q}^2\right) - w\left(\tfrac{9}{2}\mathrm{Tr}\,\underline{Q}^3\right) + u\left(\tfrac{3}{2}\mathrm{Tr}\,\underline{Q}^2\right)^2 \\ &= \tfrac{1}{2}r\,S^2 - w\,S^3 + u\,S^4.\end{aligned} \qquad (6.5)$$

The numerical factors are just a matter of choice; what matters is that a cubic term in S is allowed by symmetry considerations and hence must be assumed to be present. But this automatically entails that the transition from the isotropic $S = 0$ phase to the nematic $S \neq 0$ phase is first order. To see this, let us first consider the case $w = 0$. In Landau theory, a second order phase transition occurs for $w = 0$ at the temperature T_c where r changes sign (see box 2, in section 2.14). Indeed, very much as we saw in figure 2.19, when r is positive the free energy has a minimum at $S = 0$, while for r negative the minimum is at a nonzero value of S, indicating the ordered phase.

Now consider the case $w \neq 0$. In Landau theory it is assumed that r goes smoothly through zero, i.e., as $r = r'(T - T_c)$ with $r' > 0$. As sketched in figure 6.7, no matter how small w is, and no matter what its sign is, the presence of a nonzero cubic term always gives a two-well structure in the free energy close to the temperature T_c where r vanishes. As illustrated in the figure, this happens in the temperature range $T_c < T < T^*$ (we determine T^* in problem 6.2). Moreover, at some temperature T_{NI} between T_c and T^*, the free energies corresponding to the two minima are the same. Within Landau theory this marks the temperature at which a first order transition from the isotropic $S = 0$ phase to the nematic $S \neq 0$ phase

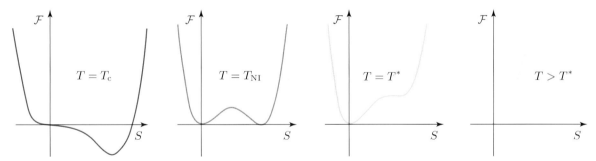

Figure 6.7. Illustration of the behavior of the Landau–de Gennes free energy $\mathcal{F}$ as a function of the order parameter S given by equation (6.5) for w small and positive. As usual in the Landau theory, it is assumed $r = r'(T - T_c)$, with T_c the critical temperature of the second order phase transition for $w = 0$. For $w \neq 0$ the free energy has two minima in the range $T_c < T < T^*$. A first order nematic-isotropic phase transition occurs at the temperature T_{NI}, at which the nematic minimum at $S > 0$ has the same free energy as the isotropic $S = 0$ state. See problem 6.2 for more details.

occurs. The smaller w is, the weaker the first order nature of the transition is. See problem 6.2 for more details.

b. Experimental confirmation

The prediction that the nematic-isotropic phase transition has to be weakly first order because we expect there to be a nonzero w term is confirmed by various types of experiments on a variety of liquid crystals. Let us discuss the behavior of the light-scattering intensity. As we show in section 3.7, $S(q)$ is proportional to the Fourier transform of the density-density correlation function; see (3.52). Now, in Landau theory the fluctuations of the order parameter fall off as $\exp(-r/\xi)/r$ with distance r, where ξ is the correlation length (see problem 6.3). From this it is easy to show that $S(q) \sim \xi^2$ in the small wavenumber limit $q\xi \ll 1$. As $\xi \sim (T - T_c)^{-1/2}$ in Landau mean-field theory, this gives for the scattering intensity $1/S(q) \sim \xi^{-2} \sim (T - T_c)$ near a second-order phase transition.

In figure 6.8 we present data of the inverse of the light-scattering intensity $S(q)$ in the isotropic phase of the liquid crystal 8CB. In agreement with the Landau theory, this inverse varies linearly in the temperature. However, just about a degree above the temperature T_c where $1/S(q)$ extrapolates to zero, deviations from the linear behavior become visible: a jump to the low-temperature nematic phase occurs, in full agreement with the above argument that the transition should be weakly first order. Data on other quantities, such as the susceptibility, the magnetic birefringence, and the critical slowing down of the dynamics, are also in agreement with a weakly first order nature of the transition in 8CB and other liquid crystals. Together such experiments allow us to determine the various parameters in the Landau–de Gennes expansion, and to confirm their mutual consistency.

An interesting aspect of the experimental data is that they imply that at the phase transition the nematic order is significant;

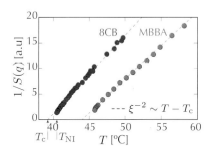

Figure 6.8. Inverse of the light-scattering intensity as a function of temperature for the liquid crystal molecules 8CB and MBBA. The inverse scattering intensity is proportional to ξ^{-2}, which in mean-field theory goes as $T - T_c$, where T_c is the point where r vanishes and where a second order phase transition occurs for $w = 0$. The data show that just about a degree above T_c, where ξ would diverge, the actual transition takes place. The difference between T_c and T_{NI} is only about a degree, so the transition is weakly first order. Adapted from Stinson and Litster, 1970 and Gramsbergen et al., 1986.[10]

The analysis of Onsager also predicts a first order phase transition for long, thin needles; see problem 6.1. The amount of nematic order at the transition is also quite large for needles, $S \approx 0.84$ according to the Onsager analysis.

That the nematic-to-smectic-A transition is weakly first order due to director fluctuations was first shown by Halperin et al., 1974 and Halperin and Lubensky, 1974 based on the analogy of superconductors and liquid crystals demonstrated by de Gennes, 1972a. For a discussion of the experimental evidence of the weak first order nature on the basis of a detailed comparison of the Landau expansion with experimental data, see Anisimov et al., 1990.

In the absence of the symmetry $\hat{n} \to -\hat{n}$, a term $(\hat{n} \cdot \vec{\nabla} \times \hat{n})(\vec{\nabla} \cdot \hat{n})$ would be allowed.[12]

In practical calculations it is often simpler to take $K_1 = K_2 = K_3$. This is called the one-constant approximation. The Frank energy is then $\mathcal{F}_d = 1/2\, K[(\vec{\nabla} \cdot \hat{n})^2 + (\vec{\nabla} \times \hat{n})^2]$. This expression is not as innocuous as it may appear at first sight, because the constraint that $\hat{n}$ is a unit vector leads to an implicit nonlinear coupling of the components of $\hat{n}$. The one-constant approximation is actually equivalent to the O(3) nonlinear sigma model of field theory.

for instance, the order parameter $S(T_{\mathrm{NI}}) \approx 0.4$ in 8CB. Moreover, within a degree or so below T_{NI}, S rises to a value pretty close to 1. Hence for all practical purposes the nematic order can be considered well established throughout the nematic phase.[11]

Note that also computer simulations of spherocylinders confirm the prediction that the transition is weakly first order: in the phase diagram shown in figure 6.2 a very thin biphase region is indicated. This marks a tiny region of phase separation associated with a weakly first order transition.

In closing this section, we note that, as is visible in figure 6.2, the smectic-A-to-nematic transition is weakly first order too. The origin of this is somewhat different, though: as shown in problem 6.4, fluctuations in the director generate an effective free energy which has a cubic term in the smectic order parameter amplitude, similar to the wS^3 term in (6.5).

6.3 Frank energy expression for the nematic director field

Now that we know the nature of the nematic order, and that the order is close to complete throughout the nematic phase, let us study the Frank energy for a spatially varying director field.

6.3.1 The Frank free energy

In the presence of a broken symmetry, one generally arrives at the appropriate Landau coarse-grained free energy expression in terms of the order parameter by expanding the free energy in terms of the lowest order gradients of the order parameter which are allowed by symmetry. For a nematic in the absence of a field, the free energy cannot depend on the direction of the orientation $\hat{n}$ itself, so the lowest order free energy contribution is indeed associated with gradients in $\hat{n}$. The appropriate gradient expansion which is invariant under the transformation $\hat{n} \leftrightarrow -\hat{n}$ is

$$\mathcal{F}_d = \frac{1}{2} K_1 (\vec{\nabla} \cdot \hat{n})^2 + \frac{1}{2} K_2 (\hat{n} \cdot \vec{\nabla} \times \hat{n})^2 + \frac{1}{2} K_3 (\hat{n} \times (\vec{\nabla} \times \hat{n}))^2.$$

(6.6)

The subscript d of $\mathcal{F}_d$ is here to remind us that this is a free energy density for *distortions* of the director field. This form of the free energy for a nematic is also referred to as the Frank free energy, after Sir Frank, who developed it in 1958.

In line with the discussion above, you should think of the terms in (6.6) as the first nontrivial terms in a gradient expansion of the variations that are consistent with the $\hat{n} \leftrightarrow -\hat{n}$ symmetry. It is accurate for spatial variations on scales much larger than the length of the molecules; since this length scale is typically small, around a few nanometers, expression equation (6.6) is essentially accurate in all situations of practical interest for molecular liquid crystals.

We will explore the three terms in (6.6) in more detail below. The Ks are called elastic constants. They are three independent effective constants, which depend on the precise molecular details. They are usually somewhat temperature-dependent and typically differ from each other only by a factor of 2 to 3 or so.[13] Since $\mathcal{F}_d$ has the dimension of an energy density (energy per volume), their dimension is energy per length, i.e., force. One expects them to be of order U/a, where $U \simeq k_B T$ is a typical molecular interaction energy, and $a \simeq 1$ nm a typical molecular length. This estimate yields a value of about 4 piconewtons, which is indeed the order of magnitude of typical elastic constants.

Given the order of magnitude and the fact that the director typically varies on scales much larger than a, the Frank distortion energy is small compared to the total energy in the system. Just as bending distortions are the small-energy distortions of thin rods, director distortions are the low-energy distortions of the nematic phase.

Typical values for the Ks are on the order of piconewtons. Values for MBBA and PAA at various temperatures are given by de Gennes and Prost, 1993. Keep in mind that, close to the other phase transitions, some elastic constants can exhibit nontrivial temperature dependence. For instance, since smectic layering suppresses bending, the bending coefficient K_3 diverges upon approaching the nematic-to-smectic-A transition, while K_1 and K_2 remain finite.[14]

6.3.2 Splay, twist, and bend distortions

The three fundamental deformations of a nematic, namely splay, twist, and bend, are illustrated in figure 6.9. We analyze below cases of pure splay, twist, and bend deformations, defined as the deformations for which only one of the three terms in the Frank elastic energy $\mathcal{F}_d$ is nonzero. On the left we illustrate them with caricatures of molecules; on the right we illustrate the corresponding idealized director field solutions which we will discuss below. Note by the way that in order to realize the deformations between plates, one needs to manipulate the boundary conditions. We will come back to these soon.

a. Splay

Pure splay distortions are such that the director essentially fans out, as sketched between the plates in figure 6.9. For the ideal splay case with cylindrical symmetry sketched at the bottom, the director field is given in Cartesian coordinates by

$$\hat{n} = (\cos\phi, \sin\phi, 0) \qquad (\hat{n} = \hat{r} \text{ in cylindrical coord.}), \qquad (6.7)$$

where $\vec{r}$ is the two-dimensional radial vector of cylindrical coordinates and ϕ the polar angle relative to the x axis. The divergence

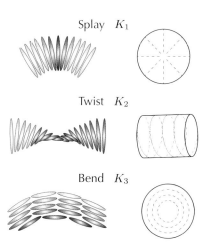

Figure 6.9. Illustration of the three basic distortions of a nematic liquid crystal: splay, twist, and bend. The lines and dashes are parallel to the director, except for the cylinder illustrating the twist distortion, where the spiral indicates the chiral-like twist of the director angle.

The expressions for the splay, twist, and bend energy densities are derived explicitly in problem 6.5.

and rotation are easily worked out and we get for this case

$$\mathcal{F}_{\rm d} = \frac{K_1}{2\,r^2}, \qquad (6.8)$$

confirming that this is a case of pure splay. K_1 is therefore called the splay elastic constant.

That the splay energy density diverges in the origin for this configuration is a result of the fact that there is a defect in the origin.[15] We will discuss defects in more detail in section 6.6.

b. Twist

In pure twist configurations the director rotates in a chiral fashion (as the cholesteric does spontaneously) along an axis perpendicular to $\hat{n}$. For the case sketched in the middle right of figure 6.9, the director is everywhere pointing in the plane perpendicular to the cylinder axis, while the twist angle $\theta(z)$ winds in the vertical direction z (it could be linear in z),[16]

$$\hat{n} = (\cos\theta(z), \sin\theta(z), 0), \qquad (6.9)$$

with $\vec{\nabla} \cdot \hat{n} = 0$ and $\vec{\nabla} \times \hat{n} = -\theta'(z)\hat{n}$; the only nonzero term is

$$\mathcal{F}_{\rm d} = \tfrac{1}{2} K_2 \left(\theta'(z)\right)^2, \qquad (6.10)$$

justifying the *twist elastic constant* name for K_2. Note that this term is finite—there is no defect and hence no singularity.

c. Bend

Pure bend distortions are sketched on the bottom right of figure 6.9. In this case, the director is perpendicular to the radial vector from the center, so we have

$$\hat{n} = (-\sin\phi, \cos\phi, 0) \qquad (\hat{n} = \hat{\phi} \text{ in cylindrical coord.}). \quad (6.11)$$

In this case $\vec{\nabla} \cdot \hat{n} = 0$ while $\vec{\nabla} \times \hat{n} = \hat{z}/r$, so that the only nonzero term in the Frank energy (6.6) is

$$\mathcal{F}_{\rm d} = \frac{K_3}{2\,r^2}, \qquad (6.12)$$

with K_3 the *bend elastic constant*. As in the example of splay, the energy density for this configuration diverges.

6.3.3 Boundary conditions

What are the boundary conditions for the director at a surface where the nematic phase is in contact with another material?

If the surface is solid, the director orientation can often be manipulated by appropriate surface treatment. The so-called strong anchoring regime is based on the possibility of preparing a surface so that the director points perpendicularly to the surface (so-called homeotropic boundary conditions) or is parallel to it (so-called planar or homogeneous boundary conditions)—see figure 6.10. In the latter case it is even possible, sometimes by simply rubbing the surface in one direction, to prescribe the director to point in a particular direction along the surface.

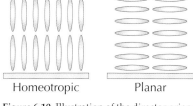

Homeotropic   Planar

Figure 6.10. Illustration of the director orientation at a boundary for the two most common types of boundary conditions.

Controlling the boundary conditions at an interface with a liquid is also important for many applications. An interesting example occurs when liquid crystal droplets are suspended in a liquid—as we shall see later on, suspended droplets offer a special opportunity to study liquid crystal defects. Anchoring transitions at an interface with a liquid can sometimes be triggered with surfactant molecules or by playing with temperature or electrostatic effects.[17]

6.4 Analysis of equilibrium solutions

The Frank energy density is zero when the director is uniform in a sample, so that there are no gradients. In practice, the boundary conditions are such that a uniform configuration is impossible. The problem of determining the equilibrium director configuration then becomes one of minimizing the total free energy

$$F = \int \mathrm{d}^3\vec{r}\, \mathcal{F}_{\mathrm{d}}, \qquad (6.13)$$

subject to the constraint that $\hat{n}$ is a unit vector. In the energy minimum, the variation of the energy is zero for all types of variations, so the equations determining the equilibrium configuration are in principle determined from the equation

$$\delta F = 0 \quad \text{for variations } \delta\vec{n} \text{ such that } |\vec{n} + \delta\vec{n}|^2 = 1. \qquad (6.14)$$

In some simple cases it is possible to do this conveniently by expressing $\hat{n}$ in polar angles, and then analyzing the variations. But this is usually inconvenient for a numerical analysis with complicated boundaries. A way to treat general situations is to allow the variations $\delta\vec{n}$ to be arbitrary, while imposing the constraint $|\vec{n}|^2 = 1$ by a Lagrange multiplier. This approach naturally leads to the introduction of the so-called molecular field, which also helps to formulate the hydrodynamic equations for liquid crystals. We therefore treat this analysis briefly here.

The analysis of the Fréedericksz transition in section 6.5.1 is an example of a calculation done in terms of polar angles.

Imposing the constraint via a Lagrange multiplier means that we allow for arbitrary variations $\delta\vec{n}$ of the augmented variational problem

$$\delta \int d^3\vec{r} \left[\mathcal{F}_d(\vec{n}) + \lambda(\vec{n} \cdot \vec{n} - 1) \right] = 0. \tag{6.15}$$

Variations $\delta\lambda(\vec{r})$ of λ immediately lead to the desired constraint that $|\vec{n}|^2 = 1$ everywhere. For general variations $\delta\vec{n}$, the equation becomes

$$\int d^3\vec{r} \left(\frac{\delta F_d}{\delta\vec{n}} + 2\lambda\vec{n} \right) \cdot \delta\vec{n}(\vec{r}) = 0. \tag{6.16}$$

Here $\delta F / \delta\vec{n}$ is a functional derivative, since F depends on the whole function $\vec{n}(r)$. Along standard lines of analysis, it can be written as

$$\int d^3\vec{r} \frac{\delta F_d}{\delta\vec{n}} \cdot \delta\vec{n}(\vec{r}) = \int d^3\vec{r} \left[\frac{\partial \mathcal{F}_d}{\partial n_j} \delta n_j + \frac{\partial \mathcal{F}_d}{\partial(\nabla_i n_j)} \nabla_i \delta n_j \right]$$

$$= \int d^3\vec{r} \left[\frac{\partial \mathcal{F}_d}{\partial n_j} - \nabla_i \left(\frac{\partial \mathcal{F}_d}{\partial(\nabla_i n_j)} \right) \right] \delta n_j. \tag{6.17}$$

Here the implicit summation convention is used for the indices i and j, and the second step is based on integration by parts. It is customary to call the functional derivative term of the Frank energy the *molecular field* $\vec{h}$,

$$\vec{h} = -\frac{\delta F}{\delta\vec{n}}, \quad \rightarrow \quad h_j = -\frac{\partial \mathcal{F}_d}{\partial n_j} + \nabla_i \left(\frac{\partial \mathcal{F}_d}{\partial(\nabla_i n_j)} \right). \tag{6.18}$$

With this notation, the equilibrium equation (6.16) becomes simply

$$\vec{h} = 2\lambda\,\vec{n}. \tag{6.19}$$

In three dimensions, this vector equation amounts to a set of three equations for the three components. The equation shows that the 'molecular field' $\vec{h}$ is always parallel to $\vec{n}$, and by projecting the equation onto $\vec{n}$ we see that λ is simply given by

$$\lambda = \tfrac{1}{2}\,\vec{n} \cdot \vec{h}, \tag{6.20}$$

so that the equilibrium equations can be summarized in the form

$$\vec{h} - \vec{n}(\vec{n} \cdot \vec{h}) = 0. \tag{6.21}$$

As an illustration, the Fréedericksz transition discussed in the next section with polar coordinates is treated with the molecular field in problem 6.6. The molecular field will also be an important ingredient of the hydrodynamical equations of nematics discussed in section 6.7.

6.5 Switching the director with a field: The Fréedericksz transition and LCDs

In this section we will discuss the Fréedericksz transition, the basic instability that describes the change of an undistorted nematic state between plates to a distorted one as a magnetic or electric field exceeds a threshold value. The possibility of switching the 'state' (director orientation) of a liquid crystal with a magnetic field was discovered at the beginning of the previous century by Fréedericksz,[18] and is nowadays a crucial element of almost any liquid crystal–based device. Although actual liquid crystal displays (LCDs) do not quite use the Fréedericksz instability in its pure form, in essence the same physics underlies their operation.

The Fréedericksz transition also provides an instructive illustration of the concepts discussed above; in addition it is a simple route to measuring the elastic constants.

6.5.1 The Fréedericksz transition

Liquid crystal molecules typically have an anisotropic magnetic susceptibility and electric polarizability. The origin is not the same for the two effects, but both invariably originate from the strongly anisotropic properties of the molecules. The origin of dielectric anisotropy usually traces to the long shape of the molecule, or the existence of easily polarizable end groups (such as the CN end group of 8CB shown in figure 6.1); most molecules are indeed more easily polarized along their long axis. The origin of the anisotropic magnetic coupling is usually associated with the presence of benzene-type rings in liquid crystal molecules, like the 8CB shown in figure 6.1.

Whatever the molecular origin and size of the effect, it should come as no surprise that the average orientation of the molecules couples to the magnetic or electrical field. We leave out the isotropic contribution, and only focus on the anisotropic one. In order to stress the similarity of the two cases, we write both resulting terms, the magnetic and electric contributions to the free energy,[20]

$$\text{Magnetic case:} \quad \mathcal{F}_M = -\tfrac{1}{2}\mu_0 \chi_a (\hat{n} \cdot \vec{H})^2, \qquad (6.22)$$

$$\text{Electrical case:} \quad \mathcal{F}_E = -\tfrac{1}{2}\epsilon_0 \epsilon_a (\hat{n} \cdot \vec{E})^2. \qquad (6.23)$$

Here χ_a and ϵ_a measure the magnetic and dielectric anisotropies of the molecules. We assume both of them to be positive, as is usually the case, so that the lowest energy state corresponds to the director pointing parallel to the field. While the electrical case is of more interest from the point of view of technical applications,

There are many more applications based on switching the liquid crystal state with a field. For instance, switchable glass, which can be switched from opaque to transparent, uses glass with many small nematic droplets inside. In the 'off' state, without a voltage, the director orientations of the droplets are random. This results in strong light scattering, and hence opaque glass. When a small voltage is applied, the director orientations of all droplets align, and the glass becomes transparent.

A magnetic field pointing through the ring induces a small current in it, which in turn induces a magnetic field that counteracts the imposed external field. This leads to a diamagnetic effect.[19] A field in the plane of the ring, on the other hand, has no component through the ring and hence gives no such diamagnetic effect.

Note again the $\hat{n} \leftrightarrow -\hat{n}$ symmetry of these expressions. A term linear in $\hat{n}$ would violate this symmetry.

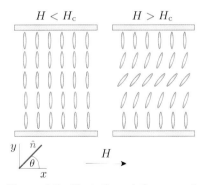

Figure 6.11. Illustration of the magnetic field Fréedericksz transition. For small fields $H < H_c$ the director remains undistorted and perpendicular to the plates. For $H > H_c$ there is a bending distortion, with the director oriented toward the magnetic field away from the plates.

we will from here on analyze the magnetic case. This actually was the one investigated by Fréedericksz. The electrical case has the added complication that in the nonlinear regime, the rotation of the director affects the electrical field. Moreover, in a DC, field ions can accumulate at the boundaries and change the field.[21] This is why in practice one uses AC fields. The magnetic case is not hampered by such effects. As the above expressions show, conceptually the two cases are essentially the same.

We consider the setup sketched in figure 6.11: a liquid crystal is confined between two plates a distance d apart, with strong anchoring homeotropic boundary conditions, so that $\hat{n}$ points normal to the plates at $y = 0$ and $y = d$. The magnetic field H is oriented parallel to the plates, in the x direction. The Fréedericksz transition is the transition from the undistorted state to the one where the director becomes distorted, because the molecules prefer to orient parallel to the field.

Due to the coupling term (6.22), the magnetic field favors the director field to orient in the x direction, but this is resisted by the Frank distortion energy. For not too large fields, the undistorted solution with $\hat{n}$ pointing in the y direction is the stable low-energy solution, but for large enough fields there is a transition to a distorted state where $\hat{n}$ tilts toward the x axis in the center. As the figure illustrates, the distortion is initially mostly of bending type. We will indeed see that the threshold is determined by the bend elastic constant K_3 only for this particular setup. However, at larger fields the deformations are a combination of splay and bend.

We will content ourselves here with determining the threshold for the instability. We do so by writing the director in terms of a polar angle so that it is automatically of unit length. The alternative route via the molecular field as well as a more extensive weakly nonlinear analysis are treated in problems 6.6 and 6.7.

For the setup and coordinates indicated in figure 6.11, we have

$$\hat{n} = (\cos\theta(y), \sin\theta(y), 0), \quad \vec{H} = (H, 0, 0), \quad (6.24)$$

and the fact that the director is anchored in the normal direction at the plates implies the boundary conditions

$$\theta(y = 0) = \theta(y = d) = \pi/2. \quad (6.25)$$

Check for yourself that $\vec{\nabla} \cdot \hat{n} = \theta'(y) \cos\theta$, $\vec{\nabla} \times \hat{n} = \theta'(y) \sin\theta \, \hat{z}$, so that $\hat{n} \cdot (\vec{\nabla} \times \hat{n}) = 0$ and $\hat{n} \times (\vec{\nabla} \times \hat{n}) = \theta'(y) \sin\theta(\sin\theta \, \hat{x} - \cos\theta \, \hat{y})$.

For this case the total free energy per unit area, the sum of the Frank distortion energy and the magnetic term, becomes

$$\frac{F}{\text{area}} = \frac{1}{2} \int_0^d dy \Big[\big(K_1 \cos^2\theta + K_3 \sin^2\theta\big) \left(\frac{d\theta}{dy}\right)^2 \\ - \mu_0 \chi_a H^2 \cos^2\theta \Big]. \quad (6.26)$$

The energy expression (6.26) consists of elastic terms which raise the energy if the director changes orientation and a magnetic term which lowers the energy if the molecules orient with the field. We also see that for small distortions ($\theta \approx \pi/2$), bend effects dominate the elastic terms energy, while for large distortions (θ small) splay deformations dominate.

The threshold of the Fréedericksz transition can be determined along lines reminiscent of our analysis of the onset of the transition to a buckled beam in section 2.8.2. We start by writing $\theta = \pi/2 + \Delta\theta$ and noting that just above threshold, the deviations $\Delta\theta$ in the angle from $\pi/2$ will be small. Thus, for determining the onset it suffices to analyze the energy to quadratic order in $\Delta\theta$. To this order, the energy is

$$\frac{F}{\text{area}} = \frac{1}{2} \int_0^d dy \left[K_3 \left(\frac{d\Delta\theta}{dy} \right)^2 - \mu_0 \chi_a H^2 (\Delta\theta)^2 \right]. \tag{6.27}$$

As in the buckled beam case, we can analyze the onset by allowing for a sum of Fourier modes that satisfy the boundary conditions $\Delta\theta(0) = \Delta\theta(d) = 0$. Based on what we find there, we can anticipate that the mode which becomes unstable first is a simple sine mode,

$$\Delta\theta(y) = A_1 \sin\left(\frac{\pi y}{d} \right). \tag{6.28}$$

Upon substitution of this in the quadratic expression (6.27) for the free energy, we get

$$\frac{F}{\text{area}} = \frac{d}{4} \left[K_3 \left(\frac{\pi}{d} \right)^2 - \mu_0 \chi_a H^2 \right] A_1^2. \tag{6.29}$$

For small fields, the elastic term dominates, so that the energy of this mode is higher than the energy $\mathcal{F} = 0$ of the undistorted state with $\theta = \pi/2$. However, as H increases, beyond a critical value

$$H_c = \frac{\pi}{d} \left(\frac{K_3}{\mu_0 \chi_a} \right)^{\frac{1}{2}}, \tag{6.30}$$

the energy of the distorted mode is negative, so lower than that of the undistorted state. In other words, H_c is the critical field for the magnetic Fréedericksz transition in this particular configuration. To determine the value of the amplitude A_1 corresponding to the minimum of the free energy above threshold, one has to expand the free energy to a higher order. This is done in problem 6.7.

The analysis illustrates that it is possible to measure the Frank elastic constants from the critical field—the above expression is valid for this particular configuration, but by playing with the field orientation and boundary conditions it is possible to design other setups that allow us to probe all three elastic constants.

You can check for yourself that the nth harmonic mode indeed becomes unstable at a field H which is n times larger than the $n = 1$ mode. Compare expression (2.85) of problem 2.3.

Because of the similarity between the magnetic and electrical Fréedericksz transitions, expressed by equations (6.22) and (6.23), the threshold E_c for the electrical field can in principle be obtained by transcribing the interaction term to the electrical case. In practice one has to make sure that the spurious charge effects mentioned earlier are avoided by using an AC field.

6.5.2 Liquid crystal displays

One important application of liquid crystals you are surely familiar with is the LCD, or liquid crystal display. In essence, its operation is based on switching the director orientation with an electric field, in combination with the fact that liquid crystals are optically anisotropic, hence birefringent.

How these effects are used is illustrated in figure 6.12. Each pixel consists of a (uniaxial) nematic liquid crystal between two small planar transparent electrodes, attached to two crossed polarizers. The surface preparation of the electrodes is such that the director is everywhere parallel to the plates, but it makes a 90-degree twist from left to right when no field is applied.[22] As indicated in the top panel, in the field-off rest state the unpolarized light coming from the left gets polarized, and it enters the nematic with the polarization parallel to the director. Because of the birefringence, the polarization of the light rotates 90 degrees with the twisted director as it propagates through the sample, to arrive at the right plate with its polarization parallel to the right polarizer. Hence it passes the analyzer on the right and exits: the pixel emits light.

As sketched in the lower panel, in the field-on state with a voltage applied across the electrodes, the electric field makes the director point sideward along the normal to the planes. In this configuration the polarization of the light remains unchanged as it propagates from left to right through the sample. Hence the light gets blocked by the right polarizer, so the pixel does not emit light—it remains dark. Effectively, when we view it from the right we have a pixel which switches from light to dark by our putting a voltage across it. Combination with suitable color filters gives the color pixels necessary for a color display.

Clearly, both LCDs and the Fréedericksz transition are manifestations of the ability to manipulate the director orientation with an electric field. The main difference is that the Fréedericksz transition starts from a undistorted state where the director is everywhere parallel. As a result, there is a clear threshold for the transition. The base state in the LCD, on the other hand, is already twisted. As a result, there is no sharp threshold, and switching is faster than when there is a pure Fréedericksz transition.

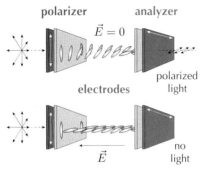

Figure 6.12. An LCD pixel consists of a nematic liquid crystal between two electrodes and crossed polarizers (polarizers with the polarization direction, indicated by a white arrow, oriented in perpendicular directions). In the field-off state the polarization of light coming from the left follows the 90-degree twist rotation of the director, so that it passes through the right polarizer unchanged. In the field-on state the light polarization remains unchanged, so that the light is blocked by the right polarizer.

Actually, the switching to the dark field-on state is often fast, while the relaxation back to the twisted field-off state is typically quite a bit slower, but still fast enough, typically 10 ms for normal displays.[23]

6.6 Topological defects in the director orientation

We noted before that the pure splay and pure bend configurations illustrated in figure 6.9 are divergent in the origin, and we remarked that this divergence is associated with there being a defect in the center. It is important to realize that these defects are not simply a result of the fact that we considered idealized pure splay and bend configurations, and that they cannot simply be removed by deforming the field smoothly. They are examples of topological defects, classified by topological numbers. In this section we discuss these defects and their classification in more detail.

Through the existence of defects in the orientational order parameter, liquid crystal physics is intimately connected with many other branches of physics, from complex ordered phases in condensed matter physics to cosmology.

An elementary but somewhat dated introduction to the richness of defects in liquid crystals is given by Brinkman and Cladis, 1982, while G. P. Alexander et al., 2012 give a full review of disclinations and point defects in nematic liquid crystals. Harth and Stannarius, 2020 review modern observations of defects in quasi-two-dimensional liquid crystal films. A nice introduction to defects in a more general context is given by Nelson, 1996. For the general topological classification of the defects in liquid crystals and many other ordered phases, see Mermin, 1979.

In fact we already encountered the important role defects play in two-dimensional melting in section 2.11.

6.6.1 Defects in the director field

When we considered the cases of pure splay and bend sketched in figure 6.9, we noticed the divergence of the elastic energy terms in the origin. This resulted from the singular behavior in the core of these configurations. Both of these cases, shown again in figure 6.13, are examples of *topological defects*, which naturally occur in liquid crystals. In the context of liquid crystals, the line defects we consider (they extend as lines out of the page) are called disclinations. The word *disclination* means a discontinuity in the orientation.

See also the introduction of section 2.12 on topological mechanics.

The word 'topological' implies that the defect is a robust quantity which can be characterized by a precise topological number which is invariant under smooth transformations of the director field away from the defect—in other words, defects cannot be 'undone' by continuous transformations or deformations of the director field. Defects can annihilate, or move to a boundary and disappear, but they cannot appear or disappear out of nowhere in the bulk. They can form or annihilate in pairs though. As a result, their existence and interactions often govern most of the long time and long-range dynamics, and it is important to understand their interactions.

a. Topological number of disclination lines

What is the proper topological number of the liquid crystal defects? Let us first consider line defects, which extend as a line in the direction perpendicular to the plane we drew in the examples considered so far. When viewed within the plane perpendicular to this line, the director orientation of line defects can be characterized

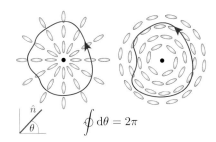

Figure 6.13. The splay and bend defects of figure 6.9 enlarged. If we follow the red closed loop around the core in the direction of increasing polar angle ϕ, the director angle makes a turn of 2π. These are therefore $s = 1$ disclinations. Compare the discussion in section 2.11 of the Burgers vector, which is the topological number characterizing a dislocation in a crystal.

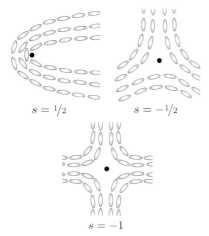

$s = \frac{1}{2}$ $s = -\frac{1}{2}$

$s = -1$

Figure 6.14. Illustration of the two half-integer defects and the one with $s = -1$.

We shall see in section 6.6.3 that defects of opposite charge attract each other, whereas those with charges of the same sign repel.

Conceptually, the situation is very much as in the core of a vortex line in a type II superconductor: if we analyze a vortex in the Ginzburg-Landau theory for a superconductor, we explicitly find that the superconductor order parameter vanishes in the core. One may imagine that when we start from the full tensor formulation of nematic ordering discussed in section 6.2, we find that the nematic order parameter S vanishes in the core.

by one angle, for instance, the angle θ the director makes with one of the axes, as indicated in figure 6.13. Whenever we follow the director orientation in the anticlockwise direction of increasing polar angle ϕ along any closed path encircling the origin, we see that the director angle makes a turn of 2π. This holds for these two cases, but there are other possibilities: because of the $\hat{n} \leftrightarrow -\hat{n}$ symmetry of a nematic, continuity of the director field requires that when we follow a closed path around a defect, the director angles at the beginning and end of the path be the same, up to a multiple of π, so that

$$\oint_{\mathcal{C}} \mathrm{d}\theta = 2\pi s, \qquad s = \pm 1/2, \pm 1, \dots \qquad (6.31)$$

for any path along a closed contour $\mathcal{C}$ encircling the disclination. We can therefore classify a defect by its topological winding number s, the number of times in units of 2π the director rotates.

The two half-integer disclinations, which are illustrated in figure 6.14 together with the one with $s = -1$, are special to liquid crystals because they result from the $\hat{n} \leftrightarrow -\hat{n}$ symmetry. We shall see that the $s = 1/2$ disclinations play an important role in active media with nematic order, as their asymmetry gives them a natural propagation direction (see chapter 9). The half-integer defects are very uncommon for order parameters in other condensed matter phases. Indeed, the symmetry of quantum mechanics for changes in the phase by 2π naturally gives rise to integer defects in ordered quantum phases.

When topological defects merge, the total topological charge is conserved, so defects of opposite topological charge can merge and annihilate. Together with the fact that the core energy of a defect goes as s^2, this is the reason that defects with charge larger than 1 normally don't occur, as they have a higher energy than the total energy of the two defects into which they can split. Depending on boundary conditions, defects can also disappear at boundaries.

b. Behavior near the core

At the core, the disclinations have a singularity. This has two consequences. First of all, we already found in our discussion in section 6.3.2 of the pure $s = 1$ splay and bend distortion that the distortion energy of disclinations diverges near the core. Of course, a divergence cannot really happen in practice—the resolution lies in the following observation. In our analysis, we have *assumed* the nematic ordering as given and fixed. But when we approach the core and the strain builds up, at some point this starts to suppress the nematic ordering: there is a small isotropic core of the defect when the energy density of the highly distorted nematic phase near the core becomes larger than that of the isotropic phase.

A second consequence of the distortion energy building up upon approaching the core is that the angular variation of the director field becomes as smooth as possible, given that the gradients are building up and that there is a defect in the core. This is because any deviation from a smooth angular dependence is penalized more and more by the $1/r^2$ divergence near the core. In other words, if we use ϕ for the polar angle relative to the horizontal x-axis, the angle θ the director makes with this axis approaches close to the defect the ideal expression,

$$\theta = s\,\phi + \theta_0. \tag{6.32}$$

This behavior is clearly visible in the experimental pictures shown below.

c. Point defects: Hedgehogs and boojums

Besides line defects, also point defects can occur in liquid crystals. Figure 6.15 illustrates how the director of the so-called hedgehog defect is pointing out radially—the name of this defect is indeed inspired by the similarity to the needles pointing outward from a hedgehog, which curls up to defend itself. Hedgehog point defects are favored in the center of small droplets of liquid crystals if the director prefers to point perpendicular to the interface of the droplet, as this boundary condition perfectly matches with that of a hedgehog in the center. Boojums, on the other hand, are surface defects; two of them are favored in droplets when the director is preferentially oriented parallel to the surface, as the figure illustrates.

d. Walls

In a large system and in the absence of external fields, the distortion energy tends to minimize the gradients, causing the director to vary smoothly away from defects. However, boundaries in thin samples or external fields can have the effect of favoring domains with a particular director orientation. As a result, the deviations from the preferred orientation are then mostly confined to thin domain walls.

6.6.2 Visualization of defects in thin samples between crossed polarizers

In thin samples, defects can be made visible very easily between crossed polarizers. As we discussed already in the context of LCDs, the nematic phase is birefringent; we assume it is uniaxial, so that directions perpendicular to the director are equivalent. This implies the following when light propagates through a thin sample sandwiched between two crossed polarizers, whose polarization

Check that $\theta_0 = 0$ for the splay defect and $\theta_0 = \pi/2$ for the bend defect of figure 6.13.

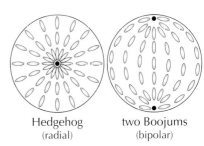

Hedgehog (radial)   two Boojums (bipolar)

Figure 6.15. Point defects are natural defects in small nematic droplets. The hedgehog defect is favored in the center of a droplet if the director likes to point perpendicular to the droplet's surface. When the director preferentially orients parallel to the surface, a low-energy configuration is with two boojums on opposite sides of the droplet.[24]

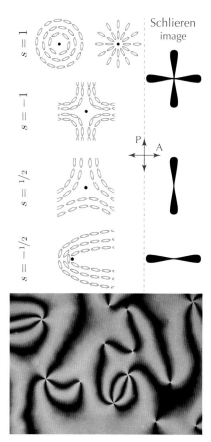

Schlieren image

Figure 6.16. Illustration of how the various defects in a nematic appear between crossed polarizers in a Schlieren image. Integer defects are visible as crosses, half-integer defects as black streaks which thin near the core of the defect. The bottom image shows an example of a nematic liquid crystal between crossed polarizers; both integer and half-integer defects are visible. Image courtesy of Oleg Lavrentovich.

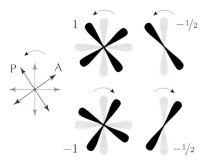

Figure 6.17. When the polarizer and analyzer are rotated, the Schlieren image of positive defects rotates in the same direction, that of a negative defect in the opposite direction.

directions are perpendicular to each other. As we illustrate in figure 6.16, in regions where the director is oriented in the same direction as the polarization direction of the incoming light or perpendicular to it, the polarization direction of the light does not rotate as it propagates through the sample. Hence it gets blocked completely by the second polarizer. In regions where the director is not lining up with any of the two polarizers, however, the polarization gets rotated by the nematic, so that the light (partially) crosses the second polarizer. Thus in the image of the light coming through the sample, black streaks mark regions where the director is aligned with one of the polarizers. This technique is often called Schlieren imaging. The figure illustrates how each of the half-integer and integer defects translates into a characteristic dark streak pattern between crossed polarizers.

Figure 6.17 shows a simple example of such a Schlieren image of a thin sample of a nematic liquid crystal between two crossed polarizers. The existence of defects is immediately clear from the black cross-type patterns. As each black streak of a cross marks a region where the nematic director is parallel to one of the polarizers, the director orientations between neighboring streaks differ by $\pi/2$. Hence these cross patterns with four streaks indicate defects with $s = 1$ or $s = -1$, where the director makes a total turn of 2π or -2π, in accord with what is indicated in figure 6.16. Note also how it is clearly visible that close to the cores, the director patterns approach the ideal form given by (6.32), in agreement with the arguments given above in section 6.6.1.b.

One cannot determine the sign of a defect in isolation from a still picture. But if in an experiment one rotates the orientation of the polarizers while keeping them crossed at 90-degree angles, this gives information on the sign of a defect. For, during the rotation of the polarizers, the streak pattern of a positive defect rotates in the same direction as that of the polarizers, while the streak patterns of the negative defects rotate in the opposite direction as illustrated in figure 6.17. This, together with the constraints from the fact that the director is continuous outside the defect cores, allows experienced researchers to quickly decode even a complicated Schlieren pattern.

A pair of surface boojums—they can only occur in pairs—is captured in the Schlieren pattern from a thin nematic sample of figure 6.18. The colors result from the added effect of optical interference due to the small thickness of the sample. That the boojums have opposite topological charges shows up in the difference between the director profiles in their neighborhood. Note that four streaks emanate from each boojum; this is consistent with the fact that if one encircles a boojum along the surface, the director rotates by 2π. At the upper left corner, the stripe pattern that is visible is a part of a domain wall.

6.6.3 Interaction of defects in two dimensions

Since disclinations are topological, their interaction and dynamics usually dominate the long time evolution of a director pattern. It is illuminating to illustrate this by specializing to thin two-dimensional films and to make contact with the statistical mechanics of other topological phases discussed in section 2.11 on two-dimensional melting.

In two dimensions, we can specify the director with the angle θ as we did in the discussion of the Fréedericksz transition. Indeed, in the one-constant approximation, so that the splay and bend elastic constants K_1 and K_3 are equal, $K_1 = K_3 = K$, we can write the total distortion free energy in the absence of fields as

$$F = \frac{K}{2} \int \mathrm{d}^2\vec{r}\,|\vec{\nabla}\theta|^2, \tag{6.33}$$

in analogy with equation (6.26) (the integral is taken over two-dimensional space and $\vec{r}$ is a two-dimensional vector here and below). This equation implies that the angle θ almost everywhere obeys the equation

$$\nabla^2\theta = 0, \tag{6.34}$$

except in the cores of a collection of N defects. These defects should enter as source terms on the right-hand side of this Laplace equation—let's investigate how.

For a single defect labeled α with topological charge s_α we have according to (6.31)

$$\oint_{\mathcal{C}} \mathrm{d}\vec{\ell}\cdot\vec{\nabla}\theta = 2\pi\,s_\alpha \tag{6.35}$$

for any path encircling the defect, with $\mathrm{d}\vec{\ell}$ the line element along the contour $\mathcal{C}$. In the presence of N vortices at positions $\vec{r}_\alpha$ we can write the 'topological charge density' n_d as

$$n_\mathrm{d}(\vec{r}) = 2\pi \sum_{\alpha=1}^{N} s_\alpha\,\delta(\vec{r}-\vec{r}_\alpha), \tag{6.36}$$

and then we have for any contour encircling these vortices the following equality

$$\oint_{\mathcal{C}} \mathrm{d}\vec{\ell}\cdot\vec{\nabla}\theta = \int \mathrm{d}^2\vec{r}\,n_\mathrm{d}(\vec{r}). \tag{6.37}$$

The solution of these equations for the angle θ in the presence of the defects can be obtained along standard lines in terms of the Green's function of the Poisson equation,

$$\nabla^2 G(\vec{r},\vec{r}') = \delta(\vec{r}-\vec{r}'), \tag{6.38}$$

See Nelson, 1980 and Nelson, 1996 for overviews of two-dimensional melting and applications in other fields such as superfluids and membranes.

Figure 6.18. Schlieren image of a pair of surface boojums. At the upper left corner, part of a domain wall is visible. Image courtesy of Oleg Lavrentovich.

whose solution in two dimensions is

$$G(\vec{r}, \vec{r}') = \frac{1}{2\pi} \ln \left(\frac{|\vec{r} - \vec{r}'|}{a} \right) + C, \tag{6.39}$$

where a is a short-range cutoff of order the isotropic core of a vortex and C is a constant. The detailed steps of the analysis are followed in problem 6.8. The result is that the director angle is given in terms of the defect density n_d as

$$\vec{\nabla}\theta(\vec{r}) = 2\pi(\hat{z} \times \vec{\nabla}) \int \mathrm{d}^2\vec{r}' \, n_\mathrm{d}(\vec{r}') G(\vec{r}, \vec{r}') + \vec{\nabla}\phi, \tag{6.40}$$

where ϕ is a smooth single-valued phase not affected by the defects (note that since (6.34) is linear, one can add any smooth solution to θ). Moreover, the free energy can be written as

$$F = F_{\mathrm{defect}} + \tfrac{1}{2} K \int \mathrm{d}^2\vec{r} \, |\vec{\nabla}\phi|^2, \tag{6.41}$$

where the defect free energy is given by

$$F_{\mathrm{defect}} = -\pi K \sum_{\alpha \neq \beta} s_\alpha s_\beta \ln \left(\frac{|\vec{r}_\alpha - \vec{r}'_\beta|}{a} \right) + E_\mathrm{c} \sum_\alpha s_\alpha^2. \tag{6.42}$$

The core energy E_c of a defect in this expression is related to the constant C in equation (6.39).

The above expression for the defect free energy is the basis for much of the work on two-dimensional systems, like the two-dimensional melting we discussed in section 2.11. It clearly shows that defects of the same sign have a logarithmic repulsion, while those of opposite signs attract. Moreover, very much as with charges, the interaction strength increases with the 'topological charge' of the defect. Finally, both the interaction and the core energy term increase with the charge s_α squared, confirming that defects with s larger than 1 can lower their energy by splitting into two defects of smaller charge, and an $s = 1$ defect can lower its energy by splitting into two $s = 1/2$ defects.

6.7 Nematohydrodynamics based on non-equilibrium thermodynamics

It is worth sketching briefly the essentials of the derivation of the nematohydrodynamic equations describing the hydrodynamic flow of the nematic phase. For, in essence, this is conceptually a straightforward though nontrivial application of the formalism of non-equilibrium thermodynamics discussed in section 1.5. Indeed, the separation of scales arguments which we gave for the rapid local equilibrium of the constituent molecules in a regular fluid hold for molecular liquid crystals as well. Moreover, typical flows of molecular liquid crystals correspond to very small Péclet numbers, as we discussed in section 6.1.2. Therefore the Onsager formalism for the linear relations between thermodynamic 'forces' and dissipative 'fluxes' applies to liquid crystals.

For nematohydrodynamics we need a dynamical equation for the director field $\hat{n}$ in addition to the usual hydrodynamic equation for the velocity field. Since the director is a unit vector, this is an equation governing the rotation of the director field in response to a force that acts like a torque. The vector $\hat{n}$ also rotates simply with the local rotational component $\vec{\omega} = 1/2 \vec{\nabla} \times \vec{v}$ of the velocity field. This is a non-dissipative effect. For dissipative processes the relevant time derivative is the comoving derivative with this (solid body–like) rotation subtracted,

$$\frac{d\vec{n}}{dt} - \vec{\omega} \times \vec{n} = \frac{\partial \vec{n}}{\partial t} + \vec{v} \cdot \vec{\nabla} \vec{n} - \vec{\omega} \times \vec{n} \equiv \vec{N}. \tag{6.43}$$

We write the director here as $\vec{n}$ since, as in our discussion of the molecular field in section 6.4, we allow it to vary in the analysis. If one traces the analysis in more detail, only the components of the molecular field $\vec{h}$ perpendicular to $\vec{n}$ give rise to a physical torque, while the component parallel to $\vec{n}$ has no physical effect.[25]

In the framework of the theory of irreversible processes with forces and fluxes, the stress tensor $\underline{\sigma}^d$ is the force conjugate to the shear rate (the flux), and the molecular field $\vec{h}$ defined in equation (6.18) is the force conjugate to the flux $\vec{N}$. Applied to the case of nematics, the linear relations (1.19) between forces and fluxes then become

$$\sigma_{ij}^d = L_{ijkl} A_{kl} + M_{ijk} N_k,$$
$$h_k = M_{ijk} A_{ij} + P_{kl} N_l. \tag{6.44}$$

Here the $A_{ij} = (\partial v_i / \partial x_j + \partial v_j / \partial x_i)/2$ are the symmetrized velocity gradient components defined in (1.3) and we have used the implicit summation convention. Note also that we have already taken advantage of the Onsager reciprocity relation (1.21) by writing the same cross-coefficient M_{ijk} in both equations.

While the stress and velocity are even functions of $\vec{n}$, the vectors $\vec{h}$ and $\vec{N}$ are odd in $\vec{n}$. The most general expression which is invariant under change of $\vec{n}$ into $-\vec{n}$ and which respects the Onsager reciprocity relations is then

$$\sigma_{ij}^d = \rho_1 \delta_{ij} A_{ll} + \rho_2 n_i n_j A_{ll} + \rho_3 \delta_{ij} n_k n_l A_{kl} + \alpha_4 A_{ij}$$
$$+ \alpha_1 n_i n_j n_k n_l A_{kl} + \tfrac{1}{2}(\alpha_5 + \alpha_6)(n_i A_{kj} + n_j A_{ki}) n_k$$
$$+ \tfrac{1}{2} \gamma_2 (n_i N_j + n_j N_i), \tag{6.45}$$
$$h_k = \gamma_2 n_i A_{ik} + \gamma_1 N_k. \tag{6.46}$$

Here the eight (!) coefficients ρ, α, and γ are phenomenological coefficients with the dimension of the well-known dynamic viscosity η of a fluid.

Imposing that the flow is incompressible simplifies life a little bit: it eliminates the terms with ρ_1 and ρ_2, while the term with ρ_3 is diagonal and can be absorbed in the pressure. But this still leaves five terms.

Even though the derivation and structure of these somewhat overwhelming equations and the resulting dynamical equations for the flow velocity and director field may be straightforward from a conceptual point of view, it should be clear that their practical use is limited to the most simple cases. On top of that, it is far from trivial to measure all the coefficients independently.

Luckily, the practical consequences of these limitations are moderate: there appear to be very few if any new effects that are inherently associated with this intricate structure. This is not very surprising. After all, the first equation essentially expresses that in some directions measured relative to the director, the effective viscosity is somewhat different than in others, while the second one states that the component of the dissipative shear gradient parallel to the director gives rise to a torque on the director. This flow-induced torque is actually the main new effect. Moreover, for applications it is often more important to control the director and reorient or switch it sufficiently quickly than to understand how exactly the dynamics precisely happens—in applied sciences, the best way forward is often to completely control the system and avoid complications, rather than to study and solve these!

6.8 Playing with the molecular shape

During the emergence of liquid crystals as a field, the focus was mostly on rod-shaped molecules. It gradually became possible to make molecules with other shapes that prefer different types of ordering, like the discotics illustrated in figure 6.3. With the ever-increasing possibilities of synthesizing molecules with specific shapes, it has become feasible in more recent years to play with these shapes to an extent that new phases and applications have become possible. We give an idea of this with some examples.

a. Splay-induced stripe domains

Figure 6.19 gives an artist's impression of a new modulated phase that has been observed for the polar wedge-shaped molecule RM734. This molecule defies the classical argument illustrated in figure 6.6 about why both polar interactions and a wedge shape of a molecule tend to favor the $\hat{n} \leftrightarrow -\hat{n}$ symmetry of a nematic. Instead, the molecule RM734 has been found to form alternating domains. Within a domain the molecules show nematic-like order with the up-down symmetry broken, so each domain is slightly

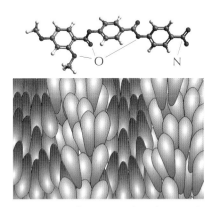

Figure 6.19. Illustration of a phase made up of polar wedge-shaped molecules, which has been found to form slightly splayed stripes of domains pointing alternatingly up and down. The experimental molecule is the so-called RM734 molecule whose structure is shown at the top. Adapted from Miller, 2020.[26]

splayed and has a polarization—the formation of a striped phase prevents an overall polarization and splay.

b. Coupling of splay and bend to polarization

That for asymmetric polar molecules splay and bend deformations are strongly coupled to polarization has actually been long known from studies of flexoelectricity.[27] This phenomenon is illustrated in figure 6.20. As illustrated in the upper panel, when bent molecules have a dipole oriented along their short axis, then the nematic phase has the $\hat{n} \leftrightarrow -\hat{n}$ symmetry with, on average, as many molecules oriented in one direction as in the opposite one in the absence of distortions. But, as illustrated, a bending deformation tends to favor a polarization of the molecules, hence the name flexoelectricity. Likewise, wedge-shaped molecules show a splay-induced flexoelectric effect. The discovery of molecules which naturally form alternating domains is a new twist to this story.

c. Bent-shaped molecules with flexible backbone

Bent-shaped molecules—think of them as banana-shaped molecules—have been found to offer a wealth of new opportunities, including the discovery of new liquid crystalline phases. It is found that the flexibility of the central part of banana-shaped molecules plays an important role in determining what types of mesophases such molecules tend to form. If their backbone is rather rigid, they prefer to form smectic and sometimes columnar phases, while the formation of nematic phases is suppressed. More flexible bent-shaped molecules, on the other hand, are more prone to forming nematic phases. One of the newly discovered phases of bent molecules with a flexible backbone is a twist-bend nematic phase, a modulated phase in which the director exhibits a certain amount of helical (chiral) order. Also, this phase has a splay-type analogue.

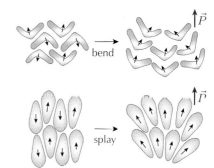

Figure 6.20. Illustration of the flexoelectric effects in liquid crystals consisting of asymmetric molecules with a dipole: a bend distortion of banana-shaped molecules or a splay distortion of wedge-shaped molecules leads to an electric polarization. Adapted from Miller, 2020.

For a review of the rich physics of liquid crystals formed by bent-shaped molecules, see Jákli et al., 2018.

6.9 Opportunities and challenges at interfaces with other fields

6.9.1 Biological liquid crystals

In section 6.1.2 we reflected on the fact that conceptually there is no sharp distinction between typical molecular liquid crystals and rod-shaped colloidal particles, apart from their size. They both tend to form a nematic phase in equilibrium. The most crucial difference is that the time it takes for an equilibrium phase to establish increases rapidly with the length of the rod. Correspondingly, the Péclet number (6.1) characterizing shear increases with the rod

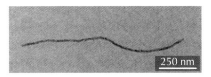

Figure 6.21. Transmission electron microscope image of the mutant fd virus, which exhibits both nematic and smectic phases. The mutation causes the persistence length of the rod to be 3.5 times larger than the persistence length of the wild-type virus (about 7.7 μm), while the aspect ratio is similar (the virus is on average 917 nm long and 6.8 nm wide). The increased stiffness causes the isotropic-nematic transition to be located exactly at the value predicted by Onsager's theory. Image courtesy of Eric Grelet.[28]

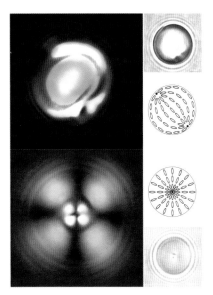

Figure 6.22. Brightfield (right) and crossed-polarizers images (left) of liquid crystal droplets with a diameter of about 8 μm in contact with water without (upper row) and with (lower row) endotoxin from *E. coli* (endotoxin is a toxin inside a bacterium which is released when the bacterium disintegrates). The nematic state switches from two boojums at the surface to a hedgehog in the center for very small changes in the endotoxin concentration in the water (on the order of picograms per milliliter), illustrating the ability to use liquid crystal droplets for detection of small concentrations (Lin et al., 2011). Images courtesy of Xin Wang and Nicholas Abbott.

length cubed. This means that sheared colloidal liquid crystals are in the strongly non-equilibrium large shear rate regime for any reasonable imposed shear, where effects like shear thinning and jamming are important. Flowing molecular liquid crystals, on the other hand, are at very small Péclet numbers. Then the material remains everywhere locally in equilibrium so that nematohydrodynamics based on the non-equilibrium thermodynamics formalism for irreversible processes applies.

Biological liquid crystals,[29] mesophases formed by biological molecules, often span the intermediate size range between synthetic molecular and colloidal liquid crystals. They have the added complication that the molecules are not completely rigid if their persistence length is less than or of order their length. Flexibility typically suppresses the tendency to form nematic order. Moreover, as we have seen, biomolecules are commonly polyelectrolytes whose stiffness and effective diameter are affected by charges, and hence by the ionic strength of the solution. An example is the fd virus shown in figure 6.21, whose persistence length is about twice its length. At sufficiently high concentrations a solution with this virus exhibits an isotropic-to-cholesteric transition, although it takes several days for the cholesteric order to develop; apparently the energy difference with the nematic phase is small. Where the transition occurs also depends sensitively on the ionic strength of the solvent.

Additional complications with biomolecules can arise from structural transitions of the biomolecule itself as a function of temperature or ionic concentration, so biomolecular liquid crystal is a subfield with its own set of challenges. Clearly, many of these are strongly affected by those posed by biopolymers.

6.9.2 Liquid crystals in droplets and other confined geometries

We already mentioned in section 6.6.1 that small liquid crystal droplets provide a nice platform to create point defects in liquid crystals. As figure 6.22 illustrates, the sensitivity of the director pattern to the surface properties provides even an opportunity to see a transition from a hedgehog defect in the center of the droplet to a pair of boojum defects on opposite sides of the surface.

Since minute changes in the deformation state of a liquid crystal in a microdroplet are easily visible with a microscope, small droplets are also used as microscopic biosensors and for drug release.[30] Likewise, three-dimensional line defects have been stabilized at the center of microtubes, and these can be used for imaging techniques.[31]

6.9.3 Colloidal liquid crystals and beyond

Related effects occur when we study colloidal particles immersed in a nematic liquid crystal. First of all, the director field causes effective anisotropic and long-range interactions. Secondly the boundary conditions at the surface naturally give rise to liquid crystal defects at or near the surface. This opens up a plethora of new possibilities to explore.[32]

Consider for instance the case of the Janus-type particle shown in figure 4.18 which is used as an active swimmer. As illustrated in figure 6.23, the swimming direction becomes coupled to the director field. So you can steer the motion of such a swimmer: we now have a building block for active media with directionally dependent dynamical interactions. Just imagine what becomes possible if you in addition play with the shapes of the colloids or if you explore the opportunities by combining liquid crystal mediated interactions with active matter behavior.

Variations on this theme are also virtually unlimited. If one replaces the Janus particle with a droplet with bacteria which, like the colloid, is immersed in a liquid crystal, one gets, along similar lines, an active droplet which propels itself in the direction of the hedgehog just in front of it.[34] In fact, as we shall discuss in chapter 9, active media of elongated particles inherit various characteristic liquid crystal properties.

A very accessible and complete overview of the field of liquid crystal colloids is given by Muševič, 2017.

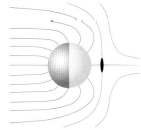

Figure 6.23. Illustration of a Janus colloidal particle in a nematic liquid crystal with normal boundary condition at the surface (so that the director is normal to the surface). This induces a hedgehog-type defect just in front of the particle. The configuration allows us to steer the propagation direction of the Janus particle with the nematic field orientation.[33]

6.9.4 Mesophases of lipid molecules relevant to pharmaceutics, cosmetics, and food

At the interface with drug research there is an active research field focused on understanding the structure and behavior of liquid crystal mesophases with the aim of optimizing drug delivery. An ideal drug delivery platform should be based on harmless molecules which are biodegradable and biocompatible, and which easily incorporate the relevant pharmaceutical ingredients without losing their effect or altering their activity. Clearly, similar requirements also play an important role in the food and cosmetics industries. Especially in drug research, regulating the delivery and speed of uptake of drugs is an additional important issue.

Lipid molecules which are amphiphilic play an important role in this. Lipids are molecules which play an important role in living organisms: they are fatty molecules which generally do not dissolve well in water. As a result, beyond some critical micelle concentration, lipid molecules self-organize into micelles. We shall focus here on lipids which have some amphiphilic properties, as they have on one side a polar group which likes to be in touch with

A very accessible physics-oriented introduction to the mesophases of drug delivery systems is given by Aleandri and Mezzenga, 2020.

Amphiphile is the generic name for a molecule with a hydrophilic head which likes water and a hydrophobic tail which prefers to avoid water.

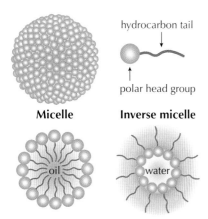

Micelle   **Inverse micelle**

hydrocarbon tail

polar head group

oil

water

Figure 6.24. Illustration of micelles and inverse micelles, formed by amphiphilic molecules with a polar hydrophilic head and a hydrophobic tail. The general term for such molecules is *amphiphilic molecules*, but also the name *lipid molecules* is used.

Ramanathan et al., 2013 discuss the relation between the shape of an amphiphile and the preferred aggregate structure.

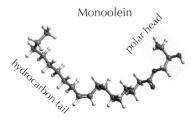

Monoolein

polar head

hydrocarbon tail

Figure 6.25. The lipid monoolein which is used everywhere in the food, pharmaceutical, and cosmetics industries.

water. As illustrated in figure 6.24, micelles formed of such lipids are small spheres consisting of the lipid molecules with the polar head groups on the outside in contact with the water, and the hydrocarbon tails on the inside. Also, steric considerations affect the preferred curvature of the interface, and this translates into a preference for certain structures over others. For instance, if the polar head group is wider than the tail, the lipids are conducive to micelle formation, while molecules with wider tails than heads prefer to form inverse micelles in hydrocarbon solvents. Amphiphiles that do not like a strongly curved interface tend to form structures with mean curvature zero. This includes three-dimensional structures where everywhere one radius of curvature is negative, and one positive, like those shown in figure 6.26, so that the mean curvature is close to zero.

In the presence of water and oil, amphiphiles strongly favor micro phase separation, with the minority phase in the core of the micelles. The amphiphile effectively reduces the surface tension of the water-oil interface to almost zero. An example of an amphiphilic lipid which is important in industry is monoolein, shown in figure 6.25. It has been called a 'magic lipid,' as over 2,200 patents in the food, pharmaceutical, and cosmetics industries are associated with it.[35]

Since within the layers the lipid molecules behave as two-dimensional disordered liquids, they are liquid crystal like mesophases. And very much as we saw for diblock copolymers in section 5.7.2, they can form ordered structures depending on their concentration and the composition of the oil and water mixture. Some of the phases which monoolein and related molecules form are illustrated in figure 6.26. Several of the block copolymer mesophases of figure 5.24 are observed with lipids like monoolein as well: the cubic micellar phases, the lamellar phase, the hexagonal phases, and the bicontinuous three-dimensional phases such as the gyroid and double diamond phases.

Clearly, the effective diffusion rate of a drug dissolved in the water phase varies strongly with the connectedness and effective dimensionality of the water-rich regions (besides, it depends on the size of the diffusing molecule relative to the size of the structure). The diffusion rate is extremely small for a molecule encapsulated in the inverse micelles, and the rate is maximal in the bicontinuous three-dimensional phases. There are even noticeable differences reported in the diffusion rates between the simple cubic, the double gyroid, and double diamond structures shown in the figure, because of their different degrees of connectedness. It should come as no surprise that also the characteristic relaxation time and the (viscoelastic) response depend heavily on the particular phase which is formed.

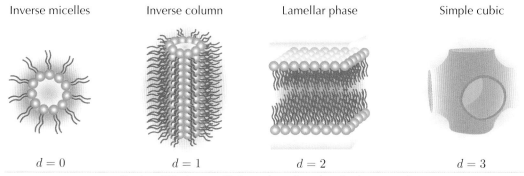

Inverse micelles $\qquad$ Inverse column $\qquad$ Lamellar phase $\qquad$ Simple cubic

$$d = 0 \qquad d = 1 \qquad d = 2 \qquad d = 3$$

effective diffusion coefficient of small molecules in water phase

Figure 6.26. Lipid molecules with a hydrophilic polar head and a hydrophobic tail can form various ordered structures. These are liquid crystalline–type mesophases, as the lipid molecules are disordered liquids within the interface. Illustrated here are some structures formed by lipids like monoolein, shown in figure 6.25. The micellar structures shown on the left are inverse micelles, with the tails pointing outward, so with the water inside the micelles. Such molecules can also form inverse hexagonal phases (with the water inside the columns), lamellar phases, and even several three-dimensional structures with a continuous water phase. Note the similarity with the phase diagram of block copolymers shown in figure 5.24. For drug delivery, the effective diffusion of drugs through the water phase is important. As indicated, the diffusion coefficient increases dramatically with the effective dimensionality of the water phase. Adapted from Aleandri and Mezzenga, 2020.

6.9.5 Epithelial cells die and disappear near +½ defects

We close this overview by pointing very briefly at exciting findings that bridge the gap between soft matter and the life sciences in an unexpected way: there is a growing body of evidence of a feedback between physical effects, mechanical forces, and biological behavior. As an example, epithelial cells in the outer layer of organs or of the skin are often somewhat elongated. Therefore one can associate a director-type order parameter with their orientation.

Now, figure 6.27 illustrates that if one analyzes the orientation patterns of epithelial kidney cells from this perspective, one discerns half-integer-type defects of both signs similar to the defects we discussed in section 6.6. This actually turns out to be more than just a curious finding: if one follows the dynamics of these cells, it turns out that their death (apoptosis) and extrusion commonly occur near $s = 1/2$ defects, but not near $s = -1/2$ ones. Whether this is simply due to the mechanical stress fields near the defect or whether other effects are dominant, is presently not known. Since also $s = 1/2$ defects turn out to dominate most of the dynamics of active matter of elongated particles—as we will discuss in section 9.5.3, in active nematics there is a strong asymmetry between plus and minus defects: $s = 1/2$ defects move while $s = -1/2$ are static[36]—one may safely assume that in the near future there will be many more intriguing findings and developments at this interface of the life sciences with liquid crystals and active matter.

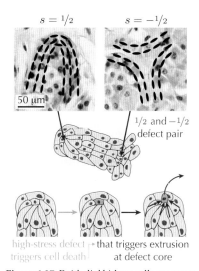

$s = 1/2 \qquad s = -1/2$

50 μm

1/2 and −1/2 defect pair

high-stress defect that triggers extrusion triggers cell death at defect core

Figure 6.27. Epithelial kidney cells are somewhat elongated. The images at the top illustrate how when their nematic order is indicated with the black lines, one immediately distinguishes both $+1/2$ and $-1/2$ defects. The sketch of a typical cell structure below highlights this further. Saw et al., 2017 have shown how death and extrusion of cells is found to occur mostly near $+1/2$ defects, as illustrated at the bottom. A nice perspective on these results by Saw et al., 2017 is given by Hirst and Charras, 2017. In section 9.9.2 we discuss how defects also play a role in morphogenesis. The epithelial kidney cell images are courtesy of Gorana Nikolic.

the increased presence of other defects in the system effectively 'screens' the Coulomb interaction between any two defects. As long as we remain below the critical point, free, unbound defects are energetically unfavorable, so any two defects of opposite charge should attract one another, regardless of how far they are from one another. The resulting effective interaction potential $V_{\text{eff}}(r)$ should therefore still be expressible as a long-range Coulomb interaction, with a renormalized elastic constant,

$$\beta V_{\text{eff}}(\vec{r}_+ - \vec{r}_-) = 4\pi^2 K_{\text{eff}} C(\vec{r}_+ - \vec{r}_-), \tag{6.49}$$

very much as the Coulomb interaction between two electrostatic charges in vacuum is rescaled by the inverse of the dielectric constant when the same charges are placed in a dielectric medium. Physically, we should expect that as the temperature increases, increased screening should decrease the magnitude of long-range interactions between defects. This should result in K_{eff} decreasing with increasing temperature. The first step of the RG calculation is to integrate out small scale structure. In our case, this amounts to increasing the minimum separation a between the vortices to ba (with b greater than but close to 1) and determining the effect of the presence of tight vortex pairs with a separation in that small range on the interactions of other, more widely separated, vortices; see figure 6.28.

6.10.3 Setting up the RG calculation

How do we actually calculate K_{eff}? Imagine placing two test defects of opposite topological charge at positions $\vec{r}_+$ and $\vec{r}_-$ within a neutral Coulomb gas below its critical temperature; see figure 6.28. The full partition function for this system can be written as an integral over all possible test charge positions, and over all possible configurations of the 'internal' charges of the Coulomb gas, including an appropriate Boltzmann factor accounting for interactions between and among the internal and test charges. We can derive an effective interaction between the two test defects by integrating over the possible internal charge configurations. This leaves us with a coarse-grained Boltzmann weight that depends only on the positions $\vec{r}_+$ and $\vec{r}_-$ of the test defect charges. In the limit of low fugacity y_0, i.e., where the temperature of the system is low compared to the energy required to nucleate defects, we can make the approximation of including only factors up to $\mathcal{O}(y_0^2)$. Physically, this means that the leading order contribution to screening is due to internal charge dipoles (as opposed to higher multipole moments), so we integrate over all possible configurations of a dipole of internal defects. In problem 6.9, we guide you step-by-step through this procedure. By comparing the explicit answer we obtain for $V_{\text{eff}}(\vec{r}_+ - \vec{r}_-)$ with equation (6.49), we determine the

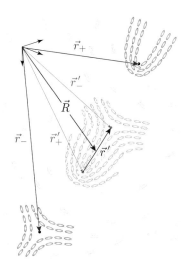

Figure 6.28. A schematic of the coarse-graining procedure for determining the long-range effective elastic constant K_{eff}, illustrated for the case of a nematic with half-integer defects. Two test defects are placed at positions $\vec{r}_+$ and $\vec{r}_-$, and an internal dipole of defects with separation vector $\vec{r}' \equiv \vec{r}_- - \vec{r}_+$ is placed between the test ones. The effective interaction is determined by integrating over all physically allowable values of the center-of-mass coordinates of the internal dipole defects, $\vec{r}'$ and $\vec{R} \equiv \frac{\vec{r}'_+ + \vec{r}'_-}{2}$. See problem 6.9.

effective elastic constant K_{eff},

$$K_{\text{eff}} = K - 4\pi^3 y_0^2 K^2 a^{2\pi K} \int_a^\infty \mathrm{d}r \, r^{3-2\pi K} + \mathcal{O}(y_0^4). \qquad (6.50)$$

This perturbative expression integrates over internal dipoles of all possible separations $(a < r < \infty)$. This is only useful, however, as long as the integral converges, which is only true for $K \leq 2/\pi$. This happens to be the value of K at which the free energy associated with an isolated vortex changes sign, suggesting that the phase transition should occur around $K \approx 2/\pi$.

In order to tame the divergent integrals for $K \geq 2/\pi$, we resort to the renormalization group procedure explained below. The partition function for this system contains two physical parameters, K and y_0, along with a lower cutoff length scale a. The value of the core size a is somewhat arbitrarily decided and can be rescaled by a factor $b > 1$, i.e., $a \to ba$. In doing so, we coarse-grain over all possible dipoles of separation between a and ba, which should in turn rescale, i.e. renormalize, the elastic interaction constant $K \to \tilde{K}$. Note that equation (6.50) gives the effective elastic constant between two defects if we coarse-grain over internal dipoles of all possible sizes. Integrating instead only over internal dipoles of separation less than ba, the renormalized elastic constant reads

$$\tilde{K} = K - 4\pi^3 y_0^2 K^2 a^{2\pi K} \int_a^{ba} \mathrm{d}r \, r^{3-2\pi K} + \mathcal{O}(y_0^4). \qquad (6.51)$$

This coarse-graining procedure should also renormalize the fugacity $y_0 \to \tilde{y}_0$, since increasing the core size decreases the configurational entropy of a single defect. The scaling dependence of y_0 is derived in problem 6.10 and yields

$$\tilde{y}_0 = b^{2-\pi K} y_0. \qquad (6.52)$$

We now iterate this procedure over and over to account for progressive rounds of coarse-graining using a mathematical construction called renormalization group flow relations.

6.10.4 How to derive the renormalization group flow relations

Considering an infinitesimal scaling factor $b = e^l \approx 1 + l$, we can derive the so-called renormalization group flow equations

$$\begin{aligned}
\frac{\mathrm{d}K}{\mathrm{d}l} &= -4\pi^3 a^4 K^2 y_0^2 + \mathcal{O}(y_0^4), \\
\frac{\mathrm{d}y_0}{\mathrm{d}l} &= (2 - \pi K) y_0 + \mathcal{O}(y_0^3)
\end{aligned} \qquad (6.53)$$

Inspection of equation (6.50) reveals that the integral scales as $r^{4-2\pi K}$.

The difficulty associated with the divergence at small K can be overcome by employing the following mathematical trick introduced by José et al., 1977: we split the integral in equation (6.50) in two parts:

$$\int_a^\infty \to \int_a^{ba} + \int_{ba}^\infty .$$

For small values of r, the integral converges, and hence its finite contribution to K_{eff} can be directly evaluated. The key point here is that this trick of splitting the integrals can be carried out order by order using a perturbation series in y_0 even if the coefficient multiplying y_0^2 diverges.

This step, which is detailed in problem 6.10, is analogous to the coarse-graining procedure implemented by the Kadanoff block construction in the percolation analysis of problem 6.12, where the parameter being renormalized is the bond probability p.

6.12 Problems

Problem 6.1* *The Onsager calculation of the isotropic-nematic transition*

In this exercise you will show how a solution of hard rods undergoes a phase transition from an isotropic to a nematic phase as the concentration is increased. As we stressed in the text, for hard core objects the transition is driven by entropy: above a critical concentration it is entropically favorable for the rods to align.

We consider a solution of hard cylinders of length L and diameter D. The rods are very long (needles with $L \gg D$) and the volume fraction ϕ is small, much smaller than unity. As the rods are hard, the only interaction between the rods is steric repulsion, i.e., they cannot interpenetrate each other. The analysis below is exact in the limit $D/L \to 0$.

a. Let c be the concentration of rods. Write down an expression for the volume fraction ϕ. Argue that in the limit $D/L \to 0$ we can ignore interactions between more than two rods.

b. As usual, $\hat{n}$ is the director. As in section 6.2, let $\hat{m}^\alpha$ be the orientation vector of a given rod α ($\hat{m}$ and $-\hat{m}$ are equivalent here) and consider the tensor order parameter Q_{ij}^{m} defined in equation (6.2). Write down an expression for the scalar order parameter $S \equiv \frac{3}{2}\langle \hat{n} \cdot \underline{Q}^{\mathrm{m}} \cdot \hat{n} \rangle$ which measures the strength of the orientational order.

c. Let $p(\hat{m})$ denote the normalized orientational distribution function of the rods, such that, for example, $\langle \hat{m}\hat{m} \rangle = \int \mathrm{d}\hat{m}\, p(\hat{m})\hat{m}\hat{m}$. Compute S in the limit of total alignment and in the isotropic state, and check the consistency of your result with the remarks made after equation (6.4).

d. In the isotropic state the free energy has two contributions. The first is an ideal gas contribution from the center of mass degrees of freedom. The second is a contribution due to the excluded volume because the rods cannot overlap. Show with the help of figure 5.13 that the excluded volume is given by $2DL^2|\sin \gamma|$, where γ is the angle between two rods.

e. Using the above result, compute the average excluded volume per particle, and then show that this leads to a free energy per particle in the isotropic state given by

$$\beta F_{\mathrm{iso}} = \ln(c\lambda^3) - 1 + \frac{\pi a L^2}{4}c, \tag{6.58}$$

where λ is the De Broglie wavelength.

f. If the system is in an anisotropic state there is the additional orientational contribution

---

Relevant coding problems and solutions for this chapter can be found on the book's website www.softmatterbook.online under Chapter 6/Coding problems.

We essentially trace the steps of the analysis of Onsager, 1949.

You should find $S = (3\langle \cos^2 \theta \rangle - 1)/2$, with θ the angle between $\hat{n}$ and $\hat{m}$.

Keep track of the fact that the excluded volume is shared by two rods!

$$\beta F_{\text{or}} = \langle \ln(4\pi p(\hat{m})) \rangle = \int d\hat{m}\, p(\hat{m}) \ln(4\pi p(\hat{m})). \qquad (6.59)$$

Check that this is the type of expression one expects. Minimizing the total free energy turns out to be a difficult problem. To make progress, Onsager proposed the ansatz

The discussion in section 5.7.1 and problem 5.6 may help you understand this expression.

$$p_\alpha(\theta) = \frac{\alpha}{4\pi} \frac{\cosh(\alpha \cos\theta)}{\sinh\alpha}, \qquad (6.60)$$

with a single free parameter α. Sketch this function. Compute an expression for the order parameter using (6.60) and sketch $S(\alpha)$.

g. Upon minimizing the total free energy using the trial function, Onsager found a function $f(c)$ showing a first order phase transition from an isotropic ($\alpha = 0$) to a nematic ($\alpha \geq 18.6$) phase. The critical volume fraction in this case is $\phi_c \simeq D/L$. Compute the order parameter at the transition point and comment on the result. Why was a first order transition to be expected and what does this mean physically?

Make a connection with the discussion in section 6.2.

Problem 6.2 *Analysis of the Landau–de Gennes free energy for the isotropic-nematic transition*

In this exercise we analyze the Landau–de Gennes expansion (6.5)

$$\mathcal{F} = \tfrac{1}{2}r(T)\,S^2 - w\,S^3 + u\,S^4 \qquad (6.61)$$

in more detail. Remember that in Landau theory, r is supposed to go linearly through zero at T_c, $r(T) = r'(T - T_c)$, where T_c is the temperature where a second order phase transition occurs for $w = 0$ (see box 2, in section 2.14). In this problem we consider the case $w \neq 0$.

a. Show that the temperature T^*, where, as indicated in figure 6.7, the free energy develops an inflection point, is given by $r = r^* = 9w^2/(16u)$.

b. Show that the first order phase transition happens at the point where $r(T_{NI}) = w^2/(2u)$. Note that this value is close to the point where the free energy as a function of S develops an inflection point, as determined in step a: $T_{NI} - T_c = \tfrac{8}{9}(T^* - T_c)$.

c. Show that the nematic order parameter at the transition is given by $S_{NI} = w/(2u)$.

d. Use the thermodynamic relation $S = -(\partial F/\partial T)_V$ for the entropy to show that at the transition the entropy difference ΔS (which gives the latent heat at the transition) between the nematic and isotropic phases is given by $\Delta S = S_{NI}^2\, r'/2$.

The analysis in this exercise can be extended to determine the susceptibility and correlation length in terms of the parameters in

as an integral over all Fourier modes of the fluctuating components $\delta \hat{n}_x$ and $\delta \hat{n}_y$.

d. Your result in the previous step should be of the form

$$\langle (\delta \hat{n})^2 \rangle_{|\psi|} \sim \int d^3 \vec{k} \, \frac{1}{Kk^2 + 2\gamma_\perp q_0^2 |\psi|^2}. \qquad (6.70)$$

Note that the integral is divergent. Introduce a cutoff k_c, and show that the integral is of the form

$$\langle (\delta \hat{n})^2 \rangle_{|\psi|} \sim C_1 k_c - C_2 |\psi|, \qquad (6.71)$$

where C_1 and C_2 are positive constants. Determine C_1 and C_2 explicitly.

e. Use the previous result in the expression (6.69)) for the free energy as a function of $|\psi|$ to show that this leads to a cubic contribution

$$\Delta \mathcal{F} = \gamma_\perp q_0^2 |\psi|^2 \langle (\delta \hat{n})^2 \rangle_{|\psi|} = -\frac{k_B T}{4\pi} \left(\frac{2\gamma_\perp q_0^2}{K} \right)^{3/2} |\psi|^3 \qquad (6.72)$$

to the free energy density. What is the effect of the term proportional to k_c in (6.70) (which gives a contribution $k_B T k_c \gamma_\perp q_0^2 |\psi|^2 / (\pi^2 K)$ to the free energy)?

f. Retrace the steps of section 6.2 to show that the term (6.72) implies that the nematic-to-smectic-A transition is (weakly) first order.

Problem 6.5 *Evaluation of the splay, twist, and bend terms for the three cases discussed in section 6.3.2*

In this problem we evaluate the terms in the Frank free energy for the three basic distortions considered in section 6.3.2. The splay and bend cases are most easily done in cylindrical coordinates. For an arbitrary vector $\vec{A}$, one has in cylindrical coordinates

$$\vec{\nabla} \cdot \vec{A} = \frac{1}{r} \partial_r (r A_r) + \frac{1}{r} \partial_\phi A_\phi + \partial_z n_z \qquad (6.73)$$

and

$$\vec{\nabla} \times \vec{A} = \left(\frac{1}{r} \partial_\phi A_z - \partial_z A_\phi \right) \hat{r} + (\partial_z n_r - \partial_r n_z) \hat{\phi}$$
$$+ \left(\frac{1}{r} \partial_r (r A_\phi) - \frac{1}{r} \partial_\phi A_r \right) \hat{z}. \qquad (6.74)$$

a. Show for the case of ideal splay specified by (6.7) that $\vec{\nabla} \cdot \hat{n} = 1/r$ while $\vec{\nabla} \times \hat{n} = 0$, and that this yields (6.8).

b. Show for the case of ideal twist specified by (6.9) that $\vec{\nabla} \cdot \hat{n} = 0$ and $\vec{\nabla} \times \hat{n} = -\theta'(z)\,\hat{n}$ and that this yields (6.10).

c. Show for the case ideal splay specified by (6.11) that $\vec{\nabla} \cdot \hat{n} = r^{-1}\partial_\phi n_\phi = 0$ and $\vec{\nabla} \times \hat{n} = r^{-1}\partial_r(r\,n_\phi)\,\hat{z} = r^{-1}\,\hat{z}$, and that these yield (6.12).

Problem 6.6 *Fréedericksz transition with a molecular field*

In section 6.4 we discussed how there are two ways to calculate the director field: one can either use polar coordinates which ensure automatically that $\hat{n}$ is a unit vector, or allow variations in $\vec{n}$ to be arbitrary, while introducing a Lagrange multiplier to ensure that the solutions are such that $\vec{n}$ is a unit vector. The latter route leads to the introduction of the concept of a molecular field $\vec{h}$ parallel to $\vec{n}$. Our analysis of the Fréedericksz transition in section 6.5 was done via the first route. In this exercise we follow the second route. You will first calculate the molecular field from the free energy, which, together with the equilibrium equations, gives rise to the equilibrium equation for θ. Then, we will use an analogy frequently encountered in physics to investigate the Fréedericksz transition.

For simplicity we consider the one-constant approximation $K_1 = K_2 = K_3 = K$ mentioned in the note in the margin next to equation (6.6). The setup we consider is sketched in figure 6.11.

a. Calculate the contribution to the molecular field $\vec{h}$ resulting from $\mathcal{F}_d$ using equation (6.18).

b. Consider the magnetic case of the Fréedericksz transition, and calculate the magnetic contribution to the molecular field $\vec{h}$.

c. The equilibrium equation in the presence of a molecular field is given by the vector equation (6.21). Show that the component parallel to $\vec{n}$ of this equation is trivial, and that the component transverse to $\vec{n}$ can be written as

$$\vec{n}^\perp \cdot \vec{h} = 0, \qquad (6.75)$$

where $\vec{n}^\perp$ is an arbitrary vector perpendicular to $\vec{n}$, that is, $\vec{n}^\perp \cdot \vec{n} = 0$.

d. Write out the equation $\vec{n}^\perp \cdot \vec{h} = 0$ using the two contributions to the molecular field, from the elastic energy and the magnetic energy.

e. We now consider the setup of figure 6.11 and anticipate that $\vec{n}$ will lie in the xy plane, as we did in section 6.5, by writing $\hat{n} = (\cos\theta, \sin\theta, 0)$. Write $\vec{n}^\perp$ in terms of θ.

f. Work out $\nabla^2 \hat{n} = d^2\hat{n}/dz^2$ for $\hat{n} = (\cos\theta(z), \sin\theta(z), 0)$ and show that it has two terms, one parallel to $\hat{n}$ and a different one along $\vec{n}^\perp$.

g. Analyze the above equation (6.75) with the molecular field which you worked out before, and show that this leads to the equation

$$K\frac{\mathrm{d}^2\theta}{\mathrm{d}y^2} - \mu_0\chi_a H^2 \sin\theta\cos\theta = 0. \qquad (6.76)$$

h. Note that equation (6.76) is similar to the equation for a particle moving in a potential, with θ playing the role of the coordinate of the particle, and y playing the role of time. What is the potential in which the particle moves, in this analogy?

i. For the case of a particle moving in a potential, the energy (the sum of potential and kinetic energies) is conserved. What does the Fréedericksz transition problem (with $\theta(0) = \theta(d) = \pi/2$) translate into in the particle in the potential language?

j. Use the analogy discussed above to write, near threshold, the equation of motion as if it were a harmonic oscillator by expanding the energy to lowest order near its minimum. Determine the oscillation frequency and deduce the threshold of the transition by requiring that the particle be back at the bottom of the well at 'time' $y = d$. Likewise, explore the similarity with the analysis of the threshold of a buckling rod discussed in section 2.8.2 and problems 2.1 and 2.2.

Problem 6.7 *Weakly nonlinear analysis of the Fréedericksz transition*

In this exercise we determine for the Fréedericksz transition the expansion of the free energy (6.26) and of the corresponding solutions to lowest nontrivial order. For simplicity, we take $K_1 = K_3 = K$. We start from (6.26), write $\theta = \pi/2 + \Delta\theta$, and make the ansatz

$$\Delta\theta = \sum_{n=1,3,5,\ldots} A_n \sin(n\pi y/d). \qquad (6.77)$$

a. Why is it safe to leave out even integer terms as $\sin(2\pi y/d)$?

b. In line with the discussion in section 2.8.2 we anticipate that just above threshold we will have $A_1 \sim \sqrt{H - H_c}$ where H_c is the critical field given by (6.30), and that $A_3 = \mathcal{O}(A_1^3)$. Convince yourself that, more generally, $A_n = \mathcal{O}(A_1^n)$ for n odd.

c. Expand the free energy expression (6.26) to fourth order in $\Delta\theta$, substitute (6.77), and show that to fourth order in A_1 we get

$$\mathcal{F} = \frac{\mu_0\chi_a d}{4}\left[(H_c^2 - H^2)A_1^2 + \tfrac{1}{2}H^2 A_1^4\right]. \qquad (6.78)$$

d. Draw the free energy as a function of A_1 and show that it conforms to the behavior sketched in figure 2.19 both above

---

If you did it right, you might wonder why the energy in the particle analogy is different from the free energy (6.26) of the Fréedericksz transition: the potential energy term is similar to the magnetic term of the free energy but has the opposite sign. This is because the energy in the particle in a potential analogy is a 'constant of motion,' whereas in the physical free energy density is lowest in regions where the director is parallel to the field.

Note that these odd integer modes are orthogonal modes on the interval $[0, d]$ for y, which is like half a period of the lowest order mode. As a result, products of two different modes are orthogonal on this interval. This is the reason that expression (6.78) is easily derived. In sixth order, though, terms of order A_3^2 do come in.

and below the transition. Derive the expression for the amplitude A_1 corresponding to the minimum of the free energy and show that it exhibits the expected square root growth just above the transition.

e. The free energy is an integrated quantity, so the consistency of the expansion and the solution itself cannot be determined. For this, we have to solve the full equation for the static solution. In the one-constant approximation, the equation to be solved is given in 6.76. Take this equation, write it in terms of $\Delta\theta$, and expand the magnetic field term to cubic order in $\Delta\theta$.

f. Take the equation derived above in e, substitute the expansion (6.77), and verify that the terms proportional to $\sin(\pi y/d)$ yield the result already derived under c for A_1.

g. Now collect the terms proportional to $\sin(3\pi y/d)$ and show that this yields

$$A_3 = -\frac{H^2}{6(9H_c^2 - H^2)} A_1^3.$$
(6.79)

This is a simple example of the fact that, near a bifurcation point, higher order modes are typically slaved to the dominant mode associated with the bifurcation. (6.79) also confirms that $A_3 = \mathcal{O}(A_1^3)$, which we anticipated when writing the expansion.

Problem 6.8* *Derivation of the effective defect interaction*

In this problem we derive that effective defect interaction (6.42) starting from (6.37).

a. Use Stokes' theorem to rewrite the left-hand side of (6.37) in terms of an area integral. Show from the previous step that since this holds for any area, this amounts to solving the equation $(\partial_x \partial_y - \partial_y \partial_x)\theta(\vec{r}) = n_d(\vec{r})$.

b. Introduce the Cauchy conjugate to the field $\partial_{x_i}\theta$, $\tilde{\theta}_i(\vec{r}) = \epsilon_{ij}\partial_{x_j}\theta(\vec{r})$, where ϵ_{ij} is the antisymmetric Levi-Civita tensor in two dimensions, $\epsilon_{xy} = -\epsilon_{yx} = 1$, $\epsilon_{xx} = \epsilon_{yy} = 0$. Give the equation obeyed by $\tilde{\theta}(\vec{r})$.

c. Write the solution for $\tilde{\theta}(\vec{r})$ using the Green's function (6.38).

d. Derive the solution (6.40) from the previous result and the expression of $\theta(\vec{r})$ in terms of $\tilde{\theta}(\vec{r})$.

e. Use (6.40) to derive the expression (6.42) for the defect interaction term in the free energy. How does the core energy depend on the constant C in (6.39)?

Hint: Use integration by parts, neglecting boundary terms, to be able to use the defining equation for one of the Green's functions in your expression.

This problem is based on the analysis by Nelson, 1996, who also gives a nice pedagogical introduction to and overview of defects in two-dimensional systems.

Problem 6.9** *Derivation of the coarse-grained effective elastic constant for a 2D Coulomb gas of defects*

Consider a system of two opposite test charges located inside of a Coulomb gas at $\vec{r}_+$ and $\vec{r}_-$. We assume these two charges will always be present, and we can therefore ignore the constant contribution to the energy due to their cores, focusing only on their interaction energy $\beta V(\vec{r}_+ - \vec{r}_-) = 4\pi^2 K C(\vec{r}_+ - \vec{r}_-)$. The full partition function for this system can be written as

$$Z_2 = \int d^2\vec{r}_+ d^2\vec{r}_- \, e^{-\beta V(\vec{r}_+ - \vec{r}_-)} P_Q(\vec{r}_+, \vec{r}_-), \qquad (6.80)$$

where P_Q is the probability associated with finding the two test charges at $\vec{r}_+$ and $\vec{r}_-$, given all possible configurations of the Coulomb gas. We are interested in understanding how the presence of other charges in a Coulomb gas 'screens' the Coulomb interaction, effectively renormalizing the elastic constant $K \to K_{\text{eff}}$. The goal of this exercise is to derive the expression for K_{eff} given in (6.50), for which we will use a perturbative procedure by coarse-graining over all possible internal dipole charge configurations.

a. To find K_{eff}, we must derive a coarse-grained partition function

$$Z_2' = \int d^2\vec{r}_+ d^2\vec{r}_- e^{-\beta V_{\text{eff}}(\vec{r}_+ - \vec{r}_-)}, \qquad (6.81)$$

where $\beta V_{\text{eff}}(\vec{r}_+ - \vec{r}_-) = 4\pi^2 K_{\text{eff}} C(\vec{r}_+ - \vec{r}_-)$; see section 6.10.2. Argue that it is useful to write the coarse-grained partition function weight as

$$e^{-\beta V_{\text{eff}}(\vec{r}_+ - \vec{r}_-)} = e^{-\beta V(\vec{r}_+ - \vec{r}_-)} P_Q(\vec{r}_+, \vec{r}_-). \qquad (6.82)$$

b. Show that $P_Q(\vec{r}_+, \vec{r}_-)$ to lowest nontrivial order in y_0 reads

$$P_Q(\vec{r}_+, \vec{r}_-) = 1 + y_0^2 \int d^2\vec{r}_+' d^2\vec{r}_-' e^{-4\pi^2 K C(\vec{r}_+' - \vec{r}_-')}$$
$$\times \left(e^{4\pi^2 K D(\vec{r}_+, \vec{r}_-; \vec{r}_+', \vec{r}_-')} - 1\right) + \mathcal{O}(y_0^4), \qquad (6.83)$$

where we define the dipole-dipole interaction as

$$D(\vec{r}_+, \vec{r}_-; \vec{r}_+', \vec{r}_-') \equiv C(\vec{r}_+ - \vec{r}_+') - C(\vec{r}_+ - \vec{r}_-')$$
$$- C(\vec{r}_- - \vec{r}_+') + C(\vec{r}_- - \vec{r}_-'). \qquad (6.84)$$

Hint: See section 6.10.1 for the expression of the grand canonical partition function Z_Q. Keep in mind that charges come in pairs.

c. Rewrite the integral in expression (6.83) in terms of the center-of-mass coordinates of the two internal charges of the Coulomb gas: $\vec{R} \equiv \frac{\vec{r}_+' + \vec{r}_-'}{2}$ and $\vec{r}' \equiv \vec{r}_-' - \vec{r}_+'$. Show how we can approximate the dipole-dipole interaction as

$$D(\vec{r}_+, \vec{r}_-; \vec{R}, \vec{r}') \approx \vec{r}' \cdot \vec{\nabla}_{\vec{R}}[C(\vec{r}_+ - \vec{R}) - C(\vec{r}_- - \vec{R})]. \qquad (6.85)$$

Hint: Take the separation between the two defects to be small, and expand to first order in $\vec{r}'$.

d. Next, perform the angular integration over $\vec{r}'$ to arrive at the new expression

$$P_Q(\vec{r}_+, \vec{r}_-) = 1 + 8\pi^5 y_0^2 K^2 \int \mathrm{d}r' \, e^{-4\pi^2 KC(r')} r'^3$$
$$\times \int \mathrm{d}^2\vec{R} \, |\vec{\nabla}_{\vec{R}}[C(\vec{r}_+ - \vec{R}) - C(\vec{r}_- - \vec{R})]|^2 + \mathcal{O}(y_0^4). \qquad (6.86)$$

e. Show that

$$\int \mathrm{d}^2\vec{R} \, |\vec{\nabla}_{\vec{R}}[C(\vec{r}_+ - \vec{R}) - C(\vec{r}_- - \vec{R})]|^2$$
$$= 2C(\vec{r}_+ - \vec{r}_-) - 2C(0), \qquad (6.87)$$

and explain why it is physically reasonable to set $C(0) = 0$.

Hint: You may find equations (6.38) and (6.39) helpful for this step.

f. Substitute your final expression for P_Q into the equation found in step a. Verify that the effective elastic constant is

$$K_{\mathrm{eff}} = K - 4\pi^3 y_0^2 K^2 a^{2\pi K} \int_a^\infty \mathrm{d}r \, r^{3-2\pi K} + \mathcal{O}(y_0^4), \qquad (6.88)$$

as was stated in (6.50).

Problem 6.10* *Renormalization group flow trajectories for the KT transition*

In this problem, we will modify the result from the previous problem and use it in a renormalization group analysis of the KT transition in a two-dimensional Coulomb gas of defects. The coarse-graining procedure here involves rescaling the lower length scale cutoff (i.e., the defect core size) from a to ba, and determining the renormalized elastic constant $\tilde{K}$ by performing the procedure from the previous problem, except now integrating only over dipoles of separation less than ba. We therefore directly write our renormalized elastic constant

$$\tilde{K} = K - 4\pi^3 y_0^2 K^2 a^{2\pi K} \int_a^{ba} \mathrm{d}r \, r^{3-2\pi K} + \mathcal{O}(y_0^4), \qquad (6.89)$$

as was given in (6.51).

a. By rescaling the defect core size to ba, we reduce individual defects' configurational entropy (see section 2.11), resulting in the scaling dependence for the fugacity (6.52),

$$\tilde{y}_0 = b^{2-\pi K} y_0. \qquad (6.90)$$

Derive this expression from the partition function for a single defect in a domain of area L^2.

b. Considering infinitesimal scaling $b = e^l \approx 1 + l$, use equations (6.89) and (6.90) to derive the following renormalization group flow recursion relations.

$$\frac{dK}{dl} = -4\pi^3 a^4 K^2 y_0^2 + \mathcal{O}(y_0^4),$$
$$\frac{dy_0}{dl} = (2 - \pi K)y_0 + \mathcal{O}(y_0^3). \qquad (6.91)$$

c. Using the recursion relations derived in step b, show that around the critical point parameterized by l along the renormalization group flow, $K(l)$ and $y_0(l)$ satisfy the equation

$$\left(K^{-1}(l) - \frac{\pi}{2} \right)^2 - \pi^4 a^4 y_0^2(l) = C, \qquad (6.92)$$

where C is an arbitrary constant.

Hint: Try rewriting the recursion relations in terms of the variables $x \equiv K^{-1} - \pi/2$ and $y \equiv a^2 y_0$, which are small near the critical point.

Problem 6.11 *Escape into the third dimension*

Consider a nematic liquid crystal in a circular capillary tube, as shown in figure 6.30. The walls are treated to keep the director normal to the cylindrical boundary. The director lines are allowed to escape along the z-axis, so that the director can be parameterized in cylindrical coordinates as

$$\hat{n} = (\cos \chi(r), 0, \sin \chi(r))^{\mathrm{T}} \qquad (6.93)$$

where χ is the angle between the director and the (r, ϕ) plane, as depicted in figure 6.30. For the escaped configuration, the boundary conditions are

$$\chi(r = R) = 0 \quad \text{and} \quad \chi(r = 0) = \frac{\pi}{2}. \qquad (6.94)$$

Consider the form of the free energy density from section 6.3.1,

$$\mathcal{F} = \frac{K_1}{2}(\vec{\nabla} \cdot \hat{n})^2 + \frac{K_2}{2}(\hat{n} \cdot \vec{\nabla} \times \hat{n})^2 + \frac{K_3}{2}(\hat{n} \times (\vec{\nabla} \times \hat{n}))^2. \qquad (6.95)$$

See section 6.10.4 and figure 6.29, where we depict renormalization group flows for the two-dimensional Coulomb gas.

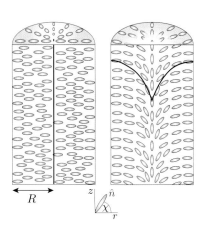

R

Figure 6.30. The defect at the center of a vertical column of nematic (left) escapes into the third dimension (right).

a. Argue why the twist term is zero for both geometries in figure 6.30.

b. Unless stated otherwise, we consider the case $K_1 = K_3 = K$. Substitute equation (6.93) into (6.95) and show that the elastic energy per unit length is equal to

$$W = \pi K \int_{-\infty}^{\ln R} d\xi \left[(\chi')^2 + \cos^2 \chi - \chi' \sin 2\chi \right], \qquad (6.96)$$

where we used the coordinate transformation $r = e^\xi$.

c. Show that the Euler-Lagrange equation, obtained by requiring that the minimum energy solution be the one which is stable to all linear variations of the director angle, is

$$\chi'' = -\cos \chi \sin \chi. \qquad (6.97)$$

d. By solving the previous equation, show that the solution to the equilibrium director distribution of the configuration has the form

$$\chi = 2 \arctan \left(\frac{R - r}{R + r} \right). \qquad (6.98)$$

e. Show that the total energy of the configuration when $K_1 = K_3$ is

$$F = 3\pi K h. \qquad (6.99)$$

f. Compare the equilibrium energy of step b to the energy of the configuration with a disclination at the center $x = y = 0$ along the entire height of the column, h. See figure 6.30 for a drawing of the two configurations that you are comparing.

g. Find the condition for which escape into the third dimension is preferred.

h. Considering now the case where $K_1 < K_3$, show that the energy of the escaped configuration is

$$F = \pi K_1 h \left(2 + \frac{\arcsin \beta}{\beta \sqrt{1 - \beta^2}} \right), \qquad (6.100)$$

where $\beta \equiv \sqrt{1 - K_1/K_3}$.

Problem 6.12 *Orientation fluctuations and correlations in a nematic liquid crystal*

In this problem, you will analyze the fluctuations of a nematic in an external magnetic field.

a. A magnetic field H oriented along z acts on the nematic. Show from the discussion in section 6.5.1 that the interaction energy can be written as

$$F_{\mathrm{M}} = \frac{1}{2} \int \mathrm{d}^3 \vec{r} \, \mu_0 \chi_{\mathrm{a}} H^2 (n_x^2 + n_y^2) + \text{const.} \qquad (6.101)$$

b. The elastic free energy density of the nematic is given by equation (6.6). Now, we consider a situation where the average director $\hat{n}_0$ is oriented along $+\hat{z}$. The fluctuations of the optical axis at any point $\vec{r}$ can then be specified by small, nonzero components $n_x(\vec{r})$, $n_y(\vec{r})$. Write equation (6.6) under this assumption in terms of derivatives of n_x and n_y.

c. Substitute the Fourier components

$$n_i(\vec{q}) = \int \mathrm{d}^3 \vec{r} \, n_i(\vec{r}) e^{i\vec{q}\cdot\vec{r}}, \quad \text{where } i = \{x, y, z\}, \qquad (6.102)$$

into the total free energy of the system to obtain

$$F = F_0 + \frac{1}{2\Omega} \sum_q [K_1 |n_x(\vec{q})q_x + n_y(\vec{q})q_y|^2 + K_2 |n_x(\vec{q})q_y$$
$$- n_y(\vec{q})q_x|^2 + (K_3 q_z^2 + \mu_0 \chi_{\mathrm{a}} H^2)(|n_x(\vec{q})|^2 + |n_y(\vec{q})|^2)], \qquad (6.103)$$

where Ω is the sample volume.

d. Now set $K_1 = K_2 = K_3$ for simplicity. The four commandments of box 3, in section 3.9, detail how to calculate equilibrium fluctuations from such a quadratic expression. Use these to show that

$$\langle |n_x(\vec{q})|^2 \rangle = \langle |n_y(\vec{q})|^2 \rangle = \frac{\Omega k_{\mathrm{B}} T}{K q^2 + \chi_{\mathrm{a}} \mu_0 H^2}, \qquad (6.104)$$

where $\langle \cdot \rangle$ denotes a thermal average. This is the central formula of fluctuation theory for nematics.

e. Now Fourier transform back the result to determine the two-point correlation function $\langle n_x(\vec{r}_1) n_x(\vec{r}_2) \rangle$, which is equal to $\langle n_y(\vec{r}_1) n_y(\vec{r}_2) \rangle$.

Hint: Use the fact that

$$\frac{1}{(2\pi)^3} \int \mathrm{d}^3 \vec{q} \, \frac{1}{q^2 + \xi^{-2}} e^{-i\vec{q}\cdot\vec{r}} = \frac{1}{4\pi|\vec{r}|} e^{-|\vec{r}|/\xi}. \qquad (6.105)$$

f. In step e, the calculation of the two-point *correlation* function of the nematic director due to thermal fluctuations gives a characteristic length in the exponential. Compare this length to the characteristic length obtained in the Fréedericksz transition from the *response* to an applied magnetic field.

Problem 6.13* *Poor man's elasticity and fluctuations of smectic and columnar phases*

Small deformations of crystals and liquid crystals can be described by the displacement field $\vec{u}$ of the lattice which represents their ground state. The vector field $\vec{u}$ is:

> three-dimensional for a 3-dimensional crystal;
> two-dimensional for columnar phases;
> one-dimensional for smectic or laminar phases.

See figure 6.31.

For a simple model of elasticity, we consider the energy of deformation of the liquid crystal under a scalar displacement u. Uniform u corresponds to uniform translation of the system, and so it does not change the deformation energy of the system. Hence, the energy can depend only on $\vec{\nabla} u$. The deformation or elastic energy of the crystal F_{cr}, columnar phase F_{co}, and smectic phase F_s are, respectively:

$$F_{cr} = \frac{C}{2} \int d^3\vec{r} \, (\vec{\nabla} u)^2,$$

$$F_{co} = \frac{C}{2} \int d^3\vec{r} \, (\vec{\nabla}_\perp u)^2 + \lambda^2 (\nabla_z^2 u)^2, \qquad (6.106)$$

$$F_s = \frac{C}{2} \int d^3\vec{r} \, (\nabla_z u)^2 + \lambda^2 (\nabla_\perp^2 u)^2,$$

where C is an elastic modulus for the material and λ a penetration length over which curvature energy is comparable to compression energy. $\vec{\nabla}_\perp$ is $\vec{\nabla}$ restricted to the xy plane so that $\vec{\nabla}_\perp = \hat{x}\partial_x + \hat{y}\partial_y$ and $\nabla_\perp^2 = \partial_x^2 + \partial_y^2$.

a. Taking $u(\vec{r}) = \frac{L^3}{(2\pi)^3} \int d^3\vec{q} \, u(\vec{q}) e^{i\vec{q}\cdot\vec{r}}$, write down expressions for F_{cr}, F_{co}, and F_s in Fourier space in terms of $u(q)$, the wavenumbers $q, q_\perp (q_\perp = \hat{x}q_x + \hat{y}q_y)$, and q_z, the sample size L, and the lattice spacing a.

b. Show using the four commandments of section 3.9 that

$$\text{crystal:} \quad \langle |u(q)|^2 \rangle = \frac{k_B T}{CL^3 q^2},$$

$$\text{columnar:} \quad \langle |u(q)|^2 \rangle = \frac{k_B T}{CL^3 (q_\perp^2 + \lambda^2 q_z^4)}, \qquad (6.107)$$

$$\text{smectic:} \quad \langle |u(q)|^2 \rangle = \frac{k_B T}{CL^3 (q_z^2 + \lambda^2 q_\perp^4)}.$$

c. Let q_c be the cutoff wavenumber above which the continuum theory ceases to hold and let L be the linear size of the

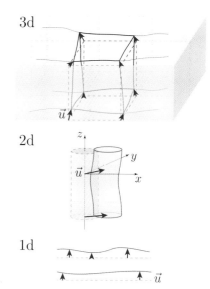

3d

2d

1d

Figure 6.31. Schematic illustrating a crystal (top), a columnar phase (middle), and a smectic phase (bottom). In each of these three cases there is a displacement field $\vec{u}$ that is three-dimensional for crystals, two-dimensional for columnar phases and one-dimensional for smectic layers. Adapted from de Gennes and Prost, 1993.

sample. Derive the following approximations for the mean square displacements in the large q_c limit, commenting on the approach you use to take the limits:

$$\text{crystal: } \frac{1}{L^3} \int d^3\vec{r} \, \langle u^2(r) \rangle \approx \frac{k_B T q_c}{2\pi^2 C},$$

$$\text{columnar: } \frac{1}{L^3} \int d^3\vec{r} \, \langle u^2(r) \rangle \approx \frac{k_B T \sqrt{q_c \lambda^{-1}}}{2\sqrt{2}\pi C}, \tag{6.108}$$

$$\text{smectic: } \frac{1}{L^3} \int d^3\vec{r} \, \langle u^2(r) \rangle \approx \frac{k_B T \log(q_c L/\pi)}{8\pi\lambda C}.$$

From this you can see that fluctuations in crystals and columnar phases are finite as $L \to \infty$, and so they are stable. For smectics, though, fluctuations are logarithmically divergent. This is known as the Peierls-Landau instability.

Hint: You may use the following identities:

For $y > \frac{1}{2}$,

$$\int_{-1}^{1} \frac{dx}{1 - x^2 + y^2 x^4}$$

$$= \frac{2\sqrt{2}y}{\sqrt{1 - 4y^2}} \left[\frac{\tan^{-1}\left(\frac{\sqrt{2}y}{\sqrt{-1-\sqrt{1-4y^2}}}\right)}{\sqrt{-1 - \sqrt{1 - 4y^2}}} \right. \tag{6.109}$$

$$\left. - \frac{\tan^{-1}\left(\frac{\sqrt{2}y}{\sqrt{-1+\sqrt{1-4y^2}}}\right)}{\sqrt{-1 + \sqrt{1 - 4y^2}}} \right].$$

For $x, \alpha \in \mathbb{R}$, in the limit $x \to \infty$,

$$\tan^{-1}(xe^{i\alpha}) \to \begin{cases} \frac{\pi}{2} & \text{if } \text{Re}[e^{i\alpha}] > 0 \\ -\frac{\pi}{2} & \text{if } \text{Re}[e^{i\alpha}] < 0, \end{cases} \tag{6.110}$$

$$\int_{-1}^{1} \frac{dx}{x^2 + y^2(1 - x^2)^2} \approx \frac{\pi}{2y} \text{ as } y \to \infty. \tag{6.111}$$

d. Estimate the root mean square displacement for each phase taking $L = 10$ mm, $k_B T = 4 \times 10^{-21}$ J, $\lambda = 3$ nm, $C_{cr} = 10^9$ Jm^{-3}, $C_{co} = C_s = 10^7$ Jm^{-3}, and $q_c = 10^{10}$ m^{-1} for crystals and $q_c = 2 \times 10^9$ m^{-1} for columnar and smectic phases.

Interfaces, Surfaces, and Membranes | 7

The previous three chapters took material building blocks of soft matter—colloidal particles, polymers and liquid crystals—as the focal point of the discussion. Even though we concentrated on bulk properties, we invariably encountered examples of how interfaces, surfaces, and membranes form an integral part of soft matter science. Think for instance of the interfacial surface tension which determines spreading properties of fluids on surfaces, of droplet dynamics driven by Marangoni effects, and of the tuning of colloidal interactions by grafting polymers to their interface. Other examples are the formation of lipid membranes, which play a central role in biology as constituting the boundary of a cell (like the red blood cell of figure 7.1) or of a compartment within a cell.

In this chapter we therefore take interfaces, surfaces, and membranes as the central focus, and we highlight some additional salient features that play a role in soft matter. Topics included are the Helfrich model for shapes of membranes, the development of icosahedral shape deformations of viruses with size, crumpling, and a soft matter experiment probing the predictions of the so-called KPZ equation.

7.1 Fluid interfaces

A crucial property of fluid interfaces is their surface tension. In our overview of fluid dynamics in chapter 1 we already treated in passing the salient features of fluid interfaces associated with the surface tension, so we simply point to those here.

For a droplet in contact with a surface, the relative values of the three surface tensions involved determine whether the fluid will spread onto the surface or form a finite contact angle (section 1.13.1). Likewise, in a concentration or temperature gradient the induced surface tension gradients give rise to flow. Such effects, which come under the name of Marangoni flow, drive much of the dynamics of small droplets in non-equilibrium situations (section 1.13.3). Moreover, due to the surface tension, a curved interface gives rise to a Laplace pressure; we have seen how this pressure tunes the resonance frequency of small droplets (section 1.14) and is an important driving force for the dynamics of thin films (section 1.12). We extend the discussion in problems 7.1 and 7.2, where we analyze the surface tension–induced instability of a cylindrical fluid layer, which gives rise to pinching and the formation of droplets, as well

Surfaces of crystals can also exhibit a roughening transition between a macroscopically flat faceted phase and a rough interface with height fluctuations not unlike those of a liquid interface in the absence of gravity. The transition is much less relevant to soft matter, because soft matter is typically disordered. Nevertheless, it is conceivable that the transition might be observable in colloidal crystals.[1]

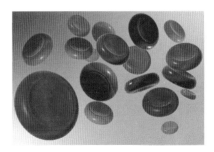

Figure 7.1. Artist's impression of red blood cells, whose outer membrane consists of a lipid bilayer. The cells' shape can be analyzed with the Helfrich model discussed in section 7.2, as figure 7.5 illustrates.[2]

as the interface profile during spreading using the lubrication approximation. The dispersion relation of capillary waves is treated in problem 1.11.

7.2 Helfrich free energy for membranes

Membranes are ubiquitous in biomatter, as they separate a living cell from its environment or enclose a particular compartment of a cell. Important examples are vesicles, the structures formed by a lipid bilayer encapsulating a fluid or cytoplasm, like the red blood cell illustrated in figure 7.1. The primary constituent of a membrane is a bilayer that forms in a water-based environment because of the hydrophilic nature of the lipid head and the hydrophobic nature of the tail. Most biomembranes are formed by bilayers of phospholipids, illustrated in figure 7.2. Phospholipids have two hydrophobic tails which want to avoid water, while their negatively charged phosphate head group is hydrophilic—i.e., likes to be in contact with water. This is the reason that in the water-based environment of cells the phospholipids naturally form bilayers with the hydrophobic tails arranged tail-to-tail on the inside, as sketched in the figure.

a. The Helfrich membrane model

We already touched on membrane formation of lipids in section 6.9.4. As we mentioned there, such membranes can be thought of as an example of liquid crystal mesophases: the molecules order in layers, but within the layers the lipids are disordered and behave like fluids. This means that within the membrane the molecules can easily flow and that the membrane properties are isotropic within the layers. Based on this observation and on what we learned from the discusion of the low-energy bending deformation of thin rods and sheets, it is thus natural to describe such membranes with an effective free energy in terms of the curvature only, the so-called Helfrich free energy (sometimes also referred to as the Helfrich Hamiltonian),

$$F_{\text{Helfrich}} = \int \mathrm{d}^2 S \left(\tfrac{1}{2} k (2H - c_0)^2 + \bar{k} K \right). \tag{7.1}$$

Here the integral is taken over the surface of the membrane. Furthermore, H is the total mean curvature, and K the so-called Gaussian curvature,

$$H = \tfrac{1}{2} \left(\frac{1}{R_1} + \frac{1}{R_2} \right), \qquad K = \frac{1}{R_1 R_2}, \tag{7.2}$$

Readers who want to have a first introduction to biomatter and the physical principles underlying living matter may enjoy consulting the book by Parthasarathy, 2022.

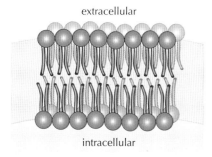

extracellular

intracellular

Figure 7.2. Phospholipids are (together with cholesterol) the main components of biological membranes. Phospholipids are amphiphilic lipids with a phosphate as the polar hydrophilic head, indicated here in orange, and two hydrocarbon tails which are hydrophobic. Lipids can also form other structures besides bilayers; see figure 6.26.

Helfrich introduced his expression for the membrane free energy in a famous paper, Helfrich, 1973, and tested and refined his approach over several decades. See Seifert, 1997 for an overview of the early work based on the Helfrich model and its extensions, and for references to Helfrich's papers. The overview by Leibler, 2004 also provides a nice introduction to membranes and historical developments.

where R_1 and R_2 are the two principal radii of curvature. The parameters k and $\bar{k}$ play the role of bending constants for mean and Gaussian curvature, while c_0 is introduced to model membranes which have a spontaneous mean curvature $H = c_0/2$. We will come back to this parameter below.

Note that the two radii of curvature have a sign: as illustrated in figure 7.3, a paraboloid-type surface has radii of the same sign, while at places where the radii have opposite signs, the surface looks locally like a saddle. Such saddle-type surfaces can have small or even zero total mean curvature, even though the two principle curvatures are nonzero.

Before discussing the Helfrich free energy in more detail, we note the similarity in spirit to our discussion of thin sheets in section 2.7, where we noted that bending deformations were the small-energy deformations. For membranes, the Helfrich model is similarly based on the idea that bending deformations are the relevant low-energy deformations. It is important to stress also that in order for this expression (7.1) to be of practical use for membranes, it has to be accompanied by two constraints, namely that the total surface area is constant and that the volume enclosed by the membrane is constant. These constraints are good approximations for large enough vesicles. For very small vesicles, it is necessary to extend the Helfrich energy with a surface energy–type term (surface area times surface tension) allowing the surface area to change, and an osmotic pressure–type term allowing the volume to change.[3]

The Helfrich free energy expression is especially useful for analyzing the shapes of closed surfaces as models for vesicles. For closed surfaces, the integrated Gaussian curvature term actually becomes independent of the precise shape, as according to the so-called Gauss-Bonnet theorem the integral

$$\oiint \mathrm{d}^2 S \, \frac{1}{R_1 \, R_2} = 4\pi \left(1 - g\right) \tag{7.3}$$

over a closed surface is a topological invariant g, the 'genus' of the surface. As illustrated in figure 7.4, the genus is an integer. Roughly speaking, the genus is the number of holes of the surface. Although the Gaussian curvature term can change, for instance, when budding leads to the splitting into two separate parts, large classes of deformations leave the genus invariant and hence do not affect the total Helfrich free energy. For such deformations, the energy explicitly contains only two parameters, the effective bending modulus k and c_0.[4]

The constant c_0 in equation (7.1) gives rise to a spontaneous curvature: if it is nonzero, the minimum of F_{Helfrich} is obtained for nonzero mean curvature H. As we remarked already in section 6.9.4, spontaneous membrane curvature is promoted by lipids or,

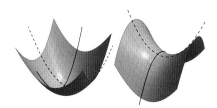

Figure 7.3. Illustration of surfaces with different curvatures. The parabolic surface on the left has two radii of the same sign and large mean curvature, while the saddle-like surface on the right has radii of opposite signs. If $R_1 = -R_2$ the mean curvature is zero.

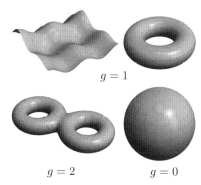

$g = 1$

$g = 2$ $g = 0$

Figure 7.4. Illustration of the genus g of closed surfaces and of the plane which extends to infinity. For a closed finite surface, the mathematical definition of genus amounts to g, the number of holes.

The modulus k can, for instance, be determined in a noninvasive way from the spectrum of thermal fluctuations, as illustrated in figure 3.13.b. For a discussion of this and of other ways to measure the parameters in the Helfrich free energy, see the review of Bassereau et al., 2014.

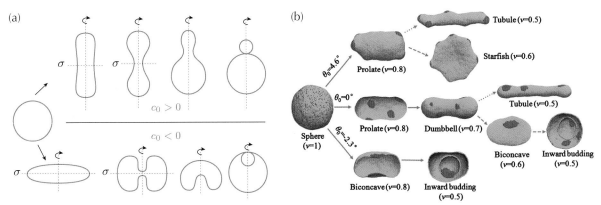

Figure 7.5. Membrane shapes predicted by the Helfrich model. Panel (a) shows the evolution of the minimum-energy shapes in the Helfrich model with parameters, both for positive spontaneous curvature parameter c_0 and for negative values of c_0. The symmetries of the various types of solutions are indicated: σ denotes reflection symmetry about the horizontal plane and the arrow indicates rotational symmetry about the vertical axis. Figure adapted from Seifert, 1997, who also discusses the underlying bifurcations. Panel (b) shows snapshots of simulations by C. Huang et al., 2011 of a particle-based membrane model with 11,054 particles as a function of changes in the parameters. The angle θ_0 is the preferred angle between the normals of neighboring membrane particles; for $\theta_0 \neq 0$ the membrane has a spontaneous curvature ($c_0 \approx 2 \sin \theta_0 / d$, where d is the diameter of the particles). The parameter ν is equal to the ratio of the actual volume enclosed to the volume of a sphere with the same area. Stomatocytes are bowl-shaped red blood cells which in blood smears appear as cells with a pale wide slit or stoma (mouth-like) area, and which are indicative of certain diseases.[6]

more generally, amphiphiles whose shape is conical, for instance, because their head is wider than the tails.[5]

b. Minimum-energy shapes of the Helfrich model

Given the Helfrich energy, the calculation of minimum-energy shapes is conceptually straightforward, although in practice one has to resort to numerical techniques as no nontrivial shapes are known analytically. Minimum-energy shapes depend not only on the two constants k and c_0, but also implicitly on the surface-to-volume ratio of the enclosed shape. The latter is usually expressed in terms of the parameter ν, the ratio of the volume of the enclosed shape to the volume of a sphere with the same surface area (defined this way, $\nu \leq 1$). The parameter space is thus rather large. Panel a of figure 7.5 shows the typical evolution and symmetry bifurcations of the shapes minimizing the Helfrich energy as parameters are changed. Schematically, the vertical direction indicates the variation of the curvature parameter c_0, while in the horizontal direction the parameter ν decreases to the right. Almost all of these shapes compare well with vesicle shapes encountered in the life sciences.

Panel b of the figure shows snapshots of simulations of a particle-based membrane model,[7] where the spontaneous membrane curvature is built in by having the axes of neighboring membrane particles make a small angle θ_0 relative to each other. Clearly the overall average shapes of the fluctuating closed membranes in this model conform very well to the minimum-energy shapes of the continuum Helfrich model.

We refer you to the review by Seifert, 1997 for a detailed discussion of the shapes determined by the Helfrich model and for a comparison of these with those of real vesicles.

Based on such studies in combination with sophisticated measurements of the parameters and detailed comparisons with actual vesicle shapes, the Helfrich model is nowadays well established as the starting point for the discussion and analysis of biomembranes and vesicles. Because of the importance and ubiquity of membranes in the life sciences, extensions have been explored in almost every imaginable direction. Modern developments include fusion and fission, shapes in phase-separated multicomponent membranes, calculation of the elastic moduli from the lipid properties, detailed calculation of the bending forces, the effect of flow on membrane shapes, wetting, budding, growth of necklace-like tubes from buds (tubulation), and the interaction of inclusions in lipid membranes.

For entries to these very diverse developments, see Komura and Andelman, 2014 for extensions to multicomponent lipid membranes; Campelo et al., 2014 for the relation between the lipid shapes and elastic moduli; Abreu et al., 2014 and Barthès-Biesel, 2016 for overviews of flow-induced deformations; Dimova and Lipowsky, 2016 for a discussion of wetting, budding, and tubulation; and Idema and Kraft, 2019 for a review of interactions between inclusions. A forward looking roadmap is given by Bassereau et al., 2018.

7.3 Virus shapes and buckling transitions in spherical shells

In chapter 2 on elasticity we discussed how bending is the lowest long-wavelength energy deformation of rods and thin sheets or shells. As we already remarked in section 2.7.3, however, when we consider deformations at increasingly smaller length scales, at some point the bending and stretching energy become of the same order of magnitude. Stated differently, there is an 'inner scale' of folds and cusps whose scaling properties are determined by a balance of stretching and bending. The membranes we discussed in the previous section are different as they are fluid-like within the membrane layer. Even so, for very small membranes one has to supplement the Helfrich bending energy terms with a surface energy term, as noted already. Mathematically, one can understand the fact that such problems have a small ('inner') scale from the observation that the bending terms involve the highest order derivative terms,[8] so that there is always a small scale where the bending terms become comparable to stretching or surface energy terms.

In the language of matched asymptotic expansions (Bender and Orszag, 1978; Van Dyke, 1975) the scale where stretching terms also become important is the 'inner' scale of such problems. Indeed, the bending modulus k in section 7.2 has a dimension $[k] = $ energy, while the surface tension σ has dimension $[\sigma] = $ energy/length^2, while according to equation (2.31) the effective Young's modulus $Y = E_Y d$ for a sheet of thickness d similarly has a dimension energy/length^2. So, for sheets the scale $\sqrt{k/Y}$, and likewise for membranes the scale $\sqrt{k/\sigma}$, sets the small length scale where stretching or surface tension becomes comparable to bending—compare the threshold expression (7.6) below.

a. Scenario for icosahedral shape deformations of viruses

In this section we give a nice demonstration of this crossover visible in the shapes of viruses, which we will think of as thin solid shells. This crossover, in combination with packing considerations that imply the presence of defects, determines most of the evolution of the shape of viruses with size. In passing, the analysis allows us to make contact with several concepts discussed before.

This section is based on the paper by Lidmar et al., 2003, which you may enjoy reading for further details and entries to the literature. For a more complete introduction to and overview of elasticity of thin shells, see Audoly and Pomeau, 2018.

Viruses are hollow structures composed of a viral shell, called the capsid, consisting of proteins, called capsomers. Most smaller viruses are spherical, like the rotavirus whose structure is shown in figure 7.6.a, while the shape of larger ones becomes more like

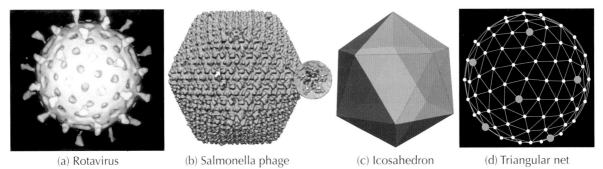

(a) Rotavirus   (b) Salmonella phage   (c) Icosahedron   (d) Triangular net

Figure 7.6. Structure of the rotavirus (a) and salmonella phage virus (b), as determined by cryo-TEM[9] measurements (a phage is a virus that kills bacteria; the tail of the salmonella phage virus is not shown). The rotavirus, with a diameter of about 80 nm, is close to spherical, while the salmonella phage, with a somewhat smaller diameter of about 60 nm, already has significant icosahedral deformations. The inset shows the fivefold coordinated packing at one of the 12 corner points.[10] An ideal icosahedron is shown in (c). As discussed in the text, the crossover from a spherical to an icosahedral shape with increasing size of the virus is induced by a buckling-like transition due to having 12 fivefold coordinated capsomers. Panel (d) shows an example of a triangulated net (an icosadeltahedron) of particles on a sphere (adapted and reproduced with permission from Lidmar et al., 2003). Vertices indicated with a large dot are fivefold coordinated. Euler's formula shows that there must be 12 such vertices for a triangulated net on a sphere.[11]

An icosahedron is the regular structure shown in figure 7.6.c, which has 20 triangular faces and 12 vertices.

Keep in mind that other shapes are found too: as figure 4.1 illustrated, the tobacco mosaic virus is rod-shaped. The figure also shows the structure of this virus with its capsomers and the DNA inside.

that of an icosahedron, shown in panel c. Although artist's impressions of some of these larger viruses are often based on the ideal isocahedron with sharp edges, one should keep in mind that in practice the ones of intermediate size are closer to icosahedrally deformed spheres. The salmonella phage virus shown in panel b provides a good illustration of this.

Many viruses, like the Covid-19 virus and the rotavirus, have protein spikes sticking out of the capsid—the Covid-19 virus pictured as a foam ball with spikes sticking out became the icon of the Covid pandemic in 2020. These spikes are critically important for receptor recognition and for the fusion process with the membrane of the cell under attack. We will not go into this here, but we will focus on the shape change from spherical to icosahedral with size of the virus.

We saw already in section 2.11 that when disk-shaped particles are packed in two dimensions, they form a triangular lattice in which all disks have six neighbors in the absence of defects. But when they are packed on a sphere, such an ideal triangulation is not possible. To see this, let us draw the capsomers forming the capsid of the virus as vertices in a network, and connect all the neigboring ones by a line. An example of this is shown in panel d of figure 7.6. The packing is thus represented by a network of vertices and edges, which triangulate the sphere. According to Euler's formula

$$F + V - E = 2 \qquad (7.4)$$

See problem 7.3 on Euler's formula and for a derivation of the fact that there are exactly 12 fivefold coordinated vertices on a triangulated net on a sphere.

for the number of vertices V, the number of edges E and the number of faces F of such a net on a sphere,[12] it is easy to derive that there must be exactly 12 vertices (particles or capsomers of the virus) which have five instead of six neighbors.[13] These are indicated by the heavy dots in figure 7.6.d. If you look closely at the vertices of the triangles superposed on the two viruses shown in panels a and b, you will indeed see that the capsomers at those vertices are fivefold coordinated, while the other ones have six neighbors.

To understand why the 12 fivefold coordinated capsomers lead to icosahedral deformations away from the sphere, note that, as illustrated in figure 7.7, a flat triangular lattice with a single fivefold disclination is highly strained. In fact, as the total deformation grows linearly with the distance from the center, the strain away from the center is roughly constant, so the total strain energy grows with the area of the system, hence quadratically with the system size. The energy for a large circular shell of radius R is therefore dominated by the area away from the center, and we can calculate this using continuum theory. Indeed, if one considers the distorted capsid as a thin shell with total two-dimensional Young's modulus Y,[15] one finds for the total strain energy E_5 resulting from a fivefold disclination defect

$$E_5 = \frac{1}{32\pi} s^2 Y R^2, \tag{7.5}$$

confirming that the energy grows with the area of the shell. Here $s = 2\pi/6$ is the disclination strength—compare our discussion in section 6.6 of the strength of defects.

The conclusion that the energy grows with the area is true if the lattice remains flat. However, bending can release most of the strain, as illustrated on the right in figure 7.7. Indeed, since for a conical shape the radius of curvature grows linearly with the distance r from the center, it is easy to see from the bending energy (7.1) with bending modulus k that the total bending energy of a bent cone goes as $\int d^2 S \frac{1}{2} k (1/r)^2 = (\pi k) \int^R dr \, r^{-1}$, which grows logarithmically with the size R of the sheet. So for sufficiently large sheets, the bent cone has a lower energy—see figure 7.8.

These considerations can be backed up by more explicit calculations which show that a flat circular sheet of radius R with a fivefold defect in the center is unstable to a buckled conical shape for radii R satisfying

$$\frac{Y R^2}{k} \geq 154. \tag{7.6}$$

For radii above the threshold radius $R_\mathrm{b} = \sqrt{154\,k/Y}$ the energy of the buckled state then grows logarithmically with size as

$$E_5 = \frac{1}{32\pi} s^2 Y R_\mathrm{b}^2 + \frac{\pi}{3} k \ln(R/R_\mathrm{b}), \quad (R \geq R_\mathrm{b}). \tag{7.7}$$

Based on these estimates and on viewing the viral capsid as a thin shell with effective Young's modulus Y and bending rigidity k, one expects the following scenario. The virus always has 12 capsomers

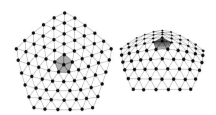

Figure 7.7. A fivefold disclination in a flat triangular lattice. The highly strained lattice leads to a total strain energy which grows as the area squared, while buckling releases some of the strain and leads to an energy growing logarithmically with size. For a virus, the lattice vertices correspond to the positions of the capsomers. Adapted and reproduced with permission from Lidmar et al., 2003.[14]

Details of the calculations are given in the papers referenced by Lidmar et al., 2003 and in problem 2.8.

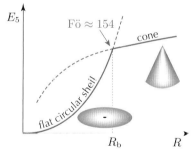

Figure 7.8. The behavior of the energy E_5 of a flat circular shell of radius R with a disclination defect in the center, and that of a cone. For $R > R_\mathrm{b}$ the cone has the lowest elastic energy.[16] In terms of the Föppl–Von Kármán number (7.8), the crossover corresponds to Fö ≈ 154.

Hamiltonian consists of the simple interaction term

$$\mathcal{H}/k_{\mathrm{B}}T = -W \sum_{\langle \alpha,\beta \rangle} \hat{n}_\alpha \cdot \hat{n}_\beta, \qquad (7.9)$$

where the sum runs over neighboring triangles and where $\hat{n}_\alpha$ denotes the normal of triangle α. This interaction term thus favors neighboring triangles to be roughly parallel. Such triangular net–based surface models are usually called tethered surfaces; in the absence of excluded volume interactions a surface can intersect itself—such artifical surfaces are sometimes referred to as phantom surfaces.

Snapshots of three phantom surface configurations obtained in Monte Carlo simulations for various values of W shown in figure 7.11 clearly demonstrate that surfaces exhibit two different phases: a crumpled phase where the surface coils up, and a flat phase where the surface on large scales behaves as a flat sheet with small fluctuations on top. Detailed analysis of the scaling of the radius of gyration R_{g} with the linear size L of the net shows that in the crumpled phase R_{g} grows very slowly with size, $R_{\mathrm{g}} \sim \ln L$, while in the flat phase one has $R_{\mathrm{g}} \sim L$, the same scaling as that of a flat sheet of paper.

The fact that R_{g} grows only as $\ln L$ means that in the crumpled phase many vertices lie almost on top of each other and that the phantom surface intersects itself a lot. After all, with excluded volume interactions, R_{g} must grow at least as $N^{1/3} \sim L^{2/3}$, since the number of vertices N of the sheet grows as L^2. So it is obvious that excluded volume interactions must change the behavior dramatically.[20] In fact, from a large number of analytical and numerical studies of generalized D-dimensional sheets embedded in d space dimensions, it has transpired that when self-avoidance effects are included, such tethered surfaces exhibit only a flat phase.[21] In other words, there is no crumpling transition.

b. Elastic sheets: Stiffening due to coupling of strain and height

An interesting case is also provided by thin elastic sheets, sheets whose elastic strain deformation has to be taken into account. That there is a strong coupling between shape fluctuations and strain is already intuitively clear from figure 7.7 and the analysis in the previous section, which showed that the strain from a fivefold disclination can be relieved by bending of the sheet. By the same token, bending induces long-range strain; this in turn enhances the bending energy.

This argument can be made more precise by writing the free energy F for a nearly flat elastic shell. For small height deformations h out of the two-dimensional reference plane, we expand the curvature term to the lowest order in h; together with the two-dimensional

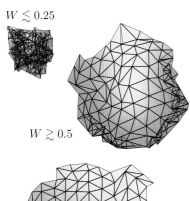

$W \lesssim 0.25$

$W \gtrsim 0.5$

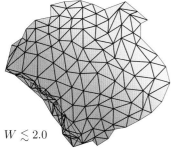

$W \lesssim 2.0$

Figure 7.11. Three representative configurations in Monte Carlo simulations of the tethered surface model with Hamiltonian (7.9) and without excluded volume interactions, for parameter values of W as indicated. The crumpling transition of this model is at $W_c \approx 0.33$. In the flat phase for $W > W_c$ the radius of gyration scales as $R_{\mathrm{g}} \sim L$ with the linear size L of the net, while in the crumpled phase for $W < W_c$, one finds $R_{\mathrm{g}} \sim \ln L$. Adapted from Kantor, 2004.

strain fields u_{ij}, we then have

$$F[h, u] = \frac{1}{2} \int d^2 \vec{r} \left[k(\nabla^2 h)^2 + 2\mu u_{ij}^2 + \lambda u_{kk}^2 \right]. \qquad (7.10)$$

At first sight, it may appear from this expression that the strain fields and height deformations are decoupled; the coupling originates, however, from the effect of an out of plane deformation on the strain. Indeed, if we go back to the full nonlinear expression (2.3) for the strain, the vertical deformations h feed back on the two-dimensional strain field through this nonlinear term,[22]

$$u_{ij} \approx \frac{1}{2} \left(\frac{\partial u_i}{\partial x_j} + \frac{\partial u_j}{\partial x_i} \right) + \frac{1}{2} \left(\frac{\partial h}{\partial x_i} \frac{\partial h}{\partial x_j} \right). \qquad (7.11)$$

When the strain fields u_{ij} are integrated out to get an effective free energy $F_{\text{eff}}[h]$ as a function of the height h only, the quadratic term in h on the right-hand side of (7.11) acts as a source term for the strain fields u. It can be shown that as a result of the long-range nature of elastic deformations, the effective energy acquires an effective nonlocal fourth order interaction term in h. Its form implies that regions where the Gaussian curvature is nonzero lead to an effective strain-induced long-range interaction of the h field which suppresses fluctuations.

In other words, the counterpart of the strain-induced buckling of viruses discussed in section 7.3 is that local cone-like bulges of a nearly flat surface lead to large long-range strain deformations, and hence they are strongly suppressed. As a result of this coupling to the shear modes, the effective bending modulus of graphene sheets illustrated in figure 2.6 is easily enhanced by three or more orders of magnitude over the bare value calculated from atomic force calculations.[23]

In conclusion, therefore, the weight of the evidence is that elastic sheets do not exhibit a thermally driven crumpling transition; even very thin sheets like graphenes have a very large bending modulus, due to the coupling of out of plane deformations to the strain modes.

c. Fluctuating membranes

As we discussed in section 7.2, the amphiphilic molecules of biological membranes behave in a fluid-like manner within the membrane layer. The overall membrane shape fluctuations are governed by the bending terms of the Helfrich energy. As the amphiphilic molecules within the membrane do not respond elastically, shape fluctuations are not suppressed by the strain deformations of elastic sheets that we just discussed. In principle such membranes are hence predicted to exhibit a crumpling transition; unfortunately, however, the practical implications of this result are limited.[24]

To get this result, we combine expressions (2.4) and (2.64) for the bending energy and the elastic energy, and we expand the curvature term to the lowest order in h. You'll remember that μ and λ are the Lamé coefficients.

The term 'integrated out' refers to a statistical description, in which the free energy $F[h, u]$ is used as the microscopic energy in a partition function, and the fluctuating height variable h and strain variables u_{ij} are to be integrated over to determine the total free energy of the fluctuating system. Integrating out the strain variables means that these variables in the partition function are integrated over for fixed height variable h. For a detailed pedagogical treatment of these arguments we refer you to Nelson, 2004, who explicitly discusses how integrating out the strain fields gives rise to a long-range effective h interaction which penalizes local deformations with nonzero Gaussian curvature. In fact, the result is so strong that, due to the long-range nature of the effective interaction, the scale-dependent bending modulus increases with length scale ℓ as ℓ^η, where $\eta \approx 0.82$. For a review of the various implications of the coupling between elasticity and shape fluctuations of elastic membranes, see Le Doussal and Radzihovsky, 2018.

d. Exploration of possible experimental realizations

Experimentally, polymerized membranes have been explored as possible realizations of tethered surfaces. It turns out that typical bending constants of most materials are much larger than $k_{\mathrm{B}}T$, so the chances to explore the temperature range where crumpling effects might become visible are few.[25] A new development is the realization that the effective bending constant and Young's modulus of thin crystalline sheets, such as layers of graphene, can be lowered by creating arrays of holes in them.[26] Using isotropically compressed thin sheets might also yield a route toward observing a thermal crumpling transition.[27]

7.4.2 Athermal crumpling by compression

Let us now turn to the crumpling of macroscopic sheets which, like the crumpled paper ball of figure 7.10, are thick and large enough that thermodynamic fluctuations play no role. Clearly, while on the one hand disorder and excluded volume effects play an important role in the formation of new creases and wrinkles, at the same time existing ones are likely to affect where and how new ones are formed. Hysteresis is sometimes also argued to play a role. The challenge is to develop a suitable statistical description of this nonlinear and far from equilibrium process.

a. Increase of the bending modulus in regular folding of a sheet

To get a feel for why a crumpled piece of paper can resist relatively large forces, even though a typical crumpled paper ball is about 80% air, let us first consider briefly regular folding. You'll certainly have experienced that you are able to fold a sheet of printing or kraft paper in half only about six or seven times. This is basically because the bending modulus grows with the cubed power d^3 of the thickness of the sheet, so every time you fold the paper, i.e., double the effective thickness, the bending modulus increases by a factor of 8. Clearly, in the folding process you are running up against the rapid exponential increase of the bending modulus as 2^{3n} with the number of folds n. This is basically the reason that six or seven is about the maximum number of folds an ordinary person is able to make.[31] Crumpling by compression is more subtle, as it is a random process that does not lead to regular folds. Nevertheless, the scaling behavior of folds does play a role.

b. Scaling behavior of the response during crumpling

Systematic studies of the process of crumpling by compressing a paper or Mylar sheet have recently explored new avenues—as we have noted at various occasions in this book, recent progress on numerical simulations and in data acquisition and processing make it possible to investigate what was out of the question just a

For a review of the stress focusing at creases and folds, and of the various open issues, see Witten, 2007.[28]

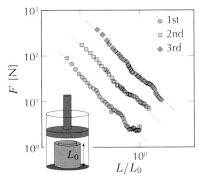

Figure 7.12. Compression experiments on crumpling paper. The inset shows the setup used by Deboeuf et al., 2013 to compress sheets of kraft paper of linear size L_0 by lowering the piston to a height L. The graph shows the measured force F in newtons as a function of compression ratio L/L_0 on a log-log scale, upon the first (blue circles), second (green squares), and third (red lozenges) compressions. Be aware that the curves have been shifted vertically for clarity of presentation: unlike what the curves appear to suggest, the crumpled paper softens with each cycle. The measured exponent[29] α defined in (7.12) decreases from 0.9 for the first compression to 0.71, and 0.70 for the second and third compressions, while $F_0 = 3.74, 1.9$, and $1.43\,\mathrm{N}$ in the first, second and third compressions. Image courtesy of Daniel Bonn.[30]

few years back. As a result our understanding of the process is still in its infancy, but growing.

Figure 7.12 shows an example of a simple experiment in which a piece of kraft paper of linear size L_0 is compressed by lowering the piston to a height L and measuring the force F on the piston during this unidirectional compression. Although the range of the data is limited to just over one decade, within this range the data are reasonably well fitted by a power law,

$$\frac{L}{L_0} = \left(\frac{F_0}{F} \right)^{\alpha} . \qquad (7.12)$$

The exponent α decreases upon each compression cycle and is about 0.91 during the first compression and about 0.7 during the third one.[32] The force needed to compress the sheet by a given amount decreases during successive cycles. These values are quite a bit larger than those measured earlier in compression experiments with Mylar,[33] which gave an exponent of about 0.53, and with the numerical simulations discussed below—it is not clear whether this is due to the experimental protocol or the material.

Numerical simulations of a triangulated surface model of elastic sheets allow one to vary parameters, investigate the effect of excluded volume interactions, and study the statistics and contributions of stretching and bending to the energy separately. Figure 7.13 shows data from these with the Föppl–Von Kármán number Fö defined in (7.8) varying over three orders of magnitude. In these simulations, sheets of radius R_0 are compressed three-dimensionally to a sphere of radius R. The data show good collapse to a power law suggested by dimensional arguments applied to stress focusing,[34]

$$\frac{R}{R_0} = C \, \text{Fö}^{\beta} \left(\frac{k}{F R_0} \right)^{\alpha} = C \left(\frac{Y R_0^2}{k} \right)^{\beta} \left(\frac{k}{F R_0} \right)^{\alpha} . \qquad (7.13)$$

Here the combination k/R_0 is essentially the buckling force of a sheet which is compressed from the side.[35]

In these numerical simulations there is a noticeable difference in the force response between phantom sheets without excluded volume interactions and elastic sheets with these interactions. Indeed, for the latter the scaling exponent $\alpha \approx 1/4$, while $\alpha \approx 3/8$ is found for phantom sheets. A value $\alpha = 3/8$ is in accord with what one expects for such sheets on the basis of the energy scaling of ridges. Furthermore, the numerical data both with and without excluded volume interactions are consistent with values of β close to the value 1/16 predicted from such scaling arguments.[36]

Simulations allow one to also study the scaling for two-dimensional and one-dimensional compressions and to extend (7.13) to these

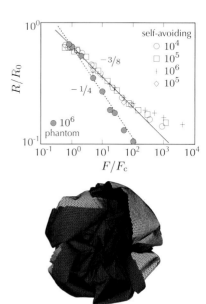

Figure 7.13. Results of a numerical study of crumpling. The lower panel shows a snapshot of a compressed sheet in the numerical simulations of Vliegenthart and Gompper, 2006. Upper panel: scaling behavior of the circular elastic sheets of radius R_0 in these simulations as a function of compression force, for various R_0. The sheets are compressed three-dimensionally to a ball of radius R. Values of Fö defined in (7.8) range from $3.4 \cdot 10^4$ to $4.7 \cdot 10^7$. The filled circles are data for a phantom sheet for Fö $= 0.9 \cdot 10^6$ without excluded volume interactions. There is good data collapse. Note that the scaling exponent α of the phantom sheets differs from the ones with excluded volume interactions. Image courtesy of Gerrit Vliegenthart.

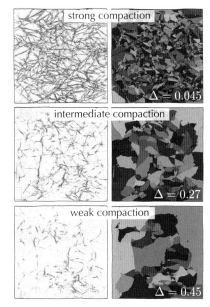

Figure 7.14. Experimental results of Andrejevic et al., 2021 on square Mylar sheets of linear size L_0, which have been compressed with a setup comparable to the one shown in figure 7.12. After compaction, the sheets have been unfolded and scanned with a laser profilometer. The amount of compaction, measured by $\Delta = L/L_0$, increases from bottom to top. The left panels show the mean curvature profiles, with red and blue denoting folds in opposite directions, the right panels the corresponding 'facets' which are colored randomly, to facilitate discerning them visually. Image courtesy of Jovana Andrejevic and Chris Rycroft.

This section is inspired by the results and analysis of Andrejevic et al., 2021. We refer you to their paper for more details and references to the literature.

results. It is found that the effective exponent α increases to $\alpha = 0.5$ for one-dimensional compression. This is still quite a bit below the value observed in the experiments of figure 7.12.[37] It should be clear from this discussion that our understanding of the scaling of crumpled sheets is still in its infancy.

For polymers and thermally crumpled sheets one of the central predictions is the scaling $R \sim L^\nu$ of the radius R with linear system size L. For the athermal case, one can also investigate the scaling of the radius of sheets crumpled by a compression force of a fixed magnitude. From the relation (7.13), one has, for fixed force,

$$R \sim R_0^\nu, \quad \text{with} \quad \nu = 1 + 2\beta - \alpha. \tag{7.14}$$

Inserting the values for α and β stated above gives $\nu \approx 0.87$. This is not very far from the rough estimate you can make by crumpling sheets of various sizes by hand and measuring their radius.[38]

c. Crumpling as a self-similar fragmentation process

Even though scaling ideas capture the overall scaling properties reasonably well, they address neither the statistical nature of the crumples and the creases, nor the process through which they form. Luckily, modern techniques make it possible to sample experimental crumples with high accuracy, and to process the data for probing the statistical properties of crumpling. Figure 7.14 shows an example of the local curvature (left column) and of the 'facets' (right column) of unfolded crumpled Mylar sheets after both weak and strong compaction. From such detailed data on wrinkles during successive compaction cycles, it is possible to follow the crumpling process statistically and to interpret it in terms of a framework developed for fragmentation of objects into smaller ones.

We briefly sketch here the main steps, which suggest that it pays to approach crumpling in terms of a self-similar process which obeys scaling forms of the type discussed in section 3.3.5.

Let us think of the crumpling as a fragmentation process in which the basic objects, the facets, have a certain probability r of breaking up into smaller pieces. If we assume a mean-field-type description in which we ignore the influence of other facets on this fragmentation process, we can write the following rate equation for the fraction $c(v, t)$ of facets of area size v at time t:

$$\frac{\partial c(v,t)}{\partial t} = -r(v)c(v,t) + \int_v^\infty \mathrm{d}w\, p(v|w)r(w)c(w,t). \tag{7.15}$$

Here $r(v)$ is the rate at which a facet of area v fragments and $p(v|w)$ the conditional probability that a facet of size $v \leq w$ is produced from the breakup of the facet of size w. The 'time' t is taken as an effective measure of the 'maturity' of the fragmentation process—it increases with the compaction cycle number and depth of the compaction. Clearly the first term on the right gives the rate at which facets of size v disappear as they are splitting into smaller ones, while the second term describes the rate at which new facets are created by the breakup of larger ones. Since the total area is conserved when a facet splits into two by creating a new ridge, one needs to impose

$$\int_0^w \mathrm{d}v\, v p(v|w) = w. \tag{7.16}$$

Note also that by writing the rate equation (7.15) we assume that the crumpling depends only on the present distribution of facets. In other words, there is no aging or memory in the process.

The advantage of this approach is that the functional form of the breakup rate $r(v)$ and the conditional probability $p(v|w)$ can be estimated from detailed analysis of successive crumpling cycles. These fragmentation studies suggest power law functional forms

$$r(v) = v^\lambda, \quad p(v|w) = \frac{1}{w}(\mu + 2)\left(\frac{v}{w}\right)^\mu \tag{7.17}$$

with $\lambda \approx 1/2$. These power law expressions naturally suggest looking for similarity-type solutions of the form $c(v) = \phi(\xi)t^{2/\lambda}$, where $\xi = vt^{1/\lambda}$ is the similarity variable (see section 3.3.5). Interestingly, the similarity variable does not depend on the exponent μ of the breakup probability p. With this self-similar scaling ansatz, the form of the facet distribution can be calculated explicitly from the rate equation (7.15). The resulting expressions then provide a convenient framework to compare in detail with the distribution functions of various quantities characterizing the crumpled Mylar during successive cycles.

The main steps of the similarity analysis are treated in problem 7.4.

The fact that the detailed data of these crumpling experiments are consistent with self-similar behavior throws on the crumpling problem a new light which is likely to bring more new insights. For example, indications of a logarithmic increase of the crease length with the number of compaction cycles had earlier been associated with creep or rare microscopic events in disordered systems. According to the above approach, however, it may naturally emerge from the self-similarity of the fragmentation process.

7.5 A soft matter realization of the one-dimensional KPZ equation

Although we cannot do justice here to the enormous literature on the dynamical scaling behavior of fluctuating interfaces, it is worth giving an idea of the types of scaling results and of the field's

KPZ refers to Kardar, Parisi, and Zhang (Kardar et al., 1986), who introduced the equation. In 2021, Giorgio Parisi received the Nobel Prize in Physics for "the discovery of the interplay of disorder and fluctuations in physical systems from atomic to planetary scale."

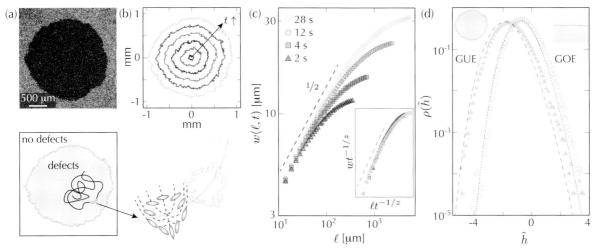

Figure 7.15. Summary of a series of experiments by Takeuchi and Sano, 2012 on growing interfaces between a defect-free turbulent region and a defect-driven turbulent region in voltage-driven convection of a nematic in a thin cell. (a) Snapshot of a circularly growing defect-driven chaotic convection phase. The experimental cells are only 12 microns thick, so they are quasi-two-dimensional. The black area is the defect-dominated turbulent domain, where defect lines constantly elongate and split while they are convected (as also discussed in section 6.6.1 and in chapter 9 on active matter). This process and the nature of the $s = 1/2$ defect is illustrated below the image. Panel (b) shows traces of the growing interface between 3 and 28 seconds for the circular case as shown in panel (a). In the experiments, also the fluctuations of growing fast interfaces are probed. (c) Data for the interface width $w(\ell, t)$ defined in (7.20) plotted as a function of ℓ for various times. The inset shows the excellent data collapse when both length and width are scaled in accord with (7.21) with the 1d-KPZ exponents $\beta = 1/3$, $z = 3/2$. (d) The probability distribution of rescaled height fluctuations $\tilde{h} = (h - v_\infty t)/(\Gamma t)^{1/3}$ for both the circular interface at $t = 10$ s (filled triangles) and the flat interface at $t = 60$ s (open circles). The dashed and dotted curves show the Gaussian Unitary Ensemble (GUE) and Gaussian Orthogonal Ensemble (GOE) distributions of random matrix theory. The experimental data for the two cases compare well to these two ensembles, without adjustable parameters and in agreement with what has been predicted for the KPZ equation. Note that the distributions are somewhat asymmetric, due to the fact that the equation is not reflection-symmetric in h. Data and images courtesy of Kazumasa Takeuchi.[45]

We see here another example—compare section 6.9.5—of the strong effect of $s = +1/2$ defects on the dynamics, which is so important in active matter.

Moreover, once this phase has nucleated, the defect-dominated domain expands into the domain without defects, and the two domains are separated by a well-defined interface. Since the only difference between the domains is the presence or absence of the defects, and since the density of defects in the defect-tangle domain is quite high, there are no obvious long-range fields affecting the dynamics of the one-dimensional interface, while the system is intrinsically noisy. This is probably why this system has proven to be such a good system for studying the interface fluctuations and comparing with KPZ scaling. Panel b shows traces of circular realizations of the interface during growth. Similar experiments are done on growing flat interfaces of the defect tangle domain.

In panel c we present the data for the width $w(\ell, t)$ defined in (7.20) as a function of the averaging length ℓ for various times. The scaling $w \sim \ell^{1/2}$ is not so surprising, since equilibrium one-dimensional interfaces also scale with this exponent (margin note below (7.21)). The scaling collapse is a more critical test of the theory, since it depends on the temporal scaling. As the inset shows, the data convincingly collapse onto a single scaling form consistent with (7.21) and the particular 1d-KPZ exponents $\beta = 1/3$ and $z = 3/2$.

Testing for scaling behavior of standard quantities like the interface width is regularly done in such types of studies, but these experiments allow one to go much further. A nontrivial example of this is presented in figure 7.15.d. It has been shown rigorously that the probability distribution of the scaled height variations in the KPZ equation conform to the so-called Gaussian Unitary Ensemble of random matrix theory for circular interfaces, while the height distribution is given by the Gaussian Orthogonal Ensemble for flat interfaces in certain growth models.[46] The height measurements in these experiments are accurate enough to test this; as panel d shows, the consistency of theory and experiment is impressive. The differences between circular and flat interfaces are subtle, but measurable. To our knowledge, these remarkable experiments are the most detailed experimental tests of the predictions of the 1d-KPZ equation to date, and the agreement is comforting.

Random matrix theory is based on analysis of eigenvalues of matrices whose elements are random variables. Depending on the symmetries of the matrix and the nature of the elements (real or complex valued), the eigenvalues and level spacings obey a particular distribution function, whose analytical form is known. For a general introduction to random matrix theory, see, e.g., Livan et al., 2018.

7.6 What have we learned

Surfaces, interfaces and membranes constitute important elements of soft matter. The surface tensions of interfaces between liquid phases or between two fluids and a substrate determine the contact angle as well as the wetting properties. Moreover, concentration and temperature gradients induce surface tension gradients which give rise to flow.

Membranes are important in the life sciences at the cellular scale. The Helfrich bending energy describes their lowest energy shape deformations, and it can be used to analyze the large variety of membrane shapes.

The shape of elastic surfaces and shells is affected by the competition between elastic deformations and bending, and, in the case of viruses, the packing constraints imposed by topology. This allows us to follow the evolution from a nearly spherical form to a more icosahedral viral shape as the size of a virus increases.

For elastic sheets, the coupling between height fluctuations and elastic fields results in a strong suppression of thermal shape fluctuations: elastic sheets are much flatter than one might initially expect. Sheets that are crumpled by force, by compressing them in a container or by applying a large external force, exhibit a number of scaling properties, which are reminiscent of self-similar scaling properties of fragmentation processes.

An interface in liquid crystal electroconvection between a defect-driven turbulent phase and a nonturbulent state provides a surprising experimental realization of a growing one-dimensional interface with fluctuations, as described by the KPZ equation.

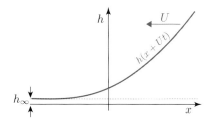

Figure 7.17. Sketch of a wetting front $h(x + Ut)$ moving with constant velocity U to the left. Ahead of the front there is already a thin layer of thickness h_∞ due to Van der Waals forces.

where $\mathrm{Ca} = \eta U / \gamma$ is the capillary number for this case (with η the dynamic viscosity as in chapter 1). Here we use a shorthand notation h_{xxx} for $\frac{\partial^3 h}{\partial x^3}$. In practice, spreading velocities for water droplets are small enough that the capillary number is small, $\mathrm{Ca} \ll 1$.

b. Show that equation (7.28) implies

$$3\,\mathrm{Ca}\,(h - h_\infty) + h^3 h_{xxx} = 0. \tag{7.29}$$

The above equation can easily be analyzed numerically; it gives a profile as in figure 7.17. Below we analyze the asymptotic behavior on both sides for large positive and negative x.

c. First consider the behavior as $x \to -\infty$. Linearize the equation in $\Delta h = h - h_\infty$, substitute $\Delta h \sim \exp(ikx)$, and show that this gives

$$k^3 = -i\,\frac{3\,\mathrm{Ca}}{h_\infty^3}. \tag{7.30}$$

d. Argue that you need solutions with $\mathrm{Im}(k) < 0$. Show that there are two roots which satisfy this criterion, and that these imply that to the left the profile approaches h_∞ exponentially with spatial oscillations.

e. We now analyze the equation for large positive x, where $h \gg h_\infty$, so that (7.29) reduces to

$$3\,\mathrm{Ca} + h^2 h_{xxx} = 0. \tag{7.31}$$

Initially, one would expect $h \sim x$ for $x \gg 1$, but clearly this does not solve the equation. Substitute the ansatz $h = Ax(\ln x/x_0)^\alpha$ into this equation and analyze the equation for large x to dominant order (i.e., up to corrections which vanish logarithmically), and show that $\alpha = 1/3$ and $A^3 = 9\,\mathrm{Ca}$, so that the solution for large positive x is

$$h \approx x\,[9\,\mathrm{Ca}\,\ln(x/x_0)]^{1/3}. \tag{7.32}$$

The constant x_0 has to be obtained from matching the solution to the behavior analyzed in (c) when $h \to h_\infty$. Note that according to this equation there are weak logarithmic corrections to the dynamic contact angle $\theta = h_x$.

Problem 7.3 *Derivation of Euler's formula (7.4) and application to viruses*

In this problem we derive Euler's formula (7.4) for an object with the topology of a cube or sphere.

a. Start from the cube in the left panel of figure 7.18. Count the number of faces F, the number of edges E, and the number of vertices V and show that (7.4) is obeyed.

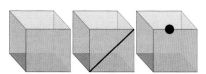

Figure 7.18. The derivation of the Euler formula can be given by starting from a cube (left panel) and considering the effect of adding an edge (middle panel) or a vertex (the dot in the right panel).

b. Show that if an edge connecting to vertices is added, as in the middle panel of figure 7.18, the change in number is such that (7.4) continues to be obeyed.

c. Show that if a vertex is added, as in the right panel of figure 7.18, the change in number is such that (7.4) continues to be obeyed. This demonstrates that the Euler formula continues to hold for any imaginable change.

The Euler formula can be generalized to $F + V - E = 2(1 - g)$, where g is the genus of the surface introduced in equation (7.3).

d. We now apply Euler's formula to viruses. Introduce, for a triangulated net on a sphere as sketched in figure 7.6.d, the number V_5 of fivefold coordinated vertices and V_6 for the number of sixfold coordinated vertices. Show that $F = (5V_5 + 6V_6)/3$ for the number of faces F and $E = 3F/2$ for the number of edges.

e. Use the results from Euler's formula (7.4) in step d to show that $V_5 = 12$, in other words, that there are exactly 12 fivefold coordinated vertices of the type indicated with a heavy dot in figure 7.6.d.

Problem 7.4* *Similarity analysis of the rate equation (7.15)*
In this exercise we perform the main steps of the similarity analysis for equation (7.15). While the analysis below is specific to the rate equation for crumpling introduced in section 7.7, it is a good illustration of how a similarity ansatz can simplify the analysis of a complicated problem.

a. Write the scaling ansatz for the fraction of facets of area v at time $tc(v, t)$ in the form $c = s^{-2}\phi(v/s)$ with $s = s(t)$ and introduce the scaling variable $\xi = v/s$. Substitute this into the rate equation (7.15) with the forms (7.17) and show that this yields the equation

$$
\begin{aligned}
&- \dot{s}s^{-(\lambda+1)} \left(2\phi(\xi) + \xi\phi'(\xi)\right) \\
&= -\xi^\lambda \phi(\xi) + (\mu + 2) \int_\xi^\infty d\eta\, \phi(\eta)\eta^{\lambda-1} \left(\frac{\xi}{\eta}\right)^\mu,
\end{aligned} \tag{7.33}
$$

where $\dot{s}$ indicates the time derivative of s and ϕ' the derivative of ϕ.

b. Next, argue that the expression (7.33) implies two following equations,

$$
-\dot{s}s^{-(\lambda+1)} = G/\lambda \tag{7.34}
$$

and

$$
\frac{-\xi^\lambda \phi(\xi) + (\mu + 2) \int_\xi^\infty d\eta\, \phi(\eta)\eta^{\lambda-1} \left(\frac{\xi}{\eta}\right)^\mu}{2\phi(\xi) + \xi\phi'(\xi)} = G/\lambda, \tag{7.35}
$$

where G is a constant.

c. Show that (7.34) implies the scaling behavior

$$s = \left(Gt + s_0^{-\lambda}\right)^{-1/\lambda}. \tag{7.36}$$

d. Next, consider equation (7.35). We will verify that the scaling function $\phi(\xi)$ is

$$\phi(\xi) = C\xi^\mu \exp(-\xi^\lambda/G), \tag{7.37}$$

with C a normalization constant. As a first step show that with this solution the integral in the numerator of (7.35) is simply equal to $(G/\lambda)\phi(\xi)$.

e. Show that

$$\xi\phi'(\xi) = \mu\phi - \frac{\lambda}{G}\xi^\lambda\phi. \tag{7.38}$$

See Andrejevic et al., 2021 for the final expression for G. In comparing, note that our G differs slightly from the one used in that paper.

f. Using the results from the previous two steps show that the numerator and denominator of the left-hand side of (7.35) both contain a term proportional to ϕ and to $\xi^\lambda\phi$, with prefactors which involve G/λ, and hence that the condition that the ratio on the left-hand side is constant finally yields an algebraic equation for the constant G in the solution (7.37).

Problem 7.5* *Invariance, scaling, and exponents of the 1d-KPZ equation*

In this exercise you will calculate the critical exponents of the 1d-KPZ equation (7.18).

For more details, see section 9.6 of Kardar, 2007.

a. Show (in arbitrary dimension) that the KPZ equation is invariant under the tilt transformation

$$h'(\vec{r}', t) = h(\vec{r} - \vec{u}t, t) - \frac{1}{\lambda}\vec{u}\cdot\vec{r} + \frac{1}{2\lambda}\vec{u}^2 t, \tag{7.39}$$

with a shifted noise term $\eta'(\vec{r}', t) = \eta(\vec{r}' + \vec{u}t, t)$, in other words, that h' obeys the same KPZ equation as h, with the noise η replaced by η'. The KPZ equation thus has a rotational invariance.

b. Write $\vec{v} = -\lambda\vec{\nabla}h$ and show that the KPZ equation is equivalent to the noisy Burgers equation

$$\partial_t\vec{v} + \vec{v}\cdot\vec{\nabla}\vec{v} = \nu\nabla^2\vec{v} + \vec{\nabla}\eta, \tag{7.40}$$

which is an equation for a potential flow velocity $\vec{v}$. Note that in this context the nonlinear term ensures Galilean invariance: the equation is the same in a frame $\vec{r}' = \vec{r} - \vec{u}t$ moving with a constant velocity $\vec{u}$.

c. Perform a scaling $\vec{r} \to b\vec{r}$, $t \to b^z t$, $h = b^\chi$, show that the KPZ equation transforms into

$$b^{\chi-z} \partial_t h = \nu b^{\chi-2} \nabla^2 h + \frac{\lambda}{2} b^{2\chi-2} (\vec{\nabla} h)^2 + \eta(b\vec{r}, b^z t), \qquad (7.41)$$

and that the correlations of the transformed noise $\eta'(\vec{r}, t) = b^{z-\chi} \eta$ $(b\vec{r}, b^z t)$ scale as

$$\overline{\eta'(\vec{r}, t) \eta'(\vec{r}', t')} = 2D\, b^{z-d-2\chi}\, \delta(\vec{r} - \vec{r}') \delta(t - t'). \qquad (7.42)$$

d. Show that the parameters are consequently transformed as

$$\nu \to b^{z-2} \nu, \qquad \lambda \to b^{\chi+z-2} \lambda, \qquad D \to b^{z-d-2\chi} D. \qquad (7.43)$$

e. Discuss how for the linear case $\lambda = 0$ the equation is made scale-invariant with the choice $z_0 = 2$, $\chi_0 = (z-d)/2$, but that in the language of critical phenomena close to this noninteracting fixed point, λ scales as $\lambda = b^{(2-d)/2}$, so that the coupling λ is relevant for $d < 2$.

f. Argue that the tilting/Galilean symmetry discussed in steps a and b should be preserved under rescaling, and that this implies that the coupling λ should not change under rescaling, and hence that

$$\chi + z = 2. \qquad (7.44)$$

This exponent relation holds generally, in arbitrary dimensions.

g. Now specify to the 1d-KPZ equation. In the discussion following (7.18) it is mentioned that the Gaussian equilibrium distribution $\mathcal{P}[h] \sim \exp[-(\nu/2D) \int \mathrm{d}x (\nabla h)^2]$ of the linear model is also the equilibrium distribution for the one-dimensional model with $\lambda \neq 0$. Show that this implies $\chi = 1/2$ and hence with (7.44) $z = 3/2$ for the 1d-KPZ equation.

Problem 7.6 *Capillary waves*

Consider a flat $d-1$ hypersurface in d dimensions whose height is described by a function of the remaining spatial coordinates $h = h(\vec{r})$, with $\vec{r} = (x_1, \ldots, x_{d-1})$.

This problem is inspired by chapter 3 of the book by Kardar, 2007.

a. The generalized 'area' is given by

$$\mathcal{A} = \int \mathrm{d}^{d-1} \vec{r} \sqrt{1 + |\nabla h(\vec{r})|^2}. \qquad (7.45)$$

Provide a geometrical argument for this formula for $d = 2$.

b. For the rest of this problem, we will consider a Hamiltonian describing a surface tension σ via $\mathcal{H} = \sigma \mathcal{A}$. For low temperatures, h is a slowly varying function. Expand the Hamiltonian to quadratic order of h and write the partition function as a functional (drop the h-independent constant).

c. Diagonalize this quadratic Hamiltonian by applying a Fourier transformation $h(\vec{r}) = \int \frac{d^{d-1}\vec{q}}{(2\pi)^{d-1}} h(\vec{q}) e^{i\vec{q}\cdot\vec{r}}$. The normal modes $h(\vec{q})$ are referred to as capillary waves.

d. These capillary waves are the hydrodynamic modes associated with a certain continuous symmetry of this system. Which continuous symmetry is broken by these hydrodynamic modes?

e. Calculate the height-height correlation $\langle [h(\vec{r}) - h(\vec{r}')]^2 \rangle$ by transforming the relevant expression into momentum space using the results from step c and the 'four commandments' of box 3.

f. Comment on the asymptotic behavior of $\langle [h(\vec{r}) - h(\vec{r}')]^2 \rangle$ for $|\vec{r} - \vec{r}'| \gg 1$. How does the result depend on dimension?

g. In this problem, we've only considered the order $\mathcal{O}(|\nabla h|^2)$. Estimate typical values of ∇h by using the results from step f, and determine when this approximation is justified.

Problem 7.7* *Fluid membrane: Normal vector correlation function*

Consider an infinitesimal sheet in the xy plane of linear size L. If the fluctuations of the sheet are small, the sheet's height above the xy plane can be modeled by a single-valued function $z(x, y)$. The free energy associated with the curvature of the sheet with the bending rigidity κ is, to lowest order in derivatives of z:

$$F = \frac{\kappa}{2} \int d^2\vec{r}\, |\nabla^2 z(\vec{r})|^2. \qquad (7.46)$$

a. Writing $z(\vec{r}) = \frac{1}{L} \sum_{\vec{k}} z_{\vec{k}} e^{i\vec{k}\cdot\vec{r}}$, derive an expression for F in terms of the $z_{\vec{k}}$.

b. Defining the position vector of a point on the surface through $\vec{R} = (x, y, z(x, y))$ we can find the normal vector $\hat{n}(x, y)$ by requiring that $\vec{n} \cdot d\vec{R} = 0$, yielding

$$\hat{n} = \frac{1}{(1 + |\vec{\nabla} z|^2)^{1/2}} (-\partial z/\partial x, -\partial z/\partial y, 1)^{\mathrm{T}}. \qquad (7.47)$$

Expand $\hat{n}(\vec{r}') \cdot \hat{n}(\vec{r}' + \vec{r})$ in powers of ∇z (to 2nd order) to show

$$\int d^2\vec{r}'\, \hat{n}(\vec{r}') \cdot \hat{n}(\vec{r}' + \vec{r}) \approx L^2 + \int d^2\vec{r}'\, \vec{\nabla} z(\vec{r}' + \vec{r}) \cdot \vec{\nabla} z(\vec{r}')$$

$$- \int d^2\vec{r}'\, \frac{1}{2} \left[\vec{\nabla} z(\vec{r}' + \vec{r})^2 + |\vec{\nabla} z(\vec{r}')|^2 \right]. \qquad (7.48)$$

c. Taking $\langle \hat{n}(\vec{r}) \cdot \hat{n}(\vec{0}) \rangle = \frac{1}{L^2} \int d^2 \vec{r}' \langle \hat{n}(\vec{r}') \cdot \hat{n}(\vec{r}' + \vec{r}) \rangle$, show that the normal vector correlation function

$$\langle \hat{n}(\vec{r}) \cdot \hat{n}(\vec{0}) \rangle \equiv \Gamma(r) = 1 - \frac{k_B T}{\kappa L^2} \sum_{\vec{k}} \frac{1 - \exp i\vec{k} \cdot \vec{r}}{k^2}. \tag{7.49}$$

d. By introducing a short distance cutoff a (or, equivalently, a high-k cutoff π/a), derive the form of $\Gamma(\vec{r})$ for $r/a \gg 1$.

Hint: The function $\Gamma(r)$ is essentially the Bessel function $J_0(r)$, from which the asymptotic behavior can be obtained.

e. Defining the persistence length ξ as the distance over which $\Gamma(\vec{r}) \geq 0$, show that

$$\xi = \frac{a}{\pi} \exp \left(\frac{2\pi\kappa}{k_B T} \right). \tag{7.50}$$

Problem 7.8 *Phantom membrane: Radius of gyration*

In the model for tethered membranes, particles occupy the vertices of a triangular network and are joined by inextensible edges of length a; see section .a. For phantom membranes (shown in figure 7.11), only the tethering constraint controls its statistical mechanics. The discrete network of tethers is approximated by a two-dimensional sheet and we assume that the phantom membrane can be described at large length scales by an effective dimensionless Hamiltonian containing only the two-dimensional version of the elasticity,

$$\mathcal{H} = K \int \sum_{i,j} d^2 \vec{x} \left(\frac{\partial r_i}{\partial x_j} \right)^2 = K \int d^2 \vec{x} \left(\frac{\partial \vec{r}}{\partial \vec{x}} \right)^2, \tag{7.51}$$

where $\vec{r}$ is the vector in the d-dimensional space in which the membrane is embedded, while the two-dimensional vector $\vec{x}$ comprises the coordinates of the points on the sheet. The position $\vec{r}(\vec{x})$ can be represented by a Fourier series

$$\vec{r}(\vec{x}) = \frac{1}{L} \sum_{\vec{k}} \vec{A}_{\vec{k}} e^{i\vec{k} \cdot \vec{x}}, \tag{7.52}$$

where $\vec{A}_{\vec{k}}$ is a d-dimensional vector, $\vec{k}$ is a two-dimensional wave-vector, and L is the linear dimension of the membrane.

a. Express $\mathcal{H}$ in terms of the $\vec{A}_{\vec{k}}$.

b. Calculate $\langle |(\vec{A}_{\vec{k}})_i|^2 \rangle$, where there is an implicit summation over the index i.

c. The size of fluctuations in the membrane in d-dimensional space is characterized by its radius of gyration, R_g. Taking the

Underlying the similarity is the fact that the Coulomb gas problem and the roughening problem are related by a duality transformation. See, e.g., Weeks, 1980.

e. Using the similarity of equations (7.63) and (7.64) with the expressions of the Coulomb gas formulation of topological defects, and essentially the same technique as in problem 6.9, show that equation (7.64) can be cast in the following form:

$$
G_k(\vec{x}-\vec{y}) = e^{-\frac{k^2}{K}C(\vec{x}-\vec{y})} \left[1 + \frac{\pi^3 k^2}{K^2} y_0^2 C(\vec{x}-\vec{y}) \right.
$$
$$
\left. \times \int dr\, r^3 e^{-\frac{2\pi \ln(r/a)}{K}} \right].
$$
(7.66)

f. Write the perturbation result in terms of an effective interaction K_{eff} given by

$$
\frac{1}{K_{\text{eff}}} = \frac{1}{K} - \frac{\pi^3}{K^2} a^{2\pi/K} y_0^2 \int_a^{\infty} dr\, r^{3-2\pi/K}.
$$
(7.67)

Show that the perturbation theory in equation (7.67) fails for K larger than a critical value K_c, which you should determine.

g. Recast the perturbation result in step f into renormalization group equations for K and y_0 by changing the lower bound in the integral from a to ba, where $b \geq 1$, following the logic we used in section 6.10.3 to derive the renormalization group equations.

Show that, to order y_0^2, the coarse-grained system you obtained is described by an effective stiffness:

$$
\frac{1}{K_{\text{eff}}} = \frac{1}{\tilde{K}} - \frac{\pi^3}{\tilde{K}^2} a^{2\pi/\tilde{K}} \tilde{y}_0^2 \int_a^{\infty} dr\, r^{3-2\pi/\tilde{K}},
$$
(7.68)

where $\tilde{K}$ and $\tilde{y}_0$ are defined by the equations

$$
\frac{1}{\tilde{K}} = \frac{1}{K} - \frac{\pi^3}{K^2} a^{2\pi/K} y_0^2 \int_a^{ba} dr\, r^{3-2\pi/K}
$$
(7.69)

$$
\tilde{y}_0^2 = b^{4-2\pi/K} y_0^2.
$$
(7.70)

Hint: First, divide the integral in (7.67) into two parts: the first part from a to ab and the second part from ab to ∞. Then, rescale the variable of integration in the second part of the integral, so that you retrieve the usual limits of integration.

h. Show that the renormalization group equations (7.69)–(7.70) can be cast as differential equations (with $b = e^{\ell} \approx 1 + \ell$):

$$
\frac{dK}{d\ell} = \pi^3 a^4 y_0^2 + \mathcal{O}(y_0^4)
$$
(7.71)

$$
\frac{dy_0}{d\ell} = \left(2 - \frac{\pi}{K} \right) y_0 + \mathcal{O}(y_0^3).
$$
(7.72)

i. Use the recursion relations in equations (7.71) and (7.72) to draw the renormalization group flow diagram. What happens to K and y_0 under coarse-graining for $K > K_c$ and for $K \leq K_c$? Briefly, discuss the physical interpretation of the two different phases in terms of the smoothness of the surface.

j. Using the result derived in equation (7.62), show that, at the transition, there is a discontinuous jump in $\left\langle \left| h(\vec{x}) - h(\vec{y}) \right|^2 \right\rangle$ whose magnitude is equal to $\frac{2}{\pi^2} \ln \frac{|\vec{x}-\vec{y}|}{a}$ for large separations $|\vec{x} - \vec{y}|$.

Hint: Note that $y_0 \to 0$ under coarse-graining implies $G_k(\vec{x} - \vec{y}) \to \langle \mathcal{G}_k \rangle_U$, which you evaluated explicitly in step b. Use also the similarity with our analysis of the Kosterlitz-Thouless transition in section 6.10.4.

Problem 7.10* *Tension-induced wrinkling*

You may have noticed that if you pull hard enough on a sheet of plastic, it will wrinkle. In this problem we investigate the height and wavelength of the observed wrinkles.

For more details on the topic of this problem see Cerda and Mahadevan, 2003.

a. Before thinking about tension-induced wrinkling, let us do a warm-up: consider first an inextensible plastic sheet lying on the surface of water. The sheet has width W in the x direction and length L in the y direction. When an experimentalist compresses the sheet by an amount Δ in the x direction, it is observed that the sheet forms sinusoidal wrinkles of the form $h(x, y) = A \sin(kx)$. Here, $h(x, y)$ is the height field, A is the amplitude and k is the wavenumber. Our goal is to determine A and k.

Our first step will be to determine the value of kA using the fact that the plastic sheet is approximately inextensible. Convince yourself that inextensibility implies the following relationship:

$$\int_0^{W-\Delta} \mathrm{d}x \sqrt{1 + \left(\frac{\mathrm{d}h}{\mathrm{d}x}\right)^2} = W. \tag{7.73}$$

Using the form $h = A \sin(kx)$, show that for $\Delta/W, kA \ll 1$, equation (7.73) implies

$$kA \propto \left(\frac{\Delta}{W}\right)^{1/2}. \tag{7.74}$$

b. Next, we are going to use energetic arguments to determine the wavenumber k. The mechanical energy takes the form

$$U = \frac{B}{2} \int \mathrm{d}^2\vec{x} \, (\nabla^2 h)^2 + \frac{\rho g}{2} \int \mathrm{d}^2\vec{x} \, h^2, \tag{7.75}$$

where B is the bending stiffness, ρ is the density of water, and g is the strength of gravity. The first term in equation (7.75) is the

d. The cross terms in (7.89) can be written as

$$\zeta(\vec{r}_\alpha) = -Y \int dS\, \rho(\vec{r}) \int dS'\, \Gamma_B(\vec{r}, \vec{r}') K(\vec{r}'). \tag{7.90}$$

$\zeta(\vec{r}_\alpha)$ is known as the *geometric potential* at the defect position $\vec{r}_\alpha$. It quantifies the energy caused by interactions between topological defects and surface curvature. Show that if we introduce an auxiliary function $V(\vec{r})$ satisfying

$$\nabla^2 V(\vec{r}) = K(\vec{r}), \tag{7.91}$$

the geometric potential can be rewritten as

$$\zeta(\vec{r}) = -Y \int dS\, \rho(\vec{r}) \int dS'\, \Gamma_L(\vec{r}, \vec{r}') V(\vec{r}'), \tag{7.92}$$

where Γ_L is the Green's function of the Laplacian.

e. Let's now consider, as an example, the energy due to a single dislocation at position $\vec{r}$ with the Burgers vector $\vec{b}$ on a Gaussian bump, specified by a height function

$$h(\vec{r}) = \alpha r_0 \exp\left(-\frac{r^2}{2r_0^2}\right). \tag{7.93}$$

Calculate the Gaussian curvature $K(\vec{r})$, and use Gauss's theorem to find that the auxiliary function is

$$V(\vec{r}) = -\frac{1}{4}\alpha^2 \exp\left(-\frac{r^2}{r_0^2}\right). \tag{7.94}$$

f. Finally, substitute the above expression for the auxiliary function into (7.92), the formula for the geometric potential, with a defect source corresponding to a single dislocation at position $\vec{r}$. Solve this integral by assuming a fairly shallow bump, so that we can make the approximation $dS \approx d^2\vec{r}$. Integrate by parts and use Gauss's theorem as you would in an electrostatic problem. Just remember we are working in two dimensions. You will end up with an expression for the geometric potential $D(r, \theta)$ which is a function of the dislocation's radial distance from the center of the bump r and the angle θ between its Burgers vector $\vec{b}$ and the radial unit vector $\hat{r}$. You should find

$$D(r, \theta) \approx Y b r_0 \frac{\alpha^2}{8} \sin\theta \left(\frac{e^{-r^2/r_0^2} - 1}{r/r_0}\right). \tag{7.95}$$

What orientation of the Burgers vector θ is energetically favored? What is the minimum-energy location for the dislocation on the bump?

You can use the Green's function of the Laplacian on a flat plane (i.e., $\Gamma_L(\vec{r}, \vec{r}')$ satisfying $\nabla^2 \Gamma_L = \delta(\vec{r} - \vec{r}')$):

$$\Gamma_L(\vec{r}, \vec{r}') = \frac{1}{4\pi} \ln|\vec{r} - \vec{r}'|^2.$$

We suggest plotting the expression you find and comparing your answer with the results of numerical simulations reported in figure 1 of Vitelli et al., 2006.

Problem 7.12 *Scaling laws for coarsening dynamics*

Phase separation is a phenomenon where two fluids spontaneously demix into domains mostly composed of one of the two fluids. An everyday example of this is seen in oil-water emulsions. Here we answer a simple question: how quickly do oil droplets grow or shrink? Surprisingly, there are quite general scaling laws for these *coarsening dynamics*. We will see that the typical size of a single domain, $L(t)$, grows with a power law,

$$L(t) \sim t^n, \tag{7.96}$$

where n depends solely on the conservation laws of the underlying dynamics.

In principle, we could work with the density fields for the two fluids, $\rho_a(\vec{x}, t)$ and $\rho_b(\vec{x}, t)$. However, for simplicity, we will assume that $\rho_a + \rho_b = \rho$ is conserved across space and time. This allows us to describe the dynamics of the system using a single order parameter,

$$\phi(\vec{x}, t) = \frac{\rho_a(\vec{x}, t) - \rho_b(\vec{x}, t)}{\rho}. \tag{7.97}$$

We see that $\phi = +1$ means that that region is composed purely of fluid a, and $\phi = -1$ that it is composed purely of fluid b. These two values of ϕ define the two different phases that separate from each other.

The free energy of the system is taken to be of the form

$$F([\phi]) = \int \mathrm{d}^d \vec{x} \, f(\phi) = \int \mathrm{d}^d \vec{x} \left[\frac{1}{2} |\vec{\nabla} \phi|^2 + V(\phi) \right]. \tag{7.98}$$

We assume that the potential $V(\phi)$ has a double well structure, for example, $V(\phi) = (1 - \phi^2)^2$, which has minima at $\phi = \pm 1$, corresponding to the two equilibrium states, or *phases*. The gradient term in $F([\phi])$ further penalizes interfaces between the two phases. The order parameter $\phi(\vec{x}, t)$ will obey either of the following dynamics, which we name using the terminology of dynamical critical phenomena:

$$\frac{\partial \phi(\vec{x}, t)}{\partial t} = \begin{cases} -\dfrac{\delta F}{\delta \phi} & \text{model A} \\[2mm] -\vec{\nabla} \cdot \vec{j} = \nabla^2 \dfrac{\delta F}{\delta \phi} & \text{model B,} \end{cases} \tag{7.99}$$

where $\dfrac{\delta F}{\delta \phi} = \dfrac{\mathrm{d} V}{\mathrm{d} \phi} - \nabla^2 \phi$. Model A corresponds to the case when ϕ, hence the relative amounts of a and b, is not conserved; in other words, a can change into b and vice versa. This describes, for example, coarsening for Ising dynamics, where the spin up and spin down states can be used to model the magnetization or to denote

For a more general introduction to and overview of phase separation and coarsening dynamics, see Bray, 1994 and Bray, 2003 or the book by Onuki, 2002.

The nomenclature of model A and model B used below is that of the theory of dynamical critical phenomena of Hohenberg and Halperin, 1977.

Note that we are missing some important constants in these equations that make the dimensions make sense. For example, with ϕ being dimensionless, model A should read

$$\frac{\partial \phi}{\partial t} = -\Gamma \frac{\delta F}{\delta \phi},$$

where Γ has units of $L^2 T^{-1}$. Similarly, model B should read

$$\frac{\partial \phi}{\partial t} = \lambda \nabla^2 \frac{\delta F}{\delta \phi},$$

where λ has units of $L^4 T^{-1}$. However, the constants we ignore are not time-dependent, so we will cautiously set them to 1 for the sake of our dynamical scaling arguments.

Use dimensional analysis on (7.110) to show

$$L(t) \sim (\gamma t)^{1/3}. \tag{7.111}$$

Noting that γ is a time-independent quantity, the relevant dynamical scaling behavior is simply $L \sim t^{1/3}$.

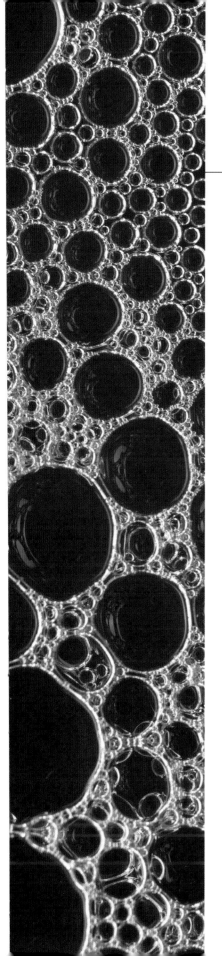

Part III

ADVANCED TOPICS

A pointwise summary of the general insights the analysis of this chapter leads to is provided in box 4, in section 8.9.1.

Nice introductions and overviews of pattern formation and dynamics outside of equilibrium are given by Hoyle, 2006, Cross and Greenside, 2009, and Manneville, 2014, while the book by Pismen, 2006 is a good follow-up as it treats more topics in detail. The second volume of J. D. Murray, 2003 provides an introduction from the perspective of (mathematical) biology.[5]

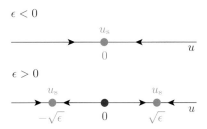

Figure 8.3. Illustration of the dynamics of u with time for the simple differential equation (8.1). In the theory of dynamical systems it is customary to think of the dynamics as a flow in the phase space u, and of the stationary points u_s as fixed points of the flow. The arrows indicate the flow direction, the green dot marks a stable (attractive) fixed point, and the red dot marks an unstable (repelling) one.

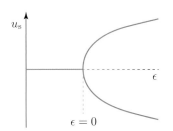

Figure 8.4. The pitchfork bifurcation. The green lines indicate stable stationary states (fixed points) u_s as a function of ϵ, while the state $u_s = 0$ is drawn with a red dashed line for $\epsilon > 0$ to indicate that it is unstable. Although arrived at for equation (8.1), the pitchfork bifurcation is a generic bifurcation.

amplitude expansion. As illustrated in figure 8.2 and discussed in detail in this chapter, the amplitude in this approach describes the slow modulation of the pattern in space and time just above the threshold of the instability that gives rise to the patterns. Even though the amplitude description is valid only close to threshold,[4] it helps us organize the way we approach questions concerning patterned states far from threshold.

The concepts developed for understanding the spontaneous formation of patterns in non-equilibrium systems are of relevance to many fields other than soft matter: we encounter examples in biology, fluid mechanics, chemistry, materials science, plasma science, ecology, population dynamics, neuroscience, astronomy, etc.[6] The organization of this chapter will reflect this—we will focus less than in earlier chapters on soft matter applications per se, and we will focus more on the general framework and concepts.

8.2 Gearing up for studying patterns in spatially extended systems

Before turning to patterns in physical systems, let us first recall an important result from dynamical systems theory and get inspiration from a model equation.

8.2.1 The pitchfork bifurcation of dynamical systems

We focus in this chapter on spatially extended systems, but to set the scene, it is useful to start with considering the following dynamical equation for a single variable u:

$$\dot{u} = \epsilon u - u^3. \tag{8.1}$$

If we think of the equation as determining a flow in the phase space u, then we can indicate the dynamics by drawing arrows indicating the change of u with time. This is drawn in figure 8.3 for the two cases, positive and negative ϵ. The stationary solutions u_s which satisfy $\dot{u}_s = 0$ correspond to fixed points of the flow. Now, for $\epsilon < 0$ the only stationary solution is $u_s = 0$. This is a stable fixed point: the flow is directed into it, implying that in time u approaches $u_s = 0$. When ϵ changes sign, however, the situation changes: for $\epsilon > 0$, the fixed point corresponding to $u_s = 0$ is unstable, while two new stationary solutions at $u_s = \pm\sqrt{\epsilon}$ exist; both correspond to stable fixed points.

The resulting stationary solutions u_s and their stabilities are plotted as a function of ϵ in figure 8.4. We see that new steady states branch or 'bifurcate' from the steady state $u_s = 0$. Because of the form of the diagram, reminiscent of a pitchfork, this type of bifurcation is called a pitchfork bifurcation. In such a diagram, the solid lines in green denote stable (attractive) states, the dashed red one an unstable state.

We arrived directly at the pitchfork bifurcation from the simple dynamical equation (8.1). But the concept is valid more generally, for higher order dynamical systems and for those involving several coupled variables: at a point where a stationary solution loses stability, new branches of steady-state solutions typically bifurcate, reminiscent of the pitchfork diagram.

Even though spatially extended pattern forming systems are much more complicated, one basic ingredient remains. Namely, the fact that a change of stability implies the existence of a new branch of solutions, and that these are, typically, the physically relevant types of solutions. In particular, when the instability is a finite wavelength instability, the new branch of solutions consists of patterned states characterized by a wavelength whose scale is set by the instability.

Studying the existence and stability as well as the competition and dynamics of these states just above the instability threshold is the main goal of this chapter. Loosely speaking, the near-threshold results developed in this chapter can be thought of as the extension of bifurcation theory to the spatiotemporal dynamics of extended systems.

8.2.2 The Swift-Hohenberg model equation

To develop some intuition for the main questions and for the challenges of spatially extended systems, let us first discuss briefly the so-called Swift-Hohenberg model equation. The equation, together with its many extensions, is easy to simulate and is often used in exploratory studies of pattern formation. It is defined by the first order partial differential equation

$$\partial_t u = \epsilon u - \left(\vec{\nabla}^2 + q_c^2 \right)^2 u - g u^3. \qquad (g > 0), \qquad (8.2)$$

for the real field $u(\vec{r}, t)$.

The linear term ϵu and nonlinear term $-g u^3$ are reminiscent of equation (8.1), but the spatial derivative terms change the behavior dramatically. Let us check the stability of the stationary state $u = 0$ by linearizing the equation about it (this amounts to ignoring the nonlinear term $-g u^3$) and substituting Fourier modes $e^{\sigma t + i q x}$. Substitution of this ansatz in the linear terms of (8.2) gives the

Technically, this is because when all eigenvalues are nonzero, then the solutions depend smoothly (analytically) on the parameters,[7] like our ϵ. But when one of the eigenvalues is zero, indicating change of stability of the fixed point, then the condition for continuing the solution upon changing a parameter becomes quadratic. This allows nonanalytic emergence of new solutions.

The amplitude equation description of pattern formation can be viewed as the extension of bifurcation analysis to spatially extended systems.[8]

Swift and Hohenberg, 1977 originally introduced the equation to study the similarities of a non-equilibrium transition with fluctuations with an equilibrium phase transition. Its richness and use as a model equation for pattern formation only became gradually clear later. In fact, Swift and Hohenberg's original motivation to introduce this equation was very similar to that of Brazovskii (Brazovskii, 1975), who had already introduced this equation to study the effect of fluctuations on a finite wavelength equilibrium transition, for problems like the one of block copolymers in figure 5.25. For this reason the free energy (8.4) is often referred to as the Brazovskii model in the polymer literature.[9]

Such symmetry arguments are not unlike those that we encountered earlier for block copolymers (figure 5.25) and the liquid crystal nematic-to-smectic-A transition (problem 6.4), where fluctuation effects generate small terms in a Landau free energy expression that drive the transition weakly first order. Similar symmetry arguments also played a role in showing that the isotropic-to-nematic transition has to be weakly first order (section 6.2).

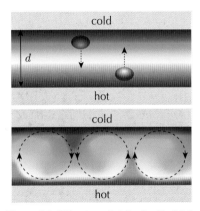

Figure 8.8. Side-view sketch of a Rayleigh-Bénard cell with a fluid contained between two plates a distance d apart, kept at fixed temperatures, with the bottom one hotter than the top one. The upper panel sketches the base state without convection, so that there is a temperature gradient in the vertical direction. The arrows indicate that the lower blob of fluid, which is a bit hotter and hence less dense than its surroundings experiences an effective buoyancy force upward, while the blob which is a bit colder and more dense than its surroundings experiences a force downward. When the effect is large enough, the base state is unstable to convection. The lower panel illustrates the resulting convective flow beyond the instability threshold. The closed flow structures have a width comparable to the cell thickness d.

Boussinesq approximation cause the Rayleigh-Bénard transition to be very weakly subcritical, and how this results in the hexagonal convection patterns close to the instability threshold. The analysis will also convince us why hexagonal patterns are the rule rather than the exception in systems with a strong symmetry breaking.

8.3 Inspiration: Rayleigh-Bénard convection and Turing patterns

In this section we will start to explore some of the main questions of pattern formation by introducing two examples which often feature as reference points in research on pattern formation. The Rayleigh-Bénard instability in fluids heated from below exemplifies how nontrivial convection patterns emerge when a featureless conductive state becomes unstable. Experimentally this system has also served as an important anchor point in the development of the field. We then turn to an analysis of the possible instabilities in two coupled reaction-diffusion equations, following Turing's work on morphogenesis. This will give us a glimpse of the richness of behavior that the interplay of diffusion and nonlinear interactions can give rise to.

8.3.1 The Rayleigh-Bénard instability

The Rayleigh-Bénard instability refers to the instability of a fluid in a flat cell heated from below, as illustrated in figure 8.8. We consider the lateral dimension L of the cell to be much larger than the distance d between the two plates: the aspect ratio $\Gamma = L/d \gg 1$. The top and bottom plates are each kept at a fixed temperature, and the temperature difference across them is ΔT—essentially, for a given fluid and cell, ΔT is the only knob the experimentalists can turn. When ΔT is not too large, the fluid will conduct heat but will not convect.

Since the colder, more dense fluid is at the top, this situation is inherently unstable because of the tendency of the fluid at the top to sink and the tendency of the fluid at the bottom to rise. But two effects oppose this buoyancy-driven instability. To see this, imagine a fluid element at the top to be slightly colder and hence heavier than its surroundings, as indicated by the blue blob near the top plate in figure 8.8. It tends to sink due to buoyancy. Likewise, the red blob near the lower plate is warmer than the surrounding fluid, so buoyancy gives an upward force. For both of these forces to result in flow, the driving needs to overcome the damping due to viscosity. In addition, these temperature fluctuations should survive sufficiently long: heat diffusion suppresses the effect by smoothing out the temperature inhomogeneity. The point where ΔT is large enough for the buoyancy to overcome these two stabilizing effects is determined by the linear stability analysis below.

a. The Boussinesq equations

In the base state, there is a temperature gradient across the fluid, but there is no convection. We want to demonstrate explicitly how to perform the linear stability calculation of this state. The result will show that there is a well-defined threshold beyond which the state is linearly unstable.

Conceptually, the stability analysis is similar to the one we did for the Rayleigh-Taylor and Kelvin-Helmholtz instabilities in problem 1.10.

To perform any calculation we first have to write the dynamical equations for the density ρ, velocity $\vec{v}$, and temperature T, together with the appropriate boundary conditions. In principle these are the full fluid dynamic equations summarized in section 1.6.2. But to make the stability analysis manageable we make a few additional well-defined approximations to arrive at a simpler form, the so-called Boussinesq equations. These equations are often also used for numerical simulations of Rayleigh-Bénard convection.

The driving force for the instability is the buoyancy term $\vec{g}\rho$, i.e., the gravitational force which through the temperature dependence of the density makes hot and hence lighter fluid elements want to rise and makes cold and heavier ones want to sink. Since, in addition, typical flow speeds are small, the Boussinesq approximation is based on two simplifying approximations. The first one amounts to writing the fluid density ρ simply as a function of the local temperature, $\rho = \rho(T)$, and assuming that temperature variations are small enough that they can be taken to be linear in the local temperature. In other words, we write

$$\rho(\vec{r}, t) = \rho_0[1 - \alpha(T(\vec{r}, t) - T_0)], \tag{8.6}$$

where T_0 is the temperature halfway between the plates at $z = 0$ in the absence of convection, and $\alpha > 0$ is the thermal expansion coefficient of the fluid. This approximation implies that small variations in the density due to the flow and pressure are ignored.

Compare our discussion in section 1.8 of when fluid flow can be treated as incompressible.

The second assumption is that since typical velocities are small, it is legitimate to take the variation of the density into account in *only* the buoyancy force $\vec{g}\rho$, that is, the flow itself is taken as incompressible: $\vec{\nabla} \cdot \vec{v} = 0$ (compare equation (1.49) for incompressible flow). As a result, variations in the density enter the equations only through the buoyancy term $\vec{g}\rho$ in equation (1.12) for the fluid velocity.

In the base state with heat conduction only, the temperature profile is linear in the height coordinate z,

$$T_{\text{cond}} = T_0 - (z/d)\Delta T, \qquad -d/2 \le z \le d/2. \tag{8.7}$$

As stated earlier, ΔT is the imposed temperature difference between the plates.

Upon writing the temperature variation about the linear conductive profile as $\theta = T - T_{\text{cond}}$ and introducing $\delta p = p - p_{\text{cond}}$, where p_{cond} is the hydrostatic pressure in the base state with density varying with height, the dynamical equations for the flow are shown in problem 8.1 to reduce in the Boussinesq approximation to

$$\vec{\nabla} \cdot \vec{v} = 0, \tag{8.8}$$

$$\frac{1}{\text{Pr}} \left(\frac{\partial \vec{v}}{\partial t} + \vec{v} \cdot \vec{\nabla} \vec{v} \right) = -\vec{\nabla} \delta p + \theta \hat{z} + \nabla^2 \vec{v}, \tag{8.9}$$

$$\frac{\partial \theta}{\partial t} + \vec{v} \cdot \vec{\nabla} \theta = \text{Ra} \, v_z + \nabla^2 \theta. \tag{8.10}$$

See problem 8.1 for a step-by-step derivation of the Boussinesq equations (8.8)–(8.10) from the full hydrodynamic equations with the aid of the approximations described above. Distances are written in units of d, times in units of d^2/D_T (with D_T the thermal diffusivity; see equation (1.28)), temperatures in units of $D_T \nu/(\alpha g d^3)$, and the pressure in terms of $\rho_0 D_T \nu/d^2$. On the left-hand side of (8.10) we recognize the convective derivative (1.2), the total time derivative of a moving fluid element. The term $\text{Ra} \, v_z$ on the right comes from the term $\vec{v} \cdot \vec{\nabla} T_{\text{cond}}$.

Here we have introduced suitable dimensionless units which are such that distances are measured in units of d. Furthermore, $\text{Pr} = \nu/D_T$ is the Prandtl number, the dimensionless number defined as the ratio of the kinematic viscosity ν and the thermal diffusivity D_T of equation (1.28).[15] Ra is the dimensionless Rayleigh number

$$\text{Ra} = \frac{\alpha g d^3 \, \Delta T}{\nu D_T}. \tag{8.11}$$

The occurrence of the various terms in the expression for Ra can nicely be understood intuitively: increasing the expansion rate α, the gravity g, and the temperature difference ΔT enhances the driving, and therefore all appear in the numerator, while viscous damping and thermal diffusion both enter in the denominator as they oppose the instability. An increase of the distance d also makes the damping from viscosity and temperature diffusion less effective; this is why d enters in the numerator. The fact that it does so as d^3 is because it is the only way to make the Rayleigh number dimensionless.[16]

The equations (8.8)–(8.10) for buoyancy-induced flow driven by a temperature-dependent density are called the Boussinesq equations. The first term on the right-hand side of (8.10), which is proportional to the Rayleigh number, clearly has the form of a source term: it enhances the temperature variations that drive the flow. Experimentalists can change the Rayleigh number simply by increasing the temperature difference ΔT—the Rayleigh number is the dimensionless control parameter of the equations.

The equations have to be supplemented with boundary conditions at the top and bottom plates. These are the conditions that the fluid velocity is zero at the plates (*no-slip* boundary conditions) and that the temperature there is equal to the one imposed experimentally. In the dimensionless units with lengths measured in units of d, this means

$$\text{no-slip b.c.:} \quad \vec{v} = 0, \quad \theta = 0, \quad \text{at } z = \pm 1/2. \tag{8.12}$$

b. Linear stability analysis

To perform a linear stability analysis of the convectionless base state with $\vec{v} = 0$, $\theta = 0$, we linearize the above equations in $\vec{v}$ and θ; the resulting equations are

$$\vec{\nabla} \cdot \vec{v} = 0,$$
$$\frac{1}{\text{Pr}} \frac{\partial \vec{v}}{\partial t} = -\vec{\nabla} \delta p + \theta \hat{z} + \nabla^2 \vec{v}, \tag{8.13}$$
$$\frac{\partial \theta}{\partial t} = \text{Ra} \, v_z + \nabla^2 \theta.$$

As we aim to analyze the stability in the large aspect ratio limit $\Gamma \to \infty$, lateral boundary conditions at the sides play no role. Moreover, translation invariance implies that the eigenmodes in the lateral direction are Fourier modes. In addition, the system is rotationally invariant in the xy plane; it is convenient however to first consider simply modes $\cos qx$ and $\sin qx$, which vary in the x direction only, and which have $v_y = 0$, and to come back to modes in other directions later.

The eigenmodes of the above linear equations also have a non-trivial z dependence. It turns out that with the no-slip boundary conditions (8.12), the z dependence of the fields is rather complicated. In order not to get sidetracked by unnecessary technicalities, we follow here Rayleigh's own insight that if we use instead free-slip boundary conditions

$$\text{slip b.c.:} \quad v_z = 0, \quad \partial v_x / \partial z = 0, \quad \theta = 0, \quad \text{at } z = \pm 1/2, \quad (8.14)$$

we can solve for the linear stability modes with the ansatz

$$
\begin{aligned}
v_z &= v_0 \, e^{\sigma_q t} \cos(qx) \cos(\pi z), \\
v_x &= v_0 \, e^{\sigma_q t} (\pi/q) \sin(qx) \sin(\pi z), \quad (8.15)\\
\theta &= \theta_0 \, e^{\sigma_q t} \cos(qx) \cos(\pi z).
\end{aligned}
$$

As detailed in problem 8.2, substitution of these solutions into (8.13) gives the following quadratic equation for $\sigma_q = \sigma(q)$:

$$(\text{Pr}^{-1}\sigma_q + \pi^2 + q^2)(\sigma_q + \pi^2 + q^2) - \frac{\text{Ra}\, q^2}{\pi^2 + q^2} = 0. \quad (8.16)$$

This equation determines the dispersion relation $\sigma(q)$. As it is a quadratic equation in $\sigma(q)$, it is easy to check from the solution that σ is always real; when $\sigma(q) < 0$ for all wavenumbers q, then the base state is linearly stable, as $e^{\sigma(q)t}$ is a decaying exponential for any q. However, whenever $\sigma(q) > 0$ for some q, then the state is unstable to these modes.

The quadratic equation gives two branches of $\sigma(q)$, one of which is always stable. In figure 8.9 we plot the solutions corresponding to the two branches of $\sigma(q)$. We see that as the Rayleigh number Ra increases, beyond some value Ra_c a band of modes is unstable, i.e., has positive values of $\sigma(q)$ for some range of wavenumbers q. Note the similarity with the behavior of $\sigma(q)$ of the Swift-Hohenberg equation sketched in figure 8.5: as the Rayleigh number increases, at some value Ra_c a band of unstable modes appears around q_c.

The analysis and plot in figure 8.9 are based on using the slip boundary conditions, for which the analysis can be done analytically. But the no-slip boundary conditions are the experimentally relevant ones. The full analysis for the no-slip case is presented

We follow here the customary presentation of the Rayleigh-Bénard problem to use sines and cosines and the growth rate σ in (8.15); alternatively, we can follow the route discussed in section 1.7.2 to use complex fields and a frequency ω, and to take the imaginary part later. This amounts to taking $\sigma = -i\omega$, so $\text{Im}\,\omega$ then corresponds to the growth rate $\text{Re}\,\sigma$.

Slip boundary conditions are the boundary conditions which allow the tangential components of the velocity field to be arbitrary, while the normal derivatives of the tangential components are zero. The latter condition expresses that the surface does not sustain shear forces, so that the components $\sigma_{xz} = \sigma_{zx} = \partial v_x / \partial z$ of the shear stress in the fluid vanish at the boundary. The expressions (8.15) satisfy this at the boundaries.

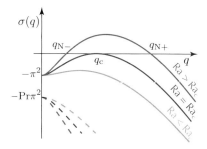

Figure 8.9. The dispersion relation of the Rayleigh-Bénard problem. The curves are drawn for the case of slip boundary conditions at the top and bottom plates, where $\sigma(q)$ is given by equation (8.16). In this case $\text{Ra}_c = 27\pi^4/4 \approx 657$ and $q_c = \pi/\sqrt{2}$ (see problem 8.2). The other two curves are drawn for Ra = 1,300 and Ra = 200, and Pr = 2. For the experimentally relevant case of no-slip boundary conditions the overall behavior is similar, but $\text{Ra}_c = 1,707.76$ and $q_c = 3.11632$ (see problem 8.3). In both cases, for Ra > Ra_c there is a band of wavenumbers for which the linear modes are unstable, i.e., for which $\sigma(q) > 0$. The dashed curves indicate the second branch of strongly damped modes.

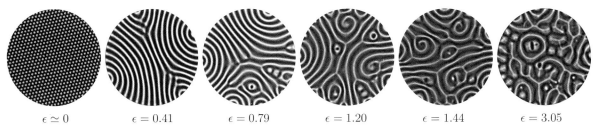

$$\epsilon \simeq 0 \qquad \epsilon = 0.41 \qquad \epsilon = 0.79 \qquad \epsilon = 1.20 \qquad \epsilon = 1.44 \qquad \epsilon = 3.05$$

Figure 8.10. Top views of convection patterns observed in Rayleigh-Bénard experiments at various values of the control parameter $\epsilon = (\mathrm{Ra} - \mathrm{Ra_c})/\mathrm{Ra_c}$, which is a measure of the distance of the Rayleigh number Ra from the value $\mathrm{Ra_c}$ where the convectionless base state becomes unstable. Except very close to threshold (left image), the convection patterns tend to form stripes, but as the control parameter ϵ increases more and more defects occur. The patterns viewed from the top are made visible with the so-called shadowgraph technique. Left image at $\epsilon \simeq 0$: courtesy of Eberhard Bodenschatz. Other images courtesy of Michael Schatz and Balachandra Suri.[17] The Prandtl number Pr in these experiments is 0.84.

as problem 8.3; the scenario and the overall form of the dispersion relation are found to be similar to the ones for the case of slip boundary, but the numerical values change. Again, the conductive state is found to be stable for small enough Ra: all modes then have $\sigma(q) < 0$ and so are damped. The value of $\mathrm{Ra_c}$ where the instability sets in is found to be 1707.76 for the realistic case of no-slip boundary conditions, and the wavenumber q_c where this happens is $q_c = 3.11632$. As this is so close to π, this means that the resulting flow structures sketched in the lower panel of figure 8.8 have a width which is very close to the cell thickness.

c. Convection patterns emerge above threshold

If a spatially extended system has a range of modes which are unstable, then the small fluctuations which generally will be present are sufficient to excite these unstable modes and have them grow out: the featureless base state is unstable so cannot persist. Moreover, since the unstable modes have a finite wavelength, it is natural to expect that a pattern with a wavelength roughly equal to $2\pi/q_c$ will emerge above threshold, very similar to what we argued for the Swift-Hohenberg equation.

That this is precisely what happens above threshold is illustrated in figure 8.10, which shows top views of convection patterns at various relative Rayleigh numbers whose distance from the threshold value $\mathrm{Ra_c}$ is measured by the ratio

$$\epsilon = \frac{\mathrm{Ra} - \mathrm{Ra_c}}{\mathrm{Ra_c}}, \tag{8.17}$$

which is a measure of the relative distance from the instability threshold. We will use ϵ from here on as the generic dimensionless *control parameter* that marks the distance from threshold in our discussion of patterns that emerge from an instability at $\epsilon = 0$.

Figure 8.10 illustrates that the flow patterns, viewed from the top and made visible with the shadowgraph technique, evolve with distance from threshold. Following the images of the circular cells from left to right starting from the second one at $\epsilon = 0.41$, the observations indicate that initially, just above threshold for small ϵ, the convection patterns tend to form 'stripes.' Inspection of the images also shows that the stripes prefer to orient themselves perpendicular to the sidewalls of the cell. As a result, there are a few defects in the flow, but overall the pattern tends to be quite coherent over long distances. As ϵ increases, however, the 'coherence length' decreases, while the number of defects goes up. Although the stills can't show this, close to threshold the pattern is typically found to adjust slowly due to the interaction of these defects. Farther above threshold, however, in the figure at $\epsilon = 1.20$ and above, the convection pattern is intrinsically time-dependent due to the existence of rotating spirals.

While the image for $\epsilon = 0.41$ illustrates that stripes are normally seen first somewhat above threshold, the image on the very left in figure 8.10 demonstrates that extremely close to threshold at $\epsilon \simeq 0$, hexagonal convection patterns are typically observed. We will demonstrate later, in connection with figure 8.23, that hexagonal patterns exist close to threshold because of symmetry-breaking terms which are left out of the Boussinesq approximation. The discussion below is focused on the regime a bit farther above threshold, where stripes are dominant.

d. Above-threshold patterns with a range of wavelengths exist

The linear stability analysis shows that when the Rayleigh number exceeds the critical value Ra_c, the convectionless base state is linearly unstable to eigenmodes over a range of wavenumbers around q_c. The width of this unstable band increases with ϵ (increasing Ra). As discussed before, this is the most common scenario in pattern forming systems, and the one the Swift-Hohenberg equation provides a simple model for. With this in mind, it may not come as a complete surprise that quite generally—not just in Rayleigh-Bénard convection—above a finite wavelength instability, pattern forming systems exhibit periodic pattern solutions over a range of wavenumbers. In other words, there is not just one unique periodic pattern state, but a whole continuous spectrum of periodic pattern states, characterized by their wavelength. Moreover, just as the band of unstable modes widens with increase of the control parameter ϵ, so does the width of the band of periodic patterns.

At this point, these are just unproven assertions—only when we develop the amplitude analysis later will we be able to back up these statements. But let us for now simply accept these statements, and illustrate them by showing results for the existence and stability of stripe patterns in Rayleigh-Bénard convection.

Figure 8.11 illustrates both the existence and stability of a regular array of stripe convection patterns, as determined from full scale

In the shadowgraph visualization technique, light is sent in, for instance, through the top of the cell, and then it is collected below (or, after reflection at the bottom, above the cell). Because the light is bent sideways toward the denser regions in the convective state (the blue regions with downflow in the lower panel of figure 8.8), the intensity of the light emerging from the cell yields a precise image of the flow pattern in the cell, with the brighter areas marking the denser downflow.

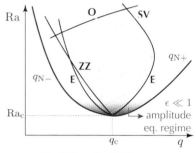

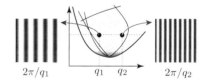

E   Eckhaus instability
O   Oscillatory instability
ZZ Zigzag instability
SV Skew-varicose instability

Figure 8.11. Typical sketch of the 'stability balloon' showing the range of existence of stable static stripe convection patterns in large Rayleigh-Bénard cells. The figure illustrates that such patterns exist for a range of Rayleigh numbers above threshold, and that for a given value of Ra, stripe patterns exist over a range of wavenumbers/wavelengths (lower panel). The domain of existence of stable stripe states, indicated by dashes, is limited by four different instabilities which can drive the stripes unstable. The Eckhaus and zigzag instabilities extend to small ϵ and are discussed in section 8.5.3, while the so-called oscillatory and skew-varicose instabilities occur only quite far beyond threshold.[18] The precise location on the stability boundaries depends on the Prandtl number $Pr = \nu/D_T$ introduced in equation (8.9).[19] The red line marks the neutral stability values of q_{N-} and q_{N+} where the growth rate of the linear modes is zero, $\sigma(q_{N\pm}) = 0$. For Rayleigh-Bénard convection the stability balloon is sometimes referred to as the Busse balloon, in honor of Fritz Busse, who played a key role in establishing it.[20]

numerical studies. The red line marks the neutral wavenumbers q_{N-} and q_{N+} for which $\sigma(q_{N-}) = \sigma(q_{N+}) = 0$. Hence for $q_{N-} < q < q_{N+}$ the linear modes from the stability calculation of the convectionless state are unstable for a given Ra (plotted along the vertical axis in figure 8.9). The dashed region indicates the domain where periodic stripe convection patterns exist *and* are stable. One sees that when Ra increases beyond Ra_c, the band of solutions increases with increasing Ra, but it lies within the red curve: it does not span the whole range from q_{N-} to q_{N+} of unstable modes.

The range of existence of stable patterns is limited by several stability boundaries, which are indicated by the black lines. Crossing one of these lines from within the dashed region in the diagram means that the striped convection pattern becomes unstable to a particular type of instability. For instance, when the upper line labeled O is crossed, stripe patterns lose stability as they start oscillating, while if the line labeled ZZ is crossed, the stripes lose stability to a so-called zigzag type of deformation. A plot like the one in figure 8.11, where the domain of stable periodic solutions and the various stability boundaries are indicated, is often referred to as a 'stability balloon.'

As indicated in the figure, the shaded region near threshold, where $\epsilon = (Ra - Ra_c)/Ra_c \ll 1$, marks the region where the pattern dynamics can be described with the amplitude equation approach, which we will develop in section 8.5. The black line labeled E (for the so-called Eckhaus instability) and the zigzag instability boundary labeled ZZ extend down to this near-threshold regime. Both the nature of these instabilities and the location of these boundaries close to threshold will be determined later with the help of the amplitude description. These stability boundaries are therefore generic for pattern forming systems; they are not specific to Rayleigh-Bénard convection. The other stability boundaries farther above threshold are specific to Rayleigh-Bénard convection, and they depend strongly on the Prandtl number Pr.

8.3.2 Turing instabilities

In Turing's own words:[21] "It is suggested that a system of chemical substances, called morphogens, reacting together and diffusing through a tissue, is adequate to account for the main phenomena of morphogenesis. Such a system, although it may originally be quite homogeneous, may later develop a pattern or structure due to an instability of the homogeneous equilibrium, which is triggered off by random disturbances. Such reaction-diffusion systems are considered in some detail in the case of an isolated ring of cells, a mathematically convenient, though

In 1952, Alan Turing published a remarkable paper in which he showed that coupled reaction diffusion equations can exhibit finite wavelength instabilities, and in which he proposed that these may lie at the basis of morphogenesis. Morphogenesis is the biological process of how a tissue or organ develops its form or shape by controlling the spatial distribution of cells during embryonic development.

We now know that structural biological information is encoded in DNA, in terms of base sequences that code for the production of proteins with a rate controlled by other proteins—these can be thought of as the 'morphogens,' the (chemical) agents which determine morphogenesis. How this information at the molecular level gets 'expressed' at the meso- or macroscopic level is still being actively investigated. Whether Turing instabilities do actually play a role in morphogenesis, and if so to what extent, is still an open question. Nevertheless, Turing's analysis of the possible instabilities of two coupled reaction-diffusion systems has become an important cornerstone and reference model of pattern formation, in particular in the life sciences. As it also further amplifies the previous discussion, it pays to summarize its main steps here.

a. The Turing model and its stability analysis

The essence of Turing's analysis can be captured by considering two coupled reaction-diffusion equations of the type

$$
\begin{aligned}
\partial_t u_1 &= D_1 \nabla^2 u_1 + f_1(u_1, u_2), \\
\partial_t u_2 &= D_2 \nabla^2 u_2 + f_2(u_1, u_2).
\end{aligned}
\tag{8.18}
$$

It is convenient to think of u_1 and u_2 as concentrations of two chemical species, but, as we shall see, coupled equations of this type also commonly emerge in other situations with a different interpretation of the us. In the chemical interpretation, the first terms on the right describe diffusion of the chemical substances, whereas f_1 and f_2 are suitable reaction terms. We can think of these functions as suitable polynomials in u_1 and u_2, which are determined by the underlying reactions.

A stability analysis proceeds again along the path threaded before: we linearize the equations around the steady state, and then we determine the dispersion relation $\sigma(q)$ of Fourier modes e^{iqx} from the linearized dynamical equations.

To implement this program, we start by assuming that the functions f_1 and f_2 allow for some uniform stationary base state u_{1s}, u_{2s} for which

$$
f_1(u_{1s}, u_{2s}) = 0, \qquad f_2(u_{1s}, u_{2s}) = 0.
\tag{8.19}
$$

By writing $u_1 = u_{1s} + \delta u_1$ and $u_2 = u_{2s} + \delta u_2$ we then linearize the equations about this stationary state by expanding the fs to first order in the δu. This yields the set of equations

$$
\begin{aligned}
\partial_t \delta u_1 &= D_1 \nabla^2 \delta u_1 + a_{11}\,\delta u_1 + a_{12}\,\delta u_2, \\
\partial_t \delta u_2 &= D_2 \nabla^2 \delta u_2 + a_{21}\,\delta u_1 + a_{22}\,\delta u_2.
\end{aligned}
\qquad a_{ij} = \left.\frac{\partial f_i}{\partial u_j}\right|_{u_s}, \tag{8.20}
$$

For the stability analysis we consider a mode with wavenumber q varying in the x direction and write, as usual,

$$
\begin{pmatrix} \delta u_1(x, t) \\ \delta u_2(x, t) \end{pmatrix} = \begin{pmatrix} \delta u_{1q} \\ \delta u_{2q} \end{pmatrix} e^{\sigma_q t} e^{iqx},
\tag{8.21}
$$

biologically unusual system. The investigation is chiefly concerned with the onset of instability." We clearly recognize Turing's insight of how patterns or structure can develop from a finite wavelength instability. For a modern perspective on the impact of Turing's work see P. Ball, 2015.

For instance, according to the law of mass action for chemical reactions, for a reaction of the type $n_A A + n_B B \to n_C C + n_D D$ for species A and B, the reaction rate is proportional to $k_{AB} c_A^{n_A} c_B^{n_B}$, where c_A and c_B are the concentrations of A and B and where k_{AB} is a rate constant.

These equations contain six parameters, the two diffusion coefficients D_1 and D_2 and the four coefficients a_{ij}. By rescaling the distances, the time, and one of the components δu_i, one sees that the linear dynamics is in fact characterized by three independent combinations only, for instance, the ratios D_1 / D_2 and a_{11} / a_{22} and the ratio of cross-couplings a_{12} / a_{21}. As figure 8.12 illustrates, in the end only two particular combinations determine the nature of the Turing instability.

Remember our discussion of complexification of real variables in section 1.7.2.

keeping in mind that in the end the physical fields are obtained by taking the real part. When this ansatz (8.21) is substituted into the linear equations (8.20), the eigenvalue equation becomes

$$\sigma_q \, \delta u_{iq} = \sum_{j=1,2} A_{ij} \, \delta u_{jq}, \qquad (i = 1, 2), \qquad (8.22)$$

where from equations (8.20) the matrix $\underline{A}$ is found to be

$$\underline{A}_q = \begin{pmatrix} a_{11} - D_1 q^2 & a_{12} \\ a_{21} & a_{22} - D_2 q^2 \end{pmatrix}. \qquad (8.23)$$

The condition that nontrivial solutions of (8.20) exist is expressed by the determinant condition

$$\left| \underline{A}_q - \sigma_q \underline{1} \right| = \begin{vmatrix} a_{11} - D_1 q^2 - \sigma_q & a_{12} \\ a_{21} & a_{22} - D_2 q^2 - \sigma_q \end{vmatrix} = 0, \quad (8.24)$$

yielding a quadratic equation for σ_q with solutions σ_{q+} and σ_{q-} given by

$$\sigma_{q\pm} = \tfrac{1}{2} \mathrm{Tr} \underline{A}_q \pm \tfrac{1}{2} \sqrt{(\mathrm{Tr} \underline{A}_q)^2 - 4 \det \underline{A}_q}, \qquad (8.25)$$

with $\mathrm{Tr} \underline{A}_q$ the trace and $\det \underline{A}_q$ the determinant of the matrix $\underline{A}_q$.

b. Implication of the stability analysis

The stability of the solutions follows straightforwardly from the above expression. Associated with the two roots of the dispersion relation are two branches, with the sign of the root with the largest real part, σ_+, determining the stability of the base state. As figure 8.12 illustrates, when plotted as a function of $\mathrm{Tr} \underline{A}_q$ and $\det \underline{A}_q$, the base state is stable only in the second quadrant. Moreover, as the negative horizontal axis is crossed downward in parameter space, σ_q remains real so the homogeneous base state shows an instability to a *stationary* mode of the type $\mathrm{Re}\, e^{iqx} = \cos(qx)$. When we cross this line in parameter space, we expect stationary periodic patterns to emerge, very much as in Rayleigh-Bénard convection or the Swift-Hohenberg equation. This is the scenario which Turing considered most interesting and explored in particular.

The diagram also shows that when we cross the upper vertical axis in parameter space toward the right, $\mathrm{Re}\, \sigma_{q\pm} > 0$ while $\mathrm{Im}\, \sigma_{q\pm} \neq 0$. This means that the base state becomes unstable to traveling modes of the type $\mathrm{Re}\, e^{i(\pm\omega_c t + qx)} = \cos(\pm\omega_c t + qx)$, where $\omega_c = |\mathrm{Im}\, \sigma_{q+}|$. We will come back to traveling wave patterns in section 8.7.

The formulation of the stability in terms of the properties of $\underline{A}_q$ actually conceals two somewhat important underlying results. First of all, for $D_1 = D_2 = D$, the determinant equation (8.24), which determines the dispersion relation, becomes an equation for the combination $\sigma_q + Dq^2$. Hence for this case the dispersion relation

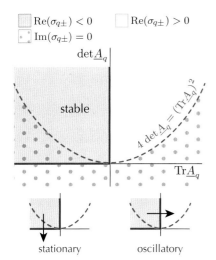

Figure 8.12. Summary of the results of the linear stability calculation of the stationary base state of the Turing model defined by the coupled reaction-diffusion equations (8.18). Results are plotted as a function of the trace and determinant of the matrix $\underline{A}_q$ which contains the various coefficients of the linear equations (8.20). Only in the second quadrant (blue) is the uniform base solution stable. When crossing the red left horizontal axis downward, the base state exhibits an instability to a stationary mode with $\mathrm{Im}\, \sigma_q = 0$, while upon crossing the red upper vertical axis the instability is to oscillatory modes with $\mathrm{Im}\, \sigma_q \neq 0$.

is simply of the form $\sigma_q = \sigma_{q=0} - Dq^2$: the growth rate is always maximal for $q = 0$. So we conclude that for finite wavelength Turing instabilities to occur, the diffusion coefficients need to be sufficiently different.

Secondly, for a finite wavelength instability to stationary patterns to occur, the diagram illustrates that $\det \underline{A}_q$ has to be negative over some finite interval of wavenumbers. If we assume, without loss of generality, that the diffusion coefficient D_2 is much larger than D_1, then we can show (see problem 8.4) from the expression for the determinant that this is possible by having $a_{11} > 0$ and $a_{22} < 0$, while the cross-coupling coefficients a_{12} and a_{21} have to have opposite signs in order for modes with sufficiently small q to be stable. In the context of the discussion of Turing patterns, the field u_1, which in the absence of coupling to u_2 tends to grow as $a_{11} > 0$, is normally called the activator, while the stabilizing field u_2 is called the inhibitor. The condition $D_1 \ll D_2$ while $a_{11} > 0$, $a_{22} < 0$, and $a_{12}\, a_{21} < 0$ for stationary Turing patterns to form is therefore commonly summarized as 'for Turing patterns to arise, one needs local activation with long-range inhibition.' Since $D_1 \ll D_2$, u_1 can grow out locally in a small spot, while the inhibitor u_2 keeps the avarage variation limited.

A specific form of the nonlinear functions f_i is needed to study finite amplitude patterns. There is a large freedom to play with these functions, and a number of popular models exist in the literature.[22] In many cases, only some generic features of these functions are needed in order to describe a certain class of phenomena—we will encounter one example of observation when we discuss excitable media in section 8.10.

In problem 8.5 we apply the above stability calculation results to the so-called Brusselator model.

c. Experimental examples of Turing patterns

Figure 8.13 shows two examples of real-life realizations of Turing patterns, the stationary patterns in a well-controlled chemical reaction cell, and the oscillations in the Min protein concentration in bacteria. Both of these cases illustrate that in practice Turing patterns emerge in out-of-equilibrium systems which are fed externally, a feature that may not be immediately clear from the theoretical model (8.18).

For chemical reactions, the terms in the Turing model have a clear physical interpretation in terms of molecular diffusion and chemical reactions. Moreover, for more complicated chemical reaction schemes, a set of Turing-type equations can often be derived in particular limits by focusing on the slowest components of the reaction. But it is good to be aware that the status of the model transcends the interpretation: the richness of the instabilities and the phenomena the Turing model can give rise to, in combination with its relative simplicity, has given the Turing equations the status of a model appropriate for capturing the essence of a large variety of complex

An example is the Belousov-Zhabotinsky reaction, an example of which is shown in figure 8.35. It involves some 18 underlying reactions.

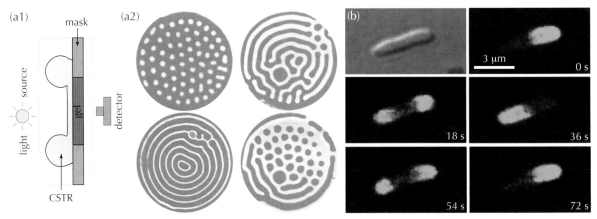

Figure 8.13. Experimental examples of Turing patterns. (a1) Sketch of the experimental cell used to study stationary Turing patterns in chemical reactions, as shown in panel (a2). The reaction chamber is made of a gel, so as to suppress any convection. Reagents are supplied and products are removed on one side with the continuously stirred tank reactor labeled CSTR. The inlet and outlet of this CSTR are not shown. Panel (a2) shows some of the patterns observed in such a cell; hexagonal, stripe, and coexisting Turing patterns are observed. The diameter of the cells is 20 mm. Images from Horváth et al., 2018.[23] (b) Observation of oscillations in the MinD protein system in live *E. coli* bacteria. The first frame shows the bacterium imaged by interference contrast microscopy; the other images show the MinD concentration measured with light intensity by labeling it with GFP (green fluorescent protein). The dynamics of the Min protein system is well described by a set of reaction-diffusion equations for MinD and MinE, their interaction with ATP, and the binding and unbinding of MinD to the membrane wall. The concentration oscillations are an example of an oscillatory Turing instability. Interestingly, these Turing oscillations of the Min protein system play a role in how a bacterium determines where the middle is during cell division (Howard et al., 2001). See Wettmann and Kruse, 2018 for an introduction and overview, and Kretschmer and Schwille, 2016 for a perspective on the role of pattern formation in cell division. Image courtesy of Karsten Kruse.[24]

problems. Examples are the vegetation patterns shown in figure 8.31, the excitable media model discussed in section 8.10, and some of the gross features of turbulent pipe flow in terms of an excitable medium in section 8.10.2.

8.4 Three types of linear instabilities

We follow the classification of Cross and Greenside, 2009 and Cross and Hohenberg, 1993 by denoting the three scenarios as type I, type II, and type III. Be aware though that these names are not widely used in the literature.

In the above examples, we encountered two different scenarios for the instability of a spatially extended system. The most common case for a pattern forming system is when the non-equilibrium system develops a *finite wavelength instability*. This case is labeled *type I* in figure 8.14: as a function of some parameter p describing the system, the dispersion relation develops a band of unstable modes with Re $\sigma_q > 0$ centered around q_c for $p > p_c$. Above the instability threshold we expect stationary patterns to exist when Im $\sigma_q = 0$—this was the scenario we found for the Rayleigh-Bénard instability and in some parameter regimes of the Turing instability. It is the case modeled by the Swift-Hohenberg equation. When Im $\sigma_q \neq 0$, however, the linear instability modes are traveling waves. Therefore, above threshold we expect traveling wave patterns or possibly standing waves, depending on the coupling between two counterpropagating waves. The amplitude description which we

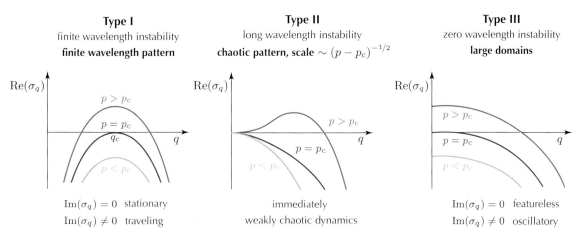

Figure 8.14. The three possible instability scenarios through which an instability can occur in a spatially extended system, as a function of some parameter p. In scenario I the system develops a finite wavelength instability; above threshold the system typically exhibits finite wavelength patterns with wavelength near $2\pi/q_c$. Examples are the Rayleigh-Bénard instability and the Turing instability, with parameters near the left horizontal axis of figure 8.12. When Im $\sigma_q \neq 0$, the patterns near threshold are traveling or standing waves. Scenario II concerns cases where a symmetry dictates $\sigma_{q=0} = 0$; an example is the phase equation discussed in section 8.5.2. In this case the prefactor of the expansion of σ_q for small q can change sign, so just above threshold, modes with very small q are unstable. Immediately above threshold, the dynamics is weakly chaotic. In scenario III the instability sets in at $q = 0$. This typically leads to featureless domains when σ_q is real and to oscillatory domains for Im $\sigma_q \neq 0$. In the latter case the phase degree of freedom associated with the oscillations can give rise to spirals or chaotic dynamics. The claims about the results for the various scenarios are substantiated in section 8.5 through section 8.7 on amplitude equations.

develop below describes the existence, stability, and dynamics of such patterns just above threshold.

We saw that when the two diffusion constants in the coupled reaction diffusion equations (8.18), whose stability we studied following Turing, are (roughly) the same, the growth rate Re σ_q is always maximal at $q = 0$. If there is an instability, it is a zero wavelength instability, indicated in figure 8.14 as *type III*. As there is then no intrinsic tendency to form finite wavelength patterns, one expects a transition to homogeneous domains of a new state to form. Whether domains form and, if so, what their size is, will depend on the problem. However, when the type III $q = 0$ instability still has Im $\sigma_{q=0} \neq 0$, the new phase will be oscillatory. In this case, different regions of the domain may get out of phase, and this can then still give rise to patterned states (just peek ahead at figure 8.30 to get an idea). If this happens, their wavelengths are not simply determined by the properties of the linear instability modes. As we will discuss later, their properties are then determined by the nonlinear behavior.

In the middle panel of figure 8.14, a third scenario is indicated. In some cases a symmetry dictates that $\sigma_{q=0} = 0$: one important example which we will encounter below is for the phase variable which enters the amplitude description. Another common example is that of an interface, where translation invariance in the direction into which the interface propagates implies $\sigma_{q=0} = 0$. In such cases it can happen that σ'' in the expansion $\sigma_q = \frac{1}{2}\sigma'' q^2 + \mathcal{O}(q^4)$ becomes positive above p_c. We will show in section 8.6 how immediately above the threshold of such a long-wavelength instability the dynamics of the relevant variable is weakly chaotic.

8.5 Amplitude equations for stationary type I instabilities

We will develop the amplitude equation description in a number of steps. In order to get a good intuitive understanding of the conceptual basis of an amplitude expansion, we will warm up with an elementary introduction to the amplitude equation approach for patterns which emerge just above the threshold of a finite wavelength instability of type I for the case that σ_q is real. As already sketched in figure 8.14, this concerns the case in which the homogeneous base state becomes unstable to stationary Fourier modes with wavenumber q_c at the point where the system parameter reaches a value p_c. This leads to a bifurcation to stationary periodic pattern states; the amplitude equation description is able to capture the dynamics and interactions of nonideal patterns in the regime close to threshold. We will first consider the one-dimensional case, and then we will discuss the various generalizations to two (and higher) dimensions, where mode interactions are crucial. Because an example of a type II instability will naturally emerge during our discussion, we take a quick detour in section 8.6 to discuss such instabilities briefly. After that we pick up the discussion of type I instabilities by analyzing traveling waves in section 8.7.

Underlying such a discussion is the assumption that the bifurcation to a patterned state corresponds to the supercritical case of figure 8.7, so that an expansion in the pattern amplitude can be made. The analysis will show what this implies for the nonlinear interaction coefficients in the amplitude description. Whether these conditions are obeyed in a particular concrete example can only be determined by calculating them explicitly from the relevant equations for the problem at hand. Often the results from experiments or numerical simulations give a good indication, and one sometimes simply chooses the coefficients to reproduce the observed behavior.

8.5.1 Inspiration from a simple perturbative calculation for the Swift-Hohenberg equation

We introduced the Swift-Hohenberg equation (8.2) as a simple model equation that exhibits a finite wavelength instability as ϵ becomes positive. Let us take advantage of this equation to first develop some intuition by constructing one-dimensional station-ary solutions of it for small ϵ. Such solutions should satisfy

Note that periodic solutions which vary only in one direction correspond to stripe solutions when extended to two dimensions.

$$0 = \epsilon u - \left(\partial_x^2 + q_c^2\right)^2 u - g u^3. \tag{8.26}$$

For $\epsilon \ll 1$, we expect stationary solutions to be very close to the mode $\cos(q_c x)$ which becomes unstable at $\epsilon = 0$. This suggests an expansion for stationary solutions u of the form[25]

$$u = a_1 \cos(qx) + a_3 \cos(3qx) + \cdots \qquad (q \approx q_c). \tag{8.27}$$

where for small ϵ we imagine $a_3 \ll a_1$. Now, if we substitute this ansatz into the equation (8.26), we get

$$0 = \left[\epsilon - (q_c^2 - q^2)^2 - \tfrac{3}{4} g a_1^2\right] a_1 \cos(qx)$$
$$+ \left[\{\epsilon - (q_c^2 - 9q^2)^2\} a_3 - \tfrac{1}{4} g a_1^3\right] \cos(3qx) + \cdots . \tag{8.28}$$

Here we used the fact that $\cos^3(qx) = \tfrac{3}{4}\cos(qx) + \tfrac{1}{4}\cos(3qx)$

For this equation to be satisfied for all x, both terms between square brackets need to be zero. The first condition implies that

$$a_1 = \sqrt{\frac{4\left[\epsilon - (q_c^2 - q^2)^2\right]}{3g}} \leq \sqrt{\frac{4\epsilon}{3g}} = \mathcal{O}(\epsilon^{1/2}). \tag{8.29}$$

This equation has two important implications. First of all, it shows that solutions exist for all values of q in the band where the growth rate $\sigma_q = \epsilon - (q_c^2 - q^2)^2$ of linear modes is positive, in other words, for all q in the linear instability band. Secondly, the mode amplitude grows as $\sqrt{\epsilon}$, is maximal in the center of the band at $q = q_c$, and vanishes at the edges of the unstable band.

Note that these conclusions hold only for $g > 0$; for this case $g < 0$, the transition is subcritical and the analysis yields in perturbation theory the first terms of the branch of unstable solutions indicated in the lower panel of figure 8.7 for $\epsilon < 0$ with a dashed line.

For the mode amplitude of the third harmonic, the requirement that the prefactor of the $\cos(3qx)$ term in (8.28) vanishes gives

$$a_3 = \frac{g}{4[\epsilon - (q_c^2 - 9q^2)^2]} a_1^3 \approx \frac{g}{256\, q_c^4} a_1^3 = \mathcal{O}(\epsilon^{3/2}), \tag{8.30}$$

where we have used that q is close to q_c.

This second result illustrates for this simple example that for $\epsilon \ll 1$, the mode with wavenumber near q_c is the dominant one, as the amplitude of the third harmonic is of order $\epsilon^{3/2}$. Extension to higher order shows that the mth harmonic is of order $\epsilon^{m/2}$.

8.5.2 Amplitude equation in one dimension for σ_q real

a. Scaling of the time and the length scales

The previous analysis showed for the specific case of the Swift-Hohenberg equation that in an expansion of the static solutions in

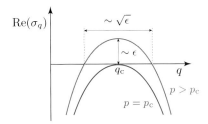

Figure 8.15. For a finite wavelength instability, just above threshold the maximum of the growth rate Re (σ_q) will grow linearly in ϵ, while the parabolic nature of the dispersion relation near q_c implies that the width of the unstable band grows as $\sqrt{\epsilon}$. We say there is a *separation of scales* for $\epsilon \ll 1$.

Note that in the case of the Rayleigh-Bénard problem these eigenvectors are z-dependent functions—compare expression (8.15)—while in the case of the Turing instability, these eigenvectors are simply the ratio of the two fields, determined by the determinant condition (8.24).

In a multiple scales approach, X_ϵ and T_ϵ are treated formally as separate scales by writing $\partial_t \to \partial_t + \epsilon \partial_{T_\epsilon}$ and $\partial_x \to \partial_x + \epsilon^{1/2} \partial_{X_\epsilon}$, while the amplitudes are expanded in powers of $\epsilon^{1/2}$, like $A(X_\epsilon, T_\epsilon) = \epsilon^{1/2} A_0 + \epsilon A_1 + \epsilon^{3/2} A_2 + \cdots$.[26] One should be aware that the multiple scales expansion method is a so-called asymptotic expansion, which 'adiabatically' completely separates the underlying pattern scale from the long scale on which the amplitude varies. 'Non-adiabatic effects,' which are intrinsically due to the coupling of the two scales, are not captured by the approach.[27]

terms of modes, their amplitudes are also nicely ordered in powers of ϵ (remember we use ϵ as a dimensionless control parameter, e.g., $\epsilon = (p - p_c)/p_c$; compare the expression (8.17) for the case of Rayleigh-Bénard convection). We now extend this observation to the dynamical behavior as a function of space and time for one-dimensional patterns, for the case in which σ_q is real.

As illustrated in figure 8.15 the growth rate just above threshold is of order ϵ; this means that the time scale for the (growth) dynamics of the amplitude of the dominant modes goes as ϵ^{-1}. Likewise, the width of the band of unstable modes scales as $\epsilon^{1/2}$, and this implies that the length scale for spatial variations of the amplitude goes as $\epsilon^{-1/2}$ (very much as the beating period of two signals with frequencies ω_1 and ω_2 is $2\pi/[\omega_1 - \omega_2]$).

b. The amplitude expansion: Basic ansatz

These observations suggest the following ansatz, which is the basis of the amplitude expansion description for the space-time dynamics of the physical fields $v_1, \cdots v_n$ of a general problem just above the instability threshold: we write

$$
\begin{pmatrix} v_1 \\ \vdots \\ v_n \end{pmatrix} = A \begin{pmatrix} v_1 \\ \vdots \\ v_n \end{pmatrix}_{ev \, q_c} e^{iq_c x} + B \begin{pmatrix} v_1 \\ \vdots \\ v_n \end{pmatrix}_{ev \, 2q_c} e^{i2q_c x} + \cdots + c.c.
$$

(8.31)

Here the terms with brackets denote the eigenvectors at wavenumbers q_c, $2q_c$, etc. (as indicated by the subscripts $ev \, q_c$, $ev \, 2q_c$) of the set of linear equations. These equations are obtained by linearizing the full set of dynamical equations characterizing the problem at hand around the stationary base state. Note that the amplitudes A, B, C, etc. are complex fields, so the real physical fields are obtained by adding the complex conjugate, as indicated by the $+c.c.$ Finally, the fact that the dominant amplitude A varies on long length scales of order $\epsilon^{-1/2}$ and long time scales of order ϵ^{-1} is expressed by

$$
A = A(X_\epsilon, T_\epsilon), \qquad A = \mathcal{O}(\epsilon^{1/2}), \quad X_\epsilon = \epsilon^{1/2} x, \; T_\epsilon = \epsilon t. \quad (8.32)
$$

The formal derivation of amplitude equations for a particular set of dynamical equations is based on the method of multiple scales. We will not follow this route explicitly here, as we will content ourselves with discussing the lowest order amplitude equation for A, and with understanding the generic form of this lowest order equation from physical principles and symmetry arguments. Nevertheless, it is good to be aware that a complete machinery exists to systematically derive such terms order by order in perturbation theory. Luckily, apart from a few exceptional cases,[28] the lowest order equations are typically sufficient to discuss the generic near-threshold behavior.

c. The lowest order amplitude equation

When focusing on the lowest order amplitude equation, it is helpful to have in mind the following simple identification of the physical fields with the amplitude:

$$\text{physical fields} \sim A e^{i q_c x} + c.c. \qquad (8.33)$$

If the precise ratios of the various physical fields are needed, these are determined by the elements of the eigenvector of the linear stability problem, as indicated in the full expression (8.31).

Now comes the important point: for the particular case we consider here, with σ_q real and variation of the pattern in the x direction only, the lowest order amplitude equation for A *cannot take any other form than*

$$\tau_0 \partial_t A = \epsilon A + \xi_0^2 \, \partial_x^2 A - g_0 |A|^2 A. \qquad (8.34)$$

To be able to play with amplitude equations and to lay the basis for exploring the necessary extensions for analyzing two-dimensional patterns and traveling waves, let us take a moment to trace the origin of the various terms and to discuss how the form of the equation is dictated by symmetry and by the properties of the dispersion relation of the linear stability modes.

Let us start by ignoring the nonlinear term $g_0 |A|^2 A$ on the right. If we retain only the linear terms in the equation and substitute

In linear order, we should just reproduce the original result of the linear stability calculation!

$$A = e^{\sigma t + i k x} \qquad (8.35)$$

into the linear terms of (8.34), we get $\sigma = (\epsilon - \xi_0^2 k^2)/\tau_0$. Since in view of (8.33) k in the above expression measures the deviation of q from q_c, the wavenumber where the instability sets in, the coefficients ξ_0 and τ_0 are simply determined by the requirement that, for the linear problem under consideration, we recover to lowest order in ϵ the expression for $\sigma(q)$ of the full problem when expanded about $\sigma(q_c, \epsilon = 0)$. By equating the two, we thus get

From here on, we use k for the wavenumber of the amplitude, so k measures the wavenumber of the pattern relative to q_c: $k = q - q_c$.

$$\tau_0^{-1} = \left. \frac{\partial \sigma_q}{\partial \epsilon} \right|_{q_c}, \qquad \frac{2 \xi_0^2}{\tau_0} = - \left. \frac{\partial^2 \sigma_q}{\partial q^2} \right|_{q_c}. \qquad (8.36)$$

In summary, the two parameters in the amplitude equation (8.34) which set the length and time scale for the variations of the amplitude can be read off from the linear dispersion relation of the problem under consideration. The coefficient g_0 in (8.34), which determines the magnitude of the amplitude above threshold and which we assume to be positive so that it leads to saturation of the pattern amplitude, is the only coefficient in the amplitude equation (8.34) which needs to be determined by a nonlinear calculation. For instance, for the Swift-Hohenberg equation, comparison with the result (8.29) shows that $g_0 = 3g$.

So the coefficients in (8.34) only set the scales. Indeed, in the rescaled variables $\tilde{X}_\epsilon = X_\epsilon / \xi_0$ and $\tilde{T}_\epsilon = T_\epsilon / \tau_0$, with X_ϵ and T_ϵ defined in equation (8.32), the equation (8.34) can be written in the scale free form $\partial_{\tilde{T}_\epsilon} \tilde{A} = \tilde{A} + \partial_{\tilde{X}_\epsilon}^2 \tilde{A} - |\tilde{A}|^2 \tilde{A}$, where $\tilde{A} = \sqrt{\epsilon / g_0} A$.

d. Phase symmetry due to translation invariance

Translation invariance in the lateral direction actually dictates that the lowest order nonlinear term in (8.34) *has* to be of the form $|A|^2 A$. Indeed, translation invariance in the x direction means that the equation describing the pattern should be invariant when we shift x by an arbitrary amount ϕ/q_c: $x \to x + \phi/q_c$. But with physical fields of the form $e^{iq_c x}$ according to (8.33), such a shift simply amounts to a phase change of the amplitude A: $A \to Ae^{i\phi}$. In short: for translation-invariant systems, the amplitude equation in one dimension should be invariant under a change of phase of A, as this boils down to a shift of the pattern. The term $|A|^2 A$ in (8.34) is the lowest order nonlinear term that obeys this requirement.

You may wonder why the lowest order nonlinearity is then not of the form $A|A|$, as this term would respect the phase symmetry of the equation. The answer is that the ansatz (8.33) leads to an expansion in powers of A and its complex conjugate A^*. Therefore, terms $|A|^2 A = A^* A^2$ *can* come in, but terms $A|A| = A\sqrt{A^* A}$ *cannot* be generated, as these are nonanalytic in A and A^*.

In summary, translation symmetry, together with the nature of the expansion, implies that the lowest order nonlinearity must be of the form $|A|^2 A = A^* A^2$.

e. Relation with Ginzburg-Landau theory

If you are familiar with the Ginzburg-Landau theory for superconductivity, you may have recognized the similarity with the Ginzburg-Landau equation if we associate A with the superconductor order parameter ψ in the absence of a magnetic field.

In fact, just as the Ginzburg-Landau equation derives from a potential, so does the amplitude equation (8.34). Indeed we can write the dynamical equation as

$$
\tau_0 \partial_t A = -\frac{\delta F}{\delta A^*},
$$
$$
F = \int \mathrm{d}x \left[-\epsilon |A|^2 + \tfrac{1}{2} g |A|^4 + \xi_0^2 \, |\partial_x A|^2 \right],
\tag{8.37}
$$

from which it is easy to show, similarly to (8.5) for the special case of the Swift-Hohenberg equation, that the dynamics is always downhill, $\mathrm{d}\mathcal{F}/\mathrm{d}t \leq 0$. We conclude that, quite generally *close to threshold, the dominant features of the dynamics of one-dimensional stationary patterns is governed by the gradient nature of the amplitude equation.* This ceases to be the case if we extend the expansion to higher order.

Moreover, the lowest order amplitude equation for traveling wave systems will turn out to be not of gradient type (see section 8.7).

In superconductivity, ξ_0 is referred to as the *coherence length*, a term which also befits the role of this parameter in pattern formation. Very much as in Ginzburg-Landau theory, the effective coherence length as a function of ϵ scales as $\xi_0/\epsilon^{1/2}$ (as the rescaled variable

$X = \epsilon^{1/2} x$ in (8.32) also illustrated), so it is larger the closer one is to the transition. This trend is visible in figure 8.10.

f. Phase winding solutions $A = a_k e^{ikx}$

Solutions of the type

$$A = a_k \, e^{ikx} \qquad (8.38)$$

with a_k and k constant are often referred to as phase winding solutions, as the phase of A winds with x at a constant rate. In view of the identification (8.33), these solutions describe patterns just above threshold with a total wavenumber $q_c + k$. If one substitutes this form into the amplitude equation (8.34), we get

$$a_k = \sqrt{(\epsilon - \xi_0^2 k^2)/g_0}, \qquad (\epsilon - \xi_0^2 k^2 > 0). \qquad (8.39)$$

Within the context of the amplitude equation, this is the equivalent of (8.29) for the Swift-Hohenberg equation. But it now shows for general pattern forming systems that, just above threshold, static pattern states exist everywhere in the band of wavenumbers where $\sigma_q > 0$, in other words, at all wavenumbers where the base state has an unstable linear mode. This result is illustrated in figure 8.16. We will study the stability of these solutions below.

g. The phase equation and stability of phase winding solutions

Let us now study the stability of the phase winding solutions by writing

$$A = (a_k + \delta a)e^{ikx + i\phi}, \qquad (8.40)$$

where a_k and k are related by (8.39). Upon substitution of this in the amplitude equation (8.34) and linearizing in δa and ϕ, we get from the real and imaginary parts

$$
\begin{aligned}
\tau_0 \, \partial_t \delta a &= -2g_0 a_k^2 \, \delta a - 2\xi_0^2 \, k a_k \partial_x \phi + \xi_0^2 \, \partial_x^2 \delta a, \\
\tau_0 a_k \, \partial_t \phi &= 2\xi_0^2 k \, \partial_x \delta a + \xi_0^2 \, a_k \partial_x^2 \phi.
\end{aligned}
\qquad (8.41)
$$

The complete stability analysis is done in problem 8.7, but the essence is captured by noting that the damping term $-2g_0 a_k^2 \, \delta a$ on the right-hand side of the first equation drives δa to a value where the terms on the right-hand side vanish, in other words, to the value given by

$$g_0 a_k \, \delta a \approx -\xi_0^2 \, k \, \partial_x \phi. \qquad (8.42)$$

This equation expresses that the variation of the amplitude follows the phase changes adiabatically. If we use this in the second equation for the variation of the phase, we get with (8.39) the

$$\text{phase diffusion equation:} \qquad \partial_t \phi = D_{||} \partial_x^2 \phi, \qquad (8.43)$$

Keep in mind that the wavenumber k of the phase winding solution $A = a_k e^{ikx}$ always measures the difference between the actual wavenumber q of the pattern and q_c: $q = q_c + k$.

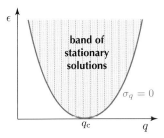

Figure 8.16. The lowest order amplitude equation shows that to lowest order in ϵ a band of stripe periodic solutions exists throughout the band where $\sigma_q > 0$. Not all of these solutions are stable. We shall see in figure 8.17 that in one dimension the solutions are unstable to phase variations near the edges of the band, while in two dimensions all solutions with $q < q_c$ are unstable to the so-called zigzag instability—see figure 8.19.

This relationship is sometimes expressed as the amplitude being 'slaved' to the phase.

where

$$D_{||} = \frac{\xi_0^2}{\tau_0} \left[\frac{-2\xi_0^2 \, k^2 + g_0 a_k^2}{g_0 a_k^2} \right] = \frac{\xi_0^2}{\tau_0} \left[\frac{\epsilon - 3\xi_0^2 \, k^2}{\epsilon - \xi_0^2 \, k^2} \right]. \tag{8.44}$$

We see that the phase ϕ obeys a simple diffusion equation. The ratio ξ_0^2 / τ_0 plays the role of a bare diffusion constant, but the most important factor is the numerator $\epsilon - 3\xi_0^2 k^2$ in the term for $D_{||}$ between square brackets, which is positive in the central portion of the band but negative near the outer edges for $\epsilon/3 \le \xi_0^2 k^2 \le \epsilon$. A negative diffusion coefficient $D_{||}$ means that small inhomogeneities in the phase *increase in time*, instead of being smoothed out, so the phase winding solutions at the edges of the band experience a phase instability. This situation is sketched in figure 8.17. A change of sign of $D_{||}$ is actually an example of a type II instability of figure 8.14; we discuss the dynamics just above this instability in section 8.6.

If one follows the dynamics from the full nonlinear amplitude equation, starting from an initial condition close to a phase winding solution with k in this unstable band, one finds that the variations grow, in accord with the phase diffusion coefficient being negative. Once the gradients become large when k is deeply into the unstable band, the modulus $|A|$ gets suppressed more and more, until at some point $|A|$ vanishes and what in superconductivity is called a phase slip event[29] occurs: a winding of ϕ by 2π is lost, and so the overall number of windings of the amplitude is reduced. Indeed, as long as the modulus is nonzero, the phase is continuous, but a winding can be lost at points in space-time where $|A| = 0$. These are the phase slip events. In the end, the dynamics typically converge to some phase winding solution in the stable band. In the language of the original pattern, this dynamics means that extra periods are created, or periods are lost, until the pattern settles to a periodic state in the stable band.

h. Phase diffusion—Symmetry, slowness, and defect interactions

The fact that, for patterns close to periodic stable states, the phase obeys a diffusion equation can be understood from the phase symmetry noted above: since the amplitude equation is invariant under a change of phase of A, the dynamical equation for ϕ can involve only spatial gradients, and since the system is reflection-symmetric, the lowest order term which can emerge in a gradient expansion is $\partial_x^2 \phi$. This conclusion is not limited to the near-threshold regime: it also holds farther away from threshold deeply into the nonlinear regime. Moreover, because diffusion is slow, the long time dynamics of patterns is dominated by phase diffusion.

This observation generalizes to patterns in two and higher dimensions, although for stripe patterns the diffusion coefficient will be

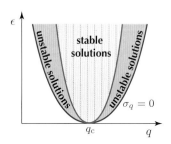

Figure 8.17. The stability analysis of the phase winding solutions of the amplitude equation in one dimension shows that the solutions are unstable at the edges of the band, for $\epsilon/3 \le \xi_0^2 k^2 \le \epsilon$. In this range the phase diffusion coefficient given by (8.43) is negative. In the context of pattern formation the phase instability in the sidebands is often referred to as the Eckhaus instability.

different for the directions perpendicular to and along the stripes. Nevertheless, the fact that the phase obeys a diffusion-type equation shows that the interaction of pattern defects, such as those in figure 8.10, is governed by long-range diffusion fields.

The expression of phase diffusion of two-dimensional stripe patterns is given in equation (8.49).

8.5.3 Two-dimensional patterns

a. Extension of amplitude equations to two-dimensional stripes

We now extend our amplitude analysis to two-dimensional patterns. The above analysis of one-dimensional patterns included the behavior of perfectly straight stripe patterns—which correspond to phase winding solutions $\sim e^{ikx}$ of the amplitude—but did not test the stability of stripes to perturbations in the y direction along the stripes. Let us first extend the previous analysis to include these effects, and let us then discuss the competition with other types of regular patterns, like squares and hexagons.

Patterns in two dimensions are described by a wavevector $\vec{q}$ with components q_x and q_y. For a rotationally invariant two-dimensional system, the growth rate σ of linear modes does not depend on the direction of $\vec{q}$, but only on its modulus $|\vec{q}|$, as illustrated in figure 8.18. When we consider stripes in two dimensions and follow (8.33) with the amplitude ansatz $Ae^{iq_c x}$ for the stripe pattern and then subsequently substitute $A \sim e^{\sigma t + i \vec{k} \cdot \vec{r}}$ in probing the linear stability of the base state, we effectively consider the stability of a mode with wavevector $\vec{q} = q_c \hat{x} + \vec{k}$. For this mode with $|\vec{k}| \ll q_c$, we have

$$|\vec{q}| - q_c = \sqrt{(q_c + k_x)^2 + k_y^2} - q_c \approx k_x + \frac{k_y^2}{2q_c}. \quad (8.45)$$

Now, our discussion of (8.36) showed that the spatial derivative terms of the amplitude equation simply encode the expansion of the growth rate about its maximum,

$$\tau_0 \sigma = \epsilon - \xi_0^2 \left(|\vec{q}| - q_c \right)^2 + h.o.t. \quad (8.46)$$

Combining these two results for $\vec{q} = q_c \hat{x} + \vec{k}$ with $k \ll q_c$, we have

$$\tau_0 \sigma = \epsilon - \xi_0^2 \left(k_x + \frac{k_y^2}{q_c} \right)^2. \quad (8.47)$$

The proper extension of the amplitude equation (8.34) can be obtained from the requirement that when we substitute $A \sim e^{\sigma t + i \vec{k} \cdot \vec{r}}$ in the linear terms in the amplitude equation, these terms should reproduce expression (8.47). The correct version is

$$\tau_0 \, \partial_t A = \epsilon A + \xi_0^2 \left(\partial_x - \frac{i}{2q_c} \partial_y^2 \right)^2 A - g_0 |A|^2 A. \quad (8.48)$$

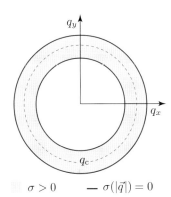

Figure 8.18. Sketch of $\sigma(q)$ for a rotationally symmetric two-dimensional system with a finite wavelength instability. The growth rate depends only on $|\vec{q}|$ and is positive within the blue ring and maximum on the dashed circle with radius q_c. It is negative elsewhere.

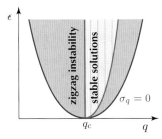

Figure 8.19. Result of the stability calculation of stripe patterns in two dimensions, from the lowest order amplitude expansion. All solutions with $k = q - q_c < 0$ are unstable to the so-called zigzag instability, as for these $D_\perp < 0$ according to (8.50). The right stability boundary is the Eckhaus stability boundary already identified in figure 8.17.

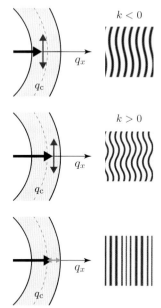

Figure 8.20. Illustration of the stability properties of a stripe phase in two dimensions as determined by phase variations. Part of the q_x, q_y space sketched in figure 8.18 is drawn for three cases. The heavy black arrow indicates the total wavevector $(q_c + k)\hat{x}$ of the periodic stripe state whose stability is probed, and the colored arrow indicates the local variations in the wavevector. The top panel shows how for $k < 0$ slow variation of the phase ϕ in the y direction effectively changes the local wavevector to regions near the dashed line where the linear growth is maximal. Hence these variations have a growth rate which is larger than the one associated with the pattern whose stability is probed. As a result the original state is unstable to these variations. The middle panel illustrates how for $k > 0$ such y modulations of the phase always lead to lower linear growth rates, so patterns with $q > q_c$ are stable to these phase modulations. The bottom panel, finally, illustrates why modulations of the phase in the x direction lead to local values of the wavevector, which in some ranges enhance linear growth rates, and in others reduce these. For small enough k, the net effect is that the pattern state is stable, but for large enough k (to be precise, for $k\xi_0 > \sqrt{\epsilon/3}$) the nonlinear coupling via the amplitude leads to instability. On the right the corresponding modulated pattern is illustrated for each of the three situations.

Do check for yourself that by substituting $A \sim e^{\sigma t + i\vec{k}\cdot\vec{r}}$ you do indeed obtain equation (8.47) for σ from the terms linear in A.

b. Stability of stripe states in two dimensions

The linear stability of stripe states can be determined directly from this amplitude equation along the same lines as before. If we start again from (8.40) and linearize the equation in δa and ϕ, the y derivative terms give an additional term to the phase diffusion equation (8.43),

$$\partial_t \phi = D_{||}\partial_x^2\phi + D_\perp\partial_y^2\phi, \qquad (8.49)$$

where $D_{||}$ is unchanged from (8.44) and

$$D_\perp = \frac{\xi_0^2}{\tau_0}\frac{k}{q_c}. \qquad (8.50)$$

As the diffusion coefficient is negative for $k < 0$, we find that stripe solutions are unstable to small modulations along the stripe direction for all $q = q_c + k < q_c$! The resulting stability balloon for stripes as it thus emerges to the lowest order in ϵ is sketched in figure 8.19—the left half of the band of periodic solutions, which earlier was found to be stable to expansion- and compression-type variations in the stripe positions in the direction perpendicular to the stripes, is unstable to a zigzag type of modulation of the pattern along the stripe direction. The origin of this zigzag instability is illustrated in figure 8.20.

c. The stability diagram for two-dimensional stripes

Figure 8.19 summarizes the existence and stability of stripe states as they follow from the lowest order amplitude equations, hence to the lowest order in ϵ. We stress that the behavior displayed in the figure holds for the general case of a system exhibiting a finite wavelength instability to stationary striped patterns. The figure shows that generally above threshold a family of stable striped states exist, characterized by their wavelength.

The calculation can be extended to higher order in ϵ by including higher order amplitude equations. However, the results start to depend quickly on the system under consideration, via the nonlinear coupling terms involved. For instance, as the stability balloon of striped convection states shown in figure 8.11 illustrates, the zigzag instability boundary, which we found above to be at $k = 0$ (so $q = q_c$) to order $\sqrt{\epsilon}$, is found to have a finite slope in order ϵ. This effect can still be captured perturbatively. However, the other instabilities that mark the various stability boundaries occur for values of ϵ of order unity, so these can only be determined numerically. Also, the convection patterns shown in figure 8.10 illustrate that in circular cells, spirals and other intrinsically dynamical states take over once ϵ is not really small anymore.

The extent to which such effects are within reach of amplitude expansions depends on the specific question asked. For instance, the precise locations of the stability boundaries, especially those associated with other types of instabilities, are hard to calculate analytically. On the other hand, several properties of various types of defects which often dominate the patterns can be captured with an amplitude approach.

d. Other regular patterns: Interaction of modes

So far, we have only considered stripe patterns: these are governed by one wavevector $\vec{q}$, together with (because of the complex conjugate) its twin $-\vec{q}$. Other regular pattern states—squares, hexagons and even more complicated patterns—can be thought of as consisting of a sum over modes with different wavevectors, consistent with the symmetry of the lattice (very much as one can read off the symmetry of a crystal lattice of atoms from the Bragg spots seen in X-ray scattering). Just above threshold, obviously the modes on the circle $|\vec{q}| = q_c$ dominate, as indicated in figure 8.21. So we can understand their near-threshold behavior by studying the interaction between these modes only.

e. Square patterns

Let us first consider the amplitude equations for square patterns. Because of rotational invariance, the linear equations for the modes A_x and A_y are the same. Furthermore, remember that translation invariance of the equation in one dimension implied that the amplitude equation should be invariant under a change of phase of A. Along the same lines, the nonlinear equation should now be invariant under a change of phase of A_x and A_y separately—a change of phase of A_y amounts to a shift of the pattern in the y direction, a change of phase of A_x to a shift in the x direction. For a translation-invariant system, such shifts should not matter.

For the set of equations to be invariant under such phase changes, the nonlinear terms should transform like the linear terms in the equation. Imposing these requirements implies that the lowest order amplitude equations have to be of the form

In other words, these equations form the proper equivariant dynamical system, i.e., the dynamical equations obeying the proper symmetries.

$$\tau_0 \partial_t A_x = \epsilon A_x + \xi_0^2 \partial_x^2 A_x - g_0 |A_x|^2 A_x - g_{90} |A_y|^2 A_x,$$
$$\tau_0 \partial_t A_y = \epsilon A_y + \xi_0^2 \partial_y^2 A_y - g_0 |A_y|^2 A_y - g_{90} |A_x|^2 A_y. \tag{8.51}$$

For simplicity we have included only the lowest order spatial derivative terms (ignoring terms in the transverse direction). As in the one-dimensional case, these equations derive from a potential, and from this it is easy to map out the competition of the square pattern with $|A_x| = |A_y|$ with a stripe solution where one of the amplitudes vanishes.[30] For $g_{90} > g_0$ the stripe solution has the lowest energy, and for $g_{90} < g_0$ the square pattern is favored.

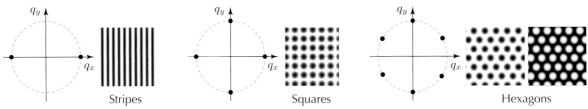

Figure 8.21. Stripe, square, and hexagon patterns and the dominant modes they are built from. Just above the instability threshold the dominant wavevectors lie on the dashed circle with radius q_c in the $q_x q_y$ plane, as these correspond to the wavenumbers where the linear growth is maximal. Vertically oriented stripe patterns (left) consist of modes with wavenumber $q_x = \pm q_c$, square patterns (middle) involve modes at $q_x = \pm q_c$ and $q_y = \pm q_c$, and hexagonal patterns (right) involve modes which make a $120°$ angle and their inversions. The hexagonal patterns actually come in two variants, depending on the sign of the amplitudes. For convection, these are distinguished by the flow in the center being down (bright spots) or up (dark spots). Adapted from Cross and Greenside, 2009.

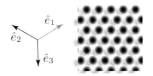

Figure 8.22. The three unit vectors used to describe hexagonal patterns in the amplitude expression (8.52). Since $\hat{e}_1 + \hat{e}_2 + \hat{e}_3 = 0$ and $\hat{e}_1 \cdot \hat{e}_{2,3} = -1/2$, the form (8.52) in terms of the amplitudes implies that a translation of the hexagonal pattern a distance a in the $\hat{e}_1$ direction, as sketched on the right, is equivalent to a phase change $e^{iq_c a}$ in A_1 and a phase change $e^{-iq_c a/2}$ in A_2 and A_3. Shifts of the pattern in the $\hat{e}_2$ and $\hat{e}_3$ directions lead to similar expressions with the indices permuted.

Interestingly, in rotating Rayleigh-Bénard cells the chiral symmetry is broken, so $g_{120} \neq g_{-120}$. In this case, which does not derive from an energy, interesting dynamics can occur: domains where mode A_1 dominates are invaded by domains dominated by mode A_2, which in turn are invaded by domains dominated by mode A_3, etc. A continuously dynamic state ensues, characterized by domains each overtaking one other type while being overtaken by the third type (Tu and Cross, 1992).

f. Hexagonal patterns

The case of hexagonal patterns is more interesting, since the phase degree of freedom allows for a quadratic term. To see this, note that in this case we write in analogy to (8.33) the physical fields as

$$\text{physical fields} \sim A_1 e^{iq_c \hat{e}_1 \cdot \vec{r}} + A_2 e^{iq_c \hat{e}_2 \cdot \vec{r}} + A_3 e^{iq_c \hat{e}_3 \cdot \vec{r}} + c.c., \tag{8.52}$$

where the three unit vectors $\hat{e}_i$ all make a 120-degree angle, as indicated in figure 8.22. Because of this form, and since $\hat{e}_1 = -(\hat{e}_2 + \hat{e}_3)$, translation invariance of the pattern in the $\hat{e}_1$ direction implies that the equations should be invariant under the combined change of phase

$$A_1 \to A_1 e^{i\phi}, \quad A_2 \to A_2 e^{-i\phi/2}, \quad A_3 \to A_3 e^{-i\phi/2}, \tag{8.53}$$

and similarly for permutations of the indices. The lowest order amplitude equations consistent with this phase symmetry are

$$
\begin{aligned}
\tau_0 \, \partial_t A_1 &= \epsilon A_1 + \xi_0^2 \, (\hat{e}_1 \cdot \vec{\nabla})^2 A_1 - g_0 |A_1|^2 A_1 \\
&\quad - g_{120} |A_2|^2 A_1 - g_{-120} |A_3|^2 A_1 + \gamma A_2^* A_3^*, \\
\tau_0 \, \partial_t A_2 &= \epsilon A_2 + \xi_0^2 \, (\hat{e}_2 \cdot \vec{\nabla})^2 A_2 - g_0 |A_2|^2 A_2 \\
&\quad - g_{120} |A_3|^2 A_2 - g_{-120} |A_1|^2 A_2 + \gamma A_3^* A_1^*, \\
\tau_0 \, \partial_t A_3 &= \epsilon A_3 + \xi_0^2 \, (\hat{e}_2 \cdot \vec{\nabla})^2 A_3 - g_0 |A_3|^2 A_3 \\
&\quad - g_{120} |A_1|^2 A_3 - g_{-120} |A_2|^2 A_3 + \gamma A_1^* A_2^*.
\end{aligned}
\tag{8.54}
$$

For stationary cells the coefficients g_{120} and g_{-120} are equal because of rotation symmetry: if they were different, the suppression of mode 1 by mode 2 would differ from the suppression of mode 2 by mode 1, and similarly for other pairs, so then the equations would have a chirality. When the two coefficients are equal, the equations can be derived from an energy function.

The quadratic terms proportional to γ are new and have important consequences. They break the $A_i \to -A_i$ symmetry of the equation, which all the other terms respect. To understand their relevance, consider for instance the Swift-Hohenberg equation (8.2). The equation is unchanged under a change of sign of u, so when studied in two dimensions, also the corresponding amplitude equation should respect this symmetry. So, for this case, $\gamma = 0$. However, if a symmetry-breaking term bu^2 is added to the Swift-Hohenberg equation, this symmetry is broken, and γ will be nonzero (and small for b small).

The situation is similar though somewhat more subtle for the convection case. In the Boussinesq approximation discussed in section 8.3.1, the flow fields in the downward flow region are precisely the mirror image of the flow in the upward region (see problem 8.8). This implies that in the Boussinesq approximation, the resulting amplitude equations will respect the $A_i \to -A_i$ symmetry, and hence that $\gamma = 0$. In this case, stripes are the favored (lowest energy) patterns. However, the Boussinesq approximation is excellent but not exact, and terms like the nonlinear temperature dependence of the density, which go beyond the linear relation (8.6), break the symmetry of the Boussinesq approximation. Hence in practice γ will be nonzero but very small.

It can be shown[31] that a nonzero value of γ results in a subcritical transition to a hexagon state with all three moduli $|A_i|$ being equal. Intuitively, one may understand the effect by noting that a quadratic term breaks the invariance under a change of sign of the amplitudes, so that a quadratic term will always enhance the growth of the amplitudes for a suitable combination of relative signs. In summary, the upshot of these arguments is that *real fluids always break the Boussinesq symmetry weakly*, resulting in a very small nonzero value of γ and hence in a weakly subcritical transition to a hexagon state.

Figure 8.23 and the leftmost image in figure 8.10 illustrate that hexagon convection patterns are indeed observed slightly below and above the transition. Pretty soon above threshold though, the stripe states are favored and seen experimentally. Interestingly, also the observation that the hexagonal domain looks striped at the edge can be understood from equations (8.54)—see the figure caption. Note also that the sign of γ does matter: it determines the relative phases of the amplitudes in the subcritical state, and it distinguishes between the two states shown in the right panel of figure 8.21, the one with downflow in the centers (right image, bright spots) or upflow in the centers (left panel, dark spots).

While the weakly subcritical nature of the transition in convection experiments at extremely small ϵ may appear to be an academic issue, we emphasize that the general conclusion, namely that systems with strong up-down symmetry breaking have a strong tendency to exhibit hexagonal patterns rather than stripes, is a nontrivial and relevant result. This is why chemical Turing patterns are

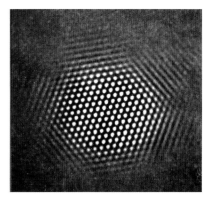

Figure 8.23. Hexagonal convection pattern observed in experiments by Bodenschatz et al., 1992 just below threshold, for $\epsilon = -1.92 \cdot 10^{-3}$. This is due to non-Boussinesq effects, which, as explained in the text, give rise to a weakly subcritical transition to hexagonal patterns. Note that at the edge of the domains, the pattern looks striped. This can be understood immediately from (8.54): at an interface where the fields vary in the $\hat{e}_1$ direction, we have $(\hat{e}_2 \cdot \vec{\nabla})^2 = (\hat{e}_3 \cdot \vec{\nabla})^2 = \frac{1}{4}(\hat{e}_1 \cdot \vec{\nabla})^2$. Hence the amplitude A_1 effectively has a larger coherence length than A_2 and A_3, so the pattern in the interface region is dominated by A_1 and looks striped (Pismen and Nepomnyashchy, 1994). This is a general phenomenon. Image courtesy of Eberhard Bodenschatz.[32]

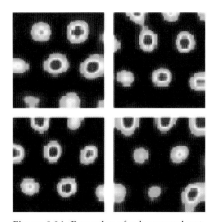

Figure 8.24. Examples of a hexagonal pattern in the spatial response correlation function of a neural model. Such hexagonal-type response patterns are especially found in realistic neural models that are based on signals of positive sign only, so that their updating dynamics involves nonnegativity constraints of the signal. As pointed out by Sorscher et al., 2019 with an expansion reminiscent of our analysis, the strong symmetry breaking which a nonnegativity constraint entails naturally gives rise to hexagonal patterns. In essence, this happens for the same reason that the symmetry-breaking non-Boussinesq effects give rise to hexagonal convection patterns in Rayleigh-Bénard cells, as figure 8.23 illustrates. Images are from the github page github.com/ganguli-lab of the Lab of Surya Ganguli at Stanford University. Creative Commons license.

often hexagonal—chemical concentrations can't be negative, so the equations do not have a symmetry under a change of sign!—and why, similarly, in neural networks response correlations are also often hexagonal, as figure 8.24 illustrates.

8.6 Dynamics just above a type II instability

Before we turn to oscillatory instabilities, we take a moment to discuss briefly the behavior just above a long-wavelength instability of type II, illustrated in figure 8.14. This is because, as we already mentioned after equation (8.44), the long-wavelength instability near the edges of the band of solutions (the so-called Eckhaus instability) is an example of such a type II instability. Moreover, as we shall see below, this phase instability also plays a significant role in oscillatory systems.

As already mentioned, a type II instability occurs when the second derivative $\sigma'' = \mathrm{d}^2\sigma/\mathrm{d}q^2|_{q=0}$ becomes positive, and this situation most naturally occurs in cases where a symmetry dictates $\sigma(q=0) = 0$. A good example is the phase equation where the phase symmetry due to translation invariance implies $\sigma = 0$ for $q = 0$. When we translate, just beyond the instability threshold, the expression $\sigma(q) \approx \frac{1}{2}\sigma''q^2 + \frac{1}{4!}\sigma''''q^4 + \cdots$ back into real space, we get

$$\partial_t \phi = -\frac{1}{2}\sigma'' \nabla^2 \phi + \frac{1}{4!}\sigma'''' \nabla^4 \phi + \cdots, \tag{8.55}$$

where just above the threshold $0 < \sigma'' \ll 1$, assuming $\sigma'''' < 0$.[33]

What are the proper nonlinear terms entering above threshold? Because of the phase symmetry, nonlinear terms cannot depend on ϕ; they can only depend on gradients of ϕ. And since we consider a long-wavelength instability for $0 < \sigma'' \ll 1$, the lowest order nonlinear gradient term is $(\vec{\nabla}\phi)^2$. After a proper rescaling of length, time, and phase,[34] we therefore conclude that in rescaled variables, indicated by tildes, the lowest order dynamical equation that governs the dynamics just above the instability threshold is

$$\partial_t \tilde{\phi} = -\tilde{\nabla}^2 \tilde{\phi} - \tilde{\nabla}^4 \tilde{\phi} - \frac{1}{2}(\tilde{\vec{\nabla}}\tilde{\phi})^2. \tag{8.56}$$

If we take a gradient of the equation, we see that this equation is equivalent to the following equation for $\tilde{q} = \tilde{\vec{\nabla}}\tilde{\phi}$:

$$\partial_t \tilde{q} = -\tilde{\nabla}^2 \tilde{q} - \tilde{\nabla}^4 \tilde{q} - \tilde{q} \cdot \tilde{\vec{\nabla}}\tilde{q}. \tag{8.57}$$

Both equivalent forms of the equation are referred to in the literature as the Kuramoto-Sivashinsky equation, after the authors who introduced it for the phase.[35] The equation is interesting in itself as a dynamical equation. First of all, note that in these scaled variables, the equations are parameter-free; the only remaining parameter is the system size, the length L in one dimension. Moreover, while the equations do admit periodic solutions, it is found that as the system size increases, the relative basin of attraction of these states shrinks, and the dominant dynamics is weakly chaotic (and sometimes intermittent).[36] Figure 8.25 illustrates the resulting chaotic dynamics of the one-dimensional equation.

We will not go into the detailed dynamics here. For us it suffices to underline the general finding that in sufficiently large systems, the dynamics of the equation is chaotic: the chance that under the dynamics the field ϕ or $\vec{q}$ converges to a well-defined regular patterns state rapidly vanishes with increasing systems size. This is an important result, as the Kuramoto-Sivashinsky equation (8.56) is *the proper generic lowest order equation* for the phase variable just above the threshold of a type II instability. Since its dynamics is chaotic for large enough systems, we conclude that in large systems *the natural dynamics just above a type II instability is weakly chaotic.*

Whether the specific Kuramoto-Sivashinsky-type behavior dominates the dynamics of an unstable phase winding solution of the amplitude equation typically depends on how deeply into the unstable regime one is, i.e., how far the base state is from the Eckhaus stability boundary. Deep into unstable regime, the coupling to the modulus of the amplitude becomes important, and the phase slip events discussed after equation (8.44) become more frequent.

8.7 Amplitude equations for oscillatory type I instabilities

So far, we have focused on the amplitude equations near the bifurcation to stationary patterns, which arise for finite wavelength instabilities with $\sigma(q)$ real. When σ has a nonzero imaginary part at the instability threshold, then the basic modes which go unstable are of the form $e^{i(q_c x \pm \omega_c t)}$, where $\omega_c = |\mathrm{Im}\,(\sigma(q_c))|$ at $\epsilon = 0$. These are traveling waves moving to the left (+) or to the right (-).

8.7.1 Amplitude equations for one-dimensional traveling waves

Conceptually, the derivation of the amplitude equation for a one-dimensional system follows essentially the same steps as before: in analogy with (8.33), we write, for the right-moving mode with

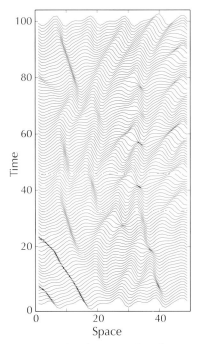

Figure 8.25. Simulation results of Tajima and Greenside, 2002 of the one-dimensional Kuramoto-Sivashinsky equation (8.57) on an interval of length 50 with rigid boundary conditions. Shown is a space-time plot, where the solution q is drawn at intervals of $\Delta t = 1$ with a small shift upward, so that the space-time dynamics becomes visible. For this case the dynamics is chaotic. Figure courtesy of Henry Greenside.[37]

We write the equation for a right-moving pattern only. The equation for A_L is similar, with obvious sign changes for the imaginary parts. For analyzing the interaction of left- and right-traveling modes, allowing for the possibility of standing waves with both amplitudes equal, the amplitude equation (8.59) for A_R has to be extended by an interaction term $-g_{LR}(1 - i c_{LR})|A_L|^2 A_R$. In most practical traveling wave systems, each mode strongly suppresses the other so that it is safe to analyze a single equation in separate domains.

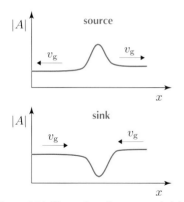

Figure 8.28. Illustration of sources and sinks in traveling wave systems. A source is a coherent structure in the amplitude with the group velocity v_g on both sides pointing outward (it sends out waves); a sink is a structure between waves with the group velocity pointing inward. The picture is drawn in the frame moving with the source or sink. After van Saarloos and Hohenberg, 1992.

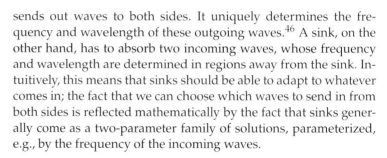

sends out waves to both sides. It uniquely determines the frequency and wavelength of these outgoing waves.[46] A sink, on the other hand, has to absorb two incoming waves, whose frequency and wavelength are determined in regions away from the sink. Intuitively, this means that sinks should be able to adapt to whatever comes in; the fact that we can choose which waves to send in from both sides is reflected mathematically by the fact that sinks generally come as a two-parameter family of solutions, parameterized, e.g., by the frequency of the incoming waves.

The upshot of this analysis is that in one-dimensional traveling wave systems, sources are the dominant localized structures, and these uniquely determine the patterns emerging in the domains. Such sources are truly nonlinear solutions, whose properties (such as the frequency or wavenumber of the waves they emit) depend on the various parameters. Traveling waves from different domains can collide at sinks. Their motion and properties are determined by the incoming waves.

8.7.3 What about two and higher dimensions?

For a formal description near threshold of a finite wavelength instability to traveling waves in two and higher dimensions, we can in principle follow the steps of section 8.5.3 to consider sums of modes with different wavevectors on the circle $|\vec{q}| = q_c$ for analyzing regular lattices. However, as mentioned in the introduction, in practical situations one traveling wave strongly suppresses all others, so the pattern becomes dominated by large domains of waves spreading out circularly in all directions from some localized wave-generating center. A single amplitude equation, which is based on perturbing around a particular finite wavevector $\vec{q}$, cannot capture this. For oscillatory type III instabilities, though, an extension of the CGLE to a higher dimension is relevant close to threshold, as we discuss below.

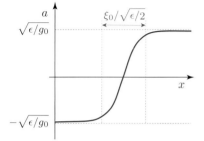

Figure 8.29. Illustration of the solution $a = \sqrt{\epsilon/g_0}\tanh(x/\xi)$ with $\xi = \xi_0/\sqrt{\epsilon/2}$, which is a stationary domain wall solution of equation (8.62), as can easily be checked by direct substitution.

8.8 Amplitude equations for type III instabilities

As we illustrated in figure 8.14, instabilities of type III are ones for which the growth rate is maximal at $q = 0$. Since there is no associated intrinsic length scale from the linear dispersion relation, systems exhibiting such an instability tend to form large homogeneous domains, not patterns with a characteristic length scale.

When $\sigma(q)$ is real, the eigenvectors are real as well, and the new basis states to which the system bifurcates are homogeneous states with real amplitude a. If the system still obeys an $a \to -a$ symmetry, the appropriate lowest order equation is the real version of (8.34),

$$\tau_0 \, \partial_t a = \epsilon a + \xi_0^2 \, \partial_x^2 a - g_0 a^3, \qquad (8.62)$$

with, as before, τ_0 and ξ_0 following from the dispersion relation according to (8.36). The most important nontrivial solution of this equation, which arises in many corners of physics, is the stationary tanh domain wall solution sketched in figure 8.29, which connects the positive and negative steady stable states $a = \pm(\epsilon/g_0)^{1/2}$.

The case of oscillatory type III instabilities is much richer since different parts of the domain can get out of phase. In this case, the eigenvalue σ and the eigenvectors of the linear operator are complex, and the system bifurcates to an oscillatory state characterized by the frequency $\omega_c = |\mathrm{Im}\, \sigma(q=0)|$. So we now have

$$\text{physical fields} \sim A e^{-i\omega_c t} + c.c. \qquad (8.63)$$

Even though the maximum growth rate is at zero wavenumber, and so does not have an intrinsic pattern length scale associated with it, the oscillations in different regions can get out of phase; for an oscillatory chemical reaction, for instance, this translates into a spatial pattern set by composition variations.

For deriving the amplitude equation close to threshold, we can follow the same steps as before except that now we need to take $q_c = 0$. The main difference with the case of finite q_c is that the state of the system is governed by a single amplitude A and that the amplitude expansion is rotationally invariant around $\vec{q} = 0$. The relevant amplitude equation is thus (compare equation (8.58))

$$\tau_0 \, \partial_t A = \epsilon(1 + ic_0)A + \xi_0^2 \, (1 + ic_1)\nabla^2 A - g_0(1 - ic_3)|A|^2 A. \qquad (8.64)$$

The study of this equation has become a field in itself,[50] as it exhibits a bewildering range of behavior. Figure 8.30 illustrates this. Of course, the equation contains all the chaotic dynamics reported briefly above for the one-dimensional CGLE. On top of it, the equation in two dimensions allows for target- and spiral- type solutions. Such targets and spirals act as the two-dimensional equivalents of sources. Their interactions are usually mediated by the waves sent out by them.[51]

These studies are useful for exploring all possible dynamics and interactions, but one should keep in mind that the validity of the amplitude equation is limited to the regime close to threshold. In real-life applications, one is often soon far into the nonlinear regime, so that it is necessary to return to the underlying equations—for instance, the reaction-diffusion equations of the Turing model (8.18). In many realistic cases with chemical reaction waves, sharp gradients occur, due to strong differences in diffusivities and strong

The equation has also served as the starting point of many studies of front propagation into an unstable state (in this case the state $a = 0$), which is a rich field in itself.[47]

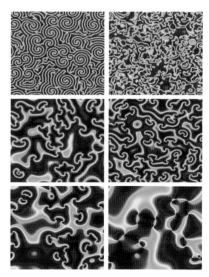

Figure 8.30. Illustration of the varied dynamics of the CGLE (8.64) in two dimensions for various values of c_1 and c_3, color coded according to the phase.[48] Note the formation of spirals in the upper left panel, and of interacting defects reminiscent of liquid crystal defects in the lower right corner. Mocenni et al., 2010 explore various statistical characterizations of the dynamics. Image courtesy of Chiara Mocenni.[49]

nonlinearities of the equation. We will encounter an example of this in our discussion of excitable media—just look ahead at the experimental waves of figure 8.35. Nevertheless, in regimes where the dynamics is incoherent or even chaotic, it may well be that the statistical characterizations of the dynamics are quite universal and that the CGLE captures these sufficiently well. The extent to which this is the case is presently not clear.

8.9 Taking stock of pattern formation and the amplitude description

In this chapter, we have introduced various examples of pattern formation in extended out-of-equilibrium systems and discussed in detail what the amplitude equation description can tell us about the existence, stability, and dynamics of patterns just above the instability threshold. This is a good place to summarize what we have learned and to discuss the issue of pattern selection.

8.9.1 Box 4: Summary of insights from amplitude equation approach

The amplitude equation approach is a powerful formalism that allows us to understand many of the main features of pattern forming systems near threshold. It also helps to organize our thinking farther above threshold or if we aim to understand a new type of problem that we are confronted with. It took quite a few steps to arrive at the overall understanding, and the body of knowledge that we aimed to sketch may have appeared overwhelming. We therefore summarize the main results of the analysis in box 4, with pointers to where the issues are treated.

8.9.2 Pattern selection?

It has emerged from our analysis that in general above the threshold of a finite wavelength instability, a family of stable pattern solutions exists. As the stability balloon for stripe convection patterns and the corresponding one from amplitude expansions showed, these pattern states span a band of wavenumbers or, equivalently, wavelengths. This picture, which extends to other regular patterns, such as squares and hexagons, naturally brings up the question of 'pattern selection': is there a mechanism or a principle according to which a pattern with a well-defined wavelength is selected or favored in an actual experimental situation?[53]

We emphasize that within pattern formation we typically focus on well-defined patterns and dynamics which exist for not too large driving in cells with large aspect ratio, $\Gamma \gg 1$. In convection cells with $\Gamma \approx 1$ and Ra not too far above threshold, one typically observes a single convection roll, whose dynamics can be thought of in terms of a low-dimensional dynamical system—it can even exhibit the period doubling route to chaos specific to low-dimensional dynamical systems (Gollub et al., 1980). For very large driving Ra $\gg$ Ra$_c$ one observes well-defined turbulence with many length scales involved, and analyses and simulations in this regime are focused on understanding the scaling behavior with Ra.[52]

Compare figure 8.11 and figure 8.19.

Box 4. Summary of what amplitude equations teach us about patterns in out-of-equilibrium systems

1. The amplitude equation approach describes pattern behavior just above threshold of systems exhibiting a finite wavelength instability. It applies to systems which exhibit a forward (supercritical) bifurcation to pattern states, or a weakly subcritical transition. (Section 8.2.3, section 8.5.)

2. As a function of the control parameter ϵ, the time scale with which the pattern adjusts goes as ϵ^{-1} and the coherence length of the pattern as $\epsilon^{-1/2}$. Prefactors determining the time and length scales can be calculated from the dispersion relation of stability modes at onset. (See figure 8.15.)

3. Above the threshold of a finite wavelength instability at wavenumber q_c, the system exhibits a continuum of periodic patterns with a different wavelength. The wavelength is close to $2\pi/q_c$, and the width of the band of solutions grows as $\epsilon^{1/2}$. Periodic pattern states need not necessarily be stable. (See figure 8.16.)

4. Above a transition to stationary patterns, there is a range of periodic states which are stable. In one dimension, the band is symmetric and limited by the Eckhaus instability; for two-dimensional striped patterns only a range of periodic solutions with $q > q_c$ is stable. Periodic solutions with $q < q_c$ exhibit a zigzag instability. (See figure 8.17 for the one-dimensional case, figure 8.19 for two-dimensional stripes.)

5. The lowest order amplitude equation in one dimension is relaxational, and its dynamics is governed by downhill dynamics of an energy function. The form of the equation is reminiscent of the Ginzburg-Landau equation. (See equation (8.37) and the adjacent discussion.)

6. Farther above threshold, the 'stability balloon' that indicates the range of existence of stable periodic solutions is limited by other stability boundaries that are system-specific and usually have to be determined numerically. (See figure 8.11 for Rayleigh-Bénard, and figure 8.31 for a Turing-type model.)

7. The dynamics of inhomogeneous patterns is governed by phase diffusion. The concept of phase diffusion also extends farther above threshold. (See equation (8.43) for one and (8.49) for two dimensions.)

8. Ordered patterns other than stripes, like squares and hexagons, can be studied by taking mode interactions into account. The competition between ordered patterns depends on the nonlinear terms in the amplitude equation. Hexagon states are favored in systems without reflection symmetry. (Figure 8.21 and figure 8.23.)

9. The Turing model, consisting of two coupled reaction-diffusion systems, can exhibit both stationary and oscillatory instabilities. It serves as a basic reference model for many types of instabilities. (Section 8.3.2.)

10. Systems exhibiting a transition to traveling wave states exhibit a rich behavior. The stability and dynamics of the traveling wave states near onset depend sensitively on two parameters that model the dependence of the wave frequency on wavenumber and amplitude. (Section 8.7, especially text following equation (8.60).)

11. In some regions of parameter space, all periodic traveling wave states are unstable. Various types of chaotic dynamics can occur in these parameter ranges. (Section 8.7, figure 8.26.)

12. Oscillatory systems exhibiting a zero wavenumber instability can still exhibit an interesting pattern dynamics due to different regions of the system being out of phase. The dynamics can be modeled by the complex Ginzburg-Landau equation; patterns tend to be dominated by spirals and target patterns. (Section 8.8.)

13. Localized structures—sources in one dimension, spirals and target patterns in two dimensions—typically dominate the overall dynamics of traveling wave and oscillatory systems. This is because they send out waves which sway the surrounding domain. Domains controlled by different sources or spirals meet at boundaries, where waves coming in from both sides get annihilated (sink-type patterns). (Section 8.7.2, Section 8.7.3, Figure 8.35.)

14. The dynamics just above threshold of a long-wavelength instability of type II generally exhibits chaotic behavior immediately above threshold provided the system is large enough. (Section 8.6 and Figure 8.25.)

a. Stationary patterns

It turns out that there is neither a simple nor a definite answer to this legitimate question. Let us first consider the issue for stationary patterns. In a well-controlled but finite system without inhomogeneities the answer often depends on the constraints provided by the boundary conditions—for example, for the convection cell of figure 8.10, striped convection patterns prefer to orient perpendicularly to the walls, and so in a circular cell the pattern is bound to contain some defects.

We refer you to Cross and Hohenberg, 1993 for an in-depth discussion of the various selection mechanisms and their relevance as functions of distance from the threshold.

For stationary patterns, a number of possible selection mechanisms have indeed been identified, and it depends on the situation at hand which one, if any, is operative. For instance, a sharp wavelength can be selected by a smooth boundary in the form of a ramp where the control parameter is gradually reduced to below threshold (hence $\epsilon < 0$). Other known sharp selection mechanisms include the wavelength generated by a front propagating into the featureless base state, or by a circular target pattern. Dislocation defects and grain boundaries can also affect the pattern wavelength, but for these there are often various competing effects. As a result, selection is less sharp and there are significant crossovers because the competition between the various ingredients depends on the value of ϵ, as inspection of figure 8.10 may illustrate.

b. Is there always a pattern selection mechanism at work?

It is appropriate to further contemplate the question concerning the mechanism of pattern selection itself. There certainly are cases where appropriate boundary or initial conditions give rise to a sharply defined pattern state with a specific wavelength. But in the end, all periodic patterns within the stable band *are* linearly stable, and so they will have a certain basin of attraction. This basin of attraction—the collection of nearby states which are attracted to it under the dynamics—is expected to be bigger for states near the center of the band than for those near the instability boundaries. But all states can be realized in principle and it is not necessary, especially in nonideal systems, that there always be a clear selection mechanism at work. Indeed, it may well be that in a system where underlying disturbances play a role, all patterns in the available band occur in different patches, but those near the center occur morefrequently than those near the edges.

See Meron, 2015 and Meron, 2019 for introductions to pattern formation in ecosystems and discussions of the basis of the model (8.65) for vegetation patterns. The notation and presentation in this section follow those of Bastiaansen et al., 2018.

An actual realization of this scenario from a very different corner of science is illustrated in figure 8.31. In dryland ecosystems in Africa, vegetation often forms banded or striped patterns with various wavelengths, depending on the slope of the terrain. From an aerial picture like the one from Somalia in the upper panel, one can extract the wavenumber q of these various patterns.[54] The middle panel shows the data from this region as a function of ground slope, where the color code illustrates the frequency with which a particular pattern wavenumber occurs. One clearly finds a band for each slope value, with realizations near the edges occurring less frequently than those near the center.

Interestingly, these and observations from different regions in Africa can be organized and understood nicely in terms of a simple phenomenological Turing model for the change in local wetness w and the change in plant biomass density n,

$$\partial_t w = e\,\partial_x^2 w + \partial_x(sw) + a - w - wn^2,$$
$$\partial_t n = \partial_x^2 n - mn + wn^2,$$

(8.65)

where x is the coordinate along the slope. The reaction terms on the right in the equation for the wetness include the effects of rainfall ($+a$), evaporation ($-w$), and uptake by plants ($-wn^2$). Besides a diffusion it includes a downhill advection term proportional to the slope s. The change of plant biomass n is modeled by a mortality term ($-mn$) and one expressing plant growth ($+wn^2$).

We have already seen that Turing models consisting of two coupled reaction-diffusion terms can exhibit pattern forming instabilities, and variant (8.65) provides a nice example. The stability balloon as a function of the slope parameter is shown as the gray zone in the lower panel of figure 8.31. Given that this Turing model is an effective phenomenological model, it is remarkable how well it helps organize the data of the observed vegetation patterns in various regions with only a few parameters. In addition, it illustrates that there are cases where it is appropriate to think of the stability band as indicating accessible states, whose probability of occurrence is sampled by the realizations in non-ideal circumstances.

c. Oscillatory systems and localized structures

As we have already mentioned, in oscillatory systems patterns are typically dominated by structures which emit waves—sources in one dimension, spirals and targets in two dimensions or even their analogue in three dimensions, the so-called scroll waves. When these waves are stable, the coherent structure determines ('selects') the wavelength of the pattern in the region around it. Although there is no rigorous proof of this, sources typically tend to be unique objects, and we expect the same for spirals. Their structure and the waves they emit are determined by the full nonlinear solution of the equations for the appropriate coherent structure.

When these structures emit waves which themselves are unstable, and or in a regime where the CGLE does not emit any stable phase winding solutions, the pattern becomes chaotic. It is then less appropriate to think in terms of pattern selection: the resulting dynamical phase is more suitably characterized by its statistical properties and by concepts from chaotic dynamics.

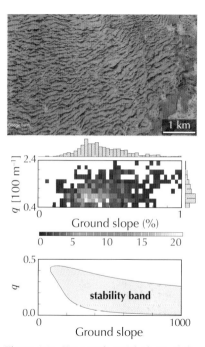

Figure 8.31. Top panel: aerial photo of the vegetation in the rather wet region of Haud, Somalia. Domains with striped vegetation patterns with various wavelengths are clearly visible. Middle panel: plot of the wavenumber q of the observed patterns as a function of ground slope as determined by Bastiaansen et al., 2018. The color coding and histograms along the axes indicate the frequency of the observed values. Note that patterns with a large range of wavenumbers are observed. The frequency of occurrence of the various values is also indicated on the right, and it is lower near the edges of the band. Lower panel: the stability balloon for the model (8.65), in gray. Note the similarity with the observed data. The analysis has also been applied to regions in Kenya, Mali, and Sudan with different wetnesses.[55] Images courtesy of Robbin Bastiaansen and Arjen Doelman.[56]

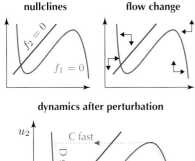

nullclines   flow change

dynamics after perturbation

Figure 8.32. Typical behavior of the null-clines $f_1(u_1, u_2) = 0$ (blue-green curve) and $f_2(u_1, u_2) = 0$ (red line) of the reaction functions f in the $u_1 u_2$ plane which give rise to excitable dynamics in the Turing model (8.66). In the upper right panel, the black arrows indicate for each segment the direction of the flow, the change in time of u_1 and u_2 under the dynamics. The black dot labeled S marks the only fixed point (stationary solution) of the equations. This is a stable fixed point, but when the solution is perturbed to the point X, the dynamics follows that indicated by the dashed lines. Since the dynamics of the change of u_1 is assumed to be much faster than that of u_2, u_1 is driven fast along the dashed line labeled A to the stable branch of the nullcline $f_1 = 0$. Subsequently, on the time scale set by u_2, the state point slowly slides up along B, following closely the nullcline f_1. When it reaches the maximum, u_1 decreases fast along C to a value close to the left branch of the nullcline, after which it relaxes slowly back via the dashed line D to the stable fixed point S. You may like to convince yourself that you can get the behavior sketched in figure 8.32 by taking f_1 to be a cubic polynomial in u_1 and linear in u_2, and f_2 linear in u_1 and u_2. In fact, the model with $f_1 = 3u_1 - u_1^3 - u_2$ and $f_2 = u_1 - a - bu_2$ and $D_2 = 0$ is the so-called FitzHugh-Nagumo model, which is one of the basic models for studying excitable medium behavior.[57]

8.10 Excitable media

Up to now, we have focused on patterns which emerge above the threshold of a linear finite wavelength instability of a featureless base state. But not all patterns originate this way. For instance, pattern dynamics may be dominated by propagating fronts or nonlinear waves separating different states of a system, or by a propagating localized solution. In this last section we sketch the main ingredients of the interesting and particularly relevant mechanism through which pulse propagation and waves occur in excitable media. Excitable media are media with a stable rest state; however, a sufficiently large trigger signal can excite the state and lead to a well-defined transient dynamics in the form of a propagating pulse or front. The study of excitable media in fact emerged from the study of the propagation of signals in nerves, but the mechanism arises in many other situations as well.

8.10.1 The basic mechanism of excitable media

The excitable medium scenario is captured most succinctly in terms of the Turing equations, which we write again for convenience,

$$\partial_t u_1 = D_1 \nabla^2 u_1 + \eta^{-1} f_1(u_1, u_2),$$
$$\partial_t u_2 = D_2 \nabla^2 u_2 + f_2(u_1, u_2). \tag{8.66}$$

These equations are essentially the Turing model equations (8.18), except that we have introduced a small parameter $\eta \ll 1$ in front of the reaction term f_1. This prefactor could be absorbed into f_1, but it is convenient to introduce it here as a measure of the time scale of the response of u_1 relative to that of u_2. When we take the functions f_1 and f_2 of the same order of magnitude, then u_1 responds on a time scale which is a factor η shorter than u_2. In other words, for $\eta \ll 1$, u_1 responds fast, u_2 slowly.

a. Reaction dynamics in the absence of diffusion

Let us first focus on the reaction dynamics by disregarding diffusion by taking $D_1 = D_2 = 0$, and imagine that f_1 and f_2 are simple polynomials in u_1 and u_2, with coefficients of order unity. We take $\eta \ll 1$, so that u_1 responds much more rapidly than u_2.

Excitable media are captured by these equations if the so-called nullclines, the lines in the $u_1 u_2$ plane where $f_i(u_1, u_2) = 0$, are of the form sketched in figure 8.32. Outside the nullclines, the direction of the flow in phase space expressing the change of u_1 and u_2 with time is determined by the signs of the functions f_i. We take these such that the flow is as indicated by the black arrows in the figure.

The intersection of the two nullclines is the point where both $f_1 = 0$ and $f_2 = 0$; it is the only fixed point of the flow where both u_1 and u_2 remain unchanged: the arrows imply that this fixed point (labeled S in the figure) is linearly stable. What makes the situation sketched in figure 8.32 into a prototype of excitable media is the fact that the fixed point does lie not too far from the minimum of the nullcline f_1, together with the separation of scales for $\eta \ll 1$.

Indeed, imagine that the system is brought into the state marked by the nearby point X below this minimum. The fast response of u_1 relative to that of u_2 for $\eta \ll 1$ implies that u_1 is then driven rapidly to the right branch of the nullcline f_1, while u_2 remains essentially unchanged. This is indicated by the dashed horizontal line labeled A in figure 8.32, while figure 8.33 shows the behavior of u_1 and u_2 separately as a function of time. This rapid response is followed by a phase where u_2 slowly increases since $f_2 > 0$ along the right branch of the nullcline of f_1; while u_2 slowly increases, u_1 adjusts adiabatically[58] so as to keep the state close to the nullcline f_1—see figure 8.33. It is as if the state of the system characterized by the point u_1, u_2 in the diagram slides upward along the curve B till it reaches the maximum. At that point the stationary branch of f_1 disappears, and then u_1 is attracted rapidly to the left branch of the nullcline as indicated by the upper dashed line C—u_1 thus decreases rapidly. After that, the state point slides down along D and returns to the stable fixed point. Figure 8.33 illustrates how after this large excursion, the state of the system returns to the stable fixed point S.

This behavior is the hallmark of an excitable medium: the system is in a stable rest state, but a not too large perturbation can excite the state sufficiently that its characteristic state variable makes a single large excursion in phase space, before it returns to its rest state.

b. With diffusion: Propagating pulses and confined waves

Let us now analyze the resulting spatiotemporal dynamics when the diffusive terms in (8.66) are taken into account. In the one-dimensional case, localized excitations of the system typically lead to propagating pulses of the type sketched in figure 8.34. This figure illustrates a pulse solution propagating to the right; the four parts labeled A–D correspond to the four branches of the reaction dynamics marked in figure 8.32.

The left flank of the pulse solution can be thought of as a front propagating into the stable state S and giving rise to values of u_1 and u_2 on the right branch of the nullcline—due to the steep gradients in this region, the u_1 field diffuses out to the left, and pushes the field as it were over the excitability barrier. Likewise, the back side of the pulse corresponds to a retracting front, while the regions B and D correspond to the slow relaxation of the u_2 field along the two branches of the nullcline of f_1, as indicated in figure

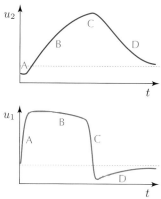

Figure 8.33. Time dependence of the two variables u_1 and u_2 for an excitable medium with nullclines as sketched in figure 8.32, in the absence of diffusion. The initial state is marked by point X in figure 8.32. Note that u_1 varies rapidly during the phases A and C, whereas u_2 varies slowly during B and D. The dashed horizontal lines mark the values associated with the stable fixed point S in figure 8.32.

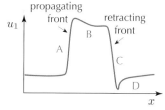

Figure 8.34. Sketch of a left-moving pulse solution in the excitable medium corresponding to figure 8.32. The labels A–D refer to the branches of the dynamics marked with these letters in this figure.[59] The flanks A and C can be viewed as propagating and retracting front solutions of the model in the limit of large separation of scales, $\eta \ll 1$. The picture illustrates that in excitable media waves the leading front is typically steeper than the retracting part: in the leading front the steep diffusion gradient is needed to help overcome the excitation barrier. The picture of a pulse solution consisting of an advancing and a retracting front connected by a plateau region with small spatial gradients can be made precise using a multiple scales analysis.

For example, for the FitzHugh-Nagumo model defined in the caption of figure 8.32, the coupled ordinary differential equations in the comoving frame $\xi = x - vt$ become

$$-v\, d_\xi u_1 = D_1 d_\xi^2 u_1 + \eta^{-1}(3u_1 - u_1^3 - u_2),$$
$$-v\, d_\xi u_2 = u_1 - a - bu_2.$$

Excitation waves like those sketched in figure 8.34 can be analyzed from this equation.

8.32. The slow relaxation in time translates into small gradients in space, just as the fast temporal relaxation for $\eta \ll 1$ gives rise to large spatial gradients.

Mathematically, such pulse solutions can be studied most easily by investigating coherent structure-type solutions of the form $u_{1,2} = u_{1,2}(\xi)$, where $\xi = x - vt$ is a comoving coordinate. By substituting this ansatz into the equations, one obtains a set of coupled ordinary differential equations, in which the velocity v enters as a parameter. As we saw for sources and sinks in the CGLE, the propagation velocity is a parameter which has to be determined from a full nonlinear analysis of the equations.[60]

8.10.2 Excitable waves in chemical systems, nerves, and beyond

A classic and well-studied example of excitable waves in chemical reaction systems is the so-called Belousov-Zhabotinsky reaction. The reaction scheme is very complex and is thought to involve around 18 different chemical reaction steps, so we refer you to the literature for details.[61] The example of the reaction in a two-dimensional cell shown in figure 8.35 illustrates that target patterns, and the interaction of waves sent out by these, dominate the dynamics. Note also that the front side of each wave is much sharper than the back, where the background state recovers. As explained in figure 8.34, this is a typical feature of excited medium waves. Another feature of waves visible in the figure is that when they meet, they annihilate, very much like the behavior at sinks in the CGLE discussed in section 8.7. The three-dimensional analogues of such target patterns are called scroll waves.[62]

Qualitatively, the behavior of such patterns in two dimensions is very reminiscent of the spiral and target patterns one finds in the CGLE (8.64) in the stable parameter range, as illustrated in figure 8.30. This explains why the latter equation has been studied so extensively as a model equation for chemical waves.

Our basic understanding of excitable media actually arose from the study of the propagation of pulses through nerves (axons).[64] In the base state of a nerve, the membrane potential, i.e., the potential difference between the outside and the inside of the axon, is negative. This potential difference is the net result of an enhanced concentration of positively charged sodium ions (Na^+) outside the membrane and an enhanced concentration of potassium (K^+) ions inside the membrane. When part of the axon is activated (e.g., via synaptic input), sodium channels in this area will open first, driving the membrane potential to positive values—see figure 8.36. This depolarization passively propagates along the membrane, quickly reaching nearby sodium channels which open and reinforce the

Figure 8.35. Snapshot of the Belousov-Zhabotinsky reaction in a thin two-dimensional cell. Image credits Michael Rogers and Stephan Morris, University of Toronto.[63]

depolarization, forming the leading edge of the pulse. The depolarization is immediately followed by the opening of the voltage-gated potassium channels which repolarize the membrane.

As we have mentioned before, the Turing equations (8.66) are often used as a simple set of model equations to capture the essential dynamics of a complex system. An intriguing observation is that the competition of the linearly stable laminar base flow and turbulent domains in pipe flow of a Newtonian fluid can be thought of in terms of an excitable medium. This allows for a phenomenological modeling of the behavior in terms of an excitable medium variant of the Turing equations.[65]

8.11 What have we learned

In spatially extended dynamical systems, many patterns occur due to a finite wavelength instability. In such cases, patterns with a range of wavenumbers emerge beyond the instability threshold.

The formation and interaction of patterns just above threshold can be analyzed with amplitude expansions. The form of the lowest order amplitude equation is dictated by the possible terms that can arise in such an expansion and by the symmetries of the underlying system, such as translation invariance. For stationary patterns, the lowest order amplitude equations are relaxational; they can be written in terms of an energy whose value decreases under the dynamics. The translation invariance of the underlying system gives rise to a phase symmetry of the pattern, and it gives rise to a slow long-time dynamics of patterns governed by a diffusion equation for the phase. The competition between various ordered patterns depends on the value of a few interaction parameters that can be calculated for a given system.

Patterns which arise near an instability to homogeneous oscillatory states or to traveling waves are not relaxational. Their dynamics, even close to threshold, is extremely rich and depends sensitively on the values of two parameters determined by the underlying system. Often the pattern dynamics in such systems becomes dominated by localized objects which send out waves. In some parameter ranges, chaotic behavior exists just above threshold.

A summary of what the amplitude equations teach us about patterns in out-of-equilibrium systems is given in section 8.9.1.

Excitable media form a different class of pattern forming systems. They are based on stable states, which can be excited by a sufficiently large perturbation that results in the medium making an excursion in its state space before returning to its stable state. Nerve pulse propagation is based on excitable medium behavior.

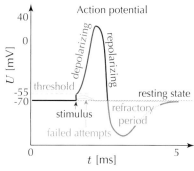

Figure 8.36. Schematic behavior of the membrane potential as a function of time, during the passage of a nerve pulse. Note the short duration of the pulse (about 1 millisecond) allowing many pulses per second to be transmitted through the axon. The speed of nerve pulses depends on the type of neuron, but it is typically several tens of meters per second.

g. The critical Rayleigh number at which the base state becomes unstable is determined by the point where the minimum of $H(q)$ becomes negative. Write $r = \pi^2 + q^2$ and show that the conditions $H(r) = 0$ and $dH/dr = 0$ give, for the critical Rayleigh number Ra_c and critical wavenumber q_c,

$$\text{Ra}_c = \frac{27\pi^4}{4}, \quad q_c = \frac{\pi}{\sqrt{2}}. \qquad \text{(slip boundary conditions)} \quad (8.75)$$

Problem 8.3** *Essence of the linear stability analysis of Rayleigh-Bénard convection with no-slip boundary conditions*

The full stability calculation is also treated by Lautrup, 2005.

In this exercise we summarize the main steps of the stability analysis of the Rayleigh-Bénard problem with no-slip boundary conditions (8.12). For simplicity, we immediately search for solutions for which $\sigma(q) = 0$. As figure 8.9 illustrates, the critical Rayleigh number is the value Ra_c where one solution appears. Above Ra_c there are two solutions; below it there are none.

a. The analysis is most easily done in terms of the stream function ψ, defined so that $v_x = -\partial_z \psi$, $v_z = \partial_x \psi$. Show that with these definitions, the incompressibility condition is automatically satisfied, and that $\vec{\nabla} \times \vec{v} = -\hat{y}\nabla^2\psi$. Convince yourself also that the name stream function is appropriate as ψ is constant along streamlines defined by equations (1.78).

b. Show that the linear problem for $\sigma_q = 0$ reduces to the two equations

$$0 = \partial_x \theta + \nabla^4 \psi, \qquad 0 = \text{Ra}\, \partial_x \psi + \nabla^2 \theta \qquad (\sigma_q = 0). \quad (8.76)$$

c. Substitute $\psi = \psi_q(z)e^{iqx}$ and $\theta = \theta_q(z)e^{iqx}$. First express θ_q in terms of ψ_q with the help of the first equation of (8.76) and then show that the second equation then reduces to

$$\left[(\partial_z^2 - q^2)^3 + \text{Ra}\, q^2\right]\psi_q(z) = 0, \qquad (\sigma_q = 0). \quad (8.77)$$

d. Equation (8.77) is a linear ordinary differential equation with constant coefficients; we can solve it with a sum of exponentials,

$$\psi_q(z) = \sum_{j=1}^{3} a_j \text{Re}\, e^{iQ_j z} = \sum_{j=1}^{3} a_j \cos(Q_j z), \quad (8.78)$$

where the a_j are real. Show from (8.77) that the Q_j are the roots

$$Q_j = \sqrt{-q^2 + (\text{Ra}\, q^2)^{1/3} e_j}, \quad (8.79)$$

where the e_j are the three cubed roots of 1, $e_1 = 1$, $e_2 = (-1 + i\sqrt{3})/2$, and $e_3 = (-1 - i\sqrt{3})/2$.

e. Show that the no-slip boundary conditions (8.12) translate into the boundary conditions $\psi_q(z) = 0$, $\partial_z \psi_q(z) = 0$, and $\theta_q(z) = 0$ at $z = \pm\frac{1}{2}$.

Actually, the stream function is defined up to a constant; for the requirement that v_z vanishes at the plates it is sufficient that ψ be constant at $z = \pm\frac{1}{2}$.

j. Each of the above boundary conditions leads to a constraint that an appropriately weighted sum of the three mode amplitudes a_j vanishes. Write out these conditions and show that the three relations have a nontrivial (i.e., nonzero) solution provided the determinant

$$
\begin{vmatrix}
\cos(\frac{Q_1}{2}) & \cos(\frac{Q_2}{2}) & \cos(\frac{Q_3}{2}) \\
Q_1 \sin(\frac{Q_1}{2}) & Q_2 \sin(\frac{Q_2}{2}) & Q_3 \sin(\frac{Q_3}{2}) \\
(Q_1^2 + q^2)^2 \cos(\frac{Q_1}{2}) & (Q_2^2 + q^2)^2 \cos(\frac{Q_2}{2}) & (Q_3^2 + q^2)^2 \cos(\frac{Q_3}{2})
\end{vmatrix}
$$

vanishes. Note that all Q_j are explicit functions of q, given by the result (8.79).

This determinant condition amounts to a transcendental equation which cannot be solved analytically, but which is easy to solve numerically. When the solutions are analyzed, the scenario is completely analogous to the case of slip boundary conditions sketched in figure 8.9. For small enough $\mathrm{Ra_c}$ the determinant equation has no solutions, while above the critical value $\mathrm{Ra_c} \approx 1707.76$ there are two values of q which solve this equation. These two values mark the edges of the band of wavenumbers within which the modes are unstable. The critical wavenumber is found to be $q_c \approx 3.11632$.

We suggest that you numerically solve the equation, obtained by requiring that the determinant vanishes, and then verify the behavior reported in the text.

Problem 8.4 *Analysis of the finite wavelength stationary Turing instability*

In this exercise we analyze the conditions for crossing the left horizontal axis in figure 8.12, so that the Turing system exhibits a finite wavelength instability to stationary patterns.

a. As a first step, demonstrate that crossing the left horizontal axis in figure 8.12 is not possible if both a_{11} and a_{22} are positive. In other words, either both coefficients are negative or one of them is positive (with their sum negative in this case).

b. Show from expression (8.25) that the occurrence of a finite wavelength instability due to crossing the left horizontal axis in parameter space from above is associated with $\det \underline{A}_q$ becoming negative for some range of values of q (remember that $\underline{A}_q$ is the matrix defined in equation (8.23)).

c. Show that the determinant $\det \underline{A}_q = |\underline{A}_q|$ is given by

$$
\begin{aligned}
|\underline{A}_q| =\, & [a_{11}a_{22} - a_{12}a_{21}] \\
& - (D_1 a_{22} + D_2 a_{11})q^2 + D_1 D_2 q^4.
\end{aligned}
\tag{8.80}
$$

d. The conclusion obtained under step b implies that we need to analyze under which conditions det $\underline{A}_q$ can become negative for a finite range of q values away from $q = 0$. Therefore investigate the behavior of det $\underline{A}_q$ by analyzing the term proportional to q^2. Verify that for the determinant to be negative, a_{11} and a_{22} must have opposite signs, and that the instability is favored for the choice $D_2 \gg D_1$ made in section 8.3.2 if $a_{11} > 1$ and $a_{22} < 0$.

e. Argue from (8.80) that for a finite wavelength instability to occur, the term between square brackets needs to be positive, and that this, together with the earlier results, implies that also a_{12} and a_{21} need to have opposite signs. Thus, the criteria derived so far which express that the base state has no zero wavelength instability can be summarized as

$$a_{11} + a_{22} < 0, \qquad a_{11}a_{22} < 0, \qquad a_{11}a_{22} - a_{12}a_{21} > 0. \qquad (8.81)$$

f. By analyzing (8.80) determine the conditions under which a finite wavelength instability exists and show that this occurs for

$$D_1 a_{22} + D_2 a_{11} > 2\sqrt{D_1 D_2 (a_{11}a_{22} - a_{12}a_{21})}, \qquad (8.82)$$

with the critical wavenumber q_c at which the instability sets in given by

$$q_c^2 = \frac{1}{2}\left(\frac{a_{11}}{D_1} + \frac{a_{22}}{D_2} \right). \qquad (8.83)$$

This result can be viewed as a qualification of the statement that the instability requires 'local activation with long-range inhibition,' as $\ell_1 \ll \ell_2$ is required.

g. Define diffusion lengths $\ell_i = \sqrt{D_1/|a_{ii}|}$ and show that the above instability criterion can be written as

$$q_c^2 = \frac{1}{2}\left(\frac{1}{\ell_1^2} - \frac{1}{\ell_2^2} \right) > \sqrt{\frac{a_{11}a_{22} - a_{12}a_{21}}{D_1 D_2}}. \qquad (8.84)$$

Problem 8.5 *Turing instability in the Brusselator*

In this problem we apply the analysis of the previous exercise to the Brusselator, a system of chemical reaction diffusion equations defined by the following model equations:

$$\begin{aligned}
\partial_t u_1 &= D_1 \nabla^2 u_1 + a - (b+1) + u_1^2 u_2, \\
\partial_t u_2 &= D_2 \nabla^2 u_2 + bu_1 - u_1^2 u_2.
\end{aligned} \qquad (8.85)$$

It is convenient to view a as fixed, and b as the control parameter which is varied.

a. Show that the equations admit a uniform solution $u_1 = a$, $u_2 = b/a$.

b. Use what you learned in the previous problem to analyze the stability of this uniform solution, and show that the state exhibits a finite wavelength instability for $b \geq b_c$ with

$$b_c = \left(1 + a\sqrt{\frac{D_1}{D_2}}\right)^2, \qquad q_c^2 = \frac{a}{\sqrt{D_1 D_2}}. \qquad (8.86)$$

Problem 8.6* *Expansion of stationary patterns in the Swift-Hohenberg equation with a symmetry-breaking quadratic term added*

In this exercise we extend the expansion of section 8.5.1 for stationary patterns of the Swift-Hohenberg equation to the case in which a symmetry-breaking quadratic term $+bu^2$ is added to the right-hand side of the equation, yielding

$$\partial_t u = \epsilon u - \left(\partial_x^2 + q_c^2\right)^2 u + bu^2 - gu^3, \qquad (g > 0). \qquad (8.87)$$

The term bu^2 breaks the $u \leftrightarrow -u$ symmetry of the equation, implying that stationary patterns are not up-down symmetric. As a result, even harmonics arise in the expansion. For positive b this biases u to positive values: it enhances the growth of positive values of u while suppressing negative values. To lowest order the result averages out, but the analysis below will illustrate that for large enough b the bifurcation to patterned states is subcritical. The critical value is determined explicitly.

We will limit the analysis to an expansion of stationary patterns which satisfy

$$0 = \epsilon u - \left(\partial_x^2 + q_c^2\right)^2 u + bu^2 - gu^3, \qquad (8.88)$$

and only carry the expansion to lowest order. In analogy with equation (8.27) we write the expansion of stationary solutions in the form

$$u = a_0 + a_1 \cos(qx) + a_2 \cos(2qx) + a_3 \cos(3qx) + \cdots. \qquad (8.89)$$

For simplicity, we take $q = q_c$ in all the steps below.

a. Substitute the expansion (8.89) for $q = q_c$ into (8.88) and anticipate that the dominant mode is still $a_1 \cos(q_c x)$ and that $a_1 = \mathcal{O}(\epsilon^{1/2})$. Show that to lowest order

$$a_0 = \frac{ba_1^2}{2q_c^4} = \mathcal{O}(\epsilon). \qquad (8.90)$$

b. Analyze the lowest order terms of the second harmonic $\cos(2q_c x)$ and show that

$$a_2 = \frac{ba_1^2}{18q_c^4} = \mathcal{O}(\epsilon). \qquad (8.91)$$

Equation (8.87) is not only a nice illustration of how the bifurcation to patterned states can become subcritical in the presence of symmetry-breaking terms; the equation has also played an important role in the study of the problem of front propagation into unstable states. In the regime where the bifurcation to patterns is supercritical, such fronts are called pulled fronts, whose speed is determined by the linear spreading speed of perturbations about the unstable state $u = 0$. The rate of approach to this asymptotic speed is power law–like and universal (Ebert and van Saarloos, 2000; Storm et al., 2000). In the regime where the bifurcation is subcritical, fronts are called pushed fronts, and their speed is determined by a nonlinear solution. See van Saarloos, 2003 for an introduction and review.

Use the identity $2\cos^2(q_c x) = \cos(2q_c x) + 1$ and use similar identities below to convert powers of cosines into harmonics of the cosine.

c. Now analyze the terms proportional to $\cos(3q_c x)$ in order $\epsilon^{3/2}$. Show that both terms analyzed in the previous two steps contribute to this order, and that this leads to

$$a_1 = \epsilon^{1/2} \left(\frac{3g}{4} - \frac{19b^2}{18q_c^4} \right)^{-1/2}. \tag{8.92}$$

d. Argue that the last result is a good indication that the bifurcation to patterns is subcritical for

$$b \geq \left(\frac{27gq_c^4}{38} \right)^{1/2}. \tag{8.93}$$

Problem 8.7* *Full stability calculation of the phase winding solutions*

In this exercise, we perform the full linear stability calculation of phase winding solutions starting from the linear equations (8.41). We keep all the parameters that set the scales—see the discussion following equation (8.36)—but you may decide to make your life easier by setting $\tau_0 = \xi_0 = \epsilon = g_0 = 1$. This simply amounts to working in rescaled coordinates, as mentioned in the note in the margin after expressions (8.36).

a. The calculation follows the usual steps to determine a dispersion relation of linear modes. Starting from equations (8.41) we probe the stability against modes where δa and ϕ vary with space and time as $e^{\sigma t + iQx}$—note that we are now probing the stability of phase winding solutions with a given wavenumber k against sinusoidal variations of amplitude and phase with wavenumber Q. Substitute the ansatz into the equations and show that the determinant equation for σ is

$$\begin{vmatrix} -2g_0 a_k^2 - \xi_0^2 Q^2 - \tau_0 \sigma & -2i\xi_0^2 k a_k Q \\ 2i\xi_0^2 k Q & -\xi_0^2 a_k Q^2 - \tau_0 a_k \sigma \end{vmatrix} = 0. \tag{8.94}$$

b. Show from this result that the two eigenvalues for σ are

$$\tau_0 \sigma_\pm = (g_0 a_k^2 + \xi_0^2 Q^2) \pm \sqrt{g_0^2 a_k^4 + 4\xi_0^4 k^2 Q^2}. \tag{8.95}$$

c. Verify that values σ_- on the minus-branch are always negative and gapped away from zero, and that they correspond to the rapid response of the amplitude. Next, show that σ_+ vanishes in the limit $Q \to 0$ and that σ_+ can be expanded for small Q as

$$\sigma_+ = - \left(\frac{\epsilon - 3\xi_0^2 k^2}{\epsilon - \xi_0^2 k^2} \right) \left(\frac{\xi_0^2}{\tau_0} \right) Q^2 - \left(\frac{2\xi_0^4 k^4}{\left(\epsilon - \xi_0^2 k^2 \right)^3} \right) \left(\frac{\xi_0^4}{\tau_0} \right) Q^4 + \cdots. \tag{8.96}$$

d. Verify that the preceding result is consistent with the phase diffusion equation (8.43) together with the expression (8.44) for the phase diffusion coefficient.

e. Verify that the form of this expression is such that the Eckhaus instability that occurs for $k^2\xi_0^2 > \epsilon/3$ is indeed a type II instability, as defined in figure 8.14 and discussed in section 8.6.

Problem 8.8 *Symmetry of the Boussinesq equations*

The flow simulation shown in the lower panel of figure 8.8, which is based on the Boussinesq equations (8.8)–(8.10), shows a remarkable symmetry: the profiles of the flow and temperature fields in the cold, downward-flowing regions look exactly the same as those in the hot, upward-flowing areas, apart from an obvious change in sign. This is indeed a consequence of the underlying symmetry of the equations, which we investigate in this problem. We shall see that the symmetry is specific to the Boussinesq approximation, and discuss what type of effects weakly break this symmetry.

a. Put the origin of the coordinate system in the center of a convection cell as indicated in figure 8.37, and let $\vec{r}$ denote the coordinate in this system. Argue that the symmetry suggested by this figure is expressed by

$$\vec{v}(\vec{r}) = -\vec{v}(-\vec{r}), \qquad \theta(\vec{r}) = -\theta(-\vec{r}). \qquad (8.97)$$

b. Show that the Boussinesq equations (8.8)–(8.10) are indeed invariant under the reflection symmetry (8.97).

c. Argue that in an amplitude equation description, the symmetry (8.97) of the Boussinesq equations implies that the amplitude equation should be invariant under change of sign of the amplitude, $A \to -A$.

d. Use this symmetry to argue that in the Boussinesq approximation, the term γ in the amplitude equations (8.54) has to be identically zero.

e. In the Boussinesq approximation, the variation of the density is taken as linear in the temperature variation; see equation (8.6). Trace how quadratic corrections to the variation of the density with the temperature enter equation (8.8) and show that the resulting equation does not obey the symmetry (8.97) between cold downflow and hot upflow. In other words, non-Boussinesq terms break the symmetry. Would other small effects, such as a weak temperature dependence of the viscosity, also contribute to this symmetry breaking?

f. Argue that, as a result of the symmetry breaking, $\gamma \neq 0$ in (8.54) due to non-Boussinesq effects, and that this term will be small if non-Boussinesq effects are small.

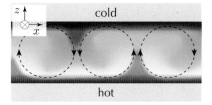

Figure 8.37. In the Boussinesq approximation the flow equations are invariant under the symmetry (8.97) relative to the origin in the center of a convection roll. Stated differently: when the figure is turned upside down, and blue is changed into red and vice versa, the flow pattern looks exactly the same.

Problem 8.9 *The group velocity $v_g = \mathrm{d}\omega/\mathrm{d}k$*

In this problem we illustrate how the group velocity v_g defined in equation (8.61) emerges for a packet of waves with dispersion relation $\omega(k)$. We do so for a signal ψ which results from a Gaussian wave packet centered around $k = k_\mathrm{m}$,

$$\psi = \int_{-\infty}^{\infty} \mathrm{d}k \, e^{i(kx - \omega(k)t)} e^{-(k-k_\mathrm{m})^2/2W}. \tag{8.98}$$

a. For not too large W, the wave packet is strongly peaked around k_m. Expand the terms in the exponent to show that expression (8.98) can be written as

$$\psi = e^{i(k_\mathrm{m}x - \omega(k_\mathrm{m})t)} \int_{-\infty}^{\infty} \mathrm{d}\Delta k \, e^{i\Delta k(x - \mathrm{d}\omega(k)/\mathrm{d}k|_{k_\mathrm{m}}t)} e^{-(\Delta k)^2/2W}, \tag{8.99}$$

where we have taken $\Delta k = k - k_\mathrm{m}$.

b. Complete the square in the exponent and perform the resulting Gaussian integral in Δk to show that

$$\psi \sim e^{i(k_\mathrm{m}x - \omega(k_\mathrm{m})t)} e^{-\frac{W}{2}(x - v_\mathrm{g}t)^2}. \tag{8.100}$$

c. Discuss how the last expression describes a Gaussian pulse in real space, propagating with velocity $v_\mathrm{g} = \mathrm{d}\omega/\mathrm{d}k|_{k_\mathrm{m}}$. As v_g is also the velocity with which information or a small perturbation travels when the wave is modulated, the name *group velocity* is appropriate. As discussed in section 8.7, the group velocity distinguishes between sources and sinks. In the comoving frame a sink is sandwiched between two incoming waves, while sources send out waves to both directions—see figure 8.28.

Problem 8.10 *'Poor man's renormalization' of the noisy pitchfork bifurcation*

As you will see, the renormalization technique we will use does not rely on the Feynman diagrams one typically uses, but it will nevertheless yield the same result to first order in the noise. We shall see how the noise shifts the bifurcation that occurs in (8.101) at $r = 0$ in the absence of noise.

See Martin et al., 2021a for a detailed explanation of this method and its application to more complex hydrodynamic equations.

In this problem, we'll study the effect of the noise on the following pitchfork bifurcation:

$$\dot{w} = rw - \gamma w^3 + \sqrt{2\sigma}\eta, \tag{8.101}$$

where η is Gaussian white noise of unit variance $\langle \eta(t)\eta(t')\rangle = \delta(t - t')$.

To perturbatively compute the impact of noise on the field $\tilde{w} = \langle w \rangle$, where $\langle \cdot \rangle$ denotes the noise average, we consider the case $r > 0$, where $\tilde{w}$ should have a nonzero value. In this regime, we introduce the field $z = w - \sqrt{r/\gamma}$ in (8.101), leading to the evolution equation,

$$\dot{z} = -2rz - \gamma z^3 - 3\sqrt{\gamma r}z^2 + \sqrt{2\sigma}\eta. \tag{8.102}$$

We further define, for convenience, $f(z) = -2rz - \gamma z^3 - 3\sqrt{\gamma r}z^2$ so that

$$\dot{z} = f(z) + \sqrt{2\sigma}\eta. \tag{8.103}$$

a. By expanding z as $z = z_0 + \sqrt{\sigma}z_1 + \sigma z_2$ and assuming σ to be small in equation (8.103), show that z_0, z_1, and z_2 evolve according to

$$\dot{z}_0 = f(z_0), \tag{8.104}$$

$$\dot{z}_1 = f'(z_0)z_1 + \sqrt{2}\eta, \tag{8.105}$$

$$\dot{z}_2 = f'(z_0)z_2 + \frac{1}{2}f''(z_0)z_1^2. \tag{8.106}$$

b. Using Itô's lemma in equation (8.105), assuming that z_0 remains fixed, show that in the steady state

See problem 3.10 for a summary of Itô's lemma.

$$\langle z_1^2 \rangle = -\frac{1}{f'(z_0)}. \tag{8.107}$$

c. By summing together equation (8.104), equation (8.105) multiplied by $\sqrt{\sigma}$, and equation (8.106) multiplied by σ, show that, to $\mathcal{O}(\sigma)$,

$$\dot{\tilde{z}} = f(z_0) + \sqrt{\sigma}f'(z_0)\langle z_1\rangle + \sigma f'(z_0)\langle z_2\rangle + \frac{\sigma}{2}f''(z_0)\langle z_1^2\rangle, \tag{8.108}$$

where $\tilde{z} = z_0 + \sqrt{\sigma}\langle z_1\rangle + \sigma\langle z_2\rangle$.

d. Next, show that

$$f(\tilde{z}) = f(z_0) + \sqrt{\sigma}f'(z_0)\langle z_1\rangle + \frac{1}{2}\sigma f''(z_0)\langle z_1\rangle^2 + \sigma f'(z_0)\langle z_2\rangle, \tag{8.109}$$

where $\tilde{z} = z_0 + \sqrt{\sigma}\langle z_1\rangle + \sigma\langle z_2\rangle$.

e. By substituting equation (8.109) into (8.108), show that

$$\dot{\tilde{z}} = f(\tilde{z}) + \frac{1}{2}\sigma f''(z_0)\left(\langle z_1^2\rangle - \langle z_1\rangle^2\right). \tag{8.110}$$

f. Noting that $\langle z_1\rangle = 0$ in the steady state and further using equation (8.107) in (8.110), show that the evolution of $\tilde{z}$ is given to $\mathcal{O}(\sigma)$ by

$$\dot{\tilde{z}} = f(\tilde{z}) - \frac{1}{2}\sigma f''(\tilde{z})\frac{1}{f'(\tilde{z})}. \tag{8.111}$$

g. Show that

$$\tilde{z} = -\sigma\frac{3\sqrt{\gamma r}}{4r^2} \tag{8.112}$$

is a solution of equation (8.111) to $\mathcal{O}(\sigma)$.

In the regime $r < 0$ the mean-field value is zero, while the noise gives rise to an average value of z. You may wish to retrace the steps in this problem to evaluate the effect of noise in this regime.

h. Coming back to the definition of z, deduce that, in the presence of noise, the field w in the $r > 0$ regime is given to first order in σ by

$$\tilde{w} = \sqrt{\frac{r}{\gamma}} - \sigma \frac{3\sqrt{\gamma r}}{4r^2}. \tag{8.113}$$

As we might intuitively expect, the addition of noise has decreased w with respect to its mean-field value.

Active Matter | 9

This chapter provides an introduction to active matter, an umbrella term that encompasses several out-of-equilibrium systems composed of individual components that convert energy into nonconservative forces and motion at the microscale. This can be achieved in situ via chemical reactions that happen on individual building blocks, as is the case for colloids powered by autocatalytic reactions and for biofilaments with attached ATP-burning molecular motors. Alternatively, external fields can be applied to drive each particle, which blurs the distinction between active and driven matter, but often leads to similar macroscopic phenomena.

Our perspective is to focus on emergent phenomena arising from the out-of-equilibrium character of the basic constituents. Experimental inspiration comes from biological systems like swarming bacteria, bird flocks, cell tissues, and synthetic matter, like motorized colloids that self-propel or rotate, and collections of robots.

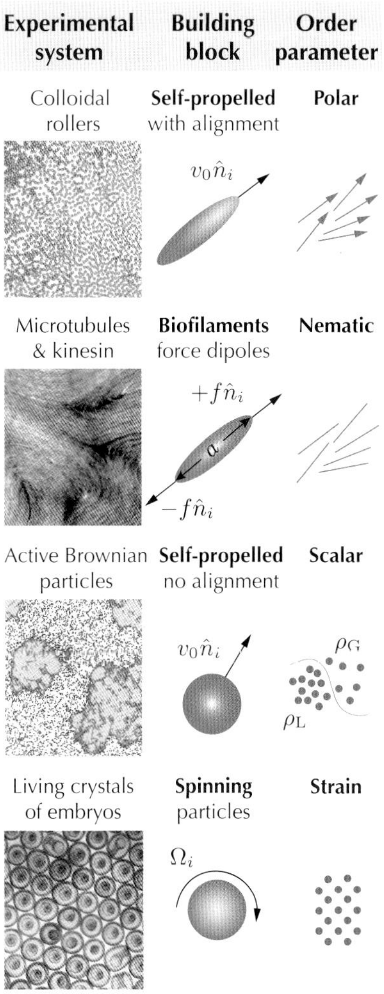

9.1 Hydrodynamic theories of active matter

Our emphasis will be on introducing some of the most common classes of active matter behavior through the lens of simplified hydrodynamic theories supported by microscopic models and experimental results that motivate them. Such approaches extend the continuum formulations presented in the previous chapters to situations where detailed balance is broken at the microscopic scale. The present chapter builds on interesting parallels that exist at the mathematical, conceptual, and phenomenological levels between active matter and the spatially extended dynamical systems we already encountered when studying pattern formation out of equilibrium in chapter 8. In both cases, describing systems out of equilibrium using continuum equations, familiar from our intuitive understanding of passive fluids or solids, requires a careful re-examination of the symmetries and conservation laws that are present (or absent!) in each active system under study.

Energy injection at the microscale is responsible for unusual mechanical states, phase transitions, and transport properties that are absent at equilibrium. Typically the unusual properties are robust phenomena which do not depend sensitively on the details of the model; we shall see that they can often be described by including the lowest order terms that are allowed by the symmetry breaking in a phenomenological description. Within this approach, one

For a more detailed introduction to active matter, see the review paper by Marchetti et al., 2013 or the book by Pismen, 2021.

Figure 9.1. A gallery of active matter: polar fluids, active nematics, suspension of active Brownian particles, and crystals of spinning embryos. All these examples of active matter are classified by the symmetries of their building blocks and the nature of the active forces which together determine the symmetries of the resulting order parameters. The experimental examples will all return later in the chapter.

Such so-called Janus particles are typically made by coating one side of the colloid with a catalyst for a reaction that takes place in the fluid; see figure 4.18. Figure 4.19 shows various examples of rod-like active particles which tend to align.

Well-known examples of molecular motors are kinesin and myosin. For more information, see section 3.6.2 and the illustrations of molecular motors in figure 3.11 and figure 3.12.

identifies the symmetries of the building blocks and the appropriate order parameter for the relevant collective phases, very much as in passive matter. Figure 9.1 illustrates this with four examples that we will encounter in this chapter.

The first experimental system, illustrated in the top row of figure 9.1, is composed of colloids that can self-propel in a given direction at a certain speed. In the presence of orientation-aligning interactions the colloids tend to move along a common direction; the proper order parameter is therefore a polar vector.

The second example illustrated in the figure concerns biopolymers, such as microtubules or actin filaments, with molecular motors that apply force dipoles along the direction in which they are oriented. On large scales, this mixture can exhibit nematic order described by the vector order parameter of liquid crystals, the director $\hat{n}$ introduced in section 6.1.4. If the particles are, however, self-propelled in a direction that undergoes a random walk without alignment (similarly to bacteria that run and tumble), density is the appropriate *scalar* order parameter, as illustrated in the third example.

The final example concerns so-called living crystals made of embryos, volvox algae, or driven colloids that constantly spin instead of self-propel. In this case, a tensor order parameter describing strains is needed to describe the system, as in ordinary solids. In addition, one could also add a field describing the local spinning rate of the particles $\Omega(r, t)$ in the hydrodynamic description.

Throughout this chapter we will use the order parameter description to discuss the hydrodynamics equations that capture the microscopic source of activity relevant to all of these models.

9.2 Flocking

Flocking is perhaps the most paradigmatic example of collective behavior in active matter systems. Active constituents, like flying birds, self-organize into complex patterns over length scales which are very large compared to the size of a bird, see figure 9.2. Within the flock, nearby birds fly in the same average direction, while on large scales the average flight orientation typically varies. In this section, we will see how traditional physics approaches can be used to construct a rich explanatory framework to model flocking.

9.2.1 The Vicsek model

Flocking can be described within a simplified setting usually referred to as the Vicsek model.[3] This model makes a drastic

Figure 9.2. A flock of birds. Albert Beukhof/Shutterstock. Schools of fish similarly exhibit coordinated movement.

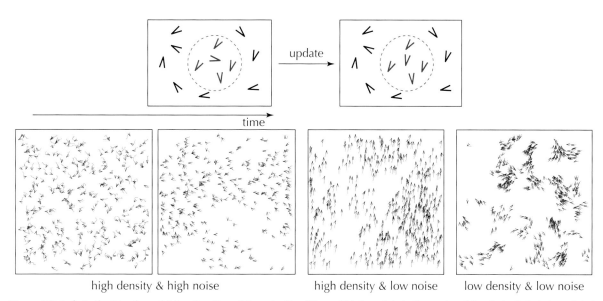

high density & high noise   high density & low noise   low density & low noise

Figure 9.3. Left: in the Vicsek model the direction of the velocity of the red bird particle in the center of the dashed circle is updated to be the average orientation of all particles within that circle (the red and blue ones), plus a random noise term. The four snapshots of simulations of the Vicsek model for various densities and noise strengths on the right illustrate that flocking occurs for low noise strength.[1] For small densities and noise the particles form flocks which move in random directions, while for large densities and small noise an ordered phase emerges with all particles moving on average in the same direction. Adapted from Vicsek et al., 1995[2].

simplification and views a flock of birds as a set of particles which in each time step move a fixed distance forward in the direction in which they are oriented. However, their flight direction is updated in each time step to become equal to the average orientation of the birds within some finite range R plus noise, as illustrated on the left side of figure 9.3.

In the two-dimensional Vicsek model, each particle is characterized by the position $\vec{r}_i$ and θ_i, where θ_i is the orientation angle of particle i with respect to some fixed axis. The update rule for the positions and angles of the particles is

$$\vec{r}_i(t + \Delta t) = \vec{r}_i(t) + v_0 \Delta t \begin{pmatrix} \cos \theta_i(t) \\ \sin \theta_i(t) \end{pmatrix},$$

$$\theta_i(t + \Delta t) = \langle \theta(t) \rangle_{R,i} + \eta_i(t),$$

(9.1)

with v_0 denoting a characteristic self-propulsion speed, Δt the characteristic time between steps, and the average angle defined as

$$\langle \theta(t) \rangle_{R,i} \equiv \arg \left(\sum_j e^{i\theta_j(t)} \right),$$

(9.2)

where the index j runs over all particles within some distance R from the particle i. For simplicity we take $\Delta t = 1$. The term η_i represents a stochastic noise and is assumed to be uncorrelated

At www.complexity-explorables.org/ you can play with simulations of the Vicsek model. The website contains many other instructive examples of complex systems, including one called *Flock'n Roll*.

between particles and between different update times,[4]

$$\overline{\eta_i(t)} = 0, \qquad \overline{\eta_i(t)\eta_j(t')} = 2\gamma\,\delta_{ij}\delta_{tt'}. \qquad (9.3)$$

Figure 9.3(b) illustrates the behavior of the model for various densities and noise strengths. For large noise strength, the birds tend to fly mostly in random directions. But for small noise strength, when they have a strong tendency to align with their neighbors, they tend to form flocks. For small densities (upper right snapshot) the birds cluster in flocks which move in random directions, while for large densities all birds tend to fly on average in the same direction. These snapshots are indicative of what has emerged from many detailed studies, namely that the model exhibits three phases, a disordered, an ordered homogeneous, and an ordered banded phase in which the bird density is spatially modulated.

Intuitively, it is rather easy to understand why the model has a tendency to form flocks with relatively sharp boundaries. The update rule favors birds in a dense region flying on average in the same direction, making escaping a flock difficult. Moreover, once a lonely bird in an empty region happens to fly into a flock, it tends to adjust its flight orientation almost immediately to join the flock. In other words, there is a strong tendency in the model for the particles to separate into dense flocks and relatively low density regions.

Flocking behavior of the type seen in such simulations has been observed and studied in detail in a variety of active matter systems. Figure 9.4 shows this for the case of active colloids driven by the Quincke rotation mechanism illustrated in figure 4.19.d.

We will not go into the details of phase diagram and scaling properties of the Vicsek model here[6] but will concentrate on the continuum formulations which capture the essence of the behavior, while providing a good starting point for including additional effects.

9.2.2 Flocking and the Mermin-Wagner theorem

Before we embark on our discussion of phenomenological equations for active media, it is good to pause for a moment to stress that in a system at equilibrium, flocking would not happen in two dimensions. Let us show why.

Suppose we do not let the 'agents' move (we fix them in space, as $v_0 = 0$), but update their orientation angles θ_i according to (9.1). Rather than a model for flying birds, we now have a stochastic spin model of the type familiar from equilibrium physics. Indeed, the update rule of our fixed agents is reminiscent of relaxational dynamics of spins, where at each time step the orientation of each angle is driven toward the average orientation of the spin

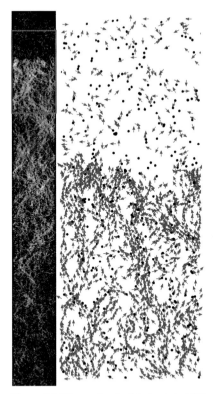

Figure 9.4. Observation by Morin et al., 2016 of collective motion of colloidal particles in a long channel with small random obstacles. Left: global view. The channel contains 8,500 colloids, and it is 7 mm long and about 850 microns wide. The active motion is due to Quincke rotation, illustrated in figure 4.19(e). The right-hand image shows an enlargement of the flocking front in the area indicated by a red box in the left-hand image, with arrows indicating the velocity orientation. Black dots indicate the obstacles used in these experiments to probe the behavior of flocking fronts in a random environment. A sharp density front is clearly visible. Quincke rotation experiments provide a versatile soft matter realization for studying flocking as well as studying front, band, and vortex formation.[5] Image courtesy of Denis Bartolo.

The argument in this section is inspired by the lecture notes of Toner, 2018; see also the review by Vicsek and Zafeiris, 2012.

and its neighbors. In addition the update rule is affected by noise. On a regular two-dimensional lattice with spacing a, this update rule is in fact the discrete form of a noisy diffusion equation,

$$\partial_t \theta = D\nabla^2 \theta + \eta(\vec{r}, t), \quad \left(D = \frac{a^2}{5} \right), \qquad (9.4)$$

where $\eta(\vec{r}, t)$ denotes white noise with zero mean and D the diffusion constant. Let us now suppose there is orientational order of these spins and that one of the spins gets oriented in a sufficiently different direction θ_0. What are the consequences of introducing this 'error' within the static version of the Vicsek model? According to (9.4) the angular variables are given by a noisy diffusion equation, with the dynamics causing angles to spread diffusively in time over a distance (see section 3.3.4 on diffusion)

$$r \sim \sqrt{t}. \qquad (9.5)$$

In the absence of noise the total angle θ is conserved according to (9.4), meaning that any error or fluctuation will slowly spread out across the system. For an initial error θ_0, at time t $N(t)$ spins will be affected. In d dimensions, we then have

$$N(t) \sim r(t)^d \sim t^{d/2}. \qquad (9.6)$$

As the error profile spreads out as a Gaussian, the typical size of the error $\Delta\theta(t)$ of the affected spins is of order

$$\Delta\theta(t) \simeq \frac{\theta_0}{N(t)} \sim \frac{\theta_0}{t^{d/2}}. \qquad (9.7)$$

From this analysis we conclude that if a local perturbation is introduced to agents which were originally ordered, in the absence of noise, the error decays so the order is reestablished as $\Delta\theta(t) \to 0$. We now ask ourselves whether the buildup of perturbations due to the noise term in equation (9.4) can overcome this decay of errors—i.e., lead to $\Delta\theta$ growing in time—if so, we interpret this to imply that the fluctuations destroy the orientational order and hence prevent the emergence of a flock.

The noise η on a given spin will spread to others as time progresses. In fact a typical spin will accumulate $t \sim r^2$ errors, and the total collection of spins affected will accumulate a total of $N(t)t \sim r^{d+2}$ errors during this time. Now, because of the law of large numbers, the width of the distribution of angles, which gives the magnitude of the typical deviation of the spin angle from the average value, is

$$\sqrt{\langle(\Delta\theta)^2\rangle} \simeq \frac{\sqrt{\text{total errors}}}{N(t)} \sim \frac{\sqrt{r^{d+2}}}{r^d} \sim r^{1-d/2}, \qquad (9.8)$$

The derivation, which is similar to the derivation of a finite-difference method for solving Laplace's equation $\nabla^2 \phi = 0$ in electrostatics, is presented as problem 9.1. Of course, diffusion-type spreading of the angular variables is not limited to regular lattices, but on a disordered lattice the effective diffusion coefficient will depend on the local lattice connectivity.

The diffusion equation (9.4) implies that the error spreads out as a Gaussian; see section 3.3.4.

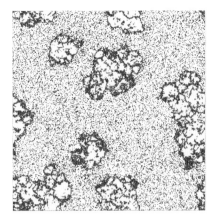

Figure 9.6. Simulations of self-propelled particles interacting via a pairwise Lennard-Jones radial potential in a periodic domain in simulations by Martin et al., 2021b.[9] The snapshot shows the occurrence of motility-induced phase separation. The two average densities corresponding to the dense and dilute clusters ($\rho_0 = 0.9$ and $\rho_0 = 0.5$) are indicated in yellow and pink, respectively.

For a review of motility-induced phase separation, see Cates and Tailleur, 2015.

The mechanism of motility-induced phase separation is reminiscent of traffic jams on highways: for sufficiently high density of cars ρ, the average driving speed $v(\rho)$ decreases with density, $\partial v / \partial \rho < 0$. This behavior of $v(\rho)$ shares some similarities with that leading to the formation of shocks, fronts, and domains in the nonlinear wave equation $\partial_t \rho + \partial_x [v(\rho)\rho] = D\partial_x^2 \rho$, which is a simplified model for phenomena like traffic flows; see, e.g., Whitham, 1974. In the regime $\partial v / \partial \rho < 0$, traffic is prone to the spontaneous emergence of traffic jams due to variations ('fluctuations') in the speeds of individual cars.

9.3 Motility-induced phase separation

Motility-induced phase separation is a manifestation of the phenomenology of self-propelled particles that, unlike flocking, does not necessitate the presence of interactions that align the (flight) orientation of individual particles or agents. In essence this phenomenon relies on two ingredients.

First, particles naturally accumulate where the position-dependent propulsion speed $v(\vec{r})$ is low. Second, the self-propulsion speed $v(\vec{r})$ decreases with the local density $\rho(\vec{r})$, i.e., $\frac{\partial v}{\partial \rho} < 0$. Once these two conditions are met, a feedback loop is created that makes a uniform suspension unstable. Instead, a phase-separated state emerges in which a dilute active gas coexists with a dense liquid with reduced motility; see figure 9.6.

9.3.1 Active Brownian particles

Motility-induced phase separation happens quite generally, but it is instructive to consider a concrete model of self-propelled particles that, unlike the Vicsek model, does not display orientation-aligning interactions. In this model, self-propelled particles, called active Brownian particles, satisfy the following Langevin dynamics:

$$\dot{\vec{r}}(t) = v(\vec{r}(t))\hat{m}(\theta(t)), \quad \dot{\theta}(t) = \sqrt{2D_\theta}\,\eta(t), \quad (9.13)$$

where $\vec{r}$ is the position of the particle, η is a Gaussian white noise, D_θ is the angular diffusion constant, and $\hat{m}(\theta) = [\cos\theta, \sin\theta]$ is the unitary vector with orientation θ. The corresponding Fokker-Planck equation for the probability distribution $P(\vec{r}, \theta)$ reads

$$\partial_t P(\vec{r}, \theta) = -\hat{n}(\theta) \cdot \vec{\nabla}\left[v(\vec{r})P(\vec{r}, \theta)\right] + D_\theta \partial_{\theta\theta} P(\vec{r}, \theta), \quad (9.14)$$

where $\vec{\nabla}$ denotes the spatial gradient and $\hat{n}$ is the average orientation. In simple terms, these equations describe a gas of non-interacting particles that self-propel with a position-dependent speed while the orientation angle θ of their velocity undergoes a random walk.

9.3.2 The mechanism behind the instability

To study the mechanism of instability that leads to motility-induced phase separation, we first write an approximate closed-form equation for the density,

$$\rho(\vec{r}, t) = \int d\theta P(\vec{r}, \theta, t). \quad (9.15)$$

In problem 9.5, you will be guided to prove that the Fokker-Planck equation (9.14) reduces to

$$\partial_t \rho = \frac{1}{2D_\theta} \vec{\nabla} \cdot \left[v(\rho) \vec{\nabla} \left(v(\rho)\rho \right) \right], \qquad (9.16)$$

where we have assumed that the mean self-propulsion speed $v(\rho)$ depends only on density. We can then perform a linear stability analysis using the methodology explained in chapter 8. We linearize equation (9.16) around a uniform density state by writing $\rho(\vec{r}, t) = \rho_0 + \delta\rho(\vec{r}, t)$, leading to

$$\partial_t \delta\rho = \frac{v_{\mathrm{u}}(v_{\mathrm{u}} + \rho_0 v')}{2D_\theta} \nabla^2 \delta\rho + \mathcal{O}(\delta\rho^2), \qquad (9.17)$$

where $v' = \partial v / \partial \rho$ and where v_{u} is the velocity in the state with uniform density. This equation is formally equivalent to a diffusion equation for the density. Crucially, the diffusion coefficient becomes negative when

$$\frac{1}{v_{\mathrm{u}}} \frac{\partial v}{\partial \rho} < -\frac{1}{\rho_0}. \qquad (9.18)$$

This is the condition for the amplification of density perturbations that yields the region of parameter space for which the instability leading to motility-induced phase separation occurs. Indeed, the Fourier transform of equation (9.17) is

$$\partial_t \delta\rho = -\frac{v_{\mathrm{u}}(v_{\mathrm{u}} + \rho_0 v')}{2D_\theta} q^2 \delta\rho, \qquad (9.19)$$

where q is the wavenumber. Therefore, perturbations with $q \neq 0$ are amplified whenever the condition (9.18) is satisfied. This result, derived using heuristic reasoning, agrees with more detailed mathematical analysis.[10] The fate of the system when perturbations grow is controlled by nonlinearities not considered at the level of linear stability. Recent experiments carried out with Quincke rollers in the dense regime have verified the existence of similar phenomena in motility-induced phase separation; see figure 9.7.

The hydrodynamic equation we used to discuss motility-induced phase separation is an illustration of an active model where the order parameter, the density of self-propelled particles, is a scalar. (see the third row of the diagram in figure 9.1).

9.4 Bacterial suspensions

As shown in figure 1.7, bacteria are typically elongated and about a micron in size. So when they are dead, a dense suspension of

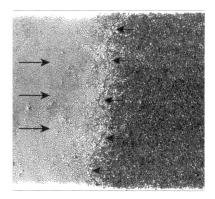

Figure 9.7. Active solidification occurring at high density in assemblies of Quincke rollers in the experiments by Geyer et al., 2019. The disordered arrested solid appears dark, while the less dense flocking fluid phase is light gray. The rollers in the solid are arrested but the interface between the two phases propagates to the left into the flocking phase, as indicated by the red arrows. Conservation of particles at the interface zone implies that there is a flow toward the interface in the fluid flocking phase, as indicated by the blue arrows. Courtesy of Delphine Geyer and Denis Bartolo.

it contains stresses σ_{ij}, which include elastic and active stresses, which will be discussed in the next paragraph. The second equation describes the evolution of the nematic order parameter Q_{ij}. According to (9.22), the evolution of this order parameter is mainly influenced by the molecular field

$$H_{ij} = -\frac{\delta F}{\delta Q_{ij}}, \tag{9.23}$$

defined as the functional derivative of the Landau–de Gennes free energy

$$F = \frac{1}{2} \int d^2\vec{r}\, \left[K_Q \,|\nabla Q|^2 + C \operatorname{Tr} \underline{Q}^2 (\operatorname{Tr} \underline{Q}^2 - 1) \right], \tag{9.24}$$

where K, K_Q and C are material constants. The expression involving the trace of Q^2 in this expression is a simplified version of equation (6.5), while a gradient term penalizing inhomogeneities in the order parameter has been added, in the spirit of the Frank free energy discussed in section 6.3.1. The term $\lambda S u_{ij}$ describes the alignment of the nematic director $\hat{n}$ with the flow, with S denoting the magnitude of the Q tensor (see equation (6.4)) and λ a parameter describing the strength of the alignment. Finally, the term $\omega_{ik} Q_{kj} - Q_{ik}\omega_{kj}$ on the right-hand side is required to take the tensorial nature of the order parameter in the material derivative into account (it is part of the so-called corotational derivative[17]).

According to (6.45), the stress $\sigma_{ij}^{\text{elastic}}$ also contains terms proportional to $n_i n_j$. For passive nematics these terms are proportional to gradients of the velocity. The active nematic has nonzero stress terms in the absence of velocity gradients. The argument in the text for the physical origin of this term is from Ramaswamy, 2010, who gives a more detailed discussion.

Compare our discussion of equation (6.43) in section 6.7 on nematohydrodynamics.

Molecular motors introduce active forces in active nematics. These active forces are distinct from and compete with the elastic stresses that tend to minimize the Frank free energy for elastic distortions of a passive nematic liquid crystal (see section 6.3.1). Therefore, the stress tensor can be written as

$$\sigma_{ij} = \sigma_{ij}^{\text{elastic}} + \sigma_{ij}^{\text{active}}, \tag{9.25}$$

where the elastic stress $\sigma_{ij}^{\text{elastic}} = -\lambda H_{ij} + Q_{ik} H_{kj} - H_{ik} Q_{kj}$ is the same as in the case of passive nematic liquid crystal. The leading-order contribution of the molecular motors adds an active term,

$$\sigma_{ij}^{\text{active}} = \alpha\, Q_{ij} \sim S\alpha\, n_i n_j. \tag{9.26}$$

The form of this active stress term can be understood through a simple argument.[18] Because the motors pull filaments along each other, the direction of any motor force will be along the direction of $\hat{n}$. Here we assume for simplicity that the stress σ_{ij} is symmetric, i.e., the motors do not exert net torques. Hence, the leading-order active stress should be the simplest symmetric tensor depending on the director (or on the Q tensor), leading to (9.26). The coefficient α is called the activity parameter and it is related to the density and pulling strength of the molecular motors.

This single active term turns out to be effective at modeling the observed hydrodynamic behavior of active nematic liquid crystals (see problem 9.6).

9.5.2 Transition to chaos and topological defects

The flow in an active nematic is governed by competing active and elastic stresses, as shown in equation (9.25). The elastic stress attempts to align the director, while the active stress tries to destroy this alignment. Their interplay determines the degree of activity-induced disorder in the material. We can determine which effect dominates using dimensional arguments. The strength of elastic stresses is characterized by a modulus K which appears in the Frank free energy

$$F = \frac{K}{2} \int \mathrm{d}^3 \vec{r} \, (\partial_i n_j)^2 . \tag{9.27}$$

The right-hand side of this equation has dimensions $[K]$ times length, therefore $[K] = \text{energy}/\text{length} = \text{force}$. The active stresses are controlled by the activity parameter α, which has dimension $[\alpha] = [\sigma^a] = \text{force}/\text{length}^2$. For a system of size L, the competing forces denoted by f are thus approximately

$$\begin{aligned} f^{\text{active}} &\sim \alpha L^2, \\ f^{\text{elastic}} &\sim K. \end{aligned} \tag{9.28}$$

Equating these two forces gives a critical activity α_c

$$\alpha_c \sim K/L^2. \tag{9.29}$$

For fixed elastic constant K and size L and $\alpha < \alpha_c$, the elastic stresses dominate and the biofilaments remain mostly aligned in a *quiescent* state. If $\alpha > \alpha_c$, activity wins the competition and the system enters a chaotic state sometimes referred to as active turbulence. In this case, the biofilaments are constantly moving and their local alignment is disrupted by the molecular motors, leading to the proliferation of topological defects. In contrast to the Kosterlitz Thouless transition studied in section 2.11, topological defects in active nematics behave like self-propelled objects (see next section) leading to a time-dependent state with chaotic dynamics.

For α and K fixed, one can similarly define an active length $L_\alpha \sim \sqrt{K/\alpha}$. This sets a critical system size below which elasticity dominates and above which chaotic active behavior will appear. Experiments and numerical simulations[20] have shown that many characteristic length scales in active nematics, such as the mean defect distance and the director correlation length, are proportional to L_α.

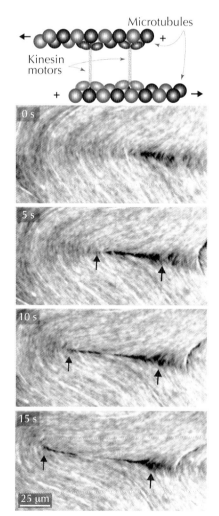

Figure 9.10. A sequence of images showing buckling, folding, and the ensuing dynamics of a $-1/2$ defect (blue) and a $+1/2$ defect (red) in an active nematic confined to a two-dimensional fluid interface. The initial fracture ends in the two defects and then heals. Note that the $+1/2$ defect is much more dynamic than the $-1/2$ one, in accord with the analysis of section 9.5.3. The motion of $+1/2$ defects is an important driver for the active turbulence.[19] Courtesy of Sattvic Ray and Zvonimir Dogic (C. Joshi et al., 2022; Sanchez et al., 2012).

For a review on active turbulence we refer you to Alert et al., 2022.

obtain the following dispersion relation

$$\sigma - i\omega = -\kappa q^2 + i\beta q, \qquad (9.33)$$

where σ is the growth rate of small disturbances and ω is their vibration frequency. We see that $\omega(q) = -\beta q = -\omega(-q)$. This dispersion relation can be contrasted to that of normal elastic waves in an underdamped solid (ruled by the equation $\partial_t^2 u = \kappa \partial_x^2 u$) for which $\omega(q) = \sqrt{\kappa}|q| = \omega(-q)$ (see section 2.6).

Continuing with figure 9.12, bottom panel shows another type of active solid, composed of moving objects. Here, however, self-propelled particles (similar to the ones we already encountered in section 9.2) occupy nodes of a two-dimensional lattice made of elastic medium. For describing this system at the continuum level, it is therefore natural to combine an order parameter describing deformations $u_i(\vec{r}, t)$ (as in elasticity; see chapter 2) with an order parameter describing the polarization $\vec{p}(\vec{r}, t)$ of the particles (as in flocking; see section 9.2). This hydrodynamic theory allows us to predict the onset of the collective rotational actuation arising from the combination of self-propulsion and pinning of the system at the edge.

9.6.2 Odd elasticity

Figure 9.13 illustrates an in vitro biological realization of chiral crystals of rotating embryos in which anomalous mechanical behavior has been reported and attributed to an active generalization of elasticity known as odd elasticity. Nonliving matter examples of active robotic metamaterials are shown in figure 9.14. Although we will not study these systems in detail, our analysis of odd elasticity applies to them. Elasticity usually assumes conservation of energy. Odd elasticity is a generalization of traditional elasticity with this assumption lifted—forces in active solids are typically nonconservative, and hence cannot be derived from a potential.

Let us start with the general constitutive relation (compare the discussion following equation (2.3))

$$\sigma_{ij} = \sigma_{ij}^{\text{pre}} + K_{ijkl}\, e_{kl}. \qquad (9.34)$$

Here, σ_{ij} is the stress tensor, $e_{kl} \equiv \partial_l v_k$ is the (unsymmetrized) deformation gradient tensor, and σ_{ij}^{pre} are stresses present even when the solid is undeformed (they are called prestresses). The tensor K_{ijkl} contains elastic moduli. If all forces are conservative, then the stress tensor can be written as the derivative

$$\sigma_{ij} = \frac{\partial \mathcal{F}}{\partial e_{ij}} \qquad (9.35)$$

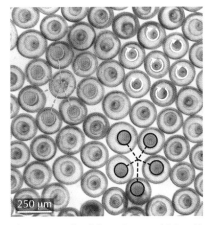

Figure 9.13. Starfish embryos which self-organize into living chiral crystals perform a global collective rotation. From Tan et al., 2022. The image shows a magnified snapshot of the lattice showing topological defects called disclinations. A fivefold defect in the lattice is marked in purple, and a sevenfold defect in orange. The yellow arrows indicate the spinning direction of the embryos. Image courtesy of Nikta Fakhri. Measurements of odd-elastic moduli for this system have been reported by Tan et al., 2022 by comparing experimental measurements of the strain field around defects with theoretical predictions by Braverman et al., 2021.

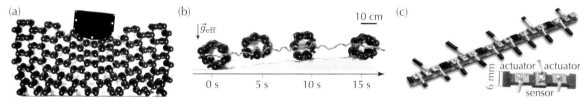

Figure 9.14. Active solids with odd elasticity are relevant to robotic metamaterials, as the energy cycles can power robotic functionalities such as propulsion (see also chapter 10). The figure shows three examples. (a) A robotic metamaterial with active hinges; each one of the elements is a minimal robot that is able to sense the behavior of its neighbors and act consequently. This type of system exhibits anomalous impact response: a bullet hitting the odd-elastic wall is deflected at an angle controlled by the sign of the odd modulus K^{o}. From Brandenbourger et al., 2021. (b) Four successive snapshots illustrating locomotion of a chiral robotic element with nonlinear work cycle. The elements consist of chiral elements connected with passive bands (blue) providing bending stiffness, and with sensors providing feedback. From Brandenbourger et al., 2021. Images a and b courtesy of Martin Brandenbourger. (c) A freestanding active metabeam with piezoelectric elements and electronic feed-forward control that displays odd elasticity. This results in a direction-dependent bending modulus and unidirectional wave amplification. Adapted from Y. Chen et al., 2021.

of some strain energy density

$$\mathcal{F} = \frac{1}{2} C_{ijkl} e_{ij} e_{kl}, \qquad (9.36)$$

so that $\sigma_{ij} = \frac{1}{2} \left(C_{ijkl} + C_{klij} \right) e_{kl}$ and therefore

$$K_{ijkl} = K_{klij}. \qquad (9.37)$$

The symmetry $K_{ijkl} = K_{klij}$ of the stiffness tensor is known as Maxwell-Betti reciprocity.

Odd elasticity emerges from nonconservative forces which cannot be represented as a derivative of an energy. Now, in general the elastic stiffness tensor can be decomposed into even and odd parts,

$$K_{ijkl} = K^{\mathrm{e}}_{ijkl} + K^{\mathrm{o}}_{ijkl}, \qquad (9.38)$$

with

$$K^{\mathrm{e}}_{ijkl} = K^{\mathrm{e}}_{klij} \qquad \text{and} \qquad K^{\mathrm{o}}_{ijkl} = -K^{\mathrm{o}}_{klij}. \qquad (9.39)$$

We refer you to Fruchart et al., 2023 for an introduction to odd elasticity and to Scheibner et al., 2020 for a more detailed discussion.

The tensor K^{e}_{ijkl} contains the usual (even) moduli arising from conservative forces, while K^{o}_{ijkl} holds odd-elastic moduli arising from nonconservative forces. Odd elasticity results from the influence of the additional elastic moduli contained in K^{o}_{ijkl}.

We will focus on odd elasticity in the special case of a two-dimensional solid. In such a material, the modulus tensor K_{ijkl} has 16 independent components while e_{kl} has four components. Though e_{kl} is a rank 2 tensor, we can represent it as a column vector e_α using as a basis the matrices given in (9.40). These correspond to the four modes of deformation, namely dilation, rotation, and two different shears. They form a complete basis, which allows us to write any matrix e_{ij} as a column vector e_α such that $e_{ij} = \sum_\alpha e_\alpha \left(\tau_\alpha \right)_{ij}$.

We can similarly write the stress components as a column vector $\sigma_{ij} = \sigma_\gamma \left(\tau_\gamma \right)_{ij}$, where the basis matrices now correspond to pressure, torque, and two shear stresses. The elastic moduli tensor

The four real basis matrices are given by

$$\tau_0 = \begin{pmatrix} 1 & 0 \\ 0 & 1 \end{pmatrix}, \quad \tau_1 = \begin{pmatrix} 0 & 1 \\ -1 & 0 \end{pmatrix}, \\ \tau_2 = \begin{pmatrix} 1 & 0 \\ 0 & -1 \end{pmatrix}, \quad \tau_3 = \begin{pmatrix} 0 & 1 \\ 1 & 0 \end{pmatrix}. \qquad (9.40)$$

When viewed as vectors in the space of 2×2 matrices, these matrices are normed and orthogonal to each other. One can show (see problem 9.8) that they obey the identity

$$\frac{1}{2} \sum_{ij} (\tau_\alpha)_{ij} (\tau_\beta)_{ij} = \delta_{\alpha\beta}. \qquad (9.41)$$

Any 2×2 matrix X can uniquely be written as $X = x_\alpha \tau_\alpha$. In this section, the components of quantities with respect to this basis are indicated with Greek indices, while the spatial components of matrices (tensors) are indicated with Roman indices.

Table 9.1. Irreducible components of rank 2 tensors in two dimensions. A pure shear (rate) corresponds to a (rate of) change in shape without a change in volume or orientation. Shear 1 describes a horizontal elongation and vertical compression while shear 2 describes elongation along the $45°$ direction and compression along the $-45°$ direction. They are mathematically orthogonal. We note that $\dot{e}_0 = \nabla \cdot v$ and $\dot{e}_1 = \omega$ is vorticity. The stresses are the conjugate forces to these deformations, and they have similar interpretations. In particular, σ_0 includes pressure and the antisymmetric stress is σ_1. Adapted from Fruchart et al., 2023.

deformation	deformation rate	stress	geometric meaning
$e_0 = \text{▦} = \partial_x u_x + \partial_y u_y$	$\dot{e}_0 = \dot{\text{▦}} = \partial_x v_x + \partial_y v_y$	$\sigma_0 = \oplus = [\sigma_{xx} + \sigma_{yy}]/2$	isotropic area change
$e_1 = \text{◩} = \partial_x u_y - \partial_y u_x$	$\dot{e}_1 = \dot{\text{◩}} = \partial_x v_y - \partial_y v_x$	$\sigma_1 = \circlearrowleft = [\sigma_{yx} - \sigma_{xy}]/2$	rotation
$e_2 = \text{▦} = \partial_x u_x - \partial_y u_y$	$\dot{e}_2 = \dot{\text{▦}} = \partial_x v_x - \partial_y v_y$	$\sigma_2 = \text{✛} = [\sigma_{xx} - \sigma_{yy}]/2$	pure shear 1
$e_3 = \text{▱} = \partial_x u_y + \partial_y u_x$	$\dot{e}_3 = \dot{\text{▱}} = \partial_x v_y + \partial_y v_x$	$\sigma_3 = \otimes = [\sigma_{xy} + \sigma_{yx}]/2$	pure shear 2

becomes a matrix in this basis, allowing the constitutive relation (9.34) to be written compactly as (see problem 9.8 for the intermediate steps and more details)

$$\sigma_\alpha = K_{\alpha\beta} e_\beta, \tag{9.42}$$

where

$$K_{\alpha\beta} = \frac{1}{2}(\tau_\alpha)_{ij}\, K_{ijkl}\, (\tau_\beta)_{kl}. \tag{9.43}$$

Using the graphical notation summarized in table 9.1, the column vectors for stress and strain read

$$\sigma_\alpha = \begin{pmatrix} \oplus \\ \circlearrowleft \\ \text{✛} \\ \otimes \end{pmatrix} \quad \text{and} \quad e_\beta = \begin{pmatrix} \text{▦} \\ \text{◩} \\ \text{▦} \\ \text{▱} \end{pmatrix}, \tag{9.44}$$

where similar to the components of σ_α the components of e_β represent dilation, rotation, and the two shears (see problem 9.8).

In addition to the work on spinning embryos, see Bililign et al., 2021 for a study of spinning colloids, where experimental behaviors compatible with odd elasticity are reported. See also Shankar and Mahadevan, 2022 and Zahalak, 1996 for a discussion of anisotropic odd elasticity in the context of muscles.

Assuming that the material under consideration is invariant under rotations, the most general matrix $K_{\alpha\beta}$ then reads

$$\underbrace{\begin{pmatrix} \oplus \\ \circlearrowleft \\ \oplus \\ \otimes \end{pmatrix}}_{\sigma_\alpha} = \underbrace{\begin{pmatrix} -p^{(\mathrm{pre})} \\ -\tau^{(\mathrm{pre})} \\ 0 \\ 0 \end{pmatrix}}_{\sigma_\alpha^{(\mathrm{pre})}} + \underbrace{\begin{pmatrix} B & \Lambda & 0 & 0 \\ A & \Gamma & 0 & 0 \\ 0 & 0 & \mu & K^{\mathrm{o}} \\ 0 & 0 & -K^{\mathrm{o}} & \mu \end{pmatrix}}_{K_{\alpha\beta}} \underbrace{\begin{pmatrix} \text{▦} \\ \text{◩} \\ \text{▦} \\ \text{▱} \end{pmatrix}}_{e_\beta}. \tag{9.45}$$

We refer you to Poncet and Bartolo, 2022 for a systematic treatment of active solids with effective violations of Newton's third law and to Baek et al., 2018; Chajwa et al., 2020; Ivlev et al., 2015; Meredith et al., 2020; Saha et al., 2014; and Soto and Golestanian, 2014 for examples of systems in which such forces are generated.

In this equation, the prestresses include a hydrostatic pressure p^{pre} and a hydrostatic torque τ^{pre} present even when the system is undeformed. The hydrostatic torque is therefore a zeroth order effect that drives a large part of the phenomenology of systems with broken mirror symmetry. The coefficients B and μ are the familiar compression and shear moduli from passive elasticity.[25] One often assumes that solid-body rotations do not induce stresses, implying $\Lambda = \Gamma = 0$. This assumption can, however, be violated, for instance, when the solid is on a substrate.

The remaining terms A and K^{o} are odd-elastic moduli. They are allowed when the symmetry constraint $K_{ijkl} = K_{klij}$ (equivalently, $K_{\alpha\beta} = K_{\beta\alpha}$) is lifted. A represents a coupling of dilation to torques while K^{o} describes a shear stress response which is rotated with respect to an applied strain.

As a result of these new moduli, the work done on the system is no longer a state function but in fact depends on the path taken in strain space. If odd elasticity is present, work can be extracted or lost from the system by performing quasi-static strain cycles. Elastic forces are given by $f_i = \partial_j \sigma_{ij}$, so the power per unit area exerted by elastic forces when the solid is deformed is $f_i \dot{u}_i$. Consider now a cyclic deformation $u_i(t) = u_i(t + T)$ imposed from the outside. The work done over one cycle of deformation is then

$$\Delta W = \int \mathrm{d}t\, f_i \dot{u}_i = -\oint \mathrm{d}e_{ij} \sigma_{ij} = -\oint \mathrm{d}e_{ij} K^{\mathrm{o}}_{ijkl} e_{kl} \neq 0. \quad (9.46)$$

To see that this quantity is in general not zero, let us take the dilation-torque coupling $A = 0$. The work done over some cyclic path in strain space is then

$$\Delta W = \oint \mathrm{d}e_{ij} \sigma_{ij} = \oint \mathrm{d}e_\beta \sigma_\beta = \int \mathrm{d}^2 S\, \epsilon_{\alpha\beta} \frac{\partial \sigma_\beta}{\partial e_\alpha}, \quad (9.47)$$

where in the rightmost term the integral is over the area enclosed by the path in strain space and $\underline{\epsilon}$ is the two-dimensional Levi-Civita tensor (see equation (2.124)). Only the odd components $K_{\alpha\beta} = K^{\mathrm{o}} \epsilon_{\alpha\beta}$ will remain, leading to

$$\Delta W = \int \mathrm{d}^2 S\, \epsilon_{\alpha\beta} \epsilon_{\alpha\beta} K^{\mathrm{o}} = 2 K^{\mathrm{o}} \times \text{Area}. \quad (9.48)$$

Thus, the work extracted from an odd-elastic engine is proportional to the odd-elastic modulus and the area of the cycle in strain space. We can visualize this using an example of a microscopic model with odd-elastic bonds, depicted in figure 9.15. The force law of the linear spring with odd elasticity is

$$\vec{f} = -(k\hat{r} + k^{\mathrm{o}}\hat{\phi})\delta r, \quad (9.49)$$

where we use polar coordinates, so that $\hat{r}$ is the unit vector pointing along the spring, $\hat{\phi}$ the unit vector orthogonal to $\hat{r}$, and δr the spring extension, while k^{o} is the odd and k the regular microscopic spring constant.

In problem 9.9, we calculate the net work done over the cycle shown in figure 9.15 directly, and we confirm that it is proportional

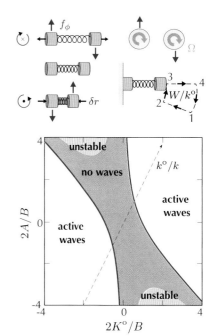

Figure 9.15. Upper panel right: schematic illustration of two disks of diameter d, separated by a distance r and spinning in a fluid at angular speed Ω. They experience non-central hydrodynamic forces (i.e., forces that have a component f_ϕ transverse to the vector joining the centers of the particles). Upper panel left: A microscopic odd-elastic spring. In addition to the standard compression/extension force, the spring exerts a force $f_\phi = k^{\mathrm{o}} \delta r$ transverse to its radial displacement δr. Bottom right of upper panel: an odd-elastic energy cycle. The spring is stretched through the path shown and the work is proportional to the area of the path times k^{o}. Lower panel: schematic phase diagram of the odd-elastic media as a function of the odd moduli A and K^{o} normalized by the bulk modulus B. The red lines indicate the transition between an overdamped solid without wave propagation and active waves (it coincides with a line of exceptional points; see text). The dashed line represents a trajectory of increasing odd microscopic spring constant k^{o}. Note that the medium becomes unstable in the yellow areas. Adapted from Scheibner et al., 2020.

For a study of the hydrodynamics of swimming algae we refer you to Drescher et al., 2009.

to both k^o and the area enclosed by the path. This ability to extract work enables self-sustained waves powered by odd-elastic energy cycles in overdamped media that contain the noncentral springs. An experimentally viable way of generating noncentral interaction forces is by having colloidal particles, swimming algae, or embryos that constantly spin in a fluid so that they experience friction-like transverse forces. In fact, it is possible to obtain explicit formulas for the odd moduli by coarse-graining these microscopic forces.[26]

9.6.3 Odd elastodynamics

An example of transverse hydrodynamic interaction is the so-called lubrication force $f_\phi \propto \Omega \log \left(\frac{r-d}{d} \right)$ between two disks whose diameter is equal to d, at distance r from each other, spinning with fixed angular speed Ω (Happel and H. Brenner, 1982). Note, however, that these hydrodynamic forces are not pairwise additive.

We now discuss the effect of odd elasticity on dynamics. As we shall see, odd elasticity can lead to active waves in overdamped systems and even to instabilities. For overdamped systems, the dynamics of the displacement field can be described by taking

$$\gamma \partial_t u_i = f_i = \partial_j \sigma_{ij}, \qquad (9.50)$$

where γ is a friction coefficient. Using the constitutive relation in equation (9.45) along with table 9.1, we find that the odd-elastodynamics equations explicitly read

$$\gamma \partial_t \vec{u} = B \vec{\nabla}(\vec{\nabla} \cdot \vec{u}) + \mu \Delta \vec{u} + K^o \underline{\epsilon} \cdot \Delta \vec{u} - A \underline{\epsilon} \cdot \vec{\nabla}(\vec{\nabla} \cdot \vec{u}). \qquad (9.51)$$

As in (9.47), $\underline{\epsilon}$ is the Levi-Civita tensor.

Performing a Fourier transform of equation (9.51), we find

$$i\gamma\omega \begin{pmatrix} u_\parallel \\ u_\perp \end{pmatrix} = q^2 \begin{pmatrix} B+\mu & K^o \\ -K^o + A & \mu \end{pmatrix} \begin{pmatrix} u_\parallel \\ u_\perp \end{pmatrix}, \qquad (9.52)$$

In rotating Rayleigh-Bénard convection, the fluid is put under rotation, leading to Coriolis forces (see chapter 8). The relaxation of the pattern is then described by an equation of motion formally identical to that of odd elastodynamics, even if the interpretation in terms of stress does not carry through; see Fruchart et al., 2023 and references therein.

in which we have defined the longitudinal $u_\parallel = \hat{q} \cdot u$ and transverse $\vec{u}_\perp = \vec{u} - u_\parallel \hat{q}$ components of the displacement. In an overdamped system, passive elastodynamics ($A = K^o = 0$) is diffusive: $\omega = -iq^2 \frac{B+\mu}{\gamma}$ and $\omega = -iq^2 \frac{\mu}{\gamma}$ (with our conventions, a negative imaginary frequency implies that a wave is attenuated). In the case of an overdamped odd-elastic solid with $A, K^o \neq 0$, we obtain

$$\omega = -iq^2 \frac{B/2 + \mu \pm \sqrt{(B/2)^2 - K^o(K^o - A)}}{\gamma}. \qquad (9.53)$$

Experimental observations of such elastic waves have been reported in the systems shown in figure 9.14.

When $K^o(K^o - A) > (B/2)^2$, the frequency has a real part: there are oscillations even though the system is overdamped. The transition between exponential relaxation and damped oscillations, marked by an exceptional point of the matrix in equation (9.52), is the point where the matrix is not diagonalizable (red lines in figure 9.15), in a similar way as in a damped harmonic oscillator. The system can even become unstable (yellow region in figure 9.15) when the imaginary part of the frequency becomes positive.

In this case, linear elasticity cannot accurately describe the system and nonlinearities have to be considered.

9.7 Chiral active fluids

Consider an active granular fluid whose particles are constantly spinning in a plane. In a passive granular system, the particles lose energy in collisions through friction, while in an active one, the particles can gain energy when their rotation speed is reset after a collision by microscopic torques or external fields. Active matter whose particles are all spinning in the same direction is known as a chiral active fluid. We already showed a couple of examples of chiral active fluids in figure 9.14 (collections of robots) and figure 9.13 (spinning embryos), but chiral active behavior has also been observed in magnetic colloids; see figure 9.16. In all these systems there is a central force that models a soft repulsion between particles and a noncentral force that captures interparticle friction. In addition to this interaction force, each particle experiences an active torque which tends to maintain a constant angular velocity. This active torque imparts a fixed chirality to the system. The macroscopic effects of these microscopic ingredients can be captured by the hydrodynamic approach summarized below.

9.7.1 Hydrodynamics of self-spinning particles

The Navier-Stokes equations that we introduced in chapter 1 describe the conservation of mass and linear momentum (and possibly of energy). In order to describe a collection of spinning objects, one has to also consider the angular momentum of the particles at the continuum level. For identical particles, this can be done by introducing the angular velocity field $\vec{\Omega}(\vec{r}, t)$ as the coarse-grained version of the spinning speed of individual particles, very much as the velocity field $\vec{v}(\vec{r}, t)$ is a coarse-grained version of the translation speeds of individual particles.[27]

For concreteness, we consider a two-dimensional fluid of spinners, where the angular velocity field is a pseudoscalar $\Omega(\vec{r}, t)$. The resulting continuum equations read

$$(\partial_t + \vec{v} \cdot \vec{\nabla})\rho = -\rho \vec{\nabla} \cdot \vec{v},$$

$$\rho(\partial_t + \vec{v} \cdot \vec{\nabla})\vec{v} = \vec{\nabla} \cdot \underline{\sigma} + \vec{f}_{\text{vol}}, \qquad (9.54)$$

$$I(\partial_t + \vec{v} \cdot \vec{\nabla})\Omega = \tau + \epsilon_{ij}\sigma_{ij} - \Gamma_\Omega \Omega + \eta_\Omega \nabla^2 \Omega,$$

where $\omega = (\vec{\nabla} \times \vec{v})_z$ is the vorticity and $\vec{f}_{\text{vol}}$ are body forces, and τ are body torques. Besides, Γ_Ω is a phenomenological damping

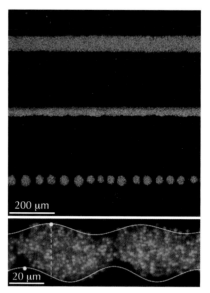

Figure 9.16. Layers of a chiral active fluid consisting of spinning colloidal magnets. The particles are made to spin in a magnetic field. Upper panel: various snapshots of layers of varying thickness. Below a thickness of 32 μm the layers exhibit a hydrodynamic instability. Lower panel: a chiral fluid strip approaching instability. Experiments on this instability allow us to test minimal hydrodynamic models of chiral fluids in detail. Adapted from Soni et al., 2019.

Even in passive fluids, it is sometimes necessary to add a continuum equation to describe the conservation of angular momentum. This equation is in principle required as soon as the particles are not point-like, and even more so if they are not spherical. However, it is often the case that the angular velocity field relaxes more quickly than other fields, and it can therefore be ignored or eliminated.

term describing the damping of the spinning, while Γ describes the coupling between spinning and vorticity, and η_Ω is a rotational viscosity. Finally, the stress is

$$\sigma_{ij} = \epsilon_{ij}\frac{\Gamma}{2}(\Omega - \omega) - P\delta_{ij} + \eta_{ijkl}\partial_l v_k + \frac{I\Omega}{2}\left(\partial_i \epsilon_{jl}v_l + \epsilon_{il}\partial_l v_j\right).$$
(9.55)

While these equations have several moving parts, note that the angular velocity field and the velocity field are coupled by the stress σ_{ij}. When the driving performed by external torques τ is strong enough, a steady state with almost uniform and almost constant angular velocity $\Omega \approx \tau/\Gamma_\Omega$ can be achieved. In this regime, the angular velocity field can be integrated out to obtain an effective hydrodynamic equation for the velocity and the density fields. At this level, the main effect of a maintained rotation of the particles is the appearance of additional terms in the hydrodynamic response—in the viscosity tensor—as we discuss in the next section.

Outside of the regime, where Ω is approximately uniform and constant, more complex behaviors can occur, where the full continuum equations (9.54) are needed.

9.7.2 Odd viscosity

See Fruchart et al., 2023 for an introduction to odd viscosity.

The evolution of the velocity field $\vec{v}$ of the chiral active fluid is described by the Navier-Stokes equation for an incompressible fluid,

$$\rho(\partial_t + \vec{v}\cdot\vec{\nabla})\vec{v} = \vec{\nabla}\cdot\underline{\sigma} + \vec{f}_{\text{vol}},$$
(9.56)

where the term on the left is the material derivative (1.1), $\underline{\sigma}$ is the stress tensor, f_{vol} are external body forces, and $\rho = nm$ is the mass density (n is the number density), which is taken as constant (see section 1.9). The stress tensor $\underline{\sigma}$ in equation (9.56) is composed of a reversible part $\underline{\sigma}^{\text{ss}}$ (the pressure for ordinary fluids; see section 1.5), and a dissipative or viscous part $\underline{\sigma}^{\text{vis}}$ that describes surface forces between fluid layers that arise in response to velocity gradients.

The viscous stress is given by $\sigma_{ij}^{\text{vis}} = \eta_{ijkl}\,\partial_l v_k$, where η_{ijkl} is the viscosity tensor of the fluid. In the same way as in section 9.6.2, we decompose the viscosity tensor

Just as in section 9.6.2 on odd elasticity, it is convenient to express σ_{ij}^{vis} and the unsymmetrized shear rate $\dot{e}_{kl} \equiv \partial_l v_k$ as column vectors $\sigma_\alpha^{\text{vis}}$ and $\dot{e}_\beta$ using as basis the matrices introduced in equation (9.40). Then η_{ijkl} can be represented as a matrix $\eta_{\alpha\beta}$.

$$\eta_{ijk\ell} = \eta_{ijk\ell}^{\text{e}} + \eta_{ijk\ell}^{\text{o}}$$
(9.57)

into symmetric (even) $\eta_{ijk\ell}^{\text{e}} = \eta_{k\ell ij}^{\text{e}}$ and antisymmetric (odd) $\eta_{ijk\ell}^{\text{o}} = -\eta_{k\ell ij}^{\text{o}}$ parts. Odd viscosities are contained in $\eta_{ijk\ell}^{\text{o}}$, i.e., they are those that violate the symmetry $\eta_{ijk\ell} = \eta_{k\ell ij}$ (or, equivalently, $\eta_{\alpha\beta} = \eta_{\beta\alpha}$; see the margin note next to (9.57)).

Usual viscosities are dissipative. The rate of loss of mechanical energy by viscous dissipation per unit volume is[28]

$$\dot{w} = \sigma_{ij}^{\mathrm{vis}} \partial_j v_i = \eta_{ijk\ell}(\partial_j v_i)(\partial_\ell v_k) = \eta_{ijk\ell}^{\mathrm{e}}(\partial_j v_i)(\partial_\ell v_k). \qquad (9.58)$$

Therefore, only the symmetric part of the viscosity tensor contributes to viscous dissipation. Odd viscosities correspond to the non-dissipative part of the viscosity tensor.

As an example, let us consider the constitutive relation for rotation invariant two-dimensional fluids,

$$
\underbrace{\begin{pmatrix} \oplus \\ \circlearrowright \\ \oplus \\ \otimes \end{pmatrix}}_{\sigma_\alpha} - \underbrace{\begin{pmatrix} -p \\ -\tau \\ 0 \\ 0 \end{pmatrix}}_{\sigma_\alpha^{\mathrm{h}}} + \underbrace{\begin{pmatrix} \zeta & \eta^{\mathrm{B}} & 0 & 0 \\ \eta^{\mathrm{A}} & \eta^{\mathrm{R}} & 0 & 0 \\ 0 & 0 & \eta & \eta^{\mathrm{o}} \\ 0 & 0 & -\eta^{\mathrm{o}} & \eta \end{pmatrix}}_{\eta_{\alpha\beta}} \underbrace{\begin{pmatrix} \bullet \\ \diamond \\ \blacksquare \\ \square \end{pmatrix}}_{\dot{e}_\beta}.
$$
$$\qquad (9.59)$$

See table 9.1 for the deformation rates and stresses σ_α. The viscosity matrix includes the standard bulk ζ and shear η viscosity coefficients, and several new ones which are allowed due to broken time-reversal and parity (i.e., mirror) symmetry. The *odd viscosity* η^{o} couples shear strains and stresses, and additional parity-violating viscosities η^{A} and η^{B} couple compression and rotation.

Using the constitutive relation in equation (9.59) along with table 9.1, we find that the full Navier-Stokes equation reads

$$
\begin{aligned}
\rho \mathrm{d}_t \vec{v} = {}& \vec{\nabla} \cdot \underline{\sigma}^{\mathrm{h}} + \zeta\, \vec{\nabla}(\vec{\nabla} \cdot \vec{v}) + \eta\, \Delta \vec{v} + \eta^{\mathrm{o}}\, \underline{\epsilon} \cdot \Delta \vec{v} \\
& - \eta^{\mathrm{A}}\, \underline{\epsilon} \cdot \vec{\nabla}(\vec{\nabla} \cdot \vec{v}) + \eta^{\mathrm{B}}\, \vec{\nabla}(\vec{\nabla} \times \vec{v}) - \eta^{\mathrm{R}}\, \underline{\epsilon} \cdot \vec{\nabla}(\vec{\nabla} \times \vec{v}),
\end{aligned}
$$
$$\qquad (9.60)$$

in which $\underline{\sigma}^{\mathrm{h}}$ is the hydrostatic stress tensor $(\vec{\nabla} \times \vec{v})_z = \epsilon_{ij}\partial_i v_j$ in two dimensions, and d_t denotes the material derivative (1.2).

The appearance of odd viscosity can be understood from the more general hydrodynamics discussed in section 9.7.1 by focusing on the last term in the stress equation (9.55). Assuming incompressibility ($\vec{\nabla} \cdot \vec{v} = 0$) and using the identity $\vec{\nabla}\omega = \underline{\epsilon} \cdot \nabla^2 \vec{v}$ for incompressible two-dimensional fluids, one can show that

$$\partial_j \sigma_{ij} = \frac{I\Omega}{2}\partial_j\left(\epsilon_{jl}\partial_i v_l + \epsilon_{il}\partial_l v_j\right) + \cdots = \frac{I\Omega}{2}\epsilon_{ij}\nabla^2 v_j + \cdots.$$
$$\qquad (9.61)$$

Comparison between equations (9.60) and (9.61) shows that this system exhibits odd viscosity, with

$$\eta^{\mathrm{o}} = \frac{I\Omega}{2}. \qquad (9.62)$$

The odd viscosity is a hydrodynamic manifestation of the chiral collisions in a microscopic model. When two self-spinning particles collide, their outgoing velocities will be rotated compared to the

In the context of non-equilibrium thermodynamics, the symmetry $\eta_{ijk\ell} = \eta_{k\ell ij}$ of the viscosity tensor is known as Onsager reciprocity; see section 1.5.

Odd viscosity has been measured in a number of systems, including chiral fluids composed of self-spinning colloids (Soni et al., 2019)—see figure 9.16—as well as magnetized polyatomic gases—see Beenakker and McCourt, 1970 and references therein (table 2)—and electrons in the presence of an external magnetic field—see Berdyugin et al., 2019.

The relation between odd viscosity and microscopic violations of mirror symmetry can be quantified in certain microscopic models. For each collision, one computes the twisting angle $\theta \equiv \mathrm{angle}(\vec{v}, \Delta\vec{v})$ between the initial velocity $\vec{v}$ and the change in velocity during the collision $\Delta\vec{v}$. The ratio η^{o}/η is then proportional to $\langle\theta\rangle$, the average twisting over the ensemble of all particles (M. Han et al., 2021).

incoming ones in a manner set by their spinning direction. Odd viscosity is not related to energy dissipation and, unlike standard (even) viscosity, cannot be derived from an entropy production rate equation, very much as an odd-elastic coefficient cannot be derived from variations of an elastic potential energy.

Hydrodynamic theories of active fluids capture several striking phenomena observed in experiments, including the instability shown in figure 9.16.

9.8 Nonreciprocal phase transitions

Non-equilibrium systems are typically modeled by stochastic processes that violate detailed balance. As a result, the steady states of these systems are characterized by nonvanishing probability currents between microstates, and they exhibit entropy production. A simple example is a system composed of three states with cyclic clockwise transition rates. The steady state is reached when the probabilities of being in each state are equal. This system is not at equilibrium even if it possesses a Boltzmann distribution with constant energy because it does not obey detailed balance. Similarly, physical systems with absorbing states—states out of which transitions have zero probability—clearly violate detailed balance.[30] Flocking states are a non-equilibrium example of such behavior. Nonreciprocal phase transitions describe the transitions from and to these non-equilibrium steady states.

When we discussed the flocking transition in terms of the Toner-Tu theory of section 9.2.3, we associated the transition with a pitchfork bifurcation arising from minimizing a quartic potential (see section 8.2.1 for a brief summary of the pitchfork bifurcation). The analysis of chapter 8 showed that bifurcations to time-dependent states (such as traveling waves) are typically non-potential. We now consider an example of this in active matter, in which nonreciprocal interactions lead to time-dependent phases that spontaneously break parity (mirror symmetry).[31]

9.8.1 Chiral phases in nonreciprocal active matter

We can illustrate the main features of nonreciprocal active matter with the following model:

$$\partial_t \theta_m = \omega_m + \sum_n J_{mn} \sin(\theta_n - \theta_m) + \eta_m(t), \qquad (9.63)$$

which can be thought of as a simple extension of the Vicsek model. When the agents are at *fixed* positions, this model is known as the

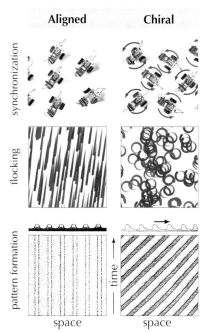

Aligned   **Chiral**

synchronization

flocking

pattern formation

time

space   space

Figure 9.17. Nonreciprocal interactions between two species, R (Red) and B (Blue), induce a phase transition from static alignment to a chiral motion that spontaneously breaks parity. Top: nonreciprocal synchronization. Angular variables with nonreciprocal interactions drawn as robots spontaneously rotate either clockwise or counterclockwise, despite no average natural frequency ($\omega_m = 0$ in equation (9.63)). Middle row: non-reciprocal flocking. Self-propelled particles run in circles despite the absence of external torques. Bottom: space-time plots of an example of pattern formation with nonreciprocal interaction. A one-dimensional stationary pattern starts traveling, either to the left or to the right as in the chiral case, when nonreciprocal interactions are turned on. The figure represents an experimental observation of a moving oil-air interface (so-called viscous fingering). Adapted and reproduced with permission from Fruchart et al., 2021.[29]

Kuramoto model, which was introduced to study the synchronization of coupled oscillators with phase angles θ_m. It qualitatively describes the collective behavior of clocks ticking, neurons firing, or fireflies flashing.[32] With strong enough coupling, a synchronized state emerges where all oscillators evolve in phase with the same frequency.

In our case, the variable θ_m describe the angle in the plane of the velocities with which the agents move, so that the positions $\vec{r}_m$ in the plane are given by

$$\partial_t \vec{r}_m = v_0 \begin{pmatrix} \cos \theta_m \\ \sin \theta_m \end{pmatrix}. \tag{9.64}$$

An agent m tends to align with an agent n when $J_{mn} > 0$, or to antialign when $J_{mn} < 0$. The standard Vicsek flocking model corresponds to $J_{mn} > 0$. In the absence of interactions and noise the agents all rotate independently with their own frequency ω_m. For positive J_{mn}, there is a critical coupling at which a transition takes place, from incoherent rotations or motion to synchronized rotation (when the positions are fixed while $\omega_m \neq 0$) or to flocking when the particles move according to equation (9.64).

Now consider two copies of the Vicsek model describing two species, labeled 1 and 2. Without interaction between the two species, each has its own order parameter (average velocity) describing the flocking. The behavior of the model becomes especially interesting, however, when there are interactions between the species. When the interactions are reciprocal, $J_{12} = J_{21}$, we find, in addition to a disordered phase, two *static* phases where $\vec{v}_1$ and $\vec{v}_2$ are (anti)aligned, in analogy with (anti)ferromagnetism. When the interactions are nonreciprocal, $J_{12} \neq J_{21}$, a time-dependent chiral phase with no equilibrium analogue emerges between the static phases. In this chiral phase, parity is spontaneously broken: $\vec{v}_1$ and $\vec{v}_2$ (the two species are represented in red and blue in figure 9.17) rotate with a fixed relative angle $\Delta\phi$, either clockwise or counterclockwise, at a constant rotation rate $\Omega_{ss} \equiv \partial_t \bar{\phi}$, where $\bar{\phi}$ is the angle between $(\vec{v}_1 + \vec{v}_2)/2$ and a fixed direction. The chiral phase is caused by the frustration experienced by agents with opposite goals: species 1 (red) wants to align with species 2 (blue) but not vice versa. This dynamical frustration results in a 'chase and run away' motion of the order parameters $\vec{v}_1$ and $\vec{v}_2$.

Figure 9.17 illustrates the aligned-to-chiral transition in flocking as well as in synchronization and pattern formation.[33] The bottom row of the figure illustrates a strong link with the formation of patterns as discussed in chapter 8: in pattern-forming non-equilibrium systems, nonreciprocal interactions can similarly lead to a transition from stationary to moving patterns, and the methods of the theory of pattern formation and dynamical systems can be used to understand general features of the phases and phase transitions.

Interestingly, Rayleigh-Bénard convection in rotating cells is described by a nonreciprocal model, in which the nonreciprocal effects affect the nonlinear terms describing mode interactions. See the note in the margin of equation (8.54).

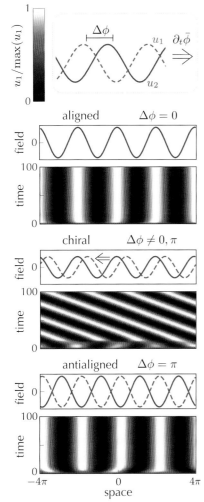

Figure 9.18. Space-time density plots of the field $u_2(x,t)$ shown with snapshots of the fields $u_1(x,t)$ and $u_2(x,t)$ at the top. In the chiral phase with a finite phase difference $\Delta\phi$, the patterns move at constant velocity, either to the left or to the right (spontaneously breaking parity). In the antialigned case $\Delta\phi = \pi$, the patterns are stationary again. Image courtesy of Michel Fruchart.

In problem 9.6, you will show explicitly how the amplitude equations reduce to (9.67)–(9.68) and derive the explicit expressions for the coefficients $\alpha = 2[\epsilon_-^2 - \epsilon_+\epsilon_0]/\epsilon_0$ and $\gamma = 2\epsilon_-$. Note also that we have expanded in these equations the right-hand sides in $\Delta\phi$, to bring out the bifurcation structure of (9.67). The full equations are periodic in $\Delta\phi$. This reflects the fact that when one of the patterns is shifted by one wavelength relative to the other, the pattern does not change.

To illustrate the cross-fertilization of the two fields, we will show below how such transitions can be studied with the methods developed in chapter 8.

9.8.2 Nonreciprocal pattern formation: A case study

Let us start by considering a classic model of pattern formation that we already encountered in section 8.2.2, the Swift-Hohenberg model, as it is the simplest type of spatiotemporal model that exhibits finite wavelength periodic patterns. We generalize to two fields u_1 and u_2 obeying a set of two Swift-Hohenberg equations with nonreciprocal interactions, namely

$$\partial_t u_i = \epsilon_{ij} u_j - (1 + \nabla^2)^2 u_i - g u_i^3, \qquad (9.65)$$

where we have introduced the nonsymmetric coupling matrix $\epsilon_{12} \neq \epsilon_{21}$. Unlike the standard Swift-Hohenberg model that can be derived from a potential, the nonreciprocal ones cannot.

Following the amplitude equations approach developed in the chapter on pattern formation (see section 8.5), we can write $u_i(x) = A_i(x)e^{ik_c x} + \text{c.c.}$, where k_c is the critical wavevector where the instability sets in (here $k_c = 1$). Note that shifting the phase of the amplitude A_i by $\delta\phi$ ($A_i \to A_i e^{i\delta\phi}$) amounts to a translation of the periodic one-dimensional pattern $u_i(x,t)$ by a distance equal to $\delta\phi/k_c$ (see section 8.5.2.d.) Using the symmetry considerations discussed in the previous chapter (or a full-length explicit derivation), the set of nonreciprocal Swift-Hohenberg equations reduces, just above the instability threshold, to the following set of nonreciprocally coupled amplitude equations:

$$\begin{aligned} \partial_t A_1 &= \epsilon_0 A_1 + \epsilon_{12} A_2 - g|A_1|^2 A_1, \\ \partial_t A_2 &= \epsilon_0 A_2 + \epsilon_{21} A_1 - g|A_2|^2 A_2, \end{aligned} \qquad (9.66)$$

where we have introduced the nonsymmetric coupling among the amplitudes $\epsilon_{12} = \epsilon_+ + \epsilon_-$ and $\epsilon_{21} = \epsilon_+ - \epsilon_-$, and ignored the gradient terms.

We now write the two complex amplitudes (that act as order parameters in this pattern formation problem) as $A_i = a_i e^{i\phi_i}$ ($a_i = |A_i|$) and assume that the two fields are primarily coupled antisymmetrically, so that $\epsilon_0 \gg \epsilon_- \gg \epsilon_+$. The amplitude equations reduce to

$$\partial_t \Delta\phi = \alpha\Delta\phi - \beta\Delta\phi^3, \qquad (\beta > 0) \qquad (9.67)$$

$$\partial_t \bar\phi = \gamma\Delta\phi, \qquad (9.68)$$

where $\Delta\phi = \phi_2 - \phi_1$ and $\bar\phi = \phi_2 + \phi_1$. The stationary state of the first equation is $\Delta\phi = 0$ if $\alpha < 0$ and $\Delta\phi = \pm\sqrt{\alpha/\beta}$ if $\alpha > 0$. The case $\alpha < 0$ results in two *static* patterns that are completely in phase, while the case $\alpha > 0$ with $\Delta\phi \neq 0$ corresponds to two periodic patterns offset by a constant phase difference $\Delta\phi$ both *moving* (because the sum of their phases increases indefinitely) at a speed proportional to $\gamma\Delta\phi$, as illustrated in figure 9.18. The resulting phase diagram from this perturbative analysis is shown in figure 9.19, while figure 9.20.a shows the full phase diagram. By itself, equation (9.67) represents a standard pitchfork bifurcation, but in conjunction with equation (9.68) it forms a system of dynamical equations that cannot be expressed in terms of the gradient of a potential, and that describes a class of bifurcations known as parity-breaking bifurcations or drift instabilities.[34]

This pattern formation problem exemplifies a transition from a state described by static order parameters to a non-equilibrium steady state described by time-varying order parameters.

9.8.3 Exceptional points and parity-breaking bifurcations

To understand the mechanism underlying the transition, we linearize the above equations about $\Delta\phi = 0$ and an arbitrary fixed $\bar{\phi}$, to obtain

$$\partial_t \begin{pmatrix} \Delta\phi \\ \bar{\phi} \end{pmatrix} = \begin{pmatrix} \alpha & 0 \\ \gamma & 0 \end{pmatrix} \begin{pmatrix} \Delta\phi \\ \bar{\phi} \end{pmatrix}. \tag{9.69}$$

At the transition point $\alpha = 0$, the matrix on the right-hand side is not diagonalizable since it has two eigenvalues equal to zero; at that point

$$\partial_t \begin{pmatrix} \Delta\phi \\ \bar{\phi} \end{pmatrix} = \gamma \begin{pmatrix} 0 & 0 \\ 1 & 0 \end{pmatrix} \begin{pmatrix} \Delta\phi \\ \bar{\phi} \end{pmatrix}. \tag{9.70}$$

The point where two eigenvalues of a matrix vanish is called an exceptional point of the matrixes and it is characteristic of non-reciprocal phase transitions.

In the Jacobian matrix above, one eigenvalue always vanishes because it corresponds to translation invariance of the pattern (and to rotation invariance of the $\vec{v}_i$ in the flocking model): it is known as the Goldstone mode of broken translation (or rotation) invariance, as indicated by the green line in figure 9.20.b. At the bifurcation, the remaining eigenmode (orange line in the figure) coalesces with the Goldstone mode (green line) at exceptional points, in red. In addition to having the same eigenvalue (zero), the two eigenmodes become parallel at this point.

The structure of exceptional points leads to a pictorial description of the phase transition to the chiral phase. Let us represent the order parameter as a ball constantly kicked by noise at the bottom of a wine bottle–shaped potential. Because of the nonreciprocal couplings, there are transverse nonconservative forces in addition

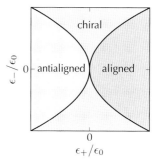

Figure 9.19. The perturbative phase diagram of the exceptional transition computed in the main text and problem 9.6 as a function of the ratios ϵ_-/ϵ_0 and ϵ_+/ϵ_0.[35]

See section 8.2.1 for an introduction to the pitchfork bifurcation.

See problem 9.10 for an explicit derivation of these equations. The matrix in (9.69) is called a Jacobian. Generally, the Jacobian of a vector field is the matrix of its derivatives.

While we have derived equations (9.67), (9.68), and (9.70) in the context of a specific example, they apply more generally to non-reciprocal phase transitions in various systems. For instance, nonreciprocal flocking can be analyzed in the same way by considering two Toner-Tu equations (section 9.2.3) with asymmetric couplings.

The same approach applies to order parameters associated with conservation laws, e.g., continuum mechanics models with odd-elasticity and odd-viscosity that conserve linear momentum (see previous sections) and nonreciprocal models of phase separation that conserve mass, discussed for example by You et al., 2020 and Saha et al., 2020. In all these problems one obtains (nonlinear) diffusion-like equations with additional diffusion coefficients accounting for nonreciprocal couplings between the components of the relevant fields as illustrated in (9.51).

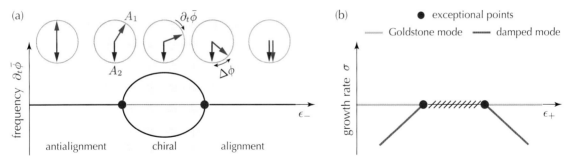

Figure 9.20. (a) Schematic bifurcation diagram of the exceptional transition for two coupled vector models (9.63), showing the frequency of the steady state $\Omega_{ss} \equiv \partial_t \bar{\phi}$. Between the static (anti)aligned phases with $\Omega_{ss} = 0$, an intermediate chiral phase spontaneously breaks parity. Two equivalent steady states (clockwise and counterclockwise, corresponding to opposite values of Ω_{ss}) are present in this time-dependent phase. The chiral phase continuously interpolates between the antialigned and aligned phases, both through $|\Omega_{ss}|$ and the angle between the order parameters $\vec{v}_A$ and $\vec{v}_B$. (b) The transition between (anti)aligned and chiral phases occurs through the coalescence of a damped (orange) and a Goldstone mode (green) at an exceptional point (red dot). Adapted from Fruchart et al., 2021.

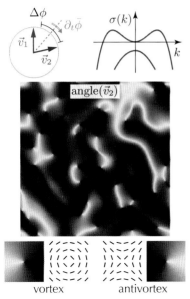

Figure 9.21. As explained in the text, a type II instability discussed in section 8.6 occurs near exceptional points. We consider here nonreciprocal flocking (top left), where the two velocity vectors rotate at a constant speed while keeping their relative orientation fixed. The growth rate of perturbations (top right) is shown signaling a finite momentum instability. The middle panel shows a snapshot of the phase angle of one of the velocities in two-dimensional hydrodynamic simulations in the unstable regime: the resulting chaotic pattern is dominated by vortices and antivortices (bottom panel). In time-dependent simulations these are found to constantly annihilate and unbind. Adapted from Fruchart et al., 2021.

to the potential energy landscape. When you kick the ball uphill, it moves perpendicular to the direction of the height gradient along the bottom of the potential, but a kick in the direction along the bottom does not drive the ball up the gradient. This arises because of the non-orthogonality of the eigenmodes of the Jacobian near the exceptional point. At the exceptional point, the ball moves only along the bottom of the potential, irrespective of how it is kicked: this is the onset of the chiral phase.

9.8.4 Exceptional points-induced instabilities

In a spatially extended system, one has to consider gradient terms in the field theory. These depend on the physical system under consideration: for instance, there can be convective terms in nonreciprocal flocking, which would be absent in nonreciprocal magnets. Let us therefore consider the generic equation of motion,

$$\partial_t \begin{pmatrix} \Delta\phi \\ \bar{\phi} \end{pmatrix} = \left[\begin{pmatrix} 0 & 0 \\ 1 & 0 \end{pmatrix} + M\vec{v}_0 \cdot \vec{\nabla} + N\nabla^2 \right] \begin{pmatrix} \Delta\phi \\ \bar{\phi} \end{pmatrix}, \qquad (9.71)$$

where M and N are 2×2 matrices that are not necessarily symmetric. Equation (9.71) is obtained by including gradient terms in equation (9.66) and repeating the analysis, leading to (9.70).

In problem 9.11, you will take a Fourier transform and diagonalize the matrix equation to obtain the complex growth rates $\sigma(k)$. When $\sigma(k) > 0$ with a maximum at finite k, a type II instability is triggered, following the classification scheme introduced in the previous chapter—compare figure 8.14 to the spectrum in figure 9.21 (top panel, right). Equation (9.71) with $M = 0$ describes the soft modes of the Kuramoto-Sivashinsky equation discussed in section 8.6.[36] As the lower panels of figure 9.21 illustrate, in the case of nonreciprocal flocking the non-equilibrium steady state is characterized by continuous unbinding and annihilation of vortices and antivortices in the vector order parameters, reminiscent of active nematic turbulence.

9.9 Applications to biological problems

While the study of active matter was initially stimulated by problems in the life sciences, the field is now increasingly having an impact on these disciplines, as the methods and insights which have been developed throw new light on biological problems. We close this chapter by sketching some examples of the avenues which are opening up. As we shall see, insights from non-equilibrium pattern formation and active matter often go hand in hand here.

9.9.1 Active gels

An active gel is a material composed of a network of crosslinked biopolymer filaments and molecular motors, as illustrated in figure 9.22. These active media are driven out of equilibrium by two distinct phenomena: the action of the molecular motors on the filaments and the spontaneous polymerization and depolymerization of the monomers within each filament. Both behaviors are energy transduction mechanisms which generate mechanical work from chemical energy. The relevant hydrodynamic theory reflects the complexity of the system, but in schematic form it can be rationalized by putting together key equations we already encountered in our treatment of (passive) polymer viscoelasticity, polar and active nematic fluids, and intuitive symmetry-based reasoning.

a. Modeling of active gels

To model active gels at the hydrodynamic level we identify the relevant conserved quantities: the overall mass, the number of monomers, the number of motors and momentum. Thus, the hydrodynamic equations will describe how the densities of these conserved quantities evolve. Since the filaments are on average parallel to each other, we need to include an additional broken-symmetry variable that describes their local orientation. Depending on whether there is nematic or polar order, this will be the director $\hat{n}$ or the polarization vector $\vec{p}$, respectively. The relevant hydrodynamic equations were already written in section 9.5.1 for active nematics.

The momentum equation includes a total stress tensor given by

$$\sigma_{ij} = \sigma_{ij}^{\text{vis}} + \sigma_{ij}^{\text{el}} + \sigma_{ij}^{\text{a}}. \tag{9.72}$$

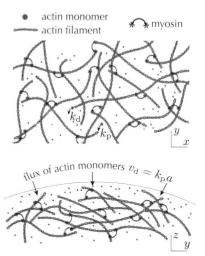

Figure 9.22. Schematic picture of an active gel composed of actin filaments and myosin motors which crosslink these filaments. The polymerization processes with rate constant k_{p} and depolarization processes with rate constant k_{d} are indicated.

The first term $\underline{\sigma}^{\mathrm{vis}}$ on the right-hand side is the viscous stress term of a (viscoelastic) fluid, while $\underline{\sigma}^{\mathrm{el}}$ models, as in a nematic liquid crystal, elastic terms associated with the broken-symmetry variables (the director or polarization vector). Finally, the constitutive relation for the total stress acquires an *active stress* term $\underline{\sigma}^{\mathrm{a}}$ given by

$$\sigma_{ij}^{\mathrm{a}} = \zeta \, Q_{ij} + \bar{\zeta} \, \delta_{ij} \qquad \left(Q_{ij} = n_i n_j - \tfrac{1}{3}\delta_{ij}\right). \qquad (9.73)$$

The equation for the active stress $\underline{\sigma}^{\mathrm{a}}$ looks very similar to the one for active nematics. Furthermore, $Q_{ij} = (n_i n_j - \frac{1}{3}\delta_{ij})$ is the traceless symmetric order parameter tensor of a nematic—compare our discussion in section 6.2 of the isotropic-nematic. There we used essentially the same tensor in equation (6.4), except that we included the strength S of the nematic order in the definition of Q_{ij}.

Here, ζ and $\bar{\zeta}$ are phenomenological coefficients related to the motor and filament densities; they determine the shear and pressure component of the active stress, respectively.

Crosslinkers binding filaments together introduce a new elastic regime at intermediate time scales. We can account for this by writing a dynamical equation for $\sigma_{ij}^{\mathrm{vis}}$ in (9.72) in the spirit of our previous discussion of viscoelasticity. Recall that in section 2.5 we introduced the Maxwell model of viscoelasticity by considering a spring in series with a dashpot. Here, the Maxwell relaxation time τ_{M} corresponds to the time scale over which crosslinkers remain bound. Following our treatment of polymer viscoelasticity in section 5.10.3, we can write a dynamical equation for $\sigma_{ij}^{\mathrm{vis}}$, which in schematic form reads

In this equation, $\mathrm{d}/\mathrm{d}t$ denotes the material time derivative (1.2).

$$\left(1 + \tau_{\mathrm{M}} \frac{\mathrm{d}}{\mathrm{d}t}\right) \sigma_{ij}^{\mathrm{vis}} = \eta \left(\partial_i v_j + \partial_j v_i - \tfrac{2}{3}\partial_k v_k \delta_{ij}\right) + \bar{\eta} \, \partial_k v_k \delta_{ij}. \qquad (9.74)$$

On time scales shorter than τ_{M} the derivative dominates the left-hand side and the behavior is that of an elastic medium with shear modulus η/τ_{M}. This elastic regime breaks down at longer time scales, where the time derivative can be neglected.

The final step is to suitably modify the mass conservation equation to model the continuous process of monomer detachment and polymerization of the filaments, which in turn depend on active stresses and polymer conformation. The conservation equation for total monomer mass can be separated into parts describing the density ρ_{f} of monomers polymerized in filaments and the density ρ_{m} of unbound monomers. While the total mass is conserved, the quantities ρ_{f} and ρ_{m} can vary according to

For a detailed derivation of these equations, see Callan-Jones and Jülicher, 2011.

$$\begin{aligned} \partial_t \rho_{\mathrm{f}} + \vec{\nabla} \cdot \vec{J}^{\mathrm{f}} &= k_{\mathrm{p}} \rho_{\mathrm{m}} - k_{\mathrm{d}} \rho_{\mathrm{f}}, \\ \partial_t \rho_{\mathrm{m}} + \vec{\nabla} \cdot \vec{J}^{\mathrm{m}} &= -k_{\mathrm{p}} \rho_{\mathrm{m}} + k_{\mathrm{d}} \rho_{\mathrm{f}}. \end{aligned} \qquad (9.75)$$

The polymerization and depolymerization rates k_{p}, k_{d} do not obey detailed balance as a result of the different chemical processes involved. The fluxes of polymerized and unbound monomers $\vec{J}^{\mathrm{f}}$, $\vec{J}^{\mathrm{m}}$ are given by phenomenologically determined constitutive relations. For example, the monomers can spontaneously flow due to

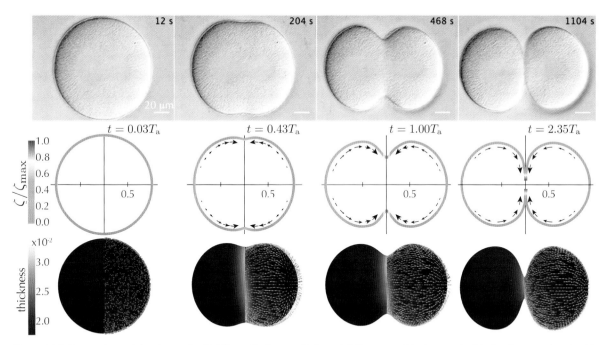

Figure 9.23. Comparison of the shape of a dividing cell of a so-called sand dollar sea urchin (top row) with the shape analyzed with the aid of the active gel theory (middle row) in the work of Turlier et al., 2014. The color code in the calculated shapes indicates the active stress level, as indicated by the color bar on the left, while the arrows indicate gel flows. The characteristic time T_a is the ratio of the viscosity η and the contractility ζ of the gel. The bottom row shows the corresponding shapes and flows in three dimensions (Borja da Rocha et al., 2022). Images courtesy of Hudson Borja da Rocha.[37] For a review of active gel physics, see Prost et al., 2015.

filament distortion caused by motors and in some cases due to local alignment itself. We can obtain equations governing molecular motor density by following the same procedure.

b. Active mechanics of cell division

Active gel theory can describe aspects of cell behavior and motion. As figure 9.23 illustrates, the evolution of the cell during cell division (cytokinesis) can be modeled quite suitably by adapting the above approach to this particular case. During cytokinesis the cell membrane contracts over time until the cell divides into two. This behavior is controlled by a so-called actomyosin cortex which is attached to and pulls on the membrane, and which can be modeled as a thin axisymmetric shell described by active gel physics. The membrane dynamics are driven by the competition between the cortex tension, which causes it to contract, and the cytoplasmic pressure, which attempts to prevent that contraction. The former can be obtained from active gel theory and the latter is assumed to be a uniform hydrostatic pressure. As we shall see in the next section, active hydrodynamic theories are not only successful in describing how a single cell divides, they also shed light on the biomechanics of cell tissues and organ formation.

In problem 9.12, we work out a simple energetic argument for the steady-state radius of the cytokinetic ring.

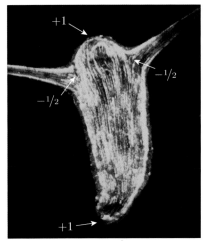

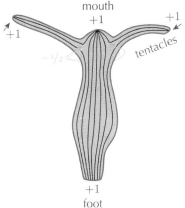

mouth

tentacles

foot

Figure 9.24. The development of a *Hydra* illustrates how the topological defects in the ectoderm fibers in the epithelial layer track the formation of the distinct body parts (the epithelium is a class of tissue forming a continuous, thin protective layer, appearing for example at the outer surfaces of organs). The upper panel shows the nematic actin fiber in a small regenerated *Hydra* specimen.[38] The lower panel shows a schematic with topological defects labeled. The nematic topological defects coincide with the morphological features of *Hydra*. Defects localized at the mouth and foot have charge $+1$. A variable number N of tentacles exhibit one defect with charge $+1$ at the tip and two defects with charge $-1/2$. See Maroudas-Sacks et al., 2021. Image courtesy of Yonit Maroudas-Sa, Kinneret Keren, and Erez Braun.

9.9.2 Active matter effects during morphogenesis

Morphogenesis is the biological process in which a tissue or organ develops its form or shape by controlling the spatial distribution of cells during embryonic development. In section 8.3.2, we discussed Turing's seminal work on pattern formation which was motivated by the desire to model morphogenesis. Turing's original suggestion[39] *'that a system of chemical substances, called morphogens, reacting together and diffusing through a tissue, is adequate to account for the main phenomena of morphogenesis'* was made before the discovery of DNA! This work laid the conceptual basis for realizing how patterns can arise from the simple combination of reaction and diffusion of chemical species; reaction and diffusion are obviously core ingredients of the early development of tissues and organs.

While an embryo is developing, flow, mechano-elastic effects, and changing geometry are important, too. This extends the pattern formation of chemical substances into the realm of physics. Turing's question about morphogenesis prompts us to ask *How did biology learn to control the physics of active matter?* Recent advances in the field of active media create new opportunities and give a road map to tackle this question. We illustrate this with two examples.

a. *Hydra* morphogenesis

Hydra, a genus of small freshwater organisms capable of regenerating damaged tissue, has long captivated the imagination of laymen and scientists alike. Figure 9.24 shows an image of a regenerated *Hydra* during its development. The epithelial layers in the organism contain actin filaments intermingled with myosin molecular motors. The filaments exhibit orientational order characteristic of nematic systems while the myosin motors generate contractile stresses. As a result, it is believed that some aspects of *Hydra* morphogenesis are governed by active nematic hydrodynamics (see section 9.5.1). The body of a *Hydra* is a closed spherical shell, so topological constraints require that the orientation field of the actin filaments exhibit defects with a net charge of $+2$.[40] This statement follows from two mathematical steps. The first relies on the Gauss-Bonnet theorem (see section 7.2) that stipulates that the so-called Euler characteristic of a surface χ is given by

$$\chi = \frac{1}{2\pi} \int d^2S \, K = 2(1-g), \qquad (9.76)$$

where K denotes the Gaussian curvature of the surface and g denotes the genus or number of handles (see figure 7.4). One can readily check the validity of the equation (9.76) for a sphere of radius R for which $K = 1/R^2$ and $g = 0$. The second step relies on the so-called Poincaré-Hopf theorem that stipulates that the sum of the topological charges s_i of the defects labeled by i is equal to the Euler characteristic χ:

$$\chi = \sum_i s_i. \tag{9.77}$$

Surprisingly, the body parts (head, foot, tentacles, etc.) of a *Hydra* develop at locations highly correlated with the presence of nematic defects at early stages. Which body part is formed depends on the topological charge of the defect. The $+1$ defects, which form spontaneously from disordered regions, are typically located at the future site of the organism's mouth, as indicated in figure 9.24. Motile $+1/2$ defects merge pairwise at sites which later become the organism's foot. The locations of $-1/2$ defects are correlated with the development of an additional body axis if they appear early from disordered regions, or with the future sites of tentacle formation if they form later through spontaneous defect unbinding.[41]

In the case of *Hydra*, $\chi = 2$ and equation (9.77) can be expressed in terms of the *Hydra* body parts as

$$\underbrace{+1}_{\text{mouth}} + \underbrace{+1}_{\text{foot}} + \underbrace{N}_{\text{tentacles}} \left(\underbrace{+1}_{\text{tip}} + 2 \times \left(\underbrace{-1/2}_{\text{base}} \right) \right) = +2. \tag{9.78}$$

This is an example of topology in action in biological systems.

b. Fruit fly embryogenesis

Genetic patterning controls the physics of living matter in order to give it shape. Model systems allow us to probe the *dynamics* of this control at the level of cells and tissues. At the same time, they enable us to study the effect of genetic perturbations. The embryo of the fruit fly (*Drosophila melanogaster*) is a model system that provides highly reproducible conditions for the biochemical pathways that underlie morphogenesis.[42] Thanks to its relatively small size, involving only a few thousand cells, the fruit fly embryo is compatible with live-cell visualization techniques, where the system under study is alive. This enables a quantitative analysis of morphodynamic processes.

During embryogenesis, feature development arises through the dynamic rearrangement of mechanically coupled cells in tissue layers. The shape and mechanical resistance of cells is mainly due to the cytoskeleton, a network of interlinking polymers. Active internal stresses are generated in the tissue through myosin motors which pull on the actin filaments in the cytoskeleton. Quantitative analysis of live imaging data showing the cellular flow of embryogenesis can be coarse-grained into a velocity field to reveal the instantaneous motion of the cells (see figure 9.25).

Changes in cell shape are driven by mechanochemical energy transduction. Therefore, the resultant tissue flow is inherently out of

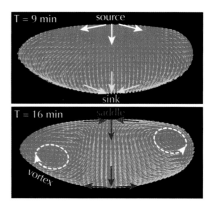

Figure 9.25. Robust features of the myosin flow on the surface of a *Drosophila* embryo (represented as a flattened map) track its shape changes during a developmental process called gastrulation (Streichan et al., 2018). Different time points exhibit different patterns of sources and sinks, with a characteristic vortex pattern that accompanies the formation of the ventral furrow. Image courtesy of Sebastian Streichan.

equilibrium and can be treated in a similar manner to the active continuum theories discussed in previous sections. We first focus on a static (instantaneous) description, where the tissues can be modeled as an active viscoelastic material. Their flow is captured[43] by a compressible Stokes equation driven by forces proportional to the spatial pattern of myosin (motor) concentration,

$$\eta \nabla^2 \vec{v} + \zeta \vec{\nabla}(\vec{\nabla} \cdot \vec{v}) = -\vec{F}^{\text{a}}. \qquad (9.79)$$

Here, η and ζ are effective shear and bulk viscosities while $\vec{F}^{\text{a}}$ is the force modeling the active drive of the system. Quantitatively, the latter is related to the force dipoles which are due to myosin pulling on actin filaments anchored within cells.

Such theoretical approaches have been tested experimentally by simultaneously measuring the tissue flow and imaging the distribution of myosin within the embryo with a technique known as light-sheet microscopy. From this data, a tensor $m_{\alpha\beta} = I_{\text{m}} n_{\alpha} n_{\beta}$ can be constructed that describes the local intensity and anisotropy of the myosin distribution. Here, I_{m} is the bare intensity of the signal, while $\hat{n}$ is the local anisotropy direction. This tensor, in turn, models the active stress due to myosin, and hence $\vec{F}^{\text{a}} = \vec{\nabla} \cdot \underline{m}$.

Despite the molecular complexity of fruit fly morphogenesis, the simple description based on Stokes' equation is sufficient to capture the instantaneous tissue flow with a remarkable degree of accuracy. Key macroscopic features of cellular flow—such as vortices related to the deformation of the embryo—can be quantitatively related to the *instantaneous* distribution and anisotropy of a single protein: myosin (see figure 9.25).

Deciphering the dynamics of embryonic development involves uncovering the subtle interplay between biochemical patterning and active continuum theories. Complex animals such as flies have thousands of protein coding genes. Many different proteins within the embryo serve distinct mechanical functions. For example, cadherin mediates adhesion between individual cells, while the actomyosin cytoskeleton generates active contractile forces. A priori it is not clear that myosin alone should be sufficient to describe the kinematics or that it should affect the long time dynamic behavior of the embryo, as certain genes not related to force generation are crucial to developmental processes.[44] The surprising result that myosin alone is sufficient to describe *any* mechanical behavior in the embryo can be used to develop a road-map to its dynamics.

How can one learn the dynamic rules of morphogenesis? Machine learning methods described in section 10.3 open up an *agnostic* (free of physical assumptions) pathway to understanding the combined action of molecular players on tissue behavior, starting from images revealing their spatiotemporal dynamics.[45] The rationale is as

$\Delta t = 15$ min

■ Exp ■ ML

Figure 9.26. Top: map of a protein known as cadherin distributed on the surface of a fruit fly embryo. Bottom: flattened projection of the cell flow made by mapping the surface of the embryo to a square. As indicated by the color bars at the bottom, the image shows integrated cell trajectories both experimentally measured and predicted by a machine learning model trained on myosin and cadherin maps like the one shown in the top panel. Note the agreement over a period of 15 minutes when gastrulation occurs. Adapted from Colen et al., 2022. Image courtesy of Jonathan Colen.

follows. If a specific protein is predictive of a given observed behavior, then a suitably designed machine learning algorithm should be able to retrieve a relationship between the spatiotemporal pattern of this protein and the associated behavior. In addition, by using machine learning tools on distinct combinations of biological inputs, one can narrow down the list of relevant factors underpinning tissue dynamics. Recent results summarized in figure 9.26 indicate that a combination of myosin motor and cadherin proteins forms a closed set of variables capable of predicting tissue flow dynamics through the formation of the so-called ventral furrow. The ventral furrow is where cells begin to invaginate, leading to a dramatic change in the qualitative structure of the deformation (encoded in the fixed point character of the flow) and its spatial distribution.

9.9.3 Tissue mechanics and vertex models

Morphogenetic processes, like those discussed in section 9.9.2, involve large scale rearrangements of collections of cells in an organism. So far, we have looked at continuum descriptions of the tissue. Ultimately, though, a tissue is composed of cells which are mechanically coupled to and are interacting with their neighbors.

In a vertex model, the tissue is described by a set of vertices at which multiple cells intersect. The boundaries between cells can be represented by edges between vertices and the shape and size of each cell can be inferred from the collection of vertices and edges. A network of vertices and edges is called a graph; see figure 9.27 (top panel). Inspired by models of foams, vertex models have been developed to explain the behavior of the epithelium.[46] In their simplest incarnation, these models consider epithelial monolayers to be networks of cells with total elastic energy E associated with the area A_i and perimeter P_i of each cell:

$$E = \sum_{i=1}^{N} [k_A (A_i - A_0)^2 + k_P (P_i - P_0)^2]. \tag{9.80}$$

Here, the model parameter A_0 specifies the preferred area which represents the internal pressure of the cell, and P_0 specifies the preferred perimeter of the cell, the two representing a competition between contractility (cortical tension) and adhesion at the cell surface. The energetic penalty of deviations from the preferred values A_0 and P_0 can be modeled by two quadratic energy terms with associated elastic constants k_A and k_P.

Active versions of the vertex model can be constructed, for example, by adding a self-propulsion term, which captures the active cell motility:

$$\frac{d\vec{r}_i}{dt} = \mu \vec{\nabla}_i E + v_0 \hat{n}_i + \eta_i, \tag{9.81}$$

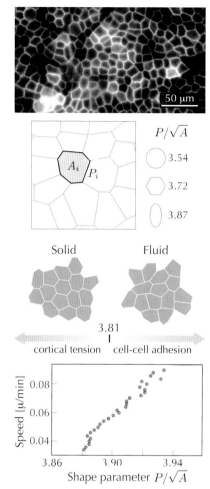

Figure 9.27. Top panel: Madin-Darby canine kidney cells, comprising a model cell line for studying epithelial tissues, form a confluent monolayer. The image shows the fluorescently labeled cell membranes. Adapted from Devany et al., 2021. Second panel: schematic illustration of a tissue as a graph composed of vertices and edges. The area and perimeter of each cell i are denoted by P_i and A_i, respectively. The shape parameter $P/\sqrt{A}$ is shown for three representative values. Third panel: schematic phase diagram where a density-independent rigidity transition from a solid to a fluid-like state is shown at a critical shape parameter value $P/\sqrt{A} \approx 3.81$. Adapted from Bi et al., 2015. Bottom panel: experimental relationship between cell speed averaged over the whole tissue and average shape parameter for the cells in epithelium. Adapted from Devany et al., 2021.

where $\vec{r}_i$ is the center of the cell i, μ is the mobility of the cells, $\hat{n}_i = (\cos\theta_i, \sin\theta_i)$ represents the direction of self-propulsion for cell i at speed v_0, very much as in the case of active Brownian particles, and η_i is a noise term.[47]

Numerical studies of vertex models indicate that there is a rigidity transition between a solid-like and a liquid-like state of the tissue as the shape parameter is raised;[48] see the middle panel of figure 9.27. Unlike the jamming transition we studied in section 2.10 this rigidity transition is density-independent. In cellular tissues, the packing fraction is always close to unity and cells typically have six neighbors on average. Hence, one does not expect the density to change much. Instead, in vertex models the point at which the phase transition occurs depends on the cell geometry parameterized by the average shape $P/\sqrt{A}$, averaged over the whole tissue. As tension increases and active stress decreases, the value of the shape parameter decreases until we reach the point of the rigidity transition, when it attains a value equal to 3.81.

Rheological measurements of embryonic tissues have shown that large changes in tissue fluidity are important in shaping the tissues during development.[49] Across many recent experiments a correlation between cell shape and displacement has been observed. For example, recent work showed such a correlation across a range of experimental conditions—see the bottom panel of figure 9.27. In qualitative agreement with the model, arrested motion occurs when the cell shape parameter is minimal. Note that the displacements measured in the fluid-like state are small (compared to the cell size) at experimental time scales because the active stress itself is very low. Additional rheological measurements as a function of cell shape are needed to probe the character of this rigidity transition.

9.10 What have we learned

In this chapter we have shown several examples of active matter. In each case, energy injection at the microscopic scale induces emergent collective motion at the macroscale. This pattern of behavior appears often in biological contexts such as cell dynamics, bacterial swarms, epithelial tissues, flocking of birds, and schools of fishes. It also arises in synthetic systems such as driven colloids, active metamaterials, and collections of robots.

We can construct effective field theories of these active media by considering what symmetry constraints have been lifted compared to their equilibrium counterparts. In all the materials considered in this chapter, energy is no longer conserved, leading to new terms appearing in constitutive relations or hydrodynamic equations.

9.11 Problems

Relevant coding problems and solutions for this chapter can be found on the book's website www.softmatterbook.online under Chapter 9/Coding problems.

Problem 9.1 *The Vicsek model as a noisy diffusion equation*

In this problem we illustrate how the Vicsek model (9.1) for fixed agents, i.e., $\vec{v}_i = 0$, on a two-dimensional lattice reduces to the noisy diffusion equation (9.4).

a. A sketch of the system is shown in figure 9.28, where we have represented agents as fixed spins on a square lattice. Show that the update rule (9.1) reduces in the case where $a < R < a\sqrt{2}$ to

$$\theta(\vec{r}, t+1) = \frac{1}{5}\left(\theta(\vec{r}, t) + \sum_{\Delta\vec{r}=\pm a\hat{e}_{x,y}} \theta(\vec{r}+\Delta\vec{r}, t)\right) + \eta(\vec{r}, t),$$

$$(9.82)$$

where the sum runs over all four near neighbor lattice directions and the vector $\vec{r}$ marks the lattice points.

b. Make the assumption that $\theta(\vec{r}, t)$ varies smoothly as a function of space and time, and perform an expansion to lowest order in the gradients. Show that this leads to equation (9.4) with a diffusion coefficient $D = a^2/5$.

c. You may wonder why truncating the above expansion to the lowest order in the gradients is justified. Compare with problem 3.4 part d, and argue that the above procedure does give the exact asymptotic long time diffusion result.

d. Consider the case in which $\sqrt{2}a < R < 2a$ and show that this gives the diffusion constant $D = a^2/3$.

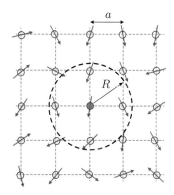

Figure 9.28. The Vicsek model on a square lattice with agents (spins) fixed on the lattice sites. The circle of radius R marks the area with spins that are included in the updating of the central spin, the dark dot. As you can see, only nearest neighbors are included in the updating in the case drawn.

Problem 9.2** *The Dean equation: Coarse-graining the Vicsek model to obtain Toner-Tu equations*

In problem 3.10 we discussed the Dean equation for fluctuating variables. In this problem we show how the method can be used to derive the coarse-grained Toner-Tu equations from the Vicsek model. Let us consider N active particles moving in a plane. Each particle is described by a position r_i and an angle θ_i, with $i = 1, \ldots, N$. The dynamics of the population is described by the set of equations

This problem is an application of the analysis by Dean discussed in problem 3.10.

$$\dot{\vec{r}}_i(t) = v_0\hat{n}[\theta_i(t)], \qquad (9.83)$$

$$\dot{\theta}_i(t) = \eta_i(t) + \sum_{j=1}^{N} J_{ij}\sin[\theta_j(t) - \theta_i(t)], \qquad (9.84)$$

where

$$\hat{n}(\theta) = (\cos(\theta), \sin(\theta))^T \qquad (9.85)$$

gives the local orientation of the agent, and $\eta_i(t)$ are Gaussian white noises with $\langle\eta_i(t)\rangle = 0$ and

$$\langle \eta_i(t)\eta_j(t')\rangle = 2\eta\delta_{ij}\delta(t-t'). \tag{9.86}$$

We set

$$J_{ij} = JH(R_0 - |\vec{r}_i - \vec{r}_j|), \tag{9.87}$$

where H is the Heaviside step function and J a coupling constant.

It will be convenient to write equations (9.83) and (9.84) in the form

$$\dot{\vec{r}}_i = \vec{A}_r(\vec{r}_i, \theta_i) + \sum_{j=1}^{N} \vec{B}_r(\vec{r}_i, \theta_i, \vec{r}_j, \theta_j), \tag{9.88}$$

$$\dot{\theta}_i = A_\theta(\vec{r}_i, \theta_i) + \sum_{j=1}^{N} B_\theta(\vec{r}_i, \theta_i, \vec{r}_j, \theta_j) + \eta_i(t). \tag{9.89}$$

a. First, show that

$$\vec{A}_r(\vec{r}_i, \theta_i) = v_0 \hat{n}(\theta_i), \qquad A_\theta(\vec{r}_i, \theta_i) = 0, \tag{9.90}$$

while

$$\begin{aligned}
\vec{B}_r(\vec{r}_i, \theta_i, \vec{r}_j, \theta_j) &= 0, \\
B_\theta(\vec{r}_i, \theta_i, \vec{r}_j, \theta_j) &= JH(R_0 - |\vec{r}_i - \vec{r}_j|)\sin(\theta_j - \theta_i).
\end{aligned} \tag{9.91}$$

b. Next, we define the stochastic single-particle distributions

$$c(\vec{r}, \theta, t) = \frac{1}{N}\sum_{i=1}^{N} \delta(\vec{r} - \vec{r}_i(t))\delta(\theta - \theta_i(t)). \tag{9.92}$$

See problem 3.10 for a summary of Itô's lemma.

By using Itô's lemma and integrating by parts, show that

$$\begin{aligned}
\frac{\partial}{\partial t}c(\vec{r}, \theta, t) = -\vec{\nabla}_r \cdot &\left[\left(\vec{A}_r(\vec{r}, \theta)\right.\right. \\
&+ \left.\int d^2\vec{r}'d\theta' \vec{B}_r(\vec{r}, \theta, \vec{r}', \theta')c(\vec{r}', \theta', t)\right)c(\vec{r}, \theta, t)\bigg] \\
- \nabla_\theta &\left[\left(A_\theta(\vec{r}, \theta)\right.\right. \\
&+ \left.\int d^2\vec{r}'d\theta' B_\theta(\vec{r}, \theta, \vec{r}', \theta')c(\vec{r}', \theta', t)\right)c(\vec{r}, \theta, t)\bigg] \\
- \nabla_\theta &\left[\sum_i \eta_i(t)c_i(\vec{r}, \theta, t)\right] + \eta\nabla_\theta^2 c(\vec{r}, \theta, t).
\end{aligned} \tag{9.93}$$

We will consider the noise-averaged version of this equation, with a mean-field approximation consisting of simply removing the noise from the equation. As the name indicates, this procedure is an approximation, and it can miss a lot of important physics! When needed, the effect of the noise can be taken into account more carefully, by increasing the complexity of the calculations. You should keep that in mind when applying this technique to new problems. Nevertheless, the mean-field approximation that we use here can be useful to quickly give us a qualitative idea of the behavior of the system.

Note that we get a diffusion term proportional to the noise strength, as happens generally when deriving a Fokker-Planck or Dean equation from a Langevin-type equation.

c. Replace in equation (9.93) the As and Bs with their expressions given in equations (9.90) and (9.91), and show that it reduces to

$$(\partial_t + v_0 \hat{n}(\theta) \cdot \vec{\nabla}_r) c(\vec{r}, \theta, t) = \eta \nabla_\theta^2 c(\vec{r}, \theta, t) - J \nabla_\theta$$

$$\times \left[\int d^2\vec{r}' d\theta' H(R_0 - |\vec{r} - \vec{r}'|) \sin(\theta' - \theta) c(\vec{r}, \theta, t) c(\vec{r}', \theta', t) \right].$$

d. To simplify the analysis, we make a second approximation by replacing $H(R_0 - |\vec{r} - \vec{r}'|)$ with $2\pi R_0^2 \delta(\vec{r} - \vec{r}')$. Show that we then have

$$(\partial_t + v_0 \hat{n}(\theta) \cdot \vec{\nabla}_r) c(\vec{r}, \theta, t) = \eta \nabla_\theta^2 c(\vec{r}, \theta, t)$$

$$- 2\pi R_0^2 J \nabla_\theta \left[\int d\theta' \sin(\theta' - \theta) c(\vec{r}, \theta, t) c(\vec{r}, \theta', t) \right]. \quad (9.94)$$

e. Show that

$$\hat{n}(\theta) \cdot \vec{\nabla}_r = \cos(\theta) \partial_x + \sin(\theta) \partial_y = \frac{1}{2} \left[e^{+i\theta} \partial_z + e^{-i\theta} \partial_{\bar{z}} \right], \quad (9.95)$$

where $\partial_z = \partial_x - i\partial_y$ and $\partial_{\bar{z}} = \partial_x + i\partial_y$.

f. Show that the definition

$$f_n(\vec{r}, t) = \int d\theta \, e^{in\theta} c(\vec{r}, \theta, t) \quad (9.96)$$

of the angular moments of $c(\vec{r}, \theta, t)$ implies that

$$c(\vec{r}, \theta, t) = \frac{1}{2\pi} \sum_n e^{-in\theta} f_n(\vec{r}, t). \quad (9.97)$$

Remember that $f_{-n}(\vec{r}, t) = \overline{f_n(\vec{r}, t)}$ because the original function is real-valued; the overline denotes complex conjugation.

g. Use the expansion (9.97) in the equation (9.94), perform the integration over θ' and simplify the sums using the Dirac deltas, to obtain:

$$\sum_n e^{-in\theta} \partial_t f_n(\vec{r}, t) + \frac{v_0}{2} \sum_n e^{-in\theta} \partial_{\bar{z}} f_{n-1}(\vec{r}, t)$$

$$+ \frac{v_0}{2} \sum_n e^{-in\theta} \partial_z f_{n+1}(\vec{r}, t) = \eta \sum_n (-in)^2 e^{-in\theta} f_n(\vec{r}, t)$$

$$- 2\pi R_0^2 J \left[\frac{1}{2i} \sum_n (-in) e^{-in\theta} \left[f_{n-1}(\vec{r}, t) f_1(\vec{r}, t) \right. \right.$$

$$\left. \left. - f_{n+1}(\vec{r}, t) f_{-1}(\vec{r}, t) \right] \right].$$

$$(9.98)$$

h. Deduce that

$$\partial_t f_n(\vec{r}, t) + \frac{v_0}{2} \left[\partial_{\bar{z}} f_{n-1}(\vec{r}, t) + \partial_z f_{n+1}(\vec{r}, t) \right] = -\eta n^2 f_n(\vec{r}, t)$$

$$+ \pi R_0^2 J n \left[f_{n-1}(\vec{r}, t) f_1(\vec{r}, t) - f_{n+1}(\vec{r}, t) f_{-1}(\vec{r}, t) \right].$$

$$(9.99)$$

Hence, expanding (9.97) in (9.94) finally yields

$$\partial_t f_n + \frac{v_0}{2} \left(\partial_{\bar{z}} f_{n-1} + \partial_z f_{n+1} \right)$$

$$= -n^2 \eta f_n + \pi R_0^2 J n \left[f_{n-1} f_1 - f_{n+1} f_{-1} \right].$$

$$(9.100)$$

i. Write down explicitly the resulting equations for $n = 0, 1, 2$. You will note that $\partial_t f_0$ involves f_1, $\partial_t f_1$ involves f_2, etc. This infinite sequence is called a hierarchy of equations. In order to make progress, one has to truncate the infinite sequence. The way of doing so is called a closure, and choosing an appropriate closure is usually a difficult task that requires a detailed analysis of the equation. We close the hierarchy of moment equations by considering the last equation (giving $\partial_t f_2$) with the assumptions $f_3 = 0$ and $\partial_t f_2 = 0$. Show that this leads to

Our analysis follows the work by Bertin et al., 2006, Bertin et al., 2009, Marchetti et al., 2013, and Patelli et al., 2019.

$$f_2 = \frac{1}{4\eta} \left[-\frac{v_0}{2} (\partial_{\bar{z}} f_1) + 2\pi J R_0^2 f_1 f_1 \right]. \qquad (9.101)$$

j. Use the replacement (9.101) in your equation for $n = 1$ to show that

$$\partial_t f_1 + \frac{v_0}{2} \partial_z f_0 - \frac{(v_0)^2}{16\eta} (\partial_{\bar{z}} \partial_z f_1) + \frac{J R_0^2 v_0}{8\eta} \partial_{\bar{z}} (f_1 f_1) =$$

$$- \eta f_1 + \frac{J R_0^2}{2} f_0 f_1 + \frac{J R_0^2 v_0}{16\eta} \overline{f_1} (\partial_z f_1) - \frac{J^2 R_0^4}{8\eta} f_1 \overline{f_1} f_1.$$

$$(9.102)$$

k. By identifying the density ρ and polarization $\vec{p} = (p_x, p_y)^T$ as

$$f_0 = \rho \qquad \text{and} \qquad f_1 = p_x + i p_y, \qquad (9.103)$$

show that

$$\partial_t \rho + v_0 \vec{\nabla} \cdot (\vec{p}) = 0, \tag{9.104}$$

and that

$$\partial_t \vec{p} + \lambda_1 (\vec{p} \cdot \vec{\nabla}) \vec{p} + \lambda_2 \vec{p} \vec{\nabla} \cdot (\vec{p}) + \lambda_3 \vec{\nabla} (\vec{p}^2) =$$
$$- \left[-\alpha(\rho) + \beta |\vec{p}|^2 \right] \vec{p} - \frac{v_0}{2} \vec{\nabla} \rho + D \nabla^2 \vec{p}, \tag{9.105}$$

where $\lambda_1 = 3\lambda_0$, $\lambda_2 = 5\lambda_0$, $\lambda_3 = -\frac{5}{2}\lambda_0$, and

$$\alpha(\rho) = -\eta j \rho, \qquad \beta = \frac{j^2}{2\eta}, \qquad D = \frac{(v_0)^2}{16\eta},$$
$$\lambda_0 = \frac{j v_0}{8\eta}, \qquad j = \pi J R_0^2. \tag{9.106}$$

Equations (9.104) and (9.105) are the Toner-Tu equations described in section 9.2.3, describing the hydrodynamics of a two-dimensional fluid of active polar particles.

Problem 9.3 *Sound modes in flocks*

In this exercise, you'll derive the dispersion relation of sound waves in a two-dimensional fluid of active polar particles described by the Toner-Tu equations (9.104) and (9.105). We will consider small-amplitude fluctuations in ρ and $\vec{p}$ so that we can linearize the Toner-Tu equations about their homogeneous steady-state solutions $(\rho_0, \vec{p}_0)$ by writing $\rho = \rho_0 + \delta\rho$ and $\vec{p} = \vec{p}_0 + \delta\vec{p}$.

a. We first look for homogeneous steady-state solutions ρ_0 and $\vec{p}_0$ to the Toner-Tu equations. These can be found by setting all derivatives in equations (9.104) and (9.105) to zero. Show that, for some arbitrary steady-state density ρ_0, the system will end up in one of two states, characterized by

$$|\vec{p}_0| = 0 \qquad \text{or} \qquad |\vec{p}_0| = \sqrt{\frac{-\alpha_0}{\beta}}, \tag{9.107}$$

where we define $\alpha_0 \equiv \alpha(\rho_0) = \eta - j\rho_0$. The $|\vec{p}_0| = 0$ solution is known as the *isotropic* state, while $|\vec{p}_0| \neq 0$ corresponds to the *ordered* state.

b. For now, we restrict our attention to the isotropic state for $\alpha < 0$ (we will come back to the ordered state later). Substitute $\rho = \rho_0 + \delta\rho$ and $\vec{p} = \delta\vec{p}$ into the Toner-Tu equations (9.104) and (9.105), linearize the equations by keeping only terms up to $\mathcal{O}(\delta\rho, \delta\vec{p})$, to arrive at

$$\partial_t \delta\rho = -v_0 \vec{\nabla} \cdot \delta\vec{p}, \tag{9.108}$$
$$\partial_t \delta\vec{p} = -\alpha_0 \delta\vec{p} - \frac{v_0}{2} \vec{\nabla} \delta\rho + D\nabla^2 \delta\vec{p}. \tag{9.109}$$

Compare your theoretical prediction with the plots in figure 2 of Geyer et al., 2018. Do they match the experimental result?

i. Substitute the definitions of the various parameters found at the end of problem 9.2 in the above expression for the sound speeds. You should end up with an expression with only four independent parameters: v_0, ρ_0, j, and η. Create a polar plot of $c_\pm(\theta)$ with, for instance, $v_0 = j = 1$, $\rho_0 = 0.75$, and $\eta = 0.5$, visualizing the speed of sound waves as they travel through the ordered state at various angles from $\vec{p}_0$.

Problem 9.4 *** Flocking bands**

Instead of retaining the hydrodynamic field $\vec{v}$, the Toner-Tu equations presented in section 9.2.3 can be written in terms of the field $\vec{W} = \rho\vec{v}$. As $\vec{v}$ and $\vec{W}$ are both vectors sharing the same symmetries, the time evolution of $\vec{W}$ should be quantified by the same types of terms as those for $\vec{v}$. We can therefore write the Toner-Tu equations in terms of ρ and $\vec{W}$ as

$$\partial_t \rho = -\vec{\nabla} \cdot \vec{W}, \tag{9.121}$$

$$\partial_t \vec{W} + \lambda_1 (\vec{W} \cdot \vec{\nabla})\vec{W} = \\ -\vec{\nabla}P + \eta\nabla^2\vec{W} + \alpha\vec{W} - \beta|W|^2\vec{W}. \tag{9.122}$$

The goal of this exercise will be to look for traveling wave solutions of (9.123) and (9.124), defined below—i.e., those of the form $\rho = f(x - ct)$ and $W = g(x - ct)$, where c is the traveling speed. This corresponds physically to states characterized by high density bands of flocking particles all traveling in roughly the same direction. We'll therefore use the one-dimensional version of (9.121) and (9.122) with $P = \zeta\rho$ and $\alpha = \rho - \rho_c$. In this case, $\vec{W}$ becomes a scalar and we obtain

$$\partial_t \rho = -\partial_x W, \tag{9.123}$$

$$\partial_t W + \lambda_1 W \partial_x W = \\ -\zeta\partial_x \rho + \eta\partial_{xx}W + (\rho - \rho_c)W - \beta W^3. \tag{9.124}$$

Note that in (9.123) and (9.124), all parameters $\lambda_1, \zeta, \eta, \rho_c, \beta$ are constants, to be considered independent of ρ and W for this exercise.

a. Inserting $\rho = f(x - ct)$ and $W = h(x - ct)$ into (9.123), show that

$$f(z) = \rho_g + h(z)/c, \tag{9.125}$$

where ρ_g is a constant.

b. Next, substituting $\rho = f(x - ct)$, $W = h(x - ct)$ into (9.124) and using (9.125), show that h verifies

$$\eta h'' + \left(c - \frac{\zeta}{c} - \lambda_1 h\right)h' - (\rho_c - \rho_g)h + \frac{1}{c}h^2 - \beta h^3 = 0, \tag{9.126}$$

where we denote ordinary differentiation with a prime.

c. By substitution of $h(z) = m(1 + \tanh(kz))/2$ into (9.125), show that the following equation must hold:

$$a_0 + a_1 \tanh(kz) + a_2 \tanh(kz)^2 + a_3 \tanh(kz)^3 = 0, \quad (9.127)$$

where a_0, a_1, a_2 and a_3 read

$$a_0 = -\frac{m}{8c}\left(\beta c m^2 - 4c^2 k + 4c(\rho_c - \rho_g) + 2ckm\lambda_1 + 4k\zeta - 2m\right),$$

$$a_1 = -\frac{m}{8c}\left(c\left(3\beta m^2 + 4(\rho_c - \rho_g) + 8\eta k^2 + 2km\lambda_1\right) - 4m\right),$$

$$a_2 = \frac{m}{8c}\left(cm(2k\lambda_1 - 3\beta m) - 4c^2 k + 2(2k\zeta + m)\right),$$

$$a_3 = \frac{m}{8}\left(m(2k\lambda_1 - \beta m) + 8\eta k^2\right). \quad (9.128)$$

d. Verify that the system

$$a_0 = 0, \quad a_1 = 0, \quad a_2 = 0, \quad a_3 = 0$$

has a solution in the form of

$$c^\star = \frac{\sqrt{3\beta\zeta + \lambda_1}}{\sqrt{3\beta}}, \quad \rho_g^\star = \rho_c - \frac{2}{9\beta\zeta + 3\lambda_1},$$

$$m^\star = \frac{2}{\sqrt{3\beta(3\beta\zeta + \lambda_1)}}, \quad k^\star = \frac{\sqrt{8\beta\eta + \lambda_1^2} - \lambda_1}{4\eta\sqrt{3\beta(3\beta\zeta + \lambda_1)}}. \quad (9.129)$$

We have just shown that

$$\rho(x - c^\star t) = \rho_g^\star + \frac{m^\star}{2c^\star}\left[1 + \tanh(k^\star(x - c^\star t))\right], \quad (9.130)$$

$$W(x - c^\star t) = \frac{m^\star}{2}\left[1 + \tanh(k^\star(x - c^\star t))\right] \quad (9.131)$$

is a traveling solution of (9.123) and (9.124). Note that as $k^\star > 0$ and $m^\star > 0$, this solution corresponds to the traveling front going upward. One can check that the system (9.129) has another solution in the form of

$$c^\star = \frac{\sqrt{3\beta\zeta + \lambda_1}}{\sqrt{3\beta}}, \quad \rho_g^\star = \rho_c - \frac{2}{9\beta\zeta + 3\lambda_1},$$

$$m^\star = \frac{2}{\sqrt{3\beta(3\beta\zeta + \lambda_1)}}, \quad \bar{k} = \frac{-\sqrt{8\beta\eta + \lambda_1^2} - \lambda_1}{4\eta\sqrt{3\beta(3\beta\zeta + \lambda_1)}}, \quad (9.132)$$

where $\bar{k}$ is negative; this solution corresponds to the traveling front going downward. Having found traveling solutions for both the upward and downward fronts, we have shown that traveling waves can exist and sustain in the steady states of (9.123) and (9.124). These traveling waves correspond to the propagating flocks observed in simulations.

Problem 9.5* *Continuum theory for active Brownian particles and motility-induced phase separation*

The goal of this problem is to derive the instability criterion for motility-induced phase separation, discussed in section 9.3.2, systematically from the equation describing the probability distribution of a system of active Brownian particles. Let us define $P(\vec{r}, \theta, t)$, the probability of finding an active Brownian particle with spatially varying self-propulsion at position $\vec{r}$, angle θ, and time t. The Fokker-Planck equation for $P(\vec{r}, \theta, t)$ reads (see also equation (9.14))

$$\partial_t P(\vec{r}, \theta, t) = -\hat{n} \cdot \vec{\nabla} \left[v(\vec{r}) P(\vec{r}, \theta, t) \right] + D_\theta \partial_{\theta\theta} P(\vec{r}, \theta, t), \quad (9.133)$$

where $\hat{n}$ is the unit vector with orientation angle θ, $v(\vec{r})$ is the position-dependent self-propulsion, and D_θ is the angular diffusion constant. We define the density $\rho(\vec{r}, t)$ and magnetization $\vec{w}(\vec{r}, t)$ as the first and second moments of the probability distribution: $\rho(\vec{r}, t) = \int d\theta\, P(\vec{r}, \theta)$ and $\vec{w}(\vec{r}, t) = \int d\theta\, \hat{n} P(\vec{r}, \theta)$.

a. Show that integrating (9.133) over θ gives the conservation equation

$$\partial_t \rho(\vec{r}, t) + \vec{\nabla} \cdot \left[v(\vec{r}) \vec{w}(\vec{r}, t) \right] = 0. \quad (9.134)$$

b. Show that multiplying (9.133) by $\hat{n}$ and integrating over θ yields

$$\partial_t \vec{w}(\vec{r}, t) = -D_\theta \vec{w}(\vec{r}, t) - \frac{1}{2} \vec{\nabla} \left[v(\vec{r}) \rho(\vec{r}, t) \right] + \vec{B}, \quad (9.135)$$

where $\vec{B}$ contains terms of higher moments in θ depending on $\int d\theta\, \cos(2\theta) P(\vec{r}, \theta)$ and $\int d\theta\, \sin(2\theta) P(\vec{r}, \theta)$.

c. Neglecting the term $\vec{B}$ in (9.135), show that the steady-state value of $\vec{w}$ reads

$$\vec{w}(\vec{r}, t) = -\frac{1}{2D_\theta} \vec{\nabla} \left[v(\vec{r}) \rho(\vec{r}, t) \right]. \quad (9.136)$$

d. Use (9.136) in (9.134) to show that evolution of the density $\rho(\vec{r}, t)$ is given by

$$\partial_t \rho(\vec{r}, t) = \frac{1}{2D_\theta} \vec{\nabla} \cdot \left[v(\vec{r}) \vec{\nabla} \left(v(\vec{r}) \rho(\vec{r}, t) \right) \right]. \quad (9.137)$$

Taking (9.135) and assuming that the mean propulsion speed depends only on the density, i.e., $v(\vec{r}) = v(\rho(\vec{r}, t))$, verify that the Fokker-Planck equation for the density becomes

$$\partial_t \rho(\vec{r}, t) = \frac{1}{2D_\theta} \vec{\nabla} \cdot \left[v(\rho(\vec{r}, t)) \vec{\nabla} \left(v(\rho(\vec{r}, t)) \rho(\vec{r}, t) \right) \right]. \quad (9.138)$$

e. To perform the linear stability analysis on this last equation, assume that $\rho(\vec{r}, t) = \rho_0 + \delta\rho(\vec{r}, t)$ with $\delta\rho(\vec{r}, t) \ll \rho_0$, and linearize (9.137) around ρ_0 to show that

$$\partial_t \delta\rho(\vec{r}, t) = \frac{v(\rho_0)}{2D_\theta} \left[v(\rho_0) + v'(\rho_0)\rho_0 \right] \nabla^2 \delta\rho(\vec{r}, t). \quad (9.139)$$

f. Finally, transform (9.139) into Fourier space to obtain the motility-induced phase separation instability criterion

$$v(\rho_0) + v'(\rho_0)\rho_0 < 0. \quad (9.140)$$

See section 9.3.2 for further discussion of this criterion.

Problem 9.6 *Spontaneous flow in active nematics*

In this problem we study a simple example of director-driven flow in an active nematic. Figure 9.29 illustrates the cell geometry experiencing a spontaneous flow in the $\hat{y}$ direction. We make the approximation that the nematic director is frozen, that is, fixed by the boundary conditions and not affected by the flow. For simplicity we consider the case of equal elastic constants ($K_1 = K_2 = K_3 = K$). The boundary conditions at the two walls are $\theta(x = 0) = 0$ and $\theta(x = L) = \pi/2$. It is useful to express the nematic director in the following form: $\hat{n} = \hat{x} \cos\theta + \hat{y} \sin\theta$ (the director field is z-independent).

a. Use the analysis of section 6.4 to determine the director profile. As a first step, work out equation (6.19) in the one-constant approximation, and show that this gives

$$-(\theta')^2 \begin{pmatrix} \cos\theta \\ \sin\theta \end{pmatrix} + \theta'' \begin{pmatrix} -\sin\theta \\ \cos\theta \end{pmatrix} = \frac{2\lambda}{K} \begin{pmatrix} \cos\theta \\ \sin\theta \end{pmatrix}, \quad (9.141)$$

where the prime means differentation with respect to x.

b. Eliminate λ/K from the above equations and show that the resulting equation implies that $\theta = kx + c$, where c and k are constants of integration.

c. Determine the integration constants c and k in step b from the boundary conditions.

d. Show with these results for the director that the active nematic force $f_j^a = \alpha\partial_i(n_i n_j)$ (see equation (9.26)) results in a nematic force density

$$\vec{f}^a = \alpha k(-\hat{x}\sin 2kx + \hat{y}\cos 2kx). \quad (9.142)$$

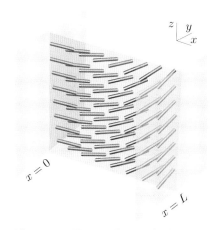

Figure 9.29. Nematic director field ground state and associated two-dimensional flow generated with mixed boundary conditions for the isotropic case ($K_1 = K_3$).

e. Now solve for the velocity field in the non-inertial limit, starting from equation (9.21). Argue that due to incompressibility, the force in the x direction will be balanced by the gradient in pressures, resulting in no flow in the x direction. Moreover, argue that there can be no pressure gradient in the y direction because of the symmetry of the problem in the flow direction.

f. Show with the results of the previous step that the y component of the hydrodynamic equation reduces to $-\eta \nabla^2 v_y = f_y^a$, which reads

$$-\eta(\partial_x^2 + \partial_y^2)v_y = \alpha k \cos 2kx. \tag{9.143}$$

g. Use the fact that v_y cannot depend on y by symmetry, so the above equation gives a simple second order differential equation for $v_y(x)$. Solve the equation to show that

$$\vec{v} = \hat{y}\frac{\alpha L}{2\pi\eta}\left(\cos\frac{\pi x}{L} + 2\frac{x}{L} - 1\right). \tag{9.144}$$

Problem 9.7** *Self-propulsion of disclinations in active nematics*
In this problem, we will first walk through the derivation of the active self-propulsion velocity of a $+1/2$ topological defect in an active nematic. We will then repeat the calculation for a $-1/2$ defect, proving its lack of self-propulsion. Consider placing a single defect with core radius a in the center of a two-dimensional circular domain of radius R. Let $r = \sqrt{x^2 + y^2}$ and $\phi = \arctan(y/x)$ be polar coordinates in the plane.

a. Write an expression for the equilibrium nematic director field $\hat{n}(\vec{r})$ around a $+1/2$ defect whose tail is oriented along the $+\hat{x}$ axis.

b. The ordering of a nematic liquid crystal is most naturally expressed in terms of the tensor order parameter Q, written in two dimensions as

$$Q_{ij} = S\left(n_i n_j - \frac{1}{2}\delta_{ij}\right). \tag{9.145}$$

Assuming the nematic is perfectly ordered outside of the defect core (i.e., setting $S = 1$), use the expression for the director $\hat{n}$ from step a to calculate Q. You should find its two independent components to be

$$Q_{xx}(\vec{r}) = \frac{1}{2}\cos\phi, \qquad Q_{xy}(\vec{r}) = \frac{1}{2}\sin\phi. \tag{9.146}$$

c. The active stress is given by $\sigma^{\text{active}} = \alpha Q$. Show that the corresponding body force on each fluid element is given by

$$\vec{f}^{\,\text{active}}(\vec{r}) = \frac{\alpha}{2r}\hat{x}. \tag{9.147}$$

d. We will approximate the nematic flow as non-inertial, so we can reduce the full nonlinear Navier-Stokes equations to Stokes flow, obeying the equations

$$0 = -\partial_i P + \eta \nabla^2 v_i + f_i,$$
$$0 = \vec{\nabla} \cdot \vec{v}. \tag{9.148}$$

The Stokes equations are linear, allowing us to split up the total body force density into active and elastic components: $\vec{f}(\vec{r}) = \vec{f}^{\text{active}}(\vec{r}) + \vec{f}^{\text{elastic}}(\vec{r}) + \dots$. We can then independently calculate the flow fields due to each of these contributions. We are interested in active self-propulsion, so we will assume the active flow field v^{active} is the primary driver of advection, and we will ignore all other contributions (this is a realistic assumption when $\eta/\gamma \gg 1$). Linearity then allows us to calculate $\vec{v}(\vec{r})$ from $\vec{f}(\vec{r})$ using the method of Green's functions

$$v_i(\vec{r}) = \int \mathrm{d}^2 A' \, G_{ij}(\vec{r} - \vec{r'}) f_j(\vec{r'}). \tag{9.149}$$

G_{ij} is the Green's function for a two-dimensional Stokes flow, also known as the two-dimensional Oseen tensor. By Fourier transforming equations (9.148) and (9.149) and applying the convolution theorem, derive the following expression for the Oseen tensor in reciprocal space:

$$G_{ij}(\vec{q}) = \frac{1}{\eta q^2} \left(\delta_{ij} - \frac{q_i q_j}{q^2} \right). \tag{9.150}$$

e. Transforming back into real space, the two-dimensional Oseen tensor reads

$$G_{ij}(\vec{r}) = \frac{1}{4\pi\eta} \left[\left(\ln \frac{\mathcal{L}}{r} - 1 \right) \delta_{ij} + \frac{r_i r_j}{r^2} \right], \tag{9.151}$$

where $\mathcal{L}$ is an arbitrary length scale which is fixed to determine the flow behavior at the boundaries. Substitute this, along with the result of step c, back into (9.149) to arrive at the following expression for the active flow field.

$$\vec{v}_+^{\text{active}} = \frac{\alpha}{12\eta} \left([3(R - r) + r \cos 2\phi] \, \hat{x} + r \sin 2\phi \hat{y} \right), \tag{9.152}$$

where we have set $\mathcal{L} = R\sqrt{e}$. (Why? Look at your final expression containing $\mathcal{L}$ and think about net flow at the boundaries.) The integrals required to solve this step are a bit tricky; you may find the following identity helpful.

$$\log |\vec{r} - \vec{r'}| = \log r_> - \sum_{n=1}^{\infty} \left(\frac{r_<}{r_>} \right)^n \cos(n(\phi - \phi')), \tag{9.153}$$

where we define $r_> = \max(|\vec{r}|, |\vec{r'}|)$ and $r_< = \min(|\vec{r}|, |\vec{r'}|)$.

f. By evaluating equation (9.152) at the core (i.e., taking $r \to 0$), recover the self-propulsion velocity of the defect. Calculate the vorticity $\omega(\vec{r}) = \hat{z} \cdot (\vec{\nabla} \times \vec{v}(\vec{r}))$, and plot it as a function of the azimuthal angle ϕ, holding r constant. Use this to help you draw a rough sketch of the active flow around the defect core.

g. Repeat the calculation above, now for an $s = -1/2$ topological defect. You should find the following expression for the velocity field:

$$
\vec{v}_-^{\,\text{active}} = \frac{\alpha r}{12\eta R} \left(\left[\left(\frac{3}{4}r - R \right) \cos 2\phi - \frac{R}{5} \cos 4\phi \right] \hat{x} \right.
$$
$$
\left. + \left[\left(\frac{3}{4}r - R \right) \sin 2\phi + \frac{R}{5} \sin 4\phi \right] \hat{y} \right). \tag{9.154}
$$

Problem 9.8 *Odd elasticity: Matrix relations*

Using the basis matrices τ_{ij}^α, we can write the elastic stress and strain tensors as column vectors σ^α, u^α defined by

$$
\sigma_{ij} = \sigma_\alpha (\tau_\alpha)_{ij}, \qquad u_{ij} = u_\alpha (\tau_\alpha)_{ij}. \tag{9.155}
$$

a. Prove the identity (9.41), $(\tau_\alpha)_{ij}(\tau_\beta)_{ij} = 2\delta_{\alpha\beta}$.

b. Show that the stress-strain relation $\sigma_{ij} = K_{ijkl}u_{kl}$ can be written as a matrix equation $\sigma_\alpha = K_{\alpha\beta}u_\beta$ and write the definition of $K_{\alpha\beta}$ that allows this.

Hint: Make use of the orthogonality condition derived in step a.

c. The basis matrices define four modes of deformation or stress that can occur in a two-dimensional solid. We will obtain a pictorial understanding of these modes. Start by drawing an arbitrary shape such as a square or rectangle.

c1. Draw the same shape after a small deformation by each basis matrix.

c2. Draw arrows on the shape corresponding to each stress mode.

c3. Check the consistency of your result with equation (9.44).

Problem 9.9 *The work done by an odd-elastic network during a cycle*

We showed in section 9.6.2 that one can use odd-elastic energy cycles to extract work from an active solid. In this problem we show this explicitly for the microscopic model with force law (9.49).

a. Consider moving the end of the spring through the cyclic path indicated in figure 9.15. Show that the energy needed to compress the spring during the first phase of the cycle is released again during the third phase when the spring extends.

b. Analyze the energy needed to move the spring during the second and fourth phases of the cycle and show that the network during the cycle is given in accord with (9.48) by $W = k^{\circ} \times Area$.

Problem 9.10** *Nonreciprocal phase transitions: Parity-breaking bifurcations and exceptional points*

In this exercise we derive equations (9.67) and (9.68) for the coupled amplitude equations (9.66).

a. Substitute the ansatz $A_i = a_i e^{i\phi_i}$ into equations (9.66). We take a_1 and a_2 to be positive. Ignore the time dependence of the a_i—as explained in section 8.5.2.g, this is allowed since the a_i relax on the fast time scale to the steady-state values consistent with the phases. Show that the real parts of the equations then yield

$$
\begin{aligned}
\epsilon_0 - g a_1^2 + \epsilon_{12} \cos(\Delta\phi) \frac{a_2}{a_1} &= 0, \\
\epsilon_0 - g a_2^2 + \epsilon_{21} \cos(\Delta\phi) \frac{a_1}{a_2} &= 0,
\end{aligned}
\tag{9.156}
$$

where $\Delta\phi = \phi_2 - \phi_1$.

b. These two equations determine a_1 and a_2 as functions of $\Delta\phi$. They cannot be solved analytically in full generality. However, in the limit in which the cross-coefficients are small,

$$
|\epsilon_{12}| \ll \epsilon_0, \qquad |\epsilon_{21}| \ll \epsilon_0,
\tag{9.157}
$$

the equations can be analyzed perturbatively. To do so, first show that it follows from (9.156) that

$$
\frac{a_1}{a_2} = \frac{\left[1 + \left(\frac{\epsilon_{12}}{\epsilon_0}\right)\left(\frac{a_2}{a_1}\right)\cos\Delta\phi\right]^{1/2}}{\left[1 + \left(\frac{\epsilon_{21}}{\epsilon_0}\right)\left(\frac{a_1}{a_2}\right)\cos\Delta\phi\right]^{1/2}}.
\tag{9.158}
$$

c. Show that in the limit (9.157) the ratio a_1/a_2 is close to 1, and that we can therefore expand the square root terms in (9.158) to get

$$
\frac{a_1}{a_2} \approx 1 + \frac{\epsilon_-}{\epsilon_0}\cos\Delta\phi + \cdots,
\tag{9.159}
$$

where, as in section 9.8.2,

$$
\epsilon_{12} = \epsilon_+ + \epsilon_-, \qquad \epsilon_{21} = \epsilon_+ - \epsilon_-.
\tag{9.160}
$$

d. Next we turn to the temporal evolution of the phases. Show that the imaginary part of the equations, obtained by substituting

Note that we are free to take a_1 and a_2 to be positive because we can incorporate a change of sign in a change of the phases by π.

the ansatz $A_i = a_i e^{i\phi_i}$, yields

$$\frac{\partial \phi_1}{\partial t} = \epsilon_{12} \left(\frac{a_2}{a_1} \right) \sin \Delta\phi,$$

$$\frac{\partial \phi_2}{\partial t} = -\epsilon_{21} \left(\frac{a_1}{a_2} \right) \sin \Delta\phi. \tag{9.161}$$

e. Note that both these equations and the equations for the evolution of a_1 and a_2 are periodic in $\Delta\phi$. What does this periodicity reflect?

Hint: Think about what a change of phase by π or 2π would amount to.

f. Show from equations (9.161) that

$$\frac{\partial \bar{\phi}}{\partial t} = \left[\epsilon_{12} \left(\frac{a_2}{a_1} \right) - \epsilon_{21} \left(\frac{a_1}{a_2} \right) \right] \sin \Delta\phi, \tag{9.162}$$

and discuss how this equation shows that when $\Delta\phi \neq 0$ the two coupled patterns move with a constant speed to the left or right.

g. Expand the term between square brackets in (9.158) in the regime (9.157) to show that

$$\frac{\partial \bar{\phi}}{\partial t} \approx 2\epsilon_- \sin \Delta\phi \approx 2\epsilon_- \Delta\phi \qquad (\Delta\phi \ll 1). \tag{9.163}$$

This gives the coefficient γ in equation (9.68).

h. Show from equations (9.161) that

$$\frac{\partial \Delta\phi}{\partial t} = - \left[\epsilon_{12} \left(\frac{a_2}{a_1} \right) + \epsilon_{21} \left(\frac{a_1}{a_2} \right) \right] \sin \Delta\phi. \tag{9.164}$$

Note that the terms between square brackets depend on $\Delta\phi$ through the dependence of the ratio a_1/a_2 on $\Delta\phi$; see equation (9.158). From this equation we can derive the coefficients on the right-hand side of (9.67) by expanding in $\Delta\phi$. We will do this only for the linear coefficient α in the limit (9.157).

i. Expand the term between square brackets in (9.164) for (9.157), to show that

$$\frac{\partial \Delta\phi}{\partial t} = 2 \frac{\epsilon_-^2 - \epsilon_+ \epsilon_0}{\epsilon_0} \Delta\phi + \cdots . \qquad (\Delta\phi \ll 1). \tag{9.165}$$

The prefactor of the $\Delta\phi$ term gives the coefficient α in (9.67). Check that this term can change sign in the regime

$$\epsilon_0 \gg \epsilon_- \gg \epsilon_+. \tag{9.166}$$

j. Perform a linear stability analysis of the equation (9.67) by linearizing about $\Delta\phi = 0$ and an arbitrary value of $\bar{\phi}$. Show that the Jacobian matrix is

$$J = \begin{pmatrix} \alpha & 0 \\ \gamma & 0 \end{pmatrix}. \tag{9.167}$$

k. Determine the eigenvalues and eigenvectors of J as a function of α. What happens to the eigenvectors when $\alpha \to 0$? Show that the matrix cannot be diagonalized when $\alpha = 0$.

Problem 9.11* *Exceptional point-induced instabilities*

In this problem, we analyze pattern formation induced by the presence of a non-diagonalizable matrix in the Jacobian of a spatially extended dynamical system.

For more details on the physical consequences, we refer you to Shraiman, 1986 and Fruchart et al., 2021 as well as the references mentioned in the endnotes in section 9.8.2.

We start with a version of (9.71) that reinstates the parameter α giving the distance to the exceptional point in the uniform system:

$$\partial_t \begin{pmatrix} \Delta\phi \\ \bar{\phi} \end{pmatrix} = \left[\begin{pmatrix} \alpha & 0 \\ 1 & 0 \end{pmatrix} + M\vec{v}_0 \cdot \vec{\nabla} + N\nabla^2 \right] \begin{pmatrix} \Delta\phi \\ \bar{\phi} \end{pmatrix}. \tag{9.168}$$

a. Perform a Fourier transform to obtain the momentum-space equation of motion

$$\partial_t \psi(q) = J(q)\psi(q), \tag{9.169}$$

where $\psi(q) = \begin{pmatrix} \Delta\phi \\ \bar{\phi} \end{pmatrix}$ and where the Jacobian is

$$J(q) = \left[\begin{pmatrix} \alpha & 0 \\ 1 & 0 \end{pmatrix} + iM\vec{v}_0 \cdot \vec{q} - Nq^2 \right]. \tag{9.170}$$

b. We first set $\alpha = 0$ and $N = D\mathbb{1}$, where $\mathbb{1}$ is the identity matrix and where the diffusion coefficient $D > 0$ is positive. Diagonalize the Jacobian $J(q)$ and plot the growth rate of small perturbations, defined as the real part of the eigenvalues. We suggest using a computer, although the calculation can be done using pen and paper. By trial and error, show that when α is small enough, it is possible to find a matrix M such that the system exhibits a type II instability discussed in section 8.6.

c. Repeat step b, still with $\alpha = 0$ but now with $M = 0$ and an arbitrary, nonsymmetric matrix N.

d. What happens when $\alpha \neq 0$?

Problem 9.12** *Cell division as a bifurcation*

In this problem, we consider the process of cell division in the geometry shown in figure 9.23. Rather than solving for the full

This exercise is based on the analysis by Turlier et al., 2014. The geometry shown in figure 9.30 was first proposed by Yoneda and Dan, 1972.

dynamics, we construct a simple energetic argument that gives the steady-state radius of the cytokinetic ring.

We consider a dividing cell with a simplified geometry, seen in figure 9.30. The actomyosin cortex is modeled as a thin spherical shell with radius R_0 and thickness $e \ll R_0$, and with a constant active surface tension N_0^a. A contractile ring is pulled inward by a line tension of strength γ, starting the creation of two daughter cells, modeled as two overlapping spheres with radius R as the contractile ring shrinks to radius $r_f = R \sin(\theta)$. Note that $0 \leq \theta \leq \pi/2$, with $\theta = \pi/2$ corresponding to an undeformed cell and $\theta = 0$ corresponding to a fully divided cell.

a. Assuming that volume is preserved, show that R_0 and R are related by

$$R_0^3 = R^3 F(\theta), \tag{9.171}$$

where $F(\theta) \equiv \dfrac{1}{2} \left(2 + 3\cos(\theta) - \cos^3(\theta) \right)$.

b. We will write the energy in terms of the cell's geometry. Calculate the circumference of the contractile ring, L, and the surface area of the cell, $A = 2R^2 \int_\theta^\pi d\theta' \int_0^{2\pi} d\phi \sin \theta'$, in the geometry shown on the bottom of figure 9.30. The energy is then written as

$$E = L\gamma + A N_0^a. \tag{9.172}$$

c. Non-dimensionalizing the membrane energy using the energy of the undeformed sphere, $E_0 = A_0 N_0^a$, show

$$\varepsilon(\theta, \kappa) \equiv \frac{E(\theta)}{E_0} = \frac{\kappa \sin(\theta)}{F^{1/3}(\theta)} + \frac{1 + \cos(\theta)}{F^{2/3}(\theta)}, \tag{9.173}$$

where the non-dimensional parameter $\kappa = \dfrac{\gamma}{2R_0 N_0^a}$ gives the relative strength of the line tension and surface tension.

d. Now we will try to find the values of $\kappa(\theta)$ that extremize ε. Show that $\partial_\theta \varepsilon = 0$ if

$$\kappa^3 = \frac{\sin^3(2\theta)}{8 + 9\cos(\theta) - \cos(3\theta)} \equiv f(\theta). \tag{9.174}$$

Argue that this condition leads to a *saddle-node bifurcation*, where for $\kappa^3 > \kappa_c^3$ there is no solution, while for $\kappa^3 < \kappa_c^3$ two solutions are found.

Hint: $f(\theta)$ is a unimodal function on $\theta \in \left(0, \frac{\pi}{2}\right)$. Find the intersections of κ^3 some constant with $f(\theta)$.

e. First, let us find a quick and dirty estimate of the critical value κ_c^3. Assume that the denominator in $f(\theta)$ is constant

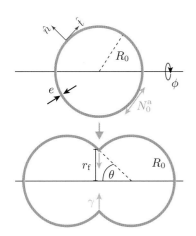

Figure 9.30. Simplified geometry of a dividing cell. A cell is initially considered a sphere with radius R_0 with cortical thickness $e \ll R_0$ and active surface tension N_0^a (top). Two portions of spheres of radius R and active surface tension N_0^a are pinched by an equatorial ring of radius r_f, of width w, and of line tension γ. The opening angle θ characterizes the state of the cell constriction.

and find θ^* that maximizes the numerator of $f(\theta)$. Deduce that $\kappa_c = f^{1/3}(\theta^*) \approx 0.404842 \ldots$.

f. One can find an exact analytical expression of the critical value κ_c but this requires some work. As $f(\theta)$ is a positive unimodal function, $\kappa_c^3 = \max(f(\theta))$, so we only need to find the maximum of $f(\theta)$. Show that, for $0 < \theta \le \pi/2$, $\partial_\theta f = 0$ when

You can use a computer algebra software to avoid tedious manipulations.

$$g(\theta) \equiv 5\cos(\theta) + 4\cos(2\theta) - \cos(3\theta) - 4 = 0. \qquad (9.175)$$

Remark on why this result requires we consider θ strictly larger than zero.

g. Use the following trigonometric identities $\cos(2\theta) = 2\cos^2(\theta) - 1$ and $\cos(3\theta) = 4\cos^3(\theta) - 3\cos(\theta)$ to show that

$$g(\theta) = -4h(\cos(\theta)) \quad \text{with} \quad h(x) = x^3 - 2x^2 - 2x + 2. \qquad (9.176)$$

Now, we are left to find $h(x) = 0$. Obtain the reduced cubic equation by the change of variables, $t = x - \frac{2}{3}$, to find

This can always be done for a cubic equation. Let

$$a_3 x^3 + a_2 x^2 + a_1 x + a_0 = 0.$$

The change of variables $x = t - \dfrac{a_2}{3a_3}$ leads to the equation

$$h(t) = t^3 - \frac{10}{3}t + \frac{2}{27}. \qquad (9.177)$$

$$t^3 + pt + q = 0,$$

where

$$p = \frac{3a_3 a_0 - a_2^2}{3a_3^2},$$

h. Now, use the 16th century result of François Viéte for cubic polynomials of the form $t^3 + pt + q = 0$, with the three real, irrational roots given by

and

$$q = \frac{2a_2^3 - 9a_3 a_2 a_1 + 27a_3^2 a_0}{27a_3^3}.$$

$$t_k = 2\sqrt{\frac{-p}{3}} \cos\left[\frac{1}{3}\arccos\left(\frac{3q}{2p}\sqrt{\frac{-3}{p}}\right) - \frac{2\pi k}{3}\right] \quad \text{for } k = 0, 1, 2. \qquad (9.178)$$

It can be easily seen that $x_k = t_k + 2/3$. Given that $x = \cos(\theta)$, and hence $|x| < 1$, find which of the three solutions gives our x^*. Use this result to find κ_c, and verify that $\kappa_c \approx 0.405325\ldots$

i. Surprisingly, there is a much easier way to get a very close estimate for κ_c. Assume that the denominator in $f(\theta)$ is constant and find θ^* that maximizes the numerator of $f(\theta)$. Then approximate $\kappa_c = f^{1/3}(\theta^*) \approx 0.404842\ldots$

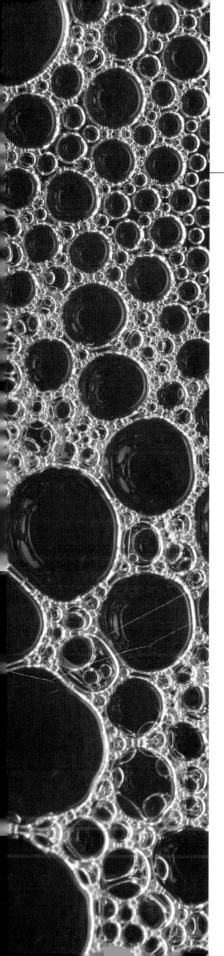

Part IV

PERSPECTIVE

NEW FRONTIERS OF SOFT MATTER

From Designing Matter to Mimicking Life

10

In this book our aim has been to present basic phenomena and concepts of soft matter, together with a selection of interesting and instructive applications. Along this journey, we have also included examples that are a step 'beyond' classical soft matter. Hopefully these examples have intrigued you.

In the last few years, several new directions have emerged at the interface with other disciplines. As the title of this chapter indicates, these developments range from designing forms of nonliving matter with new functionalities all the way to exploring how we can contribute to understanding the origin of life by mimicking some of the essential steps with nonliving matter as well as to exploring life as it could be. Some of these developments have only been made possible by bringing machine learning to bear on them.

In this final chapter we aim to give you a glimpse of these fields and opportunities that are literally exploding as we write this. Our presentation is therefore different in spirit from that in previous chapters, as we feel a textbook introduction cannot be given yet. Nonetheless, we hope to share with you our fascination with what we see happening.

The presentation of each of the topics which we will cover will start with a short introduction of the main aims and ideas, followed by an overview of some of the most important concepts that characterize the developments and that we expect to continue to play a role in the coming years. We then illustrate the topic with a number of representative examples.

We will first address *designer matter*, an emergent, fast developing, and growing field that has design of the next generations of advanced materials with unprecedented functionalities at its core. We then introduce another fast growing field, *memory formation in matter*, by discussing several types of memories that can play a role in soft matter systems. After that we will discuss *artificial intelligence* both through the tools the field is providing for material research and design and as a new paradigm for advanced learning functionalities a material can have. In the end we will turn to *artificial life*, which is in essence a synthetic approach to biology. Artificial intelligence and artificial life both venture into the study of fundamental processes of living systems and complex behavior, and even though there are overlaps, some particular questions, approaches, and methods used by the fields are markedly different. We will explore these fields independently through examples relevant to soft and biomatter, while pointing out important connections.

10.1 Designer matter

10.1.1 What we mean by designer matter

People have been designing and engineering materials for centuries: at different scales from micro to macro, within different fields from chemistry, to physics, to materials science, to biology, to engineering, to architecture. Moreover, in each chapter we have encountered examples of soft matter materials which are designed specifically to create new possibilities or functions. It is therefore useful to reflect on why we set *designer matter* apart as a field or distinct research line. Is there a single distinguishing feature unique to this endeavor?

In our view, designer matter refers to a new amalgam of focus, inspiration, and perspectives made possible by the advancement of materials control, technology, and computing power, together with the blurring of boundaries between disciplines. This change in perspective often allows us to pose our research questions from different angles and develop new concepts and approaches. Moreover, it gives us access to the heretofore unreachable parts of parameter and design spaces, potentially leading to paradigm changes in materials design, synthesis, and discovery.

Think for instance about different processes used to remove dislocations from metals, to grow defect-free semiconductor wafers, to reduce the signal loss in optical fibers, etc.

We used to want materials that were flawless, sturdy, and resilient, for a single or narrow use. The aspiration of designer matter is different. We want, or daydream about, materials that are able to transform, adapt, heal, learn, replicate, or switch between many different functions at our whim. Many of these functionalities are not far-fetched since we can see biological materials that are doing exactly that. In the end, the inspiration often comes from nature, as many living creatures easily adapt to changing circumstances. Even science fiction and art can be sources of inspiration.

With hindsight, one can even see traces of these modern developments anticipated in the visions of Feynman[1] and Lehn.[2] The field of designer matter initially emerged within robotics under the heading *programmable matter*, following the influential 'manifesto, tutorial and call for experiments' by Toffoli and Margolus, 1991.[3] The more modern perspective which underlies this section is attributable in particular to Reis et al., 2015, who promoted the name *designer matter*.

Designer matter emerges at the interface of disciplines, combining concepts, approaches, and techniques, melding them for its use. Because of that, our survey and the concepts we introduce below are not so precise and formalized (yet): they should be viewed mostly as guidelines.

10.1.2 Basic concepts and definitions

It is useful to first collect here a few basic concepts of this developing field, which are defined broadly enough to unify a vast range of systems. Most of the concepts will already be familiar to you from previous chapters or intuition but it is useful to bring them together here and to discuss their role in the context of designer matter.

Figure 10.1. Building blocks of designer matter over scales: from DNA at the nano all the way to the origami sheet at the macro scale.

a. Building block

A building block is the smallest unit composing the material. This can be a particle, a polymer, or a motif or unit cell with some size, shape, composition, and type of interaction, as depicted in figure 10.1. A colloidal particle is an example of a microscopic building block, while a piece of an elastomer (see section 2.9) with two holes of different sizes can be considered as a motif or a unit cell patterning the material.

b. Fabrication method

If building blocks are small enough, one can normally rely on thermal fluctuations to grow the material. The process in which the building blocks spontaneously come together and form materials is called self-assembly. This is a *bottom-up* approach that nature uses to make things. It is also inexpensive and hence of great interest for industry. For building block sizes above a few microns,[4] one usually has to resort to a *top-down* technique to make materials. These encompass *subtractive* manufacturing, such as laser-cutting, and *additive* fabrication techniques like 3D printing (figure 10.2).

c. Composition

Starting our discussion from the right side in figure 10.1, designer materials are usually made by introducing structure in a typical soft matter system, i.e., designer materials have typically *one* composition, as they are made of an elastomer or a particular type of granular particle. As we go down in size, however, at the micro scale we can take advantage of chemistry or building blocks of *many different types*. Why would we want thousands of different types of building blocks to make a material? Nature provides the inspiration: most natural materials are highly functional—they can adapt, repair, replicate, evolve—and this high functionality comes from the *high specificity* of short-range interactions between many different building blocks. In other words, every building block 'knows' where to go in the assembly process, and this feature can be realized in artificial systems at the micro scale as well.

d. Structure or blueprint

Structure is essential, irrespective of the scale of designer materials. The term 'blueprint' is often used for positions of different types

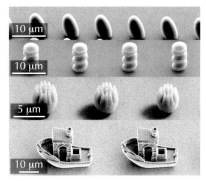

Figure 10.2. Colloidal particles of different shapes made with 3D printing. Images courtesy of Daniela Kraft.

3D printing has given rise to new opportunities and applications in many fields of science. This is a fortiori true for the field of designer materials, as it has made it possible to make forms of matter like metamaterials with heretofore unthinkable structures and shapes and which exhibit truly new properties. Examples of new types of colloids made with 3D printing are shown in figure 10.2. For an introduction to the increased role of 3D printing in physics, see Matthews, 2011, and for reviews of the applications to soft robotic systems, Wallin et al., 2018, to chemistry, Hartings and Ahmed, 2019, and to pharmaceutical and biological systems, Capel et al., 2018.

An illustration of this is DNA nanotechnology: its unprecedented possibilities are based on the combination of the staggering specificity of DNA interactions and the programmability of these interactions (Seeman and Sleiman, 2017).

of building blocks and their interactions in self-assembling structures. In the next section this will be made clear through concrete examples.

e. Designing property

The designing property is the characteristic we are actually designing when making a functional material. Depending on the scale we can essentially distinguish two classes. At the small scale, we design the building block itself—its shape, size, and type of interaction—i.e., an intrinsic property, whereas at the large scale the designing property is often the response of the whole system to external stimuli. The boundary is by no means sharp, as we will see in the examples. Whatever the property is, it defines a set of control knobs that we use to tune the system.

f. Designing approach

The way we actually search for design solutions can be through *inverse, forward, rational*, or *evolutionary* design, or a combination of these. The inverse design approach is based on defining the target property or function of the material and subsequently searching (most often numerically) for the building blocks or sets of interactions, patterns, or folds that yield the desired property. Forward design is the classical way materials are designed. However, in recent years, machine learning methods are used more and more as forward solvers to predict properties of materials.[5] Within the rational design (usually computer aided), the outcome is consistent with the knowledge and preferences of the designer. In evolutionary design one incrementally changes the system to modify or improve the performance, or change a feature.[6]

g. Function or functionality

This concept is hard to define in general, and it depends very much on the system and the observer. The way we think about it is as *the purpose of a thing* or *the relation between a set of inputs and a set of possible outputs.*

h. Memory

Memory has been an emergent organizing principle for studying and designing properties and dynamics of matter. Concepts like training, learning, storing, retrieving, and erasing also appear in relation to materials design. We will introduce in more detail various types of memories in matter in section 10.2.

i. Symmetries and Dualities

Symmetries and their breaking are powerful tools for understanding material properties, and group theory is their universal mathematical description. A symmetry is a transformation that maps

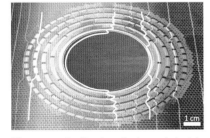

Figure 10.3. Realization of a thermal cloak in the experiments of Schittny et al., 2013. The image shows the patterned disk designed to give thermal cloaking with the measured temperature distribution around the disk in the center superimposed. Isotherms are shown in white, while color indicates the measured temperature. Image courtesy of Martin Wegener.

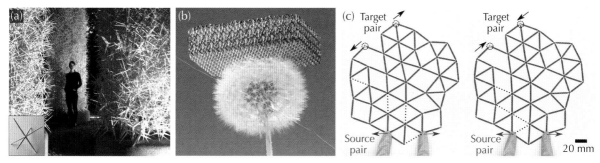

Figure 10.4. Various examples of designer matter. (a) An example of a granular architecture, an aggregate pavilion made of 3D printed hexapod particles (inset bottom left). The particle shape is designed such that the overall structure keeps its shape and can support significant weight. From Dierichs and Menges, 2021.[7] (b) Photograph of an ultra-lightweight microlattice consisting of periodic metallic hollow tubes, as studied by Schaedler et al., 2011. As pointed out by these authors, the Eiffel Tower provides an inspiration: as a structure, its density is similar to that of low-density aerogels, and its structural robustness is beyond doubt. This example also illustrates the importance of hierarchy in structural features for the design. Image courtesy of Tobias Schaedler. (c) Illustration of the fact that the local response of granular media can depend sensitively on a few bonds, even if the response is determined far from the sites used as sources of the tuning. The images show two realizations obtained from the network, identifying the bonds of a two-dimensional packing. From each of them four bonds, indicated with dashed lines, have been removed. When the pair of nodes at the bottom used as *source* are moved apart, the response of the *target pair* of nodes at the upper side is opposite in the two cases. Rocks et al., 2017 discuss several realizations of structures that exhibit such large tunability. Image courtesy of Nidhi Pashine and Sid Nagel.

a system onto itself. A duality, on the other hand, relates distinct models or structures. In self-dual systems the distinction between dualities and symmetries is blurred.

10.1.3 Examples

Let us consider various examples of what we view as designer matter, going from the macro scale to the micro scale.

a. Thermal cloaking

In section 2.9 we introduced metamaterials and mentioned the intruiging property of the optical cloaking à la Harry Potter. Figure 10.3 shows a thermal cloak, where the heat flow is steered around an object as if there is just a homogeneous medium without any obstacle: the isothermal lines to the left and right of the object in the center are straight. The cloak itself is a copper plate that was machined with a prescribed arrangement of holes that were then filled with PDMS, which has a lower heat conductivity compared to copper. The designing property here is the thermal conductivity tensor of the patterned material.

In section 1.6.1 the thermal conductivity constant was introduced for isotropic media. Anisotropy naturally leads to a tensor generalization of the thermal conductivity.

b. Inverse design of granular matter

In granular matter research, the impact of the shape of the particles on the properties of the resulting material is an important research topic.[8] Increasingly, computers are being employed for inverse design aimed at finding the particle shape that leads to the desired material response. As granular materials can be easily

A simple example of the result of such an inverse design approach with the aid of an evolutionary algorithm is shown in figure 2.35 in section 2.10.7, on granular matter.

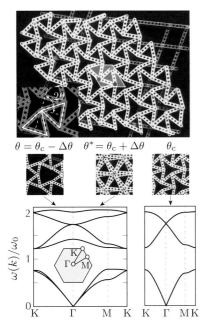

$\theta = \theta_c - \Delta\theta \quad \theta^* = \theta_c + \Delta\theta \quad \theta_c$

Figure 10.5. Twisted Kagome lattices and their band structures. Top: a Lego bricks realization of the twisted Kagome lattice. Inset left bottom: visualization of the twisting angle θ. The white diamond marks the unit cell of the mechanical structure. The three inequivalent points (i.e., not related by Bravais lattice translations) are marked with a red, a green, and a blue dot. Bottom: band structures and realizations of the mechanical structures at different twisting angles.[11]

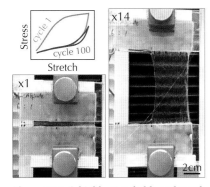

Figure 10.6. A highly stretchable and tough hydrogel designed by Sun et al., 2012. The left picture shows the gel at its original length; the right picture shows this gel stretched to 14 times its original length. The upper left panel illustrates the aging of the stress-strain curves. After a few stretches, the response follows the blue line. Image courtesy of Jeong-Yun Sun.

recycled and reused, and are extremely versatile, applications in architectural design abound. Figure 10.4.a illustrates an architectural tunnel made from hexapod particles optimized to build large structures that can support significant weight.

c. Designing ultra-lightweight materials

Fabricating lightweight[9] materials with a large Young's modulus E_Y opens up a path to many useful applications. Figure 10.4.b illustrates a metallic material with a density less than that of air, made of hollow cylinders. The resulting structure is able to recover from more than 50% compression while it is capable of large energy absorption upon cyclic loading.

d. Designing the local response of disordered networks

The examples we have seen so far discussed design of the global response of the system. But we can also tune the design so as to give a local response to a local perturbation some distance away. This type of behavior is reminiscent of allostery in proteins—a process by which local binding of a ligand, for example, to one site of the protein complex, affects the behavior at a distant site. Disordered elastic networks, for which allostery is not a common behavior, can also be designed to have this response. In figure 10.4.c a two-dimensional realization of such designed networks is shown. It is even possible to design these networks to have more than one response, thus making *multifunctional* allosteric metamaterials.[10]

e. Making use of dualities in design

To illustrate how duality can be used to understand material properties when symmetry fails, we discuss an example of a family of mechanical metamaterials, twisted Kagome lattices, and their corresponding phononic band diagrams shown in figure 10.5. These lattices change shape by means of a collapse mechanism. One can observe that pairs of distinct configurations exhibit the same vibrational spectrum and related elastic moduli, arising from a duality between the pairs with twisting angles $\theta_c \pm \Delta\theta$ (left and bottom panels). The critical point, i.e., the self-dual lattice with twisting angle θ_c (right bottom panel) has isotropic elasticity and a twofold degenerate band structure. These results suggest that dualities and their breaking may play as crucial a role in the design of metamaterials as symmetries do.

f. Endowing designer matter with memory

A whole spectrum of designer matter systems can be endowed with memory, i.e., systems can be trained to store states or memories that can be retrieved at will. We will give a more systematic overview of different types of memories in section 10.2.

g. Smart gels

Hydrogels are polymeric materials that are capable of absorbing large amounts of water. They are often called smart gels, as they can be programmed to respond to various signals from the environment, like changes in temperature, external stress, pH, etc. Applications range from drug delivery and diapers to actuators for optics. A lot of research focuses on the design of the mechanical properties of hydrogels, as most are quite brittle. The example shown in figure 10.6 combines polymers capable of forming ionically and covalently crosslinked networks, resulting in a gel that can be stretched by a factor of 14 without tearing.

h. Reconfigurable structures based on hydrogels

Hydrogels can also be used to make reconfigurable structures. Figure 10.7 illustrates how when a hydrogel layer is sandwiched between two rigid (micropatterned) layers, the temperature-dependent swelling and deswelling of the hydrogel can cause the small origami pattern to fold and unfold reversibly by itself when the temperature is changed over a range of about 30 degrees.

i. Inverse design: Assembly of target structures

Stating a problem in inverse—asking what the parameters or conditions are that give rise to a target or desired observable (be it a structure or a material, property, or function)—and solving it[12] is a way to find parameter solutions that are not obvious, or cannot be observed or envisioned in advance. Although powerful, inverse problems are difficult to solve and usually even ill-posed. Still, this approach has been used extensively in designing matter. A striking example, with inverse design at its core, is that of 'digital alchemy,' where the goal is to find an optimal particle shape for the assembly of target structures by designing entropy. Figure 10.8 shows an example of this approach yielding a new crystal structure.

j. Creating new material properties with nanoparticles

Playing with nanoparticles opens up a whole new vista on the nanoworld and the possibility of creating new materials and properties. We enter here the realm between chemistry and designer matter. We start with the example of adhesives, which are normally made from polymers. Unlike most other materials, polymers provide good contact with surfaces by covering irregularities and preventing fracture by energy dissipation under stress. Polymers, however, cannot simply be used to glue polymer gels together. Interestingly, this impediment can be overcome[13] by using a droplet of a suitable nanoparticle solution; these adsorb onto polymer gels and act as connectors helping to glue hydrogel polymer chains from different samples together. This approach can be used to create biocompatible glues, removing the necessity of sutures when closing wounds in surgery.[14]

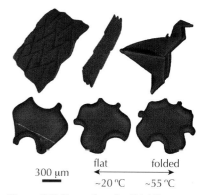

Figure 10.7. Examples of self-folding structures, made by Na et al., 2014 from a hydrogel polymer layer sandwiched between two rigid layers. Thermal actuation causes the hydrogel layer to swell, causing material to fold. Note the small scale and the fact that the material deforms reversibly upon heating and cooling. Image courtesy of Ryan Hayward.

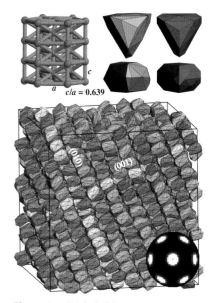

Figure 10.8. Digital alchemy is an approach that treats particle shapes as thermodynamic variables. For example, by sampling more than 10^{11} particle shapes differing only in their convexity, this algorithm and accompanying alchemy Monte Carlo simulations yielded a novel crystal structure that has no known atomic equivalent. In the top, a part of this 'distorted HCP' structure with only eight neighbors is shown. The bottom panel shows the spontaneous self-assembly of a crystal from a single particle type shown above (from different views). Adapted from Geng et al., 2019.

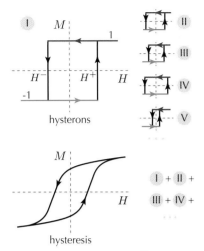

hysterons

hysteresis

$I + II +$
$III + IV +$
$\cdots$

Figure 10.13. In 1935 Preisach[21] proposed a model of hysteresis. The central idea is to think of a material as a network of small domains that have their own individual response to a change in a scalar field. All combined, the distribution of these different hysterons captures the hysteretic response of the material. For an example of hysteretic response encountered in this book, see the response of the metamaterial in figure 2.26, and see figure 5.38 for hysteresis in the transition to viscoelastic turbulence.

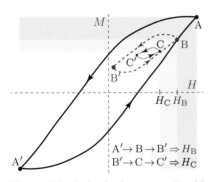

$A' \to B \to B' \Rightarrow H_B$
$B' \to C \to C' \Rightarrow H_C$

Figure 10.14. A sketch of magnetization M of a Preisach ferromagnet model as the magnetic field H is changed. Varying H in the sequence $A' \to B \to B'$ creates a memory at $H = H_B$. As long as $H_{A'} \le H \le H_B$, returning to H_B will restore the system to the same state B every time, regardless of intervening events (like the excursion to B'). The sequence $B' \to C \to C'$ creates a new subloop and encodes a second memory at $H = H_C$. Encoded memories can be read out by sweeping H from H_C to H_A and following the change of the slope dM/dH. Each exit from a subloop (points C and B) erases history with the slope of the curve changing. The readout will have jumps at H_C and H_B, indicating stored memories.

and aperture. The shell interior is functionalized with hepatitis B virus-specific parts.

10.2 Memory formation in matter

Memory is a crucial aspect of our consciousness and we daily build on what we remember. In recent years, however, the concept of memory has been recognized as an organizing principle for studying the properties and dynamics of matter. Memories may be stored in a myriad of different systems, from solids to fluids. They can be encoded in time (spin echoes),[20] in temperature (rejuvenation, aging, and memory in glassy systems), in chemical bonding (colloids with designed interparticle interactions), or in position (sheared suspensions). For each of these forms of memory there are specific training protocols, some systems needing only one training pulse and others requiring repeated cyclic training before a signal can be read out reliably.

Many materials exhibit memory of past conditions by some form of history dependence. We will mention here types of memory where a readout method also exists that recovers the encoded information with some fidelity. The field of *memory formation in matter* is young and evolving.[22] For that reason the categories we single out here, influenced by the content of this book, should not be considered as exhaustive and final.

10.2.1 Types of memories in matter

a. Hysteresis and return-point memory

Consider a system that responds to a scalar field H and which can be in a '+1' or '−1' state. The two values H^+ and H^- are properties of the system that specify when it switches states, with $H^+ > H^-$ for a dissipative system. The system is always in the +1 state when $H > H^+$, and in the −1 state when $H < H^-$, but in between, the state depends on the driving history. Systems of this type are basic elements of hysteresis, termed hysterons.

When they make up a larger system, the hysterons can give rise to a rich behavior called return-point memory, which describes the system's ability to recall a previous state when H is returned to a previous value. To illustrate this we imagine a ferromagnet in which each hysteron represents a magnetic domain that is coupled only to the applied magnetic field H, and there are distributions of H^+ and H^- to represent the material's disorder as in figure 10.13. Return-point memory in magnetization M (the average state of the hysterons) versus H for a model of a ferromagnet is shown in figure 10.14.

As we follow the evolution of M along some trajectory (H varied along arrows), we can identify time intervals during which H is bounded above and below. For example, let us focus on moving along the hysteresis branch from the state A$'$ to the state A. If we stop when reaching the state B and reverse the field (for example, by moving along the red hysteresis sub-loop), this state will last as long as $H'_A \leq H \leq H_B$. Return-point memory means that during this interval, returning to $H = H_B$ will always restore the system to state B. For $H < H_B$ the state of the system depends on the history, which is erased when we return to H_B. Return-point memory also gives us an example of multiple memories. We can see this by applying the memory definition recursively.

Many types of matter can be modeled as collections of subsystems that are individually hysteretic. It should therefore come as no surprise that return-point memory is observed in a wide range of systems beyond ferromagnets.[23] Interactions within such systems can give rise to cooperative and even critical phenomena such as avalanches, so the actual dynamics are usually dramatically different from the ferromagnetic one we described above.

b. Memories from cyclic driving

The memories we discussed above may be 'written' by applying a deformation or changing a field just once. But repeated cyclic driving is also ubiquitous: buildings and bridges are repeatedly loaded and unloaded, temperatures change between day and night, and we practice a skill repeatedly in the hope of learning it. These forms of driving may create memories.

Driving that lasts for multiple cycles may also be used to store multiple values, by varying its parameters (e.g., strain amplitude) from cycle to cycle. In the previous memory class, we encountered one way that a system can remember multiple values of a single variable. By considering the case of cyclic driving, we will be able to talk about these various behaviors using similar language.

When some systems are driven repeatedly, e.g., by shear, electrical pulses, or temperature, they (may) eventually reach a steady state in which the system is left virtually unchanged by further repetitions. This we call a memory of an amplitude. It tends to occur when the amplitude of the driving is large enough to change the material (e.g., by particle rearrangements, as shown in figure 10.15), but too small to disrupt its state completely. The steady state contains a memory of the driving that formed it over many cycles. Often there is also an amplitude beyond which the system cannot reach a steady state.

Memories formed over many cycles of driving are sometimes called self-organized. This refers to the somewhat efficient way that the system evolves its many degrees of freedom to conform to the

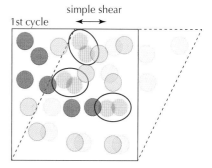

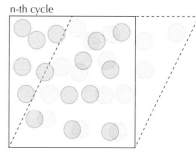

Figure 10.15. Repeated shearing of a particle-based system causes particles to rearrange until a steady state is reached, identified by the lack of particle collisions (the maximum amplitude is kept fixed for all the cycles). The red ovals in the upper panel mark pairs of particles which collide and hence rearrange in the first shear cycle. After a certain number of periodic shears, all particles have found a position where they do not collide with others anymore during a shear cycle. This behavior is common among non-equilibrium systems, including granular matter, crystalline and amorphous solids made of colloids, bubbles, or molecules, colloidal suspensions and gels, liquid crystals, vortices in superconductors, charge-density wave conductors, and even crumpled sheets of plastic.[24]

driving. When the driving is varied from one cycle to the next, some of these systems are known to retain memories of multiple values, but only before the transient self-organization has finished. This behavior, called multiple transient memories, was first seen in charge-density waves when the duration of the pulses was varied, and afterward in suspension systems, examples of which we will discuss in section 10.2.2.

In principle, for large enough systems, we can store arbitrarily many memories in this way. There is no restriction on the order in which we apply these amplitudes; this is in contrast to the case of return-point memories. However, once self-organization is completed for the amplitude interval zero to max amplitude, the system's response is uniform. There is no way to read or write the memory of any intermediate amplitude, and only the memories at zero and max amplitude remain.

Remarkably, this long-term memory loss can be avoided. Adding noise to the charge density wave and suspension systems[25] has the effect of prolonging the transient self-organization indefinitely, preserving the ability to retain multiple memories and allowing the memory contents to evolve as inputs change. This is a concrete example of 'memory plasticity' whereby a system has the ability to continue storing new memories.

c. Memory through path reversal—echoes

A shout across a mountain valley often results in an acoustic echo as the sound is returned to its source a few moments later. The sound waves reflecting from the far side of the valley follow in reverse the path along which they propagated in the first half of their journey. This is as close as one can get to time reversal—the velocities of the wave packets are reversed and they follow the identical path both to and from the valley's far side. The sound waves that return to their source contain a memory of what was shouted, including the timing between different syllables.

Another situation in which this type of behavior shows up is in a class of relaxation phenomena. Consider a system that has a set of modes by which relaxation can take place. Each of these modes has a specified relaxation time (or spectrum of times). The origin of these relaxation times is a detail of the system that is often unclear, though in some cases they are simply the relaxations of a material's subsystems individually. Independently of where the system started, it will approach an equilibrium state via these relaxation modes. The ones with short time constants will equilibrate first and the ones with longer time constants will equilibrate more slowly. It is important to specify that the time constants themselves do not change in response to a perturbation of the system. If the system is perturbed so that it relaxes toward a new state, all the modes will relax. However, if the system is held in this new position only for some waiting time (training time) before it is returned closer to its starting configuration, then only the modes that have relaxation times smaller than this time will contribute appreciably to that relaxation. The modes with longer relaxation times will not have had a chance to evolve much. The memory may be seen in the behavior of the system after the set of two perturbations has been applied. We will illustrate this class of memory through various examples in section 10.2.2.

This effect is named after André Kovacs, who discovered it in amorphous polymers.[26]

d. Associative memory

A memory that we all have firsthand experience with is associative memory. At one point or another you probably had a hard time remembering the name of a person you knew or a place you once visited. Yet if you can conjure up some specific detail associated with it—a scene or an event or even a smell or taste—suddenly the name comes back to you. What makes this particular kind of memory special is that you can use a partial or approximate version of the memory to recall much more.

This experience is by its very nature subjective, and there may be many important biological factors that affect this processing of information in our brains. Nonetheless, one can formulate model systems, in the spirit of minimal models of physics, which also exhibit the phenomena of associative memories. Some of this work has contributed to the study of artificial neural networks with associative memory—systems that have growing applications in technology and scientific research;[28] see section 10.2.2.

A seminal model concerns a mathematical abstraction of neurons themselves, the so-called Hopfield neural network model. Consider a network of identical nodes, each node denoting a neuron i, and each connection ij being assigned a weight J_{ij} as in figure 10.16. Each neuron may be 'on' $(s_i = 1)$ or 'off' $(s_i = -1)$. The connection weights govern the evolution of the system: starting from some initial state, the neurons are updated by turning s_i 'on' if $\sum_j J_{ij} s_j \geq 0$ and 'off' otherwise. A memory is one particular state of the system, $s_{\text{mem}} = \{s_1^{\text{mem}}, s_2^{\text{mem}} \ldots\}$, defined by our choice of which neurons are 'on.' It is an appealing and intuitive idea that you store a memory by making certain connections stronger. Following this notion, we set the weights $J_{ij} = 1$ if $s_i^{\text{mem}} = s_j^{\text{mem}}$ (strengthening the connection between neurons which are both 'on' or 'off' in the memory) and $J_{ij} = -1$ if $s_i^{\text{mem}} \neq s_j^{\text{mem}}$. The model with such weights reproduces the desired associative memory behavior: if you are in a state s that is merely close to the memory s_{mem}, then the evolution will lead to the state $s = s_{\text{mem}}$ which is a stable fixed point unchanged by further evolution. In that sense, the stored memory is successfully retrieved by the system.

Many extensions of the Hopfield model have been studied. As an illustration, the straightforward learning rule in equation (10.1),

Although the basic behavior of the Hopfield model might be unsurprising, consider trying to store three different memories in the same network in such a way that any of them can be retrieved. Suppose the memories of three persons 1, 2, and 3 would be stored in sets of bond strengths, $J_{ij}^{(1)}$, $J_{ij}^{(2)}$, and $J_{ij}^{(3)}$. A straightforward approach would be to take the sum of the weights for each individual memory:

$$J_{ij} = J_{ij}^{(1)} + J_{ij}^{(2)} + J_{ij}^{(3)}. \qquad (10.1)$$

You may expect the system to form three stable fixed points; one for each of the memories. But trouble is right around the corner: spurious stable fixed points also arise. Some are mixtures of the desired states (e.g., a combination of 1 and 2) and others look completely unrelated. So while you set out to write a few memories in the system, you have unwittingly written many more 'false' memories—things that you can now remember but have never experienced! John Hopfield and others showed in the 1980s that although this is true, the idea can still work. Below some threshold number of stored memories, i.e., the capacity of a network, each desired memory is indeed a stable fixed point, while all spurious fixed points have a smaller basin of attraction, i.e., only very nearby states converge to them.[27] Thus, below the threshold (which is of the same order as the number of nodes) multiple memories may be simultaneously stored across many nodes without any unintended consequences.

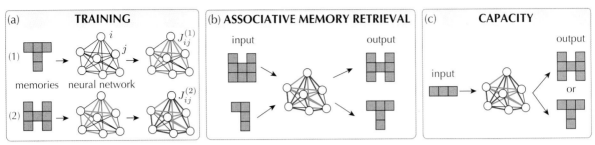

Figure 10.16. (a) Sketch of a Hopfield neural network trained to remember two patterns, the letters T and H. (b) After the training, complete memories can be retrieved if the network is presented with appropriate associations. (c) Neural networks have a capacity, a maximum number of memories that they can store and retrieve without confusion.

called Hebbian learning, may be replaced by various others; further, different dynamics have been studied by allowing simultaneous or sequential updates of neurons; and there are extensive studies of time dependences, stochasticity, and memory correlations.

Although the network model started as an abstraction of neuronal behavior, its mathematical form is reminiscent of magnetic systems in condensed matter physics, starting from the observation that each node could describe a spin state, $s_i = 1$ or -1.[29] This relationship points to the possibility of observing associative memory in a host of settings, given the pervasiveness of physical systems that may be modeled as coupled spins.

10.2.2 Examples

Let us now consider examples of memory formation in matter.

a. Memory formation and hysterons in a corrugated sheet

An interesting example of memory effects in the context of designer matter is the simple corrugated elastic sheet shown in figure 10.17. When we apply compression at the boundaries of the sheet, multiple pathways, i.e., a snapping sequence of different regions, can be observed and controlled experimentally. Each state in the pathway can be encoded by the binary state of material bits, i.e., hysterons, and the strength of their interactions plays a crucial role. If we stop the compression U once the state S_{011} is reached, while $U_{000} \leq U \leq U_{011}$, the system will retain the memory of this state. This model system opens a route to probing, manipulating, and understanding complex pathways in materials, impacting future applications in soft robotics and information processing in materials.

b. Memory effects in sheared suspensions

Memories from cyclic driving can be observed in experiments on non-Brownian suspensions. In figure 10.18 we show an example of a suspension of particles in a liquid when inertia and Brownian

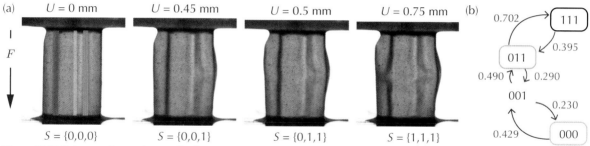

Figure 10.17. Robust pathways in a cyclically compressed corrugated sheet (Bense and van Hecke, 2021). (a) Upon compression, a sample can reach different mesostates, associated with sudden snapping of distinct regions (colored strips in the first image). Each region and its snapping can be represented as a two-state hysteron, being in '0' or '1' state depending on the history. (b) The transition graph of a sample, containing four states (nodes) labeled by the state of each hysteron. Red (blue) arrows correspond to up (down) transitions at (de)compression U as indicated in millimeters. Image courtesy of Hadrien Bense and Martin van Hecke.

motion are negligible.[30] For pure Stokes flow, cyclically shearing such a suspension back and forth will return each particle exactly to its starting point. However, pairs of particles that come too close to each other during shearing may touch and change their trajectories irreversibly.[31] Over many cycles of shearing between zero strain and a constant amplitude, the particles may move into new positions where they no longer disturb each other. When viewed stroboscopically (once per cycle), the system stops changing. But this steady state persists only if the strain stays between the extrema encountered so far (0 and the given amplitude). To read out the memory, we begin with a cycle of smaller strain amplitude which does not change the system, and then we apply cycles with larger and larger amplitudes until a change is observed. Just past the maximal amplitude, many pairs of particles that have been 'swept out' of the training interval promptly come into contact.

As already mentioned, multiple transient memories can be formed in suspension systems, associated with particles moving to positions where they no longer make any contact with their neighbors during a shear cycle. As the particle density is increased, the system may no longer be able to find such positions. Thus, one might expect that a steady state with memory of a shear-cycle amplitude can no longer be formed. However, even in this higher density regime, one can still observe memory formation with a surprisingly similar phenomenology to that of the dilute suspensions, but with crucial differences. For the example of a jammed system that we encountered in section 2.10, under cyclic shear, the systems can still reach a steady state in which subsequent cycles leave the system unchanged.[33] A suitable readout protocol can recover the strain amplitude, but the character of memory buildup is different—in dilute suspensions, the steady-state quasi-static motion was fully reversible, with particles following the same paths in forward and reverse directions in each cycle; in jammed systems the motion is periodic but not reversible, so the particles trace different paths in the forward and return parts of the cycle. Thus, each particle

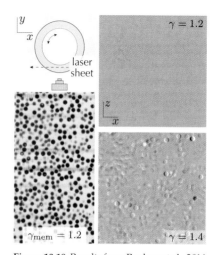

Figure 10.18. Results from Paulsen et al., 2014 on a sheared non-Brownian suspension in a cylindrical Couette cell, as sketched on the upper left. In these experiments the suspensions are sheared cyclically up to a maximum strain γ, and after each cycle the displacements of the suspended particles are compared with those after the previous cycle. The volume fraction is $\phi = 0.35$. The left panel shows part of the sheet which is imaged in the experiments. In the two right panels the images of two successive cycles are subtracted. For $\gamma = 1.2$ (top) the image is evenly gray, implying that the particles have returned to their previous position, so the flow is reversible. For $\gamma = 1.4$ many particles are displaced and the flow is irreversible. With cyclic shearing protocol, suspensions can be 'trained' with 200 strain cycles. Images courtesy of Joseph Paulsen.[32]

traverses a loop so that, at the end of a cycle, it returns to the identical position it had at the start of that cycle.[34] In some cases periodic states are found, where it takes multiple cycles to return to a previous configuration.[35]

c. Memory effects upon flow reversal at low Reynolds numbers

It is well known that a liquid of sufficiently high viscosity, i.e., low Reynolds number, is reversible if the boundaries are distorted and then returned to their original positions along precisely the reverse sequence of motions as those with which they were originally deformed.[36] We illustrated this behavior with figure 1.14. As is sketched in figure 10.19, a blob of dye is introduced into a viscous liquid occupying a narrow gap between two cylindrical walls. When the inner cylinder is rotated, the fluid is sheared and the spot of dye smears out into a sheet. After the first shear, it appears as if the fluid is completely mixed and there is no way of regaining the original conformation. But there is a subtle memory in the fluid, i.e., the blob 'remembers' where it came from because at low Reynolds numbers all fluid elements retrace their paths upon flow reversal. As a result, when the shear is reversed, the spot reappears. This memory, like the spin echo, is surprising but is based on the idea of time (or more precisely, path) reversal. If the system can be manipulated so that after a first set of deformations the dynamics can be made to repeat itself but in the opposite direction, then there is a reversal of dynamics.

d. Kovacs-type memory effects in crumpled sheets

A Kovacs-type memory effect can be observed in crumpled sheets. In this case, training is done by preparing a sheet of plastic by crumpling it many times, and then confining it to a cylinder.[37] The sheet is then compressed into a smaller volume for some training time and then allowed to expand into an intermediate type of volume. Once the sheet expands it begins to exert a force on the container. This force continues to grow for a time comparable to the training time, but then it starts to decay as the slower modes become involved—modes slow enough to effectively ignore the time spent at the smallest volume. The elapsed time until the peak force is proportional to the training time and is a memory.

e. Associative memory effects in self-assembly

Associative memory may arise in the apparently very different context of self-assembly. A simple physical model, shown in figure 10.20, captures the basic phenomenology. The model consists of many different types of building blocks (puzzle-like pieces) interacting through specific interactions. The target structures, i.e., memories, are composed of the same types of building blocks but arranged in a different manner. To encode assembly of a structure, one introduces a strong binding between any pair of particles that

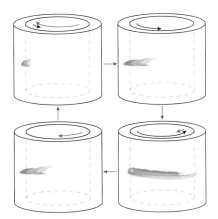

Figure 10.19. Illustration of a path reversal in a low Reynolds number liquid. The invariance of the small Reynolds number fluid dynamics equations under flow reversal is discussed in section 1.11.1 and also illustrated in figure 1.14.

See section 7.4.2 for a discussion of crumpling. Note that figure 7.12 shows an example of different response scaling during the first, second, and third crumples.

A similar behavior has been observed in measurements of the total area of microscopic contact between two solid objects, where a large normal force is applied for some duration and then reduced to a smaller value, thus demonstrating that frictional interfaces exhibit a similar type of memory of their loading history.

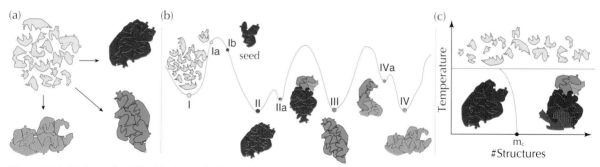

Figure 10.20. (a) A set of building blocks can be designed to have specific interactions. Three different assembled structures (red, green, and yellow) represent three memories. The memories are simultaneously stored by defining attractive interactions between blocks which are neighbors in any of the structures. (b) This self-assembling system exhibits the same phenomenology as the Hopfield neural networks (Murugan et al., 2015). A mixture of free-floating building blocks with designed interactions can be stimulated to assemble a specific structure (memory) if an appropriate seed is put in, or if the concentration of the building blocks that compose the seed is increased, or if their interaction strength is increased. These assembling mixtures also have a capacity, that is, there is a maximal number of structures that can be assembled with the same set of building blocks without confusion. (c) Schematic phase diagram of self-assembly regimes in simulations. Three behaviors (separated by heavy lines) are distinguished by the fate of an introduced seed for a desired structure, i.e., by making a memory association. Below a critical number of stored structures, m_c, and at low enough temperatures (bottom left regime), a seed for the 'red' structure (the irregular red shape shown in (a)) leads to its self-assembly. In other words: the desired red memory is successfully retrieved. Above storage capacity the same seed leads to an erroneous structure (bottom right regime). When the thermal energy sufficiently exceeds the binding energy, all assembly is prevented.[40]

appear as a nearest neighbor pair in that structure. To encode multiple structures, one adds bond strengths in the spirit of the Hebbian learning rule (10.1) in neural networks. One would intuitively think that the possibility of a building block binding any of the several different partners, each corresponding to a desired neighbor in one of the multiple stored structures, would inevitably cause errors. However, it was shown[38] that below some number of stored structures m_c (scaling with a power of the number of building blocks), successful retrieval of any of the memories through introduction of a corresponding associative piece is possible.

10.3 Artificial intelligence

Artificial intelligence (AI) is a term used for any 'thing' that mimics human intelligence, i.e., all the different ways the brain processes information, from visual perception to learning. Over the past decade one of the most important subsets of AI has been *machine learning* (ML), which focuses on allowing machines to learn from data without explicit programming; see figure 10.21.

Both machine learning and artificial intelligence have had a burgeoning role in providing new tools for modeling and characterization of soft materials as well as in creating new paradigms for making 'intelligent' matter. These leaps have been made possible mostly by recent technological advances, as computers can now carry out tasks that before seemed unrealistic.

In one of his visionary lectures, given in 1947 to the London Mathematical Society, Alan Turing, one of the founders of computer science, proposed the idea of digital computing machines that can learn from experience.[39] This lecture is considered to be the beginning of AI and ML. The actual term *artificial intelligence* was coined by John McCarthy in 1956. After the initial boom, the developments in the AI field stagnated, basically waiting for the advancements in computing power and technology of the 1980s. The main boost, though, started in the 1990s, with the use of machine learning algorithms and techniques instead of predominantly rule-based ones, coupled with access to the large amounts of data needed for training.

For reviews on the topic of how machine learning can boost materials research and discovery, see Batra et al., 2020 and Guo et al., 2021. For perspectives on the rise of intelligent matter, see Kaspar et al., 2021.

Figure 10.21. There are many ways to define the field of artificial intelligence, and these definitions evolve with new conceptual and technological advancements. In essence it is the science and engineering of making stuff mimic human intelligence. Machine learning, a subfield of AI, is the study of computer algorithms that allow computer programs to automatically improve through experience.[41] Deep learning is a part of machine learning inspired by the structure and function of the brain, where multilayered neural networks learn from huge amounts of examples.

For a recent review on material learning, see Stern and Murugan, 2023.

T. M. Mitchell, 1997, defines any machine learning problem as the problem of improving some measure of performance P when executing some task T, through some type of training experience E, where performance metric P can, for example, be the accuracy of a classifier. Likewise, the training experience might consist of objects that are labeled or are a mixture of labeled and unlabeled. In any case, once the triplet (T, P, E) is fully specified, we say that the learning problem is defined.

With data coming from experiments and simulations, machine learning algorithms such as deep neural networks can complement standard theories based, for example, on hydrodynamics and statistical mechanics. Such hybrid approaches become increasingly important as the scope of soft matter broadens. For example, the hydrodynamic approaches based on symmetries and conservation laws that we have successfully employed in this book so far can fail altogether when confronted with systems at the interface between biology and active materials. Hence, it is beneficial to augment the very process of constructing mathematical models by using machine learning.

It is not surprising that we are still far from realizations of matter that will be able to take information from the environment, and to store it, learn from it, and adapt its actions and behavior, i.e., to be a form of intelligent matter. Challenges ahead are broad in nature, from discovering design rules for adaptive matter, to developing matter that can learn as well as developing new fabrication methods. This will clearly require an interdisciplinary, concerted effort of different fields, across theories, simulations, and experiments. In this spirit, moving in tandem with machine learning, there is a strong effort underway to create soft matter platforms whose building blocks are themselves endowed with (primitive forms of) artificial intelligence, allowing them to interact with the environment, learn from the inputs, and regulate actions.

In view of these developments, in what follows we provide a glossary of key ideas which can form the basis for either machine learning algorithms or design principles for soft matter endowed with artificial intelligence. In the spirit of this chapter, we follow the organization of previous sections: after an introduction of basic concepts and definitions of different types of learning and neural networks, we provide some illustrative state-of-the-art examples of applications of machine learning to the modeling and characterization of soft materials, as well as some first steps in the direction of intelligent matter.

10.3.1 Types of machine learning

The most relevant classification of learning for this book is into supervised learning, unsupervised learning, and reinforcement learning.

a. Supervised learning

In general, supervised learning refers to a class of problems where one starts from pairs of input examples and output variables (training data), and then uses a model to learn (or extract) the relationship (or mapping) between them.[42] In soft matter research, one typically uses data generated either from simulations with known physics models or directly from experiments, and then trains neural networks that have the capability to distill the underlying physics and achieve robust predictions on these high-dimensional data.

b. Unsupervised learning

In contrast to supervised learning, unsupervised learning refers to a class of problems where one starts from input data only and then uses a model to learn relationships in the data. There is no 'teacher' (output variables) that corrects the model in unsupervised learning; it must learn without any guidance to make sense of it all, to find the underlying organization principles. This type of learning is particularly important when, for example, data classification is not known in advance.

c. Reinforcement learning

Reinforcement learning is a form of machine learning rooted in psychological theory; reinforcement selects actions that lead to favorable outcomes. It is used in many disciplines, from game theory to simulation-based optimization to swarm intelligence. It can be considered neither a type of supervised learning, as the desired outputs are not known in advance, nor a type of unsupervised learning, as there is no need to classify anything and hence the training is not limited. As reinforcement learning is becoming increasingly important in physics research, we will introduce some of its terms and concepts and then go into a bit more detail about a typical reinforcement learning scenario.

> Within AI, the term *intelligent agent* is used for anything that perceives its environment, is capable of taking autonomous actions to be able to achieve goals, and moreover can improve its performance with learning or knowledge (see figure 10.24 for a general scheme). You can think of a human being, but also of a thermostat, as an example of an intelligent agent.
>
> *Environment* is defined as the surroundings of the agent. The agent perceives (takes input from) the environment via sensors and performs actions via effectors. An environment can have many descriptors from the point of view of an agent. For example, the environment can be fully or partially observable (playing Go versus walking in a city); competitive or collaborative (running a marathon versus building a house); dynamic or static (driving a car versus solving a maze); or deterministic or stochastic (playing Go versus walking in a city).
>
> *Policy* defines how the agent behaves at any time. A policy maps the states of the environment to the corresponding actions that are taken. This map can be stochastic, and it can

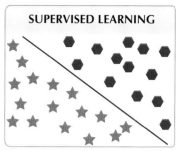

Figure 10.22. Illustration of a result from supervised machine learning: a network is trained by a set of labeled training data to learn the mapping between input and output. This allows it then, for example, to classify new data as orange stars or blue hexagons by 'fitting' the input properties to a predetermined set of criteria or formulas. The line indicates that after learning, the network is able to clearly classify the data into hexagons and stars. Learning algorithms in this class fall into classification, regression, and optimization and control categories.

In 2017, Google's AlphaGo computer program succeeded in defeating a professional Go player, a challenge that for a long time had seemed unattainable due to the game's extremely high complexity. The core of the success lay in incorporation of a reinforcement learning algorithm into the program.

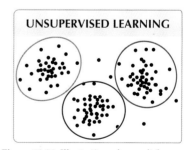

Figure 10.23. Illustration of a result from unsupervised machine learning, learning without a predetermined set of criteria or formulas. Starting from scratch, the program is able to learn relations between the data and cluster them into appropriate groups, as indicated by the colored circles. Learning algorithms in this class fall into clustering and dimensionality reduction categories.

Figure 10.24. A general scheme of an intelligent agent (orange) interacting with its environment (green). Which elements the agent uses to probe the environment and improve its performance, and how it weighs the various ingredients, depends on the particular implementation.

For an accessible introduction to reinforcement learning, see Sutton and Barto, 2020.

for instance have the form of a lookup table or a function.

At each perception step, the environment gives the agent a *reward*. The reward can be low or high depending on the decisions the agent makes before acting. The *reward signal* measures the performance of the agent with respect to the task at hand, i.e., tells the agent what were the good and what were the bad decisions. The central goal of the agent is to maximize the total reward it receives.

What is good for the agent in the long run is specified by the *value*. It accounts for the amount of reward that is likely to follow looking ahead. Values are constantly re-estimated during the lifetime of an agent.

Within the reinforcement learning framework (see figure 10.24), at each iteration step, the agent sees the environment state and decides which action to take. In the following step, the agent gets a reward and is now presented with a new state of the environment. Agents explore the environment by taking different actions, and rewards can be deterministic or random to promote exploration. There are several different methods for action selection and estimation of action values. We will mention here one of the simplest ones, the so-called action-value method.

If an agent knows the expected value for each action, the problem is solved. The agent will maximize its reward by always choosing the action with the highest expected value. During the learning process, the agent operates in an uncertain environment and does not know the reward for a given environment-action pair a priori. However, it can make estimates, for example, by averaging over the past rewards received for each specific action. Given the estimates, at any step, there will be one or several actions with the highest reward estimate. When an agent chooses among these actions, it is said to be *exploiting* the current knowledge to maximize reward and making a *greedy* action selection. If, however, an agent chooses actions that do not necessarily maximize the immediate reward, it is said to be *exploring* the environment. Exploration has benefits in the long run. While one action might be "bad" (low reward), it could open the door to new environments and actions which have very high rewards.

We now proceed with a review of the basic neural network architectures relevant to this chapter.

10.3.2 Neural network architectures

In section 10.2 when we discussed associative memory, we introduced the concept of a neural network as a nonlinear computing

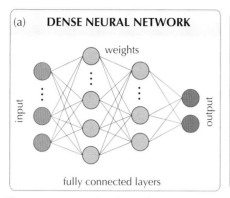
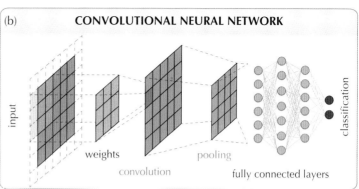

Figure 10.25. Examples of neural network architectures. (a) Dense neural network. When an input vector (blue circles) passes through the network, it is processed by multiple layers of neurons (gray circles) to generate the final predictive output (green circles). Neurons at any two adjacent layers are fully connected. Each connection line contains a weight, which will be optimized to capture the relation between output and input during the training process. (b) Convolutional neural network. It can be used for data that can be represented on a grid. The network employs neural patches (filters or weights) to scan through the input. This not only takes the spatial connectivity of the input data into account but also largely reduces the model's complexity as the weights are shared by each input component. After a convolutional layer, pooling is typically applied, where only some parts are further propagated for classification, the measure often being maximal value.

model inspired by the structure of the human brain. We also illustrated with figure 10.16 an example of a neural network that is composed of only one layer of neurons. A broader family of machine learning approaches, the so-called deep learning, employs more complex network architectures, ones that are composed of multiple interconnected layers of neurons, as depicted in figure 10.25.a. When an input passes through such a network architecture, it is processed sequentially at each neural layer and eventually exits the network as a predicted output. During this computation, the output of each layer is passed as the input to the next layer and then processed with the trainable weights on the interlayer connections to obtain the output of the next layer.

In the following paragraphs we will introduce some common neural network architectures used for deep learning.

Generally, an update rule for a neuron can be written as

$$x_j^{\ell+1} = f(w_{ij}^\ell x_i^\ell + b_j^\ell), \qquad (10.2)$$

where ℓ is the layer index, w denotes the weights between neurons, b denotes a bias parameter to be trained on each neuron, and f is some fixed nonlinear function such as a sigmoid or hyperbolic tangent which enables neural networks to have strong expressive power. We use the Einstein summation convention on the indices i. During training, the weights and biases are adjusted to minimize the difference between the predicted output and the targeted output.

a. Dense neural network (DNN)

In a fully connected or dense neural network like the one shown in figure 10.25.a, pairs of neurons in adjacent layers are connected with a unique weight. A dense neural network with a single sufficiently wide layer between inputs and outputs is enough to approximate arbitrary functions. However, when the number of neurons in a layer is large, the number of trainable weights grows rapidly and becomes intractable. This can occur, for example, when the input layer is a large image like the spatially extended nematic field we saw in figure I.14. In these cases, other architectural choices, such as convolutional or recurrent neural networks, are better suited for solving the problem. In some cases dense neural networks can be *combined* with other architectural choices, such as convolution-based ones, to compress something before the DNN is used, so that the number of weights isn't as high.

b. Convolutional neural network (CNN)

A convolutional neural network is an architectural choice originally designed for image processing. It is commonly used to extract features in pictures that can contain millions of pixels, but it is generally applicable to any input data that can be represented on a regular grid of arbitrary dimension. Very recently, deep convolutional neural networks have been enabled by advances in hardware. Graphics processing units (GPUs) and the newer 'tensor processing units' (TPUs) make CNN training fast, which before was practically impossible (in any meaningful application) using traditional CPUs. An example of a CNN architecture is illustrated in figure 10.25.b.

Unlike simple neural networks, where each input unit interacts with an output unit, convolutional neural networks have a so-called sparse interaction, that is, CNNs use filters (or weights) of smaller size to spatially scan through the values on the grid and learn about different parts of the input. This leads to fewer weights to be trained, and therefore increased efficiency in processing high-dimensional data.

CNN architecture is inspired by the hierarchical way the brain's visual cortex processes information. The visual cortex has a topographical map (representing the visual field) in which nearby cells process information from nearby visual fields.

These weights are also *shared* by each input component and, as a result, fewer parameters are stored, reducing the memory requirement of the model. Sharing weights in this way also enables CNNs to be trained on smaller-sized systems and later to be applied to arbitrarily large systems. This not only avoids the issue of retraining the model for different input sizes, which is typically required for a dense neural network, but also allows us to conveniently generate large amounts of training data using small systems. While a single convolutional layer can collect only local information up to the size of each filter, a network can obtain information from larger regions by using a stack of convolutional layers connected in sequence. By its doing so, each subsequent layer of filters gets a progressively larger field of view on the input. At the end, CNNs use *pooling* (or subsampling), where the most valued parts of the grid according to some measure (e.g., maximal value) are propagated to a fully connected network for final classification.

More formally, translational invariance can be written as

$$f(g(I)) = f(I), \qquad (10.3)$$

where I is an input image, g is an element of the translational group, and f is a feature or a function; and translational equivariance is written as

$$f(g(I)) = g(f(I)), \qquad (10.4)$$

where g is now an element of some transformation group.

Finally, we emphasize here two properties of a convolutional neural network: one property is *translational invariance*, which means that if we translate the input, the network will classify it the same way as before the translation. This is the result of the pooling operation. The second property is *translational equivariance*, which means that the translation of the input features results in an equivalent translation of the outputs. This is the result of the weight sharing operation.

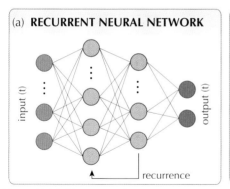
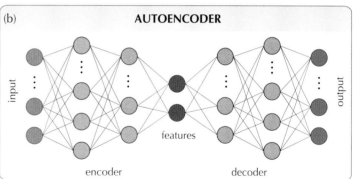

Figure 10.26. More examples of neural network architectures. (a) Recurrent neural network. RNNs are often used for analyzing sequential data. Because of the recurrence link between the inner layers of the network, memory effects can be captured along the input sequence. (b) Autoencoder. This type of architecture is useful when one wants to find low-dimensional representations of the system, or, in other words, to learn meaningful features of the system.

c. Recurrent neural network (RNN)

A recurrent neural network is an architectural choice designed to process sequential data. It exploits time invariance by scanning through an input sequence with the same neural cell, as illustrated in figure 10.26.a. Unlike the CNN filters, which are connected only with adjacent layers, a recurrent neural cell is connected with itself through a so-called recurrence connection. This allows an RNN to preserve useful information collected in the past and capture memory effects that could exist in a system. The prediction made at any point in time is based on both the current input and any information it has kept from the past.

Because the same set of weights is used at each point in time, a recurrent neural network can be used to study complex dynamical problems even with long-term memory effects without requiring an excessive number of weights.

d. Autoencoder architecture

An autoencoder is a neural network commonly used in computer vision to compress images and learn meaningful features by using an unsupervised learning approach. In physics, autoencoders are capable of achieving dimensionality reduction and identifying meaningful degrees of freedom in a system. An autoencoder is composed of two neural network units: an encoder and a decoder, as shown in figure 10.26.b. The encoder, which is typically composed of a sequence of convolutional layers, compresses an input to a set of features, also called a latent vector. The compression is the result of a filter scanning through the data grid with a stride bigger than 1, resulting in a convolution grid that is smaller than the input one. The decoder upsamples those latent vectors and attempts to accurately reconstruct the original input. The upsampling can be done in different ways, by either transpose convolution or inverse

In signal processing, downsampling is the process of reducing the sampling rate of a signal. This is usually done to reduce the data rate or the size of the data. The downsampling factor is usually an integer or a rational fraction greater than unity. Upsampling is the process of inserting 'samples' between the original ones to increase the sampling rate. The concepts are used in a more broad sense here.

Stride is a parameter of a neural network's filter that describes how much the filter moves over an image. For example, in figure 10.25.b, the filter (gray weight patch) has stride = 1 means that the patch shifts by one cell while scanning through the data grid.

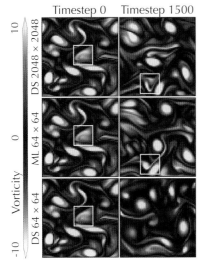

Timestep 0   Timestep 1500

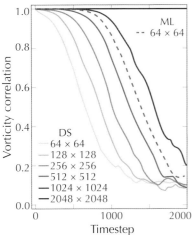

Figure 10.27. Top: evolution of predicted vorticity fields for a 2048×2048 direct simulation (DS), a 64×64 machine learning–based simulation (ML), and baseline 64×64 direct simulation (DS), starting from the same initial velocities. The left column shows the initial vortex field, the right column the final one at time 1,500. The yellow box traces the evolution of a single vortex between the initial and the final time step. Note that the results of the CNN-based ML simulation on a 64×64 grid compare well with the DS on a 2048×2048 grid: learned interpolation based on a CNN reaches the same accuracy as direct simulation, with many fewer grid points. Bottom: comparison of the vorticity correlation between predicted low-resolution flows and the reference high-resolution solution for a model with CNN and standard direct numerical simulation solvers. Adapted from Kochkov et al., 2021.

pooling. As a result, the encoder learns to capture a set of features which are essential for describing the data.

In physics contexts, the autoencoder architecture provides us an automatic way to distill a low-dimensional representation of a system. Here, the encoded latent information can be used to construct collective variables or order parameters. In addition, one can also make predictions or perform efficient dynamic forecasting in a lower-dimensional feature space.

10.3.3 Examples

Let us now consider different examples of soft matter systems that benefit from utilizing machine learning, as well as an example of matter exhibiting intelligence.

a. Computational fluid dynamics

Computational fluid dynamics recently received a boost when machine learning and deep neural networks were used to systematically derive discretizations for continuous physical systems,[43] and it was shown that learned models can surpass the performance of standard high order polynomial approximations.[44]

It is well known that direct numerical simulations of the fluid dynamics equations are prohibitively expensive, as the full Navier-Stokes equations contain many scales in space and in time.[45] Recent advances in hardware (GPUs and TPUs) helped the development of new approaches for simulating the fluid dynamics equations, based on convolutional deep neural networks.[46]

In numerical simulations of turbulence, one can try to start with a low-resolution grid and upscale to a high-resolution grid. This method leads to errors, especially when the flow is turbulent. With machine learning, it was shown that convolutional neural networks can be used to determine a correction term that allows scaling up from low- to high-resolution grids and capture the features of the very expensive high-resolution direct numerical simulations. It relies on statistics of the flows: what neighboring grid cells are doing and what the past is telling us about what the high-resolution result should be. An example is shown in figure 10.27. In the top panel, the three rows correspond to high-resolution direct simulations, low-resolution CNN-based simulations, and to low-resolution direct simulations of the vorticity field. Within each row an initial and a final time step is shown and it can be clearly seen that low-resolution CNN-based simulations impressively capture turbulence features of high-resolution simulations. The bottom panel shows how the CNN result compares to direct simulation results at different resolutions.

b. Pushing the boundaries of soft matter research and engineering

In the last few years, machine learning has pushed the boundaries of soft matter research and engineering, aiding in the design and discovery of next-generation materials. The few reviews on the topic[47] cover illustrative examples, including inverse design of self-assembling materials, nonlinear learning of protein folding landscapes, high-throughput antimicrobial peptide design, data-driven materials design, application of CNN to particle tracking, characterization of ordered arrangements of particles, finding correlations between local structure and susceptibility toward rearrangements of building blocks in glass research, and the design of composite soft materials. Here we single out two topics, one concerning the development of algorithms for inverse design of colloids self-assembling into a target structure, and a second one concerning understanding assembly pathways and their control.

We discussed in section 10.1 an example of how the inverse design can be used to find particle shapes that assemble into target structures. The vastness of the design space in terms of variety of colloidal building blocks and assembly conditions, however, limits this approach. To overcome this limitation, researchers have been using the inverse design in a different manner, i.e., by using machine learning to find the solutions rather than by systematically exploring the parameter space. Generic inverse design methods for efficient reverse-engineering of crystals, quasicrystals, and liquid crystals by targeting their diffraction patterns are already being proposed.[48]

Another very interesting direction in the field combines machine learning with exploration of self-assembly pathways, for the moment mostly focusing on understanding or uncovering the mechanisms of the formation of the critical nucleus leading to the growth of desired crystal structures.[49] Even though this phenomenon has been extensively studied, the rate at which it occurs is difficult to predict. Moreover, there is a huge discrepancy (many orders of magnitude) between what is extracted from experiments and what can be predicted with simulations.[50] The main idea is to train neural networks (architectures depend on the task at hand) to identify local structural motifs that help discriminate between a fluid and a solid phase as well as between different kinetic pathways to homogeneous nucleation of specific crystal structures.[51]

c. Active matter

In chapter 9 we introduced active matter and the challenges hydrodynamic theories face in their description, mostly due to the fact that hydrodynamic parameters are fields themselves. Here is an example of how machine learning can aid in spatiotemporal characterization of an active nematic, even in the absence of knowledge

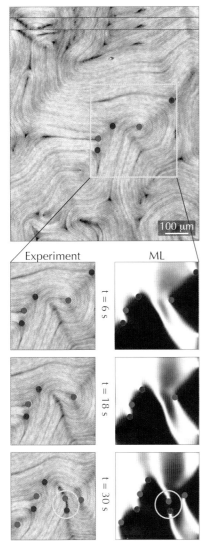

Figure 10.28. A defect nucleation event as seen in experiment and as predicted by the machine learning model trained on microtubule-kinesin experimental data. Machine learning predictions show the evolution of the magnitude of $\sin(2\theta)$, where θ is the angle of the director field. The $+1/2$ and $-1/2$ defects are marked as red and blue dots, respectively. Adapted from Colen et al., 2021.

of the underlying dynamics. Starting from simulated and experimental movies of systems of biofilament/molecular motors with microtubule/kinesin and actin/myosin complexes, a neural network was able to extract how activity and elastic moduli change as a function of time and space as well as adenosine triphosphate (ATP) or motor concentration. With this knowledge, and a combination of autoencoder and recurrent neural network architecture (see figure 10.26), the evolution of active biomatter can be predicted, using only image sequences of their past, as shown in figure 10.28. This general framework is not limited to active nematics, but can be applied to diverse physical and biological systems.

d. Relation between biochemistry and traction forces

Cells exert forces on their surroundings as a result of intricate signaling networks and the interplay of hundreds of different proteins. Finding the relationship between chemical observables like protein concentrations and traction stresses (force per unit surface area of a given orientation) using physics-based approaches is challenging: biological systems, with lots of fluctuations and intricate structure on the measurement length scale, cannot be simply modeled by hydrodynamic approaches to extract order parameters.

Neural networks are extremely flexible in terms of the functions they can approximate, and they are able to directly learn a mapping between proteins and traction stresses. In the example shown in figure 10.29.a, a so-called U-Net has been trained with experimental traction force data to extract features at multiple scales during multiple steps of coarse-graining. The particular architecture of the U-Net allows local information to propagate through the network, thus enabling fine-scale features to be represented in the output. As can be seen in the right panel of figure 10.29.a, as compared to the middle panel, the network is able to predict forces in the cell by learning from the biochemistry encoded in the zyxin protein distribution (left panel).

e. An 'intelligent fluid'

Figure 10.29.b shows an example of an 'intelligent fluid.' This is an adaptive system in which basic constituents learn to coherently move by harnessing non-equilibrium fluctuations in their local environment. The snapshots illustrate the behavior of a minimal computer model of spherical grains that learn how to move solely by adjusting their radii. As they modify their collision cross-section, these intelligent grains learn to accomplish complex tasks, such as spontaneous separation or directed motion, using only a single policy of action. Large-scale simulations reveal that, when poured on a plate, the intelligent grains perform typographic pattern formation, i.e., they sequentially type letters. The required policy, found through reinforcement learning, relies on breaking reflection symmetry in velocity space.

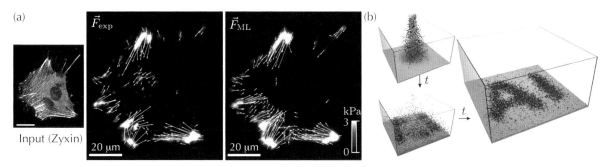

Figure 10.29. (a) Comparison of experimentally determined cell forces and those found with machine learning using a U-Net (Schmitt et al., 2022). U-Net is a type of convolutional neural network, developed for biomedical image segmentation. The fully convolutional neural network we show in figure 10.25.b is supplemented by successive layers, and the pooling operations are replaced by upsampling operators. Image courtesy of Matthew Schmitt. (b) Three successive snapshots from a simulation of intelligent grains which are capable of adjusting their radii depending on their surroundings. (VanSaders and Vitelli, 2023). Image courtesy of Bryan VanSaders.

f. DNA-based network with neuromorphic architecture

As discussed in section 5.4 and section 10.1, over the past two decades DNA has emerged as a versatile building block for designing materials. DNA is, however, also capable of storing and processing information at the molecular scale, which leads to the design of a plethora of DNA-based computational molecular devices.[54]

We mention here the work on a DNA-based network with neuromorphic architecture, which could lead to molecular decision-making devices not unlike gene regulatory networks. The lower panel of figure 10.30 shows a scheme of a molecular realization of a linear classifier, where the artificial neuron takes DNA strands as inputs, and computes their weighted sum. It then takes an ON state (green data in the upper panel) if the weighted sum exceeds a concentration threshold, and remains OFF otherwise (black data). Stated differently, this network mimics a so-called perceptron, an algorithm for supervised learning of binary classification, with a step function as the nonlinear activation function. This network was expanded to take many inputs leading to the realization of a 10-bit majority voting function (one of the central functions essential to many decision rules, the state of the art up to 2021 being three bits), and such perceptrons were further combined to create a multilayered perceptron.[55] This system can be considered an example of intelligent matter, i.e., the system is behaving as a computer.

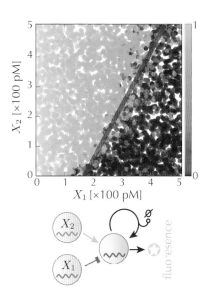

Figure 10.30. Upper panel: An example of an enzymatic neuron in action as a linear classifier.[52] Each data point is a result of a computation done in a droplet containing all the ingredients and fluorescence of the reporter strand is measured after six hours. Note the similarity of the diagram to that of supervised learning shown in figure 10.22. Lower panel: Chemical architecture of the enzymatic neuron: The autocatalytic amplification of the output DNA strand (black circular arrow) is triggered when the weighted activation (green and gray) by input strands overcomes the thresholding mechanism indicated by $\emptyset$.[53] The color indicates the fluorescence intensity, as indicated by the bar in the lower panel. Image courtesy of Guillaume Gines, adapted from Okumura et al., 2022.

10.4 Artificial Life

10.4.1 What do we mean by artificial life?

Biology is a discipline studying life as we know it: life on earth, based on carbon chemistry. Life emerged at some point in time

The name 'artificial life' was coined by the theoretical biologist Christopher Langton (Langton, 1991; Langton, 1995). Historically, John von Neumann can be seen as a fore-runner of the field. In the late 1940s he had already addressed the question of what kind of logical organization is sufficient for an object to reproduce itself. Von Neumann considered a thought experiment in which one starts from a list of unambiguously defined elementary parts that are floating around in a large container. He then imagined an object, i.e., a constructor, going around, whose essential activity is to pick up the parts and put them together into a copy of itself, together with attaching a copy of the instruction it followed, similar to Turing's tape (von Neumann and Burks, 1966). As a proof of concept, von Neumann constructed a two-dimensional lattice of coupled cellular automata,[56] each requiring 29 states in which a finite area of the lattice is able to replicate itself onto an adjacent region.[57] The so-called universal constructor together with the instruction tape supports open-ended evolution, i.e., growth of complexity reminiscent of biological organisms. Even though questions like 'What are the causes and conditions of life? Can we make living creatures?' have been asked throughout history, it is widely accepted that von Neumann's work on self-replication was a seminal step toward capturing the essence of artificial life (Aguilar et al., 2014).

under given conditions, and flora and fauna that are around us are the result of millennia of evolution through natural selection. They are by no means the only possible 'solutions,' leading to one of the fundamental obstacles in theoretical biology: how can one establish general theories of life when only one instance of it is available to us? Driven by this question, the field of *artificial life*, or *ALife*, emerged in the 1980s. As a field of study *artificial life* is in essence a synthetic approach to biology made possible by the rapid expansion and availability of computer power and by the concomitant emergence of subfields in (bio)chemisty such as supramolecular chemistry and DNA nanotechnology, which we encountered before. These developments have enabled a bottom-up approach to exploring life as it could be.

ALife research spans a dozen or so themes, focusing on properties of living systems, life at different scales, and our understanding, uses, and descriptions we have of the living.[58] Like the other fields discussed in this chapter, ALife emerges at the interface of disciplines, covering a wide range of literature and developments that go beyond this book. We will focus here on examples from softmatter and biomatter, and we start by introducing some concepts that play a role in addition to those mentioned in previous sections of this chapter.

10.4.2 Additional concepts and glossary

a. Self-organization

Self-organization is used to define processes in which a system displays global (ordered, spatiotemporal) patterns originating solely from the local interactions among the constituents of the system. You can think of the swarms or flocks, examples of which we encountered in the active matter chapter, or the Belousov-Zhabotinsky reaction system illustrated in figure 8.35. One can observe processes like these in both living and artificial systems, establishing self-organization as a basic concept in a range of disciplines from physics and biology all the way to engineering.

b. Self-assembly

Self-assembly can be viewed as a form of self-organization. It is a process whereby preexisting components come together and form patterns or structures without external intervention.[59] You encountered this bottom-up process throughout the book. It is especially important in section 10.1.

c. Self-replication

A special case of self-organization is self-replication, a process where an object, a replicator, acts as a template and organizes preexisting components into a copy of itself. Once separated from the 'parent,' each 'daughter' acts as a replicator.[60] This process is ubiquitous in nature (think of DNA replication) and realizing it in artificial systems has been a dream for many decades. Some first steps have been made over the last few decades, which we will discuss below through concrete examples.

d. Catalysis

In all living organisms, most molecular processes are controlled by catalysis, the *acceleration* of chemical reactions by molecules that are *not consumed* in the reaction. Nature's catalysts are enzymes, highly efficient machines that can enhance reaction rates by many orders of magnitude.[61]

e. Autocatalysis

Unlike in the process of self-replication, where *the reaction product* can catalyze *only its own formation* (highly specific), in the process of autocatalysis *one of the reaction products* is a catalyst for *the same or a coupled reaction.* Moreover, a set of reactions can be collectively autocatalytic if it is self-sustained, i.e., the products of some of the reactions are catalysts for other reactions in the set (nonspecific), so the entire catalytic reaction network is self-sustained.

Autocatalysis occurs in the Turing model discussed in section 8.3.2.

f. Adaptation

Another central process in all living organisms is adaptation. Very roughly it can be defined as a change or an adjustment in an agent or system to the environment in order to improve its chances of surviving in that environment, or of realizing its goal. Adaptation can be observed on different time scales. We refer to the slowest time scale as *evolution*, the medium scale as *development*, and the fastest scale as *learning*.

We mention here in passing one of the impacts ALife had on AI. Traditionally, AI was all about control and prediction. One of the big pushes for bringing more concepts from ALife to AI goes back to the 1990s,[62] when adaptability as a desirable property was more seriously considered.[63]

g. Evolution

We are all familiar with biological evolution, which refers to changes in heritable characteristics of populations over many generations. However, this process is heavily used in artificial systems as well. One of the most successful methods used in protein engineering is directed evolution. In other fields of research, evolutionary processes have been used to efficiently explore vast phase spaces, which is challenging to achieve with traditional methods. In section 10.1 we mentioned evolutionary design as one of the main design approaches.

The development of directed evolution methods to develop enzymes was recognized by the awarding of the 2018 Nobel Prize in Chemistry to Frances Arnold.[64]

Computationally, artificial evolution can be implemented with an evolutionary algorithm, like the genetic algorithm, to try to solve an optimization problem. Very roughly, this iterative process starts with a population of randomly generated candidate solutions, called a generation, with each candidate characterized by a set of properties that can be subject to change (mutation). At each

See figure 2.35 for the use of such an evolutionary algorithm in granular media research.

iteration, the fitness of each individual candidate (the value of the objective function of the optimization problem) is evaluated. Among the more fit, a subset of candidates is randomly selected and their properties are modified to form a new generation that serves as a starting point for the next iteration.

h. Learning

One can think of learning as a change in behavior of an agent or a system, brought by experience.[65] In section 10.3 we encountered artificial neural networks as an approach to learning that mimics some aspects and functionality of biological neural networks, as well as reinforcement learning, where learning occurs through interactions with the environment.

i. Behavior

Note that in figure 10.24, which sketches a general scheme of an intelligent agent, what is happening within the agent is what within the AI field is sometimes called behavior, which can be misleading.[66]

Behavior of an agent (or a system) is a result of an interaction between the agent and the environment that can be observed, and that cannot be explained from some internal mechanisms of an agent only. The complexity of a behavior is not necessarily correlated with the complexity of internal mechanisms.

j. Artificial chemistries

A nice introductory book on the topic is Banzhaf and Yamamoto, 2015, and an in-depth review published in the journal *Artificial Life* is by Dittrich et al., 2001.

In the spirit of the definition of ALife, artificial chemistries study and explore 'chemical' reactions as they could be imagined. In an artificial chemistry, one starts with reactants, i.e., atoms, that follow certain interaction rules we prescribe. This leads to a combinatorial space of possible structures, i.e., molecules, that goes beyond what we are familiar with in the chemistry we live in.

The next step is to define how the molecules in the artificial chemistry interact (this can be whatever one wishes), leading to the dynamics in a combinatorial space that is usually vast. Even though mostly computational, with the development of the field of nanotechnology, experimental realizations of artificial chemistries are no longer so far-fetched.

k. Emergence

The term 'emergent phenomenon' is sometimes used more loosely in physics for the emergence of ordered phases and their collective modes from a description in terms of constituent particles. See the last paragraph of section 1.1 and references therein.

A property of a system is said to be emergent when it comes from interactions of the building blocks of the system, but cannot be explained or predicted from a simple building block. The whole is more than the sum of its parts. For complex behavior to emerge the constituents do not have to have complex interactions. The converse statement is also true: simple behavior can emerge from complex interactions.

l. Origin of life

This is an area of research focused on uncovering how life on earth started. The two major approaches follow the 'metabolism-first'

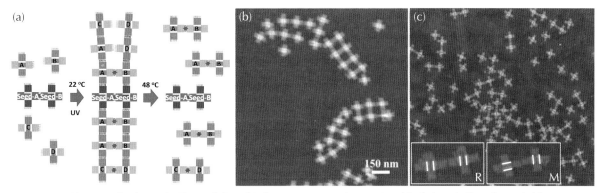

Figure 10.31. (a) Self-replication cycle. The seed dimer binds two monomers (orange A and B tiles) vertically to form a ladder structure. After exposure to UV light, the two orange tiles are chemically crosslinked, making the bond irreversible at the lower temperatures used in the experiment. After the system is heated to $48°C$, the seed dimer and the orange dimers separate. The crosslinked dimers can then act as seeds to generate further generations. (b) An AFM image of the seed AB ladders, before the heating step. (c) An AFM image of a different mixture of cross-tiles after many cycles, where 'mutations' are allowed, leading to competition between replicators for 'food.' Insets show the two types of binding, a regular one which is characterized by the two | | signs on the cross-tiles in parallel, and a mutated one binding with one of the signs rotated by $90°$. Panels (b) and (c) are adapted from Zhou et al., 2021.[69]

or the 'replicator-first' (RNA-first) hypothesis of the origin. In the former, life started with the emergence of a metabolism, i.e., self-sustained chemical reaction networks that slowly evolved in complexity.[67] In the latter, life started with the appearance of hereditary molecules, requiring preexistence of replication machinery.[68]

10.4.3 Examples

We now discuss, in the spirit of the general approach of this chapter, examples from several areas of research in ALife. In order to facilitate understanding of how they tie in with the various general concepts mentioned above, we cluster them under six headings.

If you want to find out more, good entries are provided by the topical reviews published regularly in the journal *Artificial Life*.

a. Self-replication

Over the past decade or so we witnessed several realizations of self-replication in artificial systems. Here we mention ones that used artificial building blocks and where exponential replication has been achieved.[70]

Achieving exponential growth has been one of the main obstacles to successful realizations of self-replication in artificial systems. A breakthrough came with the work of Lincoln and Joyce, 2009.

a.1 Self-replication using DNA origami

DNA origami have been used as building blocks of self-replicating systems, with replicators being quasi-one-dimensional. We discuss here only one model system where some external input is needed to drive the replication process. In other words, the system is not autonomous, though one could substitute the external driving with the day-night cycle.

In figure 10.31.a we show a replication scheme in a system with the so-called origami cross-tiles as building blocks. The structure being replicated is a dimer. A solution of free-floating orange A and B cross-tiles is seeded with blue dimers. All tiles are functionalized with DNA single strands (sticky ends) for horizontal and vertical binding to other complementary monomer tiles. When the system is cooled, the seed dimer vertically binds two pairs of monomers. Being held close together, monomers bind horizontally. After UV activation, monomers are covalently linked and represent daughter dimers. Heating the system leads to the detachment of the parent dimer and the daughter dimers. The daughter dimers can now act as parents for the next generation.

The fact that both top and bottom sides of the seed dimer are functionalized leads to a 270,000-fold amplification after six cycles (if only one side is functionalized the amplification factor is about 2). This is thanks to the formation of long ladder structures (see figure 10.31.b) that, after heating cycles, lead to the proliferation of daughter dimers.

This particular system of cross-tiles has been pushed further in the efforts to understand the fundamental processes in nature and discover next-generation materials with the introduction of interaction mutation and growth advantages. As before, one dimer type is put in a solution and replicated, leading to dimer doubling in each cycle. The solution additionally contains tiles that can form a different type of dimer, one that can lead to dimer quadrupling in each cycle, but this process is unseeded at the initial step. By introducing a small mutation rate that leads to the first dimer type infrequently templating the other type, it can be shown that the offspring of the second type takes over the system in approximately six generations in an advantageous environment. Figure 10.31.c shows an AFM image of this system.

a.2 Self-replication using DNA particle systems

DNA particle systems have also been used to demonstrate self-replication with exponential growth, though so far in simulations only. In figure 10.32 we show a scheme for self-replication of a small six-particle cluster with octahedral symmetry. The octahedron can be efficiently replicated if another cluster, a dimer, is present. These two 'parents' seed a solution of free-floating particles, all interacting via designed, specific, short-range interactions, inspired by DNA-coated colloids. Each particle in the parent cluster can bind one particle of specific type from the solution. Once bound, these attached particles can interact and form bonds.[71] In the replication scheme both parents contribute to the formation of a copy of the octahedron, and the parent octahedron makes a copy of the dimer.

This system exhibits exponential growth, and autonomous replication is achieved by programmed release of children from parents, which is easy to realize in simulations.

Note that if there were methods for self-replicating complex materials, then selection amplification cycles would lead to discovery of solutions with greatly enhanced properties.

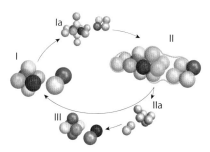

Figure 10.32. Scheme of self-replication of a small six-particle cluster.

b. Catalysis

Most biological reactions are facilitated by enzymes which, besides being extremely efficient and specific, are not energy consuming and are systematically recycled. The current challenge for making artificial catalysts is on the theory side: there is no theoretical framework for designing catalysts from scratch. At present, the dominant design rule is Pauling's principle, according to which catalysts should stabilize transition states.[72] This principle is behind the selection of catalytic antibodies (currently the most successful mimics of enzymes) as well as behind the computational design of new enzymes.[73] In either case the artificially produced enzymes fall short of matching natural ones.

b.1 Catalysis: Design of a dimer catalyst for bond cleavage

A general approach to design artificial catalysts is starting to be developed, in which a theoretical framework of catalysis integrates kinetic, geometrical, and physical constraints. Figure 10.33 shows the application of this framework to a system with DNA-coated particles, where the reaction that is being catalyzed is bond cleavage. The scheme at the top shows a proposed cycle where a catalyst, designed as a rigid dimer with precise geometry (red particles at a certain distance), cleaves a dimer bond (blue particles).

The contour plot shows that there is a parameter regime where catalysis is possible: in the regime below the dashed white line, dimer bonds break faster with a catalyst present than spontaneously due to thermal fluctuations.[74] The rate enhancement depends on the strength of the bond being cleaved. Although much still needs to be achieved to reach enhancement enzymes reach, results like those of figure 10.33 are a step toward understanding and building catalysts from the bottom up at the colloidal scale.

c. Autocatalysis

Autocatalysis as a fundamental concept is used in a wide range of disciplines, from biology all the way to economics. Over the last decades it has increasingly played a role in self-organization, artificial life, and origin of life research.

Autocatalysis can be viewed as a precondition for the emergence of life.[75] Whether any of the theoretical scenarios that have been proposed for the emergence of autocatalysts naturally occurs in practice remains to be seen. The fundamental question is to determine how likely it is for a soup of interacting catalysts to self-organize into an exponentially growing autocatalytic cycle. Moreover, how special do the interactions between the different components need to be for spontaneous exponentially growing autocatalytic cycles to emerge? And how diverse do these autocatalytic reaction networks need to be for evolution to emerge from transitions of the system between different (connected) networks?[77]

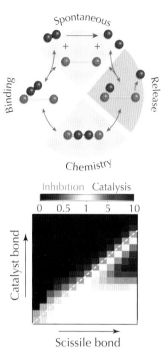

Figure 10.33. Design of a dimer catalyst that helps cleave a bond. Top: proposed scheme. Bottom: contour plot showing that there is an experimentally accessible parameter regime in which catalytic cleaving behavior can be observed. Image courtesy of Maitane Muñoz Basagoiti. See Muñoz-Basagoiti et al., 2023 for more details.

Understanding how to design interactions and catalyst geometry to cleave (or form) bonds is also an important ingredient for realization of autonomous self-replication in the artificial systems we discussed above.

Using a simple mathematical model, the so-called garbage bag model, Dyson, 1982 presented a scenario through which such a complex metabolism could arise spontaneously. His idea was that a random set of catalysts will catalyze arbitrary chemical reactions in a non-synergistic fashion. But if each catalyst is more likely to function when there are others that are synergistic with it, then there is a critical amount of cooperativity above which metabolic cycles spontaneously emerge. In a similar spirit Kauffman and collaborators have shown that if the probability of one species catalyzing the formation of another is above a threshold, then catalytic cycles naturally emerge.[76]

c.1 Autocatalysis: Design with DNA-coated colloids

In the numerical system of DNA-coated colloids discussed above, coupled autocatalytic cycles can also be realized. If one asks whether interactions can be designed such that one small structure serves as a template for a different structure, i.e., catalyzes its formation, the solution leads to the emergence of coupled autocatalytic cycles. Depending on the number of different types of building blocks used and their interactions, the autocatalytic reaction networks contain different levels of diversity in structure type and composition.[78]

d. Folding

Proteins, DNA, and RNA are biopolymers, which through the folding process reach their three-dimensional structures that contain many complicated twists and turns. What a protein does is determined by its structure, and predicting protein structure from its sequence has been one of the central goals in computational biology. The importance does not stop there: being able to predict protein structure will aid in the design of artificial enzymes, that can be used, for example, for digesting plastics, or to aid in identification of protein structures in certain diseases, which would accelerate drug development.

One of the biggest breakthroughs in this field came a few years ago, when Google's DeepMind released AlphaFold,[79] an AI deep learning–based program that can predict the protein structure from its sequence in a fraction of the time in which previous codes could, and with much higher precision. This is the current state of the art in the field of structure prediction. AlphaFold, however, does not perform well in modeling groups of proteins that interact with each other, which remains a challenge.

Other goals in the field concern understanding folding pathways themselves and the structure of the folding landscape as well as how to build functional complexes from the bottom up in a given artificial chemistry. Below, we discuss two experimental examples of artificial systems which have recently been introduced to throw light on the open questions concerning folding.

d.1 Folding: Mimicking protein folding with emulsion droplets

Colloidomers are chains of emulsion droplets that are irreversibly bound with valence 2 to form a linear backbone. They are used as model systems with which to study the fundamental process of folding, with two goals in mind. The first goal, more in the spirit of section 10.1, is to show how self-assembly of functional next-generation materials can be assisted by using folding. The second, more relevant to this section, is to mimic folding of proteins and explore the sequence-structure-function relation in artificial systems, as droplets can be endowed with DNA and therefore droplet-droplet interactions can be programmed.

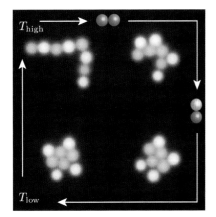

Figure 10.34. Fluorescent images that follow a temperature protocol giving rise to stepwise folding of an octamer chain into the hourglass structure. Starting from a linear configuration, first, blue-blue droplet-type interaction is turned on, followed by the yellow-blue and finally yellow-yellow, which locks the remaining bonds. The droplets are about a micron in diameter. Image courtesy of Angus McMullen. See McMullen et al., 2022 for more details.

In figure 10.34 we show a sequence of experimental images in which an octamer chain of alternating types of droplets (blue and yellow droplets have different DNA coating) folds into a so-called hourglass structure after a specific sequence of secondary interactions is turned on. Folded colloidomers resemble proteins in that they can be functional and adaptable to changes in the environment and can be evolved by manipulation of the DNA coating. Currently this system is studied in two dimensions; however, nothing prevents the two-dimensional structures from folding into the third dimension, which is one of the next steps to explore. Unlike AlphaFold, this model system visualizes folding pathways on the scale of the monomer, giving way to uncovering rules underlying folding.

d.2 Folding: Mimicking protein folding with magnetic particles

On the macroscopic scale, some first steps in studying folding have been made in the so-called magnetic handshake systems. In figure 10.35 we show two examples of elastic nets that in their vertices have disks endowed with patterns of small magnets, whose interaction can be designed. When the nets are shaken at varying amplitudes, different secondary interactions are turned on: a bond that has formed by shaking at a particular amplitude will not break during shaking at a lower amplitude. The sequence of amplitudes leads to the folding of two-dimensional structures into three-dimensional ones such as cubes and tetrahedrons.

e. Artificial cells

From the origin of life perspective, there is a strong interest in understanding what was the composition of the first protocell membranes. It is believed that fatty acids and lipids were available in the prebiotic environment. These can form different phases depending on the concentration, such as micelles, lamellae and vesicles. For example, one of the questions that is explored is the replication of RNA with the compatibility of the protocell membranes: RNA replication requires high levels of certain ions, which themselves have detrimental influence on the fatty acids composing the membranes.[80]

In recent years there has been a lot of effort put toward reconstituting a minimal, functional artificial cell from the bottom up. Starting from natural biomolecules in just the right combination to approximate different aspects of life, researchers aim at making artificial cells that can convert food into energy, use DNA to transfer information, grow and reproduce. The dream is to engineer cells that one day may produce medicines or any other needed chemicals or useful molecules or which serve as microbial factories. This is a rich field of research that combines biologists, chemists, and physicists.[81] Below we give an example of a nonbiological analogue of a cell.

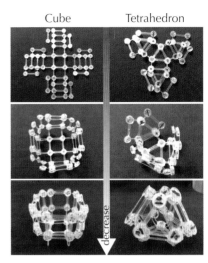

Cube   Tetrahedron

decrease

Figure 10.35. Folding of two-dimensional nets with magnets into three-dimensional structures in experiments by Niu et al., 2019. Binding energies of different magnet faces, their orientation, and the width, length, and thickness of the net elements can be controlled. The left column shows snapshots of a net folding into a cube structure and the right column shows a net folding into a tetrahedron structure. To fold the final structures, nets are placed into a shaker apparatus. The arrow indicates that the shaking amplitude decreases from top to bottom. Images courtesy of Ran Niu and Itai Cohen.

The field of artificial cell research is huge. We mention here just a few fast-developing directions within the field.

Greek symbols

Symbol	Description	Dimension	Section
α	bending stiffness of wormlike chain	$\text{kg m}^3/\text{s}^2$	5.4.1
α	thermal expansion coefficient	K^{-1}	8.3.1
γ	surface tension	kg/s^2	1.12
γ	correlation strength of 'stochastic force'	*depends*	3.2.2
γ	global affine strain	–	5.8.3
$\dot{\gamma}$	global shear rate	s^{-1}	4.6.3
Γ	relaxation rate	s^{-1}	3.2
ϵ	energy density per unit mass	m^2/s	1.3.5
ϵ	dimensionless control parameter	–	8.3.1
ϵ_0	dielectric permittivity of vacuum	$\text{A}^2\,\text{s}^4/\text{kg m}^3$	4.3.1
ε	global strain	–	2.5
ζ	bulk viscosity	$\text{kg}/\text{m s}$	1.6.1
η	dynamic viscosity	$\text{kg}/\text{m s}$ (=Pa·s)	1.6.1
κ	curvature	m^{-1}	2.7
κ	inverse screening length in Debye-Hückel theory	m^{-1}	4.4.1
κ_T	thermal conductivity	$\text{kg m}/\text{s}^3$	1.6.1
λ	Lamé coefficient	$\text{kg}/\text{m s}^2$	2.3
λ	relaxation time of UCM and Oldroyd-B model	s	5.10.3
λ_B	Bjerrum length	m	4.4.1
μ	shear modulus (also Lamé coefficient)	$\text{kg}/\text{m s}^2$ (=Pa)	2.3
ν	kinematic viscosity	m^2/s	1.7
ν	Poisson ratio	–	2.3
ν	scaling exponent	–	5.3.3

ν	enclosed volume of membrane relative to sphere	–	7.2
ξ	similarity variable	*depends*	3.3.5
ξ	correlation length	m	6.2
ξ_0	coherence length	m	8.3.1
ξ_ϕ	crossover length in polymer solutions	m	5.5.2
Π	osmotic pressure	$kg/m\,s^2$	5.5.2
ρ	fluid density	kg/m^3	1.3.2
$\underline{\sigma}$	stress tensor	$kg/m\,s^2$ (=Pa)	1.3.4
$\sigma(q)$	growth rate of stability mode	s^{-1}	8.3.1
σ_X	mean square variation of position X	m^2	3.2.3
τ	time scale	s	1.8
ϕ	volume fraction	s	5.5.2
ϕ	phase	–	8.3.1
θ	temperature	K	8.3.1
ω	angular frequency	s^{-1}	1.7.2

Roman symbols

Symbol	Description	Dimension	Section
A	amplitude in amplitude expansion	–	8.5.2
B	bending modulus of rod	$kg\,m^3/s^2$	2.7.2
$\vec{b}$	Burgers vector	m	2.11
c	concentration	m^{-3}	5.5.2
c^*	crossover concentration dilute to semi-dilute regime	m^{-3}	5.5.2
c_s	speed of sound	m /s	1.7
D	diffusion coefficient	m^2/s	3.2.4

D_T	thermal diffusivity	$\mathrm{m^2/s}$	1.6.3
d	dimension	–	2.10
d	thickness	m	2.7
E	energy	$\mathrm{kg\,m^2/s^2}$	2.3
$\mathcal{E}$	energy density	$\mathrm{kg/m\,s^2}$	2.3
$\vec{E}$	electric field	$\mathrm{kg\,/s^3\,A}$	4.3.1
E_Y	Young's modulus	$\mathrm{kg/m\,s^2}$	2.4
F	free energy	$\mathrm{kg\,m^2/s^2}$	5.3.3
$\mathcal{F}$	free energy density	$\mathrm{kg/m\,s^2}$	6.3.1
F	force	$\mathrm{kg\,m/s^2}$ (=N)	2.8.2
f	force	$\mathrm{kg\,m/s^2}$ (=N)	5.4.4
f_1, f_2	nonlinear functions in Turing equations	–	8.3.2
G	generalized modulus	$\mathrm{kg\,/m\,s^2}$	2.4
g	gravitational acceleration	$\mathrm{m/s^2}$	1.3.4
g	genus of a surface	–	7.2
$\vec{H}$	magnetic field	$\mathrm{A/m}$	6.5.1
H	mean curvature	$\mathrm{m^{-1}}$	7.2
H	Hamaker constant	$\mathrm{kg\,m^2/s^2}$	4.3.1
$\mathcal{H}$	Hamiltonian	$\mathrm{kg\,m^2/s^2}$	7.4
h	height	m	1.12
$\vec{h}$	molecular field	$\mathrm{kg/m\,s^2}$	6.4
J	current or current density	*depends*	1.12
K	compression modulus	$\mathrm{kg\,/m\,s^2}$ (=Pa)	2.3
K	Gaussian curvature	$\mathrm{m^{-2}}$	7.2
K_1, K_2, K_3	Frank elastic constants	$\mathrm{kg\,m/s^2}$	6.3.1
k	wavenumber	$\mathrm{m^{-1}}$	1.7.2

k	wavenumber, deviation from critical value	m^{-1}	8.3.1
k_B	Boltzmann constant	$\mathrm{kg\,m^2/s^2\,K}$	2.11
ℓ	characteristic length of a problem	m	1.8
ℓ_P	persistence length of WLC model	m	5.4.1
ℓ_K	Kuhn length of polymer	m	5.3.1
L	length	m	1.10.2
L	stochastic acceleration ('stochastic force')	$\mathrm{m/s^2}$	3.2
$L_{\alpha\beta}$	Onsager coefficients	–	1.5
M	mass (fluid element, Brownian particle)	kg	1.3.2
M	moment	$\mathrm{kg\,m^2/s^2}$	2.8.3
$\hat{n}$	director field of nematic order	–	6.1.4
$\hat{n}$	normal vector	–	2.8.1
N	number of monomers of polymer chain	–	5.3.1
$\mathcal{P}$	probability distribution	*depends*	3.3.1
p	pressure	$\mathrm{kg/m\,s^2}$ (=Pa)	1.5
$\vec{p}$	polarization of a flock	–	9.2.3
Q_{ij}	nematic order parameter	–	6.2
$\vec{q}$	scattering vector	m^{-1}	3.7.2
q	wavenumber of a mode	m^{-1}	8.3.1
q_c	critical wavenumber at instability threshold	m^{-1}	8.2.2
R	radius (of droplet, particle, or bubble)	m	1.14
R	radius of curvature (of sheet or rod)	m	2.7.1
R	end-to-end distance of chain or polymer	m	5.3.1
S	surface of fluid element	$\mathrm{m^2}$	1.3.2
S	entropy	$\mathrm{kg\,m^2/s^2\,K}$	5.3.1
$S(q)$	scattering intensity	–	3.7.2

s	entropy per unit mass	$\mathrm{m^2/s^2\,K}$	1.5
s	topological number of singularity	–	6.6.1
T	temperature	K	1.5
T_ϵ	threshold-dependent time scale of pattern variation	s	8.5.2
$\hat{t}$	tangent vector	–	2.8.1
U	potential energy	$\mathrm{kg\,m^2/s^2}$	3.3.3
U	internal energy	$\mathrm{kg\,m^2/s^2}$	2.11
u_{ij}	strain rate tensor	–	2.2
u_1, u_2	fields obeying Turing equations	–	8.3.2
V	volume of fluid element or material considered	$\mathrm{m^3}$	1.3.2
V	velocity of Brownian particle	m/s	3.2
$\vec{v}$	velocity field	m/s	1.3.1
v_g	group velocity	m/s	8.7
X	position of Brownian particle	m	3.2.3
X_ϵ	threshold-dependent spatial scale of pattern variation	m	8.5.2
Y	effective Young's modulus of a sheet	$\mathrm{kg/s^2}$	7.3

Notes

Introduction

1. Figure taken from figure 2 of Perrin, 1909.

2. Figure redrawn from figure 2 of Besteman et al., 2007, copyright (2007) by the American Physical Society. The setup of these experiments is sketched in section 3.6.1. As explained there, the smaller the pulling force, the larger the size of the fluctuations. This is visible in the increase of the width of the shaded band with decreasing force. The fact that the data depend strongly on the buffer illustrates that the electric charges affect the behavior of DNA in solution. This will be discussed in section 5.4.2.

3. See, e.g., the reviews by Prévost et al., 2009 or Deniz et al., 2007.

4. Figure from Wikipedia, published under CC BY-SA 3.0 License. © Kevin R. Johnson

5. This figure is inspired by figure 3 of the website wiki.anton-paar.com/tw-zh/the-influence-of-particles-on-suspension-rheology/. Important aspects of the rheology of colloids are discussed in section 4.6.

6. The topic of rheology of colloids is treated in section 4.6. The cornstarch image in the upper left is from "Structural changes of corn starch during fuel ethanol production from corn flour" by Daria Szymanowska-Powałowska et al. (BioTechnologia 93(3): 333–341) © 2012 Institute of Bioorganic Chemistry, Polish Academy of Sciences, published under a CC BY-NC-ND license, Szymanowska-Powałowska et al., 2012. The clay image in the upper right is from "The Influence of Individual Clay Minerals on Formation Damage of Reservoir Sandstones: a Critical Review with Some New Insights" by M. J. Wilson, L. Wilson, I. Patey (Clay Minerals 49, no. 2: 147–64) © 2014, The Mineralogical Society of Great Britain and Ireland, M. J. Wilson et al., 2014. Lower right and left: Zorana Zeravcic.

7. See Bonn et al., 2017 for a review of the occurrence of yield stress in soft matter.

8. For similar figures, see Schall et al., 2007. The topic is discussed further in section 4.5.4.

9. The divergence of the viscosity upon approaching the glass transition is discussed further in section 4.5.4. Figure I.10, redrawn from figure 1 of Berthier et al., 2016, shows equilibrium relaxation timescale of polydisperse (23%) hard spheres from standard Monte Carlo simulations. The red line is a power law fit from the paper. The shaded gray zone indicates the jamming range in density based on various extrapolations discussed in the paper.

10. Printed with permission from AAAS. From "Predictive Self-Assembly of Polyhedra into Complex Structures" by S. C. Glotzer, M. Engel, and P. F. Damasceno. (Science, Vol 337, Issue 6093. 2012: 453-457), Damasceno et al., 2012. We thank Karen Coulter and Sharon Glotzer for obtaining permission from *Science* to use this figure and Michael Engel for providing the high-resolution version of the image.

11. Adapted from Zeravcic and M. P. Brenner, 2017.

12. See, e.g., various articles in the collection Berthier et al., 2011.

13. For a similar picture and further discussion, see Jaeger et al., 1996.

14. For a modern perspective on bacterial swarming as an example of active matter, see section 9.4. For an overview of the earlier work on bacterial swarming and the effects of hydrodynamic interactions, see Pedley and Kessler, 1992.

15. See C. Joshi et al., 2022 for more details.

16. See the illustration of paint and mayonnaise in figure I.8. Applications in the processing and food industry are also touched upon in section 1.15 and section 6.9.4.

17. See the illustration in figure I.8, and section 6.9.4 for applications in the food and pharmaceutical industries. The industrial liquid-liquid extraction process is mentioned in section 1.15 and put in broader perspective in the review by Lohse and X. Zhang, 2020. Lubrication theory of thin layers, the topic of section 1.12, is important for the spreading and dripping of paint and in many industrial processes. See Lohse, 2022 for a review of the complex flow features of inkjet printing.

18. Supramolecular polymers are being explored for biodegradable plastics—see figure 5.2 for an illustration. For a pointer to soft self-healing electronics, see https://physicsworld.com/a/soft-electronics-self-heal/.

19. See section 6.9.4 and in particular the review by Aleandri and Mezzenga, 2020.

20. See Bourouiba et al., 2014 and Yang et al., 2021 for the spreading of aerosols and Poon et al., 2020 for a review of soft matter science and Covid-19.

21. The discussion of section 1.13 is relevant to this issue. See Bonn et al., 2009a for further discussion and references for the application to pesticide spreading.

22. See section 6.5.2 for a brief discussion of how liquid crystal displays work. In the margin note in the beginning of section 6.5 we briefly touch on the physics of switchable glass.

23. We will encounter various examples of this in chapter 10 in the context of our discussion of designer matter. See figure 2.2 and the paper by Bertoldi et al., 2017 for references.

24. See figure 2.1 and section 2.9 and in particular chapter 10 for a more systematic discussion.

Chapter 1

1. The snapshot in (a) was generated using Python example code from the book by Allen and Tildesley, 2017. All the examples that accompany that book can be found on github https://github.com/Allen-Tildesley.

2. The Boltzmann equation is treated in many statistical physics textbooks; see, e.g., Kestin and Dorfman, 1972 or Balescu, 1975.

3. There are many interesting videos about the Reynolds number dependence of flows. You will find links to some of them on the book's website www.softmatterbook.online.

4. This paper, together with all the other works of Euler, can be found in the Euler Archive at scholarlycommons.pacific.edu. See also this book's website www.softmatterbook.online.

5. The use of the term is inspired by the broken symmetries approach illustrated in figure I.15 and was popularized in particular by Nobel Prize winner Robert Laughlin (Laughlin, 2005). See R. C. Bishop, 2019 for a general approach to emergent phenomena from a physics perspective. As mentioned in section 10.4.2, use of the concept is more common in the field of *artificial life*.

6. If there is an *external* force like gravity, then the total momentum of the fluid element as a whole also changes due to this external force; likewise the energy of the fluid element in the presence of an external force changes due to the displacement of the fluid element in the force field. The terms which express this are therefore of different type: they are not the net sum of integrals over the boundary but are like 'source terms'—see, e.g., equation (1.25), where the gravity enters like a source term in the momentum equation.

7. The important property that distinguishes a tensor from the matrix formed by its components is that a tensor is a mathematical entity that lives in a structure and interacts with other entities (for instance, the stress tensor is a force on a surface whose normal is in a given direction). If one transforms the other entities in the structure in a regular way, then the tensor must obey a related transformation rule. The transformation property of a tensor is the key that distinguishes it from a mere matrix.

8. As a reminder, the Gauss divergence theorem in three dimensions states that if V is a volume in $\mathbb{R}^3$ which is compact and has a piecewise smooth boundary S (often written as $\partial V = S$), and if $\vec{F}$ is a continuously differentiable vector field defined on a neighborhood of V, then: $\int_V \mathrm{d}^3\vec{r}\,(\vec{\nabla}\cdot\vec{F}) = \int_S \mathrm{d}^2\vec{S}\cdot\vec{F}$, where $\mathrm{d}\vec{S}$ is the vector normal to the surface element.

9. For a more complete discussion of the symmetry of the stress tensor and the underlying physical reason, see, e.g., Chaikin and Lubensky, 1995, page 335 or Landau and Lifshitz, 1970, page 6.

10. Remember that the transpose of a tensor is obtained by interchanging its coefficients, $\sigma_{ij}^T = \sigma_{ji}$. The transpose of the stress tensor arises when we apply the Gauss divergence theorem, for the gradient operator is then contracted with the second index of the stress tensor.

11. For an introduction and overview of non-equilibrium thermodynamics, see de Groot and Mazur, 1964.

12. For instance, quantities of different tensorial character cannot couple. This actually dictates the form (1.22) for the stress tensor as the sum of a traceless tensor and the unit tensor. See, e.g., de Groot and Mazur, 1964 for more details on this.

13. See the customer center at www.meerstetter.ch on the Peltier effect and this book's website www.softmatterbook.online.

14. See the explanation of the Seebeck effect on tuitionphysics.com and this book's website www.softmatterbook.online.

15. The precise derivation of this equation can be found in section 15 of Landau and Lifshitz, 1987 and in Kleman and O. D. Lavrentovich, 2003; useful treatments of tensors, vectors, and scalars can be found in Lyubarskii, 1960. Here is a more formal version of the argument in a nutshell: we start from a fact that σ_{ij}^{d} is a rank 2 tensor and that friction and dissipation can arise when nearby fluid elements have

different velocities. In the limit of small gradients, we can therefore assume that σ_{ij}^{d} depends only on the first derivatives of the velocity, i.e., the elements will be linear combinations of $\nabla_i v_j$. From the Onsager reciprocity relations (1.21), we know that σ_{ij}^{d} is symmetric, and so we consider the symmetrized expression $\partial v_i / \partial x_j + \partial v_j / \partial x_i$. Going back to the flow itself (figure 1.5), we can observe that the symmetric linear combinations of flow gradients correspond to dilation and shearing motion. (Note that the antisymmetrized expression $\partial v_i / \partial x_j - \partial v_j / \partial x_i = -\varepsilon_{ijk}(\vec{\nabla} \times \vec{v})_k$ (with ε_{ijk} the Levi-Civita symbol) corresponds to a rigid body rotation in which there is no dissipation, so this term is excluded from σ_{ij}^{d} in accordance with Onsager.) Next, one observes that the trace of a tensor is scalar-invariant (does not change with the change of coordinate system) and is therefore an independent quantity. This illustrates the remark made three endnotes back that quantities of different tensorial character cannot couple. In our case, $\mathrm{Tr}\,\underline{\sigma}^{\mathrm{d}} \sim \vec{\nabla} \cdot \vec{v}$ is clearly a scalar, and it corresponds to a volume change of a fluid (see figure 1.5). Therefore we can separate our symmetric expression into the trace times the identity tensor, and its traceless part, of which the latter cannot be reduced further, naturally leading to the final expression (1.22) for $\underline{\sigma}^{\mathrm{d}}$.

16. The simplest way to arrive at these equations is to write all the expressions using indices i, j, k. Also note that we assume here that the viscosity coefficients do not change noticeably in the fluid and can therefore be taken outside the gradient operator.

17. So while it is common to consider the five hydrodynamic equations as a closed set of five equations, one could also say one has five dynamical equations plus a sixth equation of state. It is a bit of a matter of preference how one counts.

18. To convince yourself this is the case, you can change x to $x \pm x_0$ and t to $t \pm t_0$ in equation (1.34) or equation (1.33), and see that these equations do not change.

19. It is sometimes suggested (e.g., in Kleman and O. D. Lavrentovich, 2003) that incompressible fluids have $\zeta = 0$, but this is neither true (for water $\zeta \approx 3\eta$) nor actually the point. The proper question is, under which conditions is it fair to treat a flow as incompressible, which, as the analysis shows, is more related to the time and length scales on which the flow varies? Our analysis here will follow the line of Landau and Lifshitz, 1987.

20. $\mathrm{Re}_i = \Omega_i R_i d / \nu$, where Ω_i is the rotation rate of the inner cylinder, R_i is its radius, and d is the gap distance between the inner and outer cylinders—compare the definition of the Reynolds number in equation (1.52).

21. For the full figure see the review by Grossmann et al., 2016, which includes the detailed phase diagram, drawn for a ratio of inner to outer radius of the cylinders equal to 0.833. A nice introduction of the Taylor-Couette cell, aimed at soft matter scientists, is given by Fardin et al., 2014.

22. Check that the set (1.49) constitutes a set of four equations for four fields, the three-dimensional vector $\vec{v}$, and the pressure p.

23. Poiseuille pipe flow is linearly stable at all Reynolds numbers. When the Reynolds number is increased, at some point around $\mathrm{Re} \approx 2000$, turbulent domains ('puffs') appear, and the flow is an alternation of laminar and turbulent domains. Turbulent puffs can also split and decay, and in the end the transition to turbulence can be associated with a directed percolation type spreading of the turbulent domains throughout the whole pipe (Shih et al., 2015). See Barkley, 2016 for comprehensive overviews of this scenario. The driving mechanism of the instability and the relation to chaos is more stressed in the earlier paper by Eckhardt et al., 2007. The sketch in figure 1.12 is figure 13 in the paper by Reynolds, 1883.

24. The story and Taylor's log-log plot, which he made to extract E_{blast} from the data of the radius R of the blast at various times, can be found on various websites. The example is also discussed by Barenblatt, 1996—some editions of this book even have the photo of the atomic bomb explosion on the cover.

25. For fascinating experimental photographs of flow past a cylinder at various Reynolds numbers that illustrate these drawings, see the book by Van Dyke, 1982.

26. If we would have kept the gravity in the discussion, we would have obtained a dimensionless acceleration term proportional to $gL/U^2 = 1/\mathrm{Fr}^2$. Here Fr is the dimensionless Froude number, which measures the relative importance of kinetic energy over gravitational energy. It plays a role in ship design.

27. For an overview, see the book *Low Reynolds Number Hydrodynamics* by Happel and H. Brenner, 1982.

28. There are many videos which nicely illustrate this counterintuitive effect; see, e.g., the YouTube demonstrations of laminar flow from the University of New Mexico, and this book's website www.softmatterbook.online.

29. The Stokes drag is a very basic result which is derived in many books, among others the ones by Landau and Lifshitz, 1987 and Happel and H. Brenner, 1982.

30. If you want to catch up on modern work, the recent review *Fluid Dynamics at the Scale of the Cell* by Goldstein, 2016 gives a good entry to recent developments such as cytoplasmic streaming in plant cells, synchronization of eukaryotes, flagella, interactions between swimming cells and surfaces, and collective behavior in suspensions of microswimmers. Bacterial hydrodynamics is reviewed by Lauga, 2016.

31. Research on the swimming of bacteria was pioneered by Howard Berg (H. Berg, 1993). His website has many interesting videos of swimming bacteria. Videos of *E. coli* clearly show the rotating tail and the fact that bacteria change their swimming direction (tumble) by briefly counterrotating their flagella. For a review of this work, see H. Berg, 2004.

32. For Poiseuille pipe flow, the Reynolds number is defined as $\mathrm{Re} = U_{\mathrm{mean}} D / \nu$, where U_{mean} is the mean flow velocity and D the pipe diameter. With blood flow of about 5 liters/minute and and aorta diameter of about 2 cm, one estimates a mean flow velocity of about 25 cm/s. With a viscosity about three to four times that of water, this gives Re in the range 600–800, whereas the critical Reynolds number for pipe flow is usually taken to be around 2000. Besides the fact that arteries are not straight and of uniform size, it is important to note that while blood plasma behaves quite well as a Newtonian fluid, the red blood cells make blood a shear-thinning complex fluid.

33. There are many beautiful photos of the Kelvin-Helmholtz instability in cloud formation on the internet. For a review of the Kelvin-Helmholtz instability in space science, see Masson and Nykyri, 2018.

34. A famous example is the collapse of the Tacoma Narrows Bridge in 1940 due to such resonances. We show the video on our website www.softmatterbook.online, but you will also easily find versions of it on the internet or via the Wiki page. Modern bridges often have either dampers to suppress resonances, or structures around cables to suppress the formation of the vortices.

35. The nonlinear scenario was first proposed and supported by suggestive experiments by Bertola et al., 2003. The theoretical arguments to support this scenario were reviewed by Morozov and van Saarloos, 2007, and more in-depth analysis of the transition for viscoelastic channel flows can be found in the paper by Morozov and Saarloos, 2019. Experimentally, the issue was settled by a series of very nice experiments by Arratia and coworkers on long micro-channels, where the flow was disturbed near the inlet by a varying number of micro-pillars (Pan et al., 2013; Qin and Arratia, 2017). Recently, Morozov, 2022 has presented convincing evidence that the subcritical instability is associated with the existence of a branch of exact traveling wave solutions of viscoelastic polymer flow.

36. Reprinted with permission from Z. J. Wang, 2000. Copyright (2000) by the American Physical Society. See the original paper for details and Z. J. Wang, 2005 for a review of the fluid dynamics of insect flight. The lift calculated from the two-dimensional airflows in this simulation is sufficient to support the weight of a hovering dragonfly.

37. See, e.g., Eggers, 1997. For a more basic account of matched asymptotic expansions, see Van Dyke, 1975.

38. Based on the observations from Bouwhuis et al., 2012. In this paper the situation is analyzed theoretically with the lubrication approximation and also studied numerically. It is shown that there is an optimal velocity, determined by the dimensionless Stokes number, for entrainment of an air layer under a bubble.

39. Interestingly, the statistical properties of wet and dry patches on the hot plate are in the universality class of directed percolation (Chantelot and Lohse, 2021), very much as the statistical properties of turbulent domains in pipe flow can be described by directed percolation (Barkley, 2016; Shih et al., 2015).

40. An example of the analysis of the dewetting rim of films using (1.62) can be found in a paper by Snoeijer and Eggers, 2010.

41. See Rayleigh, 1917.

42. This may not be obvious at first sight, especially since the components σ_{rr}^{d} and $\sigma_{\theta\theta}^{\mathrm{d}}$ are nonzero. You can check it explicitly from the equations given in spherical coordinates in section 15 of Landau and Lifshitz, 1987.

43. Sonoluminescence is reviewed by M. P. Brenner et al., 2002 and Lohse, 2018.

44. See for details and references Lohse and X. Zhang, 2020, the source of the inspiration for the figures.

45. See Lohse and X. Zhang, 2020 for further discussion of this.

46. See, e.g., de Groot and Mazur, 1964.

47. Panels (a), (c), (d), and (f) are taken from figure 16 of Rahman and San, 2019.

48. A review of the Rayleigh-Taylor instability is given by Kull, 1991.

Chapter 2

1. The patterns in region B, which are not shown, are intermediate between those in A and C.

2. Reprinted with permission from Bowden et al., 1999. Copyright (1999) by the American Institute of Physics. The formation of such wrinkles in skin and in materials science is reviewed by Genzer and Groenewold, 2006; see also Hendricks et al., 2010.

3. For an accessible introduction to wrinkled materials, including skin, see Genzer and Groenewold, 2006.

4. Convince yourself that the symmetry $K_{ijkl} = K_{jikl} = K_{ijlk}$ still allows 36 coefficients! To give you a feel for the reduction of this number by crystalline symmetries: as discussed, e.g., by Landau and Lifshitz, 1970, the number of different elastic constants for a monoclinic crystal (with two basis vectors at 90-degree angles, and a third one making an angle less than 90 degrees) is still 11.

5. Be aware that sometimes, e.g., in the book by Kleman and O. D. Lavrentovich, 2003, the notation $u_{i,j}$ is used to denote the partial derivative $\partial u_i / \partial x_j$, so that the strain tensor (2.3) is then written as $\frac{1}{2}(u_{i,k} + u_{k,i} + u_{l,i}u_{l,k})$.

6. Since the total energy E_{elas} depends on the *function* $u_{ij}(\vec{r})$, we can also write $\sigma_{ij} = \partial \mathcal{E}_{\text{elas}} / \partial u_{ij} = \delta E_{\text{elas}} / \delta u_{ij}$, with $\delta E_{\text{elas}} / \delta u_{ij}$ a functional derivative. See section 6.4 for further discussion of the functional derivative.

7. Don't confuse the Poisson ratio ν with the kinematic viscosity of the Navier-Stokes equation (1.49)! Also note that the Poisson ratio is sometimes denoted by σ in the literature.

8. Figure inspired by the work by Bertoldi et al., 2017, who discuss cork in the context of metamaterials with patterns of holes.

9. PDMS stands for polydimethylsiloxane, with chemical formula $(CH_3)_3$-Si-$[O\text{-}Si(CH_3)_2]_n$-O-$Si(CH_3)_3$.

10. An example of a very anisotropic soft matter material, a polyurethane foam, with a Poisson ratio even larger than 1, is presented by T. Lee and R. S. Lakes, 1997. Note that for anisotropic materials the Poisson ratio will depend on the direction in which the stress is applied, so the Poisson ratio of anisotropic materials is actually not a single number for a given material.

11. Note that in the time domain equation, (2.21) reads $\sigma_{xy}(t) = \int_{-\infty}^{\infty} dt_1 \, G(t - t_1)) u_{xy}(t_1)$. The fact that causality dictates that the response cannot depend on driving or disturbances in the future, implies that $G(t) = 0$ for $t < 0$. As $G(t) = (2\pi)^{-1} \int d\omega \, G(\omega) e^{-i\omega t}$, this implies that $G(\omega)$ must be analytic in the upper half ω plane: when the contour is closed in the upper half plane, the integral along the upper semicircle vanishes for $t < 0$, and when $G(\omega)$ has no poles in the upper half plane, this then gives $G(t) = 0$ for $t < 0$. For a discussion of the Kramers-Kronig relations, see, e.g., the books by Forster, 1994; Landau and Lifshitz, 1981a; or Lovesy, 1986; or the short note by Hu, 1989.

12. The cross section is taken to remain normal to the surface, which is appropriate for homogeneous rods of constant thickness bent along one of their symmetry directions.

13. For the derivation and discussion of the Föppl–Von Kármán equations, see, for instance, Landau and Lifshitz, 1970 or Audoly and Pomeau, 2018.

14. See also the margin note to the first paragraph of section 7.3.

15. The effect is due to the mismatch in the Gaussian curvature discussed in section 7.2 between a sphere with nonzero Gaussian curvature and a plane with zero curvature. See Hure et al., 2012 for details of the experiments and the origin of these wrinkles.

16. Reproduced with permission from Hure et al., 2012. Copyright (2012) by the American Physical Society.

17. Reproduced with permission from Cerda and Mahadevan, 2003. Copyright (2003) by the American Physical Society. In this paper a detailed analysis of the balance between bending and stretching is given. See Genzer and Groenewold, 2006 for an introduction to and review of such wrinkling effects and of their use in biology and materials science.

18. This is a nice illustration of the fact that linear elasticity can give rise to strongly nonlinear problems. Something similar happens for so-called moving boundary problems. An example of a moving boundary problem is posed by the calculation of the shape of a growing crystal whose interface dynamics is governed by the diffusion of latent heat released at the interface upon growth. The diffusion equation in this case is linear, but the intricate interaction between the interface shape and the diffusion field—the diffusion field depends on the shape, while the interface evolution depends on the interface shape and the diffusion field—makes such problems highly nonlinear and nontrivial.

19. It is easy to check that the sine deformation (2.45) solves the linear equation (2.56) with $f_{\text{ext}} = 0$ and $F_{\text{el},x} = F^{\text{ext}}$. One can also obtain equation (2.56) for the stationary shape from equation (2.44) by requiring that the expression be stationary for small variations δy about the shape, provided one also imposes that the boundary terms from the integration by parts vanish. Indeed, the change in energy δE^{total} to first order in δy then gives $\delta E^{\text{total}} = \int_0^L dx \, [By'''' + F^{\text{ext}} y''] \delta y$. For a stationary state, the term between square brackets has to vanish, and this gives equation (2.57). The boundary terms from the partial integration indeed vanish for the case considered here, where no torque is applied at the boundaries.

20. For an introduction to the Landau theory of phase transitions, see, e.g., the book by Onuki, 2002 or the one by P. Tolédano and J.-C. Tolédano, 1987.

21. With the exception of saddle-node bifurcations, where two stationary states merge, bifurcations are associated with a change of stability of a state, i.e., with a stationary solution or state going from stable to unstable or the other way around. We show this for an elementary example in section 8.2.1. Bifurcations are typically formulated in terms of the behavior of ordinary differential equations; many patterns in spatially extended systems are similarly the result of an instability to finite wavelength perturbations, as we will

discuss extensively in chapter 8. For introductions to bifurcation theory, see, e.g., Strogatz, 2015, V. I. Arnold, 1973, or, from a more mathematical perspective Guckenheimer and Holmes, 1983. Short elementary introductions to the notion of bifurcation are also given in section 1.3.2 of Pismen, 2006 and section III.A.2 of Cross and Hohenberg, 1993.

22. The plot contains 9,227 data points. The website (in French) patternformation.wordpress.com by Fabian Brau of the Université Libre de Bruxelles contains a large number of interesting pattern formation examples, including the analysis of maximum tree heights. A more detailed treatment and discussion of the tree scaling problem is given in the lecture notes by Brau, which are available via dipot.ulb.ac.be and via the book's website www.softmatterbook.online.

23. Sketch from Codex Madrid I, page 137.

24. For the connection with supercoiling of DNA and the nonlinear Schrödinger equation, see, e.g., Shi and Hearst, 1994. For a general analysis of nonlinear solutions of the equations for a rod, see Scarpello and Ritelli, 2006.

25. The brighter spot is actually a so-called microtubule seed: a stabilized microtubule that is attached to the surface and from where microtubules nucleate and grow. Effectively, it thus acts as the attachment point; it is brighter since a higher ratio of labeled proteins is used for it, so it can be distinguished in the experiments from the freely fluctuating microtubule part.

26. Experiments of this type to extract the bending stiffness of microtubules started with the paper by Dogterom and Yurke, 1997. The analysis of the bending stiffness from thermal bending fluctuations is discussed by Janson and Dogterom, 2004.

27. This is, for instance, the case of fcc cystals (Milstein and K. Huang, 1979) and cubic metals (Baughman et al., 1998).

28. A foam with auxetic response was introduced by R. Lakes, 1987.

29. For reviews of electromagnetic cloaking, see, e.g., B. Zhang, 2012 or Fleury and Alu, 2014. See also chapter 10, where we show an example of a thermal cloak in figure 10.3.

30. For similar pictures and more details, see Bertoldi et al., 2017.

31. The two-dimensional version of this problem is like that of the 'ice rule' of two-dimensional vertex models in statistical physics; see, e.g., Lieb and Wu, 1972 or Lavis and Bell, 1999.

32. The sample is about 65 mm wide, so typical strains are on the order of 10% in these experiments.

33. Soft robots provide their own challenges, opportunities, and design principles (Oliveri and Overvelde, 2020; Rafsanjani et al., 2019); they are already being used a lot in the biomedical field (Cianchetti et al., 2018).

34. See the papers by Mistry et al., 2018 and Suzuki et al., 2016.

35. Left: Image courtesy of Sid Nagel. Middle: Image courtesy of Zorana Zeravcic. Right: Image courtesy of Jasna Brujic.

36. The argument in this section follows in particular O'Hern et al., 2003.

37. A similar and somewhat simpler model system is a random network of masses connected by springs. The average coordination z of the network, i.e., the average number of springs each node in the network is connected with, is the equivalent of the average number of contacts of each sphere in the packing. This node-and-spring model is easiest to explore by starting from a network with large coordination number z and by cutting bonds to reduce z.

38. See Thorpe, 1985 for an elementary introduction to rigidity percolation. This topic has long been studied. An interesting new twist to the story is the indication that for colloidal gels to acquire rigidity correlations, attractive interactions are important (S. Zhang et al., 2019). Another noteworthy new development is the recent evidence that rigidity percolation occurs in blastoderm organization in zebra fish (Petridou et al., 2021). These as well as jamming ideas are increasingly being explored for living tissue (Merkel and Manning, 2018).

39. There is a subtlety here: we should include in the counting only the particles which have contacts. In computer generated packings there is typically a small percentage of 'rattlers,' particles which have sufficient room to move around freely in a cage formed by other particles (remember that there is no external force such as gravity, and that the temperature is zero so that particles do not move). These rattlers thus have zero contacts.

40. Data for packings with so-called Hertzian forces are obtained from the response to a point force and to bulk deformations. See Ellenbroek et al., 2009 for details. The outlier data point for the smallest value of Δz is determined only from fitting the point force response for various values of the Poisson ratio, and then determining the value for which the fit is optimal. This procedure becomes quite inaccurate near the isostatic point, where $\mu / K \ll 1$.

41. The result is treated in most textbooks, such as Ashcroft and Mermin, 1976 and Kestin and Dorfman, 1972.

42. Plot based on data by the author (ZZ). The first analysis of the spectrum of particle packings in three dimensions was reported in Silbert et al., 2005. A book chapter on the topic with more details is by A. J. Liu et al., 2011.

43. Based on data by the author (ZZ). See Silbert et al., 2005 for the original illustration of the characteristic behavior of the modes using smaller packings.

44. Intuitively, the scaling $\omega \sim \Delta z$ can be understood by noting that long-wavelength elastic modes have a linear dispersion $\omega = ck$; assuming that the speed of c of these modes is independent of Δz yields $\omega^* \simeq ck^* = c(2\pi/\ell^*) \sim \Delta z$.

45. Note that Miskin and Jaeger, 2013 replace all the bonds by elastic springs with the *same* spring constant, so the individual bonds are all the same. However, as the network is random, the contributions from the bonds to the change in energy under compression or shear deformation are different. The possibility of tuning either the compression modulus K or the shear modulus μ illustrates that bonds contribute mostly to either the stiffness under compression, or the stiffness under shear deformation.

46. See Goodrich et al., 2014.

47. In terms of the bond orientational order parameter $\psi_{6,i} = 1/n \sum_{j=1}^{n} \exp(i6\theta_{ij}(t))$, the correlation function $g_6(t)$ analyzed in the experiment is defined as $g_6(t) = \text{Re}\langle[\psi_{6,i}(t) \cdot \psi_{6,i}^*(0)]/|\psi_{6,i}(0)|^2\rangle$. See Thorneywork et al., 2018 for details.

48. Reproduced with permission of *Physical Review Letters* from Thorneywork et al., 2017. Copyright (2017) by the American Physical Society.

49. See, e.g., equation (4.27) of Nelson, 1980 and section 6.10.1.

50. See Brock et al., 1989 for a discussion of the experiments on smectics. The authors conclude "The existence of two-dimensional hexatics has now been proven beyond a doubt." One of the first conclusive observations in a colloidal system was done by C. A. Murray and Winkle, 1987.

51. See Bertoldi et al., 2017.

52. See in particular Bertoldi et al., 2017 in the context of this discussion.

53. There would be no zero mode if the same structure had the topology of a ring, because there would be an additional bond.

54. See B. G. Chen et al., 2014; Kane and Lubensky, 2014.

55. The Landau theory of phase transitions is treated in many books on statistical physics and phase transitions. For a comprehensive introduction, see in particular the book by P. Tolédano and J.-C. Tolédano, 1987 devoted to Landau theory. The book on dynamics by Onuki, 2002 also gives a nice introduction.

56. The functional derivative $\delta F/\delta\phi$ is discussed below equation (6.16).

57. Static fluctuations obey the Debye-Hückel equation (4.15) discussed in chapter 4. This implies that when ϕ is changed locally, the spatial decay of ϕ is given by equation (4.17).

58. See, e.g., Goldenfeld, 1992 or Herbut, 2007 for an introduction to the renormalization group. See also section 6.10.

59. See Wiggins et al., 1998 for a detailed derivation and discussion of this equation.

Chapter 3

1. Keep in mind that the figure is only illustrative: for typical Brownian particles of order 1 micron, the difference in scale of the small particles and the large one is several orders of magnitude, and an extremely large number of small particles collide with the large one within a collision time.

2. See, e.g., Gardiner, 2004 for the generalization to three dimensions.

3. The law of large numbers states that the average of a number of random variables independently drawn from a distribution converge to the mean for large numbers and that the distribution converges to a normal distribution centered around this mean. That means: the variation about the mean becomes normally (Gaussian) distributed when the number of observations is sufficiently large. This law is sometimes used as a justification for taking the random forces to be Gaussian distributed, but it is important to keep in mind that it is an assumption which is not always justified.

4. There are actually mathematical subtleties associated with using delta-correlated noise. Loosely speaking, these are related to whether one takes the induced fluctuating quantity $V(t)$ to be correlated or uncorrelated with $L(t)$ at the same time. This issue comes up if one considers averages like $\overline{V(t)dV(t)/dt}$, as this gives a term $\overline{V(t)L(t)}$. With (3.11), this naively gives a term $\int_0^t dt_1 \exp[-\Gamma(t-t_1)]\overline{L(t_1)L(t)} = 2\gamma \int_0^t dt_1 \exp[-\Gamma(t-t_1)]\delta(t_1-t)$. The upper integration boundary is just *at* the time where the argument of the

delta function vanishes—this is a bit fishy, and it is a sign that we have to watch out. In essence, because of the delta function correlation, we have to give meaning to such types of averages, and we have the freedom to make a choice about how to interpret terms consisting of products of the variable and the noise. If one considers the delta function variance (3.8) to be a simple way to express the limiting process of having correlations over small but nonzero time—as is the customary way to think about it in the physical sciences—the so-called Stratonovich interpretation is intuitively the more natural one. In this interpretation, $V(t)$ and $L(t)$ are correlated with strength as if one were integrating 'through only half of the delta function.' Nevertheless, one does have the freedom to define how to handle such stochastic integrals with delta-correlated noise differently: the stochastic differential equation becomes a well-defined equation only in combination with an interpretation rule. In mathematics and control theory, the so-called Itô interpretation is indeed often preferred. We also use it in the discussion of the Dean equation in problem 3.10. In this interpretation, averages like $\overline{V(t)L(t)}$ are taken to be uncorrelated. The motivation often to do this is that in mathematics and control theory the variables at that moment are uncorrelated with the stochastic signal. In the language of the Langevin equation, the preferred choice is then for the stochastic force to be uncorrelated with the variables driven by it. This has some elegance, and in control theory it can be considered closer to the process one aims to mimic. We stress that both types of approaches can be used, provided one sticks with the choice made and does not overinterpret the results. Luckily, these subtleties do not play up for the types of questions that we will consider in the main text, but they do play a role in the derivation of the Dean equation in problem 3.10. For more details concerning these interpretations from a physics perspective, we refer you to Gardiner, 2004; van Kampen, 1981; and van Kampen, 2009; while L. Arnold, 1974 gives a more in-depth mathematical discussion.

5. Reproduced with permission from *American Journal of Physics*, from Newburgh et al., 2006, to which we refer you for further details.

6. Beware that when b is a function $b(X)$ of X, the subtleties of the differences between the Itô and Stratonovich interpretations, mentioned in note 4, play a role. The equivalence expressed by (3.26) extends also to cases where b is a function $b(X)$, *provided* that the Langevin equation is interpreted in the Itô sense. The Langevin equation with Stratonovich interpretation of the noise, which is equivalent to the Fokker-Planck equation in (3.26), has an additional term, $-\gamma b(X)\partial_x b(X)$, on the right-hand side. See equation (4.3.9) of Gardiner, 2004.

7. Given the full distribution $\mathcal{P}(X, V, t)$, one has $\mathcal{P}(X, t) = \int dV\, \mathcal{P}(X, V, t)$, and similarly $\mathcal{P}(V, t) = \int dX\, \mathcal{P}(X, V, t)$.

8. In cases in which the particles have short-range interactions but do not tend to form clusters, the overall density will again obey the diffusion equation. The effective diffusion coefficient then does depend on the particle density. For example, the hydrodynamic interactions of otherwise noninteracting Brownian particles in suspension makes their collective diffusion coefficient density-dependent—see, e.g., Happel and H. Brenner, 1982 or van de Ven, 1988. Also, excluded volume effects contribute to the density dependence.

9. It is easy to show, by expanding the Gaussian in the integral (3.35) in $X_1 - X_0$ with X_0 chosen so that $X_0 = \int dX_1\, \mathcal{P}_0(X_1)X_1$ is the average initial position, that the correction term to the Gaussian falls off as $-(4Dt)^{-1} \int dX_1\, \mathcal{P}_0(X_1)(X_1 - X_0)^2$.

10. The full analysis requires in-depth matching analysis. It turns out that large X expansion of the nonlinear front region imposes that the leading dynamics in the front region be given by the term (3.39), which normally is a subdominant solution. See Ebert and van Saarloos, 2000 for details.

11. Reproduced with permission from Elsevier, owner of the copyright of Kramers, 1940.

12. The behavior is named after Svante August Arrhenius, who earned the 1903 Nobel Prize in Chemistry "for the advancement of chemistry by his electrolytic theory of dissociation." Interestingly, Arrhenius was also the first one to predict the relation between increasing CO_2 concentration and rising earth temperature that causes the present climate problems (although Arrhenius himself felt that a rise in temperatures might be beneficial for the Scandinavian countries). See in particular the podcasts of the American Institute of Physics at https://www.aip.org/initialconditions for the interesting history of climate science in the early twentieth century.

13. See Weidenmüller and J.-S. Zhang, 1984.

14. The colloids are made of polystyrene and are 1 micron in diameter. The starting suspension is 2 ml of 5% w/w solution of colloids, corresponding to a volume fraction around 0.8%. Solution densities are increased by adding, from left to right, 0, 0.2, 0.5, 0.6, 1.5, and 3.0 molar sucrose solution. The resulting densities are, from left to right, 0.997, 1.023, 1.05, 1.074, 1.189, and 1.382 g/cm^3.

15. A nice time-lapse video illustrating the slow dynamics can be found at on the book's website www.softmatterbook.online (you can find it easily be searching for 'index matched colloids decalin' on YouTube). It shows an experiment on 264-nm PMMA spheres in a decalin/tetralin mixture. The shiny speckle-type pattern visible at the end indicates that the colloids form crystalline grains that Bragg-reflect the light. For nice photos illustrating the phase behavior of hard core colloids as a function of the concentration of colloids, see Pusey and van Megen, 1986.

16. We focus here on recent developments. Many interesting older examples of random walks in biology, in particular those of bacteria, can be found in the book by H. Berg, 1993.

17. We thank Nynke Dekker for providing us with a high-resolution version of the original figure.

18. The typical step sizes and time scales correspond to those in the experiments of Yanagida et al., 2008 on a single-headed myosin motor, the so-called myosin VI, which makes large 40-nm steps while pulling a bead along an actin filament, as illustrated in the upper

left part of the top panel. The steps of the myosin are so large that the bead first has to catch up through Brownian motion, before a next step can be made.

19. Figure adapted from Yanagida et al., 2008.

20. Adapted with permission from *Langmuir 29*, 14588–14594. Copyright 2013 American Chemical Society.

21. Reproduced with permission from Caragine et al., 2018. Copyright (2018) by the American Physical Society.

22. As the field is rather new, there is no easily accessible overview yet where these developments are discussed. The paper by Stuij et al., 2019 gives a good entry into the literature through 2019.

23. Adapted from Stuij et al., 2019. We thank Simon Stuij and Peter Schall for providing us with the original data that allowed us to redraw the fluctuation figure from their publication.

24. See Spruijt et al., 2013 and references therein.

25. An example of this is when one probes the statistics of a single polymer in a dense polymer solvent by substituting hydrogen by deuterium in a small fraction of the polymers. As explained in section 5.5.3, this does not change the polymers chemically, but it allows us to extract their signal from the scattering of the neutrons from the deuterium. Examples where this technique is used are shown in figure 5.20 and figure 5.17.

26. So, in static light scattering one measures all the scattered light, i.e., integrated over all frequencies (or, equivalently, energies). Note that if you integrate $S(\vec{q}, \omega)$ given by (3.53) over all ω and use the fact that $(2\pi)^{-1} \int d\omega \, exp(-i\omega t') = \delta(t')$, you see that the expression reduces to (3.52) for $S(\vec{q})$ in terms of the equal time correlation function.

27. Such spectra can be found in many publications on light scattering. For an example of such a light-scattering spectrum from water, see, e.g., J. Xu et al., 2003. The full light-scattering spectrum is derived in section 8.4.8 of Chaikin and Lubensky, 1995. See section 3.7.1 for an introduction to the essentials of scattering methods.

28. The central peak in this spectrum corresponds to density fluctuations resulting from thermal fluctuations. The intensity of this peak grows as one approaches the critical point, and the ratio of the strength of this peak to the sound-mode peaks can also be calculated explicitly from the thermodynamic fluctuations—see e.g. Landau and Lifshitz, 1980.

29. A very accessible elementary introduction and overview of critical scattering theory is given by Kociński and Wojtczak, 1978. For a review of the use of light scattering as a probe of interfaces and surfaces, see Earnshaw, 1996; of colloids and interface, see Johnson, 1993; and of polymer monolayers, see Cicuta and Hopkinson, 2004. Probing liquid crystals with light scattering is reviewed by Val'kov et al., 1994.

30. For translation-invariant systems, Fourier modes diagonalize the energy as they are eigenmodes of the translation operator. You may be familiar with this from quantum mechanics: if the energy is translation-invariant, it commutes with the momentum operator and both momentum and energy have the same eigenfunctions; and Fourier modes are eigenfunctions of the momentum operator. Compare the derivation in solid state physics of the fact that Bloch waves are eigenfunctions of the Hamiltonian of a periodic system.

31. We can illustrate this for the general case in which the total energy is written to quadratic order in the fields as $E_{tot} = \sum_q E(q)|A(q)|^2$. Here $A(q)$ is the Fourier transform of some real physical field, so that $A(-q) = A^*(q)$, and $E(-q) = E(q)$. If we now write $A(q)$ in terms of its real and imaginary parts, $E_{tot} = \sum_q E(q)[A_r^2(q) + A_i^2(q)] = \sum_{q>0} 2E(q)[A_r^2(q) + A_i^2(q)]$, we see that the real and imaginary parts are nicely Gaussian distributed real variables. Due to the correlation of modes q and $-q$, we have for these $\langle A_r^2(q)\rangle = \langle A_i^2(q)\rangle = 1/2k_BT/(2E(q))$. But when we handwavingly put $\langle|A(q)|^2\rangle = 1/2k_BT/E(q)$ by treating each wavenumber as a single independent mode, we do get the proper answer, as $\langle|A(q)|^2\rangle = \langle A_r^2(q)\rangle + \langle A_i^2(q)\rangle = 2[1/2k_BT/(2E(q))] = 1/2k_BT/E(q)$—a rare occasion where being sloppy a second time masks an earlier careless step. If you want to be even more precise, you have to pay extra attention to the $q = 0$ mode, for which this doubling does not occur. But this too does not change the answer as it is only a single term of order $1/L$.

32. For the analysis of bending fluctuations of a tubule, see Janson and Dogterom, 2004.

33. The bending shape fluctuations of the experiment in figure 3.13.b is presented in detail by Loftus et al., 2013.

34. Thermally excited capillary waves are discussed in section 3.3 of Safran, 1994.

Chapter 4

1. Source: https://ictv.global/report/chapter/virgaviridae.

2. The significance of this paper by Onsager, 1949 is nicely put in perspective by Frenkel, 2000. Besides being the first paper showing that order can arise from entropy, Onsager's analysis appears to be the first example of the use of density functional theory.

3. Note that Witten and Pincus, 2004 use cgs units, so in comparison with our discussion of the Van der Waals interaction, factors like $4\pi\epsilon_0$ are absent in their treatment.

4. As you may have noted, in our discussion we assume that the electric field responds instantaneously to the fluctuating dipole. This is a very good approximation, but at some very large distances where the potential is normally already very small, the finite speed of propagation of the field does start to play a role. These so-called retardation effects, discussed, for instance, in chapter 5 of Safran, 1994 and section 1.4.2 of Kleman and O. D. Lavrentovich, 2003, were first analyzed by Casimir and Polder, 1948, who showed that the $1/r_{12}^6$ crosses over to a $1/r_{12}^7$ behavior at large distances. It should also be noted that already from London's original derivation of the Van der Waals interaction it is clear that the interaction is an expansion in inverse distances; e.g., the quadrupole moment gives a term of order $1/r_{12}^8$. A more accurate theory for calculating the macroscopic Van der Waals force is the Lifshitz theory (which goes beyond pairwise additivity and also takes retardation effects into account). For a complete overview and historical introduction, see Israelachvili, 2011.

5. For a nice introduction to an overview of the depletion interaction, and of the phase behavior of colloids resulting from it, see Lekkerkerker and Tuinier, 2011.

6. Indeed, the depletion effect was discussed first in the context of polymers by Asakura and Oosawa, 1954, as Kleman and O. D. Lavrentovich, 2003 discuss in section 14.3.2.

7. Even though this is somewhat buried in the literature, in the field of non-equilibrium pattern formation discussed in chapter 8 it is known that while localized objects in the real Landau equation have an exponential interaction, localized objects like the source and sink solutions discussed in section 8.7 can have power law interactions if their interactions are mediated by so-called phase winding solutions. Two examples where this insight is used can be found in the papers by Riecke and Rappel, 1995 and Clerc et al., 2005. Also, in this field there are many cases of patterns with oscillatory decay of the field away from localized structures, resulting in locking of structures at particular distances. Early examples of this phenomenon can be found in the papers by Coullet et al., 1987 and Dee and van Saarloos, 1988. A brief review of the interactions of such structures in pattern formation is given by Coullet, 2002.

8. For instance, many colloids are made of silica. For silica in contact with water the silanolgroups tend to ionize, Si-OH $\rightarrow$ SiO$^-$ + H$^+$, leading to a negatively charged surface of silica colloids with positively charged counterions in solution.

9. The explanation of the long time tail effects has been advanced in particular by the mode coupling analysis in the papers by Ernst et al., 1970 and Ernst et al., 1976. Dorfman and Cohen, 1970 pursued a kinetic approach while Bedeaux and Mazur, 1975 and Mazur, 1975 focused on the renormalization of transport coefficients. The long time tail effect is actually simple to understand, even quantitatively. In addition to the spreading through Brownian motion, one has to take into account that according to linear hydrodynamics, the transverse part of the velocity field (in view of the condition $\nabla \cdot \vec{v} = 0$, in Fourier modes $\vec{v}$ is transverse to $\vec{k}$: $\vec{k} \cdot \vec{v}_{\vec{k}} = 0$) obeys the diffusion-type equation $\partial_t \vec{v} = \nu \nabla^2 \vec{v}$ for its d-1 transverse components. So a small velocity fluctuation also spreads out diffusively. When the Brownian motion and velocity relaxation are combined, one obtains the dominant $t^{-d/2}$ relaxation of the velocity correlation function (van der Hoef and Frenkel, 1995). Note by the way that this is a rare example in linear hydrodynamics in which the time derivative term $\partial_t \vec{v}$ cannot be neglected, because the fast relaxation of the velocity on the small (near-molecular) scale is actually at the origin of the long time tails.

10. Let's say one starts with a particle in the center. One then introduces a new particle far away and lets it diffuse around till it touches the first particle in the center. The diffusing particle then sticks to the first particle at the point where it touches it, so a cluster of two particles is formed. Next, a third particle is introduced far away; it is allowed to diffuse around, till it hits the cluster in the center and sticks too. By repeating this, one gets an ever-growing cluster (aggregate) in the center, which is very branched and fluffy, as in the example of the experimental fractal aggregate shown in figure I.5. The model is most easily simulated on a lattice, but one then does have to be aware of strong lattice effects. You may try to write computer code to do this yourself, but you can also find Python code on the internet. The opposite regime, when the sticking probability of a particle's joining the cluster is so small that the attachment process becomes the rate-limiting process, is referred to as the rate-limited aggregation regime. Clusters in this regime tend to be more compact, i.e., their fractal dimension is larger.

11. For an elementary introduction to regular fractals, see, for instance, sections 7.1 and 7.2 of Kleman and O. D. Lavrentovich, 2003.

12. Multifractal means that the structures are actually not characterized by a single exponent, but that they have a whole continuous distribution of fractal exponents. See Halsey et al., 1986 and Tél and Vicsek, 1987. The multifractal nature of DLA clusters is discussed by Vicsek et al., 1990. For a full discussion of DLA and other related growth models, see Meakin, 1998.

13. The q^{-4} fall-off of the scattering is usually referred to as Porod's law. At the scale of the particle, one has to go back to the form (3.51) as translation invariance is lost on the scale of the particles. The derivation of Porod's law for arbitrary shaped particles is discussed in an accessible way in Ciccariello et al., 2002 and Bray, 1994.

14. To see this, note that with $\rho(r) \sim r^{d_{\mathrm{f}}-d}$ we get $S(q) \propto \int \mathrm{d}^d \vec{r} \, r^{d_{\mathrm{f}}-d} \, \exp(i\vec{q}\cdot\vec{r})$, which upon changing to the variable $\vec{w} = q\vec{r}$ can be written as $S(q) = q^{-d_{\mathrm{f}}} \int \mathrm{d}^d \vec{w} w^{d_{\mathrm{f}}-d} \, \exp(i\hat{q}\cdot\vec{w})$.

15. Note that the value of d_f which is typically found in DLA simulations of three-dimensional clusters is around 2.2. In DLA simulations, particles do not interact till they touch. In experiments of colloidal aggregation, forces between colloidal particles typically do play a role. Indeed, careful experiments by Burns et al., 1997 have shown that the effective fractal dimension of aggregating colloidal clusters in three dimensions depends strongly on the salt concentration. At high salt concentrations, when electric charges are largely screened (see section 4.4.1), d_f is indeed observed to be around 2.2, but at low salt concentrations, when repulsive electric charges are important, d_f reduces to about 1.8 (see figure 5 of Burns et al., 1997). Note that the paper by Glover et al., 2000, on which our figure is based, also presents data on the size distribution of the clusters and measurements from sedimentation experiments. The light-scattering data are also discussed in the overview by Bushell et al., 2002.

16. In simulations of hard cubes, Smallenburg et al., 2012 observed a crystalline ordered lattice of size of size 41^3 with only 40^3 cubes. This implies a vacancy density of about 6%. The cubic lattice formed by cubic particles is comparable to a three-dimensional version of a slide puzzle, with very mobile vacancies which can easily 'shift' along the crystalline axes. Note that unlike crystalline lattices of atoms which are studied at low temperatures so that vacancies hardly move, the hard core cube system is effectively an infinite temperature system, so the order is entropic and vacancies are very mobile.

17. We thank Job Thijssen from the University of Edinburgh for permission to use this figure from his 2007 PhD thesis at Utrecht University. The original image was taken by Dannis 't Hart.

18. The idea of using self-assembled colloidal crystals to make photonic crystals was discovered by several groups more or less at the same time. See van Blaaderen, 1998 for an early commentary on these inventions and their importance, and for references to the various groups which contributed. For a more recent review of photonic crystals, see, e.g., Ruda and Matsuura, 2017.

19. $\phi = 0.76$ is close to the glass transition value in this two-dimensional system, as determined by comparison with mode coupling theory.

20. When a fit to the data is made with expression (4.24), Hunter and E. R. Weeks, 2012 find $C = 1.15$ and $\phi_0 = 0.638$.

21. Typical speeds for micron-sized particles are about 1 micron per second. Due to the rotational diffusion, the Janus particles perform a random walk with diffusion coefficient of order $0.1 \, \mu m^2/s$ for H_2O_2 concentrations up to 10%, which is of the same order of magnitude as the diffusion coefficient due to molecular collisions, given by the Einstein relation (see section 3.5.1). Such microswimmers are found to collect in a well-defined layer about half a micron from the wall (Ketzetzi et al., 2020). The values for the speed and diffusion coefficient are taken from this paper, so it is important to keep in mind that these are values for particles moving along a wall.

22. This figure is inspired by figure 1 of the review by J. Zhang et al., 2017 of the various applications of active colloids.

23. See, e.g., Paxton et al., 2006.

24. See, e.g., the experiments by Di Leonardo et al., 2010 for a micromotor with a shape similar to the one drawn which is driven by *E. coli*. In their experiment, the inner gear is about 48 μm wide, and 10 μm thick. In a bath with *E. coli*, the gear rotates at about 1 rpm.

25. In the experiments of Bricard et al., 2013, which stressed the use of the Quincke rotation mechanism for creating a model system, spheres with radius of 2.4 μm are used, and typical speeds of the beads are on the order of mm/s.

26. Gagnon et al., 2014 use a particle like the one sketched by attaching a sphere to a rod with total length of about 3 mm, suspended in a wormlike micellar solution. When the rod-like particles are made to wiggle by an oscillating magnetic field, they move.

27. See Michelin et al., 2013 and in particular Maass et al., 2016 for a review.

28. A practical rheometer is based on a 'cone and plate' geometry; as both the gap and the rotation speed increase linearly with the distance from the center, the shear rate is independent of the distance from the center. From the torque needed to rotate the cone with a given rotation rate, one infers the rheology.

29. The vorticity or local rotation of a fluid flow equals $\vec{\omega} = \vec{\nabla} \times \vec{v}$, which for the simple shear flow considered here gives $\omega_z = \dot{\gamma}$.

30. The effective viscosity is defined as the ratio of the shear stress σ to the shear rate $\dot{\gamma}$, see the discussion leading to equation (4.29).

31. In the bottom panel, the $\bar{\eta}$ denotes the viscosity of the suspension normalized by the limiting values for small shear rates/Péclet numbers. This way of plotting allows us to easily compare the trends of dispersions with various densities and interactions, as all curves approach $\bar{\eta} \approx 1$ for small Péclet numbers. See figure 4 of N. J. Wagner and Brady, 2009 for actual data.

32. We stress again that the cartoon of the polymer brushes is not to scale: the brushes typically form a layer with thickness up to 1/50th the diameter of the colloid.

33. N. J. Wagner and Brady, 2009 give a nice and accessible overview of these effects, with references to the literature.

34. So, these hydrodynamic interactions are derived by analyzing the linear low Reynolds number hydrodynamic equations for arbitrary configurations of the particles, relating their velocities to the forces. For these generalized mobilities, three- and four-particle interactions are important if one expands up to seventh order in the inverse of the interparticle distances (Mazur and van Saarloos, 1982).

35. For a more extensive discussion of these effects, which obviously are intimately connected with the emergence of a yield stress of a sample left at rest, we refer you to Bonn et al., 2017.

36. See Moller et al., 2009, who report results for a 4% bentonite in water suspension; for bentonite the viscosity decreases initially with time. Bentonite is a natural clay which is often used in civil engineering, e.g., as a component of tunnels, as bentonite is not very permeable by water at high concentrations.

37. Provided there are no concentration gradients which would give rise to Marangoni flow effects as discussed in section 1.13.3. Note that droplets flowing in emulsions will interact via relatively long-range hydrodynamic interactions. This might affect the exponents, in particular, the flow exponent Γ. In other words, the exponents for systems with hydrodynamic interactions may differ from the ones in simulations, as these have been almost exclusively done with spheres which do not interact as long as they do not touch.

38. The system studied is of a microemulsion of castor oil in water, stabilized with a surfactant. Mean droplet size is about 3.2 μm.

39. The value of ϕ_J obtained from the fitting corresponds well to the random close packing value of spheres, even though the droplets have a polydispersity of about 20%.

40. The scaling exponents resulting from the best fit are $\Delta = 2.13 \pm 0.11$, $\Gamma = 3.84 \pm 0.44$.

41. Reprinted with permission from Paredes et al., 2013. Copyright (2013) by the American Physical Society.

42. The droplets in the emulsion are reported to have a polydispersity of 20%. The polydispersity will change the random close packing value, but the work by Desmond and E. R. Weeks, 2014 indicates that the effect on the random close packing values will be immeasurably small for a 20% polydispersity.

43. Illustration based on results from figure 5a of Forterre and Pouliquen, 2008.

Chapter 5

1. These types of polymers have been developed by Leibler and coworkers; see Cordier et al., 2008.

2. See Montarnal et al., 2011 for more details and Van Zee and Nicolaÿ, 2020 for a review on the topic.

3. This work is a part of ESPCI-VUB consortium to develop soft robots, under the project name SHERO. More details can be found in Terryn et al., 2021.

4. Reproduced from Lysetska et al., 2002.

5. See Rubinstein and Colby, 2003 for a more extensive discussion of the relation between the ratio of the bond length and the Kuhn length, and the ratio between n and N. The relation between the two also involves some angular projection factors.

6. For an introduction to the theory of critical phenomena, see, e.g., Goldenfeld, 1992, Herbut, 2007, Nishimori and Ortiz, 2010, or Honig and Spalek, 2018.

7. See de Gennes, 1972b.

8. Note also that one has to be careful with how the step size is defined. In the ideal chain model, the steps of length ℓ_K are in arbitrary directions in three dimensions, and this conforms to the natural way to think in terms of steps or of the mean free path of particles that scatter. But in our discussion in one dimension in chapter 3 we used both the one-dimensional diffusion coefficient and the step size ℓ_1 in one direction in the estimate (3.21). Clearly for a particle diffusing in three dimensions the step size ℓ needed in (5.7) is the total step size in three directions, for which obviously $\ell^2 = 3\ell_1^2$. This is also the origin of the factor 3 in the diffusion expression on the right-hand side of equation (5.6).

9. In section 3.3.5 in our chapter on diffusion and Brownian motion, we made a short digression to discuss self-similarity, intermediate asymptotics, and scaling in the diffusion equation. In section 3.2.1 of their book, Witten and Pincus, 2004 approach the discussion of the end-to-end probability in the ideal chain model by starting with a scaling assumption. This analysis leads to the conclusion that the intermediate asymptotics equation for the end-to-end probability distribution is a diffusion equation. You may find it interesting to read this section and to reflect back on our discussion in section 3.3.5 and in particular the one at the end of section 3.4.

10. Here we see a nice illustration of the power of exploiting the insights from critical phenomena. As discussed by de Gennes, 1979 in chapter 10, de Gennes had already shown in 1972 that the statistical properties of polymers can be thought of as the $n \to 0$ limit of the O(n) model (de Gennes, 1972b). The critical exponents of the O(n) model in two dimensions were obtained exactly by Nienhuis, 1982. Actually, de Gennes's 1972 approximate result for ν_{SAW} in three dimensions, derived with the ϵ-expansion, was already remarkably close to the present value 0.588.

11. The results are obtained by mapping the O(n) model onto a solid-on-solid model for surface heights; see Nienhuis, 1982.

12. See also section I.3 of de Gennes, 1979 or section 3.1.2 of Rubinstein and Colby, 2003.

13. The applicability of the wormlike chain (WLC) model to biopolymers was advanced in particular in the influential paper by Marko and Siggia, 1995, who used it to analyze the stretching of DNA. Much in this section is based on their very readable paper. As we will see in figure 5.16, deviations from the wormlike chain behavior are often signals of changes in the DNA due to overstretching. The WLC model is a continuum version of a model introduced by Kratky and Porod, 1949. For this reason the model is sometimes (e.g., in the classic book on polymer dynamics by Doi and Edwards, 1986) referred to as the Kratky-Porod model.

14. These images are based on cryo-TEM (low-temperature transmission electron microscopy) imaging of the actual structures built with this DNA technology. Note that a lock-and-key binding site is clearly visible on the reconstructions.

15. These structures are reconstructed using the so-called mrDNA and oxDNA packages. OxDNA is a simulation package originally developed to implement the coarse-grained DNA model, but it is nowadays used for different types of simulations. The electron density map from the cryo-TEM reconstruction is used to generate the pseudo-atomic model.

16. For sufficiently strong interactions, the building blocks assemble into a sheet that closes into a tubule with a diameter and chirality that can be programmed.

17. The last expression for $d\theta/ds$ is derived in (2.39) with the help of figure 2.17.

18. See section 6.2 of Kardar, 2007.

19. For an introduction to polyelectrolytes and the effects of screening on their bending stiffness, see section 3.4.5 of Witten and Pincus, 2004, or section 11.2.1 of de Gennes, 1979. Accurate experimental data as a function of ion concentration for various types of ions are reported by Guilbaud et al., 2019. We have given (5.32) in the form in which it is often stated, but note that with (4.16) and (4.15) we can write $\lambda_B/(4A^2\kappa^2) = 1/(32\pi A^2 c)$, where c is the ion concentration.

20. See, e.g., Barrat and Hansen, 2003, from which this picture is adapted, for further discussion of this.

21. The molecular diameter of DNA is only about 2 nm, but in solution the effective diameter is larger due to the electrostatic repulsion: the range increases with decreasing solvent ion density, due to Debye-Hückel screening. The value we use here is appropriate for sodium ion density of 10^{-2}M; see section III.C of Marko and Siggia, 1995. The formula for the swelling as a function of z is given in equation (5) of Odijk and Houwaart, 1978 and is derived with the expansion method discussed by Yamakawa, 1971 for weak excluded volume effects. One should keep in mind that the parameter z used in this paper differs by a factor of $(3/2\pi)^3$ from the one defined in our equation (5.33), and hence it is smaller by a factor of 8. For the example given in the text, the formula of Odijk and Houwaart, 1978 gives an enhancement of the mean radius squared by about 10%–15%. The swelling of DNA for small extensions due to excluded volume effects is also discussed by Marko and Siggia, 1995 in their section III.C.

22. Parseval's theorem (5.39) follows straightforwardly from substituting the expression of $g(s)$ by its expression in terms of its Fourier components, analogous to (5.38): $L^{-1} \int_0^L ds\, g^2(s) = L^{-1} \sum_{qq'} \int_0^L ds\, g(q)g(q')e^{i(q+q')s} = \sum_{qq'} g(q)g(q')\delta_{qq'} = \sum_q g(q)g(-q) = \sum_q |g(q)|^2$.

23. Remember that $\pi^{-1} \int_{-\infty}^{\infty} dx\, (x^2+1)^{-1} = 1$.

24. The percentage error of this interpolation formula is reported to be up to about 15%. More accurate interpolation formulas, with accuracy of $\pm 1\%$ and less, are given by Bouchiat et al., 1999 and Petrosyan, 2016.

25. We thank Márcio Rocha for providing us with the data from Rocha, 2015 which are drawn in the graph.

26. In this particular experiment, the Brownian motion of the beads is used to determine the curvature of the optical well potential only once, before the actual experiment is performed—during the experiments, the pulling force on the DNA is determined simply by the average position in the optical trap. The precise procedure is described in section 2.1.5 of Gross et al., 2010.

27. See, e.g., Beaucage, 1996, who shows data over two orders of magnitude in q for various polymer blends, and who discusses in detail the fitting to extract parameters such as the R_g and the fractal dimension d_f. The range of q is often too small to distinguish convincingly between q^{-2} and $q^{-5/3}$ behavior. For data on poly(acrylic acid) consistent with $\nu = 1/d_f \approx 3/5$, see Spruijt et al., 2013.

28. Adapted from "Cross-over in polymer solutions" by B. Farnoux et al., (*J. Phys. France* 39, 77-86). © 1978 EDP Sciences.

29. An example of a study of the behavior of the crossover length ξ_ϕ with concentration is Falcão et al., 1993; the crossover is also reviewed by Mortensen, 2001; see his figure 13.

30. This way of thinking was advanced in particular by de Gennes, 1979.

31. In the data analysis of Witten and Pincus, 2004, the derivative $d\tilde{\Pi}/d\tilde{\phi}$ is normalized for each polymer sample so that this approaches 1 in the small $\tilde{\phi}$ limit, while $\tilde{\phi}$ values are normalized so that the second derivative $d^2\tilde{\Pi}/d\tilde{\phi}^2$ approaches 2 at low concentrations. The

data on polyisoprene with molecular weight $58 \cdot 10^4$ are from Adam et al., 1991 and the data on poly(α-methylstyrene), a polymer like polystyrene, from Noda et al., 1981. The shortest molecules in these experiments have molecular weight of about $7 \cdot 10^4$, the longest ones $747 \cdot 10^4$.

32. We are grateful to Tom Witten for providing us with the data underlying figure 4.3 from Witten and Pincus, 2004, which allowed us to replot the data.

33. There is good experimental evidence (Anisimov et al., 1990) that the nematic-to-smectic-C transition is also weakly first order due to the same mechanism, but here the interaction parameter cannot be tuned easily.

34. See in particular Žagar et al., 2015 for a discussion of simulations of athermal models of crosslinked fibers, and van Dillen et al., 2008 for a comparison of the force-extension curves of the thermal and athermal models, and for a derivation of the 3/2 law (5.71).

35. Actually, for F-actin, fibrin, collagen, vimentin, and polyacrylamide, the real part of the dynamic shear modulus measured at a frequency of 10 rad/s is shown (compare the discussion in section 2.5). For small enough frequencies this should be close to the static shear modulus. See Storm et al., 2005 for details.

36. Reproduced with permission from Broedersz and MacKintosh, 2014. Copyright (2014) American Physical Society.

37. There is actually a slight subtlety here. In section 5.4.4, in order to have the opportunity to expose the general approach in section 3.9, we used in (5.38) complex modes which are only independent with periodic boundary conditions, and hence with $q = \pm 2\pi/L, \pm 4\pi/L, \ldots$. For the clamped polymer, also a mode $\sin(\pi z/\ell_{\text{crlk}})$ is allowed. Allowing for this possibility and using real Fourier modes then gives a sum over positive integer values of n as in (5.59). The lesson from this is again that one always has to be careful when writing a field in Fourier modes.

38. See Broedersz and MacKintosh, 2014.

39. As discussed by Broedersz and MacKintosh, 2014, the results discussed here are exact for $\ell_{\text{crlk}} \ll \ell_{\text{p}}$, while for $\epsilon \to -1$ finite temperature buckling must be included.

40. The contribution to σ_{xy} also involves a term $\sin\theta \cos\theta \sin\phi$. Taking all terms together then implies that the angular average becomes $(1/4\pi) \int_0^{2\pi} \mathrm{d}\phi \int_0^\pi \mathrm{d}\theta [\cos^2\theta \sin^3\theta] \sin^2\phi = 1/15$. The analysis sketched by Broedersz and MacKintosh, 2014 is done in detail in a pedagogical way by Storm, 2018. Note that our angle ϕ differs by $\pi/2$ from the one of Storm.

41. See figure 15 of Broedersz and MacKintosh, 2014.

42. See figure 15 of Broedersz and MacKintosh, 2014.

43. Reprinted with permission from Žagar et al., 2011. Copyright 2011 American Chemical Society. See also the follow-up paper by Žagar et al., 2015, where the model is studied in more detail for parameters relevant to actin.

44. See figure 19 of Broedersz and MacKintosh, 2014.

45. See figures 21 and 27 of Broedersz and MacKintosh, 2014.

46. For reviews of the origin of drag reduction, which is still not completely understood, see White and Mungal, 2008 and Graham, 2014.

47. The precise conditions are not stated by Chu, 1998, but in the original work by Perkins et al., 1994b, similar pictures are taken 1.5 seconds apart, and the concentration is stated to be 7 molecules per cubic micron.

48. Reprinted with permission of *Reviews of Modern Physics* from Steven Chu's 1997 Nobel lecture, Chu, 1998. Copyright (1998) American Physical Society.

49. See, e.g., Rubinstein and Colby, 2003, Strobl, 1997, or Doi and Edwards, 1986 for a discussion of the Rouse model. Besides the fact that (5.75) is a natural extension of (3.19) for a string of beads, you can also easily convince yourself from (3.22) that the diffusion coefficient of the center of mass of N walkers is $1/N$ times the diffusion coefficient of a single Brownian walker.

50. Figure 9.5 of Rubinstein and Colby, 2003, on which our plot is based, actually reports data in terms of the molar weight M rather than N, but as the two are proportional and scaled variables are used, we have labeled the scaled variables in terms of N. See the caption of the original figure for references and values of M_{c} used in the rescaling. Hydrogenated polybutadiene refers to a particular preparation process which reduces the polydispersity.

51. See https://web.mit.edu/nnf/.

52. What we consider here is actually the so-called first normal stress difference N_1. The second normal stress difference N_2 is defined as $N_2 = \sigma_{yy} - \sigma_{zz}$ in the case of shear $v_x(y)$. This stress difference is typically (much) smaller than N_1 and is less relevant to the hoop stresses that cause most instabilities. For a velocity $v_x(y) = \dot{\gamma}y$, the constitutive equation (5.80) of the Oldroyd-B and upper convected Maxwell model actually gives $\sigma_{xx} = 2\lambda\eta_{\text{p}}\dot{\gamma}^2$, while $\sigma_{yy} = 0$. You will derive this in problem 5.9.

53. Once shear thinning effects discussed in section 4.6 become important, as is the case for most industrially relevant polymers (Pahl et al., 1991; Rauwendaal, 2001; Strobl, 1997), deviations from linearity are typically observed in the shear rate dependence of the Weissenberg number. Boger fluids (James, 2009) are polymer solutions whose rheology closely follows that of the Oldroyd-B model discussed in section 5.10.3. They are therefore popular experimental model fluids.

54. An extension of the equations which is often considered is a finitely extensible nonlinear elastic (FENE) model. FENE-P stands for a particular closure of the equations by Peterlin. See Larson, 1999, Morozov and Spagnolie, 2015, or Bird et al., 1987a and Bird et al., 1987b for details.

55. Because of its bilinear nature, in regions of high shear rate some components of $\underline{\sigma}_{\mathrm{p}}$ tend to grow exponentially. This effect makes the model prone to numerical instabilities, especially at higher Weissenberg numbers. For this reason the rapidly increasing susceptibility to numerical instabilities with increasing Weissenberg number is sometimes referred to as the high-Weissenberg number problem. See, e.g., Owens and Phillips, 2002 for a general introduction to computation rheology of non-Newtonian fluids, and Hulsen et al., 2005 for a discussion of ways to deal with the exponential divergence.

56. For the Oldroyd-B model the Weissenberg number is also linear in $\dot{\gamma}$ but the shear stress is enhanced by the solvent viscosity as well. Hence the Weissenberg number is reduced by a factor $\eta_{\mathrm{p}} / (\eta_{\mathrm{s}} + \eta_{\mathrm{p}})$ relative to (5.81).

57. Reprinted from Groisman and Steinberg, 2004, which is published under Creative Commons license. The earlier article Groisman and Steinberg, 2000 about these experiments was instrumental in identifying viscoelastic turbulence as an important field in itself; see Larson, 2000.

58. A nice example is the observation by Pahl et al., 1991 that for a wide range of industrial polymers, the so-called melt fracture instability occurs at a Weissenberg number of about 4.8. The term 'melt fracture' refers to the fact that the instability can become so violent that the extrudate fractures (Rauwendaal, 2001). As discussed by Morozov and van Saarloos, 2007, where the data from Pahl et al. are reproduced, this stimulated investigations about whether viscoelastic pipe flow would be nonlinearly unstable. Evidence for this scenario was found by Pan et al., 2013 using a cleverly designed microfluidics experiment.

59. In the literature, also the Deborah number De is used. For a flow with typical velocity U and length scale L, De $= \lambda U / L$. For the circular geometries usually employed in studies of viscoelastic instabilities, the typical length scale L is the radius of the setup, so the criterion (5.82) can also be written as Wi De $> M^2$. This is the form in which it is also given by Pakdel and McKinley, 1996. Interestingly, in geometries which are not curved, like the flow through straight channels studied by Pan et al., 2013, the criterion can also be used to estimate the threshold of the nonlinear perturbation needed to drive the flow unstable (Morozov and van Saarloos, 2007).

Chapter 6

1. The origin of the anisotropic magnetic coupling is often associated with the presence of benzene-type rings in the molecules: a magnetic field pointing through the ring produces a small Faraday current in the ring, which itself creates a magnetic field that interacts with the imposed one, while a field in the plane of the ring has no such effect. The electric polarizability, on the other hand, is often associated with specific side or end groups which are easily polarized. For instance, for the 8CB molecule shown in figure 6.1 the CN end group is mostly responsible for its electric polarizability.

2. See Freelon et al., 2011 for data on the molecule 8S5.

3. Adapted from H. N. W. Lekkerkerker and G. J. Vroege, "Liquid crystal phase transitions in suspensions of mineral colloids: new life from old roots" (*Phil. Trans. R. Soc. A.* 371, 20120263, (2013)), based on the computer simulations of Peter Bolhuis and Daan Frenkel, "Tracing the phase boundaries of hard spherocylinders" (*J. Chem. Phys:* 106(2), (1997)). Reprinted and adapted with permission by AIP Publishing.

4. See Onsager, 1949. The analysis of Onsager is retraced in problem 6.1.

5. For a very accessible overview of the diverse world of liquid crystals, see Palffy-Muhoray, 2007.

6. For several examples of nematic and smectic phases of rod-shaped colloidal particles of a length ranging from 50 nm (much less than the TMV virus) to 5 µm, see, e.g., Sharma et al., 2009 and Kuijk et al., 2011.

7. See section 4.5.3 for a brief discussion of photonic bandgap materials made with colloids.

8. Remember that a delta function $\delta(x)$ has a dimension of 1 over length, and so $\delta(\vec{r})$ has a dimension of 1 over volume. Hence the volume factor V in (6.2) nicely makes Q_{ij} dimensionless.

9. From the point of view of analogies and differences with liquid crystal defects, the books by Vollhardt and Wölfle, 1990 and G. E. Volovik, 1992 are worth consulting for an introduction and overview of the various order parameters and defects of the superfluid phases of ^{3}He.

10. Reprinted and adapted from "Pretransitional Phenomena in the Isotropic Phase of a Nematic Liquid Crystal" by T. W. Stinson, III and J. D. Litster (*Phys. Rev. Lett.* 25, 503) © 1970 American Physical Society and "Landau theory of the nematic-isotropic phase transition" by Egbert F. Gramsbergen et al. (*Phys. Rep.* Vol. 135, Iss. 4) © 1986 Elsevier. The transition temperature of MBBA is $T_N = 45.2$. The data are reproduced with an enlargement of the critical region in the review by Gramsbergen et al., 1986. The application of the Landau–de Gennes theory was in particular explored by Stinson and Litster, 1970, who reported light-scattering-intensity measurements for MBBA as well as magnetic birefringence and dynamical slowing down data. The light-scattering data of Stinson and Litster, 1970 are reproduced by de Gennes and Prost, 1993 and Chandrasekhar, 1992.

11. Gramsbergen et al., 1986 give the value $S(T_N) = 0.395$ for 8CB below their equation (4.41). The conclusion that S rapidly approaches 1 when the temperature is lowered below T_{NI} follows from their figure 16, in combination with table 6, with data for the various temperatures. Beware that in this review paper $T^\dagger$ is used for our T^*; see their figure 7 and equations (2.7) and (2.8).

12. In fact, it actually took some 30 years to establish (6.6) as the appropriate free energy expression—see section 3.1.3 of de Gennes and Prost, 1993.

13. From writing a free energy as an expansion in the tensor $\underline{Q}$, one sees that the effective elastic constants vary as S^2. If the nematic-isotropic transition were second order, this dependence would be much larger than it actually is: the fact that already $S(T_{NI})$ is significant, due to the weakly first order nature of the transition (see section 6.2), helps to limit the temperature variation.

14. See, for instance, Oswald et al., 2004 for data of the elastic coefficients of 8CB.

15. The splay configuration analyzed here is a bit artificial, in three dimensions, as it is unstable to a bending-type deformation where $\hat{n}$ gets a component in the z-direction; the phrase sometimes used in this context is that "the director escapes in the third dimension" (Brinkman and Cladis, 1982). See problem 6.11. In thin quasi-two-dimensional samples, however, such defects are common.

16. Note that since the directions $\hat{n}$ and $-\hat{n}$ are equivalent, the actual physical periodicity is half of the distance over which the angle θ changes by 2π.

17. See, for instance, Durey et al., 2020 and Mundoor et al., 2019. For a somewhat more extensive discussion of the boundary effects, see section 3.1.4 of de Gennes and Prost, 1993, and for a full account of the surface physics of liquid crystals, see Sonin, 1995.

18. Vsevolod Konstantinovich Fréedericksz (1885 - 1944) was a Russian scientist. His last name is sometimes transcribed to Frederiks.

19. See Ruda and Matsuura, 2017 for a discussion of magnetic effects in liquid crystals and the molecular origin.

20. We use standard mksA units in this book. In the cgs units used in de Gennes and Prost, 1993 and Chandrasekhar, 1992 and much of the literature, the factor μ_0 is missing from the magnetic energy expression. In the electrical case, the factor ϵ_0 is absent in cgs units, while there is an extra factor $1/4\pi$.

21. The effect is larger the higher the electrical conductivity is. For example, free ions dissolved in the liquid crystal can lead to significant conductivity, and the mobility of such ions along the director is typically significantly larger than in the orthogonal direction. Hence there are focusing effects of the electrical currents for some director configurations. As stated, an AC voltage is usually applied to avoid such effects.

22. Sometimes some chiral liquid crystal molecules are added to make sure the twist is everywhere in the same direction.

23. With other electro-optic effects, switching times as fast as 30 ns can be achieved; see Borshch et al., 2013. Also the reduction of the switching time by playing with particular molecular shapes and other phases like the smectic-A phase continues to be explored (C. Meyer et al., 2021).

24. See G. D. Volovik and O. D. Lavrentovich, 1983 and O. D. Lavrentovich, 1998 for a discussion of boojums in nematic droplets and of their topological charges.

25. See the remarks at the end of section 5.1.4.5 of de Gennes and Prost, 1993. The torque term is given in their equation (5.17).

26. See Sebastián et al., 2020 for the domain structure interpretation of the experiments of Mandle et al., 2017.

27. The origin of flexoelectricity was first put forward by R. B. Meyer, 1969.

28. The virus in the image is acutally a mutant type fd-Y21M that exhibits a persistence length 3.5 times larger than the wild-type fd-wt. See Pouget et al., 2011 for more details on the work and Dogic and Fraden, 2014 for a review of experiments on viruses and the phase diagram. The latter also shows an image of the wild-type virus.

29. See Zhao et al., 2019 for a brief review of biological liquid crystals and the challenges these pose.

30. See, for instance, Verma et al., 2017 and Kwon et al., 2016. These authors give various references to the use of liquid crystal droplets as a sensing platform for developing immunoassays and for detecting pH levels, glucose, bacterial endotoxin, bacteria and viruses, and even cancer cells.

31. See Zhong and Jang, 2015 and references therein.

32. The phase diagram of mixtures of a molecular liquid crystal and anisotropic particles like plates or rods is indeed very rich and includes several unusual liquid crystal phases, like biaxial nematics (Mundoor et al., 2018). One can even make knotted structures which form crystal phases (Tai and Smalyukh, 2019).

33. See Mušević, 2017 for an example of a Janus particle coated with gold on one side and with DMOAP on the other side (to ensure good homeotropic (normal) boundary conditions of the director) in a 10-micron-thick cell of the liquid crystal 5CB.

34. Directed motion of droplets full of swimming bacteria, which enhance the Brownian motion of the droplet, is discussed by Rajabi et al., 2020. Liquid crystals are increasingly being explored as examples of active media. For instance, in a thin liquid crystal sample, skyrmion-type topological states can be created, which upon driving with a field show collective dynamical behavior reminiscent of schools of fish (Sohn et al., 2019).

35. For a review of 'the magic lipid' monoolein and its use see Kulkarni et al., 2011. The molecule also features prominently in the introductory article by Aleandri and Mezzenga, 2020.

36. The phenomenon was first observed experimentally by Sanchez et al., 2012 and explained theoretically by Giomi et al., 2013. See Doostmohammadi et al., 2018 for a review of active nematics.

37. See D. J. Bishop and Reppy, 1978.

38. Halperin and Lubensky, 1974 show that the term actually already integrates to zero in the total free energy for a given $\delta\hat{n}$. That this has to be so can easily be seen from the fact that upon a partial integration it gives a term proportional to the overall phase ϕ. Clearly the free energy cannot depend on the phase, so the term in front of it must be zero. Halperin and Lubensky, 1974 show explicitly that this is so. But even if the term would not integrate to zero for arbitrary $\delta\hat{n}$, this term, which is odd in $\delta\hat{n}$, would average to zero.

Chapter 7

1. For an introduction to the roughening transition, see J. D. Weeks, 1980 or section 10.6.2 of Chaikin and Lubensky, 1995. To our knowledge the roughening transition has not been observed in colloidal crystals (see Li et al., 2016 for a review of transitions in colloidal crystals) although properties of grain boundaries in two-dimensional crystals have been studied (Liao et al., 2018; Skinner et al., 2010).

2. See Grzhibovskis et al., 2016 for an application to the calculation of the shape of red blood cells in contact with an artificial surface.

3. Very much like the fact that the behavior near sharp folds in a piece of paper is determined by a balance of bending and stretching elastic terms.

4. See section 14.2.1.2 of Kleman and O. D. Lavrentovich, 2003 for microscopic estimates of the curvature moduli from the underlying molecular picture.

5. See discussion in Ramanathan et al., 2013 on the relation between the steric shape of amphiphiles and the preferred structures they form.

6. Reprinted with permission from C. Huang et al., 2011. Copyright (2011) by the American Physical Society.

7. The model is similar to the one used by Seung and Nelson, 1988 in their analysis and simulations underlying the discussion of buckling transitions in the virus shapes of section 7.3.

8. The bending terms κ^2 (curvature squared) in the energy give rise to fourth order derivative terms—see, e.g., equation (2.55)—while surface tension or stretching terms lead to second order derivative terms.

9. Cryo-TEM stands for cryogenic transmission electron microscopy, a technological breakthrough that earned its developers the Nobel Prize in Chemistry in 2017.

10. Image (a) from Graham Beards/Wikimedia Commons. Image (b) from Heymann, J. et al. "The Mottled Capsid of the Salmonella Giant Phage SPN3US, a Likely Maturation Intermediate with a Novel Internal Shell" (*Viruses* 12, no. 9: 910) CC BY 4.0.

11. Image (d) is adapted and reproduced with permission from J. Lidmar, L. Mirny, and D. R. Nelson (2003). 'Virus shapes and buckling transitions in spherical shells'. (*Phys. Rev. E* 68).

12. Euler's formula as given in (7.4) is valid on surfaces with the topology of a sphere. For surfaces with genus g illustrated in figure 7.4, the Euler's formula reads $F + V - E = 2(1 - g)$.

13. The analysis of problem 7.3 implies that there must be exactly 12 fivefold coordinated vertices in the absence of any other defects. If we deform the packing further by creating pairs of five- and sevenfold coordinated vertices, the number of fivefold coordinated vertices becomes larger. But creating such defects costs a lot of energy.

14. Adapted and reproduced with permission from J. Lidmar, L. Mirny, and D. R. Nelson (2003). 'Virus shapes and buckling transitions in spherical shells'. (*Phys. Rev. E* 68).

15. For a thin sheet of thickness d, the effective two-dimensional Young's modulus is $Y = E_Y d$, where E_Y is the Young's modulus of the material of which the sheet is made; see equation (2.31).

16. The fact that the two curves cross at a finite angle would suggest a first order transition at R_b, but this may well be the result of the lowest order expressions (7.5) and (7.7). The behavior near the transition is subtle, and it is actually not completely clear what the nature of the transition is (Seung and Nelson, 1988).

17. Adapted and reproduced with permission from J. Lidmar, L. Mirny, and D. R. Nelson (2003). 'Virus shapes and buckling transitions in spherical shells'. (*Phys. Rev. E* 68).

18. If $\hat{n}_\alpha$ is the normal of the triangular face α, the bending Hamiltonian is $\mathcal{H}_b = (\tilde{k}/2) \sum_{\langle \alpha\beta \rangle} (\hat{n}_\alpha - \hat{n}_\beta)^2$, where the sum runs over neighboring triangular faces. This term is very similar to (7.9), used for simulating tethered surfaces.

19. See, e.g., the micron-sized capsid-like vesicles observed by Béalle et al., 2011, and references in this paper.

20. A Landau theory for the crumpling transition together with a Flory-type treatment for the excluded volume interactions gives $R_g \sim L^{4/5}$ in the crumpled phase (Nelson, 2004).

21. Bowick and Travesset, 2001 conclude after an extensive review of all the work, "in conclusion the bulk of accumulated evidence indicates that flatness is an intrinsic consequence of self-avoidance."

22. We focus on the lowest order terms, so we ignore the nonlinear terms $(\partial u_k/\partial x_i)(\partial u_k/\partial x_j)/2$.

23. Since the free energy F in (7.10) orders of magnitude is quadratic in the elastic strain fields, the thermally excited Gaussian strain field fluctuations u_{ij} can be integrated out to yield a long-range effective contribution to the bending energy term. See Le Doussal and Radzihovsky, 2018 for an overview of these effects. Experiments on the bending modulus of graphene are reported by Blees et al., 2015. The enhancement of the bending modulus by a factor of 5,000 is shown in their extended data figure 4.

24. More concretely, short-wavelength fluctuations have been demonstrated to renormalize the effective bending rigidity k to zero on long length scales. The renormalization of the bending rigidity $k_R(q)$ as a function of wavenumber is indeed given by $k_R = k_0 - (3k_B T/4\pi) \ln(1/qa)$ (Nelson, 2004). But this renormalization with scale is only logarithmically slow, so the suppression of the bending modulus becomes noticeable only on exponentially long length scales.

25. Even though with excluded volume effects tethered surfaces are predicted to be always flat, this does not completely exclude the possibility of seeing thermal crumpling in a system where excluded volume effects are sufficiently small, just as biopolymers with a sufficiently large persistence length exhibit ideal chain behavior as far as their scaling is concerned—see section 5.4.3.

26. See Yllanes et al., 2017. The reduction of the bending modulus and Young's modulus is reported to be significant, and to be essentially only a function of the fraction of the removed area.

27. See Shankar and Nelson, 2021.

28. There are also interesting modern developments on thin plate fracture; see Marthelot and Roman, 2021.

29. The exponent β reported by Deboeuf et al., 2013 is written here as the exponent $1/\alpha$ for ease of comparison with the data of Vliegenthart and Gompper, 2006 and Matan et al., 2002. For similar reasons we have used L_0 for the diameter D used by Deboeuf et al., 2013.

30. Reprinted with permission from Deboeuf et al., 2013. Copyright (2013) by the American Physical Society.

31. See Deboeuf et al., 2013 for further discussion of this, based on typical parameters for a paper sheet and the force a human can apply by hand.

32. This is a form of memory formation in matter, which we discuss more systematically in section 10.2.

33. See Matan et al., 2002. The authors also study the time dependence of the height when a constant weight is put on the Mylar.

34. See E. M. Kramer and Witten, 1997 and Witten, 2007.

35. When comparing with the expression (2.47) for buckling of one-dimensional rods, keep in mind that the total buckling force is applied from the side over a length of order $R_0 \simeq L$. This is why the critical buckling force for the sheet is of order k/R_0, while the buckling force for the rod in (2.47) goes as $1/L^2$.

36. See E. M. Kramer and Witten, 1997 and Witten, 2007.

37. It may well be that the discrepancy has (partly) to do with the fact that the compression is imposed differently in numerical simulations and in real experiments. In the numerics, a force is applied to the outer edges on the sheet, while in the experiments the sheet is compressed with hard walls. In experiments it is found that crumpled elastic sheets indeed become orientationally ordered at the walls (Cambou and Menon, 2015), i.e., they are folded in particular along the walls.

38. Kantor et al., 1987 estimated $\nu \approx 0.8$ by simply crumpling pieces of paper and aluminum foil by hand. The linear scale in these Friday afternoon–type experiments varied only by about a factor of 10.

39. A recent discovery, suggesting possible new mathematical connections, is that of the scaling properties of a random walker who has to walk around a lake; see Vladimirov et al., 2020.

40. We can illustrate this as follows for a one-dimensional interface growing in the z direction. If the interface has a constant growth rate v_0 in the direction normal to the *actual* interface, and if this normal makes an angle θ with the z axis, then this constant normal growth leads, by a simple projection, to a contribution $v_0 / \cos\theta \approx v_0(1 + 1/2(\nabla h)^2 + \dots)$ to the growth $\partial_t h$ of the height in the direction normal to the flat *reference surface*.

41. See, in particular, Krug, 1997 and chapter 10 of Kardar, 2007 for an accessible discussion of the mapping onto the directed polymer in a random environment problem. Corwin, 2012 focuses on connections with various mathematical breakthroughs and mentions connections with random matrix theory, random tilings, and certain combinatorial problems.

42. This is because in the Fokker-Planck equation for $\mathcal{P}$, the nonlinear dynamical term gives an extra term $\frac{\delta}{\delta h}\int \mathrm{d}x\,(\lambda/2)(\nabla h)^2\mathcal{P}[h]$. This gives a term $\mathcal{P}\int \mathrm{d}x\lambda(\nabla h)^2\nabla^2 h$. The integrand can be written as $(\lambda/3)\nabla(\nabla h)^3$, so that the integral vanishes for the case of periodic boundary conditions. See section 9.6 of Kardar, 2007 for more details.

43. For an overview of electroconvection, see Kai and Zimmermann, 1989 or the brief introduction in section 3.1.3 of van Hecke et al., 1994. In the literature, these convection states are often referred to as turbulent, but they are not turbulent in the sense of displaying scaling of correlation functions over several decades. We therefore simply refer to the states as irregular or chaotic.

44. Interestingly, superfluid helium also has turbulent states dominated by vortex tangles (Donnelly, 1993). The superfluid flow velocity is proportional to the gradient of the phase of the superfluid wave function, so in this case, too, vortices (which in ^{4}He have a core of only about 1 Ångström) play a dominant role in determining the superfluid flow and the turbulence. There are quite a few indications of the presence of relatively sharp interfaces separating domains with vortex tangle dominated flow and domains with smooth superfluid flow without vortices (Nemirovskii, 2010).

45. Reprinted with permission from Takeuchi and Sano, 2010. Copyright (2010) by the American Physical Society. We are grateful to Kazumasa Takeuchi for providing the data underlying the plots in their paper.

46. See Johansson, 2000 and Prähofer and Spohn, 2000 and the review by Corwin, 2012 for the connection between interface fluctuations and random matrix theory. Takeuchi and Sano, 2012 also give various references to the literature.

47. The increased laser intensity of the trap presumably sucked many lipid membrane molecules into the trap and thus increased the surface tension.

48. Reproduced with permission from Bar-Ziv and Moses, 1994. Copyright (1994) American Physical Society.

49. See the paper by Tchoufag et al., 2022 for an overview of the refinements of the orginal work by Powers and Goldstein, 1997 on the propagating Rayleigh instability. The authors of this paper account for the intra-membrane viscosity and for the base flow of lipids. Interestingly, the Föppl–Von Kármán number (7.8) is one of the two dimensionless numbers characterizing such membranes. The other one is the so-called Scriven-Love number which characterizes the base flow speed; its size determines whether the instability is absolutely unstable or convectively unstable (van Saarloos, 2003).

50. Adapted from Powers and Goldstein, 1997. Copyright (1997) American Physical Society.

51. In practice, U may vary in time, but it does so on much longer time scales than the inner time scale associated with the variations in height with time, at fixed positions.

Chapter 8

1. The formation of patterns in a fluid heated from below is an (imperfect) example of the Rayleigh-Bénard instability discussed in section 8.3.1, but at the free surface, Marangoni effects discussed in section 1.13.3 play a role in the convective state as well. In a closed Rayleigh-Bénard cell, such Marangoni effects do not play a role.

2. Tears of wine involve both Marangoni effects and evaporation (Hosoi and Bush, 2001).

3. Patterned states can of course also arise in static equilibrium or stationary states; we encountered several examples to support this observation in earlier chapters. For instance, the formation of static wrinkles when a plastic sheet is stretched as in figure 2.15 can be described in terms of the minimization of the elastic energy given the external constraints. Likewise, we learned that when we consider viruses of increasing size, at some point their shape becomes increasingly icosahedral, as figure 7.6 exemplifies. Again, this development of the shape can be understood from the minimization of the elastic and bending energy, given the constraints associated with packing molecules on a closed surface with the topology of a sphere. Director patterns in nematic liquid crystals are also determined by energy minimization and boundary conditions.

4. Just as happens near phase transitions, fluctuations can become noticeable extremely close to threshold in systems that form patterns on a mesoscopic or macroscopic scale. We will not go into the effect of fluctuations on patterns near the transition, a topic which is reviewed by Sagués et al., 2007.

5. The book by Walgraef, 1997 also gives a broad introduction, with attention to materials science applications like irradiated materials and plastic instabilities. The comprehensive review by Cross and Hohenberg, 1993 still serves as an important reference guide for the field, but is less accessible for the newcomer. An elementary introduction to amplitude equations is given by van Hecke et al., 1994 and a beginner's guide to the complex Ginzburg-Landau equation by van Saarloos, 1994.

6. In fact, the Turing patterns which we will discuss in section 8.3.2 are a good illustration of how historically important ideas were stimulated by problems emerging first in the life sciences. These and any other examples can be found in the books by J. D. Murray, 2003, whose first edition was published in 1989 as a single volume. A well-known example of a problem with a varied history that includes disparate fields was the front propagation problem for second order partial differential equations introduced by R. A. Fisher, 1937 and Kolmogorov et al., 1937 in the same year (J. D. Murray, 2003). The problem emerged in population dynamics and was followed up on and refined mainly in the mathematics literature on second order partial differential equations (Aronson and Weinberger, 1978). It received a new impetus when physicists in the 1980s realized that there appeared to be a more general mechanism underlying front propagation into unstable states in more general classes of equations. Eventually, merging these ideas with concepts developed in plasma science led to several surprising exact results (Ebert and van Saarloos, 2000) for such fronts in general systems. See van Saarloos, 2003 for an overview of this field with its interesting history.

7. Mathematically, this is guaranteed by the so-called implicity function theorem.

8. Bifurcation analysis is developed for 'low dimensional systems,' i.e., systems described by ordinary differential equations for a few degrees of freedom. Pattern formation focuses on spatially extended systems governed (typically) by partial differential equations in which the spatial variation of the fields is a crucial ingredient, in particular because there is a finite wavelength instability.

9. See, e.g., Carilli et al., 2018 and references therein.

10. It is easy to convince oneself of the existence of this band of stationary periodic solutions of the ordinary differential equation $\epsilon u - \left(\partial_x^2 + q_c^2\right)^2 u - gu^3 = 0$, using standard methods. For instance, in the spirit of a shooting method, it is convenient to write the equation as a set of four first order equations. If one then starts at $x = 0$ with a particular amplitude $u = u_0$ and with $\partial_x u|_0 = 0$ and $\partial_x^3 u|_0 = 0$, one has one free parameter, namely $\partial_x^2 u$ at $x = 0$. Upon integrating the set of first order equations along x till u has reached the value u_0, one can use freedom to adjust $\partial_x^2 u|_0$ to also require $\partial_x u = 0$ at this point. Because the equation is reflection-symmetric in x, one then has $\partial_x^3 u = 0$ at that point as well, so one has obtained a periodic solution. Varying the value u_0 then yields the one-parameter family of stationary periodic solutions, characterized by u_0 or, equivalently, the wavelength. For a review of the construction and properties of such periodic solutions in fourth order equations, see Peletier and Troy, 2001.

11. We are grateful to Peimeng Yin for running some additional simulations based on the methods detailed in H. Liu and Yin, 2022a and H. Liu and Yin, 2022b, and for providing us with images and supporting information.

12. In more technical terms, F plays the role of a Lyapunov function for the dynamics. Generally, a Lyapunov function is the function associated with the dynamics of a system of (partial) differential equations which proves the stability of a fixed point or the attractive nature of a certain state or class of states. Note that the fact that the dynamics is downhill does not mean that the dynamics is attracted to the absolute minimum of F, since as we shall discuss in section 8.5.2, there is a band of stable periodic solutions which can attract the dynamics. For instance, a front propagating into the unstable state $u = 0$ generates a periodic solution with a wavelength different from the one with wavenumber q_c (van Saarloos, 2003). Likewise, a 'ramp,' a region where ϵ slowly varies from a value below zero to a positive value, selects a different well-defined periodic state (Cross and Hohenberg, 1993).

13. For example, an extension inspired by questions concerning Rayleigh-Bénard convection in rotating cells is discussed in section 5.2 of Cross and Greenside, 2009. It consists of adding terms $g_2 \hat{z} \cdot \vec{\nabla} \times [(\vec{\nabla} u)^2 \vec{\nabla} u] + g_3 \vec{\nabla} \cdot [(\vec{\nabla} u)^2 \vec{\nabla} u]$ to the right-hand side of (8.2). The term with g_2 does not derive from a potential, and it breaks the symmetry between clockwise and anticlockwise directions.

14. For instance, when a symmetry-breaking term bu^2 is added to the Swift-Hohenberg equation (8.2), the bifurcation becomes subcritical for $b^2 > 27g/38$. A detailed calculation is needed to show this (van Saarloos, 1989).

15. The Prandtl number is typically just somewhat larger than 1, because for most liquids thermal diffusion is somewhat slower than momentum diffusion due to viscous transport. An exception to this rule is formed by metallic liquids, for which the electrons dominate

the heat diffusion. For instance, liquid mercury has $\text{Pr} \approx 0.015$. Be aware that in discussions of Rayleigh-Bénard convection, the thermal diffusivity is often written as κ. We reserve κ_T for the thermal conductivity; see equation (1.27).

16. One can also think of the Rayleigh number as the ratio $\tau_{\text{visc}} \tau_{\text{heat}} / \tau_{\text{buoy}}^2$, where $\tau_{\text{visc}} = d^2/\nu$ is the viscous damping time, $\tau_{\text{heat}} = d^2/D_T$ the heat diffusion time, and $\tau_{\text{buoy}} = [d/(\alpha g \, \Delta T)]^{1/2}$ the time it takes a blob with the imposed density difference to fall down a distance d because of gravity.

17. See also the website of the Schatz group at Georgia Institute of Technology, schatzlab.gatech.edu, which shows various similar images from experiments of Krishan et al., 2007 and Kurtuldu et al., 2011. A video of the convection pattern at $\epsilon \approx 2.5$ can be found at the website www.weizmann.ac.il/complex/steinberg of the Steinberg group at the Weizmann Institute. See also the book's website www.softmatterbook.online for various videos of convection patterns.

18. The oscillatory instability is a finite wavelength oscillatory instability of the stripes, while the skew-varicose instability is somewhat like the zigzag instability but involves mean flow effects, which gives rise to a component along the stripes and perpendicular to the stripes. See the book by Cross and Greenside, 2009 for further details.

19. See, e.g., figure 4.4 of Cross and Greenside, 2009, where the Busse balloon is shown for $\text{Pr} = 0.7$, relevant to air at room temperature, and for $\text{Pr} = 7$, corresponding to the Prandtl number for water.

20. From about 1965 on, Fritz Busse and coworkers were leading in studying numerically the nonlinear convection patterns in detail and in analyzing their range of existence and stability. Busse also set up an extensive experimental program to compare experimental observations with the numerical stability results. Note that in such numerical studies of the stability of ideal ordered patterns like stripes, one mimics infinite systems by using periodic boundary conditions. Cross and Greenside, 2009 mark the period of this pioneering work by Busse and coworkers as the 'second phase' of pattern formation research, in which the implications of the nonlinearities were starting to be explored. Cross and Greenside refer to the linear stability studies of Rayleigh, Taylor, Turing, and Chandrasekhar as the first phase of pattern formation. In this spirit one might call the phase starting in the late 1970s, when the amplitude equation description was developed following the seminal work by Newell and Whitehead, 1969, and when many ingenious new experiments became available, as the third phase of pattern formation.

21. The full abstract is still wonderful to read from a present perspective, as it illustrates Turing's foresight:

> "It is suggested that a system of chemical substances, called morphogens, reacting together and diffusing through a tissue, is adequate to account for the main phenomena of morphogenesis. Such a system, although it may originally be quite homogeneous, may later develop a pattern or structure due to an instability of the homogeneous equilibrium, which is triggered off by random disturbances. Such reaction-diffusion systems are considered in some detail in the case of an isolated ring of cells, a mathematically convenient, though biologically unusual system. The investigation is chiefly concerned with the onset of instability. It is found that there are six essentially different forms which this may take. In the most interesting form stationary waves appear on the ring. It is suggested that this might account, for instance, for the tentacle patterns on Hydra and for whorled leaves. A system of reactions and diffusion on a sphere is also considered. Such a system appears to account for gastrulation. Another reaction system in two dimensions gives rise to patterns reminiscent of dappling. It is also suggested that stationary waves in two dimensions could account for the phenomena of phyllotaxis. The purpose of this paper is to discuss a possible mechanism by which the genes of a zygote may determine the anatomical structure of the resulting organism. The theory does not make any new hypotheses; it merely suggests that certain well-known physical laws are sufficient to account for many of the facts. The full understanding of the paper requires a good knowledge of mathematics, some biology, and some elementary chemistry. Since readers cannot be expected to be experts in all of these subjects, a number of elementary facts are explained, which can be found in text-books, but whose omission would make the paper difficult reading" (Turing, 1952).

Also Turing's remarks on broken symmetries are still worth reading.

22. See, e.g., chapters 3 and 11 of Cross and Greenside, 2009. The books by Gray and Scott, 1990 and Epstein and Pojman, 1998 both focus on chemical systems, while J. D. Murray, 2003 focuses on applications in biology and Desai and Kapral, 2009 focus on self-organized and self-assembled structures.

23. Adapted with permission from Horváth et al., 2018. Copyright 2018 American Chemical Society.

24. Published with permission from Wettmann and Kruse, 2018, which is published under the Creative Commons license.

25. Even harmonics of the type $\cos(2qx)$ do not occur: as the equation (8.26) is invariant under $u \to -u$, static solutions u should obey the symmetry $u(x + \pi/q) = -u(x)$. This symmetry is only respected by odd harmonics. When a symmetry-breaking term u^2 is included in the Swift-Hohenberg equation, even harmonics are generated—see problem 8.6.

26. See appendix 2 of Cross and Greenside, 2009 or appendix A of Cross and Hohenberg, 1993 for an introduction to multiple scales perturbation theory and an application to the derivation of amplitude equations. Bender and Orszag, 1978 give a very accessible more general introduction to multiple scales analysis in their chapter 11.

27. An example occurs in the case of a subcritical bifurcation when one considers a front between the featureless state and the patterned state. In the amplitude description, the front is stationary only at one particular value of the control parameter, whereas in practice the interface is stationary over an exponentially small but finite parameter interval, due to its being locked to the underlying lattice of the patterned state (Bensimon et al., 1988; Pomeau, 1986).

28. A remarkable example is the accurate calculation of the nonlinear instability threshold for viscoelastic pipe flow by going to the 11th (!) order in the expansion (Morozov and Saarloos, 2019).

29. In superconductivity, the phase winding solutions carry a current, so phase slip events reduce it leading to a resistance.

30. See, e.g., section IV.A.1.a.(iii) of Cross and Hohenberg, 1993 for further details.

31. The analysis of the amplitude equations (8.54) is conceptually straightforward but technically somewhat involved—see section 7.3.3. of Cross and Greenside, 2009 for a full discussion of this issue. We use the word *subcritical* somewhat loosely in the text, to bring across that hexagons exist in a small subcritical range for $\epsilon < 0$. The quadratic term with γ in (8.54) actually leads to a transcritical bifurcation at $\epsilon = 0$, and a saddle-node bifurcation at a small negative value of ϵ. See appendix 1 of Cross and Greenside, 2009 or section 1.3.2 of Pismen, 2006 for a brief introduction to the various types of bifurcations.

32. As discussed in section 3.2 of van Saarloos, 2003, the effect can also be seen in the experiments of Limat et al., 1992 on the propagating Rayleigh-Taylor instability in thin oil films. Image courtesy of Eberthard Bodenschatz.

33. If σ'''' would be positive, the system would exhibit a finite wavelength instability of type I, at some negative value of σ''.

34. Rescale the space and time by writing $x = (\sigma''''/\sigma'')^{1/2}\,\tilde{x}$ and $t = \sigma''''/(\sigma'')^2\,\tilde{t}$, and rescale the phase so as to make the prefactor of the nonlinear term unity. The sign of the prefactor of the nonlinear term is immaterial, as this can also be chosen by a proper choice of the sign of $\tilde{\phi}$. We have made the choice so that equation (8.57) coincides with the one simulated by Tajima and Greenside, 2002 and shown in figure 8.25.

35. See Kuramoto and Tsuzuki, 1976 and Sivashinsky, 1977, and for its importance for oscillatory chemical systems in particular, the book by Kuramoto, 1984.

36. This scenario was advanced in particular by Shraiman, 1986. See the paper by Tajima and Greenside, 2002 for references to the literature on the nature of the chaos in the Kuramoto-Sivashinsky equation.

37. Reproduced with permission from Tajima and Greenside, 2002. Copyright (2002) American Physical Society.

38. Note that there are various conventions for the imaginary parts of the complex Ginzburg-Landau equation. We follow the convention of Cross and Hohenberg, 1993 and van Saarloos and Hohenberg, 1992, but Aranson and L. Kramer, 2002 use b for our c_1 and $-c$ for our c_3. The convention of Aranson and L. Kramer, 2002 is also used on the website www.cgleatlas.com, where you can simulate the equation in two dimensions with one mouse click. For an elementary introduction to the equation, see van Saarloos, 1994.

39. Compare figure 15 of Garnier et al., 2003, where grayscale images for the local wavenumber are also shown.

40. The calculation is conceptually similar to the one for the real equation, but it is technically more involved. The analysis is given explicitly in section II.D of Aranson and L. Kramer, 2002.

41. In the context of pattern formation, solutions which are stationary in a frame $\xi = x - vt$ moving with velocity v are called coherent structures. These solutions are governed by ordinary differential equations. For complex fields like the amplitude, a temporal space-independent phase winding term is allowed in the definition, in other words, a coherent structure is a solution of the form $A(x,t) = e^{-i\Omega t}\tilde{A}(\xi)$. See van Saarloos and Hohenberg, 1992.

42. See van Hecke, 1998.

43. An interesting experimental realization of a one-dimensional traveling wave system, where the parameters of the CGLE can be obtained explicitly and where sources and sinks can be studied in detail, is a heated wire suspended a few millimeters below a fluid surface. In the experiments of Pastur et al., 2003a shown in figure 8.27, $c_1 = -1.7$ and $c_3 = 0.5$ are found. These parameters are deep into the stable regime, i.e., far from the region $c_1 c_3 > 1$ where all phase winding solutions are unstable.

44. See Pastur et al., 2003a and Pastur et al., 2003b for further details.

45. For an in-depth analysis of sources and sinks (as well as fronts and pulses) in the CGLE and extensions to subcritical cases, see van Saarloos and Hohenberg, 1992.

46. More mathematically: source solutions come as a discrete set of isolated solutions. See van Saarloos and Hohenberg, 1992 and Aranson and L. Kramer, 2002 for further details. Coherent structures are solutions of the form $A_R(x,t) = e^{-i\Omega t}\tilde{A}_R(\xi)$, where $\xi = x - vt$ is a comoving coordinate. Not surprisingly, the actual picture is quite a bit more subtle than what we are sketching here. First of all, the CGLE actually admits, contrary to general expectation (van Saarloos and Hohenberg, 1992), a special one-parameter family of sources (so-called Nozaki-Bekki holes) which has been attributed to a hidden symmetry or accidental degeneracy of the equation and which does not survive when perturbations are added to the equations—see section II.A.1 of Aranson and L. Kramer,

2002. Secondly, also other types of coherent structures play a role. Homoclons, which correspond to homoclinic orbits in the ordinary differential equations governing uniformly translating coherent structures, play quite an important role in organizing the chaotic behavior. Thirdly, also fronts and (in the subcritical case) pulse solutions can organize the dynamics in some regimes.

47. See, e.g., van Saarloos, 2003 for an overview of the field of front propagation into unstable states.

48. As argued in section 8.7.1, c_1 and c_3 are the only nontrivial parameters, as all others can be eliminated by a proper rescaling of space and time, and by including a time-dependent phase factor in A.

49. Reprinted with permission from Mocenni et al., 2010. Copyright (2010) *Proceedings of the National Academy of Sciences*.

50. The review of Aranson and L. Kramer, 2002 provides a very good introduction and an overview, but research on this equation has continued since its publication. Play a bit with the simulator available on www.cgleatlas.com to get a feel for the dynamics.

51. Even when the vortex arms are stable, the resulting interaction dynamics can become so slow that the multi-vortex state has been referred to as 'glassy' (Brito et al., 2003).

52. See, e.g., Stevens et al., 2011 for simulations up to $\text{Ra} = 2 \cdot 10^{12}$, and references therein. Interestingly, in large aspect ratio simulations of Rayleigh-Bénard cells, at extremely large driving the turbulent intensity shows modulations with a rather well-defined wavelength (Stevens et al., 2018). The origin of these turbulent 'superstructure' patterns, which have also been observed experimentally in Taylor-Couette cells (Prigent et al., 2002), is not really understood, even though, quite remarkably, the behavior of such patterns conforms very well to the amplitude description discussed in section 8.5 (Prigent et al., 2002).

53. The quest for a 'pattern selection mechanism' was stimulated in the 1980s by research on front propagation and on dendritic growth. For fronts propagating into unstable states, a family of front solutions typically exists, parametrized by the front speed v. The question thus arose about which solution is typically 'selected' from this family. Whether this is the most appropriate picture is debatable—from later work it has emerged (Ebert and van Saarloos, 2000; van Saarloos, 2003) that a more powerful concept is the linear spreading speed v^*, with which perturbations about the unstable state spread according to the linear equations. This linear spreading in a large class of cases pulls the nonlinear front along—hence the term 'pulled front' for nonlinear fronts moving with velocity v^*. Likewise, for dendrites the idea of 'selection' is to some extent evaporated by later developments. Upon our ignoring surface tension, a family of needle solution shapes exists. It was initially thought that with surface tension included perturbatively, one of these solutions is 'selected' dynamically, but it became clear later on that the existence of a family of needle solutions is an artifact of initially neglecting surface tension and then doing a perturbation expansion. Instead, with surface tension included from the start only a discrete family of needle solutions exists (see, e.g., Pelcé, 2004 for a review) . The idea of 'selection' of this state out of a family of solutions, which to some extent are an artifact of an unjustified approximation, turned out to be of limited value in the end.

54. See Bastiaansen et al., 2018 for full details. In their figure 3, also a histogram of observed values of the wavenumber is plotted, for the data from Haud, Somalia, as well as for the data from the Sool Plateau pastoral area in Somalia. In both cases, the histogram appears peaked away from the edges. The authors also discuss data for the biomass and for the downhill migration speed as a function of the slope s (for which they use the symbol v).

55. See Bastiaansen et al., 2018.

56. Reprinted with permission from Bastiaansen et al., 2018. Copyright (2018) *Proceedings of the National Academy of Sciences*.

57. Cross and Greenside, 2009 discuss in section 11.1.3 how FitzHugh and independently Nagumo arrived at this model and how it can be viewed as a simplified version of the so-called Hodgkin-Huxley model for the action potential of nerves.

58. One can show this explicitly using the method of multiple scales. See, e.g., chapter 11 of Bender and Orszag, 1978. For the analysis of pulses in terms of two fronts and a plateau, elements of matched asymptotic expansions also come in.

59. Note that the shape of the propagating pulse resembles roughly that of the time dependence of u_1 sketched in figure 8.33. Indeed, a pulse solution propagating with a constant speed v to the left is a function of the coordinate $x + vt$ only, so an increase in x is like an increase in t. Of course, the propagating pulse is affected by diffusion, so the two profiles are not exactly the same.

60. Such an analysis is sometimes referred to as a nonlinear eigen value problem.

61. A nice short history of the Belousov-Zhabotinsky reaction, with references to the early literature in Russian, is given by Winfree, 1984. Easily accessible papers by the two authors whose name the reaction bears are Belousov, 1984 and Zhabotinsky, 1964 and Zhabotinsky, 1995. Two well-known reviews of many excitation patterns in chemical reactions are the book edited by Field and Burger, 1984 and the one edited by Kapral and Showalter, 1995, while Tsuji and Müller, 2019 and Zykov, 2018 review more recent developments.

62. Scroll waves were popularized in particular by Winfree, 1973 and Winfree, 1974.

63. Image reproduced from P. Ball, 2015, published under the Creative Commons license.

64. An elementary introduction to the physics of nerve fibers is given in section 11.1.1 of Cross and Greenside, 2009.

65. See the overview of Barkley, 2016 for an extensive discussion of this approach.

Chapter 9

1. The snapshots correspond to those from figure 1 of Vicsek et al., 1995. See the figure caption for the precise values of density and noise strength.

2. Reproduced with permission from Vicsek et al., 1990. Copyright (1995) American Physical Society.

3. The model is standardly referred to in the literature as the Vicsek model, in honor of Tamás Vicsek, who played a crucial role in establishing the field of active media. Note, though, that the original paper where the Vicsek model was introduced has five coauthors (Vicsek et al., 1995).

4. Note that since the time in this model is discrete, we use a Kronecker delta for the temporal correlation, instead of the delta function correlation we used in section 3.2.2 and section 7.5 for the continuum time Langevin equations.

5. The supplementary information of Bricard et al., 2013 includes several nice movies of the propagating fronts in a homogeneous cell. See Bricard et al., 2015 and K. Han et al., 2020 for an example of a study of multivortex states of Quincke rollers, and Kaiser et al., 2017 for global rotations using ferromagnetic particles. In the paper by Morin et al., 2016, from which the figure is reproduced, the destruction of flocks in a disordered environment is studied, so the coherent flocking motion in the dense phase is disordered. It is also possible to study sound (Geyer et al., 2018) and dynamic vortex glasses in disordered environments, using active colloids driven by Quincke rotation (Chardac et al., 2021).

6. See Ginelli, 2016 for a review of the Vicsek model and its behavior.

7. Mermin and H. Wagner, 1966 proved the result for Heisenberg spin models. The theorem is sometimes also called the Mermin-Wagner-Hohenberg theorem since Hohenberg proved the absence of long-range order for two-dimensional Bose and Fermi systems (Hohenberg, 1967); since the phase of the wave function is like the angle of a spin, the two results are essentially the same.

8. Note that the unit vector $\hat{m}$ that gives the orientation of each particle is the same as the one used in our discussion of the order parameter of liquid crystals in section 6.2. Be aware that Marchetti et al., 2013 denote this unit vector by $\hat{\nu}$.

9. Reproduced with permission from Martin et al., 2021b. Copyright (2021) American Physical Society.

10. And with the general criteria discussed by Whitham, 1974 for the formation of domains in the nonlinear wave equation.

11. See the review by Pedley and Kessler, 1992.

12. The dark regions are upwardly flowing, while the bright regions are high-concentration downwellings. The small grains/flakes are most likely immotile cells that are trapped in stagnation points of the flow at the upper (air-water) interface.

13. More detailed observations and analysis can be found in the paper by Jánosi et al., 1998. For a review of early studies of bacterial swarming, see Pedley and Kessler, 1992.

14. Reproduced with permission from Wensink et al., 2012. Copyright (2012) *Proceedings of the National Academy of Sciences*.

15. Reproduced with permission from López et al., 2015. Copyright (2015) American Physical Society.

16. See also Mackay et al., 2020 for a detailed discussion of the various regimes of Poiseuile flow of an active nematic fluid.

17. See, for instance, §9 in Bird et al., 1987a for details.

18. Experimental evidence suggests this picture is not quite complete. Because the motors both propel *and* connect filaments, their presence might also affect the material properties of the nematic. Recent experiments appear to suggest that modifying the chemical fuel (ATP) or the motors' physical properties affects *both* their activity *and* their elastic or viscous coefficients (see figure S2 of the supplementary information of Lemma et al., 2019). Clarifying the relationships between active and passive nematic properties as well as microscopic behavior is an area of ongoing research (Lemma et al., 2019).

19. Note, by the way, that in regular passive nematics electroconvection, 'turbulent' states also exist where likewise the chaotic dynamics is driven by $+1/2$ defects—in the experiments of figure 7.15, the boundary between this defect-driven state and regular convective states was used to test the scaling predictions of the KPZ equation.

20. See Lemma et al., 2019 and references therein.

21. Reproduced with permission from Giomi et al., 2014; published under the Creative Commons license.

22. The most prominent example of the influence of $+1/2$ defects on the dynamics is the active nematic turbulent state illustrated in figure 9.10. Other examples are the occurrence of apoptosis near $+1/2$ defects, illustrated in figure 6.27, and the defect-driven electroconvection of nematics of figure 7.15.

23. The analysis in the section follows the paper by Giomi et al., 2014, where more details can be found.

24. Reprinted with permission from Ghatak et al., 2020. Copyright (2020) *Proceedings of the National Academy of Sciences*.

25. Be aware that in chapter 2 the bulk compression modulus was written as K.

26. See Braverman et al., 2021 for explicit formulas.

27. More details can be found in Banerjee et al., 2017; Condiff and Dahler, 1964; Dahler and Scriven, 1961; Fürthauer et al., 2012; and Tsai et al., 2005.

28. See, e.g., Landau and Lifshitz, 1987.

29. Adapted from M. Fruchart, R. Hanai, P. B. Littlewood, and V. Vitelli, "Non-reciprocal phase transitions." *Nature* 592, 363–369 (2021). Reproduced with permission.

30. In many chaotic extended dynamical systems, absorbing states occur. For instance, in the transition to turbulence in Newtonian pipe flow, the laminar state acts like an absorbing state for turbulent domains. The dynamical competition between turbulent spots and the laminar state can be understood in terms of 'directed percolation'—see figure 1.12 and references in the caption.

31. Hydrodynamic models for nonreciprocal systems are, for instance, discussed in Fruchart et al., 2020 for non-conserved order parameters (model A in the classification of Hohenberg and Halperin) and in Saha et al., 2020 and You et al., 2020 for conserved order parameters (model B).

32. See, e.g., the book by Strogatz, 2015 for a discussion of this.

33. Nonreciprocal synchronization can be described in the same way as nonreciprocal flocking by setting $v_0 = 0$.

34. We refer you to Coullet and Fauve, 1985; Fauve et al., 1991; Knobloch et al., 1995; and the textbook by Hoyle, 2006 for more details.

35. Image courtesy of Vincenzo Vitelli and Michel Fruchart.

36. We refer you to Shraiman, 1986 for more details. In both cases, an exceptional point exists (first term on the right-hand side of equation (9.71)), but in the case of the Kuramoto-Sivashinsky equation, only the Laplacian terms from the third term on the right-hand side of (9.71) are present. Here, and in the case of nonreciprocal flocking, an additional gradient term $M\vec{v}_0 \cdot \vec{\nabla}$ proportional to the self-propulsion speed v_0 exists. In both cases, the instabilities are accompanied by spatiotemporal chaos.

37. Panels a and b are from Turlier et al., 2014; panel c is from Borja da Rocha et al., 2022, which is published under the Creative Commons license.

38. The image is from a transgenic *Hydra* expressing lifeact-GFP (green fluorescent protein; lifeact-GFP is a widely used actin probe for visualizing F-actin structures in various cell types and model organisms) in the ectoderm, which labels the ectodermal muscle fibers.

39. See Turing, 1952, from which the quote is taken, and the introduction of section 8.3.2.

40. To see this, take a look at the droplet with two boojums shown in figure 6.15. When you focus on the surface only, it effectively has two +1 defects.

41. Hoffmann et al., 2022 use a combination of linear stability analysis and computational fluid dynamics to demonstrate that active layers, such as confined cell monolayers, are unstable to the formation of protrusions in the presence of disclinations, and that this lies at the origin of the interplay between defects and morphogenesis.

42. See K. D. Irvine and Wieschaus, 1994 for more details.

43. See references in Streichan et al., 2018 for entries into the theoretical research.

44. See K. D. Irvine and Wieschaus, 1994.

45. A morphodynamic atlas for *Drosophila* is presented in N. P. Mitchell et al., 2022.

46. See Farhadifar et al., 2007.

47. See Bi et al., 2016.

48. See Bi et al., 2015; Bi et al., 2016.

49. See Mongera et al., 2018.

Chapter 10

1. In his famous "There's plenty of room at the bottom" paper, Feynman, 1960 writes, "biology is not simply writing information. It is doing something about it. . . . Consider the possibility that we too can make a thing very small, which does what we want!"

2. Lehn, one of the founding fathers of supramolecular chemistry, who was honored with the 1987 Nobel Prize in Chemistry, started viewing chemistry as a 'science of informed matter.' He and others put forward ideas about programmed chemical systems, where simple molecular components spontaneously, but in a controlled manner, assemble into functional supramolecular architectures; this self-organization is stored in the covalent and non-covalent interactions between components. See Aida and E. W. Meijer, 2020 for a modern personal perspective on the field.

3. The name 'programmable matter' reflects the fact that Toffoli and Margolus, 1991 shaped the field initially based on their background in computer science, as is visible in quotes like "in programmable matter, the same cubic meter of machinery can become a wind tunnel at one moment, a polymer soup at the next; it can model a sea of fermions, a genetic pool, or an epidemiology experiment at the flick of a console key. . . . Flexibility, instant reconfigurability, variable resolution . . . make such programmable matter worth a premium over ordinary matter." The authors introduced, among other things, a scalable cellular automata machine that embodies some of these ideas.

4. We discussed in section 3.5 how a micron is roughly the crossover scale separating the regime of small colloidal particles which disperse through Brownian motion from that of granular particles which are so large that they tend to settle and that their equilibration and self-organization are extremely slow or even prohibited due to gravity.

5. See, e.g., Dijkstra and Luijten, 2021 for a discussion of applications of machine learning to colloids; Colen et al., 2021 for applications to active matter; and Kochkov et al., 2021 for a demonstration of how fluid dynamics simulations can be sped up with machine learning.

6. Figure 2.35 shows an example of the use of evolutionary design in granular media research.

7. Reprinted from Dierichs and Menges, 2021 with permission on account of the Creative Commons license.

8. See section 2.10 for a discussion of granular media. Some recent work on the impact of particle shape but also friction on the material properties can be found in Azéma et al., 2017; Binaree et al., 2020; and Sarkar et al., 2019.

9. Such ultralight materials are being explored for a variety of applications, like thermal insulation, battery electrodes, catalyst supports, and shock and vibration or acoustic damping. Silica aerogels, carbon nanotubes, and metallic and polymer foams are often the materials of choice for the fabrication of lightweight structures.

10. See, in particular, Rocks et al., 2017 and Rocks et al., 2019.

11. The physical frequencies are nondimensionalized by a characteristic frequency, ω_0. Adapted and reproduced with permission from Fruchart et al., 2020.

12. As we will see in section 10.3, solving inverse problems is at the core of many machine learning algorithms.

13. See Rose et al., 2013.

14. See Meddahi-Pellé et al., 2014.

15. Reproduced from Fang et al., 2020 on account of the Creative Commons license, under which the paper has been published.

16. For instance, one can read from the interaction matrix that particle 1 attracts only particles 5, 6, and 7, while particle 2 attracts 5, 6, and 8. The interaction matrix is symmetric, since interactions are defined only between pairs of particles. The interaction matrix on the left shows that in the cluster shown at the top, particles 1 only attract particles 2, while particles of type 2 attract both particles of type 1 and those of type 2.

17. The term is from the paper by Macfarlane et al., 2011, which was one of the first to contain a whole range of architectures.

18. The simplest solution that typically works (modulo some pathological cases) is the one where every building block is made to be different, i.e., a particle is interacting favorably only with its nearest neighbors in a desired structure and unfavorably with everyone else. To find other design solutions requires knowledge of all the other possible structures that can be made of the same number of building blocks and then solving the inverse problem. For large assemblies we fight with the combinatorial explosion of the number of possible shapes, but for smaller ones enumerations are tractable.

19. See Meng et al., 2010 for more details.

20. In nuclear magnetic resonance and magnetic resonance imaging, one of the crucial components for producing a 3D representation of water- or fat-containing samples is spin echoes. The common spin echo technique quantifies the mutual interactions of spins in a sample using a sequence of resonant transitions. Magnetic resonance measurements start with a sample in a large static magnetic

field $H_0(z)$ which produces a small polarization of the spins along the $+z$ direction. When a radio frequency (RF) pulse resonantly tuned to the average Larmor frequency $<\omega>$ is applied in a direction perpendicular to H_0, the spins rotate away from $+z$ axis. In the spin echo technique one chooses the pulse duration such that the spin polarization rotates into the xy plane (called a $\pi/2$ pulse), in which it precesses clockwise around the $+z$ axis. However, the polarization decays since each individual spin precesses at its own Larmor frequency $\omega(x, y, z)$ locally determined by the slight spatial inhomogeneity of the chemical environment and of H_0. The trick with the spin echo is that after the spins have been allowed to dephase for a time τ, another RF pulse of twice the duration (called a π pulse) is applied, flipping the orientation of spins in the xy plane. Each reversed spin now precesses counterclockwise, keeping its own frequency, so by waiting for another length of time τ all the dephasing gets undone, and the realignment of spins produces a peak in the polarization signal at time 2τ although no pulses are applied at that time. This is an echo signal, representing a memory of the time delay, τ, between the two RF pulses. The decay of signal from its initial value to the echo value is hence purely an effect of spin-spin interactions, i.e., time-dependent fluctuations which do not memorize the effect of pulses. By changing the delay time τ, we can recover different information about tissues, like detecting edema and inflammation. The spin echo technique is also used to characterize interactions of spins (dynamics of electron spins) in samples of materials in condensed matter.

21. See Preisach, 1935.

22. This section on memory builds heavily on the review by Keim et al., 2019—it even contains, verbatim, parts of the text. We are grateful to ZZ's coauthors Nathan Keim, Joseph Paulsen, Srikanth Sastry, and Sid Nagel for their permission to integrate here parts of their review. Sheared suspensions are discussed in section 4.6.1, rejuvenation in section 4.6.2, colloidal glasses in section 4.5.4, and colloids with designed interparticle interactions in section 4.5.5.

23. Examples of systems in which return-point memory has been observed include spin ice and high-temperature superconductors, adsorption of gases on surfaces, solids with shape memory (not to be confused with the shape-memory effect itself), and jammed systems. Particular examples of return-point memory in mentioned systems can be found in Deutsch et al., 2004; Emmett and Cines, 1947; Ortín, 1991; and Panagopoulos et al., 2006, respectively. Mungan and Witten, 2019 have begun work toward a general framework for describing systems with return-point memory.

24. For each of the systems you can find more information in the papers by Corté et al., 2008; Fleming and Schneemeyer, 1986; Haw et al., 1998; Lahini et al., 2017; Laurson and Alava, 2012; M. H. Lee and Furst, 2008; Lundberg et al., 2008; Mangan et al., 2008; Packard et al., 2010; Sircar and Q. Wang, 2010; and Toiya et al., 2004.

25. See, for example, Paulsen et al., 2014 and Povinelli et al., 1999.

26. See Kovacs, 1963 and Kovacs et al., 1979.

27. For more details see Amit et al., 1985; Hertz et al., 1991 and Hopfield, 1982.

28. See, for example, Mehta et al., 2019.

29. We refer you to the excellent textbook by Hertz et al., 1991 for an introduction to the study of neural networks using the theory of magnetic systems and the tools of statistical physics.

30. See, for example, Corté et al., 2008; Keim and Nagel, 2011; Keim et al., 2013; and Paulsen et al., 2011.

31. For irreversible trajectory change, see Corté et al., 2008; Pine et al., 2005; and Popova et al., 2007 and Pham et al., 2015.

32. Reproduced with permission of *Physical Review Letters* from Paulsen et al., 2014. Copyright (2014) American Physical Society.

33. See, for example, Hébraud et al., 1997 and Petekidis et al., 2002.

34. See Keim and Arratia, 2014; Nagamanasa et al., 2014; and Slotterback et al., 2012.

35. See M. O. Lavrentovich et al., 2017; Mungan and Witten, 2019; Regev et al., 2013; and Royer and Chaikin, 2015.

36. It is interesting to note that in the video-recorded demonstration by G. I. Taylor for the National Committee for Fluid Mechanics Films, he begins his synopsis with: "Low Reynolds number flows are reversible when the direction of motion of the boundaries which gave rise to the flow is reversed. This may lead to some surprising situations, which might almost make one believe that the fluid has a memory of its own."

37. See Lahini et al., 2017.

38. See Murugan et al., 2015.

39. The transcript can be found at www.vordenker.de. An excerpt referring to AI is: "It has been said that computing machines can only carry out the processes that they are instructed to do. This is certainly true in the sense that if they do something other than what they were instructed then they have just made some mistake. It is also true that the intention in constructing these machines in the first instance is to treat them as slaves, giving them only jobs which have been thought out in detail, jobs such that the user of the machine fully understands what in principle is going on all the time. Up till the present machines have only been used in this way. But is it necessary that they should always be used in such a manner? Let us suppose we have set up a machine with certain initial instruction

tables, so constructed that these tables might on occasion, if good reason arose, modify those tables. . . . What we want is a machine that can learn from experience."

40. Adapted from Murugan et al., 2015.

41. This is the definition from the book by T. M. Mitchell, 1997.

42. A good introductory book on machine learning is the one by C. M. Bishop, 2007.

43. See the paper by Bar-Sinai et al., 2019 and a perspectives paper by M. P. Brenner et al., 2019.

44. Machine learning has been used in three different areas of computational fluid dynamics: First, it has been used to accelerate direct numerical simulations. Second, it has been used to accelerate turbulence closure models (large eddy and Reynolds average Navier-Stokes models). And third, machine learning has been used to improve physical understanding, scalability, and generalizability of turbulence models and classical reduced order models. For a very nice review on the use of machine learning in fluid dynamics, see Brunton et al., 2020.

45. To give an idea, for the simplest setups of flow around an airplane wing or turbine blade one would need months and months of simulation time, and that is for only one set of design parameters. This translates into decades for full fidelity simulations of flows that would help design better wings or blades, even if we take into consideration Moore's law (the doubling of the number of transistors per chip approximately every two years) of increased speed and capability of computers.

46. See the paper by Kochkov et al., 2021 for more details.

47. See Ferguson, 2017 and Clegg, 2021.

48. The algorithm proposed by Coli et al., 2022 uses an evolutionary strategy for parameter optimization, and it uses a convolutional neural network as an order parameter, resulting in solutions for colloidal interactions specifically optimized to stabilize the desired structure. We also mention here the work by the group of Sharon Glotzer at the University of Michigan (Geng et al., 2018), who explore similar lines.

49. Nucleation is a process in which a crystal nucleus spontaneously forms in the metastable fluid phase due to a statistical fluctuation.

50. One of the very first people to point this out was Daan Frenkel. An interested reader can follow the developments from the nearly 1,000 papers that cite Auer and Frenkel, 2001.

51. Interesting papers on the topic are Adorf et al., 2019 and Coli and Dijkstra, 2021.

52. This realization mimics the perceptron by Rosenblatt, 1958.

53. Building blocks of neural networks of this type are the enzymes that produce DNA (i.e., polymerase), cut DNA (i.e., nickase), and degrade DNA (i.e., exonuclease).

54. The work of Eric Winfree is a good starting point if you are interested in DNA-based computational molecular devices; see Qian et al., 2011; F. Wang et al., 2020; and Woods et al., 2019.

55. See Okumura et al., 2022 for more details.

56. Von Neumann built here on the formalism of cellular automata that he developed together with Stan Ulam in the 1940s.

57. Cellular automata-based systems continued to be developed in the 1970s and 1980s. Some of the most famous examples are https://playgameoflife.com/, John Conway's Game of Life (Gardner, 1970), and Stephen Wolfram's cellular automata (Wolfram, 1983).

58. See Aguilar et al., 2014. The structure of this paper is used as an inspiration for the concepts part of this section.

59. One of the first papers to address self-assembly on all scales was the one by Whitesides and Grzybowski, 2002.

60. A review that covers the first 50 years of research on self-replication, starting from von Neumann, was published in the journal *Artificial Life* (Sipper, 1998).

61. See Wolfenden and Snider, 2001 for examples.

62. In 1995 a volume titled *The Artificial Life Route to Artificial Intelligence: Building Embodied, Situated Agents* was published as a result of a workshop in which a number of leading researchers from the fields of artificial intelligence and biology gathered to examine whether there was any ground to assume that a new AI paradigm was forming itself and what the essential ingredients of this new paradigm were (Steels and Brooks, 1995). The aim was to clarify the common ingredients, see what had been achieved so far, and examine in what way the research could move further. The volume contained contributions which were distributed before the workshop, but it was then substantially broadened and revised to reflect the workshop discussions and more recent technical work.

63. Even though this book does not discuss the robotics and computer science axes of ALife and AI, we would like to draw your attention to the work done by the Boston Dynamics group (see www.bostondynamics.com), which spans 30 years of robotics research. The book's website www.softmatterbook.online contains a video which shows the evolution of robots built by the Boston Dynamics group since 2009. Apart from the change in the appearance of the robots, one can clearly see how processes such as adaptation and learning have been added over time.

64. Frances Arnold received half of the prize for the first development, in 1993, of enzymes through selection and evolution. George Smith and Gregory Winter received each one quarter of the prize for phage display.

65. This definition is adapted from R. A. Wilson and Keil, 2001, which you can download from The MIT Encyclopedia of the Cognitive Sciences (available via MIT's website) free of charge. It is by no means the only definition, just the one we chose to write here.

66. If we use behavior for the general scheme of an intelligent agent, this includes the whole setup of the internal processing, through sensing, mapping/modeling, and effecting. This then implies that the behavior of an agent has more to do with the inference based on the internal mappings and representations than it has to do with the interactions with the environment which is at the core of behavior.

67. See, for example, the early works Oparin, 1957; Dyson, 1982; Dyson, 1985; Kauffman, 1986; and Kauffman, 1993.

68. See, for example, Gilbert, 1986; Neveu et al., 2013; and Robertson and Joyce, 2010.

69. Reproduced from Zhou et al., 2021 on account of the Creative Commons license, under which the paper has been published.

70. For an inspiring realization in a chemical system, see the work from Sijbren Otto, starting with the paper by Carnall et al., 2010.

71. While in solution, free-floating particles can also interact, but the barrier to nucleation is sufficiently high that spontaneous self-assembly is not observed on the scale of the numerical experiment. For more details, see Zeravcic and M. P. Brenner, 2014.

72. Pauling's seminal paper is Pauling, 1946.

73. The work of Rohl et al., 2004 is one of the most successful examples of software used for design of new enzymes.

74. The mechanism of bond breaking is strain: once both particles in a dimer are bound to the catalyst, they cannot find equilibrium positions and the dimer bond is pulled to breaking (stage C_2 to C_3 indicated in the figure).

75. The works we refer to are Dyson, 1982 and Oparin, 1957.

76. See the works of Kaufmann and collaborators, such as Kauffman, 1986.

77. For review and research papers that discuss these questions in mostly chemical experimental systems or that discuss autocatalysis in general, see Ameta et al., 2021; Hanopolskyi et al., 2020; Plasson et al., 2011; and Schuster, 2019.

78. More details can be found in Zeravcic and M. P. Brenner, 2017.

79. See www.deepmind.com.

80. For more work on protocells we invite you to explore the research of Jack Szostak, who won the 2009 Nobel Prize in Physiology or Medicine for this work.

81. A recent review that discusses some aspects of this vast field is the one by Cho and Lu, 2020; also the *Nature* news article by Powell, 2018 discusses approaches on how to make artificial cells from the bottom up. One of the biggest multidisciplinary consortiums aimed at building a synthetic cell is based in the Netherlands and led by Marileen Dogterom.

82. See the work done in the lab of Hod Lipson at Columbia University (see www.creativemachineslab.com) for more information and interesting examples.

83. For video's of different evolved soft robots in motion, search YouTube for 'evolving soft robots with multiple materials.' See also the book's website www.softmatterbook.online.

Anderson, P. W. (1972). 'More is different'. In: *Science* 177, pp. 393–396 (cit. on p. 9).

Anderson, P. W. (1997). *Basic Notions Of Condensed Matter Physics*. Advanced Book Classics, CRC Press (cit. on p. 12).

Andrejevic, J., L. M. Lee, S. M. Rubinstein, and C. H. Rycroft (2021). 'A model for the fragmentation kinetics of crumpled thin sheets'. In: *Nat. Commun.* 12, p. 1470 (cit. on pp. 324, 334).

Anisimov, M. A., P. E. Cladis, E. E. Gorodetskii, D. A. Huse, V. E. Podneks, V. G. Taratuta, W. van Saarloos, and V. P. Voronov (1990). 'Experimental test of a fluctuation induced first order transition: The nematic-smectic A transition'. In: *Phys. Rev. A* 41, pp. 6749–6762 (cit. on pp. 102, 268, 298, 518).

Aranson, I. S. and L. Kramer (2002). 'The world of the complex Ginzburg-Landau equation'. In: *Rev. Mod. Phys.* 74, pp. 99–143 (cit. on pp. 383, 526, 527).

Arnold, L. (1974). *Stochastic Differential Equations: Theory and Applications*. John Wiley (cit. on p. 512).

Arnold, V. I. (1973). *Ordinary Differential Equations*. MIT Press (cit. on pp. 84, 510).

Aronson, D. G and H. F. Weinberger (1978). 'Multidimensional nonlinear diffusion arising in population genetics'. In: *Adv. Math.* 30, pp. 33–76 (cit. on p. 524).

Asakura, S. and F. Oosawa (1954). 'On interaction between two bodies immersed in a solution of macro-molecules'. In: *J. Chem. Phys.* 22, pp. 1255–1256 (cit. on p. 514).

Ashcroft, N. W. and N. D. Mermin (1976). *Solid State Physics*. Harcourt (cit. on p. 510).

Audoly, B. and Y. Pomeau (2018). *Elasticity and Geometry—From Hair Curls to the Non-linear Response of Shells*. Oxford University Press (cit. on pp. 315, 509).

Auer, S. and D. Frenkel (2001). 'Prediction of absolute crystal-nucleation rate in hard-sphere colloids'. In: *Nature* 409, pp. 1020–1023 (cit. on p. 532).

Azéma, E., N. Estrada, I. Preechawuttipong, J.-Y. Delenne, and F. Radjai (2017). 'Systematic description of the effect of particle shape on the strength properties of granular media'. In: *EPJ Web of Conferences*. Ed. by F. Radjai, S. Nezamabadi, S. Luding, and J.Y. Delenne. Vol. 140. EDP Sciences, p. 06026 (cit. on p. 530).

Baconnier, P., D. Shohat, C. Hernández-López, C. Coulais, V. Démery, G. Düring, and O. Dauchot (2022). 'Selective and collective actuation in active solids'. In: *Nat. Phys.* 18, pp. 1234–1239 (cit. on p. 419).

Baek, Y., A. P. Solon, X. Xu, N. Nikola, and Y. Kafri (2018). 'Generic long-range interactions between passive bodies in an active fluid'. In: *Phys. Rev. Lett.* 120, p. 58002 (cit. on p. 422).

Balescu, R. (1975). *Equilibrium and Non-equilibrium Statistical Mechanics*. John Wiley & Sons (cit. on p. 506).

Ball, P. (2015). 'Forging patterns and making waves from biology to geology: A commentary on Turing (1952) 'The chemical basis of morphogenesis''. In: *Phil. Trans. R. Soc. B* 370, p. 20140218 (cit. on pp. 363, 527).

Banerjee, D., A. Souslov, A. G. Abanov, and V. Vitelli (2017). 'Odd viscosity in chiral active fluids'. In: *Nat. Commun.* 8, p. 1573 (cit. on p. 529).

Banzhaf, W. and L. Yamamoto (2015). *Artificial Chemistries*. Cambridge, Mass, MIT Press (cit. on p. 492).

Bar-Sinai, Y., S. Hoyer, J. Hickey, and M. P. Brenner (2019). 'Learning data-driven discretizations for partial differential equations'. In: *Proc. Natl. Acad. Sci.* 116, pp. 15344–15349 (cit. on p. 532).

Bar-Ziv, R. and E. Moses (1994). 'Instability and "pearling" states produced in tubular membranes by competition of curvature and tension'. In: *Phys. Rev. Lett.* 73, pp. 1392–1395 (cit. on pp. 330, 523).

Barenblatt, G. I. (1987). *Dimensional Analysis*. Gordon and Breach (cit. on p. 41).

Barenblatt, G. I. (1996). *Scaling, Self-Similarity, and Intermediate Asymptotics*. Cambridge University Press (cit. on pp. 41, 131, 507).

Barkley, D. (2016). 'Theoretical perspective on the route to turbulence in a pipe'. In: *J. Fluid Mech.* 803, P1–P80 (cit. on pp. 507, 508, 527).

Barrat, J.-L. and J.-P. Hansen (2003). *Basic Concepts for Simple and Complex Liquids*. Academic (cit. on p. 517).

Barthès-Biesel, D. (2016). 'Motion and deformation of elastic capsules and vesicles in flow'. In: *Annu. Rev. Fluid Mech.* 48, pp. 25–52 (cit. on p. 315).

Bassereau, P., R. Jin, T. Baumgart, M. Deserno, R. Dimova, V. A. Frolov, P. V. Bashkirov, H. Grubmüller, R. Jahn, H. J. Risselada, L. Johannes, M. M. Kozlov, R. Lipowsky, T. J. Pucadyil, W. F. Zeno, J. C. Stachowiak, D. Stamou, A. Breuer, L. Lauritsen, C. Simon, C. Sykes, G. A .Voth, and T. R. Weikl (2018). 'The 2018 biomembrane curvature and remodeling roadmap'. In: *J. Physics D: Appl. Phys.* 51, p. 343001 (cit. on p. 315).

Bassereau, P., B. Sorre, and A. Lévy (2014). 'Bending lipid membranes: Experiments after W. Helfrich's model'. In: *Adv. Colloid Interface Sci.* 208, pp. 47–57 (cit. on p. 313).

Bastiaansen, R., O. Jaïbi, V. Deblauwe, M. B. Eppinga, K. Siteur, E. Siero, S. Mermoz, A. Bouvet, A. Doelman, and M. Rietkerk (2018). 'Multistability of model and real dryland ecosystems through spatial self-organization'. In: *Proc. Natl. Acad. Sci.* 115, pp. 11256–11261 (cit. on pp. 388, 389, 527).

Bates, F. S. and G. H. Fredrickson (1990). 'Block copolymer thermodynamics: Theory and experiment'. In: *Annu. Rev. Phys. Chem.* 41, pp. 525–557 (cit. on p. 230).

Bates, F. S., J. H. Rosedale, and G. H. Fredrickson (1990). 'Fluctuation effects in a symmetric diblock copolymer near the order-disorder transition'. In: *J. Chem. Phys.* 92, pp. 6255–6270 (cit. on p. 231).

Batra, R., L. Song, and R. Ramprasad (2020). 'Emerging materials intelligence ecosystems propelled by machine learning'. In: *Nat. Rev. Mater.* 6, pp. 655–678 (cit. on p. 480).

Baughman, R. H., J. M. Shacklette, A. A. Zakhidov, and S. Stafström (1998). 'Negative Poisson ratios as a common feature of cubic metals'. In: *Nature* 392, pp. 362–365 (cit. on p. 510).

Béalle, G., J. Jestin, and D. Carrière (2011). 'Osmotically induced deformation of capsid-like icosahedral vesicles'. In: *Soft Matter* 7, pp. 1084–1089 (cit. on p. 522).

Beatus, T., T. Tlusty, and R. Bar-Ziv (2006). 'Phonons in a one-dimensional microfluidic crystal'. In: *Nat. Phys.* 2, pp. 743–748 (cit. on p. 419).

Beaucage, G. (1996). 'Small-angle scattering from polymeric mass fractals of arbitrary mass-fractal dimension'. In: *J. Appl. Cryst.* 29, pp. 134–146 (cit. on p. 517).

Bedeaux, D. and P. Mazur (1975). 'Renormalization of the diffusion coefficient in a fluctuating fluid III'. In: *Physica A* 80, pp. 189–202 (cit. on p. 514).

Beenakker, J. J. M. and F. R. McCourt (1970). 'Magnetic and electric effects on transport properties'. In: *Annu. Rev. Phys. Chem.* 21, pp. 47–72 (cit. on p. 427).

Belousov, B. P. (1984). 'A periodic reaction and its mechanism'. In: *Oscillations and Traveling Waves in Chemical Systems*. Ed. by R. J. Field and M. Burger. Wiley, pp. 605–614 (cit. on p. 527).

Bender, C. M. and S. A. Orszag (1978). *Advanced Mathematical Methods for Scientists and Engineers I: Asymptotic Methods and Perturbation Theory*. Springer (cit. on pp. 315, 525, 527).

Bowick, M. J. and L. Giomi (2009). 'Two-dimensional matter: Order, curvature and defects'. In: *Adv. Phys.* 58, pp. 449–563 (cit. on p. 342).

Bowick, M. J. and A. Travesset (2001). 'The statistical mechanics of membranes'. In: *Phys. Rep.* 344, pp. 255–308 (cit. on pp. 319, 522).

Brady, J. F. and G. Bossis (1988). 'Stokesian dynamics'. In: *Annu. Rev. Fluid Mech.* 20, pp. 111–157 (cit. on pp. 66, 184).

Brandenbourger, M., C. Scheibner, J. Veenstra, V. Vitelli, and C. Coulais (2021). 'Limit cycles turn active matter into robots'. Preprint arXiv:2108.08837 (cit. on p. 421).

Braverman, L., C. Scheibner, B. VanSaders, and V. Vitelli (2021). 'Topological defects in solids with odd elasticity'. In: *Phys. Rev. Lett.* 127, p. 268001 (cit. on pp. 420, 529).

Bray, A. J. (1994). 'Theory of phase-ordering kinetics'. In: *Adv. Phys.* 43, pp. 357–459 (cit. on pp. 345, 514).

Bray, A. J. (2003). 'Coarsening dynamics of phase-separating systems'. In: *Philosophical Transactions of the Royal Society of London. Series A: Mathematical, Physical and Engineering Sciences* 361. Ed. by T. C. B. McLeish, M. E. Cates, J. S. Higgins, and P. D. Olmsted, pp. 781–792 (cit. on p. 345).

Brazovskii, S. A. (1975). 'Phase transition of an isotropic system to a nonuniform state'. In: *J. Exp. Theor. Phys.* 68, pp. 175–185. [*Sov. Phys. JETP* 41, pp. 85–89 (1975)] (cit. on p. 353).

Brenner, M. P., J. D. Eldredge, and J. B. Freund (2019). 'Perspective on machine learning for advancing fluid mechanics'. In: *Phys. Rev. Fluids* 4, p. 100501 (cit. on p. 532).

Brenner, M. P., S. Hilgenfeldt, and D. Lohse (2002). 'Single-bubble sonoluminescence'. In: *Rev. Mod. Phys.* 74, pp. 425–484 (cit. on p. 508).

Brenner, M. P., P. Sörensen, and D. Weitz (2020). *Science and Cooking: Physics Meets Food, From Homemade to Haute Cuisine*. W. W. Norton (cit. on p. 14).

Bricard, A., J.-B. Caussin, D. Das, C. Savoie, V. Chikkadi, K. Shitara, O. Chepizhko, F. Peruani, D. Saintillan, and D. Bartolo (2015). 'Emergent vortices in populations of colloidal rollers'. In: *Nat. Commun.* 6, p. 8470 (cit. on p. 528).

Bricard, A., J.-B. Caussin, N. Desreumaux, O. Dauchot, and D. Bartolo (2013). 'Emergence of macroscopic directed motion in populations of motile colloids'. In: *Nature* 503, pp. 95–98 (cit. on pp. 197, 515, 528).

Brinkman, W. F. and P. E. Cladis (1982). 'Defects in liquid crystals'. In: *Phys. Today* 35(5), pp. 48–54 (cit. on pp. 277, 520).

Brito, C., I. S. Aranson, and H. Chaté (2003). 'Vortex glass and vortex liquid in oscillatory media'. In: *Phys. Rev. Lett.* 90, p. 068301 (cit. on p. 527).

Brock, J. D., R. J. Birgeneau, J. D. Litster, and A. Aharony (1989). 'Liquids, crystals and liquid crystals'. In: *Phys. Today* 42(7), pp. 52–59 (cit. on p. 511).

Broedersz, C. P. and F. C. MacKintosh (2014). 'Modeling semiflexible polymer networks'. In: *Rev. Mod. Phys.* 86, pp. 995–1036 (cit. on pp. 232, 233, 235, 518).

Brunton, S. L., B. R. Noack, and P. Koumoutsakos (2020). 'Machine learning for fluid mechanics'. In: *Annu. Rev. Fluid Mech.* 52, pp. 477–508 (cit. on p. 532).

Burns, J. L., Y.-d Yan, G. J. Jameson, and S. Biggs (1997). 'A light scattering study of the fractal aggregation behavior of a model colloidal system'. In: *Langmuir* 13, pp. 6413–6420 (cit. on p. 515).

Bushby, R. J. and O. R. Lozman (2002). 'Discotic liquid crystals 25 years on'. In: *Curr. Opin. Colloid Interface* 7, pp. 343–354 (cit. on p. 262).

Bushell, G. C., D. Yan, D. Woodfield, J. Raper, and R. Amala (2002). 'On techniques for the measurement of the mass fractal dimension of aggregates'. In: *Adv. Colloid Interface Sci.* 95, pp. 1–50 (cit. on p. 515).

Callan-Jones, A. C. and F. Jülicher (2011). 'Hydrodynamics of active permeating gels'. In: *New J. Phys.* 13, p. 093027 (cit. on p. 434).

Cambou, A. D. and N. Menon (2015). 'Orientational ordering in crumpled elastic sheets'. In: *Europhys. Lett.* 112, p. 14003 (cit. on p. 523).

Campelo, F., C. Arnarez, S. J. Marrink, and M. M. Kozlov (2014). 'Helfrich model of membrane bending: From Gibbs theory of liquid interfaces to membranes as thick anisotropic elastic layers'. In: *Adv. Colloid Interface Sci.* 208, pp. 25–33 (cit. on p. 315).

Capel, A. J., R. P. Rimington, M. P. Lewis, and S. D. R. Christie (2018). '3D printing for chemical, pharmaceutical and biological applications'. In: *Nat. Rev. Chem.* 2, pp. 422–436 (cit. on p. 465).

Caragine, C. M., S. C. Haley, and A. Zidovska (2018). 'Surface fluctuations and coalescence of nucleolar droplets in the human cell nucleus'. In: *Phys. Rev. Lett.* 121, p. 148101 (cit. on pp. 140, 513).

Carilli, M. F., K. T. Delaney, and G. H. Fredrickson (2018). 'Nucleation of the lamellar phase from the disordered phase of the renormalized Landau-Brazovskii model'. In: *J. Chem. Phys.* 148, p. 054903 (cit. on p. 524).

Carnall, J. M. A., C. A. Waudby, A. M. Belenguer, M. C. A. Stuart, J. J.-P. Peyralans, and S. Otto (2010). 'Mechanosensitive self-replication driven by self-organization'. In: *Science* 327, pp. 1502–1506 (cit. on p. 533).

Carraher Jr., Ch. E. (2017). *Polymer Chemistry, (4th Edition)*. CRC Press (cit. on pp. 10, 204).

Casimir, H. B. G. and D. Polder (1948). 'The influence of retardation on the London-Van der Waals forces'. In: *Phys. Rev.* 73, pp. 360–372 (cit. on p. 514).

Cates, M. E. and J. Tailleur (2015). 'Motility-induced phase separation'. In: *Annu. Rev. Condens. Matter Phys.* 6, pp. 219–244 (cit. on p. 412).

Cerda, E. and L. Mahadevan (2003). 'Geometry and physics of wrinkling'. In: *Phys. Rev. Lett.* 90, p. 074302 (cit. on pp. 341, 509).

Chaikin, P. M. and T. C. Lubensky (1995). *Principles of Condensed Matter Physics*. Cambridge University Press (cit. on pp. 33, 143, 265, 506, 513, 521).

Chajwa, R., N. Menon, S. Ramaswamy, and R. Govindarajan (2020). 'Waves, algebraic growth, and clumping in sedimenting disk arrays'. In: *Phys. Rev. X* 10, p. 041016 (cit. on p. 422).

Chan, B., N. J. Balmforth, and A. E. Hosoi (2005). 'Building a better snail: Lubrication and adhesive locomotion'. In: *Phys. Fluids* 17, p. 113101 (cit. on p. 46).

Chandrasekhar, S. (1992). *Liquid Crystals, 2nd edition*. Cambridge University Press (cit. on pp. 260, 265, 520).

Chantelot, P. and D. Lohse (2021). 'Leidenfrost effect as a directed percolation phase transition'. In: *Phys. Rev. Lett.* 127, p. 124502 (cit. on p. 508).

Chardac, A., S. Shankar, M. C. Marchetti, and D. Bartolo (2021). 'Emergence of dynamic vortex glasses in disordered polar active fluids'. In: *Proc. Natl. Acad. Sci.* 118, e2018218118 (cit. on p. 528).

de Gennes, P.-G. (1981). 'Polymer solutions near an interface. Adsorption and depletion layers'. In: *J. Phys. (Paris)* 14, pp. 1637–1644 (cit. on p. 227).

de Gennes, P.-G. (1987). 'Polymers at an interface: A simplified view'. In: *Adv. Colloid Interface Sci.* 27, pp. 189–209 (cit. on p. 227).

de Gennes, P.-G., F. Brochard-Wyart, and D. Quéré (2004). *Capillarity and Wetting Phenomena, Drops, Bubbles, Pearls, Waves*. Springer (cit. on p. 48).

de Gennes, P.-G. and J. Prost (1993). *The Physics of Liquid Crystals*. Clarendon Press (cit. on pp. 260, 265, 269, 309, 520).

de Groot, S. R. and P. Mazur (1964). *Non-equilibrium Thermodynamics*. North-Holland (cit. on pp. 22, 26, 30, 506, 508).

Deam, R. T. and S. F. Edwards (1976). 'The theory of rubber elasticity'. In: *Philosophical Transactions of the Royal Society of London. Series A, Mathematical and Physical Sciences* 280, pp. 317–353 (cit. on p. 256).

Dean, D. S. (1996). 'Langevin equation for the density of a system of interacting Langevin processes'. In: *J. Phys. A Math.* 29, pp. L613–L617 (cit. on p. 157).

Deboeuf, S., E. Katzav, A. Boudaoud, D. Bonn, and M. Adda-Bedia (2013). 'Comparative study of crumpling and folding of thin sheets'. In: *Phys. Rev. Lett.* 110, p. 104301 (cit. on pp. 322, 522).

Dee, G. and W. van Saarloos (1988). 'Bistable systems with propagating fronts leading to pattern formation'. In: *Phys. Rev. Lett.* 60, pp. 2641–2644 (cit. on p. 514).

Deegan, R. D., O. Bakajin, T. F. Dupont, G. Huber, S. R. Nagel, and T. A. Witten (1997). 'Capillary flow as the cause of ring stains from dried liquid drops'. In: *Nature* 389, pp. 1275–1278 (cit. on p. 49).

Deniz, A. A., S. Mukhopadhyay, and E. A Lemke (2007). 'Single-molecule biophysics: At the interface of biology, physics and chemistry'. In: *J. R. Soc. Interface* 5, pp. 15–45 (cit. on p. 505).

Desai, R. C. and R. Kapral (2009). *Dynamics of Self-Organized and Self-Assembled Structures*. Cambridge University Press (cit. on p. 525).

Desmond, K. W. and E. R. Weeks (2014). 'Influence of particle size distribution on random close packing of spheres'. In: *Phys. Rev. E* 90, p. 022204 (cit. on p. 516).

Deutsch, J. M., A. Dhar, and O. Narayan (2004). 'Return to return point memory'. In: *Phys. Rev. Lett.* 92, p. 227203 (cit. on p. 531).

Devany, J., D. M. Sussman, T. Yamamoto, M. L. Manning, and M. L. Gardel (2021). 'Cell cycle-dependent active stress drives epithelia remodeling'. In: *Proc. Natl. Acad. Sci.* 118, p. 1917853118 (cit. on p. 439).

Di Leonardo, R., L. Angelani, D. Dell'Arciprete, G. Ruocco, V. Iebba, S. Schippa, M. P. Conte, F. Mecarini, F. De Angelis, and E. Di Fabrizio (2010). 'Bacterial ratchet motors'. In: *Proc. Natl. Acad. Sci. USA* 107, pp. 9541–9545 (cit. on p. 515).

Díaz, J. S. (2021). 'Explosion analysis from images: Trinity and Beirut'. In: *Eur. J. Phys.* 42, p. 035803 (cit. on p. 39).

Dierichs, K. and A. Menges (2021). 'Designing architectural materials: From granular form to functional granular material'. In: *Bioinspir. Biomim.* 16, p. 065010 (cit. on pp. 467, 530).

Dijkstra, M. and E. Luijten (2021). 'From predictive modelling to machine learning and reverse engineering of colloidal self-assembly'. In: *Nat. Mat.* 20, pp. 762–773 (cit. on pp. 167, 530).

Dimova, R. and R. Lipowsky (2016). 'Giant vesicles exposed to aqueous two-phase systems: Membrane wetting, budding processes, and spontaneous Tubulation'. In: *Adv. Mater. Interfaces* 4, p. 1600451 (cit. on p. 315).

Dittrich, P., J. Ziegler, and W. Banzhaf (2001). 'Artificial chemistries—A review'. In: *Artif. Life* 7, pp. 225–275 (cit. on p. 492).

Dogic, Z. and S. Fraden (2014). 'Phase behavior of rod-like viruses and virus-sphere mixtures'. In: *Soft Matter Vol. 2: Complex Colloidal Suspensions*. Ed. by G. Gompper and M. Schick. Wiley-VCH Verlag, pp. 1–86 (cit. on p. 520).

Dogterom, M. and B. Yurke (1997). 'Measurement of the force-velocity relation for growing microtubules'. In: *Science* 278, pp. 856–860 (cit. on p. 510).

Doi, M. (2013). *Soft Matter Physics*. Oxford University Press (cit. on p. 238).

Doi, M. and S. F. Edwards (1986). *Theory of Polymer Dynamics*. Clarendon Press (cit. on pp. 202, 517, 518).

Dollet, B., Ph. Marmottant, and V. Garbin (2019). 'Bubble dynamics in soft and biological matter'. In: *Annu. Rev. Fluid Mech.* 51, pp. 331–355 (cit. on p. 52).

Donnelly, R. J. (1993). 'Quantized vortices and turbulence in helium II'. In: *Annu. Rev. Fluid Mech.* 25, pp. 325–371 (cit. on p. 523).

Doostmohammadi, A., J. Ignés-Mullol, J. M. Yeomans, and F. Sagués (2018). 'Active nematics'. In: *Nat. Commun.* 9, p. 3246 (cit. on pp. 415, 521).

Dorfman, J. R. and E. G. D. Cohen (1970). 'Velocity correlation functions in two and three dimensions'. In: *Phys. Rev. Lett.* 25, pp. 1257–1260 (cit. on p. 514).

Drescher, K., K. C. Leptos, I. Tuval, T. Ishikawa, T. J. Pedley, and R. E. Goldstein (2009). 'Hydrodynamics bound states of swimming algae'. In: *Phys. Rev. Lett.* 102, p. 168101 (cit. on p. 424).

Durey, G., Y. Ishii, and T. Lopez-Leon (2020). 'Temperature-driven anchoring transitions at liquid crystal/water interfaces'. In: *Langmuir* 36, pp. 9368–9376 (cit. on p. 520).

Dyson, F. J. (1982). 'A model for the origin of life'. In: *J. Mol. Evol.* 18, pp. 344–350 (cit. on pp. 495, 533).

Dyson, F. J. (1985). *Origins of life*. Cambridge University Press (cit. on p. 533).

Earnshaw, J.C. (1996). 'Light scattering as a probe of liquid surfaces and interfaces'. In: *Adv. Colloid Interface Sci.* 68, pp. 1–29 (cit. on p. 513).

Ebert, U. and W. van Saarloos (2000). 'Front propagation into unstable states: Universal algebraic convergence towards uniformly translating pulled fronts'. In: *Physica D* 146, pp. 1–99 (cit. on pp. 132, 399, 512, 524, 527).

Eckhardt, B., T. M. Schneider, B. Hof, and J. Westerweel (2007). 'Turbulence transition in pipe flow'. In: *Annu. Rev. Fluid Mech.* 39, pp. 447–468 (cit. on p. 507).

Edwards, S. F. (1965). 'The statistical mechanics of polymers with excluded volume'. In: *Proc. Phys. Soc.* 85, pp. 613–624 (cit. on p. 202).

Eggers, J. (1997). 'Nonlinear dynamics and breakup of free-surface flows'. In: *Rev. Mod. Phys.* 69, pp. 865–929 (cit. on pp. 46, 508).

Ellenbroek, W. G., M. van Hecke, and W. van Saarloos (2009). 'Jammed frictionless disks: Connecting local and global response'. In: *Phys. Rev. E* 80, p. 061307 (cit. on pp. 95, 510).

Emmett, P. H. and M. Cines (1947). 'Adsorption of argon, nitrogen, and butane on porous glass.' In: *J. Phys. Chem.* 51, pp. 1248–1262 (cit. on p. 531).

Epstein, I. R. and J. A. Pojman (1998). *An Introcution to Nonlinear Chemical Dynamics*. Oxford University Press (cit. on p. 525).

Ernst, M. H., E. H. Hauge, and J. M. J. van Leeuwen (1970). 'Asymptotic time behavior of correlation functions'. In: *Phys. Rev. Lett.* 25, pp. 1254–1256 (cit. on p. 514).

Ernst, M. H., E. H. Hauge, and J. M. J. van Leeuwen (1976). 'Asymptotic time behavior of correlation functions. III. Local equilibrium and mode-coupling theory'. In: *J. Stat. Phys.* 15, p. 23 (cit. on p. 514).

Evans, D. J. and G. Morriss (2008). *Statistical Mechanics of Nonequilibrium Liquids*. Cambridge University Press (cit. on p. 126).

Falcão, A. N., J. S. Pedersen, and K. Mortensen (1993). 'Structure of randomly cross-linked poly(dimethylsiloxane) metworks produced by electron irradiation'. In: *Macromolecules* 26, pp. 5350–5364 (cit. on p. 517).

Fang, H., M. F. Hagan, and W. B. Rogers (2020). 'Two-step crystallization and solid-solid transitions in binary colloidal mixtures'. In: *Proc. Natl. Acad. Sci.* 117, pp. 27927–27933 (cit. on pp. 470, 530).

Fardin, M. A., C. Perge, and N. Taberlet (2014). 'The hydrogen atom of fluid dynamics—Introduction to the Taylor-Couette flow for soft matter scientists'. In: *Soft Matter* 10, p. 3523 (cit. on p. 507).

Farhadifar, R., J.- C. Röper, B. Aigouy, S. Eaton, and F. Jülicher (2007). 'The influence of cell mechanics, cell-cell interactions, and proliferation on epithelial packing'. In: *Curr. Biol.* 17, pp. 2095–2104 (cit. on p. 529).

Farnoux, B., F. Boue, J. P. Cotton, M. Daoud, G. Jannink, M. Nierlich, and P.-G. de Gennes (1978). 'Cross-over in polymer solutions'. In: *J. Phys. France* 39, pp. 77–86 (cit. on p. 517).

Fauve, S., S. Douady, and O. Thual (1991). 'Drift instabilities of cellular patterns'. In: *J. phys. II* 1, pp. 311–322 (cit. on p. 529).

Feng, J. and S. Weinbaum (2000). 'Lubrication theory in highly compressible porous media: The mechanics of skiing, from red cells to humans'. In: *J. Fluid Mech.* 422, pp. 281–317 (cit. on p. 46).

Ferguson, A. L. (2017). 'Machine learning and data science in soft materials engineering'. In: *J. Condens. Matter Phys.* 30, p. 043002 (cit. on p. 532).

Feynman, R. (1960). 'There's plenty of room at the bottom: An invitation to enter a new field of physics'. In: *Eng. Sci.* XXIII (cit. on p. 530).

Field, R. J. and M. Burger, eds. (1984). *Oscillations and Traveling Waves in Chemical Systems*. Wiley (cit. on p. 527).

Fisher, M. E. and P.-G. de Gennes (1978). 'Phénomènes aux parois dans un mélange binaire critique'. In: *C. R. Acad. Sci. Ser.* B287, pp. 207–209 (cit. on p. 171).

Fisher, R. A. (1937). 'The wave of advance of advantageous genes'. In: *Ann. Eugen.* 7, pp. 355–369 (cit. on p. 524).

Fleming, R. M. and L. F. Schneemeyer (1986). 'Observation of a pulse-duration memory effect in $K_{0.30}MoO_3$'. In: *Phys. Rev. B* 33, pp. 2930–2933 (cit. on p. 531).

Fleury, R. and A. Alu (2014). 'Cloaking and invisibility: A review'. In: *Electromagn. Waves (Camb)* 147, pp. 171–202 (cit. on p. 510).

Florijn, B., C. Coulais, and M. van Hecke (2016). 'Programmable mechanical metamaterials: The role of geometry'. In: *Soft Matter* 12, pp. 8736–8743 (cit. on p. 90).

Flory, P. J. (1969). *Statistical Mechanics of Chain Molecules*. John Wiley (cit. on pp. 202, 211).

Forster, D. (1994). *Hydrodynamic Fluctuations, Broken Symmetry, and Correlation Functions*. Taylor and Frances (Advanced Book Classics, reprint from the 1975 edition) (cit. on pp. 126, 509).

Forterre, Y. and O. Pouliquen (2008). 'Flows of dense granular media'. In: *Annu. Rev. Fluid Mech.* 40, pp. 1–24 (cit. on pp. 188, 516).

Fredrickson, G. H. and E. Helfand (1997). 'Fluctuation effects in the theory of microphase separation in block copolymers'. In: *J. Chem. Phys.* 87, pp. 697–705 (cit. on p. 231).

Freelon, B., M. Ramazanoglu, P. J. Chung, R. N. Page, Y.-T. Lo, P. Valdivia, C. W. Garland, and R. J. Birgeneau (2011). 'Smectic-A and smectic-C phases and phase transitions in 8̄S5 liquid-crystal-aerosil gels'. In: *Phys. Rev. E* 84, p. 031705 (cit. on p. 519).

Frenkel, D. (2000). 'Perspective on "The effect of shape on the interaction of colloidal particles"'. In: *Theor. Chem. Acc.* 103, pp. 212–213 (cit. on p. 514).

Frenkel, D. and B. Smit (2023). *Understanding Molecular Simulation: From Algorithms to Applications (3rd edition)*. Academic Press (cit. on p. 22).

Fruchart, M. and D. Carpentier (2013). 'An introduction to topological insulators'. In: *C. R. Phys.* 14, pp. 779–815 (cit. on p. 102).

Fruchart, M., R. Hanai, P. B. Littlewood, and V. Vitelli (2021). 'Non-reciprocal phase transitions'. In: *Nature* 592, pp. 363–369 (cit. on pp. 428, 432, 457).

Fruchart, M., C. Scheibner, and V. Vitelli (2023). 'Odd viscosity and odd elasticity'. In: *Annu. Rev. Condens. Matter Phys.* 14, pp. 471–510 (cit. on pp. 421, 422, 424, 426).

Fruchart, M. and V. Vitelli (2020). 'Symmetries and dualities in the theory of elasticity'. In: *Phys. Rev. Lett.* 124, p. 248001 (cit. on p. 78).

Fruchart, M., Y. Zhou, and V. Vitelli (2020). 'Dualities and non-Abelian mechanics'. In: *Nature* 577, pp. 636–640 (cit. on pp. 529, 530).

Fürthauer, S., M. Strempel, S. W. Grill, and F. Jülicher (2012). 'Active chiral fluids'. In: *Eur. Phys. J. E* 35, p. 89 (cit. on p. 529).

Gagnon, D. A., N. C. Keim, X. Shen, and P. E. Arratia (2014). 'Fluid-induced propulsion of rigid particles in wormlike micellar solutions'. In: *Phys. Fluids* 26, p. 103101 (cit. on p. 515).

Gambassi, A. (2009). 'The Casimir effect: From quantum to critical fluctuations'. In: *J. Phys. Conf. Ser.* 161, p. 012037 (cit. on p. 171).

Gardel, M. L., J. H. Shin, F. C. MacKintosh, L. Mahadevan, P. Matsudaira, and D. A. Weitz (2004). 'Elastic behavior of cross-linked and bundled actin networks'. In: *Science* 304, pp. 1301–1305 (cit. on p. 236).

Gardiner, C. W., ed. (2004). *Handbook of Stochastic Methods: For Physics, Chemistry and the Natural Sciences, 3rd ed.* Springer (cit. on pp. 123, 128, 158, 511, 512).

Gardner, M. (1970). 'The fantastic combinations of John Conway's new solitaire game "life"'. In: *Sci. Am.* 223, pp. 120–123 (cit. on p. 532).

Garnier, N., A. Chiffaudel, F. Daviaud, and A. Prigent (2003). 'Nonlinear dynamics of waves and modulated waves in 1D thermocapillary flows. I. General presentation and periodic solutions'. In: *Physica D* 174, pp. 1–29 (cit. on pp. 382, 526).

Geng, Y., G. van Anders, P. M. Dodd, J. Dshemuchadse, and S. C. Glotzer (2019). 'Engineering entropy for the inverse design of colloidal crystals from hard shapes'. In: *Sci. Adv.* 5, aaw0514 (cit. on p. 469).

Geng, Y., G. van Anders, and S. C. Glotzer (2018). 'Predicting colloidal crystals from shapes via inverse design and machine learning'. [Data set], University of Michigan—Deep Blue Data; arXiv:1801.06219 (cit. on p. 532).

Genzer, J. and J. Groenewold (2006). 'Soft matter with hard skin: From skin wrinkles to templating and material characterization'. In: *Soft Matter* 2, pp. 310–323 (cit. on pp. 508, 509).

Geyer, D., D. Martin, J. Tailleur, and D. Bartolo (2019). 'Freezing a flock: Motility-induced phase separation in polar active liquids'. In: *Phys. Rev. X* 9, p. 031043 (cit. on p. 413).

Geyer, D., A. Morin, and D. Bartolo (2018). 'Sounds and hydrodynamics of polar active fluids'. In: *Nat. Mater.* 17, pp. 789–793 (cit. on pp. 448, 528).

Ghatak, A., M. Brandenbourger, J. van Wezel, and C. Coulais (2020). 'Observation of non-Hermitian topology and its bulk-edge correspondence in an active mechanical metamaterial'. In: *Proc. Natl. Acad. Sci.* 117, pp. 29561–29568 (cit. on pp. 419, 529).

Gilbert, W. (1986). 'Origin of life: The RNA world'. In: *Nature* 319, pp. 618–618 (cit. on p. 533).

Ginelli, F. (2016). 'The physics of the Vicsek model'. In: *Eur. Phys. J. Spec. Top.* 225, pp. 2099–2117 (cit. on p. 528).

Giomi, L. (2015). 'Geometry and topology of turbulence in active nematics'. In: *Phys. Rev. X* 5, p. 031003 (cit. on p. 415).

Giomi, L., M. J. Bowick, X. Ma, and M. C. Marchetti (2013). 'Defect annihilation and proliferation in active nematics'. In: *Phys. Rev. Lett.* 110, p. 228101 (cit. on p. 521).

Giomi, L., M. J. Bowick, P. Mishra, R. Sknepnek, and M. C. Marchetti (2014). 'Defect dynamics in active nematics'. In: *Phil. Trans. Roy. Soc. A* 372, p. 20130365 (cit. on pp. 418, 528).

Glover, S. M., Y. Yan, G. J. Jameson, and S. Biggs (2000). 'Bridging flocculation studied by light scattering and settling'. In: *Chem. Eng. J.* 80, pp. 3–12 (cit. on pp. 176, 515).

Goddard, B. D., R. D. Mills-Williams, and J. Sun (2020). 'The singular hydrodynamic interactions between two spheres in Stokes flow'. In: *Phys. Fluids* 32, p. 062001 (cit. on p. 66).

Gokhale, S., A. K. Sood, and R. Ganapathy (2016). 'Deconstructing the glass transition through critical experiments on colloids'. In: *Adv. Phys.* 65, pp. 363–452 (cit. on p. 177).

Goldbart, P. M. and N. Goldenfeld (1989). 'Microscopic theory for cross-linked macromolecules. II. Replica theory of the transition to the solid state'. In: *Phys. Rev. A* 39, pp. 1412–1419 (cit. on p. 259).

Goldbart, P. M. and N. Goldenfeld (2004). 'Sam Edwards and the statistical mechanics of rubber'. In: *Stealing the gold: A celebration of the pioneering physics of Sam Edwards*. Oxford University Press, pp. 275–300 (cit. on p. 256).

Goldenfeld, N. (1992). *Lectures on Phase Transitions and the Renormalization Group (Frontiers in Physics 85)*. Taylor and Francis (cit. on pp. 199, 201, 211, 259, 290, 511, 516).

Goldenfeld, N. (2015). 'Samuel Frederick Edwards: Founder of modern polymer and soft matter theory'. In: *Proc. Natl. Acad. Sci.* 113, pp. 10–11 (cit. on p. 202).

Goldstein, R. E. (2016). 'Fluid dynamics at the scale of the cell'. In: *J. Fluid Mech.* 103, p. 168103 (cit. on p. 507).

Gollub, J. P., S. V. Benson, and J. Steinman (1980). 'A subharmonic route to turbulent convection'. In: *Ann. N.Y. Acad. Sci.* 357, pp. 22–27 (cit. on p. 386).

Goodrich, C. P., A. J. Liu, and S. R. Nagel (2014). 'Solids between the mechanical extremes of order and disorder'. In: *Nat. Phys.* 10, pp. 578–581 (cit. on p. 511).

Goodrich, C. P., A. J. Liu, and S. R. Nagel (2015). 'The principle of independent bond-level response: Tuning by pruning to exploit disorder for global behavior'. In: *Phys. Rev. Lett.* 114, p. 225501 (cit. on p. 97).

Graham, M. D. (2014). 'Drag reduction and the dynamics of turbulence in simple and complex fluids'. In: *Phys. Fluids* 26, p. 101301 (cit. on p. 518).

Gramsbergen, E. F., L. Longa, and W. H. de Jeu (1986). 'Landau theory of the nematic-isotropic phase transition'. In: *Phys. Rep.* 135, pp. 195–257 (cit. on pp. 267, 298, 520).

Gray, P. and S. K. Scott (1990). *Chemical Oscillations and Instabilities*. Oxford University Press (cit. on p. 525).

Grier, D. (2003). 'A revolution in optical manipulation'. In: *Nature* 424, pp. 810–816 (cit. on p. 136).

Groisman, A. and V. Steinberg (2000). 'Elastic turbulence in a polymer solution flow'. In: *Nature* 405, pp. 53–55 (cit. on p. 519).

Groisman, A. and V. Steinberg (2004). 'Elastic turbulence in curvilinear flows of polymer solutions'. In: *New J. Phys.* 6, p. 29 (cit. on pp. 245, 246, 519).

Gross, P., G. Farge, E. J. G. Peterman, and G. J. L. Wuite (2010). 'Combining optical tweezers, single-molecule fluorescence microscopy, and microfluidics for studies of DNA-protein interactions'. In: *Methods in Enzymology*. Ed. by N. G. Walter. Elsevier, pp. 427–453 (cit. on p. 517).

Gross, P., N. Laurens, L. B. Odershede, U. Bockelmann, E. J. G. Peterman, and G. J. L. Wuite (2011). 'Quantifying how DNA stretches, melts and changes twist under tension'. In: *Nat. Phys.* 7, pp. 731–736 (cit. on p. 222).

Grossmann, S., D. Lohse, and C. Sun (2016). 'High Reynolds number Taylor-Couette turbulence'. In: *Annu. Rev. Fluid Mech.* 48, pp. 53–80 (cit. on pp. 38, 507).

Grzhibovskis, R., E. Krämer, I. Bernhardt, B. Kemper, C. Zanden, N. V. Repin, B. V. Tkachuk, and M. V. Voinova (2016). 'Shape of red blood cells in contact with artificial surfaces'. In: *Eur. Biophys. J.* 46, pp. 141–148 (cit. on p. 521).

Guckenheimer, J. and P. Holmes (1983). *Nonlinear Oscillations, Dynamical Systems, and Bifurcations of Vector Fields*. Springer (cit. on p. 510).

Guilbaud, S., L. Salomé, N. Destainville, M. Manghi, and C. Tardin (2019). 'Dependence of DNA persistence length on ionic strength and ion type'. In: *Phys. Rev. Lett.* 122, p. 028102 (cit. on p. 517).

Guo, K., Z. Yang, C.-H. Yu, and M. J. Buehler (2021). 'Artificial intelligence and machine learning in design of mechanical materials'. In: *Mater. Horiz.* 8, pp. 1153–1172 (cit. on p. 480).

Halperin, B. I. (1979). 'Superfluidity, melting and liquid crystal phases in two dimensions'. In: *Proceeding of Kyoto Summer Institute 1979 – Physics of Low Dimensional Systems*. Ed. by Y. Nagaoka and S. Hikami. Publications Office, Progress of Theoretical Physics, p. 53 (cit. on p. 291).

Halperin, B. I. and T. C. Lubensky (1974). 'On the analogy between smectic-A liquid crystals and superconductors'. In: *Solid State Commun.* 14, pp. 997–1001 (cit. on pp. 268, 298, 521).

Halperin, B. I., T. C. Lubensky, and S.-k. Ma (1974). 'First-order phase transitions in superconductors and smectic-A liquid crystals'. In: *Phys. Rev. Lett.* 32, pp. 292–295 (cit. on pp. 268, 298).

Halpin-Healy, T. and Y.-C. Zhang (1995). 'Kinetic roughening phenomena, stochastic growth, directed polymers and all that. Aspects of multidisciplinary statistical mechanics'. In: *Phys. Rep.* 254, pp. 215–414 (cit. on p. 326).

Halsey, Th. C., M. H. Jensen, L. P. Kadanoff, I. Procaccia, and B. I. Shraiman (1986). 'Fractal measures and their singularities: The characterization of strange sets'. In: *Phys. Rev. A* 33, pp. 1141–1151 (cit. on p. 514).

Han, K., G. Kokot, O. Tovkach, A. Glatz, I. S. Aranson, and A. Snezhko (2020). 'Emergence of self-organized multivortex states in flocks of active rollers'. In: *Proc. Natl. Acad. Sci.* 117, pp. 9706–9711 (cit. on p. 528).

Han, M., M. Fruchart, C. Scheibner, S. Vaikuntanathan, J. J. de Pablo, and V. Vitelli (2021). 'Fluctuating hydrodynamics of chiral active fluids'. In: *Nat. Phys.* 17, pp. 1260–1269 (cit. on p. 427).

Hanopolskyi, A. I., V. A. Smaliak, A. I. Novichkov, and S. N. Semenov (2020). 'Autocatalysis: Kinetics, mechanisms and design'. In: *ChemSystemsChem* 3, e2000026 (cit. on p. 533).

Happel, J. and H. Brenner (1982). *Low Reynolds Number Hydrodynamics*. Springer (cit. on pp. 424, 507, 512).

Harth, K. and R. Stannarius (2020). 'Topological point defects of liquid crystals in quasi-two-dimensional geometries'. In: *Front. Phys.* 8, p. 00112 (cit. on p. 277).

Hartings, M. R. and Z. Ahmed (2019). 'Chemistry from 3D printed objects'. In: *Nat. Rev. Chem.* 3, pp. 305–314 (cit. on p. 465).

Hasan, M. Z. and C. L. Kane (2010). 'Colloquium: Topological insulators'. In: *Rev. Mod. Phys.* 82, pp. 3045–3067 (cit. on p. 102).

Haw, M. D., W. C. K. Poon, P. N. Pusey, P. Hébraud, and F. Lequeux (1998). 'Colloidal glasses under shear strain'. In: *Phys. Rev. E* 58, pp. 4673–4682 (cit. on p. 531).

Hayakawa, D., T. E. Videbaek, D. M. Hall, H. Fang, C. Sigl, E. Feigl, H. Dietz, S. Fraden, M. F. Hagan, G. M. Grason, and W. B. Rogers (2022). 'Geometrically programmed self-limited assembly of tubules using DNA origami colloids'. In: *Proc. Natl. Acad. Sci.* 119 (cit. on p. 214).

He, M., J. P. Gales, E. Ducrot, Z. Gong, G. Yi, S. Sacanna, and D. J. Pine (2020). 'Colloidal diamond'. In: *Nature* 585, pp. 524–529 (cit. on p. 470).

Hébraud, P., F. Lequeux, J. P. Munch, and D. J. Pine (1997). 'Yielding and rearrangements in disordered emulsions'. In: *Phys. Rev. Lett.* 78, pp. 4657–4660 (cit. on p. 531).

Heinrichs, V., S. Dieluweit, J. Stellbrink, W. Pyckhout-Hintzen, N. Hersch, D. Richter, and R. Merkel (2018). 'Chemically defined, ultrasoft PDMS elastomers with selectable elasticity for mechanobiology'. In: *PLoS One* 13, e0195180 (cit. on p. 75).

Helfrich, W. (1973). 'Elastic properties of lipid bilayers: Theory and possible experiments'. In: *Z. Naturforsch. C* 28, pp. 693–703 (cit. on p. 312).

Hendricks, T. R., W. Wang, and I. Lee (2010). 'Buckling in nanomechanical films'. In: *Soft Matter* 6, pp. 3701–3706 (cit. on p. 508).

Herbut, I. (2007). *A Modern Approach to Critical Phenomena*. Cambridge University Press (cit. on pp. 199, 211, 511, 516).

Hertz, J., A. Krogh, and R. G. Palmer (1991). *Introduction to the Theory of Neural Computation*. Boston, MA, USA: Addison-Wesley (cit. on p. 531).

Hirst, L. S. and G. Charras (2017). 'Liquid crystals in living tissue'. In: *Nature* 544, pp. 164–165 (cit. on p. 289).

Hoffmann, L. A., L. N. Carenza, J. Eckert, and L. Giomi (2022). 'Theory of defect-mediated morphogenesis'. In: *Sci. Ad.* 8, abk2712 (cit. on p. 529).

Hohenberg, P. C. (1967). 'Existence of long-range order in one and two dimensions'. In: *Phys. Rev.* 158, pp. 383–386 (cit. on p. 528).

Hohenberg, P. C. and B. I. Halperin (1977). 'Theory of dynamic critical phenomena'. In: *Rev. Mod. Phys.* 49, p. 435 (cit. on pp. 21, 345).

Honig, J. and J. Spalek (2018). *A Primer to the Theory of Critical Phenomena*. Elsevier (cit. on pp. 199, 211, 516).

Hopfield, J. J. (1982). 'Neural networks and physical systems with emergent collective computational abilities'. In: *Proc. Natl. Acad. Sci.* 79, pp. 2554–2558 (cit. on p. 531).

Hormoz, S. and M. P. Brenner (2011). 'Design principles for self-assembly with short-range interactions'. In: *Proc. Natl. Acad. Sci.* 108, pp. 5193–5198 (cit. on p. 470).

Horváth, J., I. Szalai, and P. De Kepper (2018). 'Designing stationary reaction-diffusion patterns in pH self-activated systems'. In: *Acc. Chem. Res.* 51, pp. 3183–3190 (cit. on pp. 366, 525).

Hosoi, A. E. and J. W. M. Bush (2001). 'Evaporative instabilities in climbing films'. In: *J. Fluid Mech.* 442, pp. 217–239 (cit. on p. 524).

Howard, M., A. D. Rutenberg, and S. de Vet (2001). 'Dynamic compartmentalization of bacteria: accurate division in *E. coli*'. In: *Phys. Rev. Lett.* 87, p. 278102 (cit. on p. 366).

Hoyle, R. (2006). *Pattern Formation: An introduction to Methods*. Cambridge University Press (cit. on pp. 352, 529).

Hu, B. Y-H. (1989). 'Kramers-Kronig in two lines'. In: *Am. J. Phys.* 57, pp. 821–821 (cit. on p. 509).

Huang, C., H. Yuan, and S. Zhang (2011). 'Coupled vesicle morphogenesis and domain organization'. In: *Appl. Phys. Lett.* 98, p. 043702 (cit. on pp. 314, 521).

Hueckel, T., G. M. Hocky, and S. Sacanna (2021). 'Total synthesis of colloidal matter'. In: *Nat. Rev. Mater.* 6, pp. 1053–1069 (cit. on p. 180).

Hulsen, M. A., R. Fattal, and R. Kupferman (2005). 'Flow of viscoelastic fluids past a cylinder at high Weissenberg number: Stabilized simulations using matrix logarithms'. In: *J. Non-Newtonian Fluid Mech.* 127, pp. 27–39 (cit. on p. 519).

Hunter, G. L. and E. R. Weeks (2012). 'The physics of the colloidal glass transition'. In: *Rep. Prog. Phys.* 75, p. 066501 (cit. on pp. 177, 179, 515).

Hure, J., B. Roman, and J. Bico (2012). 'Stamping and wrinkling of elastic plates'. In: *Phys. Rev. Lett.* 109, p. 054302 (cit. on pp. 82, 509).

Idema, T. and D. J. Kraft (2019). 'Interactions between model inclusions on closed lipid bilayer membranes'. In: *Curr. Opin. Colloid Interface Sci.* 40, pp. 58–69 (cit. on p. 315).

Irvine, K. D. and E. F. Wieschaus (1994). 'Cell intercalation during Drosophila germband extension and its regulation by pair-rule segmentation genes'. In: *Development* 120, pp. 827–841 (cit. on p. 529).

Irvine, W. T., V. Vitelli, and P. M. Chaikin (2010). 'Pleats in crystals on curved surfaces'. In: *Nature* 468, pp. 947–951 (cit. on p. 342).

Israelachvili, J. N. (2011). *Intermolecular and Surface Forces (3rd edition)*. Elsevier (cit. on pp. 173, 189, 194, 514).

Ivlev, A. V., J. Bartnick, M. Heinen, C.-R. Du, V. Nosenko, and H. Löwen (2015). 'Statistical mechanics where Newton's third law is broken'. In: *Phys. Rev. X* 5, p. 011035 (cit. on p. 422).

Jacobs, W. M., A. Reinhardt, and D. Frenkel (2015). 'Rational design of self-assembly pathways for complex multicomponent structures'. In: *Proc. Natl. Acad. Sci.* 112, pp. 6313–6318 (cit. on p. 471).

Jaeger, H. M., S. R., Nagel, and R. P. Behringer (1996). 'Granular solids, liquids, and gases'. In: *Rev. Mod. Phys.* 68, pp. 1259–1273 (cit. on p. 505).

Jákli, A., O. D. Lavrentovich, and J. V. Selinger (2018). 'Physics of liquid crystals of bent-shaped molecules'. In: *Rev. Mod. Phys.* 90, p. 045004 (cit. on p. 285).

James, D. F. (2009). 'Boger fluids'. In: *Annu. Rev. Fluid Mech.* 41, pp. 129–142 (cit. on pp. 244, 519).

Jánosi, I. M., J. O. Kessler, and V. K. Horváth (1998). 'Onset of bioconvection in suspensions of *Bacillus subtilis*'. In: *Phys. Rev. E* 58, pp. 4793–4800 (cit. on p. 528).

Janson, M. E. and M. Dogterom (2004). 'A bending mode analysis for growing microtubules: Evidence for a velocity-dependent rigidity'. In: *Biophys. J.* 87, pp. 2723–2736 (cit. on pp. 510, 513).

Johansson, K. (2000). 'Shape fluctuations and random matrices'. In: *Commun. Math. Phys.* 209, pp. 437–476 (cit. on p. 523).

Johnson, P. (1993). 'Light scattering in the study of colloidal and macromolecular systems'. In: *Int. Rev. Phys. Chem.* 12, pp. 61–87 (cit. on p. 513).

Jones, R. (2002). *Soft Condensed Matter*. Oxford University Press (cit. on p. 192).

José, J. V., L. P. Kadanoff, S. Kirkpatrick, and D. R. Nelson (1977). 'Renormalization, vortices, and symmetry-breaking perturbations in the two-dimensional planar model'. In: *Phys. Rev. B* 16, pp. 1217–1241 (cit. on p. 293).

Joseph, D. D. (1990). *Fluid Dynamics of Viscoelastic Liquids*. Springer (cit. on p. 240).

Joshi, C., S. Ray, L. M. Lemma, M. Varghese, G. Sharp, Z. Dogic, A. Baskaran, and M. F. Hagan (2022). 'Data-driven discovery of active nematic hydrodynamics'. In: *Phys. Rev. Lett.* 129, p. 258001 (cit. on pp. 417, 505).

Joshi, Y. M. (2014). 'Dynamics of colloidal glasses and gels'. In: *Annu. Rev. Chem. Biomol. Eng.* 5, pp. 181–202 (cit. on p. 177).

Kadanoff, L. P. (1966). 'Scaling laws for Ising models near T_c'. In: *Physics* 2, pp. 263–272 (cit. on p. 201).

Kadic, M., T. Bückmann, R. Schittny, and M. Wegener (2013). 'Metamaterials beyond electromagnetism'. In: *Rep. Prog. Phys.* 76, p. 126501 (cit. on p. 89).

Kai, S. and W. Zimmermann (1989). 'Pattern dynamics in the electrohydrodynamics of nematic liquid crystals'. In: *Prog. Theor. Phys. Suppl.* 99, pp. 458–492 (cit. on p. 523).

Kaiser, A., A. Snezhko, and I. S. Aranson (2017). 'Flocking ferromagnetic colloids'. In: *Sci. Adv.* 3, e1601469 (cit. on p. 528).

Kamrin, K. (2019). 'Non-locality in granular flow: Phenomenology and modeling approaches'. In: *Front. Phys.* 7, p. 00116 (cit. on p. 188).

Kane, C. L. and T. C. Lubensky (2014). 'Topological boundary modes in isostatic lattices'. In: *Nat. Phys.* 10, pp. 39–45 (cit. on pp. 104, 511).

Kantor, Y. (2004). 'Properties of tethered surface'. In: *Statistical mechanics of membranes and surfaces (2nd edition)*. Ed. by D. R. Nelson, T. Piran, and S. Weinberg. World Scientific, pp. 111–130 (cit. on p. 320).

Kantor, Y., M. Kardar, and D. R. Nelson (1987). 'Tethered surfaces: Statics and dynamics'. In: *Phys. Rev. A* 35, pp. 3056–3071 (cit. on p. 523).

Kapral, R. and K. Showalter, eds. (1995). *Chemical Waves and Patterns*. Springer (cit. on p. 527).

Kardar, M., ed. (2007). *Statistical Physics of Fields*. Cambridge University Press (cit. on pp. 290, 291, 326, 334, 338, 517, 523).

Kardar, M., G. Parisi, and Y.-C. Zhang (1986). 'Dynamic scaling of growing interfaces'. In: *Phys. Rev. Lett.* 56, pp. 889–892 (cit. on p. 325).

Kaspar, C., B. J. Ravoo, W. G. van der Wiel, S. V. Wegner, and W. H. P. Pernice (2021). 'The rise of intelligent matter'. In: *Nature* 594, pp. 345–355 (cit. on p. 480).

Kauffman, S. A. (1986). 'Autocatalytic sets of proteins'. In: *J. Theor. Biol.* 119, pp. 1–24 (cit. on p. 533).

Kauffman, S. A. (1993). *The Origins of Order: Self-Organization and Selection in Evolution*. Oxford University Press (cit. on p. 533).

Keim, N. C. and P. E. Arratia (2014). 'Mechanical and microscopic properties of the reversible plastic regime in a 2D jammed material'. In: *Phys. Rev. Lett.* 112, p. 028302 (cit. on p. 531).

Keim, N. C. and S. R. Nagel (2011). 'Generic transient memory formation in disordered systems with noise'. In: *Phys. Rev. Lett.* 107, p. 010603 (cit. on p. 531).

Keim, N. C., J. D. Paulsen, and S. R. Nagel (2013). 'Multiple transient memories in sheared suspensions: Robustness, structure, and routes to plasticity'. In: *Phys. Rev. E* 88, p. 32306 (cit. on p. 531).

Keim, N. C., J. D. Paulsen, Z. Zeravcic, S. Sastry, and S. R. Nagel (2019). 'Memory formation in matter'. In: *Rev. Mod. Phys.* 91, p. 035002 (cit. on p. 531).

Keller, E. F. and L. A. Segel (1971). 'Model for chemotaxis'. In: *J. Theor Biol.* 30, pp. 225–234 (cit. on p. 414).

Kestin, J. and J. R. Dorfman (1972). *Course in Statistical Thermodynamics*. Academic Press (cit. on pp. 506, 510).

Ketzetzi, S., J. de Graaf, and D. J. Kraft (2020). 'Diffusion-based height analysis reveals robust microswimmer-wall separation'. In: *Phys. Rev. Lett.* 125, p. 238001 (cit. on p. 515).

Kleman, M. and O. D. Lavrentovich (2003). *Soft Matter Physics*. Springer (cit. on pp. 169, 192, 506, 507, 509, 514, 521).

Knobloch, E., J. Hettel, and G. Dangelmayr (1995). 'Parity breaking bifurcation in inhomogeneous systems'. In: *Phys. Rev. Lett.* 74, pp. 4839–4842 (cit. on p. 529).

Kochkov, D., J. A. Smith, A. Alieva, Q. Wang, M. P. Brenner, and S. Hoyer (2021). 'Machine learning–accelerated computational fluid dynamics'. In: *Proc. Natl. Acad. Sci.* 118, e2101784118 (cit. on pp. 486, 530, 532).

Kociński, J. and L. Wojtczak (1978). *Critical Scattering Theory—An Introduction*. Elsevier (cit. on p. 513).

Kolmogorov, A., I. Petrovsky, and N. Piscounoff (1937). 'Étude de l'équations de la diffusion avec croissance de la quantité de matière et son application a un problème biologique'. In: *Bull. Univ. Moskou, Ser. Internat.* 1A, pp. 1–25 (cit. on p. 524).

Komura, S. and D. Andelman (2014). 'Physical aspects of heterogeneities in multi-component lipid membranes'. In: *Adv. Colloid Interface Sci.* 208, pp. 34–46 (cit. on p. 315).

Kovacs, A. J. (1963). 'Glass transition in amorphous polymers: A phenomenological study'. In: *Adv. Polym. Sci.* 3, pp. 394–507 (cit. on p. 531).

Kovacs, A. J., J. J. Aklonis, J. M. Hutchinson, and A. R. Ramos (1979). 'Isobaric volume and enthalpy recovery of glasses. II. A transparent multiparameter theory'. In: *J. Polym. Sci. B Polym. Phys.* 17, pp. 1097–1162 (cit. on p. 531).

Kramer, E. M. and T. A. Witten (1997). 'Stress condensation in crushed elastic manifolds'. In: *Phys. Rev. Lett.* 78, pp. 1303–1306 (cit. on pp. 522, 523).

Kramers, H. A. (1940). 'Brownian motion in a field of force and the diffusion model of chemical reactions'. In: *Physica* 7, pp. 284–304 (cit. on pp. 132, 512).

Kratky, O. and G. Porod (1949). 'X-ray investigation of dissolved chain molecules'. In: *Rec. Trav. Chim.* 68, pp. 1106–1123 (cit. on p. 517).

Krech, M. (1994). *The Casimir Effect In Critical Systems.* World Scientific (cit. on p. 171).

Kretschmer, S. and P. Schwille (2016). 'Pattern formation on membranes and its role in bacterial cell division'. In: *Curr. Opin. Cell Biol.* 38, pp. 52–59 (cit. on p. 366).

Krishan, K., H. Kurtuldu, M. F. Schatz, M. Gameiro, K. Mischaikow, and S. Madruga (2007). 'Homology and symmetry breaking in Rayleigh-Bénard convection: Experiments and simulations'. In: *Phys. Fluids* 19, p. 117105 (cit. on p. 525).

Krug, J. (1997). 'Origins of scale invariance in growth processes'. In: *Adv. Phys.* 46, pp. 139–282 (cit. on pp. 326, 523).

Kruse, K., J.-F. Joanny, F. Jülicher, J. Prost, and K. Sekimoto (2004). 'Asters, vortices, and rotating spirals in active gels of polar filaments'. In: *Phys. Rev. Lett.* 92, p. 078101 (cit. on p. 415).

Kuijk, A., A. van Blaaderen, and A. Imhof (2011). 'Synthesis of monodisperse, rodlike silica colloids with tunable aspect ratio'. In: *J. Am. Chem. Soc.* 133, pp. 2346–2349 (cit. on pp. 164, 519).

Kulkarni, C. V., W. Wachter, G. Iglesias-Salto, S. Engelskirchen, and S. Ahualli (2011). 'Monoolein: A magic lipid?' In: *Phys. Chem. Chem. Phys.* 13, pp. 3004–3021 (cit. on p. 521).

Kull, H. J. (1991). 'Theory of the Rayleigh-Taylor instability'. In: *Phys. Rep.* 206, pp. 197–325 (cit. on p. 508).

Kuramoto, Y. (1984). *Chemical Oscillations, Waves, and Turbulence.* Springer (cit. on p. 526).

Kuramoto, Y. and T. Tsuzuki (1976). 'Persistent propagation of concentration waves in dissipative media far from thermal equilibrium'. In: *Progr. Theor. Phys.* 55, pp. 356–369 (cit. on p. 526).

Kurtuldu, H., K. Mischaikow, and M. F. Schatz (2011). 'Extensive scaling from computational homology and Karhunen-Loève decomposition analysis of Rayleigh-Bénard convection experiments'. In: *Phys. Rev. Lett.* 107, p. 034503 (cit. on p. 525).

Kwon, J.-Y., M. Khan, and S.-Y. Park (2016). 'pH-Responsive liquid crystal double emulsion droplets prepared using microfluidics'. In: *RSC Advances* 6, pp. 55976–55983 (cit. on p. 520).

Lahini, Y., O. Gottesman, A. Amir, and S. M. Rubinstein (2017). 'Nonmonotonic aging and memory retention in disordered mechanical systems'. In: *Phys. Rev. Lett.* 118, p. 085501 (cit. on p. 531).

Lakes, R. (1987). 'Foam structures with negative Poisson's ratio'. In: *Science* 235, pp. 1038–1040 (cit. on p. 510).

Landau, L. D. and E. M. Lifshitz (1970). *Course of Theoretical Physics Volume 7: Theory of Elasticity*. Pergamon Press (cit. on pp. 69, 506, 509).

Landau, L. D. and E. M. Lifshitz (1980). *Course of Theoretical Physics Volume 5: Statistical Physics, Part I (3rd Edition)*. Pergamon Press (cit. on pp. 128, 146, 230, 513).

Landau, L. D. and E. M. Lifshitz (1981a). *Course of Theoretical Physics Volume 8: Electrodynamics of Continuous Media*. Pergamon Press (cit. on p. 509).

Landau, L. D. and E. M. Lifshitz (1981b). *Course of Theoretical Physics Volume 9: Statistical Physics, Part II (3rd Edition)*. Pergamon Press (cit. on p. 146).

Landau, L. D. and E. M. Lifshitz (1987). *Course of Theoretical Physics Volume 6: Fluid Mechanics*. Pergamon Press (cit. on pp. 28, 37, 506–508, 529).

Langlois, W. E. and M. O. Deville (2013). *Lubrication Theory*. Springer (cit. on p. 46).

Langton, C. G. (1991). 'Artificial life'. In: *1991 Lectures in Complex Systems*. Vol. IV. Santa Fe Institute, Addison-Wesley, p. 189 (cit. on p. 490).

Langton, C. G., ed. (1995). *Artificial Life: An Overview*. The MIT Press (cit. on p. 490).

Larson, R. G. (1992). 'Instabilities in viscoelastic flows'. In: *Rheol. Acta* 31, pp. 213–263 (cit. on p. 244).

Larson, R. G. (1999). *The Structure and Rheology of Complex Fluids*. Oxford University Press (cit. on pp. 238, 240, 519).

Larson, R. G. (2000). 'Turbulence without inertia'. In: *Nature* 405, pp. 27–28 (cit. on pp. 245, 519).

Larson, R. G., E. S. G. Shaqfeh, and S. J. Muller (1990). 'A purely elastic instability in Taylor-Couette flow'. In: *J. Fluid Mech.* 218, pp. 573–600 (cit. on p. 245).

Lauga, E. (2016). 'Bacterial hydrodynamics'. In: *Annu. Rev. Fluid Mech.* 48, pp. 105–130 (cit. on pp. 182, 507).

Laughlin, R. B. (2005). *A Different Universe: Reinventing Physics from the Bottom Down*. Basic Books (cit. on p. 506).

Laurson, L. and M. J. Alava (2012). 'Dynamic hysteresis in cyclic deformation of crystalline solids'. In: *Phys. Rev. Lett.* 109, p. 155504 (cit. on p. 531).

Lautrup, B. (2005). *Physics of Continuous Matter—Exotic and Everyday Phenomena in the Macroscopic World*. Institute of Physics Publishing (cit. on pp. 46, 396).

Lavis, D. A. and G. M. Bell (1999). *Statistical Mechanics of Lattice Systems*. Springer (cit. on p. 510).

Lavrentovich, M. O., A. J. Liu, and S. R. Nagel (2017). 'Period proliferation in periodic states in cyclically sheared jammed solids'. In: *Phys. Rev. E* 96, 020101(R) (cit. on p. 531).

Lavrentovich, O. D. (1998). 'Topological defects in dispersed words and worlds around liquid crystals, or liquid crystal drops'. In: *Liq. Cryst.* 24, pp. 117–126 (cit. on p. 520).

Le Doussal, P. and L. Radzihovsky (2018). 'Anomalous elasticity, fluctuations and disorder in elastic membranes'. In: *Ann. Phys.* 39, pp. 340–410 (cit. on pp. 321, 522).

Lee, M. H. and E. M. Furst (2008). 'Response of a colloidal gel to a microscopic oscillatory strain'. In: *Phys. Rev. E* 77, p. 041408 (cit. on p. 531).

Lee, T. and R. S Lakes (1997). 'Anisotropic polyurethane foam with Poisson's ratio greater than 1'. In: *J. Mat. Sci.* 32, pp. 2397–2401 (cit. on p. 509).

Leibler, S. (2004). 'Equilibrium statistical mechanics of fluctuating films and membranes'. In: *Statistical mechanics of membranes and surfaces (2nd edition)*. Ed. by D. R. Nelson, T. Piran, and S. Weinberg. World Scientific, pp. 49–101 (cit. on p. 312).

Lekkerkerker, H. N. W. and R. Tuinier (2011). *Colloids and the Depletion Interaction*. Springer (cit. on pp. 173, 193, 514).

Lekkerkerker, H. N. W. and G. J. Vroege (2013). 'Liquid crystal phase transitions in suspensions of mineral colloids: New life from old roots'. In: *Phil. Trans. R. Soc. A* 371, p. 20120263 (cit. on p. 262).

Lemma, L. M., S. J. DeCamp, Z. You, L. Giomi, and Z. Dogic (2019). 'Statistical properties of autonomous flows in 2D active nematics'. In: *Soft Matter* 15, pp. 3264–3272 (cit. on p. 528).

Li, B., D. Zhou, and Y. Han (2016). 'Assembly and phase transitions within colloidal crystals'. In: *Nat. Rev. Mater.* 1, p. 15011 (cit. on p. 521).

Liao, M., X. Xiao, S. T. Chui, and Y. Han (2018). 'Grain-boundary roughening in colloidal crystals'. In: *Phys. Rev. X* 8, p. 021045 (cit. on p. 521).

Lidmar, J., L. Mirny, and D. R. Nelson (2003). 'Virus shapes and buckling transitions in spherical shells'. In: *Phys. Rev. E* 68, p. 051910 (cit. on pp. 315–318).

Lieb, E. H. and F. Y. Wu (1972). 'Two-dimensional ferroelectric models'. In: *Phase Transitions and Critical Phenomena, Vol. 1*. Ed. by C. Domb and M. S. Green. London: Academic Press, p. 321 (cit. on p. 510).

Limat, L., P. Jenffer, B. Dagens, E. Touron, M. Fermigier, and J. E. Wesfreid (1992). 'Gravitational instabilities of thin liquid layers: Dynamics of pattern selection'. In: *Physica D* 61, pp. 166–182 (cit. on pp. 61, 526).

Lin, I.-H., D. S. Miller, P. J. Bertics, C. J. Murphy, J. J. de Pablo, and N. L. Abbott (2011). 'Endotoxin-induced structural transformations in liquid crystalline droplets'. In: *Science* 332, pp. 1297–1300 (cit. on p. 286).

Lincoln, T. A. and G. F. Joyce (2009). 'Self-sustained replication of an RNA enzyme'. In: *Science* 323, pp. 1229–1232 (cit. on p. 493).

Liu, A. J. and S. R. Nagel (1998). 'Jamming is not just cool anymore'. In: *Nature* 396, pp. 21–22 (cit. on p. 92).

Liu, A. J., M. Wyart, W. van Saarloos, and S. R. Nagel (2011). 'The jamming scenario—an introduction and outlook'. In: *Glasses, Colloids and Granular Media*. Ed. by L. Berthier, G. Biroli, J.-P. Bouchaud, L. Cipelletti, and W. van Saarloos. Oxford University Press (cit. on pp. 92, 510).

Liu, H. and P. Yin (2022a). 'High order unconditionally energy stable RKDG schemes for the Swift-Hohenberg equation'. In: *J. Comput. Appl. Math.* 407, p. 114015 (cit. on p. 524).

Liu, H. and P. Yin (2022b). 'On the SAV-DG method for a class of fourth order gradient flows'. In: *Numer. Methods Partial Differ. Equa.* 22, pp. 1–16 (cit. on p. 524).

Livan, G., M. Novaes, and P. Vivo, eds. (2018). *Introduction to Random Matrices—Theory and Practice*. Springer (cit. on p. 329).

Loftus, A. F., S. Noreng, V. L. Hsieh, and R. Parthasarathy (2013). 'Robust measurement of membrane bending moduli using light sheet fluorescence imaging of vesicle fluctuations'. In: *Langmuir* 29, pp. 14588–14594 (cit. on pp. 140, 513).

Lohse, D. (2018). 'Bubble puzzles: From fundamentals to applications'. In: *Phys. Rev. Fluids* 3, p. 110504 (cit. on pp. 50, 508).

Lohse, D. (2022). 'Fundamental fluid dynamics challenges in inkjet printing'. In: *Annu. Rev. Fluid Mech.* 54, pp. 349–382 (cit. on p. 505).

Lohse, D. and X. Zhang (2015). 'Surface nanobubbles and nanodroplets'. In: *Rev. Mod. Phys.* 87, pp. 981–1035 (cit. on p. 50).

Lohse, D. and X. Zhang (2020). 'Physicochemical hydrodynamics of droplets out of equilibrium'. In: *Nat. Rev. Phys.* 2, pp. 426–443 (cit. on pp. 52, 505, 508).

López, H. M., J. Gachelin, C. Douarche, H. Auradou, and E. Clément (2015). 'Turning bacteria suspensions into superfluids'. In: *Phys. Rev. Lett.* 115, p. 028301 (cit. on pp. 414, 528).

Louf, J.-F., J. Knoblauch, and K. H. Jensen (2020). 'Bending and stretching of soft pores enable passive control of fluid flows'. In: *Phys. Rev. Lett.* 125, p. 098101 (cit. on p. 58).

Lovesy, S. W. (1986). *Condensed Matter Physics: Dynamic Correlations, 2nd ed.* Benjamin/Cummings (cit. on pp. 126, 509).

Lowensohn, J., B. Oyarzún, G. Narváez Paliza, B. M. Mognetti, and W. B. Rogers (2019). 'Linker-mediated phase behavior of DNA-coated colloids'. In: *Phys. Rev. X* 9, p. 041054 (cit. on p. 470).

Lubchenko, V. (2015). 'Theory of the structural glass transition: A pedagogical review'. In: *Adv. Phys.* 64, pp. 283–443 (cit. on p. 177).

Lundberg, M., K. Krishan, N. Xu, C. S. O'Hern, and M. Dennin (2008). 'Reversible plastic events in amorphous materials'. In: *Phys. Rev. E* 77, p. 041505 (cit. on p. 531).

Lutsko, J. F. (1989). 'Generalized expressions for the calculation of elastic constants by computer simulation'. In: *J. Appl. Phys.* 65, pp. 2991–2997 (cit. on p. 78).

Lysetska, M., A. Knoll, D. Boehringer, T. Hey, and G. Krauss (2002). 'UV light-damaged DNA and its interaction with human replication protein A: An atomic force microscopy study'. In: *Nucleic Acids Res.* 30, pp. 2686–2691 (cit. on pp. 206, 516).

Lyubarskii, G.Ya. (1960). *The Application of Group Theory in Physics*. Pergamon (cit. on p. 506).

Maass, C. C., C. Krüger, S. Herminghaus, and S. Bahr (2016). 'Swimming droplets'. In: *Annu. Rev. Condens. Matter Phys.* 7, pp. 171–193 (cit. on pp. 53, 515).

Macfarlane, R. J., B. Lee, M. R. Jones, N. Harris, G. C. Schatz, and C. A. Mirkin (2011). 'Nanoparticle superlattice engineering with DNA'. In: *Science* 334, pp. 204–208 (cit. on p. 530).

Mackay, F., J. Toner, A. Morozov, and D. Marenduzzo (2020). 'Darcy's law without friction in active nematic rheology'. In: *Phys. Rev. Lett.* 124, p. 187801 (cit. on p. 528).

Mallory, S. A., C. Valeriani, and A. Cacciuto (2018). 'An active approach to colloidal self-assembly'. In: *Annu. Rev. Phys. Chem.* 69, pp. 59–79 (cit. on p. 182).

Mandelbrot, B. B. (1982). *The Fractal Geometry of Nature*. Freeman (cit. on p. 175).

Mandle, R. J., S. J. Cowling, and J. W. Goodby (2017). 'A nematic to nematic transformation exhibited by a rod-like liquid crystal'. In: *Phys. Chem. Chem. Phys.* 19, pp. 11429–11435 (cit. on p. 520).

Mangan, N., C. Reichhardt, and C. J. Olson Reichhardt (2008). 'Reversible to irreversible flow transition in periodically driven vortices'. In: *Phys. Rev. Lett.* 100, p. 187002 (cit. on p. 531).

Manneville, P. (2014). *Dissipative Structures and Weak Turbulence*. Academic Press (cit. on p. 352).

Mao, X. and T. C. Lubensky (2018). 'Maxwell lattices and topological mechanics'. In: *Annu. Rev. of Condens. Matter Phys* 9, pp. 413–433 (cit. on p. 120).

Mistry, D., S. D. Connell, S. L. Mickthwaite, P. B. Morgan, J. H. Clamp, and H. F. Gleeson (2018). 'Coincident molecular auxeticity and negative order parameter in a liquid crystal elastomer'. In: *Nat. Commun.* 9, p. 5095 (cit. on p. 510).

Mitchell, N. P., M. F. Lefebvre, V. Jain-Sharma, N. Claussen, M. K. Raich, H. J. Gustafson, A. R. Bausch, and S. J. Streichan (2022). 'Morphodynamic atlas for Drosophila development'. Preprint bioRxiv (cit. on p. 529).

Mitchell, T. M. (1997). *Machine Learning*. New York: McGraw-Hill (cit. on p. 532).

Mocenni, C., A. Facchini, and A. Vicino (2010). 'Identifying the dynamics of complex spatio-temporal systems by spatial recurrence properties'. In: *Proc. Natl. Acad. Sci.* 107, pp. 8097–8102 (cit. on pp. 385, 527).

Moffitt, J. R., Y. R. Chemla, S. B. Smith, and C. Bustamante (2008). 'Recent advances in optical tweezers'. In: *Annu. Rev. Biochem.* 77, pp. 205–228 (cit. on p. 136).

Moller, P., A. Fall, V. Chikkadi, D. Derks, and D. Bonn (2009). 'An attempt to categorize yield stress fluid behaviour'. In: *Phil. Trans. R. Soc. A* 367, pp. 5139–5155 (cit. on p. 516).

Mongera, A., P. Rowghanian, H. J. Gustafson, E. Shelton, D. A. Kealhofer, E. K. Carn, F. Serwane, A. A. Lucio, J. Giammona, and O. Campàs (2018). 'A fluid-to-solid jamming transition underlies vertebrate body axis elongation'. In: *Nature* 561, pp. 401–405 (cit. on p. 529).

Montarnal, D., M. Capelot, F. Tournilhac, and L. Leibler (2011). 'Silica-like malleable materials from permanent organic networks'. In: *Science* 334, pp. 965–968 (cit. on p. 516).

Mora, S., T. Phou, J.-M. Fromental, L. M. Pismen, and Y. Pomeau (2010). 'Capillarity driven instability of a soft solid'. In: *Phys. Rev. Lett.* 105, p. 214301 (cit. on p. 331).

Morin, A., N. Desreumaux, J.-B. Caussin, and D. Bartolo (2016). 'Distortion and destruction of colloidal flocks in disordered environments'. In: *Nat. Phys.* 13, pp. 63–67 (cit. on pp. 408, 528).

Morozov, A. (2022). 'Coherent structures in plane channel flow of dilute polymer solutions with vanishing inertia'. In: *Phys. Rev. Lett.* 129, p. 017801 (cit. on p. 508).

Morozov, A. and W. van Saarloos (2019). 'Subcritical instabilities in plane Poiseuille flow of an Oldroyd-B fluid'. In: *J. Stat. Phys.* 175, pp. 554–577 (cit. on pp. 508, 526).

Morozov, A. and S. Spagnolie (2015). 'Introduction to complex fluids'. In: *Complex Fluids in Biological Systems*. Ed. by S. E. Spagnolie. Springer New York (cit. on pp. 240, 519).

Morozov, A. and W. van Saarloos (2007). 'An introductory essay on subcritical instabilities and the transition to turbulence in visco-elastic parallel shear flows'. In: *Phys. Rep.* 447, pp. 112–143 (cit. on pp. 245, 508, 519).

Mortensen, K. (2001). 'Structural studies of polymer systems using small-angle neutron scattering'. In: *Advanced Functional Molecules and Polymers*. Ed. by H. S. Nalwa. Gordon & Breach, pp. 223–269 (cit. on pp. 230, 517).

Muševič, I. (2017). *Liquid Crystal Colloids*. Springer (cit. on pp. 287, 521).

Mundoor, H., S. Park, B. Senyuk, H. H. Wensink, and I. I. Smalyukh (2018). 'Hybrid molecular-colloidal liquid crystals'. In: *Science* 360, pp. 768–771 (cit. on p. 521).

Mundoor, H., B. Senyuk, M. Almansouri, S. Park, B. Fleury, and I. I. Smalyukh (2019). 'Electrostatically controlled surface boundary conditions in nematic liquid crystals and colloids'. In: *Sci. Adv.* 5, eaax4257 (cit. on p. 520).

Mungan, M. and T. A. Witten (2019). 'Cyclic annealing as an iterated random map'. In: *Phys. Rev. E* 99, p. 052132 (cit. on p. 531).

Muñoz-Basagoiti, M., O. Rivoire, and Z. Zeravcic (2023). 'Computational design of a minimal catalyst using colloidal particles with programmable interactions'. In: *Soft Matter* 19, pp. 3033–3039 (cit. on p. 495).

Murray, C. A. and D. H. Van Winkle (1987). 'Experimental observation of two-stage melting in a classical two-dimensional screened Coulomb system'. In: *Phys. Rev. Lett.* 58, pp. 1200–1203 (cit. on p. 511).

Murray, J. D. (2003). *Mathematical Biology, I: An Introduction; II: Spatial Models and Biomedical Applications.* Springer (cit. on pp. 352, 524, 525).

Murugan, A., Z. Zeravcic, M. P. Brenner, and S. Leibler (2015). 'Multifarious assembly mixtures: Systems allowing retrieval of diverse stored structures'. In: *Proc. Natl. Acad. Sci.* 112, pp. 54–59 (cit. on pp. 479, 531, 532).

Na, J.-H., A. A. Evans, J. Bae, M. C. Chiappelli, C. D. Santangelo, R. J. Lang, T. C. Hull, and R. C. Hayward (2014). 'Programming reversibly self-folding origami with micropatterned photo-crosslinkable polymer trilayers'. In: *Adv. Mater.* 27, pp. 79–85 (cit. on p. 469).

Nagamanasa, K. H., S. Gokhale, A. K. Sood, and R. Ganapathy (2014). 'Experimental signatures of a nonequilibrium phase transition governing the yielding of a soft glass'. In: *Phys. Rev. E* 89, p. 062308 (cit. on p. 531).

Nagamanasa, K. H., S. Gokhale, A. K. Sood, and R. Ganapathy (2015). 'Direct measurements of growing amorphous order and non-monotonic dynamic correlations in a colloidal glass-former'. In: *Nat. Phys.* 11, pp. 403–408 (cit. on p. 178).

Narumi, T., S. V. Franklin, K. W. Desmond, M. Tokuyama, and E. R. Weeks (2011). 'Spatial and temporal dynamical heterogeneities approaching the binary colloidal glass transition'. In: *Soft Matter* 7, pp. 1472–1482 (cit. on p. 178).

Nash, L. M., D. Kleckner, A. Read, V. Vitelli, A. M. Turner, and W. T. M. Irvine (2015). 'Topological mechanics of gyroscopic metamaterials'. In: *Proc. Natl. Acad. Sci.* 112, pp. 14495–14500 (cit. on p. 103).

Nelson, D. R. (1980). 'Two dimensional superfluidity and melting'. In: *Fundamental Problems in Statistical Mechanics V.* Ed. by E. G. D. Cohen. Amsterdam: North-Holland, pp. 53–108 (cit. on pp. 98, 281, 511).

Nelson, D. R. (1996). 'Defects in superfluids, superconductors and membranes'. In: *Fluctuating Geometries in Statistical Mechanics and Field Theory (Les Houches Summer School Proceedings, Vol 62).* Ed. by F. David, P. Ginsparg, and J. Zinn-Justin. Elsevier-North-Holland (cit. on pp. 277, 281, 303).

Nelson, D. R. (2002). *Defects and Geometry in Condensed Matter Physics.* Cambridge University Press (cit. on p. 295).

Nelson, D. R. (2004). 'The statistical mechanics of membranes and interfaces'. In: *Statistical Mechanics of Membranes and Surfaces (2nd edition).* Ed. by D. R. Nelson, T. Piran, and S. Weinberg. World Scientific, pp. 1–17 (cit. on pp. 319, 321, 522).

Nelson, D. R., T. Piran, and S. Weinberg, eds. (2004). *Statistical Mechanics of Membranes and Surfaces (2nd edition).* World Scientific (cit. on p. 319).

Nemirovskii, S. K. (2010). 'Propagation of a turbulent front in quantum fluids'. In: *J. Low Temp. Phys.* 162, pp. 347–353 (cit. on p. 523).

Neveu, M., H.-J. Kim, and S. A. Benner (2013). 'The "strong" RNA world hypothesis: Fifty years old'. In: *Astrobiology* 13, pp. 391–403 (cit. on p. 533).

Newburgh, R., J. Peidle, and W. Rueckner (2006). 'Einstein, Perrin, and the reality of atoms: 1905 revisited'. In: *Am. J. Phys.* 74, pp. 478–481 (cit. on p. 512).

Newell, A. C. and J. A. Whitehead (1969). 'Finite bandwidth, finite amplitude convection'. In: *J. Fluid Mech.* 38, pp. 279–303 (cit. on p. 525).

Nienhuis, B. (1982). 'Exact critical point and critical exponents of O(n) models in two dimensions'. In: *Phys. Rev. Lett.* 49, pp. 1062–1065 (cit. on pp. 516, 517).

Nishimori, H. and G. Ortiz (2010). *Elements of Phase Transitions and Critical Phenomena*. Oxford University Press (cit. on pp. 199, 211, 516).

Niu, R., C. X. Du, E. Esposito, J. Ng, M. P. Brenner, P. L. McEuen, and I. Cohen (2019). 'Magnetic handshake materials as a scale-invariant platform for programmed self-assembly'. In: *Proc. Natl. Acad. Sci.* 116, pp. 24402–24407 (cit. on p. 497).

Noda, I., N. Kato, T. Kitano, and M. Nagasawa (1981). 'Thermodynamic properties of moderately concentrated solutions of linear polymers'. In: *Macromolecules* 14, pp. 668–676 (cit. on p. 518).

O'Hern, C. S., L. E. Silbert, A. J. Liu, and S. R. Nagel (2003). 'Jamming at zero temperature and zero applied stress: The epitome of disorder'. In: *Phys. Rev. E* 68, p. 011306 (cit. on p. 510).

Odijk, T. (1977). 'Polyelectrolytes near the rod limit'. In: *J. Polym. Sci.* 15, pp. 477–483 (cit. on pp. 217, 249).

Odijk, T. and A. C. Houwaart (1978). 'On the theory of the excluded-volume effect of a polyelectrolyte in a 1-1 electrolyte solution'. In: *J. Polym. Sci.* 16, pp. 627–639 (cit. on p. 517).

Ohbayashi, K., T. Kohno, and H. Utiyama (1983). 'Photon correlation spectroscopy of the non-Markovian Brownian motion of spherical particles'. In: *Phys. Rev. A* 27, pp. 2632–2641 (cit. on p. 174).

Okumura, S., G. Gines, N. Lobato-Dauzier, A. Baccouche, R. Deteix, T. Fujii, Y. Rondelez, and A. J. Genot (2022). 'Nonlinear decision-making with enzymatic neural networks'. In: *Nature* 610, pp. 496–501 (cit. on pp. 489, 532).

Oliveri, G. and J. T. B. Overvelde (2020). 'Inverse design of mechanical metamaterials that undergo buckling'. In: *Adv. Funct. Mat.* 30, p. 1909033 (cit. on p. 510).

Ong, L. L., N. Hanikel, O. K. Yaghi, C. Grun, M. T. Strauss, P. Bron, J. Lai-Kee-Him, F. Schueder, B. Wang, P. Wang, J. Y. Kishi, C. Myhrvold, A. Zhu, R. Jungmann, G. Bellot, Y. Ke, and P. Yin (2017). 'Programmable self-assembly of three-dimensional nanostructures from 10,000 unique components'. In: *Nature* 552, pp. 72–77 (cit. on p. 471).

Onsager, L. (1931). 'Reciprocal relations in irreversible processes. I.' In: *Phys. Rev.* 37, pp. 405–426 (cit. on p. 30).

Onsager, L. (1949). 'The effect of shape on the interaction of colloidal particles'. In: *Ann. NY Acad. Sci.* 51, pp. 627–659 (cit. on pp. 163, 296, 514, 519).

Onuki, A. (2002). *Phase Transition Dynamics*. Cambridge University Press (cit. on pp. 345, 509, 511).

Oparin, A. I. (1957). *The Origin of Life on the Earth (3rd Ed.)* Academic Press (cit. on p. 533).

Oron, A., S. H. Davis, and S. G. Bankoff (1997). 'Long-scale evolution of thin liquid films'. In: *Rev. Mod. Phys.* 69, pp. 931–980 (cit. on p. 46).

Ortín, J. (1991). 'Preisach modeling of hysteresis for a pseudoelastic Cu-Zn-Al single crystal'. In: *J. Appl. Phys.* 71, pp. 1454–1461 (cit. on p. 531).

Oswald, P., J. Baudry, and T. Rondepierre (2004). 'Growth below and above the spinodal limit: The cholesteric-nematic front'. In: *Phys. Rev. E* 70, p. 041702 (cit. on p. 520).

Oswald, P. and P. Pieranski (2006a). *Nematic and Cholesteric Liquid Crystals—Concepts and Physical Properties Illustrated by Experiments*. Taylor & Francis (cit. on p. 260).

Oswald, P. and P. Pieranski (2006b). *Smectic and Columnar Liquid Crystals—Concepts and Physical Properties Illustrated by Experiments*. Taylor & Francis (cit. on p. 260).

Overbeek, J. Th. G. (1982). 'Monodisperse colloidal systems, fascinating and useful'. In: *Adv. Colloid Interface Sci.* 15, pp. 251–277 (cit. on p. 164).

Owens, R. G. and T. N. Phillips (2002). *Computational Rheology*. Imperial College Press (cit. on p. 519).

Packard, C. E., E. R. Homer, N. Al-Aqeeli, and C. A. Schuh (2010). 'Cyclic hardening of metallic glasses under Hertzian contacts: Experiments and STZ dynamics simulations'. In: *Philos. Mag.* 90, pp. 1373–1390 (cit. on p. 531).

Pahl, M., W. Gleissle, and H.-M. Laun (1991). *Praktische Rheologie der Kunststoffe und Elastomere*. VDI Verlag, (cit. on p. 519).

Pakdel, P. and G. H. McKinley (1996). 'Elastic instability and curved streamlines'. In: *Phys. Rev. Lett.* 77, pp. 2459–2462 (cit. on pp. 245, 246, 519).

Palffy-Muhoray, P. (2007). 'The diverse world of liquid crystals'. In: *Phys. Today* 60(9), pp. 54–60 (cit. on p. 519).

Pan, L., A. Morozov, C. Wagner, and P. E. Arratia (2013). 'Nonlinear elastic instability in channel flows at low Reynolds numbers'. In: *Phys. Rev. Lett.* 110, p. 174502 (cit. on pp. 508, 519).

Panagopoulos, C., M. Majoros, T. Nishizaki, and H. Iwasaki (2006). 'Weak magnetic order in the normal state of the high-T_c superconductor $La_{2-x}Sr_xCuO_4$'. In: *Phys. Rev. Lett.* 96, p. 047002 (cit. on p. 531).

Paredes, J., M. A. J. Michels, and D. Bonn (2013). 'Rheology across the zero-temperature jamming transition'. In: *Phys. Rev. Lett.* 111, p. 015701 (cit. on pp. 186, 516).

Parisi, G., P. Urbani, and F. Zamponi (2020). *Theory of simple glasses: exact solutions in infinite dimensions*. Cambridge University Press (cit. on p. 259).

Parmar, J. J., M. Woringer, and C. Zimmer (2019). 'How the genome folds: The biophysics of four-dimensional chromatin organization'. In: *Annu. Rev. Biophys.* 48, pp. 231–253 (cit. on p. 213).

Parthasarathy, R. (2022). *So Simple a Beginning: How Four Physical Principles Shape Our Living World*. Princeton University Press (cit. on p. 312).

Pastur, L., M.-T. Westra, D. Snouck, W. van de Water, M. van Hecke, C. Storm, and W. van Saarloos (2003a). 'Sources and holes in a one-dimensional traveling-wave convection experiment'. In: *Phys. Rev. E* 67, p. 036305 (cit. on p. 526).

Pastur, L., M.-T. Westra, and W. van de Water (2003b). 'Sources and sinks in a 1D traveling waves'. In: *Physica D* 174, pp. 71–83 (cit. on pp. 383, 526).

Patelli, A., I. Djafer-Cherif, I. S. Aranson, E. Bertin, and H. Chaté (2019). 'Understanding dense active nematics from microscopic models'. In: *Phys. Rev. Lett.* 123, p. 258001 (cit. on p. 444).

Paul, G. L. and P. N. Pusey (1981). 'Observation of a long-time tail in Brownian motion'. In: *J. Phys. A: Math. Gen.* 14, pp. 3301–3327 (cit. on p. 174).

Pauling, L. (1946). 'Molecular architecture and biological reactions'. In: *Chem. Eng. News* 24, pp. 1375–1377 (cit. on p. 533).

Paulsen, J. D., N. C. Keim, and S. R. Nagel (2014). 'Multiple transient memories in experiments on sheared non-Brownian suspensions'. In: *Phys. Rev. Lett.* 113, p. 068301 (cit. on pp. 477, 531).

Paxton, W. F., P. T. Baker, T. R. Kline, Y. Wang, T. E. Mallouk, and A. Sen (2006). 'Catalytically induced electrokinetics for motors and micropumps'. In: *J. Am. Chem. Soc.* 128, pp. 14881–14888 (cit. on p. 515).

Pécseli, H. L. P. (2000). *Fluctuations in Physical Systems*. Cambridge University Press (cit. on p. 153).

Pedley, T. J. and J. O. Kessler (1992). 'Hydrodynamic phenomena in suspensions of swimming microorganisms'. In: *Annu. Rev. Fluid Mech.* 24, pp. 313–358 (cit. on pp. 505, 528).

Pelcé, P. (2004). *New Visions on Form and Growth: Fingered Growth, Dendrites, and Flames*. Oxford University Press (cit. on p. 527).

Peletier, L. A. and W. C. Troy (2001). *Spatial Patterns—Higher Order Models in Physics and Mechanics*. Birkhäuser Verlag (cit. on p. 524).

Perkins, T., S. R. Quake, D. Smith, and S. Chu (1994a). 'Relaxation of a single DNA molecule observed by optical microscopy'. In: *Science* 264, pp. 822–826 (cit. on p. 239).

Perkins, T., D. Smith, and S. Chu (1994b). 'Direct observation of tube-like motion of a single polymer chain'. In: *Science* 264, pp. 819–822 (cit. on pp. 239, 518).

Perrin, J. (1909). 'Mouvement brownien et molécules'. In: *J. Phys. Theor. Appl.* 9, pp. 5–39 (cit. on p. 505).

Petekidis, G., A. Moussaïd, and P. N. Pusey (2002). 'Rearrangements in hard-sphere glasses under oscillatory shear strain'. In: *Phys. Rev. E* 66, p. 051402 (cit. on p. 531).

Petridou, N. I., B. Corominas-Murtra, C.-P. Heisenberg, and E. Hannezo (2021). 'Rigidity percolation uncovers a structural basis for embryonic tissue phase transitions'. In: *Cell* 184, pp. 1914–1928 (cit. on p. 510).

Petrosyan, R. (2016). 'Improved approximations for some polymer extension models'. In: *Rheol. Acta* 56, pp. 21–26 (cit. on p. 517).

Pham, P., B. Metzger, and J. E. Butler (2015). 'Particle dispersion in sheared suspensions: Crucial role of solid-solid contacts'. In: *Phys. Fluids* 27, p. 051701 (cit. on p. 531).

Pine, D. J., J. P. Gollub, J. F. Brady, and A. M. Leshansky (2005). 'Chaos and threshold for irreversibility in sheared suspensions'. In: *Nature* 438, pp. 997–1000 (cit. on p. 531).

Pismen, L. M. (2006). *Patterns and Interfaces in Dissipative Systems*. Springer (cit. on pp. 352, 510, 526).

Pismen, L. M. (2021). *Active Matter within and around Us*. Springer (cit. on p. 405).

Pismen, L. M. and A. A. Nepomnyashchy (1994). 'Propagation of the hexagonal pattern'. In: *Europhys. Lett.* 27, pp. 433–436 (cit. on p. 379).

Plasson, R., A. Brandenburg, L. Jullien, and H. Bersini (2011). 'Autocatalysis: At the root of self-replication'. In: *Artif. Life* 17, pp. 219–236 (cit. on p. 533).

Pomeau, Y. (1986). 'Front motion, metastability and subcritical bifurcations in hydrodynamics'. In: *Physica D* 23, pp. 3–11 (cit. on p. 526).

Poncet, A. and D. Bartolo (2022). 'When soft crystals defy Newton's third Law: Nonreciprocal mechanics and dislocation motility'. In: *Phys. Rev. Lett.* 128, p. 048002 (cit. on p. 422).

Poon, W. C. K., A. T. Brown, S. O. L. Direito, D. J. M. Hodgson, L. Le Nagard, A. Lips, C. E. MacPhee, D. Marenduzzo, J. R. Royer, A. F. Silva, J. H. J. Thijssen, and S. Titmuss (2020). 'Soft matter science and the COVID-19 pandemic'. In: *Soft Matter* 16, pp. 8310–8324 (cit. on p. 506).

Pope, S. B. (2000). *Turbulent Flows*. Cambridge University Press (cit. on p. 45).

Popova, M., P. Vorobieff, M. S. Ingber, and A. L. Graham (2007). 'Interaction of two particles in a shear flow'. In: *Phys. Rev. E* 75, p. 066309 (cit. on p. 531).

Pouget, E., E. Grelet, and P. M. Lettinga (2011). 'Dynamics in the smectic phase of stiff viral rods'. In: *Phys. Rev. E* 84, p. 041704 (cit. on p. 520).

Povinelli, M. L., S. N. Coppersmith, L. P. Kadanoff, S. R. Nagel, and S. C. Venkataramani (1999). 'Noise stabilization of self-organized memories'. In: *Phys. Rev. E* 59, pp. 4970–4982 (cit. on p. 531).

Powell, K. (2018). 'How biologists are creating life-like cells from scratch'. In: *Nature* 563, pp. 172–175 (cit. on p. 533).

Powers, T. R. and R. E. Goldstein (1997). 'Pearling and pinching: Propagation of Rayleigh instabilities'. In: *Phys. Rev. Lett.* 78, pp. 2555–2558 (cit. on pp. 330, 523).

Powers, T. R., D. Zhang, R. E. Goldstein, and H. A. Stone (1998). 'Propagation of a topological transition: The Rayleigh instability'. In: *Phys. Fluids* 10, pp. 1052–1057 (cit. on p. 330).

Prähofer, M. and H. Spohn (2000). 'Universal distributions for growth processes in 1+1 dimensions and random matrices'. In: *Phys. Rev. Lett.* 84, pp. 4882–4885 (cit. on p. 523).

Preisach, F. (1935). 'Über die magnetische Nachwirkung'. In: *Z. Phys.* 94, pp. 277–302 (cit. on p. 531).

Prévost, C., M. Takahashi, and R. Lavery (2009). 'Deforming DNA: From physics to biology'. In: *Chem. Phys. Chem.* 10, pp. 1399–1404 (cit. on p. 505).

Prigent, A., G. Grégoire, H. Chaté, O. Dauchot, and W. van Saarloos (2002). 'Large-scale finite-wavelength modulation within turbulent shear flows'. In: *Phys. Rev. Lett.* 89, p. 014501 (cit. on p. 527).

Pritchard, R. H., P. Lava, D. Debruyne, and E. M. Terentjev (2013). 'Precise determination of the Poisson ratio in soft materials with 2D digital image correlation'. In: *Soft Matter* 9, pp. 6037–6045 (cit. on p. 74).

Prosperetti, A. (2017). 'Vapor bubbles'. In: *Annu. Rev. Fluid Mech.* 49, pp. 221–248 (cit. on p. 51).

Prost, J., F. Jülicher, and J.-F. Joanny (2015). 'Active gel physics'. In: *Nat. Phys.* 11, pp. 111–117 (cit. on p. 435).

Purcell, E. M. (1977). 'Life at low Reynolds numbers'. In: *Am. J. Phys.* 45, pp. 3–11 (cit. on p. 42).

Pusey, P. N. (1991). 'Colloidal suspensions'. In: *Liquids, Freezing and Glass Transition: Les Houches Summer Schools of Theoretical Physics Session LI (1989)*. Ed. by J. P. Hansen, D. Levesque, and J. Zinn-Justin. Elsevier Science Publ BV, pp. 763–942 (cit. on p. 167).

Pusey, P. N. and W. van Megen (1986). 'Phase behaviour of concentrated suspensions of nearly hard colloidal spheres'. In: *Nature* 320, pp. 340–342 (cit. on p. 512).

Py, C., P. Reverdy, L. Doppler, J. Bico, B. Roman, and C. N. Baroud (2007). 'Capillary origami: spontaneous wrapping of a droplet with an elastic sheet'. In: *Phys. Rev. Lett.* 98, p. 156103 (cit. on p. 81).

Qian, L., E. Winfree, and J. Bruck (2011). 'Neural network computation with DNA strand displacement cascades'. In: *Nature* 475, pp. 368–372 (cit. on p. 532).

Qin, B. and P. E. Arratia (2017). 'Characterizing elastic turbulence in channel flows at low Reynolds number'. In: *Phys. Rev. Fluids* 2, p. 083302 (cit. on p. 508).

Rafsanjani, A., K. Bertoldi, and A. R. Studart (2019). 'Programming soft robots with flexible mechanical metamaterials'. In: *Sci. Robot.* 4, eaav7874 (cit. on p. 510).

Rahman, Sk. M. and O. San (2019). 'A relaxation filtering approach for two-dimensional Rayleigh-Taylor instability-induced flows'. In: *Fluids* 4, p. 78 (cit. on p. 508).

Rajabi, M., H. Baza, T. Turiv, and O. D. Lavrentovich (2020). 'Directional self-locomotion of active droplets enabled by nematic environment'. In: *Nat. Phys.* 17, pp. 260–266 (cit. on p. 521).

Ramanathan, M., L. K. Shrestha, T. Mori, Q. Ji, J. P. Hill, and K. Ariga (2013). 'Amphiphile nanoarchitectonics: from basic physical chemistry to advanced applications'. In: *Phys. Chem. Chem. Phys.* 15, p. 10580 (cit. on pp. 288, 521).

Ramaswamy, S. (2010). 'The mechanics and statistics of active matter'. In: *Annu. Rev. Condens. Matter Phys.* 1, pp. 323–345 (cit. on p. 416).

Rauwendaal, C. (2001). *Polymer Extrusion*. Hanser Publishers (cit. on pp. 241, 519).

Rayleigh, L. (1917). 'On the pressure developed in a liquid during the collapse of a spherical cavity'. In: *Philos. Mag.* 34, p. 94 (cit. on p. 508).

Regev, I., T. Lookman, and C. Reichhardt (2013). 'Onset of irreversibility and chaos in amorphous solids under periodic shear'. In: *Phys. Rev. E* 88, p. 062401 (cit. on p. 531).

Reis, P. M., H. M. Jaeger, and M. van Hecke (2015). 'Designer matter: A perspective'. In: *Extreme Mech. Lett.* 5, pp. 25–29 (cit. on p. 464).

Reynolds, O. (1883). 'XXIX. An experimental investigation of the circumstances which determine whether the motion of water shall be direct or sinuous, and of the law of resistance in parallel channels'. In: *Philos. Trans. R. Soc. Lond.* 174, pp. 935–982 (cit. on p. 507).

Reynolds, O. (1886). 'On the theory of lubrication and its application to Mr Tower's experiments'. In: *Philos. Trans. R. Soc. Lond.* 177, pp. 157–235 (cit. on p. 46).

Riecke, H. and W. J. Rappel (1995). 'Coexisting pulses in a model for binary-mixture convection'. In: *Phys. Rev. Lett.* 75, pp. 4035–4038 (cit. on p. 514).

Robertson, M. P. and G. F. Joyce (2010). 'The origins of the RNA world'. In: *Cold Spring Harb. Perspect. Biol.* 4, a003608 (cit. on p. 533).

Rocha, M. S. (2015). 'Extracting physical chemistry from mechanics: A new approach to investigate DNA interactions with drugs and proteins in single molecule experiments'. In: *Integr. Biol.* 7, pp. 967–986 (cit. on pp. 221, 517).

Rocks, J. W., N. Pashine, I. Bischofberger, C. P. Goodrich, A. J. Liu, and S. R. Nagel (2017). 'Designing allostery-inspired response in mechanical networks'. In: *Proc. Natl. Acad. Sci.* 114, pp. 2520–2525 (cit. on pp. 467, 530).

Rocks, J. W., H. Ronellenfitsch, A. J. Liu, S. R. Nagel, and E. Katifori (2019). 'Limits of multifunctionality in tunable networks'. In: *Proc. Natl. Acad. Sci.* 116, pp. 2506–2511 (cit. on p. 530).

Rogers, W. B. and V. N. Manoharan (2015). 'Programming colloidal phase transitions with DNA strand displacement'. In: *Science* 347, pp. 639–642 (cit. on p. 470).

Rohl, C. A., C. E. M. Strauss, K. M. S. Misura, and D. Baker (2004). 'Protein structure prediction using Rosetta'. In: *Meth. Enzymol.* Ed. by L. M. Hirks. Elsevier, pp. 66–93 (cit. on p. 533).

Rose, S., A. Prevoteau, P. Elzière, D. Hourdet, A. Marcellan, and L. Leibler (2013). 'Nanoparticle solutions as adhesives for gels and biological tissues'. In: *Nature* 505, pp. 382–385 (cit. on p. 530).

Rosenblatt, F. (1958). 'The perceptron: A probabilistic model for information storage and organization in the brain.' In: *Psychol. Rev.* 65, pp. 386–408 (cit. on p. 532).

Royer, J. R. and P. M. Chaikin (2015). 'Precisely cyclic sand: Self-organization of periodically sheared frictional grains.' In: *Proc. Natl. Acad. Sci.* 112, pp. 49–53 (cit. on p. 531).

Rubinstein, M. and R. H. Colby (2003). *Polymer Physics*. Oxford University Press (cit. on pp. 202, 204, 228, 229, 232, 238, 240, 254, 516–518).

Ruda, H. E. and N. Matsuura (2017). 'Nano-engineered tunable photonic crystals'. In: *Springer Handbook of Electronic and Photonic Materials*. Ed. by S. Kasap and P. Capper. Springer International Publishing (cit. on pp. 515, 520).

Russel, W. B., D. A. Saville, and W. R. Schowalter (1989). *Colloidal Dispersions*. Cambridge University Press (cit. on pp. 66, 167, 182).

Ryzhov, V. N., E. E. Tareyeva, Yu. D. Fomin, and E. N. Tsiok (2017). 'Berezinskii-Kosterlitz-Thouless transition and two-dimensional melting'. In: *Phys. Usp.* 60, pp. 857–885 (cit. on pp. 98, 101).

Sacanna, S., W. T. M. Irvine, P. M. Chaikin, and D. J. Pine (2010). 'Lock and key colloids'. In: *Nature* 464, pp. 575–578 (cit. on p. 193).

Sadd, M. H. (2014). *Elasticity—Theory, Applications, and Numerics*. Academic Press (cit. on p. 69).

Safran, S. A. (1994). *Statistical Thermodynamics of Surfaces, Interfaces, and Membranes*. Addison Wesley (cit. on pp. 513, 514).

Sagués, F., J. M. Sancho, and J. García-Ojalvo (2007). 'Spatiotemporal order out of noise'. In: *Rev. Mod. Phys.* 79, pp. 829–882 (cit. on p. 524).

Saha, S., J. Agudo-Canalejo, and R. Golestanian (2020). 'Scalar active mixtures: The nonreciprocal Cahn-Hilliard model'. In: *Phys. Rev. X* 10, p. 041009 (cit. on pp. 431, 529).

Saha, S., R. Golestanian, and S. Ramaswamy (2014). 'Clusters, asters, and collective oscillations in chemotactic colloids'. In: *Phys. Rev. E* 89, p. 062316 (cit. on p. 422).

Sanchez, T., D. T. N. Chen, S. J. DeCamp, M. Heymann, and Z. Dogic (2012). 'Spontaneous motion in hierarchically assembled active matter'. In: *Nature* 491, pp. 431–434 (cit. on pp. 417, 521).

Sarkar, D., M. Goudarzy, and D. König (2019). 'An interpretation of the influence of particle shape on the mechanical behavior of granular material'. In: *Granul. Matter* 21, p. 53 (cit. on p. 530).

Saw, T. B., A. Doostmohammadi, V. Nier, L. Kocgozlu, S. Thampi, Y. Toyama, P. Marcq, C. T. Lim, J. M. Yeomans, and B. Ladoux (2017). 'Topological defects in epithelia govern cell death and extrusion'. In: *Nature* 544, pp. 212–216 (cit. on p. 289).

Scarpello, G. M. and D. Ritelli (2006). 'Elliptic integral solutions of spatial elastica of a thin straight rod bent under concentrated terminal forces'. In: *Meccanica* 41, pp. 519–527 (cit. on p. 510).

Schaedler, T. A., A. J. Jacobsen, A. Torrents, A. E. Sorensen, J. Lian, J. R. Greer, L. Valdevit, and W. B. Carter (2011). 'Ultralight metallic microlattices'. In: *Science* 334, pp. 962–965 (cit. on p. 467).

Schall, P., D. A. Weitz, and F. Spaepen (2007). 'Structural rearrangements that govern flow in colloidal glasses'. In: *Science* 318, pp. 1895–1899 (cit. on p. 505).

Scheibner, C., A. Souslov, D. Banerjee, P. Surówka, W. T. M. Irvine, and V. Vitelli (2020). 'Odd elasticity'. In: *Nat. Phys.* 16, pp. 475–480 (cit. on pp. 78, 421, 423).

Schiessel, H. (2013). *Biophysics for Beginners—A Journey through the Cell Nucleus*. Routledge (cit. on p. 213).

Schittny, R., M. Kadic, S. Guenneau, and M. Wegener (2013). 'Experiments on transformation thermodynamics: Molding the flow of heat'. In: *Phys. Rev. Lett.* 110, p. 195901 (cit. on p. 466).

Schmitt, M., J. Colen, S. Sala, M. L. Gardel, P. W. Oakes, and V. Vitelli (2022). 'Machine learning models of cellular forces'. (in preparation) (cit. on p. 489).

Schmitz, K. S. (1990). *An Introduction to Dynamic Light Scattering by Macromolecules*. Academic Press (cit. on p. 143).

Schuster, P. (2019). 'What is special about autocatalysis?' In: *Monatsh. Chem.* 150, pp. 763–775 (cit. on p. 533).

Sebastián, N., L. Cmok, R. J. Mandle, M. Rosario de la Fuente, I. Drevenšek Olenik, M. Čopič, and A. Mertelj (2020). 'Ferroelectric-ferroelastic phase transition in a nematic liquid crystal'. In: *Phys. Rev. Lett.* 124, p. 037801 (cit. on p. 520).

Seeman, N. C. and H. F. Sleiman (2017). 'DNA nanotechnology'. In: *Nat. Rev. Mater.* 3, p. 17068 (cit. on p. 465).

Seifert, U. (1997). 'Configurations of fluid membranes and vesicles'. In: *Adv. Phys.* 46, pp. 13–137 (cit. on pp. 312, 314).

Seung, H. S. and D. R. Nelson (1988). 'Defects in flexible membranes with crystalline order'. In: *Phys. Rev. A* 38, pp. 1005–1018 (cit. on pp. 521, 522).

Shankar, S. and L. Mahadevan (2022). 'Active muscular hydraulics'. Preprint bioRxiv (cit. on p. 422).

Shankar, S. and D. R. Nelson (2021). 'Thermalized buckling of isotropically compressed thin sheets'. In: *Phys. Rev. E* 104, p. 054141 (cit. on p. 522).

Shankar, S., A. Souslov, M. J. Bowick, M. C. Marchetti, and V. Vitelli (2022). 'Topological active matter'. In: *Nat. Rev. Phys.* 4, pp. 380–398 (cit. on p. 102).

Shaqfeh, E. S. G. (1996). 'Purely elastic enstabilities in viscometric flows'. In: *Annu. Rev. Fluid Mech.* 28, pp. 129–185 (cit. on p. 245).

Sharma, V., K. Park, and M. Srinivasarao (2009). 'Colloidal dispersion of gold nanorods: Historical background, optical properties, seed-mediated synthesis, shape separation and self-assembly'. In: *Mater. Sci. Eng. R* 65, pp. 1–38 (cit. on p. 519).

Shaw, D. (2013). *Introduction to Colloid and Surface Chemistry (4th Edition)*. Elsevier (cit. on p. 10).

Shi, Y. and J. E. Hearst (1994). 'The Kirchhoff elastic rod, the nonlinear Schrödinger equation, and DNA supercoiling'. In: *J. Chem. Phys.* 101, pp. 5186–5200 (cit. on p. 510).

Shih, H.-Y., T.-L. Hsieh, and N. Goldenfeld (2015). 'Ecological collapse and the emergence of travelling waves at the onset of shear turbulence'. In: *Nat. Phys.* 12, pp. 245–248 (cit. on pp. 507, 508).

Shraiman, B. I. (1986). 'Order, disorder, and phase turbulence'. In: *Phys. Rev. Lett.* 57, pp. 325–328 (cit. on pp. 457, 526, 529).

Sigl, C., E. M. Willner, W. Engelen, J. A. Kretzmann, K. Sachenbacher, A. Liedl, F. Kolbe, F. Wilsch, S. A. Aghvami, U. Protzer, M. F. Hagan, S. Fraden, and H. Dietz (2021). 'Programmable icosahedral shell system for virus trapping'. In: *Nat. Mater.* 20, pp. 1281–1289 (cit. on p. 471).

Silbert, L. E., A. J. Liu, and S. R. Nagel (2005). 'Vibrations and diverging length scales near the unjamming transition'. In: *Phys. Rev. Lett.* 95, p. 098301 (cit. on pp. 510, 511).

Simon, S. H., ed. (2013). *The Oxford Solid State Basics*. Oxford University Press (cit. on p. 77).

Sims, K. (1994). 'Evolving virtual creatures'. In: *Proceedings of the 21st Annual Conference on Computer Graphics and Interactive Techniques—SIGGRAPH '94*. ACM Press (cit. on p. 498).

Sipper, M. (1998). 'Fifty years of research on self-replication: An overview'. In: *Artif. Life* 4, pp. 237–257 (cit. on p. 532).

Sircar, S. and Q. Wang (2010). 'Transient rheological responses in sheared biaxial liquid crystals'. In: *Rheol. Acta* 49, pp. 699–717 (cit. on p. 531).

Sivashinsky, G. I. (1977). 'Nonlinear analysis of hydrodynamic instability in laminar flames I. Derivation of basic equations'. In: *Acta Astronaut.* 4, pp. 1177–1206 (cit. on p. 526).

Skinner, T. O. E., D. G. A. L. Aarts, and R. P. A. Dullens (2010). 'Grain-boundary fluctuations in two-dimensional colloidal crystals'. In: *Phys. Rev. Lett.* 105, p. 168301 (cit. on p. 521).

Skolnick, J. and M. Fixman (1977). 'Electrostatic persistence length of a wormlike polyelectrolyte'. In: *Macromolecules* 10, pp. 944–948 (cit. on p. 217).

Slotterback, S., M. Mailman, K. Ronaszegi, M. van Hecke, M. Girvan, and W. Losert (2012). 'Onset of irreversibility in cyclic shear of granular packings'. In: *Phys. Rev. E* 85, p. 021309 (cit. on p. 531).

Smallenburg, F., L. Filion, M. Marechal, and M. Dijkstra (2012). 'Vacancy-stabilized crystalline order in hard cubes'. In: *Proc. Natl. Acad. Sci. USA* 109, pp. 17886–17890 (cit. on p. 515).

Smith, A. M., M. Borkovec, and G. Trefalt (2020). 'Forces between solid surfaces in aqueous electrolyte solutions'. In: *Adv. Colloid. Int. Sci.* 275, p. 102078 (cit. on p. 173).

Smits, A. J., B. J. McKeon, and I. Marusic (2011). 'High-Reynolds number wall turbulence'. In: *Annu. Rev. Fluid Mech.* 43, pp. 353–375 (cit. on p. 60).

Snoeijer, J. H. and J. Eggers (2010). 'Asymptotic analysis of the dewetting rim'. In: *Phys. Rev. E* 82, pp. 1275–1278 (cit. on p. 508).

Sohn, H. R. O., C. D. Liu, and I. I. Smalyukh (2019). 'Schools of skyrmions with electrically tunable elastic interactions'. In: *Nat. Commun.* 10, p. 4744 (cit. on p. 521).

Soni, V., E. S. Bililign, S. Magkiriadou, S. Sacanna, D. Bartolo, M. J. Shelley, and W. T. M. Irvine (2019). 'The odd free surface flows of a colloidal chiral fluid'. In: *Nat. Phys.* 15, pp. 1188–1194 (cit. on pp. 425, 427).

Sonin, A. A. (1995). *The Surface Physics of Liquid Crystals*. Gordon and Breach (cit. on p. 520).

Sorscher, B., G. C. Mel, S. Ganguli, and S. A. Ocko (2019). 'A unified theory for the origin of grid cells through the lens of pattern formation'. In: *Advances in Neural Information Processing Systems*. Ed. by H. Wallach, H. Larochelle, A. Beygelzimer, F. d'Alché-Buc, E. Fox, and R. Garnett. Vol. 32. Curran Associates, Inc. (cit. on p. 380).

Soto, R. and R. Golestanian (2014). 'Self-assembly of catalytically active colloidal molecules: Tailoring activity through surface chemistry'. In: *Phys. Rev. Lett.* 112, p. 068301 (cit. on p. 422).

Spruijt, E., F. A. M. Leermakers, R. Fokkink, R. Schweins, A. A. van Well, M. A. Cohen Stuart, and J. van der Gucht (2013). 'Structure and dynamics of polyelectrolyte complex coacervates studied by scattering of neutrons, x-rays, and light'. In: *Macromolecules* 46, pp. 4596–4605 (cit. on pp. 513, 517).

Stauffer, D. and A. Aharony (2003). *An introduction to Percolation Theory (2nd edition).* Taylor and Francis (cit. on pp. 198, 201).

Steels, L. and R. Brooks, eds. (1995). *The "Artificial Life" Route to "Artificial Intelligence": Building Situated Embodied Agents.* Routledge (cit. on p. 532).

Stern, M. and A. Murugan (2023). 'Learning without neurons in physical systems'. In: *Annu. Rev. Condens. Matter Phys.* 14, pp. 417–441 (cit. on p. 480).

Stevens, R. J. A. M., A. Blass, X. Zhu, R. Verzicco, and D. Lohse (2018). 'Turbulent thermal superstructures in Rayleigh-Bénard convection'. In: *Phys. Rev. Fluids* 3, p. 041501 (cit. on p. 527).

Stevens, R. J. A. M., D. Lohse, and R. Verzicco (2011). 'Prandtl and Rayleigh number dependence of heat transport in high Rayleigh number thermal convection'. In: *J. Fluid Mech.* 688, pp. 31–43 (cit. on p. 527).

Stinson, T. W. and J. D. Litster (1970). 'Pretransitional phenomena in the isotropic phase of a nematic liquid crystal'. In: *Phys. Rev. Lett.* 25, pp. 503–506 (cit. on pp. 265, 267, 520).

Storm, C. (2018). 'Theory of semiflexible network materials'. In: *Lecture Notes of the 49th IFF Spring School "Physics of Life" (Forschungszentrum Jülich, 2018).* Ed. by G. Gompper, J. Dhont, J. Elgeti, C. Fahlke, D. Fedosov, S. Förster, P. Lettinga, and A. Offenhäusser. Schriften des Forschungszentrums Jülich (available online), pp. C4.1–C4.15 (cit. on p. 518).

Storm, C., J. J. Pastore, F. C. MacKintosh, T. C. Lubensky, and P. A. Janmey (2005). 'Nonlinear elasticity in biological gels'. In: *Nature* 435, pp. 191–194 (cit. on pp. 233, 518).

Storm, C., W. Spruijt, U. Ebert, and W. van Saarloos (2000). 'Universal algebraic relaxation of velocity and phase in pulled fronts generating periodic or chaotic states'. In: *Phys. Rev. E* 61, R6063–R6066 (cit. on p. 399).

Strandburg, K. J. (1988). 'Two-dimensional melting'. In: *Rev. Mod. Phys.* 60, pp. 161–207 (cit. on p. 98).

Streichan, S. J., M. F. Lefebvre, N. Noll, E. F. Wieschaus, and B. I. Shraiman (2018). 'Global morphogenetic flow is accurately predicted by the spatial distribution of myosin motors'. In: *eLife* 7, e27454 (cit. on pp. 437, 529).

Strobl, G. (1997). *The Physics of Polymers (2nd edition).* Springer (cit. on pp. 204, 228, 518, 519).

Strogatz, S. H. (2015). *Nonlinear Dynamics and Chaos: With Applications to Physics, Biology, Chemistry, and Engineering, 2nd Ed.* Westview Press (cit. on pp. 84, 510, 529).

Stuij, S., J. M. van Doorn, T. Kodger, J. Sprakel, C. Coulais, and P. Schall (2019). 'Stochastic buckling of self-assembled colloidal structures'. In: *Phys. Rev. Res.* 1, p. 023033 (cit. on pp. 141, 513).

Sun, J.-Y., Z. Zhao, W. R. K. Illeperuma, O. Chaudhuri, K. H. Oh, D. J. Mooney, J. J. Vlassak, and Z. Suo (2012). 'Highly stretchable and tough hydrogels'. In: *Nature* 489, pp. 133–136 (cit. on p. 468).

Sutton, R. S. and A. G. Barto (2020). *Reinforcement Learning: An Introduction (2nd ed.)* MIT Press (cit. on p. 482).

Suzuki, Y., G. Cardone, D. Restrepo, P. D. Zavattieri, T. S. Baker, and F. A. Tezcan (2016). 'Self-assembly of coherently dynamic, auxetic, two-dimensional protein crystals'. In: *Nature* 533, pp. 369–373 (cit. on p. 510).

Swift, J. and P. C. Hohenberg (1977). 'Hydrodynamic fluctuations at the convective instability'. In: *Phys. Rev. A* 15, pp. 319–328 (cit. on p. 353).

Szymanowska-Powałowska, D., G. Lewandowicz, W. Błaszczak, and A. Szwengiel (2012). 'Structural changes of corn starch during fuel ethanol production from corn flour'. In: *BioTechnologia* 3, pp. 333–341 (cit. on p. 505).

Tai, J.-S. B. and I. I. Smalyukh (2019). 'Three-dimensional crystals of adaptive knots'. In: *Science* 365, pp. 1449–1453 (cit. on p. 521).

Tajima, S. and H. S. Greenside (2002). 'Microextensive chaos of a spatially extended system'. In: *Phys. Rev. E* 66, p. 017205 (cit. on pp. 381, 526).

Takeuchi, K. A. and M. Sano (2010). 'Universal fluctuations of growing interfaces: Evidence in turbulent liquid crystals'. In: *Phys. Rev. Lett.* 104, p. 230601 (cit. on p. 523).

Takeuchi, K. A. and M. Sano (2012). 'Evidence for geometry-dependent universal fluctuations of the Kardar-Parisi-Zhang interfaces in liquid-crystal turbulence'. In: *J. Stat. Phys.* 147, pp. 853–890 (cit. on pp. 328, 523).

Tan, T. H., A. Mietke, J. Li, Y. Chen, H. Higinbotham, P. J. Foster, S. Gokhale, J. Dunkel, and N. Fakhri (2022). 'Odd dynamics of living chiral crystals'. In: *Nature* 607, pp. 287–293 (cit. on p. 420).

Tanner, R. L. and K. Walters (1998). *Rheology: An Historical Perspective*. Elsevier (cit. on p. 240).

Tarjus, G. (2011). 'An overview of the theories of the glass transition'. In: *Glasses, Colloids and Granular Media.* Ed. by L. Berthier, G. Biroli, J.-P. Bouchaud, L. Cipelletti, and W. van Saarloos. Oxford: Oxford University Press (cit. on p. 177).

Tasaki, H. (2020). 'Hohenberg-Mermin-Wagner-Type theorems for equilibrium models of flocking'. In: *Phys. Rev. Lett.* 125, p. 220601 (cit. on p. 410).

Tchoufag, J., A. Sahu, and K. K. Mandadapu (2022). 'Absolute vs convective instabilities and front propagation in lipid membrane tubes'. In: *Phys. Rev. Lett.* 128, p. 068101 (cit. on p. 523).

Tél, T. and T. Vicsek (1987). 'Geometrical multifractality of growing structures'. In: *J. Phys. A: Math. Gen.* 20, pp. L835–L840 (cit. on p. 514).

Terryn, S., J. Langenbach, E. Roels, J. Brancart, C. Bakkali-Hassani, Q.-A. Poutrel, A. Georgopoulou, T. G. Thuruthel, A. Safaei, P. Ferrentino, T. Sebastian, S. Norvez, F. Iida, A. W. Bosman, F. Tournilhac, F. Clemens, G. Van Assche, and B. Vanderborght (2021). 'A review on self-healing polymers for soft robotics'. In: *Mater. Today* 47, pp. 187–205 (cit. on p. 516).

Thorneywork, A. L., J. L. Abbott, D. G. A. L. Aarts, and R. P. A. Dullens (2017). 'Two-dimensional melting of colloidal hard spheres'. In: *Phys. Rev. Lett.* 118, p. 158001 (cit. on pp. 101, 511).

Thorneywork, A. L., J. L. Abbott, D. G. A. L. Aarts, P. Keim, and R. P. A. Dullens (2018). 'Bond-orientational order and Frank's constant in two-dimensional colloidal hard spheres'. In: *J. Phys. Condens. Matter* 30, p. 104003 (cit. on p. 511).

Thorpe, M. F. (1985). 'Rigidity percolation'. In: *Physics of Disordered Materials*. Ed. by D. Adler, H. Fritzsche, and S. R. Ovshinsky. Springer US, pp. 55–61 (cit. on p. 510).

Toffoli, T. and N. Margolus (1991). 'Programmable matter: Concepts and realization'. In: *Physica D* 47, pp. 263–272 (cit. on pp. 464, 530).

Toiya, M., J. Stambaugh, and W. Losert (2004). 'Transient and oscillatory granular shear flow'. In: *Phys. Rev. Lett.* 93, p. 088001 (cit. on p. 531).

Tokita, M. (1989). 'Gelation mechanism and percolation'. In: *Food Hydrocoll.* 3, pp. 263–274 (cit. on p. 198).

Tolédano, P. and J.-C. Tolédano (1987). *The Landau Theory of Phase Transitions*. World-Scientific (cit. on pp. 509, 511).

Toner, J. (2018). 'Why walking is easier than pointing: Hydrodynamics of dry active matter'. *Lecture Notes Les Houches Summer School "Active matter and non-equilibrium statistical physics"*, 2018; arXiv:1812.00310 (cit. on p. 408).

Toner, J. and Y. Tu (1995). 'Long-range order in a two-dimensional dynamical xy-model: how birds fly together'. In: *Phys. Rev. Lett.* 75, pp. 4326–4329 (cit. on p. 410).

Toner, J., Y. Tu, and S. Ramaswamy (2005). 'Hydrodynamics and phases of flocks'. In: *Ann. Phys. (Amsterdam)* 318, pp. 170–244 (cit. on p. 410).

Tsai, J.-C., F. Ye, J. Rodriguez, J. P. Gollub, and T. C. Lubensky (2005). 'A chiral granular gas'. In: *Phys. Rev. Lett.* 94, p. 214301 (cit. on p. 529).

Tsuji, K. and S. C. Müller, eds. (2019). *Spirals and Vortices—In Culture, Nature, and Science*. Springer (cit. on p. 527).

Tu, Y. and M. C. Cross (1992). 'Chaotic domain structure in rotating convection'. In: *Phys. Rev. Lett.* 69, pp. 2515–2518 (cit. on p. 378).

Turing, A. M. (1952). 'The chemical basis of morphogenesis'. In: *Phil. Tran. R. Soc. Lon. B* 237, pp. 37–72 (cit. on pp. 525, 529).

Turlier, H., B. Audoly, J. Prost, and J.-F. Joanny (2014). 'Furrow constriction in animal cell cytokinesis'. In: *Biophys. J.* 106, pp. 114–123 (cit. on pp. 435, 458, 529).

Val'kov, A. Yu, V. P. Romanov, and A. N. Shalaginov (1994). 'Fluctuations and light scattering in liquid crystals'. In: *Phys. Usp.* 37, pp. 139–183 (cit. on p. 513).

van Blaaderen, A. (1998). 'Materials science: Opals in a new light'. In: *Science* 282, pp. 887–888 (cit. on p. 515).

van de Ven, T. G. M. (1988). *Colloidal Hydrodynamics*. Academic Press (cit. on pp. 123, 512).

van der Hoef, M. A. and D. Frenkel (1995). 'Computer simulations of long-time tails: What's new?' In: *Transport Theor. Stat.* 24, pp. 1227–1247 (cit. on p. 514).

van Dillen, T., P. R. Onck, and E. van der Giessen (2008). 'Models for stiffening in cross-linked biopolymer networks: A comparative study'. In: *J. Mech. Phys. Solids* 56, pp. 2240–2264 (cit. on p. 518).

Van Dyke, M. (1975). *Perturbation Methods in Fluid Mechanics*. Parabolic Press (cit. on pp. 315, 508).

Van Dyke, M. (1982). *An Album of Fluid Motion*. Parabolic Press (cit. on p. 507).

van Hecke, M. (1998). 'Building blocks of spatiotemporal intermittency'. In: *Phys. Rev. Lett.* 80, pp. 1896–1899 (cit. on p. 526).

van Hecke, M. (2010). 'Jamming of soft particles: geometry, mechanics, scaling and isostaticity'. In: *J. Phys. C.* 22, p. 033101 (cit. on p. 92).

van Hecke, M., P. C. Hohenberg, and W. van Saarloos (1994). 'Amplitude equations for pattern forming systems'. In: *Fundamental Problems in Statistical Mechanics VIII*. Ed. by H. van Beijeren and M. H. Ernst. North-Holland, pp. 245–278 (cit. on pp. 523, 524).

van Kampen, N. G. (1981). 'Itô versus Stratonovich'. In: *J. Stat. Phys.* 5, pp. 175–185 (cit. on p. 512).

van Kampen, N. G. (2009). *Stochastic Processes in Physics and Chemistry (3rd edition)*. Elsevier (cit. on pp. 123, 128, 512).

van Saarloos, W. (1989). 'Front propagation into unstable states. II. Linear versus nonlinear marginal stability and rate of convergence'. In: *Phys. Rev. A* 39, pp. 6367–6390 (cit. on p. 524).

van Saarloos, W. (1994). 'The complex Ginzburg-Landau equation for beginners'. In: *Proceedings of the Santa Fe Workshop on 'Spatio-Temporal Patterns in Nonequilibrium Complex Systems'*. Ed. by P. E. Cladis and P. Palffy-Muhoray. Addison Wesley, pp. 19–31 (cit. on pp. 524, 526).

van Saarloos, W. (2003). 'Front propagation into unstable states'. In: *Phys. Rep.* 386, pp. 29–222 (cit. on pp. 132, 330, 399, 523, 524, 526, 527).

van Saarloos, W. and P. C. Hohenberg (1992). 'Fronts, pulses, sources and sinks in generalized complex Ginzburg-Landau equations'. In: *Physica D* 56, pp. 303–367 (cit. on pp. 384, 526).

Van Zee, N. J. and R. Nicolaÿ (2020). 'Vitrimers: Permanently crosslinked polymers with dynamic network topology'. In: *Prog. Polym. Sci.* 104, p. 101233 (cit. on p. 516).

VanSaders, B. and V. Vitelli (2023). 'Informational active matter'. (Preprint arXiv:2302.07402) (cit. on p. 489).

Verma, I., S. Sidiq, and S. K. Pal (2017). 'Poly(l-lysine)-coated liquid crystal droplets for sensitive detection of DNA and their applications in controlled release of drug molecules'. In: *ACS Omega* 2, pp. 7936–7945 (cit. on p. 520).

Vicsek, T., A. Czirók, E. Ben-Jacob, I. Cohen, and O. Shochet (1995). 'Novel type of phase transition in a system of self-driven particles'. In: *Phys. Rev. Lett.* 75, pp. 1226–1229 (cit. on pp. 407, 528).

Vicsek, T., F. Family, and P. Meakin (1990). 'Multifractal geometry of diffusion-limited aggregates'. In: *Europhys. Lett.* 12, pp. 217–222 (cit. on pp. 514, 528).

Vicsek, T. and A. Zafeiris (2012). 'Collective motion'. In: *Phys. Rep.* 517, pp. 71–140 (cit. on p. 408).

Vilfan, I. D., J. Lipfert, D. A. Koster, S. G. Lemay, and N. H. Dekker (2009). 'Magnetic tweezers for single-molecule experiments'. In: *Handbook of Single-Molecule Biophysics*. Ed. by P. Hinterdorfer and A. van Oijen. Springer, pp. 372–395 (cit. on p. 137).

Vitelli, V., J. B. Lucks, and D. R. Nelson (2006). 'Crystallography on curved surfaces'. In: *Proc. Natl. Acad. Sci.* 103, pp. 12323–12328 (cit. on p. 344).

Vladimirov, A., S. Shlosman, and S. Nechaev (2020). 'Brownian flights over a circle'. In: *Phys. Rev. E* 102, p. 012124 (cit. on p. 523).

Vliegenthart, G. A. and G. Gompper (2006). 'Forced crumpling of self-avoiding elastic sheets'. In: *Nat. Mater.* 5, pp. 216–221 (cit. on pp. 323, 522).

Vollhardt, D. and P. Wölfle (1990). *The Superfluid Phases of Helium 3*. Taylor and Francis (cit. on p. 519).

Volovik, G. D. and O. D. Lavrentovich (1983). 'Topological dynamics of defects: Boojums in nematic drops'. In: *Sov. Phys. JETP* 58, pp. 1159–1166 (cit. on p. 520).

Volovik, G. E. (1992). *Exotic Properties of Superfluid* ^{3}*He*. World Scientific (cit. on p. 519).

von Neumann, J. and A. W. Burks (1966). *Theory of Self-Reproducing Automata*. University of Illinois Press (cit. on p. 490).

Wagner, N. J. and J. F. Brady (2009). 'Shear thickening in colloidal dispersions'. In: *Phys. Today* 62(10), pp. 27–32 (cit. on pp. 182, 183, 515).

Wagner, N. J. and J. Mewis, eds. (2021). *Theory and Applications of Colloidal Suspension Rheology*. Cambridge University Press (cit. on pp. 167, 182).

Walgraef, D. (1997). *Spatio-Temporal Pattern Formation*. Springer (cit. on p. 524).

Wallin, T. J., J. Pikul, and R. F. Shepherd (2018). '3D printing of soft robotic systems'. In: *Nat. Rev. Mater.* 3, pp. 84–100 (cit. on p. 465).

Wang, F., H. Lv, Q. Li, J. Li, X. Zhang, J. Shi, L. Wang, and C. Fan (2020). 'Implementing digital computing with DNA-based switching circuits'. In: *Nat. Commun.* 11, p. 121 (cit. on p. 532).

Wang, L. and J. Simmchen (2019). 'Review: Interactions of active colloids with passive tracers'. In: *Condens. Matter* 4, p. 78 (cit. on p. 182).

Wang, Y., Y. Wang, D. R. Breed, V. N. Manoharan, L. Feng, A. D. Hollingsworth, M. Weck, and D. J. Pine (2012). 'Colloids with valence and specific directional bonding'. In: *Nature* 491, pp. 51–55 (cit. on p. 180).

Wang, Z. J. (2000). 'Two-dimensional mechanism for insect hovering'. In: *Phys. Rev. Lett.* 85, pp. 2216–2219 (cit. on pp. 44, 508).

Wang, Z. J. (2005). 'Dissecting insect flight'. In: *Annu. Rev. Fluid Mech.* 37, pp. 183–210 (cit. on p. 508).

Warnez, M. T. and E. Johnsen (2015). 'Numerical modeling of bubble dynamics in viscoelastic media with relaxation'. In: *Phys. Fluids* 27, p. 063103 (cit. on p. 52).

Weaire, D. and S. Hutzler (2001). *The Physics of Foams*. Osford University Press (cit. on p. 164).

Weeks, E. R. (2016). 'Introduction to the colloidal glass transition'. In: *ACS Macro Lett.* 6, pp. 27–34 (cit. on p. 177).

Weeks, J. D. (1980). 'The roughening transition'. In: *Ordering in Strongly Fluctuating Condensed Matter Systems*. Ed. by T. Riste. Plenum, pp. 293–317 (cit. on p. 521).

Weidenmüller, H. A. and J.-S. Zhang (1984). 'Stationary diffusion over a multidimensional potential barrier: A generalization of Kramers' formula'. In: *J. Stat. Phys.* 34, pp. 191–201 (cit. on p. 512).

Wensink, H. H., J. Dunkel, S. Heidenreich, K. Drescher, R. E. Goldstein, H. Löwen, and J. M. Yeomans (2012). 'Meso-scale turbulence in living fluids'. In: *Proc. Natl. Acad. Sci.* 109, pp. 14308–14313 (cit. on pp. 414, 528).

Wettmann, L. and K. Kruse (2018). 'The Min-protein oscillations in Escherichia coli: An example of self-organized cellular protein waves'. In: *Phil. Tran. R. Soc. Lon. B* 373, p. 20170111 (cit. on pp. 366, 525).

White, C. M. and M. G. Mungal (2008). 'Mechanics and prediction of turbulent drag reduction with polymer additives'. In: *Annu. Rev. Fluid Mech.* 40, pp. 235–256 (cit. on p. 518).

Whitesides, G. M. and B. Grzybowski (2002). 'Self-assembly at all scales'. In: *Science* 295, pp. 2418–2421 (cit. on p. 532).

Whitham, G. B. (1974). *Linear and Nonlinear Waves*. Wiley (cit. on pp. 412, 528).

Wiggins, C. H., D. Riveline, A. Ott, and R. E. Goldstein (1998). 'Trapping and wiggling: Elastohydrodynamics of driven microfilaments'. In: *Biophys. J.* 74, pp. 1043–1060 (cit. on p. 511).

Wilson, M. J., L. Wilson, and I. Patey (2014). 'The influence of individual clay minerals on formation damage of reservoir sandstones: A critical review with some new insights'. In: *Clay Miner.* 49, pp. 147–164 (cit. on p. 505).

Wilson, R. A. and F. C. Keil, eds. (2001). *The MIT Encyclopedia of the Cognitive Sciences*. MIT press (cit. on p. 533).

Winfree, A. T. (1973). 'Scroll-shaped waves of chemical activity in three dimensions'. In: *Science* 181, pp. 937–939 (cit. on p. 527).

Winfree, A. T. (1974). 'Rotating chemical reactions'. In: *Sci. Amer.* 230, pp. 82–95 (cit. on p. 527).

Winfree, A. T. (1984). 'The prehistory of the Belousov-Zhabotinsky oscillator'. In: *J. Chem. Educ.* 61, pp. 661–663 (cit. on p. 527).

Witten, T. A. (2007). 'Stress focusing in elastic sheets'. In: *Rev. Mod. Phys.* 79, pp. 643–675 (cit. on pp. 81, 322, 522, 523).

Witten, T. A. and P. A. Pincus (2004). *Structured Fluids: Polymers, Colloids, Surfactants.* Oxford University Press (cit. on pp. 143, 167, 170, 204, 224, 226, 227, 514, 516–518).

Wolfenden, R. and M. J. Snider (2001). 'The depth of chemical time and the power of enzymes as catalysts'. In: *Acc. Chem. Res.* 34, pp. 938–945 (cit. on p. 532).

Wolfram, S. (1983). 'Statistical mechanics of cellular automata'. In: *Rev. Mod. Phys.* 55, pp. 601–644 (cit. on p. 532).

Woods, D., D. Doty, C. Myhrvold, J. Hui, F. Zhou, P. Yin, and E. Winfree (2019). 'Diverse and robust molecular algorithms using reprogrammable DNA self-assembly'. In: *Nature* 567, pp. 366–372 (cit. on p. 532).

Xu, J., Z. Ren, W. Gong, R. Dai, and D. Liu (2003). 'Measurement of the bulk viscosity of liquid by Brillouin scattering'. In: *Appl. Optics* 42, pp. 6704–6709 (cit. on p. 513).

Xu, Z., T. Hueckel, W. T. M. Irvine, and S. Sacanna (2021). 'Transmembrane transport in inorganic colloidal cell-mimics'. In: *Nature* 597, pp. 220–224 (cit. on p. 498).

Yamakawa, H. (1971). *Modern Theory of Polymer Solutions.* Harper and Row (cit. on p. 517).

Yanagida, T., M. Iwaki, and Y. Ishii (2008). 'Single molecule measurements and molecular motors'. In: *Phil. Trans. R. Soc. B* 363, pp. 2123–2134 (cit. on pp. 139, 512, 513).

Yang, R., C. S. Ng, K. L. Chong, R. Verzicco, and D. Lohse (2021). 'Do increased flow rates in displacement ventilation always lead to better results?' In: *J. Fluid Mech.* 932, A3 (cit. on p. 506).

Yllanes, D., S. S. Bhabesh, D. R. Nelson, and M. J. Bowick (2017). 'Thermal crumpling of perforated two-dimensional sheets'. In: *Nat. Commun.* 8, p. 1381 (cit. on p. 522).

Yoneda, M. and K. Dan (1972). 'Tension at the surface of the dividing sea-urchin egg'. In: *J. Exp. Biol.* 57, pp. 575–587 (cit. on p. 458).

You, Z., A. Baskaran, and M. C. Marchetti (2020). 'Nonreciprocity as a generic route to traveling states'. In: *Proc. Natl. Acad. Sci.* 117, pp. 19767–19772 (cit. on pp. 431, 529).

Zaccone, Alessio (2022). 'Explicit Analytical Solution for Random Close Packing in $d = 2$ and $d = 3$'. In: *Phys. Rev. Lett.* 128 (cit. on p. 93).

Žagar, G., P. R. Onck, and E. van der Giessen (2011). 'Elasticity of rigidly cross-linked networks of athermal filaments'. In: *Macromolecules* 44, pp. 7026–7033 (cit. on pp. 237, 518).

Žagar, G., P. R. Onck, and E. van der Giessen (2015). 'Two fundamental mechanisms govern the stiffening of cross-linked networks'. In: *Biophys. J.* 108, pp. 1470–1479 (cit. on p. 518).

Zahalak, G. I. (1996). 'Non-axial muscle stress and stiffness'. In: *J. Theor Biol.* 182, pp. 59–84 (cit. on p. 422).

Zeravcic, Z. and M. P. Brenner (2014). 'Self-replicating colloidal clusters'. In: *Proc. Natl. Acad. Sci.* 111, pp. 1748–1753 (cit. on p. 533).

Zeravcic, Z. and M. P. Brenner (2017). 'Spontaneous emergence of catalytic cycles with colloidal spheres'. In: *Proc. Natl. Acad. Sci.* 114, pp. 4342–4347 (cit. on pp. 505, 533).

Zhabotinsky, A. M. (1964). 'Periodical oxidation of malonic acid in solution (a study of the Belousov reaction kinetics)'. In: *Biofizika* 9, pp. 306–311 (cit. on p. 527).

Zhabotinsky, A. M. (1995). 'Wave propagation and wave pattern formation in nonuniform reaction-diffusion systems'. In: *Chemical Waves and Patterns*. Ed. by R. Kapral and K. Showalter. Springer, pp. 401–418 (cit. on p. 527).

Zhang, B. (2012). 'Electrodynamics of transformation-based invisibility cloaking'. In: *Light Sci. Appl.* 1, e32 (cit. on p. 510).

Zhang, J., E. Luijten, B. A. Grzybowski, and S. Granick (2017). 'Active colloids with collective mobility status and research opportunities'. In: *Chem. Soc. Rev.* 46, pp. 5551–5569 (cit. on pp. 182, 515).

Zhang, S., L. Zhang, M. Bouzid, D. Z. Rocklin, E. Del Gado, and X. Mao (2019). 'Correlated rigidity percolation and colloidal gels'. In: *Phys. Rev. Lett.* 123, p. 058001 (cit. on p. 510).

Zhao, J., U. Gulan, T. Horie, N. Ohmura, J. Han, C. Yang, J. Kong, S. Wang, and B. B. Xu (2019). 'Advances in biological liquid crystals'. In: *Small* 15, p. 1900019 (cit. on p. 520).

Zhong, S. and C.-H. Jang (2015). 'Nematic liquid crystals confined in microcapillaries for imaging phenomena at liquid-liquid interfaces'. In: *Soft Matter* 11, pp. 6999–7004 (cit. on p. 521).

Zhou, F., R. Sha, H. Ni, N. Seeman, and P. M. Chaikin (2021). 'Mutations in artificial self-replicating tiles: A step toward Darwinian evolution'. In: *Proc. Natl. Acad. Sci.* 118, e2111193118 (cit. on pp. 493, 533).

Zilz, J., R. J. Poole, M. A. Alves, D. Bartolo, B. Levaché, and A. Lindner (2012). 'Geometric scaling of a purely elastic flow instability in serpentine channels'. In: *J. Fluid Mech.* 712, pp. 203–218 (cit. on pp. 245, 255, 256).

Zykov, V. S. (2018). 'Spiral wave initiation in excitable media'. In: *Phil. Trans. R. Soc. A* 376, p. 20170379 (cit. on p. 527).

Image Acknowledgments

We are grateful to the colleagues listed below for providing us with high-resolution images and plots of their work.

Since Princeton University Press is a not-for-profit publisher and since we aim to keep the price of this book as low as possible so that students can afford it, we have had to avoid using images from journals which charge high fees for the right to reproduce images which appeared there. Many authors have been so kind as to delve into old files to extract images similar to published ones which were not copyright protected. We highly appreciate this: the enthusiastic response and help provided by the colleagues listed below has been a great stimulus for us and an important service to the community.

Figure	Acknowledgments and courtesy of
figure I.1	M. Fermigier, Ecole Supérieure de Physique et de Chimie Industrielle Paris, France
figure I.8	A. McMullen and J. Brujic, New York University, New York, NY, USA
figure I.9	P. Schall, University of Amsterdam, Amsterdam, the Netherlands
figure I.11	K. Coulter and S. C. Glotzer, University of Michigan, Ann Arbor, MI, USA
figure I.13	S. R. Nagel, University of Chicago, Chicago, IL, USA
figure I.14	S. Ray and Z. Dogic, University of California Santa Barbara, Santa Barbara, CA, USA
figure 1.14	S. R. Nagel, University of Chicago, Chicago, IL, USA
figure 1.17	Z. J. Wang, Cornell University, Ithaca, NY, USA
figure 1.22	O. Dauchot, Ecole Supérieure de Physique et de Chimie Industrielle, Paris, France
figure 2.1	J. Grima, L-Università ta' Malta, Malta
figure 2.2.a	P. Reis, Ecole Polytechnique Féderale de Lausanne, Lausanne, Switzerland
figure 2.2.b	W. Huck, Radboud University, Nijmegen, the Netherlands
figure 2.14	J. Bico, Ecole Supérieure de Physique et de Chimie Industrielle, Paris, France
figure 2.15.a	J. Bico, Ecole Supérieure de Physique et de Chimie Industrielle, Paris, France
figure 2.15.b	E. Cerda, Universidad de Santiago de Chile, Santiago, Chile
figure 2.16	N. Smits, Leiden, the Netherlands
figure 2.20	F. Brau, Université Libre de Bruxelles, Brussels, Belgium
figure 2.23	M. Kok, C. Alkemade, and M. Dogterom, Delft University of Technology, Delft, the Netherlands, and A. Aher and A. Akhmanova, Utrecht University, Utrecht, the Netherlands
figure 2.25.a	K. Bertoldi, Harvard University, Cambridge, MA, USA
figure 2.25.b/c	C. J. M. Coulais, University of Amsterdam, Amsterdam and M. van Hecke, AMOLF, Amsterdam, the Netherlands
figure 2.26	M. van Hecke, AMOLF, Amsterdam, the Netherlands
figure 2.28.a	S. R. Nagel, University of Chicago, Chicago, IL, USA
figure 2.28.c	J. Brujic, New York University, New York, NY, USA
figure 2.37	A. L. Thorneywork, Oxford University, Oxford, United Kingdom and R. Dullens, Radboud University, Nijmgen, the Netherlands
figure 2.38	L. M. Nash, D. Kleckner, and W. T. M. Irvine, University of Chicago, Chicago, IL, USA
figure 3.3	J. Peidle, Harvard University, Cambridge, MA, USA
figure 3.8	A. Azadbakht and D. Kraft, Leiden University, Leiden, the Netherlands
figure 3.9	N. Dekker, Delft University of Technology, Delft, the Netherlands
figure 3.13.b	R. Partharasaty, University of Oregon, Eugene, OR, USA

figure 3.13.c	A. Zidovska, New York University, New York, NY, USA
figure 3.14	S. Stuij and P. Schall, University of Amsterdam, Amsterdam, the Netherlands
figure 4.2	S. Sacanna, New York University, New York NY, USA
figure 4.12	D. Kraft, Leiden University, Leiden, the Netherlands
figure 4.13.a	D. Kraft, Leiden University, Leiden, the Netherlands
figure 4.13.b	D. 't Hart, Utrecht University, Utrecht, the Netherlands and J. T. H. Thijssen, University of Edinburgh, United Kingdom
figure 4.14.a	E. R. Week, Emory University, Atlanta, GA, USA
figure 4.14.c	P. Schall, University of Amsterdam, Amsterdam, the Netherlands
figure 4.17	Y. (Yufeng) Wang, Y. (Yu) Wang, and D. Pine, New York Universy, New York, NY, USA
figure 4.23	D. Bonn, University of Amsterdam, Amsterdam, the Netherlands
figure 5.2.a	F. Tournillac and L. Leibler, Ecole Supérieure de Physique et de Chimie Industrielle, Paris, France, and CNRS Images
figure 5.2.b	C. Fresillon, Ecole Supérieure de Physique et de Chimie Industrielle, Paris, France, and CNRS Images
figure 5.2.c	J. Langenbach, Ecole Supérieure de Physique et de Chimie Industrielle, Paris, France
figure 5.12	D. Hayakawa, T. Videbaek, and W. B. Rogers, Brandeis University, Waltham, MA, USA
figure 5.15	M. S. Rocha, Universidade Federal de Viçosa, Viçosa, Brasil
figure 5.16	G. Wuite, Vrije Universiteit, Amsterdam, the Netherlands
figure 5.21	T. A. Witten, University of Chicago, Chicago, IL, USA
figure 5.26	C. Broedersz, Vrije Universiteit, Amsterdam, the Netherlands, F. MacKintosh, Rice University, Houston, TX USA and S. Münster, University of Erlangen-Nürnberg, Erlangen, Germany
figure 5.30	P. Onck and E. van der Giessen, University of Groningen, Groningen, the Netherlands
figure 5.34	G. H. McKinley, Massachusetts Institute of Technology, Cambridge, MA, USA
figure 5.37.c	A. Lindner, Ecole Supérieure de Physique et de Chimie Industrielle, Paris, France
figure 6.16	O. D. Lavrentovich, Liquid Crystal Institute, Kent State University, Kent, OH, USA
figure 6.18	O. D. Lavrentovich, Liquid Crystal Institute, Kent State University, Kent, OH, USA
figure 6.21	E. Grelet, Centre de Recherche Paul Pascal, Bordeaux, France
figure 6.22	X. (Xin) Wang and N. L. Abbott, Cornell University, Ithaca NY, USA
figure 6.27	G. Nikolic, University of Belgrade, Belgrade, Serbia
figure 7.12	D. Bonn, University of Amsterdam, Amsterdam, the Netherlands
figure 7.13	G. Vliegenthart, Forschungszentrum Jülich, Jülich, Germany
figure 7.14	J. Andrejevic and C. Rycroft, Harvard University, Cambridge, MA, USA
figure 7.15	K. A. Takeuchi, The University of Tokyo, Tokyo, Japan
figure 7.16	E. Moses, Weizmann Institute, Rehovot, Israel
figure 7.19	W. T. M. Irvine, University of Chicago, Chicago, IL, USA
figure 8.1	D. E. Stille, Instructional Resources, Department of Physics and Astronomy, University of Iowa, Iowa City, IA, USA
figure 8.6	P. Yin, Oak Ridge National Laboratory, Oak Ridge, TN, USA
figure 8.10.a	E. Bodenschatz, Max Planck Institute for Dynamics and Self-Organisation, Göttingen, Germany
figure 8.10.b	B. Suri, Indian Institute of Science, Bengaluru, India and M. F. Schatz, Georgia Institute of Technology, Atlanta, GA, USA
figure 8.13.b	K. Kruse, University of Geneva, Geneva, Switzerland
figure 8.23	E. Bodenschatz, Max Planck Institute for Dynamics and Self-Organisation, Göttingen, Germany
figure 8.25	H. S. Greenside, Duke University, Durham, NC, USA
figure 8.26	N. Garnier, Ecole Normale Supérieure de Lyon, Lyon, France

figure 8.27	W. van de Water, Delft University of Technology, Delft, the Netherlands
figure 8.30	C. Mocenni, Università di Siena, Siena, Italy
figure 8.31	R. Bastiaansen, Utrecht University, Utrecht, the Netherlands and A. Doelman, Leiden University, Leiden, the Netherlands
figure 9.3	T. Vicsek, Eötvös Loránd University, Budapest, Hungary
figure 9.4	D. Bartolo, Ecole Normale Supérieure de Lyon, Lyon, France and A. Morin, Leiden University, Leiden, the Netherlands
figure 9.6	D. Martin, University of Chicago, Chicago, IL, USA and J. Tailleur, Massachusetts Institute of Technology, Cambridge, MA, USA
figure 9.7	D. Geyer and D. Bartolo, Ecole Normale Supérieure de Lyon, Lyon, France
figure 9.8.a/b	R. E. Goldstein, Cambridge University, Cambridge, United Kingdom
figure 9.8.c	E. Clément, Ecole Supérieure de Physique et de Chimie Industrielle, Paris, France
figure 9.10	S. Ray and Z. Dogic, University of California Santa Barbara, Santa Barbara, CA, USA
figure 9.11	L. Giomi, Leiden University, Leiden, the Netherlands
figure 9.12.a	T. Beatus and R. Bar-Ziv, Weizmann Institute of Science, Rehovot, Israel
figure 9.12.b	C. Coulais, University of Amsterdam, Amsterdam, the Netherlands
figure 9.12.c	P. Baconnier and O. Dauchot, Ecole Supérieure de Physique et de Chimie Industrielle, Paris, France
figure 9.13	N. Fahkri, Massachusetts Institute of Technology, Cambridge, MA, USA
figure 9.14.a/b	M. Brandenbourger, Aix-Marseille Université, Aix-Marseille, France
figure 9.16	W. T. M. Irvine, University of Chicago, Chicago, IL, USA
figure 9.18	M. Fruchart, University of Chicago, Chicago, IL, USA
figure 9.23	H. Borja da Rocha, H. Tullier, and J.-F. Joanny, College de France, Paris, France
figure 9.24	Y. Maroudas-Sa, K. Keren, and E. Braun, Technion, Haifa, Israel
figure 9.25	S. Streichan and B. I. Shraiman, Kavli Institute for Theoretical Physics, Santa Barbara, CA, USA
figure 9.26	J. Colen, University of Chicago, Chicago, IL, USA
figure 10.2	D. Kraft, Leiden University, Leiden, the Netherlands
figure 10.4	M. Wegener, Karlsruhe Institute of Technology, Karlsruhe, Germany
figure 10.3.b	T. A. Schaedler, HRL Laboratories, LLC, Malibu, CA, USA
figure 10.3.c	N. Pashine and S. R. Nagel, University of Chicago, Chicago, IL, USA
figure 10.6	J.-Y. Sun, Seoul National University, Seoul, Korea
figure 10.7	R. C. Hayward, University of Colorado, Boulder, CO, USA
figure 10.9.a	M. He and D. Pine, New York Universy, New York, NY, USA
figure 10.9.b/c	W. B. Rogers, Brandeis University, Waltham, MA, USA
figure 10.11	N. Liu and P. Ying, Harvard University, Cambridge, MA, USA
figure 10.12	C. Sigl and H. Dietz, Technische Universiät München, Munich, Germany
figure 10.17	H. Bense and M. van Hecke, AMOLF, Amsterdam, the Netherlands
figure 10.18	J. D. Paulsen, Syracuse University, Syracuse, NY, USA
figure 10.27	M. P. Brenner, Harvard University, Cambridge, MA, USA
figure 10.29.a	M. Schmitt, University of Chicago, Chicago, IL, USA
figure 10.29.b	B. VanSaders, University of Chicago, Chicago, IL, USA
figure 10.30	G. Gines, Ecole Supérieure de Physique et de Chimie Industrielle, Paris, France
figure 10.33	M. Muñoz Basagoiti, Institute of Science and Technology Austria, Vienna, Austria
figure 10.35	R. Niu, Huazhong University of Science and Technology, Wuhan, Hubei, China and I. Cohen, Cornell University, Ithaca, NY, USA
figure 10.36	S. Sacanna, New York University, New York, NY, USA
figure 10.37	N. Cheney, University of Vermont, USA

Index